3D Printing

3D Printing: Fundamentals to Emerging Applications discusses the fundamentals of 3D-printing technologies and their emerging applications in many important sectors such as energy, biomedicals, and sensors. Top international authors in their fields cover the fundamentals of 3D-printing technologies for batteries, supercapacitors, fuel cells, sensors, and biomedical and other emerging applications. They also address current challenges and possible solutions in 3D-printing technologies for advanced applications.

Key features:

- Addresses the state-of-the-art progress and challenges in 3D-printing technologies
- Explores the use of various materials in 3D printing for advanced applications
- Covers fundamentals of the electrochemical behavior of various materials for energy applications
- Provides new direction and enables understanding of the chemistry, electrochemical properties, and technologies for 3D printing

This is a must-have resource for students as well as researchers and industry professionals working in energy, biomedicine, materials, and nanotechnology.

3D Printing

Fundamentals to Emerging Applications

Edited by

Ram K. Gupta

CRC Press is an imprint of the
Taylor & Francis Group, an **informa** business

Cover image credit: © Shutterstock

First edition published 2023
by CRC Press
6000 Broken Sound Parkway NW, Suite 300, Boca Raton, FL 33487-2742

and by CRC Press
4 Park Square, Milton Park, Abingdon, Oxon, OX14 4RN

CRC Press is an imprint of Taylor & Francis Group, LLC

ISBN: 9781032283999 (hbk)
ISBN: 9781032284019 (pbk)
ISBN: 9781003296676 (ebk)

DOI: 10.1201/9781003296676

Typeset in Times
by Newgen Publishing UK

Dedication

I would like to dedicate this book to my parents, family members, friends, teachers, and students. Special thanks to my wife Rajani, my daughter Anjali, and my niece Payal Gupta for their love, motivation, and encouragement.

Contents

Preface

3D-printing technology is rapidly advancing and being used in many emerging areas such as energy and biomedical. The availability of many different 3D-printing technologies allows for fast prototype devices with much-reduced cost. They can be used to print electrodes, energy generation and storage devices, sensors, and devices for biomedical applications. They can print complex 3D devices and can be modified without changing the manufacturing process. However, there are still many challenges and shortcomings which need to be overcome to use this technology on a commercial scale.

The main purpose of this book is to provide fundamentals of 3D-printing technologies, current state-of-the-art knowledge, and their emerging applications in many important sectors such as energy, biomedical, and sensors. This book provides current challenges and possible solutions in 3D-printing technologies for advanced applications. This book covers the fundamentals of 3D-printing technologies for batteries, supercapacitors, fuel cells, sensors, biomedical, and other emerging applications. All the chapters are covered by experts in these areas around the world making this a suitable textbook for students and providing new guidelines to researchers and industries working in energy, biomedical, materials, and nanotechnology.

Biography

Ram K. Gupta is an Associate Professor at Pittsburg State University. Before joining Pittsburg State University, he worked as an Assistant Research Professor at Missouri State University, Springfield, MO then as a Senior Research Scientist at North Carolina A&T State University, Greensboro, NC. Gupta's research focuses on green energy production and storage using conducting polymers and composites, electrocatalysts, fuel cells, supercapacitors, batteries, nanomaterials, optoelectronics and photovoltaics devices, organic-inorganic hetero-junctions for sensors, nanomagnetism, bio-based polymers, bio-compatible nanofibers for tissue regeneration, scaffold and antibacterial applications, bio-degradable metallic implants. Gupta published over 250 peer-reviewed articles, made over 320 national/international/regional presentations, chaired many sessions at national/international meetings, wrote several book chapters (55+), edited many books (20+) for the American Chemical Society, CRC, and Elsevier publishers, and received several million dollars for research and educational activities from external agencies. He currently serves as associate editor, guest editor, and editorial board member for various journals.

Contributors

Babak Akbari
University of Tehran, Iran

Nahal Aliheidari
University of Massachusetts-Lowell, USA

Luis M. Alvarez
Uniformed Services University of Health Sciences, USA
Lung Biotechnology PBC, USA

Amir Ameli
University of Massachusetts-Lowell, USA

Anis A. Ansari
Aligarh Muslim University, India

Sara Lopez de Armentia
Universidad Pontificia Comillas, Spain

Daniel Banks
Massachusetts Institute of Technology, USA

Rasmita Barik
Indian Institute of Technology Delhi, India

Shiva Bhardwaj
Pittsburg State University, USA

Harry Bikas
Laboratory for Manufacturing Systems & Automation (LMS),
University of Patras, Greece

Juliano A. Bonacin
University of Campinas, Brazil

Jack T. Buchen
The Henry M. Jackson Foundation for the Advancement of Military Medicine, USA
Uniformed Services University of Health Sciences, USA

Laura Mendoza Cerezo
University of Extremadura, Spain

Somnath Chattopadhyaya
Indian Institute of Technology, Dhanbad, India

Heng Li Chee
Institute of Materials Research and Engineering, Singapore

Wendell K. T. Coltro
Universidade Federal de Goiás, Brazil
Instituto Nacional de Ciência e Tecnologia de Bioanalítica, Brazil

Arthur V. Cresce
US Army Research Laboratory, USA

Kwadwo Mensah Darkwa
Kwame Nkrumah University of Science and Technology, Ghana

Evandro Datti
University of Campinas, Brazil

Lorena Maria Dering
Pontifícia Universidade Católica do Paraná, Brazil

Yash Desai
Pittsburg State University, USA

Amit Rai Dixit
Indian Institute of Technology, Dhanbad, India

Lucas C. Duarte
Universidade Federal de Goiás, Brazil

Nicholas Dunne
Dublin City University, Ireland
Queen's University of Belfast, UK
Trinity College Dublin, Ireland

Touraj Ehtezazi
Liverpool John Moores University, UK

Kateryna Fatyeyeva
Normandie University, France

Beatriz Luci Fernandes
Pontifícia Universidade Católica do Paraná, Brazil

Preethichandra D. M. Gamage
Queensland University of Technology, Brisbane
Central Queensland University, Australia

Antonio Macías García
University of Extremadura, Spain

Rafael L. Germscheidt
University of Campinas, Brazil

Natalie Golota
Massachusetts Institute of Technology, USA

Andrews Nirmala Grace
VIT University, Vellore, Tamil Nadu 632014, India

Robert Griffin
Massachusetts Institute of Technology, USA

Ram K. Gupta
Pittsburg State University, USA

Pravin P. Ingole
Indian Institute of Technology Delhi, India

Farzaneh Jabbari
Materials and Energy Research Center, Iran

Shailly H. Jariwala
The Henry M. Jackson Foundation for the Advancement of Military Medicine, USA
Uniformed Services University of Health Sciences, USA

Niki Sweta Jha
National Institute of Technology, India

Sujin P Jose
Madurai Kamaraj University, India

Paras Kalra
Indian Institute of Technology Delhi, India

Mohammad Kamil
Aligarh Muslim University, India

Rudranarayan Kandi
Indian Institute of Technology, Delhi, India

A.M. Kannan
Arizona State University, USA

Yaroslav Kobzar
Normandie University, France

Shrinivas B. Kulkarni
Dr. Homi Bhabha State University, India

Nathan Lazarus
US Army Research Laboratory, USA

André Giacomelli Leal
Pontifícia Universidade Católica do Paraná, Brazil

Chong-Yong Lee
University of Wollongong, Australia

Gaurav M. Lohar
Lal Bahadur Shastri College of Arts, Science and Commerce, India

Sudhakar Reddy Madhira
Pittsburg State University, USA

K.A.U. Madhushani
Pittsburg State University, USA

Dipanwita Majumdar
Chandernagore College, India

Brian Michael
Massachusetts Institute of Technology, USA

Dipesh Kumar Mishra
AKTU Lucknow, India

Xuening Pang
Shandong University of Technology, P.R. China

Christos Papaioannou
Laboratory for Manufacturing Systems & Automation (LMS),
University of Patras, Greece

Anupam Patel
Banaras Hindu University, India

Dayanidhi Krishana Pathak
G. B. Pant Government Engineering College, India

Eva Paz
Universidad Pontificia Comillas, Spain

Matheus Kahakura Franco Pedro
Pontifícia Universidade Católica do Paraná, Brazil

A.A.P.R. Perera
Pittsburg State University, USA

Onkar C. Pore
Lal Bahadur Shastri College of Arts, Science and Commerce, India

Nikolas Porevopoulos
Laboratory for Manufacturing Systems & Automation (LMS),
University of Patras, Greece

Sirawit Pruksawan
Institute of Materials Research and Engineering, Singapore

Suppanat Puangpathumanond
National University of Singapore, Singapore

Buwanila T. Punchihewa
University of Missouri-Kansas City, USA

V. Raja
Madurai Kamaraj University, India

Juan Carlos del Real
Universidad Pontificia Comillas, Spain

Jesús M. Rodríguez Rego
University of Extremadura, Spain

Alfonso C. Marcos Romero
University of Extremadura, Spain

Digambar S. Sawant
Lal Bahadur Shastri College of Arts, Science and Commerce, India
Dr. Homi Bhabha State University, India

Vithyasaahar Sethumadhavan
Queensland University of Technology, Brisbane

Mohammed Shariq
VIT University, Vellore, Tamil Nadu, India

Padma Sharma
National Institute of Technology, Bihar, India

Pawan Sharma
Indian Institute of Technology, Varanasi, India

Mariana B. Silva
University of Campinas, Brazil

Habdias A. Silva-Neto
Universidade Federal de Goiás, Brazil

Swee Leong Sing
National University of Singapore, Singapore

Rajendra K. Singh
Banaras Hindu University, India

Gabriel L. Smith
US Army Research Laboratory, USA

Prashant Sonar
Queensland University of Technology, Brisbane

Kenan Song
Arizona State University, USA

Thanassis Souflas
University of Patras, Greece

Felipe M. de Souza
Pittsburg State University, USA

Mauren Abreu de Souza
Pontifícia Universidade Católica do Paraná, Brazil

Panagiotis Stavropoulos
Laboratory for Manufacturing Systems & Automation (LMS),
University of Patras, Greece

Richard C. Steiner
The Henry M. Jackson Foundation for the Advancement of Military Medicine, USA
Uniformed Services University of Health Sciences, USA

Vaishali Tanwar
Indian Institute of Technology Delhi, India

Lobat Tayebi
Marquette University School of Dentistry, USA

Igor Tkachenko
National Academy of Sciences of Ukraine, Ukraine

Joshua B. Tyler
Oak Ridge Associated Universities, USA
US Army Research Laboratory, USA
University of Maryland, USA

Konstantinos Tzimanis
Laboratory for Manufacturing Systems & Automation (LMS),
University of Patras, Greece

J Vigneshwaran
Madurai Kamaraj University, India

Gordon G. Wallace
University of Wollongong, Australia

FuKe Wang
Institute of Materials Research and Engineering, Singapore

Richard J. Williams
National University of Singapore, Singapore

Naitao Yang
Shandong University of Technology, P.R. China

Jinjin Zhang
Shandong University of Technology, P.R. China

1 3D Printing
An Introduction

Richard J. Williams[1] *and Swee Leong Sing*[2,*]
[1]Department of Mechanical Engineering, College of Design and Engineering, National University of Singapore
[2]9 Engineering Drive 1, Block EA, Singapore, 117575
*Corresponding author: sweeleong.sing@nus.edu.sg

CONTENTS

1.1 WHAT IS ADDITIVE MANUFACTURING?

In its essence, Additive Manufacturing (AM) refers to a group of technologies that can manufacture end-use parts directly by progressively adding material, layer upon layer. This sets it apart from subtractive manufacturing techniques, such as machining, where the material is removed or forming processes that reshape material. In both cases, some form of tooling is always required, whereas with AM parts are produced directly and no tooling is required. Several unique benefits arise on this basis, offering great potential to the manufacturing world. Complex topologies, with internal features and overhangs, which would be prohibitive to manufacture via conventional means, can be produced directly. Moving parts and fasteners can be integrated into a single design, reducing the need for downstream assembly. Short production runs become economically viable, allowing products and designs to be customized to a user. Numerous other opportunities and advantages also exist in a similar vein.

DOI: 10.1201/9781003296676-1

Beyond sharing the common principle of material addition, the field of AM contains a diverse group of processes capable of manufacturing all classes of material, including metals, polymers, ceramics, composites, and even biological matter. The different types of AM processes can broadly be divided into seven categories, which will be described in more detail in later sections. In the following section, we will examine the overall approach involved in producing a part with AM, from raw material to end product. This process chain sits independent of the specific AM method employed.

1.2 THE ADDITIVE MANUFACTURING PROCESS CHAIN

Although AM is sometimes described as a 'one-step production technique', there are in fact several steps involved in taking an object from a design concept to an end-user component. Regardless of the AM technique, many other operations are typically required before and after the manufacturing process itself. Therefore, these steps can intuitively be grouped into three stages: (a) The pre-processing stage (i.e., activities that happen before manufacture), (b) The AM process, and (c) The post-processing stage (i.e., finishing operations that take place after the part has been made). AM can therefore be thought of as a process chain. Many of the required pre-processing steps are common to all AM techniques, whereas the post-processing steps required tend to be more specific to each AM process.

1.2.1 Pre-processing

We shall now approach the AM process chain from the beginning, where one may have a particular component in mind that they wish to 3D print. The first step here would be to create a 3D computer-aided design (CAD) model of the part; this is a requirement of all AM processes. This can either be done natively in a CAD software package or by 3D scanning a physical artifact. The latter may be useful in legacy part production or medical settings, where parts are customized to fit a particular patient. The CAD model must then be exported as a surface mesh to be loaded into the build preparation software, independent of any CAD package used. By far the most common format used here is the .stl file, in which the surface of the part is expressed as a tessellation of triangles. Various other file formats have been proposed at different stages, but the .stl remains the de-facto standard. Sometimes, .stl files may require repair operations if they contain defects, such as surface voids or bad aspect ratio triangles. If this is the case, then dedicated repair software is generally required.

Following this, the surface mesh is then read into build preparation software. The purpose of the build preparation stage is to generate the tool path, or G-code, for the 3D printer to build the part. The surface mesh will first be sliced into several 2D layers and then the tool path will be created for each layer, encoding the various printing parameters used at each point. Each printer manufacturer therefore typically has its build preparation software and build files are not transferrable between different AM systems. As it is software-heavy, the preprocessing stage may also be referred to as the 'digital phase'.

1.2.2 Manufacturing

During the manufacturing stage, the part is physically built up by one of the AM processes. Typically, some form of alignment or calibration between the 3D printer and the substrate is performed manually before starting the job, depending on the technique in question. Raw materials, or feedstock, must also be supplied to the 3D printer and these can vary widely in form depending on the material and process technique used. Powder and wire or filament are the two most common forms of feedstock, particularly when working with metals or polymers. Beyond this, sheets, liquids, and inks are also used.

Most of the techniques are then automated during the manufacturing process, such that they require little to no human intervention. As such, errors or malfunction can often go undetected, and automating failure detection forms one of the great challenges in the field today. Following completion of the build, the part is removed from the 3D printer before any post-processing operations are applied. Removal of the part from the 3D printer can be non-trivial, for example with powder-bed processes where substantial powder cleaning and removal are required.

1.2.3 Post-processing

Once the part is removed from the 3D printer, various post-processing operations may be applied depending on which AM technique was used. These finishing operations are designed to ensure the part has suitable structure and properties for its intended function. Firstly, parts produced via any AM technique that necessitates support structures to be used will first require these to be removed. This is commonly achieved by either dissolving the supports in a chemical solution or mechanically removing them, by pulling or machining. This step can be difficult and time-consuming. Metal parts, produced either via Directed Energy Deposition (DED) or Powder Bed Fusion (PBF), often undergo a heat treatment process to reduce residual stresses and achieve their intended microstructural state. Hot Isostatic Pressing (HIP) can also be used to eliminate any porosity or processing defects in the part under high heat and pressure. Polymer components may undergo chemical-based surface treatment to enhance the quality of the surface finish or ensure water tightness. Finally, in the case of binder jetting, the part obtained directly after manufacture is just an intermediate or 'green' part that requires sintering in a furnace post-manufacture to achieve full density. In this case, the post-processing steps may be considered an integral part of the manufacturing process. In all, a suite of different finishing operations may be applied to 3D-printed parts depending on the material, manufacturing method, and intended application.

1.3 AM PROCESS CATEGORIES

In the following section, we will describe each of the different AM processes, categorized in terms of their processing methodology. Some of the advantages and disadvantages of each technique will be discussed, as well as identifying their key commercial suppliers. As will become clear, different material classes, such as polymers and metals, can be processed by the same technique. The categories used here follow the taxonomy defined by the ISO/ ASTM standard [1].

1.3.1 Binder Jetting

Binder Jetting (BJT) is actually a two-step manufacturing process. In the first stage, printheads are used to selectively jet droplets of a binder compound onto a thin layer of powder, in a similar fashion to classical inkjet printing. The build platform then lowers by one layer thickness and a new layer of powder is spread on top, onto which another layer of binder is jetted. This weakly bonds the powder particles and, after many layers, a so-called 'green' part is produced. The green part is highly porous and low in strength. At the end of the 3D-printing process, the part is surrounded by unbound powder which must be removed before post-processing. A schematic of the printing phase of the BJT process is presented in Figure 1.1.

After printing, the green part must undergo post-processing to achieve full density and intended material properties. For metals and ceramics, this is done by sintering. The green part is held at a high temperature in a furnace causing the binder to melt off and the powder particles to fuse, producing a fully dense part. Alternatively, full density can be achieved via an infiltration process. A liquid material, such as epoxy resin, is used to infiltrate the voids in the green part, before solidifying and forming a dense part. Infiltration is typically used when processing polymers. The final variation of

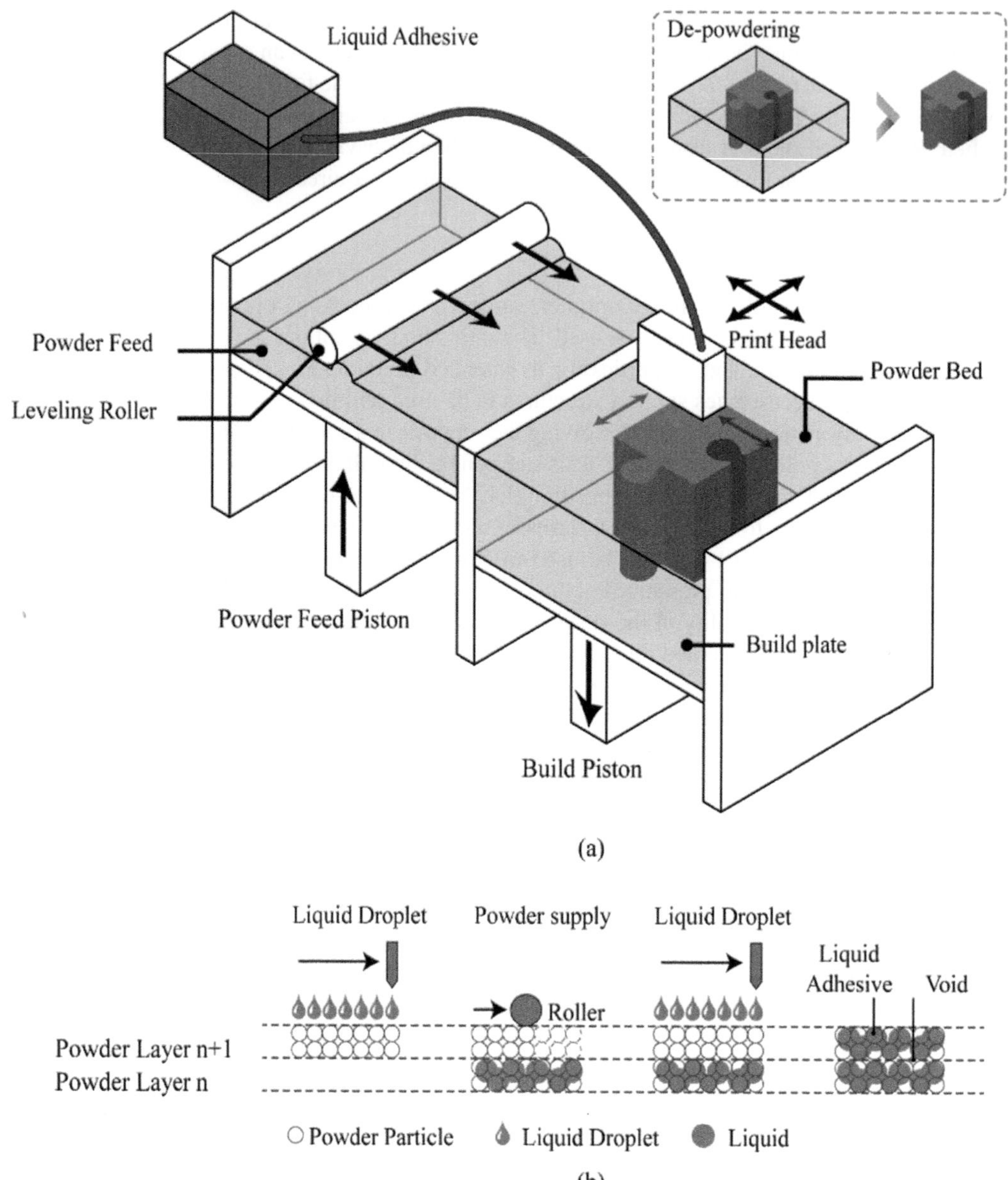

FIGURE 1.1 A schematic of the printing phase of the Binder Jetting (BJT) process, depicting (a) the processing equipment and methodology, including powder removal at the end of the process, and (b) The bonding of powder particles with the liquid binding agent. Figure reproduced with permission [2].

BJT requires no post-processing. Here, a thermally activated liquid compound is jetted during step one and the print head is surrounded by heat lamps, fully melting the powder particles in-situ and producing a dense part. Consequently, this type of process is called high-speed sintering (HSS). As the jetted compound is not a binder, it could be argued that the HSS process sits somewhere between BJT and powder bed fusion (PBF).

The BJT process can be used to process various classes of metals and non-metals, boasting arguably the broadest range of material choices of all the AM processes [3]. Regardless of the material, the feedstock must be supplied in a finely powdered form. One unique application of non-metal BJT is the 3D printing of sand to manufacture molds for metal casting. Both synthetic and organic sands can be process processed and this technology is sometimes described as 'digital sand casting'. Intricate mold shapes can be produced at low cost using BJT and new mold designs can rapidly be iterated. Similarly, ceramic powders can also be processed by BJT. These include hard, tooling ceramics, such as tungsten carbide, glasses, and alumina compounds. Polymer parts can be manufactured from certain compounds, primarily general-purpose materials such as polyamide (PA) and polypropylene (PP). Some of the non-metal feedstocks, such as sand, can be processed exclusively by BJT and are not suitable for any other AM process.

On the metals side, a wide range of alloys are available compared with other metal AM systems and processes, though still relatively few compared with conventional manufacturing processes. Ductile austenitic steels, low alloy and martensitic steels, aluminum, tungsten, titanium, and nickel alloys are all commercially available as powder feedstocks. Metals are processed at low temperatures during BJT. Therefore, unlike other metal AM processes, some of the materials challenges associated with rapid heat and cooling cycles, such as solidification cracking or chemical segregation, which affect printability, are negated. For the same reasons, minimal residual stresses are also induced in parts.

The ambient temperature printing principle of BJT gives rise to a few other advantages. Support structures are not required since residual stresses and warping are not an issue. Systems are relatively low in cost compared to laser-based processes and are also safer, thus they can be housed in office or home environments. Relatively large parts, in terms of AM, on the order of a meter in scale can be processed by BJT, particularly with the sand systems. The printing process is also relatively fast in comparison with raster scanning methods, as printheads can cover large areas with a binder quickly. On the other hand, although the printing process itself is low energy and low cost, one must potentially purchase a furnace, offsetting the low cost of the printer. The process inherently involves two steps and the post-processing requirements may be viewed as a disadvantage. Parts can often suffer from shrinkage and warping during the sintering phase. Post-sinter consolidation and density are also not guaranteed.

The original commercial exponents of BJT systems were ExOne, however, they have since been bought out by a younger competitor, Desktop Metal. Today, sand systems continue to be sold under the ExOne moniker whereas the ceramic and metal systems are sold under the Desktop Metal brand. Competition is provided by HP and Voxeljet. These two systems are the primary options for polymer binder jetting and HSS and they also produce systems capable of printing color into parts.

1.3.2 Directed Energy Deposition

Directed Energy Deposition (DED) processes employ a high-power energy source to melt and deposit feedstock material, which is continuously and locally fed into the line of the energy source. Both the energy source and the feedstock supply move with many degrees of freedom on a robotic arm to build up a part onto a substrate. The building substrate itself may also be traversed or rotated to create certain geometries, depending on the process. Beyond this, however, the form of both the energy source and the feedstock material can vary. Generally, feedstock materials fall into two classes, wire or filament, known as Wire-DED, or powder, blown locally into the melt, known as Powder-DED. The energy source is usually a high-power laser or an electrical arc, widely used in welding processes. As such, DED processes can essentially be thought of as '3D multi-pass welding'. The wire-fed, laser DED process is illustrated in Figure 1.2.

The DED process which utilizes a wire feed and an electrical arc to melt it is popularly known as wire arc additive manufacturing (WAAM). Several smaller companies offer this technology

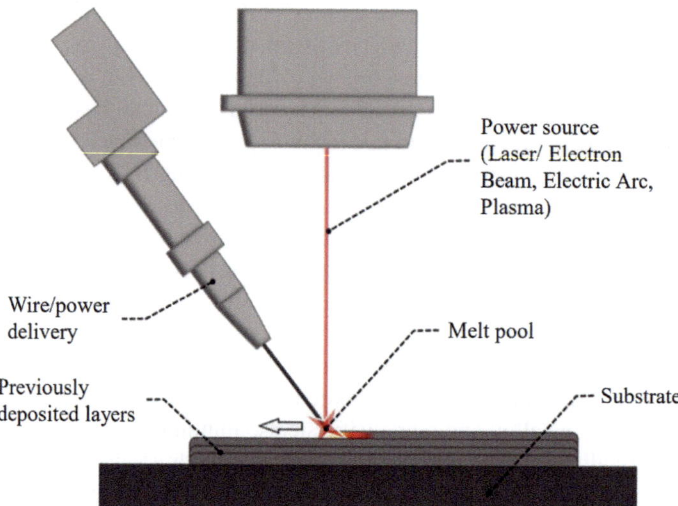

FIGURE 1.2 A schematic depicting a wire-fed DED process using a laser energy source. The processing head would move in or out of the page to deposit an adjacent track of material. Tracks are deposited with left and right traversal. Figure adapted with permission [4]. Copyright: The Authors, some rights reserved; exclusive licensee CRC Press, LLC. Distributed under a Creative Commons Attribution License 4.0 (CC BY 4.0).

commercially. The laser-driven, blown powder DED systems manufactured by Trumpf are branded as Laser Metal Deposition (LMD). These are the most established commercial DED systems in the market. As is also common in welding processes, an inert gas is used locally to shield the weld from the air and prevent the oxidation that would occur in metals at high temperatures. Although inert gas is commonly supplied locally at the energy source, some DED systems are also housed inside the large inert chamber when the lowest possible oxygen concentration is required. This imposes space constraints, restricting the size of a part that can be fabricated, whilst also being costly in terms of gas consumption.

DED processes are used exclusively to process metals. However, they enjoy a slightly broader range of alloys that can routinely be processed in comparison to PBF. High-performance nickel alloys, lightweight aerospace aluminum and titanium alloys, low-alloy structural steels, and martensitic steels are all commercially available as DED feedstocks. Beyond this, more novel material systems such as high entropy alloys [5], duplex stainless steels [6] and refractory metal alloys [7] have been processed in academic settings. When processed by DED, these alloys may exhibit drastically different microstructures and mechanical properties than their wrought counterparts, however, due to the rapid and cyclic heating and cooling [8]. This generates very large residual stresses, and it can be difficult to obtain a material's desired microstructural state in the as-built condition. Post-process heat treatments are therefore often required.

One of the unique advantages of DED is that it can be performed in an open environment and with a large robotic arm, enabling large components to be fabricated, such as pressure vessels and walls. Such components would be impossible to make by any other metal AM process. For the same reasons, DED is also uniquely suited to repair operations and several case studies have been published demonstrating the repair of turbine blades and other high-integrity metal parts with DED [9]. Despite this, the size of DED systems can act as a drawback. The diameter of the weld pool is substantially greater than that found in PBF, and the control and precision of the energy source are inferior. The resolution with which parts can be produced is, therefore, lower than PBF and parts often tend to have a poor surface finish. As a result, DED systems are rarely employed to manufacture intricate components. This demonstrates the importance of choosing the correct AM process for a given job.

1.3.3 Material Extrusion

The Material Extrusion (MEX) processes represent one of the simplest methods of producing AM parts. Here, a wire or filament is extruded through a heated nozzle to deposit layers of material. The nozzle traverses inside the build volume on a frame to deposit tracks of material selectively on each layer and produce a component at the net shape. By this virtue, overhanging features in parts will usually require support structures to be built when manufacturing via MEX, to stop the newly deposited, soft material from sinking over the void below. A schematic of the process is presented in Figure 1.3.

It is the polymer MEX process that is perhaps most synonymous with the term '3D printing' and simple, low-cost versions of these machines are widely available. They often can fit on a desktop or a table in the home. These entry-level systems can quickly and cheaply produce a 3D plastic model or part and are well-suited for rapid prototyping and hobbyists where the quality and integrity of parts are not a primary concern. Often, the parts produced by these systems are not of sufficient quality to be sold as an end-use product. They may have poorer geometric accuracy than the CAD model or contain porosity and other defects and therefore lack durability. Some of the more high-end desktop printers, however, can produce quality parts. Higher quality, industrial-grade systems also exist for producing functional components and in this context, the process is sometimes known as fused deposition modeling (FDM) and fused filament fabrication (FFF). These systems are also generally much larger in build volume, allowing larger parts to be manufactured. Here, the deposition nozzle will be controlled more finely enabling more precise and accurate material extrusion and part quality. The Stratasys FDM system is probably the most famous example in this area.

A wide range of different thermoplastic polymers is available to be processed by MEX, including acrylonitrile butadiene styrene (ABS), polylactic acid (PLA), polyamide/ Nylon (PA), and polyether

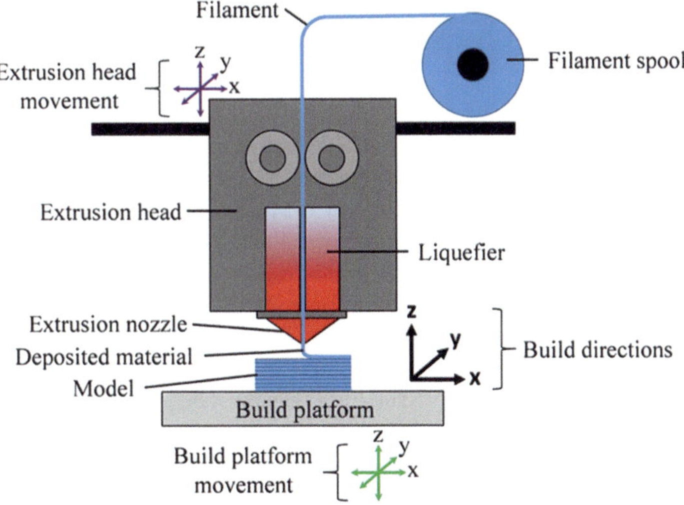

FIGURE 1.3 A schematic of the Material Extrusion (MEX) process, using a single filament of material [10]. In some cases, the second filament of a different material may also be used. Figure reproduced with permission [10]. Copyright 2019, Emerald Publishing Limited all rights reserved.

ketones (PEEK). Proprietary AM grades have also been developed by some suppliers to offer greater strength and thermal resistance than existing polymers. Therefore, as a process, it probably offers the widest choice of polymer materials out of all the AM techniques. Several MEX systems are equipped with more than one deposition nozzle, allowing multiple materials to be extruded in a single build. Hence, parts composed of more than one material can easily be built. Taking advantage of this, some material suppliers offer a water-soluble polymer filament that can be used for the manufacture of support structures. Parts can then simply be soaked in water afterward to remove the supports.

Aside from polymers, almost every other type of material beyond polymers can also be processed by MX so long as they are mixed with an extrudable liquid or polymer compound, thanks to the simplicity of the process. One can source polymer filament containing various types of wood fibers, ceramic powders, carbon fibers, and polymer fibers. Therefore, these types of materials would technically be classed as composites, though they may be described as 'wood printing' or 'ceramic printing' for example. Similarly, MEX processes can also be used for bioprinting, whereby cells, molecules, or other organic biomaterials are suspended in hydrogels and then extruded into shape [11]. The emergence of the bioprinting space is one of the most recent developments for MEX technology.

The final class of materials to be mentioned is metals. These are rarely processed by MEX, DED and PBF are by far the most common, though some options do exist. In this case, parts are not manufactured directly in a single step. A second, sintering phase is required. Powdered metal is mixed with a polymer binder and extruded in layers, as per polymer MEX, to produce a green part. The green part is then washed to remove some of the binders, leaving a porous and weakly bonded component. This is termed a 'brown' part in sintering terminology. The brown part is placed in a furnace close to its melting temperature to sinter and form a dense part. Significant shrinkage may occur during sintering which can often lead to poor geometric accuracy. The final dense parts may also still contain a degree of porosity, reducing their toughness and durability. However, metal MEX systems come at a fraction of the cost of DED and PBF systems and part production is also less expensive. They are therefore excellent for metal prototyping. In this way, metal MEX systems share some advantages with BJT systems. Desktop Metal and Markforged are the two primary commercial providers of metal MEX technology, with Desktop Metal operating in both the MEX and BJT spaces.

1.3.4 Material Jetting

In Material Jetting (MJT), droplets of photocurable polymer are jetted directly from a printhead onto a substrate to produce a component. Ultraviolet (UV) curing lamps are attached directly to the printhead, curing the layer of polymer immediately in situ. The substrate then moves down by one layer and the subsequent layer is jetted and cured onto the top surface. This procedure is illustrated in Figure 1.4. In contrast with BJT, no powder or binding agents are involved in MJT and the build material is deposited directly by the printhead. In some setups, the print head may contain a second, water-soluble polymer for printing support structures. This can easily be dissolved in water after the build, simplifying the removal process.

Exclusively polymers are processed by MJT since they must be jettable and photocurable. Hence, the feedstock is supplied in a liquid form. Different grades of polymer are available, enabling a range of material properties to be achieved. Highly flexible, rubber-like polymers are commercially available, which is unique to MJT. However, compared with the other AM processes, the range of materials available for MJT is narrow. Despite this, a full range of colors, including transparency, can be 3D printed, with detailed variation across the surface of the part. MJT excels in this area, making it well suited for prototyping and producing customized designs for consumer products. Accurate and intricate parts can be produced with an excellent surface finished, owing to the precise

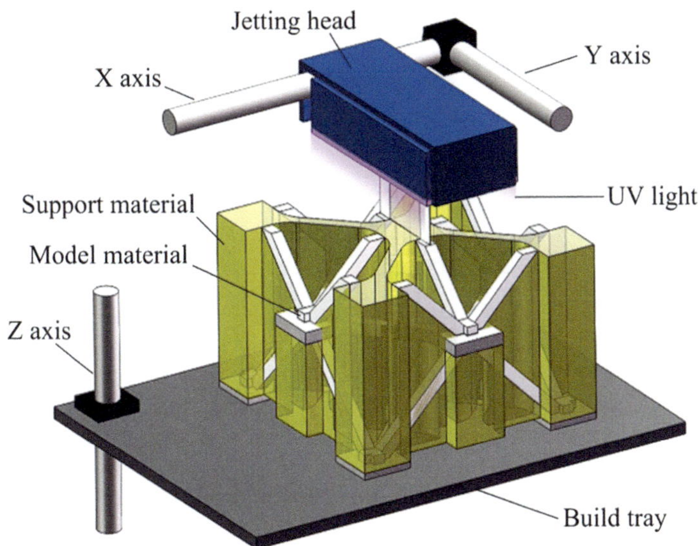

FIGURE 1.4 A schematic of the Material Jetting (MJT) process. In this case, supporting material is being deposited alongside the main building material. Figure reproduced with permission [12]. Copyright 2019, Elsevier.

and accurate nature of print head technology, which has been developed over decades for conventional printing. Support structures are usually required, however, offsetting this somewhat. The main commercial providers of MJT technology are Stratasys, with their PolyJet line of systems. This was one of the earliest printing systems on the market.

1.3.5 Powder Bed Fusion

In PBF, a high-powered energy source, typically a laser or electron beam, is used to selectively melt and fuse a thin layer of powdered material. First, a powder recoating mechanism, such as a roller or wiper blade, will coat a layer of powdered feedstock over a flat substrate – onto which the part will be built. The heat source will trace out and raster scan over the first 2D slice of the part. After this, the build platform will lower by one layer thickness and the recoating mechanism will wipe a fresh layer of powder on top before the second layer is melted. This process is repeated, layer by layer, until the part is built. The build process is depicted in a schematic in Figure 1.5. As the entire substrate must be coated with powder in each layer, the part is effectively buried at the end of the build and the excess powder must be removed before the part can be taken out of the printer. The excess powder is usually recyclable. The PBF process shares some similarities with DED, as both processes fully melt and weld metals. However, PBF processing takes place on a finer length scale, with a smaller molten pool and faster scanning speeds.

PBF is among the most popular and developed AM processes and is widely used to process both polymers and metals in an industrial setting. When processing polymers, the process is also often referred to as selective laser sintering (SLS). Most commonly, a high-power laser is used as the heat source for melting and this particular subset is sometimes designated laser-PBF or LPBF. Less commonly, an electron beam can also be used for melting metal feedstock, most notably the GE Additive system often termed electron beam melting (EBM). Interestingly, support structures are generally required when building metal PBF parts whereas, for polymers, they are not. The function of the support structures in metal processing is to anchor the part to the base plate and prevent

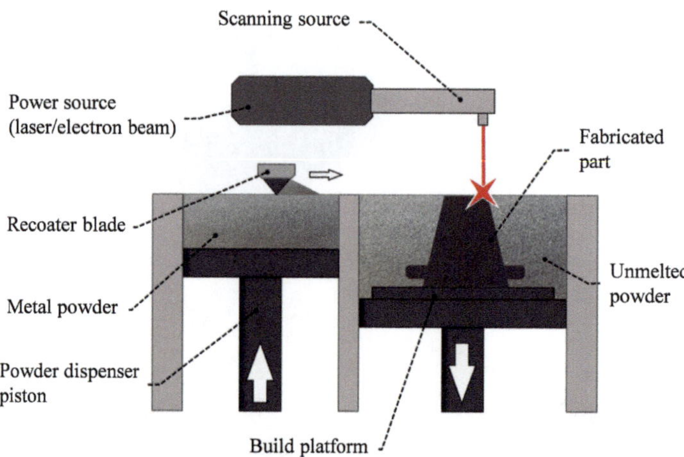

FIGURE 1.5 A schematic of the Powder Bed Fusion (PBF) process [4] using a laser of the energy source. Here, a wiper blade is used to spread a thin layer of powder before melting. Figure adapted with permission [4]. Copyright: The Authors, some rights reserved; exclusive licensee CRC Press, LLC. Distributed under a Creative Commons Attribution License 4.0 (CC BY 4.0).

warping under the severe thermal stresses that are formed, as well as provide a path for heat conduction to prevent the part from overheating. When processing polymers, the temperatures involved are lower and thus thermal expansions are less severe. Due to the chemistry of the polymers used in processing, the whole part can be held at a relatively homogenous temperature during the entire build, so support structures are not necessary. The surrounding powder provides natural support to overhanging surfaces. As a result, additional parts can also be stacked vertically in the build volume and higher machine productivity can be achieved. When processing metals, multiple parts can generally only be nested along the substrate owing to the supports.

Regardless of whether metals or polymers are being processed, the feedstock material supplied to the printer is a very fine powder, typically tens of microns in diameter. This gives PBF parts a characteristically rough, grainy surface finish. The production of such finely powdered metal feedstock via gas atomization is complex and energy intensive and is, therefore, the subject of research and development. The palette of materials readily available for processing is relatively limited, in both metals and polymers, though again the expansion of this is a topic of much ongoing academic research. The most common alloys used in metal PBF are stainless steel grade 316L, Ti-6Al-4V, Al-Si-10Mg, and Inconel alloys 625 and 718. All these alloys can easily be processed by PBF by using the manufacturer's recommended parameters. Beyond these, and a few more, the user would likely have to develop their processing parameters to build high-quality parts. In terms of polymers, the most common materials are PA 11 and 12, different PEEKs, and polypropylene (PP), comparable to those available for polymer BJT. These general-purpose materials enable the manufacturing of many everyday, functional plastic components for end-use.

Given the spot size of the laser and the precise control of the galvo scanners, PBF parts typically have excellent resolution. Fine details and lattice structures can be manufactured. Good mechanical properties can also be achieved, making parts suitable for end-use applications. However, the geometric accuracy, microstructure, and surface finish of metal PBF parts still often render them unsuitable for demanding, safety-critical applications. Systems can also be expensive, both to purchase and to run, and the process is fairly slow compared to other AM methods. Commercially, the PBF systems market is highly competitive and growing rapidly. Several new manufacturers have emerged, attempting to usurp the established companies. EOS Gmbh, GE Additive, 3D Systems, and SLM Solutions are among the most established suppliers, commanding the greatest market share.

1.3.6 Sheet Lamination

Sheet Lamination (SHL) is among the least used of the AM processes. As the name suggests, the basic principle of SHL is to join together successive premanufactured sheets of material. Therefore, in this case, the feedstock is supplied as a thin sheet of the chosen material. Beyond this common principle, several different joining methods are used to produce different classes of material.

The original SHL process, developed in the mid-1990s by Helisys Inc., was known as Laminated Object Manufacturing (LOM). This process used a heated roller to bond successive sheets of adhesive-backed paper into a 3D stack. After each layer was bonded, a laser would trace out the outline of the layer of the part being manufactured to cut away the excess paper sheet. Layer by layer, a glued 3D paper part is produced. Helisys ceased operating several years ago and other paper LOM processes based on this principle followed, with the capability to print color into parts later added. At the time of writing, no commercially produced system for paper LOM is available on the market.

Aside from paper, SHL has most frequently been used in industry to produce metallic parts. Here, thin sheets of metal are bonded together through ultrasonic vibration before the part is machined down to its net shape. This subset of SHL can be referred to as Ultrasonic Additive Manufacturing (UAM). Ultrasonic transducers, attached to a sonotrode, traverse the newly applied layer causing it to vibrate at a high frequency and bond to the surface below through plastic deformation. Machining to the net shape may happen after each layer or at the end. A schematic of the UAM process is shown in Figure 1.6.

Metal parts are manufactured at low temperatures by SHL. Internal features can also be created during the build since there is no surrounding powder feedstock. Combined with the low temperatures, this enables electronics and other components to be embedded within parts [13]. It is also relatively easy to change materials in each layer, so UAM is a good method to produce multi-material components. Joining dissimilar metals at low temperatures can deliver a favorable microstructure compared with conventional high-temperature welding processes, as well as lower residual stresses. Several alloys can be processed, including stainless steel, aluminum, copper, and titanium alloys. Large components can also be built easily, relative to PBF for example, as the process takes place in the open and large sheets of material can be joined.

1.3.7 Vat Photopolymerization

Vat Photopolymerization (VPP) processes utilize a photocurable liquid resin and a UV light source to manufacture polymer components. The most famous VPP process, and arguably the first ever AM process, is stereolithography (SLA). During SLA, a UV light source will raster scan a 2D slice of a component at the top surface of a vat of liquid photocurable resin. This will therefore solidify and fuse the first layer of the part. The part is then drawn out of the vat by a distance equal to the layer thickness in the process. This procedure is then repeated, solidifying the second layer and fusing it to the first – and so on – until the part is complete. Several developments to speed up the SLA process have been made since the first generation of systems arrived, giving rise to derivative VPP processes. A high-powered digital UV light projector has been employed to flash the whole area of a layer in the vat in one go, as opposed to raster scanning. This is usually referred to as Digital Light Projection (DLP). Schematics of both classical SLA and DLP are presented in Figure 1.7. Most recently, a 3D hologram of the part has been projected into the vat to produce the part rapidly in a single exposure [14]. Each of these methods is built around the same principle, however.

Whilst the available materials for processing via VPP are restricted to photocurable polymer liquid resins, several different grades are available including hard and soft plastics and resins mixed with non-polymers to achieve enhanced mechanical properties, similarly to the MEX filament palette. Ceramic reinforced composite resins help to promote heat resistance and fiber reinforced composites improve strength. Specific biocompatible resin grades are available for manufacturing

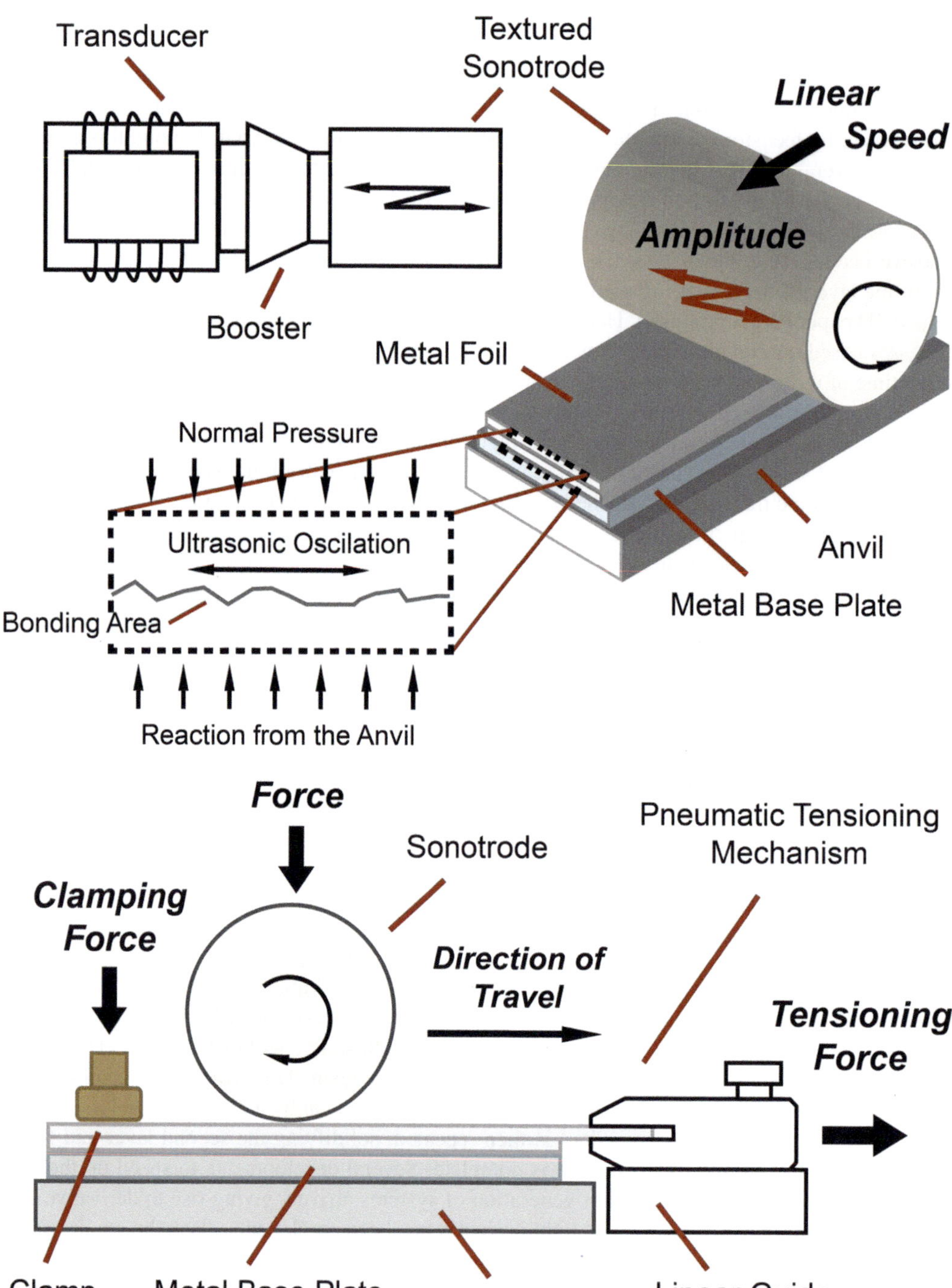

FIGURE 1.6 A schematic showing the different elements of metal SHL manufacture [13]. An ultrasonic transducer is attached to a sonotrode (top), which traverses over the metal sheet causing it to vibrate under force (middle). Successive sheets are joined in this fashion (bottom). Figure reproduced with permission [13].

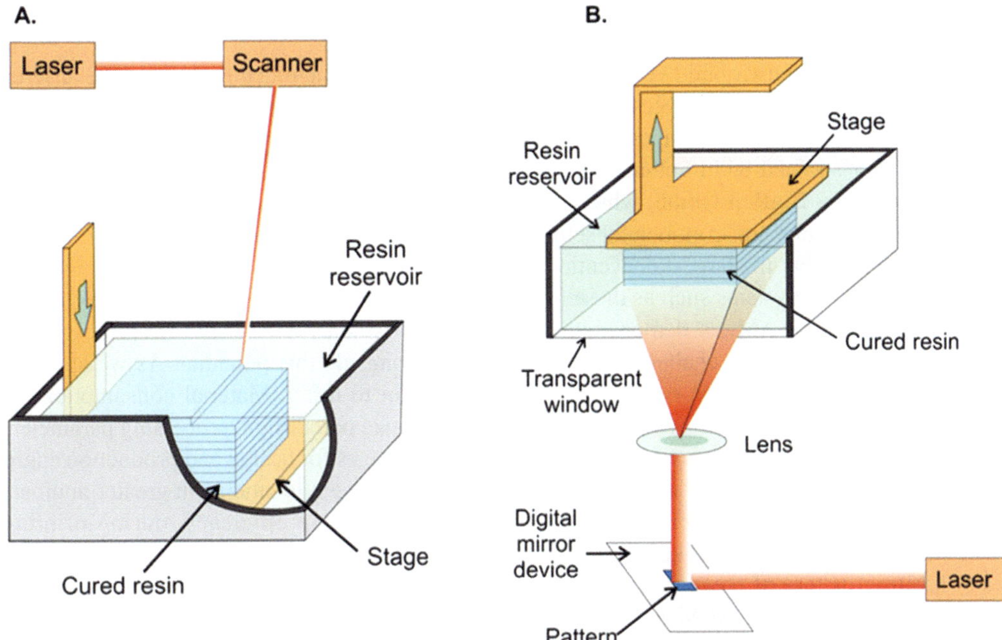

FIGURE 1.7 A schematic demonstrating two different types of VPP process [15], (a) stereolithography (SLA), and (b) digital light projection (DLP). Note that the process can take place from the top down, as in (a), or inverted, as in (b). Figure reproduced with permission [15]. Copyright 2014, American Chemical Society.

dental and medical devices. It is relatively difficult to process multiple materials in a single build by VPP as the liquid feedstock must always be contained inside the vat.

As they are produced from liquid feedstock, VPP parts typically have the best surface finish and dimensional accuracy of all the AM methods. VPP is, therefore, a good choice in complex and intricate applications and is often used to manufacture dental and medical devices. The build process is also relatively fast compared with other AM processes, particularly with the later generation VPP systems, such as DLP. This makes it also well suited to rapid prototyping. The chief drawback of VPP systems stems from the material palette. The resins processed by VPP have routinely exhibited poor mechanical properties, though this is being addressed as bespoke new resins are developed.

1.4 CURRENT DEVELOPMENTS: POTENTIAL AND CHALLENGES

In the following section, contemporary developments in the field of AM will be discussed in the context of potential opportunities for the technology and new challenges being faced.

1.4.1 Size and Productivity

Initially, it was thought that 3D printing technologies would only be economically viable for manufacturing one-off parts or very short production runs. Indeed, this is one mentioned as one of the advantages of AM. However, it was later shown that given suitable improvements to machine productivity, AM had the potential to compete on cost with processes such as injection molding at relatively high production volumes [16]. With this idea in mind, one angle of development which has been consistently followed by machine manufacturers since the emergence of AM into end-use manufacturing has been the drive to increase the scale and productivity of machines. These developments come primarily in two flavors, improving the productivity of the 3D printing process and expanding the size of the build volume, and hence, the machine.

Systems with ever larger build volumes have been released by manufacturers across the different AM process categories. This enables larger-sized parts to be manufactured as well as fitting more parts into a single build, increasing productivity. This is particularly important for those processes with heavily constrained build volumes, such as PBF. One example here is the BigRep Pro FDM system released by BigRep Gmbh, which offers a build volume of one cubic meter with a traditional FDM setup. In a non-commercial setting, a custom system developed at Oak Ridge National Laboratory is probably the largest conventional AM system presented to date, with a build volume of 4m × 2m × 2.4m. Systems such as these offer the possibility to 3D print products far larger than previously thought by AM, as demonstrated in the case of studies with chairs [17], bicycles, or automobiles. However, new challenges are encountered alongside this potential. As we approach the meter scale, the production economics of printing begin to differ. Material consumption and printing time can vary widely. It will become crucial to optimize part design, processing parameters, and build setup as cost savings are amplified. These are challenges for design and production engineers and further research will be needed as ever larger systems begin to emerge in greater numbers.

The other approach has been to increase productivity, or time efficiency, during printing. Such improvements may be essential to keep build durations in check as systems increase in scale. Melting meters of powder with a single laser scanner would likely prove prohibitive in terms of time. To this end, SLM Solutions Gmbh have recently released their flagship NXG XII 600 PBF system, boasting 12 1kW lasers and a 60cm × 60cm × 60cm build volume. The process here is identical to regular, single-laser PBF, except multiple lasers are scanning on each layer to speed up the melting process. Almost all other commercial PBF system manufacturers have likewise released multi-laser machines. In other cases, the build process has been altered more fundamentally in pursuit of enhanced productivity. An example here is the development of a volumetric SLA-type process, based on 3D holographic light projection, at Lawrence Livermore National Laboratory (LLNL). Here, three patterned light fields are projected into the vat of photopolymeric resin at orthogonal angles. At the intersection of the three projects, a 'hologram' of the part is formed, curing the resin in a single step in seconds [14]. In doing so, the layerwise paradigm of AM is bypassed entirely.

Nonetheless, additional complexities are introduced by such innovations. The introduction of multiple lasers poses questions about scanning strategies and how the lasers might interact and overlap during scanning. Porosity generation and residual stress distributions in parts are likely to be influenced by new scanning modes [18], [19]. Melting such vast quantities of material will also create challenges in the ancillary elements of processing. More complex gas flow arrangements will be required to cope with multiple sites of soot and spatter emission in multi-laser machines. Greater quantities of soot and process emissions will be generated, necessitating greater filtration of the processing atmosphere.

1.4.2 Design Considerations

AM offers the potential for unparalleled geometric freedom in parts. However, for this potential to be realized we must first be capable of designing topologically complex components with enhanced functionality over existing designs. As such, the field of 'topological optimization' has emerged with the advent of AM. Here, research is conducted into designing optimal structures to save weight, integrate assemblies or improve part performance based on a set of design constraints [20]. The design of porous, lattice structures forms a major feature of this work and these are of particular interest in the field of biomechanics. The porous nature of such structures is ideal to promote bone growth [21] and their stiffness can be tailored to mimic the behavior of bone [22]. Architected hierarchical metamaterials have also been presented, whereby different arrangements of periodic lattice structures can be arranged to enable programmable material deformation and failure [23]. A major challenge of this work is rooted in software, given the tools to mathematically represent, then perform simulations with such structures that did not previously exist. Parametric design software has emerged and has been adapted accordingly. Lattice-specific build preparation software has also been developed to produce optimal tool paths for lattice manufacture, as different processing strategies for bulk material are required.

Another avenue of great potential in the design of AM components is functionally graded materials (FGM) or structures (FGS). FGM and FGS exhibit variation, or grading, in their structure and/ or properties throughout their expanse. This variation can be programmed inherently into the design of the part or during the build setup stage, by altering the processing parameters to be used at different points in the part. For example, the periodic unit cells of lattice structures can be altered and arranged throughout a component to lead to different deformation behavior in different regions. The composition of the feedstock material can also be adjusted during the process to change the properties of the part in different regions. This is particularly straightforward during PBF and MEX processes when more than one feedstock material can easily be used. Finally, the microstructure in metallic components can be manipulated during manufacture by altering the processing parameters. A method to program and spatially vary the crystallographic grain orientation in PBF specimens by changing the angle of laser scanning has been presented in academic literature [24]. The same group also demonstrated the influence of crystallographic grain orientation on material hardness in the same alloy [25], presenting a pathway to tailoring material properties in PBF components.

As with any manufacturing process, manufacturing constraints must also be considered when designing components to be produced by AM. This has given rise to the term 'Design for Additive Manufacturing', or DfAM, born out of the original 'Design for Manufacturing'. Although many of the design constraints associated with traditional manufacturing processes are indeed removed by AM, several new considerations are introduced. The orientation in which a part is built with respect to the build direction can substantially affect its surface finish, residual stress, and microstructure and must be carefully considered during part design. Powder removal must also be considered for powder-based processes. Warpage of parts can also cause issues and some academic groups have simulated deformations before manufacture and compensated the design accordingly. A lack of knowledge of how to optimize part designs for AM remains an issue that plagues the field [26].

1.4.3 Materials

A good deal of research activity has also focused on the feedstock materials available for AM. This is a particularly fruitful area of development in the field of metals, where the first generation of alloys available was highly restrictive compared to that which is available for other manufacturing processes. Initially, conventional weldable alloys were ported over to AM and were largely all that was available. Several researchers have studied the adaptation of existing alloys to enhance their printability. Minor tweaks in composition have often resulted in drastic improvements in printability, mitigating cracking and porosity [27]–[29].

The rapid heating and cooling inherent in processing metals by PBF and DED techniques have also enabled new material systems to be created. A relatively new class of alloys has emerged as a promising candidate for AM, named High Entropy Alloys (HEA). These alloys tend to have roughly equal proportions to each constituent alloying element. Although they emerged academically in the 1980s, they were previously very challenging to process. They are straightforward to produce via AM and offer the potential for improved mechanical performance over conventional alloys in several areas. Another promising line of research investigated the use of graphene oxide as an additive to enhance the laser irradiation absorptivity of powders in PBF. This could enable different wavelengths of laser light to be used and greatly expand the range of possible feedstock materials [30].

Aside from metals, novel ceramic and polymer systems have also been developed. AM lends itself particularly well to the exploration of novel materials [11]. Different powders can rapidly be mixed and processed by PBF, for example, or new photopolymers mixed in an SLA vat. Compositions can therefore be varied more quickly than before.

1.4.4 Process Monitoring and Control

Although the accuracy and consistency of the different AM processes have been improved considerably since they were first invented, problems with the quality and consistency of the parts produced remain. Parts may frequently contain pores, defects, or other damage and be out of geometric tolerance. These problems continue to hold the technology back from wider dissemination into high-performance or safety-critical environments. Aside from this, issues often develop during the manufacturing process and can be substantial enough to cause the build to fail and require aborting. For example, metal parts can warp to such an extent that they collide with the recoater during PBF, or polymer parts can become detached from the substrate in MEX processes. The current generation of machines is relatively dumb and unable to detect such failures, costing the user time and money.

In-situ process monitoring has been identified by many as the remedy to the above issues. As such, much academic research has been conducted in this area over the past decade. Various types of sensors have been placed in the different AM systems, including cameras, photodiodes, different pyrometry setups, acoustic sensors, and ultrasonic sensors [31]–[33]. The goal here is to try and determine when a process has failed, or a defect has been induced in a part, based on a recorded signal of the sensor. Although some successes have been reported in detecting gross process failure, detecting defect formation has largely proven to be very difficult to achieve reliably, despite the degree of research conducted. Given the volume and complexity of the process data captured in-situ, machine learning (ML) methods have frequently been employed in analyses and making predictions [34]. The next step would be to implement a process control strategy, making use of a feedback loop. Some early work has been presented in this area [35]. Closed-loop process control is seen by many as critical in ensuring AM reaches its full potential. Although monitoring systems have been explored across all the different process categories, they have been a particular focus in the processing of metals by PBF and DED. This is because metal parts are often used in the most demanding, high-integrity applications and thus quality assurance is most critical in this class of materials.

1.4.5 Process Automation and Industry 4.0

Following on directly from the previous section, achieving closed-loop process control is also highly desirable in the context of process automation and Industry 4.0. Once 3D printers can correct their own mistakes and run with minimal human input, AM workflows have the potential to become highly automated. Arrays of machines could run unattended for long durations, requiring operator input only to set up and clean away the build. Corresponding developments to better automate some of the ancillary processes of AM, such as build setup, feedstock handling, and even post-processing, would also be useful. AM system manufacturers have made some progress on this front. A small number of 'one-stop shop' systems have emerged, with different processing stations joined sequentially to improve workflow. Expanding the intelligence and degree of autonomy of 3D printers also offers the potential for developing remote, deployable printers. Several authors have proposed the idea of deploying autonomous printers to build or repair structures in remote or hazardous locations, such as in space [36].

There is also tremendous potential presented by the synergy of AM and the Internet of Things (IoT). AM processes are inherently digital and once equipped with appropriate sensing and artificial intelligence capabilities have the potential to transmit and receive data and make decisions on the fly. AM systems could therefore receive logistics and inventory data or feedback directly from consumers and adapt their processes accordingly. The wealth of data captured during manufacture could likewise be used to integrate with the rest product manufacturing chain, optimize product designs and understand their lifespan and degradation better [37]. The lack of any tooling requirements in AM can also allow manufacturers to switch designs seamlessly. Alongside process automation and connectivity over IoT, this could enable agile and decentralized manufacturing supply chains to emerge. However, it must be emphasized that many of these ideas remain far on the horizon at present. Considerable technical research and development are required to turn these visions into reality and, again, the development of process monitoring systems sits central to this. The economics, and profitability, of these new manufacturing routes and supply chains, must also be established and will drive their development.

1.5 CONCLUSION

In this chapter, we have introduced the field of AM and discussed its context within the wider world of manufacturing today. The typical workflow involved in producing a component by AM has been outlined and the seven different process categories of AM were outlined. Each process has its unique operating principle, feedstock materials, and post-processing requirements. These factors largely determine which is the most suitable process to manufacture a given product.

Finally, we looked at some of the recent trends and developments across the different processes. In general, there has been a movement of increasing the size (i.e., the build volume) and productivity of AM systems. As industrial adoption of AM has increased, there has been a desire to improve part throughput and compete with conventional higher volume processes. However, larger machines introduce new challenges for part designers and process setups must be optimized to avoid escalating costs and build durations.

Further research and development of materials and design for AM are required to realize its full potential. The palette of available materials must continue to expand, particularly in the domain of metals. New AM-specific material systems should emerge. Software tools enabling straightforward design and simulation of complex structures are still lacking.

In-situ process monitoring and control systems have been identified as critical in enabling greater industrial adoption of AM. Closed-loop process control, while challenging to develop, may offer the solution to chronic issues with the quality and consistency of AM parts. Increased automation and sensor data capture will also allow closer integration of AM with the IoT ecosystem. In the future, this could allow agile, decentralized supply change to emerge, as envisioned in Industry 4.0. For now, however, this remains on the horizon.

REFERENCES

1 ISO/ASTM, "ISO/ASTM52921-13(2019) Standard Terminology for Additive Manufacturing – Coordinate Systems and Test Methodologies," 2019.

2 K. -S. Min, K. -M. Park, B. -C. Lee and Y. -S. Roh, "Chloride diffusion by build orientation of cementitious material-based binder jetting 3D printing mortar," *Materials*, vol. 14, no. 23, p. 7452, 2021.

3 M. Ziaee and N. B. Crane, "Binder jetting: A review of process, materials, and methods," *Additive Manufacturing*, vol. 28, pp. 781–801, 2019.

4 T. Childerhouse and M. Jackson, "Near net shape manufacture of titanium alloy components from powder and wire: A review of state-of-the-art process routes," *Metals*, vol. 9, no. 6, p. 689, 2019.

5 R. Savinov, Y. Wang, J. Wang and J. Shi, "Comparison of microstructure and properties of CoCrFeMnNi high-entropy alloy from selective laser melting and directed energy deposition processes," *Procedia Manufacturing*, vol. 53, pp. 435–442, 2021.

6 A. Baghdadchi, V. A. Hosseini, M. A. Valiente Bermejo, B. Axelsson, E. Harati, M. Högström and L. Karlsson, "Wire laser metal deposition of 22% Cr duplex stainless steel: As-deposited and heat-treated microstructure and mechanical properties," *Journal of Materials Science*, vol. 57, no. 21, pp. 9556–9575, 2022.

7 S. L. Sing, F. E. Wiria and W. Y. Yeong, "Selective laser melting of titanium alloy with 50 wt% tantalum: Effect of laser process parameters on part quality," *International Journal of Refractory Metals and Hard Materials*, vol. 77, pp. 120–127, 2018.

8 S. Tekumalla, R. Tosi, X. Tan and M. Seita, "Directed energy deposition and characterization of high-speed steels with high vanadium content," *Additive Manufacturing Letters*, vol. 2, p. 100029, 2022.

9 A. Saboori, A. Aversa, G. Marchese, S. Biamino, M. Lombardi and P. Fino, "Application of directed energy deposition-based additive manufacturing in repair," *Applied Sciences*, vol. 9, no. 16, p. 3316, 2019.

10 T. J. Gordelier, P. R. Thies and L. Johanning, "Optimising the FDM additive manufacturing process to achieve maximum tensile strength: A state-of-the-art review," *Rapid Prototyping Journal*, vol. 6, pp. 953–971, 2019.

11 J. M. Lee, S. L. Sing and W. Y. Yeong, "Bioprinting of multimaterials with computer-aided design/computer-aided manufacturing," *International Journal of Bioprinting*, vol. 6, no. 1, pp. 65–73, 2020.

12 W. Liu, H. Song and C. Huang, "Maximizing mechanical properties and minimizing support material of PolyJet fabricated 3D lattice structures," *Additive Manufacturing*, vol. 35, p. 101257, 2020.

13 A. Bournias-Varotsis, R. J. Friel, R. A. Harris and D. S. Engstrøm, "Ultrasonic additive manufacturing as a form-then-bond process for embedding electronic circuitry into a metal matrix," *Journal of Manufacturing Processes*, vol. 32, pp. 664–675, 2018.

14 M. Shusteff, A. E. Browar, B. E. Kelly, J. Henriksson, T. H. Weisgraber, R. M. Panas, N. X. Fang and C. M. Spadaccini, "One-step volumetric additive manufacturing of complex polymer structures," *Science Advances*, vol. 3, no. 12, p. eaao5496, 2017.

15 B. C. Gross, J. L. Erkal, S. Y. Lockwood, C. Chen and D. M. Spence, "Evaluation of 3D printing and its potential impact on biotechnology and the chemical sciences," *Analytical Chemistry*, vol. 86, no. 7, pp. 3240–3253, 2014.

16 N. Hopkinson and P. Dickens, "Analysis of rapid manufacturing – Using layer manufacturing processes for production," *Proceedings of the Institution of Mechanical Engineers, Part C: Journal of Mechanical Engineering Science*, vol. 217, no. 1, pp. 31–40, 2003.

17 J. I. Novak and J. O'Neill, "A design for additive manufacturing case study: fingerprint stool on a BigRep ONE," *Rapid Prototyping Journal*, vol. 25, no. 6, pp. 1069–1079, 2019.

18 J. Yin, D. Wang, H. Wei, L. Yang, L. Ke, M. Hu, W. Xiong, G. Wang, H. Zhu and X. Zeng, "Dual-beam laser-matter interaction at overlap region during multi-laser powder bed fusion manufacturing," *Additive Manufacturing*, vol. 46, p. 102178, 2021.

19 W. Zhang, M. Tong and N. M. Harrison, "Scanning strategies effect on temperature, residual stress and deformation by multi-laser beam powder bed fusion manufacturing," *Additive Manufacturing*, vol. 36, no. June, p. 101507, 2020.

20 J. Zhu, H. Zhou, C. Wang, L. Zhou, S. Yuan and W. Zhang, "A review of topology optimization for additive manufacturing: Status and challenges," *Chinese Journal of Aeronautics*, vol. 34, no. 1, pp. 91–110, 2021.

21 S. Lu, D. Jiang, S. Liu, H. Liang, J. Lu, H. Xu, J. Li, J. Xiao, J. Zhang and Q. Fei, "Effect of different structures fabricated by additive manufacturing on bone ingrowth," *Journal of Biomaterials Applications*, vol. 36, no. 10, pp. 1863–1872, 2022.

22 S. L. Sing, "Perspectives on additive manufacturing enabled beta-titanium alloys for biomedical applications," *International Journal of Bioprinting*, vol. 8, no. 1, pp. 1–8, 2022.

23 M. S. Pham, C. Liu, I. Todd and J. Lertthanasarn, "Damage-tolerant architected materials inspired by crystal microstructure," *Nature*, vol. 565, no. 7739, pp. 305–311, 2019.

24 K. A. Sofinowski, S. Raman, X. Wang, B. Gaskey and M. Seita, "Layer-wise engineering of grain orientation (LEGO) in laser powder bed fusion of stainless steel 316L," *Additive Manufacturing*, vol. 38, p. 101809, 2021.

25 S. Tekumalla, B. Selvarajou, S. Raman, S. Gao and M. Seita, "The role of the solidification structure on orientation-dependent hardness in stainless steel 316L produced by laser powder bed fusion," *Materials Science and Engineering A*, vol. 833, p. 142493, 2022.

26 L. E. Thomas-Seale, J. C. Kirkman-Brown, M. M. Attallah, D. M. Espino and D. E. Shepherd, "The barriers to the progression of additive manufacture: Perspectives from UK industry," *International Journal of Production Economics*, vol. 198, no. February, pp. 104–118, 2018.

27 N. J. Harrison, I. Todd and K. Mumtaz, "Reduction of micro-cracking in nickel superalloys processed by Selective Laser Melting: A fundamental alloy design approach," *Acta Materialia*, vol. 94, pp. 59–68, 2015.

28 M. L. Montero Sistiaga, R. Mertens, B. Vrancken, X. Wang, B. Van Hooreweder, J. P. Kruth and J. Van Humbeeck, "Changing the alloy composition of Al7075 for better processability by selective laser melting," *Journal of Materials Processing Technology*, vol. 238, pp. 437–445, 2016.

29 W. Yu, Z. Xiao, X. Zhang, Y. Sun, P. Xue and S. Tan, "Processing and characterization of crack-free 7075 aluminum alloys with elemental Zr modification by laser powder bed fusion," *Materials Science in Additive Manufacturing*, vol. 1, no. 1, pp. 1–11, 2022.

30 C. Lun, A. Leung, I. Elizarova, M. Isaacs, S. Marathe, E. Saiz and P. D. Lee, "Enhanced near-infrared absorption for laser powder bed fusion using reduced graphene oxide," *Applied Materials Today*, vol. 23, p. 101009, 2021.

31 R. McCann, M. A. Obeidi, C. Hughes, É. McCarthy, D. S. Egan, R. K. Vijayaraghavan, A. M. Joshi, V. Acinas Garzon, D. P. Dowling, P. J. McNally and D. Brabazon, "In-situ sensing, process monitoring and machine control in Laser Powder Bed Fusion: A review," *Additive Manufacturing*, vol. 45, p. 102058, 2021.

32 R. J. Williams, A. Piglione, T. Rønneberg, C. Jones, M. S. Pham, C. M. Davies and P. A. Hooper, "In situ thermography for laser powder bed fusion: Effects of layer temperature on porosity, microstructure and mechanical properties," *Additive Manufacturing*, vol. 30, p. 100880, 2019.

33 A. Remani, R. Williams, A. Thompson, J. Dardis, N. Jones, P. Hooper and R. Leach, "Design of a multi-sensor measurement system for in-situ defect identification in metal additive manufacturing," in *ASPE/euspen Advancing Precision in Additive Manufacturing*, St. Gallen, Switzerland, 2021.

34 S. L. Sing, C. N. Kuo, C. T. Shih, C. C. Ho and C. K. Chua, "Perspectives of using machine learning in laser powder bed fusion for metal additive manufacturing," *Virtual and Physical Prototyping*, vol. 16, no. 3, pp. 372–386, 2021.

35 D. A. Brion, M. Shen and S. W. Pattinson, "Automated recognition and correction of warp deformation in extrusion additive manufacturing," *Additive Manufacturing*, vol. 56, p. 102838, 2022.

36 M. Yashar, C. Ciardullo, M. Morris, R. Pailes-Friedman, R. Moses and D. Case, "Mars X-House: Design principles for an autonomously 3D-printed ISRU surface babitat," in *49th International Conference on Environmental System*, Boston, Massachusetts, 2019.

37 M. Khorasani, J. Loy, A. H. Ghasemi, E. Sharabian, M. Leary, H. Mirafzal, P. Cochrane, B. Rolfe and I. Gibson, "A review of Industry 4.0 and additive manufacturing synergy," *Rapid Prototyping Journal*, vol. 28, no. 8, pp. 1462–1475, 2022.

2 Dimensional Aspect of Feedstock Material Filaments for FDM 3D Printing of Continuous Fiber-Reinforced Polymer Composites

Anis A. Ansari[1] *and Mohammad Kamil*[2,*]
[1]Mechanical Engineering Section, University Polytechnic, Faculty of Engineering and Technology, Aligarh Muslim University, Aligarh 202002, India
[2]Department of Petroleum Studies, Zakir Husain College of Engineering and Technology, Aligarh Muslim University, Aligarh 202002, India
[*]Corresponding author: mkamilamu@gmail.com, smkamil@zhcet.ac.in

CONTENTS

DOI: 10.1201/9781003296676-2

2.1 INTRODUCTION

3D printing has made it possible to produce complex geometries to final shape without the use of specific tools or molds. Printing composite materials is a new area of study, with few commercially accessible printer types for making continuous fiber composites and only a few in development. 3D printing is a pretty challenging technique. Typically, the trial-and-error method is used to find the components (material, printer, process settings, post-processing) that can provide the desired quality result. Making decisions based on available knowledge from prior research projects in this field is one strategy for reducing the number of iterations involved with this experimental approach. As a result, there is a need to approach the 3D-printing process from many angles to improve understanding and knowledge in the development process of 3D-printed functional parts. The current utility of 3D printing has reached almost every trade sector ranging from the toy industry to space [1]. A brief description of the present industrial applications is listed in Table 2.1.

TABLE 2.1
Current Applications of 3D Printing

Industries	Applications
Aerospace	• Production of parts with complex geometry • Production of light-weight parts at the desired part density • Fabrication of components at space stations • Lattice truss structures
Medical/Pharmaceutical	• Accurate anatomic models for surgery • Use of printed corpses to understand the anatomy • Dental crowns • Knee implants • Patient-specific hearing aids, orthotic insoles for footwear, prosthetics • Bone, tissue, and organ 3D printing
Sports	• Production of accessories and protective equipment • prototypes for product testing
Automobile	• Integration of various parts in a complex assembly • Production of spare parts and accessories
Chemical/Petrochemical	• Structured catalysts • Metal electrodes • Directly printed catalytically functional structures • 3D printed heat exchangers
Construction	• Novel design of concretes • self-cleaning concrete • High-performance concrete • Cement free building • 3D printed houses
Education	• Models for teaching and training in various disciplines • Innovative products development
Fashion	• Personalized buttons, bangles, watches, hair clips, Jewelry, shoes, etc. • Dresses
Food industry	• Chocolates with personalized shapes and designs • Pasta, meat, etc.
Electronics	• Printed light-emitting diodes, solar cells • Integrated circuits • Flexible electronic circuits, flexible capacitors

Among the various 3D-printing techniques, fused deposition modeling (FDM) is more cost-effective than other production processes. Many researchers are focusing on FDM due to its broad applicability and affordability. The use of this method on a larger industrial scale resulted in a variety of demands that have been handled by industries. Most printer manufacturers, including Stratasys, Ultimaker, and Markforged are constantly working to improve their printing machines. However, the development of continuous fiber-reinforced composites is a pressing requirement to broaden the use of 3D-printed FDM parts for high performance [2]. The dimensional superiority of the 3D-printed part is a function of the feedstock dimensional precision, and to meet the performance criteria, dimensional accuracy and stability of the printed part are necessary for overall part quality [3], [4]. Therefore, the present study emphasizes the dimensional aspect of the different feedstock filaments used to fabricate the continuous fiber-reinforced polymer composites.

2.2 COMPOSITE 3D PRINTING

2.2.1 Evolution and Commercialization

The FDM technique is well-considered and the most widely used method for fabricating composite parts due to its cost and convenience. Because of its weak mechanical qualities, FDM printed parts are typically utilized for prototype component manufacture. Different reinforcements, including carbon black, platelets, chopped fibers, and polymer fibrils, are mixed with the thermoplastic matrix and then extruded together during printing to improve the performance of the printed polymer. The fiber orientation and fiber volume fraction have a considerable impact on the performance of these short fiber reinforced composites. However, they still outperform conventional fiber-reinforced composites in terms of mechanical performance. Compared to the conventional fabrication processes of fiber-reinforced polymers, the FDM process can provide complex, functional, and structural parts [5]. FDM machines are also available with dual extrusion nozzles with the provision to use two different materials and colors as and when required. Table 2.2 represents the factors and possible variables common to the FDM process [6].

The inception of 3D printing took place from 1981 onwards and introduced Selective Laser Sintering (SLS) and fused deposition modeling (FDM) during the first decade. The first patent titled 'apparatus for production of three-dimensional objects of stereolithography' was issued to

TABLE 2.2
Various Factors Affecting Mechanical Performance

Factors	Variables
Material	• Matrix type • Reinforcement type • Humidity
Printer	• Nozzle diameter • Envelop temperature • Bed temperature
Process parameters	• Nozzle temperature • Print speed • Layer thickness • Raster width and angle • Raster to raster air gap • Build orientation • Number of outer shells • Reinforcement orientation

Chuck Hull in 1986 for the first commercial 3D printer. Later on in 1989, Scott Crump invented a 3D desktop printer that was commercialized by Stratasys company. The cost of FDM 3D printers dropped significantly in the year 2012–2013. In 2011, Time magazine presented a list of the 50-best invention of the year, including a 3D-printed dress. In 2014, the world's first carbon fiber 3D printer changed the face of additive manufacturing forever. Carbon fiber is one of the most valuable materials available to a part designer. It is also one of the most expensive because of the time-consuming process of forming or winding it. It would be truly astonishing if carbon fiber could simply be printed. The first 3D printer for carbon fiber was introduced at the SolidWorks World 2014 conference in San Diego, USA. 3D printers can produce parts with a more excellent strength-to-weight ratio. Gregory Marks (founder of Markforged company) named his new brainchild 'Mark One' [7]. In September 2014, the world's first 3D-printed automobile was tested at the International Manufacturing Technology Show (IMTS) in Chicago. According to reports, the complete car body was created on-site in 44 hours using a 3D printer [8]. Furthermore, Matsuzaki's 2016 disclosure of 3D-printed continuous fiber-based composites was a great addition to tailoring the mechanical characteristics of printed parts with a complicated topology, which may lead to a future materials manufacturing sector [9]. In the recent past, the increased research exposure regarding 3D-printing technology caused making tools on-demand, Urbee (the first 3D-printed car), 3D-printed rocket fuel injector, lunar habitation by European Space Agency (ESA), and life-saving 3D-printed splint. The market studies for the past decade revealed a potential future scope for 3D printing and expected an increase almost exponentially in the coming years.

2.2.2 Components of FDM 3D Printer

While FDM printers use three different electromechanical systems, such as extrusion, Z motion, and XY gantry motion [10]. The operation of these systems determines the quality of the printed part. The main components of an FDM 3D printer are briefly described as follows:

- **Extrusion system**: The extrusion system heats and extrudes the polymer material through the nozzle. The print speed and accuracy of the printer are influenced by factors like nozzle size and extrusion speed.
- ***z* motion system**: Material is deposited onto the print bed via the print head and extrusion mechanism. The z motion mechanism moves the bed up and down in discrete and equal steps to produce the layers.
- ***xy* gantry motion system**: It controls the movement of the print head in x- and y-directions. It moves the tool pathways that produce each layer. The rigidity of the gantry affects the precision and accuracy of the printed part. 3D printers can build more accurate geometries on the x-y plane than on x-z or y-z planes.
- **Print head**: The print head is the component that is responsible for depositing the thermoplastic melt through a nozzle and it is the most sophisticated component in an FDM 3D printer.
- **Print bed**: This is the supporting plate that holds the extruded filament coming through the nozzle. This helps in building the part layer by layer by its displacement to the nozzle position in z-direction.

2.2.3 Continuous Fiber Reinforcement

In commonly available Markforged printers, the open-source Eiger slicing software is used. Markforged printers offer unidirectional (or isotropic) and concentric fiber patterns. Within the same layer, combinations of two patterns are also possible. With a resolution of 0.01 degrees, fibers can be placed at any angle. In the case of concentric fiber arrangement, it takes the outside geometry and offsets it inward to place the fiber. It is possible to control the number of fiber rings in each layer.

Three different fiber patterns are printed, namely concentric fiber and isotropic fiber. The isotropic fiber angle can be adjusted in any desired direction as needed. The fibers are continuous in each layer, but they do not span the layers. When creating a fiber layer, the fiber end is placed down and compressed. It causes an ironing action that alters the fiber filament's geometry slightly elliptical. After that, a blade cuts the fiber and then moves on to the next layer. The matrix can be nylon or PLA, while the fiber can be carbon, glass, or Kevlar. The number of fiber layers in each specimen can be changed to a required composition of fiber volume fraction [11].

2.3 FEEDSTOCKS FOR POLYMER COMPOSITE

2.3.1 Polymer Filaments (Continuous Phase)

Several thermoplastic materials are used in the FDM 3D-printing process. However, it is important to understand the properties of matrix material for a specific application based on the function and purpose of the printed part. Therefore, some most popular filament materials used in FDM 3D printers are described as follows:

1. **Polylactic acid (PLA)**: PLA is the most commonly used filament material and has a low melting point of 145–186 °C. It can be easily produced into filament at temperatures over 185–190 °C. PLA's biocompatibility and mechanical qualities (high strength and modulus) make it suitable for industrial and biomedical applications.
2. **Acrylonitrile butadiene styrene (ABS)**: It is an amorphous polymer based on styrene, acrylonitrile, and polybutadiene. ABS is more appealing for use in FDM because it not only has greater strength and toughness than PLA, but it also has greater resistance to corrosive substances. However, it is a little difficult to print because to the tendency to warp caused by the high shrinkage factor.
3. **Nylon 6**: It is utilized in the 3D-printing industry because of its high strength, flexibility, and durability. However, it is sensitive to moisture and should be stored in a cold, dry environment for printing high-quality items.
4. **Polycarbonate (PC)**: PC is a 3D-printing engineering thermoplastic material that is popular due to its good mechanical qualities (toughness, flexural strength, and impact resistance) and wide heat resistance range of -150 C to 140 °C. As a result, it is commonly utilized in demanding applications such as functional testing and tooling. PC, on the other hand, is a hygroscopic material that absorbs moisture from the air. To avoid filament degradation, store it in a dry or airtight environment. It can also warp between the filaments and the construction plate or between filaments. Residual stress in the polymer can be formed when the PC is printed without a warm environment. As the printing progresses, residual stress finally overcomes the bed or inter-layer adhesion, resulting in deformation.
5. **Polyether-ether-ketone (PEEK)**: It is a poly aryl ether ketone (PAEK) semicrystalline polymer. Due to the high melting temperature of roughly 343 °C, PEEK is extremely difficult to print, however since the nozzle in the FDM printer has lately developed, it has led to the development of PEEK filaments for more generic FDM printers. Furthermore, the polymer's various advantages, such as good mechanical, chemical, and thermal qualities, allow it to be used in harsh situations demanding high service temperatures or mechanical properties, such as bone, bearing, automotive parts, and aviation components.
6. **Polyether-ketone-ketone (PEKK)**: It is a semicrystalline material that belongs to the poly aryl ether ketone (PAEK) class. PEKK, like PEEK, is an emerging material in the aerospace and tooling industries due to its excellent mechanical strength and chemical resistance. PEKK has a greater melting temperature than PEEK, around 385 °C. PEKK crystallizes at a significantly slower rate than PEEK and is less sensitive to chilling in a low-temperature build chamber below 200 °C. As a result, PEKK is well-known as an advanced alternative feedstock

for FDM since it can be handled in the same way as an amorphous polymer, resulting in strong layer adhesion and dimensional stability.

7. **Nanocomposite filament**: Various forms of nanofillers, including carbon nanomaterials, are used depending on the final product's function, such as artificial bone implants.
8. **Fibers-reinforced polymer composite filaments**: Because of their strong mechanical qualities while being lightweight, fiber-reinforced polymer composites offer considerable potential in structural industries such as aerospace, automotive, and energy applications. As reinforcements for printing composite parts, both discontinuous fibers (e.g., chopped glass fibers, chopped carbon fibers, and short basalt fibers) and continuous fibers (e.g., glass fiber, carbon fiber, and aramid fiber) are utilized. Nowadays, 3D printing for fiber-reinforced composites is being intensively researched to build structural applications that make use of 3D printing's superior mechanical qualities and design flexibility. Short fiber is especially appealing due to its ease of production, low cost, and better mechanical property.

2.3.2 Filament Fabrication Process

This process uses extruders that push the material through die holes to get the product as a filament extrudate [12]. The pure polymer filament is entirely composed of polymers with no additions. Each pure polymer filament has its own set of features and mechanical qualities. However, the intrinsic qualities of pure polymers cannot always fulfill the need for mechanical properties in specific products. This dilemma necessitates the industry to constantly develop polymer filaments acceptable for commercial markets. Adding chemicals to the filament mix can help improve mechanical qualities. This procedure resulted in the polymer composite filament. As a result, several academics and enterprises have produced polymer composites as 3D-printing filament material by mixing the matrix and upgrading the components to achieve systems with structural features and functional benefits.

2.3.3 Fibers (Discontinuous Phase)

Various fibers are used as a discontinuous phase in the fabrication of polymer composites. In most applications, synthetic fibers are generally used as reinforcement. Synthetic fibers are man-made fibers produced by chemical synthesis and classified as organic or inorganic based on their content. Fiber materials are significantly stronger and stiffer than matrix materials, making them a load-bearing component in the composite structure [5], [13]. Some of the well-known fiber materials are discussed below:

1. **Glass**: Glass fibers are the most extensively used synthetic fiber reinforcement because they provide exceptional strength and durability, thermal stability, impact resistance, chemical resistance, and wear properties. However, when using typical machining processes, cutting glass fiber-reinforced polymer is relatively slow and difficult, resulting in reduced tool life [14].
2. **Carbon**: Carbon fibers are used instead of glass fibers in applications where rigidity is required. Carbon fiber-reinforced polymers have extensive uses in the aircraft, automotive, sport, and various other industries [15].
3. **Kevlar**: Because of its anisotropic nature, Kevlar fiber has lower compression strength than glass and carbon fiber counterparts [16].
4. **Graphene**: Graphene fibers, when compared to carbon fibers, are a novel type of high-performance carbonaceous fiber with superior tensile strength and electrical conductivity. Increased graphene fiber properties offer promise in applications such as lightweight conductive cables and wires, knittable supercapacitors, micromotors, solar cell textiles, actuators, and so on [17].

5. **Basalt fiber**: Basalt fiber is a material formed from extremely fine basalt fibers that contain minerals such as plagioclase, pyroxene, and olivine. Basalt fiber outperforms glass fiber in terms of physical and mechanical qualities. Furthermore, BF is much less expensive than carbon fibers.

2.3.4 Commercially Available Continuous Fibers

Continuous fiber-reinforced composites outperform discontinuous fiber-reinforced composites in terms of performance. Continuous fibers have a high aspect ratio and are typically aligned in composites.

- **Glass fiber filament**: Glass fiber (also known as Fiberglass) is the most cost-effective reinforcement material. Reinforced fiber offers high strength at an affordable price. It is widely used in day-to-day applications for vital parts. Compared to Onyx, Fiberglass has 2.5 times the strength and eight times the rigidity of Onyx. It has Flexural Strength of 200 MPa and a modulus of 22 GPa.
- **Kevlar fiber filament**: Kevlar fiber is a lightweight, bendable fiber that offers excellent durability. It is ideal for parts subjected to repeated and sudden stress. Kevlar fiber is as rigid as glass fiber but much easier to mold. It has a flexural strength of 240 MPa and a modulus of 26 GPa.
- **Carbon fiber filament**: It possesses the highest strength and thermal conductivity. It is most suitable for stiffness and strength requirements. The carbon fiber reinforcement can produce parts with six times the strength and eighteen times the stiffness of Onyx. It has a flexural strength of 540 MPa and a flexural modulus of 51 GPa. This fiber reinforcement is commonly employed in parts that replace metal sections.
- **High-strength high-temperature glass fiber filaments (HSHT) filament**: HSHT fiberglass is a particular variant of glass fiber that provides strength like aluminum. It has five times the strength and seven times the rigidity of Onyx. Its application is preferred for parts exposed to high operating temperatures. It has a flexural strength of 420 MPa and a modulus of 21 GPa.

2.4 DIMENSIONAL ASSESSMENT OF PRINTING FILAMENTS

The feedstock filaments used to fabricate continuous fiber-reinforced polymer composites by fused deposition modeling technique can be classified into three categories: polymer filament, short fiber reinforced polymer filament, and continuous fiber filaments. The filament specimens for analyzing the dimensional aspect were taken from each of the above three categories. The feedstocks covered in the present study are PLA, Nylon, Onyx, Glass fiber, Kevlar fiber, and Carbon fiber.

2.4.1 Methodology

The dimensional analysis of the feedstock filaments was performed based on the following criteria:

1. Longitudinal uniformity
2. Cross-sectional geometry

Measurements were performed to study the accuracy and consistency of the dimensions in the feedstock filaments. Since the uniformity in the cross-sectional geometry of the filament is an important quality parameter, therefore, measurement points were specified at selected locations along the longitudinal direction and a particular cross-section. The measurement was carried out with the help of micrometer MDC-25PX (Mitutoyo, Japan), having a resolution of 0.001 mm.

For longitudinal accuracy, dimensional measurement was carried out in a 50 mm long piece taken from the filament. Five equally spaced locations (P1, P2, P3, P4, and P5) were marked as measurement points to observe the longitudinal uniformity in the feedstock filaments. The cross-sectional geometry of each filament type (polymer matrix filaments, polymer composite filaments, and fiber filaments) was analyzed at four diagonal positions (AA´, BB´, CC´, and DD´) of the circular cross-section. The angular displacement between every two successive sections is kept at 45°. A schematic diagram of the filament longitudinal and cross-sectional geometry is presented in Figure 2.1.

2.4.2 Polymer Matrix Filament

Three filament types from the polymer matrix were taken polylactic acid (PLA), Nylon (PA-6), and Onyx. Each of the filaments has a 1.75 mm filament diameter. PLA and Nylon are pure polymers, while Onyx is a fiber-reinforced polymer composite filament that can be considered a matrix material for continuous fiber-reinforced 3D printing. Onyx® is a trademark of Markforged (USA) for Nylon reinforced with short carbon fibers. An SEM microscopic image showing the Onyx filament cross-section is given in Figure 2.2 which indicates variation in cross-sectional dimension in two perpendicular directions. Moreover, a magnified image of the filament cross-section indicates embedded fibers oriented along the longitudinal direction.

Dimensional analysis of the filaments was carried out for uniformity and consistency. The longitudinal geometry of the feedstock filaments is measured at specified points P1 to P5. The measured values are presented in Table 2.3.

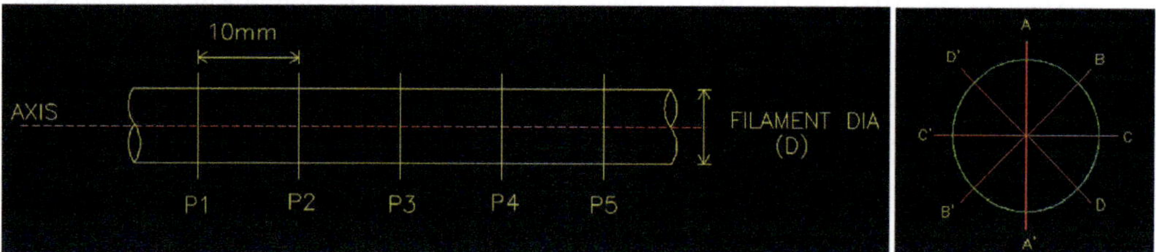

FIGURE 2.1 Schematic of the filament longitudinal geometry showing five location points P1 to P5 (left), and filament section at P1 (right).

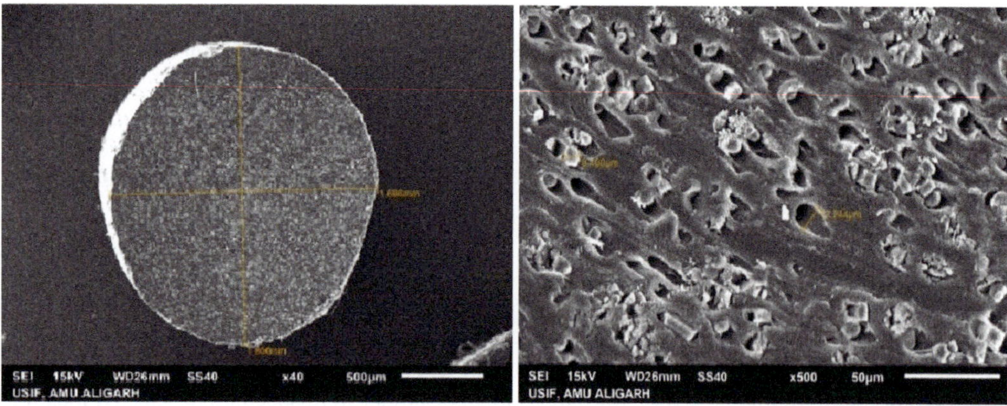

FIGURE 2.2 SEM images showing a single Onyx filament cross-section at magnification ×40 (left), and ×500 (right).

TABLE 2.3
Measured Polymer Matrix Filament Dimensions

	Diameter (mm)				
Filament Type	D_{P1}	D_{P2}	D_{P3}	D_{P4}	D_{P5}
PLA filament	1.751	1.751	1.776	1.767	1.774
Nylon filament	1.738	1.737	1.737	1.737	1.735
Onyx	1.756	1.776	1.774	1.765	1.768

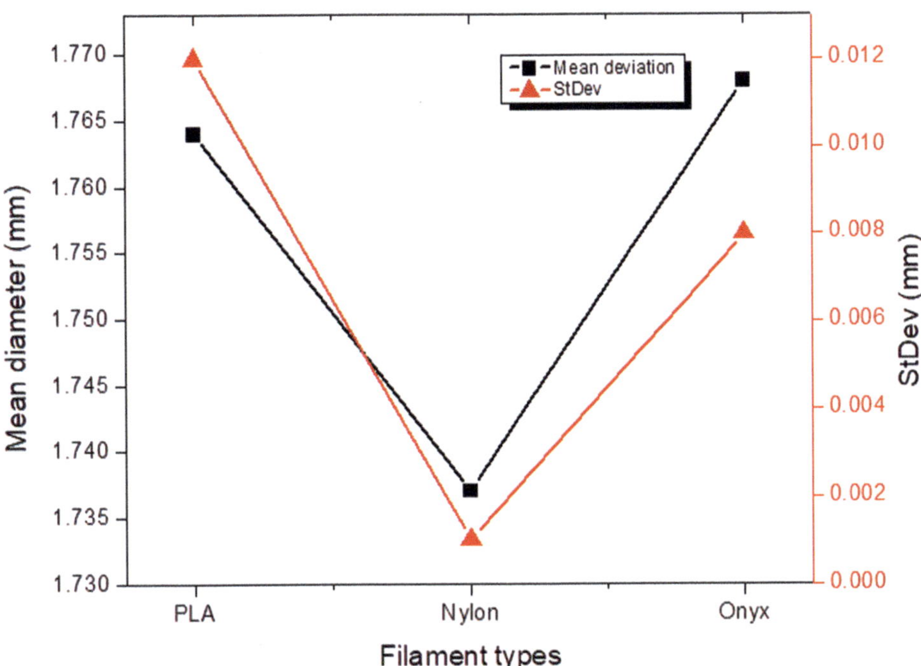

FIGURE 2.3 Comparison of filament mean diameter and longitudinal uniformity based on the standard deviation for PLA, Nylon, and Onyx.

After analysis of the obtained data, a comparison of mean diameter and standard deviation for PLA, Nylon, and Onyx filaments (as shown in Figure 2.3). The results indicate that the mean of PLA and Onyx possess higher dimensional values than the standard diameter (1.75 mm). On the other hand, Nylon filament is under dimension than the standard diameter and quite superior in terms of longitudinal uniformity as compared to the remaining filaments.

The cross-sectional geometry is analyzed through the measurement at a particular cross-section for different diagonal sections at AA´, BB´, CC´, and DD´. Based on the measured data, mean and standard deviations were evaluated for each filament. The measured dimensions along with their mean and standard deviation are given in Table 2.4.

2.4.3 Continuous Fiber Filament

Three types of continuous fibers, such as glass fiber (GF), Kevlar fiber (KF), and carbon fiber (CF) were taken for this study. A photograph of these continuous fibers is given in Figure 2.4.

TABLE 2.4
Measured Polymer Matrix Filament Cross-section, Mean, and Standard Deviation

	Diameter (mm)					
Polymer Matrix Type	$D_{AA'}$	$D_{BB'}$	$D_{CC'}$	$D_{DD'}$	Mean (mm)	StDev
PLA filament	1.771	1.762	1.775	1.79	1.775	0.012
Nylon filament	1.719	1.728	1.732	1.727	1.727	0.010
Onyx	1.618	1.657	1.745	1.715	1.684	0.057

FIGURE 2.4 Photograph of glass, Kevlar, and carbon fiber filaments.

For analyzing the longitudinal geometry of the fibers, five different locations along the longitudinal direction were assigned at an equal gap of 10 mm and marked as P1 to P5. The measured dimensional data is presented in Table 2.5.

A comparison of the longitudinal uniformity based on fiber mean diameter and the standard deviation is presented in Figure 2.5. It is observed that the mean diameter of the Kevlar fiber is slightly lesser as compared to the carbon fiber. The comparison of the standard deviation of the three fiber types indicates that carbon fiber has a minimum level of standard deviation while glass fiber has the maximum standard deviation. The diameter of the carbon fiber is significantly higher than the remaining fibers and the dimensional quality is also superior in comparison to glass and Kevlar fibers.

The cross-sectional geometry of the fibers was measured at four sectional lines as earlier. The measured dimensional data along with the mean and standard deviation is presented in Table 2.6. There is a huge difference between the standard deviation of carbon fiber, glass fiber, and Kevlar fiber. The higher standard deviation in the case of carbon fiber indicates the elliptical nature of the fiber. However, the other two fibers (KF and GF) are comparatively superior in terms of their circular geometry at the cross-section.

The obtained data for the specific sections indicate quite low dimensional variations in the fiber diameter at sections AA´, BB´, CC´, and DD´ for glass and Kevlar fibers. However, a higher standard deviation is found for carbon fiber which corresponds to a poor symmetrical geometry in comparison

TABLE 2.5
Measured Continuous Fiber Filament Dimensions

	Diameter (mm)				
Fiber Type	D_{P1}	D_{P2}	D_{P3}	D_{P4}	D_{P5}
GF	0.281	0.266	0.269	0.316	0.268
KF	0.279	0.281	0.285	0.249	0.274
CF	0.372	0.382	0.379	0.375	0.379

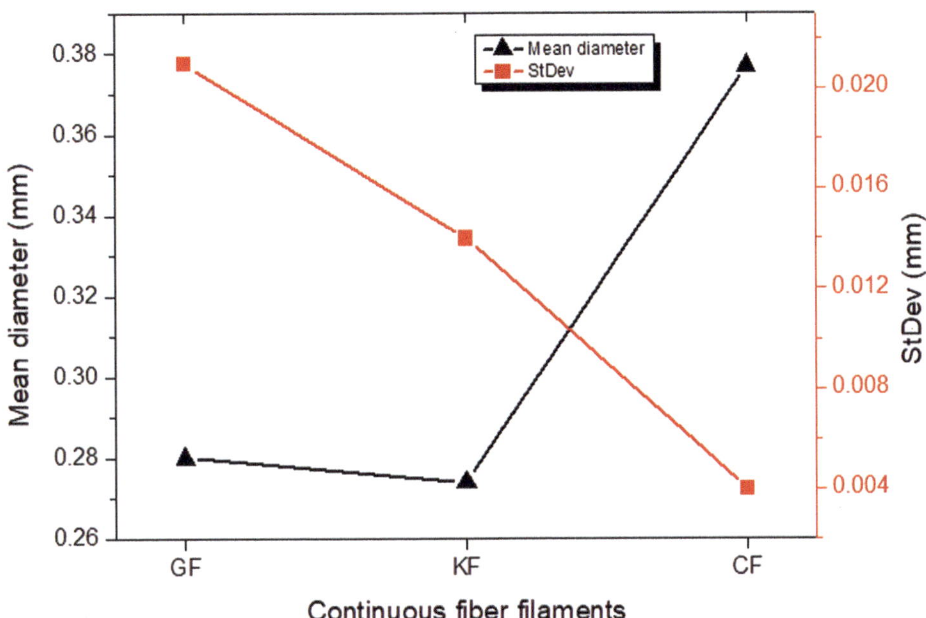

FIGURE 2.5 Comparison of mean diameter and longitudinal uniformity based on the standard deviation for GF, KF, and CF.

TABLE 2.6
Measured Fiber Cross-Section, Mean and Standard Deviation

	Diameter (mm)					
Fibers	$D_{AA'}$	$D_{BB'}$	$D_{CC'}$	$D_{DD'}$	**Mean (mm)**	**StDev**
GF cross-section	0.281	0.280	0.280	0.279	0.280	0.001
KF cross-section	0.279	0.277	0.277	0.277	0.278	0.001
CF cross-section	0.376	0.391	0.376	0.390	0.383	0.007

to glass and Kevlar fibers. Figure 2.6 shows SEM microscopic images of continuous fiber filament cross-section. The geometry of continuous fibers is seen with their actual dimensions. Also, the individual microfibers bundled within a single fiber are seen with bonding materials. Moreover, at a few locations, some void points are also observed [18].

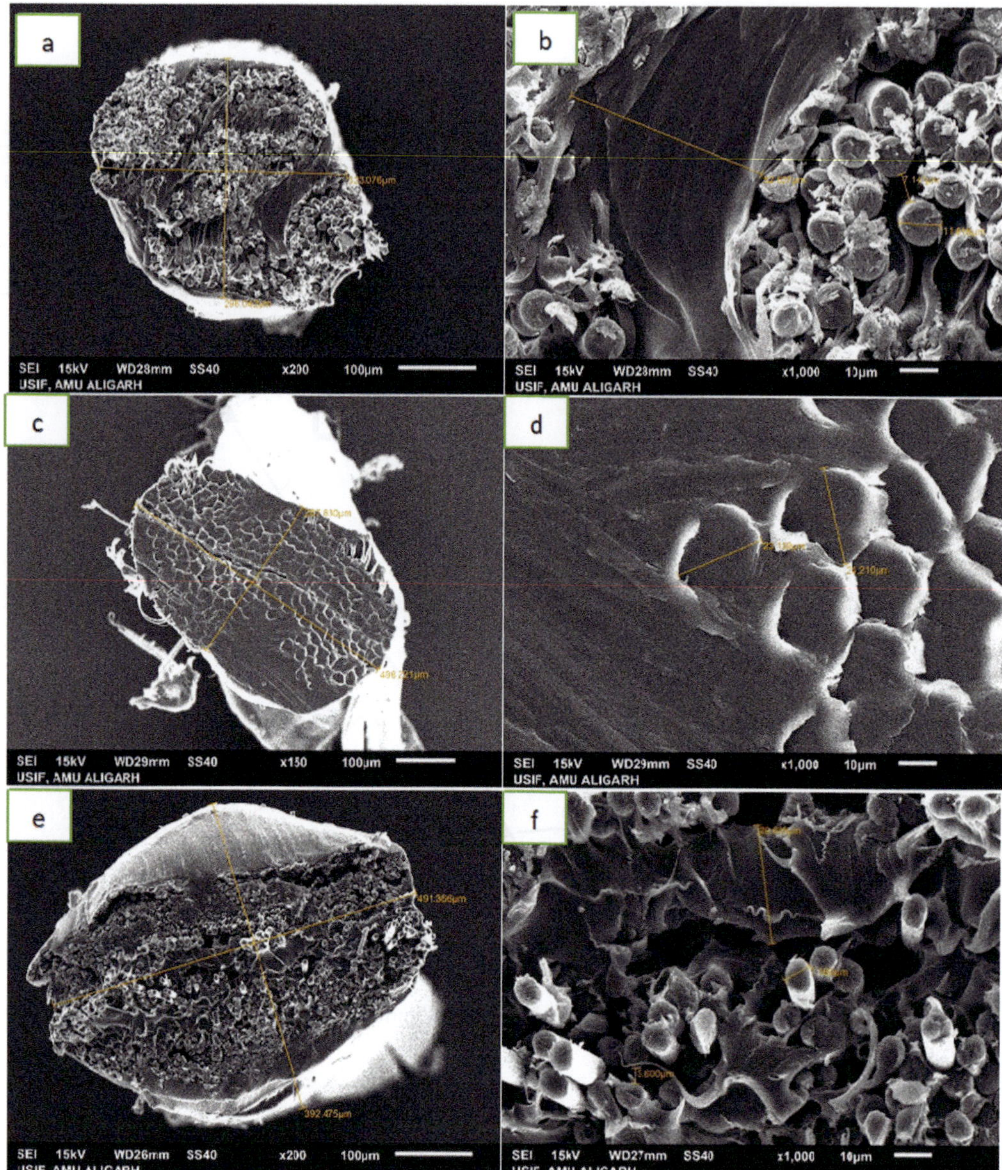

FIGURE 2.6 SEM images showing a fiber filament cross-section for (a) glass fiber at ×200, (b) glass fiber at ×1000, (c) Kevlar fiber at ×150, (d) Kevlar fiber at ×1000, (e) carbon fiber at ×200, and (f) carbon fiber at ×1000 magnification.

2.5 APPLICATIONS OF CONTINUOUS FIBER-REINFORCED POLYMER COMPOSITES

The continuous fiber-reinforced 3D-printed polymer composites have wide applications due to their higher mechanical performance. The application area covers a range of industries including aerospace, drone, sports, automobile, marine, petroleum, chemical, and other applications where lightweight materials with superior mechanical properties are required. Increased research exposure regarding this amazing technology caused the invention of 3D-printed cars, making tools on demand, rocket fuel injectors, lunar habitation, and life-saving dressings [19]–[21].

Some of the important applications are given below:

- Frame for drones
- Unmanned aerial vehicle (UAV)
- Propellers
- Structural components for airplanes
- Lightweight structures
- load-bearing applications
- high specific strength and stiffness
- Chassis for racing cars
- Medical tools and components
- Dentistry
- Wind turbines
- Satellite antennas

The applicability of composite 3D-printing technologies is enormous, but more industrial acceptance is still required so that machine manufacturers and users can adopt new materials-based technologies on composites with continuous reinforcing fibers.

2.6 SUMMARY

The present study shows an interesting aspect of the fibers used in 3D printing of continuous fiber-reinforced polymer composites. The glass and Kevlar fibers have superior cross-sectional geometry. In contrast, carbon fiber possesses superiority over glass and Kevlar fibers in terms of dimensional uniformity along their longitudinal direction. The results observed and presented are beneficial to the composite fabricators and researchers working in this area. As there was no significant study found in the literature stating the dimensional aspect of the feedstock filaments, there is tremendous scope to take forward this study by taking other useful filament materials for 3D printing. Researchers may take filaments, such as carbon fiber reinforced Polylactic acid, HSHT, and fire-resistant carbon fiber for the study of the dimensional aspect.

REFERENCES

1 A.A. Ansari and M. Kamil, 3D printed fiber-reinforced polymer composites by fused deposition modeling process – an overview, *MS Journal* 9 (2020) 957–970.

2 S.M.F. Kabir, K. Mathur, and A.F. Seyam, A critical review on 3D printed continuous fiber-reinforced composites: history, mechanism, materials and properties, *Compos. Struct.* 232 (2020) 1–24.

3 A. Bellini and S. Güçeri, Mechanical characterization of parts fabricated using fused deposition modeling, *Rapid Prototyp. J.* 9 (2003) 252–264.

4 A.A. Ansari and M. Kamil, Izod impact and hardness properties of 3D printed lightweight CF-reinforced PLA composites using design of experiment, *Int. J. Light. Mater. Manuf.* 5 (2022) 369–383.

5 C. Yang, X. Tian, T. Liu, Y. Cao, and D. Li, 3D printing for continuous fiber reinforced thermoplastic composites: mechanism and performance, *Rapid Prototyp. J.* 23 (2017) 209–215.

6 A.A. Ansari and M. Kamil, Effect of print speed and extrusion temperature on properties of 3D printed PLA using fused deposition modeling process, *Mater. Today Proc.* 45 (2021) 5462–5468.

7 www.extremetech.com/extreme/175518-worlds-first-carbon-fiber-3d-printer-demonstrated-could-change-the-face-of-additivemanufacturing-forever. (Accessed May 16, 2022).

8 A. Bandopadhyay and S. Bose, *Additive Manufacturing* (2016), Boca Raton: CRC Press, Taylor & Francis.

9 R. Matsuzaki, M. Ueda, M. Namiki, T.K. Jeong, H. Asahara, K. Horiguchi, T. Nakamura, A. Todoroki, and Y. Hirano, Three-dimensional printing of continuous-fiber composites by in-nozzle impregnation, *Sci. Rep.* 6 (2016) 1–7.
10 www.mark3d.com/nl/wp-content/uploads/sites/5/2020/10/Markforged-All-Products-3D-printing-September2020_EN_Mark3D.pdf. (Accessed May 16, 2022).
11 S.H. Sanei and D. Popescu, 3D-printed carbon fiber reinforced polymer composites: a systematic review, *J. Compos. Sci.* 4 (2020) 1–23.
12 R.B. Kristiawan, F. Imaduddin, D. Ariawan, Ubaidillah, and Z. Arifin, A review on the fused deposition modeling (FDM) 3D printing: filament processing, materials, and printing parameters, *Open Eng.* 11 (2021) 639–649.
13 K. Rajan, M. Samykano, K. Kadirgama, W.S. Harun, and M.M. Rahman, Fused deposition modeling: process, materials, parameters, properties, and applications, *Int. J. Adv. Manuf. Technol.* 120 (2022) 1531–1570.
14 P. Parandoush, Additive manufacturing of high-strength continuous fiber reinforced polymer composites, PhD Thesis (2019), Kansas State University, Manhattan, Kansas.
15 M.H. Hsueh, C.J. Lai, K.Y. Lin, C.F. Chung, S.H. Wang, C.Y. Pan, W.C. Huang, C.H. Hsieh, and Y.S. Zeng, Effects of printing temperature and filling percentage on the mechanical behavior of fused deposition molding technology components for 3D printing, *Polymers.* 13 (2021) 1–14.
16 T.J. Singh and S. Samanta, Characterization of Kevlar fiber and its composites: a review, *Mater. Today Proc.* 2 (2015) 1381–1387.
17 Z. Xu and C. Gao, Graphene fiber: a new trend in carbon fibers, *Biochem. Pharmacol.* 18 (2015) 480–492.
18 S.M.F. Kabir, K. Mathur, and A.F.M. Seyam, Maximizing the performance of 3d printed fiber-reinforced composites, *J. Compos. Sci.* 5 (2021) 1–18.
19 Q. Hu, Y. Duan, H. Zhang, D. Liu, B. Yan, and F. Peng, Manufacturing and 3D printing of continuous carbon fiber prepreg filament, *J. Mater. Sci.* 53 (2018) 1887–1898.
20 T. Yu, Z. Zhang, S. Song, Y. Bai, and D. Wu, Tensile and flexural behaviors of additively manufactured continuous carbon fiber-reinforced polymer composites, *Compos. Struct.* 225 (2020) 1–9.
21 H. Zhang, Recent progress of 3D printed continuous fiber reinforced polymer composites based on fused deposition modeling: a review, *J. Mater. Sci.* 56 (2021) 12999–13022.

3 Applications and Challenges of 3D Printing for Molecular and Atomic Scale Analytical Techniques

*Natalie Golota, Daniel Banks, Brian Michael, and Robert Griffin**
Francis Bitter Magnet Laboratory and Department of Chemistry, Massachusetts Institute of Technology, Cambridge, MA 02139, USA
Present addresses:
*Corresponding author: rgg@mit.edu

CONTENTS

3.1 INTRODUCTION

Over the past one hundred years, fundamental discoveries in the field of physical chemistry enabled the development and adoption of molecular spectroscopies based on the exposure to and subsequent absorption of electromagnetic radiation. The absorption of photons across the electromagnetic spectrum results in light–matter interactions which are then detected as a function of the frequency of irradiation. Different elements and molecules have distinct spectral signatures, enabling the identification and quantification of sample analytes ranging from small molecules to large macromolecular assemblies. These techniques have evolved to become critical, ubiquitous tools central to the development of modern science and medicine. They have enabled countless discoveries in various fields, including medicine, materials science, and chemistry.

Initially, access to spectroscopic devices was limited as spectrometers were research objectives in themselves and typically found only in research and development laboratories. However, between the 1950s and 1990s, there was rapid commercialization of scientific equipment through the founding of many large-scale scientific device manufacturers. This improved accessibility to spectroscopy and increased the rate of scientific discovery profoundly.

While the design of many spectrometers will be detailed further in the chapter, their fabrication has previously been dependent on traditional subtractive manufacturing methods requiring extensive

DOI: 10.1201/9781003296676-3

infrastructure to support. As commercial equipment began entering laboratories, machine shop facilities became less widespread and the number of laboratories developing their own customized spectrometers has decreased over time. As a result, while access to spectroscopy has increased, the ability to rapidly tailor instrumentation to experimental needs has decreased over time.

While this trend continued into the early 2000s, the commercialization and cost reduction of 3D printers has made them a standard and necessary resource in many scientific laboratories today. Technologies such as fused deposition model (FDM) and stereolithography (SLA) printers are most common in laboratories, however, 2-photon 3D printing has rapidly become a viable strategy uniquely suited for the miniaturization of scientific devices. Rapid, low-cost, and often in-house fabrication strategies using 3D printing have resulted in a rapid resurgence of customized instrumentation demonstrated by hundreds of publications each year incorporating 3D-printed components. Importantly, most published work contains open-access CAD models which further improve the accessibility of innovative instrumentation. Overall, 3D printing has not only furthered the customization of instrumentation but has also led to improved democratization of scientific equipment. For many labs, upgrades to spectrometers has become a cost-prohibitive step, however, when using 3D printing the cost is near zero, particularly when considering the availability of 3D printers on most university campuses.

Within this chapter, we discuss a small selection of the numerous creative and innovative uses of 3D printing for molecular scale analyses. Herein we focus on the application of printing to ultraviolet and visible spectrophotometry, infrared and Raman spectroscopy, mass spectrometry, and nuclear magnetic resonance.

3.2 UV/VIS SPECTROPHOTOMETRY

Ultra-violet and visible (UV/VIS) spectrophotometry is used to determine analyte compositions and concentrations of samples ranging from biomolecules to materials. UV/VIS measures the absorption or transmission of UV and visible light through the sample. The absorption of photons in this range of the electromagnetic spectrum can then be used to determine concentrations of the analyte of interest. This versatile technique is widely used in academic labs, industry, and clinical settings. Additionally, it is a foundational technique used to introduce chemistry and biology students to spectroscopic methods.

Typical spectrometers for UV/VIS in research settings cost between several hundred to several thousands of dollars and can take up a large footprint in the laboratory. All UV/VIS spectrometers use a UV/VIS light source which is then sent through a diffraction grating or monochromator to achieve spectral resolution in the UV/VIS electromagnetic region. The sample is typically contained in quartz, optical glass, or disposable plastic cuvette. The transmitted light is then recorded on a CCD detector and sample absorbance is used to determine analyte concentrations. While these simple components can become more sophisticated in high-quality research-grade equipment, the overall simplicity of the spectrometer makes it highly amenable to customization using 3D printing technology.

3D printing has accelerated the customization of UV/VIS spectrometers enabling researchers to increase the utility of these devices while decreasing the cost of spectroscopic consumables. A wide variety of printing technologies have been applied to overcome challenges related to printing air and watertight components, the use of highly transparent resins, and improved chemical resistance of printing resins. Moreover, 3D printing has accommodated the required flexibility of researchers interested in the integration of electrochemistry during UV/VIS, the study of samples in the gaseous phase or solid phases, under flow-cell conditions, and variable temperatures. While commercial suppliers have developed limited add-on components to their spectrometers to address these and other user needs, the associated high costs of such devices limit widespread adoption.

To this end, low-cost, variable volume and path length cuvettes with integrated temperature control have been fabricated primarily using FDM printing techniques. Pisaruka et al. expanded the commercially available spectrometer functionality to include an integrated water bath heating chamber surrounding the sample chamber, enabling variable temperature measurements on decreased sample volumes of amphiphile micelle concentrations to probe phospholipid aggregation properties [1]. These sample chambers were fabricated from acrylonitrile butadiene styrene (ABS) and polylactic acid (PLA) filaments via FDM printing using a MakerBot Replicator 2X printer. This was followed by a post-printing processing step in which PLA or ABS cuvettes were immersed in either chloroform or acetone to fuse layer defects and improve watertight properties.

Universal cuvette adapters have also been 3D-printed to accommodate a variety of commercial spectrometers in addition to enabling the use of common laboratory containers to serve as cuvettes directly. The development by Whitehead et al. uses a standard glass or quartz microscope slide integrated into a 3D-printed support piece, for an estimated cost of $1 or $14, respectively [2]. Furthermore, the authors demonstrate a universal adapter that can house gas-tight commercially available vials, in addition to supporting solid samples that would otherwise be difficult to study with existing devices. A total of 24 designs were printed in less than two hours each using a LulzBot Mini Printer with Black PLA filament.

Flow capabilities have been integrated into optically transparent cuvettes using a Dreamer Sygnis New Technologies printer and NinjaFlex Midnight Black Filament to fabricate rubber-like components that housed small tubing for flow studies [3]. The flow cell maintained a volume of only 50 uL and the reported total cost was ~$1.3 which the authors state as being hundred times less expensive than a commercially available flow cell. The authors validated their design with use over 20 hours on two model analytes, bromothymol blue and fluorescein.

While the previous examples used FDM printed sample chambers incorporating optical glass, Sirjani et al. used optically transparent filaments- transparent PLA (MakerBot), HD glass (Formfutura), and T-Glase (Taulman3D) to directly fabricate the entire sample cuvette [4]. Significant light scattering effects due to layer spacing in the FDM printing process were overcome using XTC-3D ®, a finishing coating for 3D-printed parts that are used to fill in print striations and imperfections to create a smooth and high gloss finish. The authors extended their application from a simple cuvette to the fabrication of a 3D-printed, optically transparent device containing three reaction chambers with embedded neodymium magnets used to immobilize enzymes conjugated to magnetic Dynabeads™. Two and three-step lactose assays were performed to validate the enzyme immobilization assay. This work represented an important step toward the rapid fabrication of optically transparent, custom reaction chambers for biosensing applications without chemical compatibility issues.

Beyond fabrication of the sample chamber, several groups have developed multiple components for UV/VIS spectroscopy which further extend its utility and applications. This includes the functionalization of printed components to serve as electrodes for combined spectro-electrochemistry applications by Vaněčková et al. [5]. This enables the study of electron transfer mechanisms and detection of electrochemical reaction products. The electrochemical electrodes were printed from PLA-carbon nanotubes (CNT) using a Prusa I3 MK3 printer followed by electrochemical activation and integration into a quartz cuvette. Importantly, the electrode design included an optical window integrated into the 3D-printed structure. The authors validated their electrodes using a standard sample and printed components behaved similarly to conventional electrodes.

3D printing has enabled the fabrication of accessible and portable spectrometers that couple inexpensive optical elements to smartphones serving as the detector. 3D-printed housings allow the integration of all components of the spectrometer. Grasse et al. offered a procedure for translating pixels from smartphone-detected images into conventional UV/VIS absorption spectra and validated their methodologies by determining Kool-Aid concentrations [6].

3D printing has also enabled the miniaturization of UV/VIS spectrometers with a total volume of only 100 μm × 100 μm × 300 μm using two-photon direct laser writing 3D printing technology, while the diffraction slit of the spectrometer was fabricated using super-fine inkjet process [7].

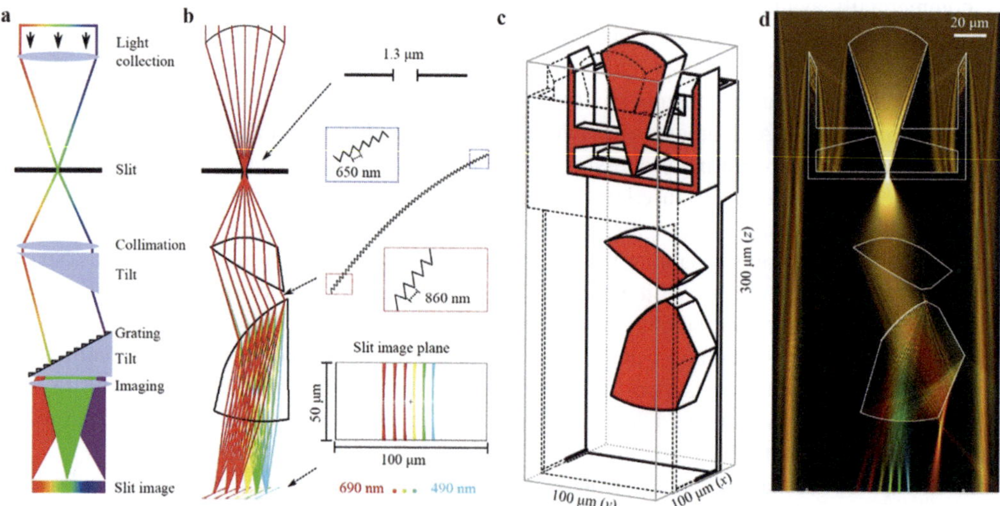

FIGURE 3.1 Optical design and simulation of the miniature spectrometer. (a) Design principle. (b) Optical design and ray-tracing simulation. Slit width blazed grating, topography, and footprint spot diagram are highlighted for the indicated surfaces. (c) Cut view of the 3D-printed volume model. Lenses, grating, slit, and mounts are printed in one step and fit in a volume of only 100 × 100 × 300 µm^3. (d) Two-dimensional wave-optical simulation (WPM) of the plane highlighted in red in (c). Around the slit, a perfectly absorbing material is assumed. Simulation wavelengths are in the visible range from 490 nm to 690 nm in 40 nm steps. Figure 3.1 is adapted with permission [7]. Copyright: The Authors, some rights reserved; exclusive licensee Light Publishing Group. Distributed under a Creative Commons Attribution License 4.0 (CC BY).

This is one of the first instances of micro-optics being fabricated using 3D-printing technology and represents a large potential opportunity for a variety of sensing and spectroscopic applications. The printed device is shown in Figure 3.1 and details the optical design and 3D-printed volume including micro-optics. The authors compared the spectrum collected with the 3D-printed miniaturized spectrometer to a commercial spectrometer and found sufficient agreement between the results, despite the footprint of the miniaturized device is eight orders of magnitude less than the commercial spectrometer.

The primary remaining challenges toward widespread adoption of 3D printing for UV/VIS spectroscopy lie in the chemical compatibility of resins and filaments used that are in contact with analyte solutions. Moreover, ensuring leak-tight characteristics of sample cuvettes and integrating optically transparent resins and filaments into designs will continue to be a challenge to researchers. However, as demonstrated by the previously mentioned studies, several authors have been able to overcome these challenges and 3D printing represents a viable method of fabricating and upgrading UV/VIS spectrometers.

3.3 FOURIER TRANSFORM INFRARED AND RAMAN SPECTROSCOPY

Fourier transform infrared (FTIR) spectroscopy and Raman spectroscopy are analytical techniques that probe the vibrational modes of molecules. These measurements provide rich information about the chemical bonds present in a sample of interest. The frequencies associated with vibrational transitions within molecules generally fall within the infrared region of the electromagnetic spectrum. FTIR measures the absorption of infrared radiation which will occur only at specific frequencies associated with particular vibrations within a sample. Raman measurements on the other hand are based on inelastic scattering. Samples are irradiated with a laser and scattered light is collected and measured. The dominant scattering effect is elastic Rayleigh scattering which preserves the frequency of the light. There are also inelastic Raman scattering effects that cause the frequency of the light to shift up or down by the frequency of vibrational transitions present in the sample.

Since FTIR is an absorption measurement it requires an infrared light source and detector that function across the entire spectral region of interest. Typically, the light source is a heated silicon carbide element or a tungsten-halogen lamp. Rather than using a monochromator or diffraction grating to produce spectral resolution (as is common practice when working with visible or ultraviolet light) FTIR employs an interferometer. The light is split into two parts that are directed to the detector via separate paths. Depending on the frequency of the light and the difference in the lengths of the two paths the two beams can positively or negatively interfere with one another. By varying the length of one of the two beam paths an interferogram can be produced that encodes the frequency and intensity of the light. This signal can then be converted into a spectrum via a Fourier transform.

Raman spectroscopy requires a more complex arrangement of optics. A laser, typically in the visible or near-infrared, is focused on a sample using a microscope objective. That same objective lens is used to collect light scattered from the sample. Great pains must be taken to avoid fluorescence which can also emit light that would be collected by the objective. Scattered light at the same frequency as the laser is produced by Rayleigh scattering and must be filtered out before detection. As the light in a Raman microscope is at higher frequencies than that in an FTIR spectrometer it generally is spectrally resolved using a monochromator or a grating paired with a CCD detector.

Most FTIR and Raman instruments that chemists and material scientists in academic and industrial labs have access to are commercial devices that are not intended to be significantly altered by the user. Constructing homebuilt instruments or modifying existing commercial instruments is generally difficult and expensive but can offer a crucial path to new and innovative spectroscopic methodologies. 3D-printing approaches have significantly lowered the barrier to experimenting with novel apparatuses needed for studying challenging samples and overcoming practical barriers that previously prevented some types of measurements.

In the most common case, FTIR measurements are performed by passing a beam of infrared light through a sample and then into an interferometer for detection. This arrangement is effective but has significant limitations. Such transmission mode measurements usually require that the sample be in the solid state and be very thin to avoid excessive attenuation of the beam. Attenuated total reflection (ATR) is a popular alternative method in which the sample is placed directly on a crystal with a high refractive index. The beam travels through the crystal and hits the interface with the sample at an oblique angle such that total internal reflection occurs, and the beam effectively bounces off the interface. A small portion of the beam, known as an evanescent wave, extends a short distance beyond the interface and interacts with the sample allowing absorption to occur. ATR attachments are available for many commercial FTIR spectrometers, but they are expensive and inflexible. 3D printing has been used to produce low-cost ATR modules that allow for variation of many aspects of the measurements, such as the type of ATR crystal [8]. By creating custom ATR modules researchers could alter the number of times the beam bounces off the interface, a parameter that generally cannot be adjusted using commercial equipment, to optimize the sensitivity of measurements of proteins in dilute solution. Additionally, because the 3D-printed modules could be made overnight at a relatively low cost, they could be used in a disposable fashion to reduce cross-contamination and study otherwise inaccessible biological samples.

3D printing has also been applied to great effect in Raman spectroscopy. Due to their increased complexity and cost, Raman instruments are often less accessible than FTIR spectrometers, especially in teaching labs. By integrating 3D-printed parts with off-the-shelf optical components, a functional Raman spectrometer was developed that cost less than $4000, even including $3000 for a 3D printer [9]. Such an approach can make it significantly cheaper to provide students with hands-on experience of Raman spectroscopy.

3D printing has also been used to build experimental apparatuses aimed at overcoming the low sensitivity of Raman measurements. One method for enhancing the signal of Raman spectroscopy makes use of a liquid core waveguide, in which the sample is placed in a Teflon-amorphous fluoropolymer (AF) tube to provide a longer path length and help capture the scattered light [10]. A major issue encountered in this approach is the difficulty of preventing bubbles in the sample. A 3D-printed negative pressure module was developed for easy removal of bubbles allowing for fast and reproducible analytical measurements.

Another commonly employed method is surface-enhanced Raman spectroscopy (SERS) in which samples adsorbed to rough metal surfaces or plasmonic nanomaterials exhibit significantly stronger Raman signals. SERS samples are frequently prepared via drop casting, by depositing a solution on a surface and allowing the solvent to evaporate. This process often produces inhomogeneous samples where the concentration of the target molecules varies with the distance from the center of deposition. 3D printing has been used to develop a set of alignment tools to precise position a pipette tip used to deposit a sample and the focal point of a Raman microscope relative to one another [11]. These simple 3D-printed parts allowed for reproducible sampling at controlled distances from the center of a drop on a gold-coated substrate. 3D printing can also offer methods to control plasmonic nanomaterials, such as silver nanoparticles suspended in a hydrogel (Figure 3.2) [12]. Extrusion-based 3D printing can alter the structure of hydrogels by causing the shear-induced alignment of nanofibers. The aligned 3D-printed hydrogel gives a significant signal enhancement compared to a hydrogel that had not gone through the printing process.

Pairing Raman spectroscopy with other modalities, such as electrochemistry can provide insight that cannot be obtained from either technique operating independently. Conducting these sorts of tandem experiments requires the development of new experimental apparatuses, which are often not commercially available, that allow electrochemical cells and electrodes to be mounted to a Raman spectrometer. The low cost and rapid iteration permitted by 3D printing have enabled the development of cheap and versatile electrochemical cells allowing in situ Raman measurements of electrochemical film deposition while avoiding expensive components such as quartz plates and platinum rings used in previously reported devices [13], [14]. These experiments were able to use Raman spectroscopy to probe the chemical changes that occurred as a function of the potential of the electrode. 3D-printed electrochemical cells have also been used to pair voltammetry of immobilized particles with Raman spectroscopy. The formation of new compounds at precise locations on the electrode surface could be directly monitored almost simultaneously by Raman providing detailed insight into the effects of the electrochemical process.

One of the major limitations of the 3D printing-based approaches discussed here is the chemical compatibility of 3D-printed parts. FTIR and Raman spectroscopy are widely applied techniques that researchers use to investigate a variety of biological, organic, and inorganic systems. If the polymers commonly used in 3D printing are incompatible with the target molecules or the solvents they are stable in, then 3D-printed parts cannot be employed. Further development of new 3D-printed materials with a variety of chemical properties could help broaden the applicability of the types of innovations described here.

3.4 MASS SPECTROMETRY

Mass spectrometry (MS) is one of the oldest and most powerful spectroscopic techniques used to analyze molecular compounds and mixtures by measuring the mass-to-charge (m/z) ratio of a molecule and its fragments. Such ratios can be used to elucidate numerous chemical properties of a compound, including its elemental composition, chemical structure, and isotopic ratios. The detection limit of modern mass spectrometers routinely allows studies on femtomole quantities of samples, making MS one of the most sensitive analytical techniques available. The information that can be

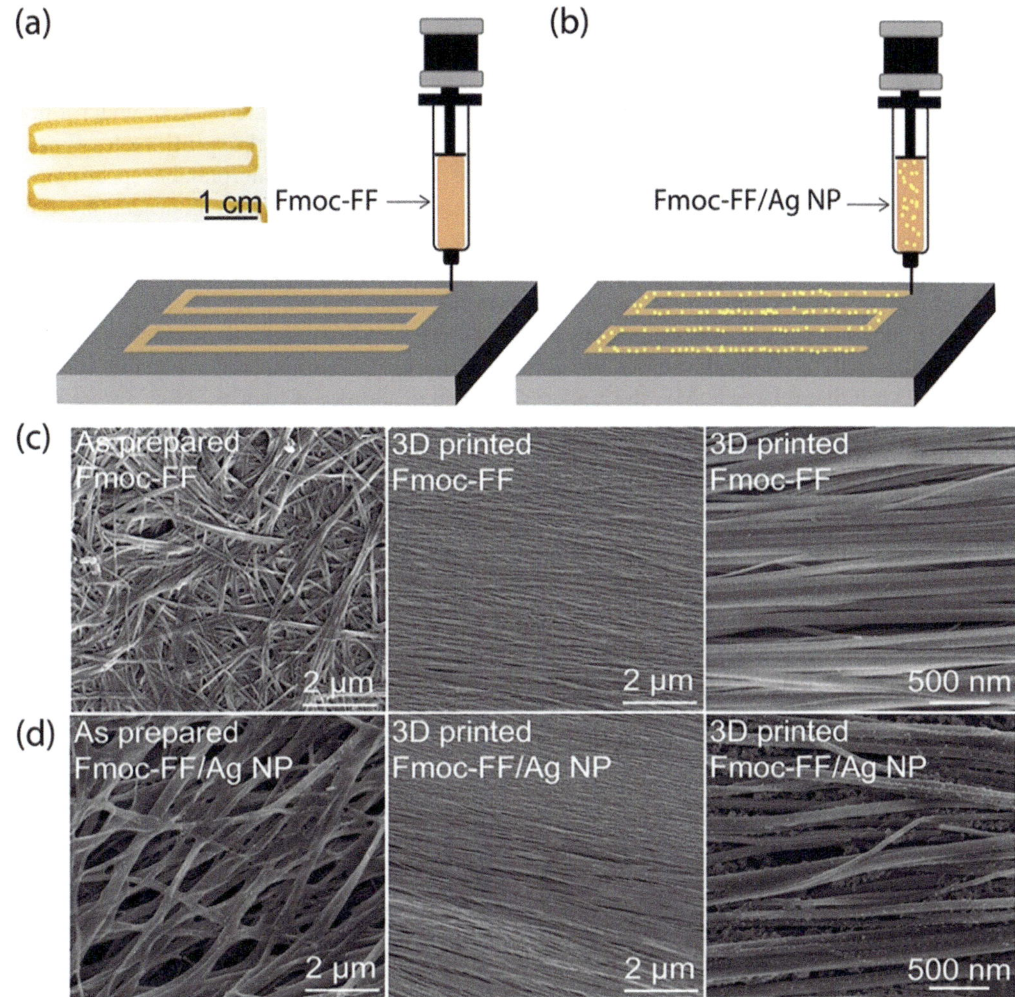

FIGURE 3.2 Schematic representation of the 3D-printed hydrogel (a) without and (b) with NPs. The inset in a is an optical image of the hydrogel (mixed with 0.06 g of orange food dye (U8-OSL0-PS8Q, Preema, UK) printed on glass. (c) SEM images of the as-prepared and 3D-printed Fmoc-FF. (d) SEM images of the as-prepared and 3D-printed Fmoc-FF with Ag NPs. Dye was used only for the inset in a. Figure 3.2 is adapted with permission [12]. Copyright 2019, American Chemical Society.

obtained from MS in conjunction with the broad availability of commercial mass spectrometers makes MS one of the most used analytical techniques for studying chemical systems. As a result, MS has had a profound impact in fields ranging from materials science to biotechnology.

The type of equipment found in a mass spectrometer can differ based on the type of sample being analyzed and the information desired to be obtained. However, in general, all mass spectrometers consist of a few essential components. First, an inlet system is required to introduce the sample of interest to the spectrometer so that it can be analyzed. Next, an ion source is used to ionize the molecules in the sample. Ion sources can be categorized as hard or soft sources based on the degree of molecular fragmentation that takes place. Depending on the initial sample phase and type of spectrometer, conversion of the sample to a gaseous phase, which is required for analysis, is accomplished either by the inlet system or the ionization source.

After ionization, the molecular ions are accelerated via an electric field through a mass analyzer, typically consisting of a magnetic or electric field, which separates the different molecular ions based on their m/z ratio. An ion detector is then used to count the number of molecular ion fragments with a given m/z ratio. The ion source, mass analyzer, and ion detector are all kept under a high vacuum to prolong the lifetime of the molecular ions as they travel through the spectrometer. Finally, a computer is used to translate the data collected via the detector into an interpretable spectrum.

To date, most MS components developed with 3D printing have focused on modifications to the inlet system and ion source. Because 3D printing allows for rapid prototyping of components for mass spectrometers, its application greatly expands the types of analysis and experiments that can be performed. When choosing materials for 3D printing numerous material properties should be considered, including heat resistance, chemical compatibility, and resistance to outgassing. Here we highlight recent MS hardware developments using 3D printing. For a more comprehensive literature review of 3D printing related to MS, the reader is referred to additional sources [15].

With respect to the inlet system, 3D printing has allowed for unique customization of both online and offline sample preparation methods. Online sample preparation methods involve direct integration with the inlet system of the spectrometer, while offline sample preparation methods prepare the sample in a separate station before interfacing with the spectrometer. An example of improvement offered by 3D-printed devices for online sample preparation is the ability to monitor chemical reactions in real-time. In one case, a 3D-printed polypropylene fluidic device allowed live reaction monitoring of the formation of a metal-salt complex using electrospray ionization MS (ESI-MS) [16]. The stoichiometry of the reaction was controlled by varying the flow rates of each reactant at the individual inlets of the 3D-printed device. In another example, 3D printing allowed for the fabrication of a polypropylene reactor with an internal magnetic stir bar and nanospray ion emitter [17]. The magnetic stir bar was incorporated into the reaction device via insertion during the FDM print, with the remainder of the housing built around the stir bar following the insertion, something not possible with traditional subtractive manufacturing methods. An image of this reactor can be seen in Figure 3.3.

One of the fastest growing areas of MS is the development of ambient ionization (AI) methods that allow the MS sample to be prepared at atmospheric pressure with minimal sample prep. One promising method of AI is paper spray ionization (PSI), which involves soaking a sheet of paper with a solvent of interest and then applying a high voltage source to ionize the sample and introduce it to the mass spectrometer. One of the disadvantages of the PSI method is the rapid depletion of the solvent, which limits analysis time. However, recent developments in 3D printing have allowed

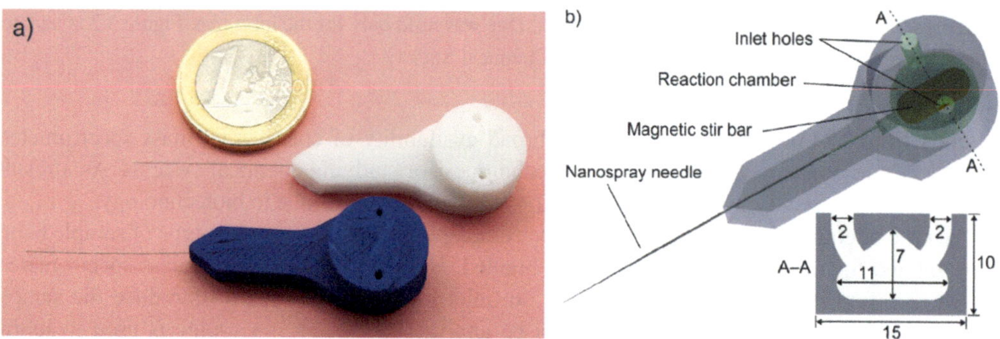

FIGURE 3.3 (a) Image of polypropylene reactor. (b) A transparent image of a polypropylene reactor with individual components is highlighted. A cross-section of the reaction can be seen, where the given dimensions are in mm. Figure 3.3 is adapted with permission [17]. Copyright: The Authors, some rights reserved; exclusive licensee Royal Society of Chemistry. Distributed under a Creative Commons Attribution License 3.0 (CC BY).

this analysis to be extended via the integration of polylactic acid (PLA) sample reservoirs. In one example, employing a PLA cartridge with a paper tip and dual hydrophilic wicks extended the sample spray to tens of minutes [18]. The design of this reservoir allowed fast wetting of the paper via one wick placed near the surface of the reservoir, while a second wick placed deeper in the reservoir allowed for an extended supply of solvent to the paper through slow capillary action. In the case of the fast-wetting wick, it was found that diffusion of the solvent through the paper occurred faster between the paper-PLA interface rather than direct wetting of the paper.

Another AI technique that has benefited from the application of 3D printing is electrospray ionization (ESI). ESI can be challenging under ambient conditions as the number of analyzable ions is greatly reduced due to the expansion of the spray before even entering the vacuum region of the spectrometer for analysis. However, this challenge of spray expansion was recently overcome through the employment of 3D-printed electrodes capable of focusing the ion beam before it enters the spectrometer [19]. The electrodes were fabricated using an FDM printer and polyethylene terephthalate glycol printing material doped with conductive CNT.

A final AI method ripe for the application of 3D printing is low-temperature plasma (LTP) ionization. In this method a plasma, often generated from helium gas, is used to generate molecular ions. The low power consumption and simplistic design of LTP probes are attractive advantages, and the customized nature of the probes (to date the majority of LTP probes have been home built) makes them well suited for production via 3D printing. In one study, an LTP probe 3D-printed from PLA and ABS was used to successfully obtain MS spectra of small molecule organic compounds [20].

One of the greatest benefits of 3D printing for MS applications is that many of the designs mentioned in this section are open source and available to researchers free of charge. However, despite these benefits, several challenges remain that impede the integration of 3D printing into MS. For example, many 3D-printing resins are soluble in organic solvents, which can limit the types of solvents that can be used in 3D-printed reaction vessels.

3.5 NUCLEAR MAGNETIC RESONANCE SPECTROSCOPY

Nuclear magnetic resonance (NMR) is an extremely powerful and versatile spectroscopic technique for the study of chemicals ranging from biomolecules to materials. NMR is a non-destructive spectroscopic method that is capable of probing individual atomic sites within a molecule at atomic resolution. Over the last few decades, NMR has proven to be an indispensable tool for the structural determination of organic compounds and biomolecules such as proteins and has found additional applications in emerging technological fields such as battery applications, among others.

NMR is most used to study chemical systems in a liquid or solid state. Generally, the type of equipment used is similar regardless of physical state. First, a superconducting magnet (typically in the range of 9–28 tesla) is used to generate an energy level splitting between the nuclear spin states in a sample. The sample itself is housed within a chamber with an RF coil surrounding it. The purpose of this coil is two-fold: it transmits RF radiation to the sample where it is absorbed and acts as a receiver to detect the corresponding RF radiation that is emitted from the sample. The RF coil and sample chamber are part of the NMR probe which sits within the bore of the superconducting magnet.

In solution state NMR the sample is typically placed in a glass NMR tube while in solid-state NMR the sample is typically placed into a ceramic NMR rotor and fitted with a microturbine cap. In the latter case, this rotor is placed into a stator and the sample is pneumatically rotated at a fixed angle (54.74°) at frequencies that can exceed 7 million RPM. This technique is referred to as magic angle spinning (MAS) and is used to substantially improve the resolution of NMR experiments on solid-state samples. Finally, the spectrometer itself is used to generate the necessary RF pulses for exciting the nuclear transitions. The spectrometer typically consists of RF transmitters, high-power amplifiers, and a receiver unit that translates the analog RF signal into a digital format that can be processed by a computer.

Over the last several years there has been a substantial increase in the application of 3D-printed materials to enhance the performance of NMR equipment. This has been driven by a need for low-cost, open-source equipment that can be rapidly customized and produced. In this section, we highlight a few of the most prominent examples of 3D printing related to NMR. For a more extensive review, the reader is referred to other literature [21].

One of the first examples of 3D printing components for NMR spectroscopy is the fabrication of 3D-printed sample exchangers for dynamic nuclear polarization (DNP) NMR [22]. DNP is a powerful technique that significantly improves the sensitivity of an NMR experiment by transferring the abundant spin polarization of electrons to the nuclei of interest via microwave irradiation. DNP experiments are most efficient when performed at low temperatures, typically <-173°C. Furthermore, because the sample must be rapidly cooled, the NMR rotor must be inserted into the probe at a low temperature. Despite these challenging temperature requirements, 3D-printed eject pipes were fabricated that allowed for seamless insertion and ejection of MAS rotors at temperatures of -173°C. The eject pipes were not only able to handle the extreme temperatures necessary for DNP experiments, but also the extreme thermal cycling that occurs between different probe cooldown cycles.

While most NMR coils in MAS NMR consist of a uniform pitch solenoid coil, non-uniform pitch coils offer improved RF homogeneity and microwave access for DNP experiments. One example of using 3D printing to create non-uniform pitch coils involves the fabrication of dissolvable coil inserts [23]. In this example, CAD software allowed the straightforward design of templates on which coils were wound with variable pitch. The coil templates were 3D printed from ABS material on an FDM printer. After winding the coil around a template, the template was then dissolved in acetone, leaving behind just the coil with the given template geometry. This approach allows easy, rapid, and reproducible prototyping of coil designs.

Stators used for MAS NMR experiments are capable of spinning rotors at extremely high rotation frequencies, currently up to ~7 million RPM. However, to accomplish this, the individual components of the stator must be fabricated to very tight tolerances, sometimes on the order of a few microns. As such, it is often difficult to manufacture these MAS stators with traditional additive manufacturing methods. However, recent work has shown that it is possible to fabricate many stator components using a combination of ceramic and plastic SLA printing, thus greatly reducing the overall fabrication cost of the stator [24]. Additionally, it was shown that the cost of certain consumables such as MAS drive caps, which function as microturbines and enable sample rotation, could be greatly reduced as well. Using the Protolabs proprietary resin microfine green, researchers were able to fabricate 3.2 mm drive caps capable of reproducibly spinning at frequencies of 1.2 million RPM for periods of weeks at a time with comparable performance to commercially machined drive caps. An image of the 3D-printed stator mounted to a spinning test station and NMR probe can be seen in Figure 3.4.

While most MAS stators are designed to spin cylindrical rotors, a recent development in MAS NMR is the implementation of spherical rotors [25]. Initial tests with 9.5 mm diameter spherical rotors found that they were able to spin to 636,000 RPM using helium gas in a 3D-printed stator. The stator design requirements for spherical rotors are significantly different than that of stators for cylindrical rotors. Therefore, when designing initial prototypes of the stator it was highly convenient to use 3D printing to rapidly iterate through stator designs with minor modifications. As the size of spherical rotors began to shrink it was found that further modifications need to be made to the stator design [26]. For 2 mm MAS spheres, the aperture for the bearing gas had to be reduced to 0.4 mm, a significant challenge for most commercial 3D printers. It was also discovered that smaller spherical rotors had difficulty remaining in the stator and would tend to eject during spinning. The issue was eliminated by 3D printing a 'cage' as part of the stator design, thus keeping the sphere in place.

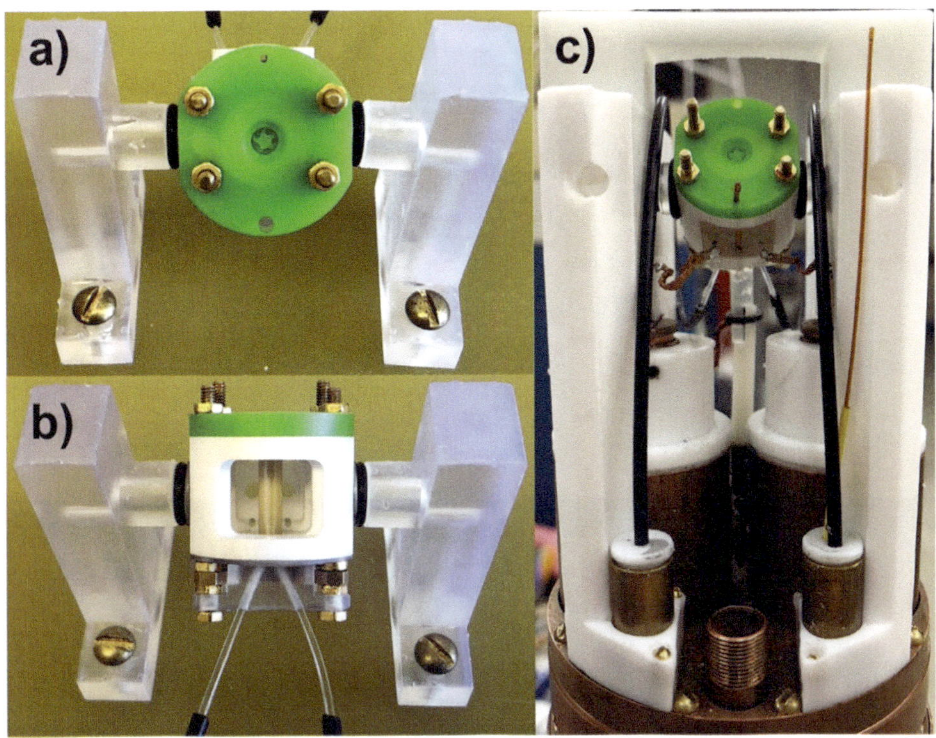

FIGURE 3.4 (a) Top and (b) front view of 3D-printed stator assembly mounted on 3D-printed spinning test station. (c) 3D-printed stator mounted on triple-resonance NMR probe. Reprinted with permission [24], Copyright 2022, Elsevier.

Using these design modifications the 3D-printed stators were capable of spinning 2 mm spherical rotors to > 4 million RPM using helium gas.

3D printing has also been used extensively to fabricate non-MAS NMR components. One example of this is using 3D printing to make custom NMR probe heads [27]. The authors utilized both FDM and SLA printing methods followed by a liquid metal injection step to fabricate high complexity, custom radio frequency coils, as shown in Figure 3.5. These inductor designs were optimized for improved sensitivity over previous designs that were limited by conventional fabrication methods. This work represents important progress toward low-cost magnetic resonance imaging (MRI) coils in addition to in situ-reaction monitoring.

In another example, researchers were able to 3D print a low-cost, easy-to-use autosampler for a benchtop NMR system [28]. The autosampler could store 30 samples at a time for solution NMR analysis, significantly increasing the throughput of the system which had previously required manual insertion of the sample. Both the CAD designs and software for controlling the autosampler are open-source and readily available to the research community.

Continued miniaturization of RF transmit and receive coils is critical for magnetic resonance devices, and recently realized using two-photon polymerization 3D printing. This printing strategy was used to fabricate sub-nanoliter (sub-nL) microfluidic channels on a single chip NMR probe with a resolution better than 1 μm^3 [29]. The high-resolution printing method enabled fabrication of the detection coil 10 μm from the sample, which allowed the detection of low concentrations (6×10^{-12} moles) of intact biomolecules at 0.1 parts per million spectral resolution.

Despite the immense progress that has been made with respect to additive manufacturing in NMR, several challenges remain. One challenge is that NMR equipment often consists of small parts

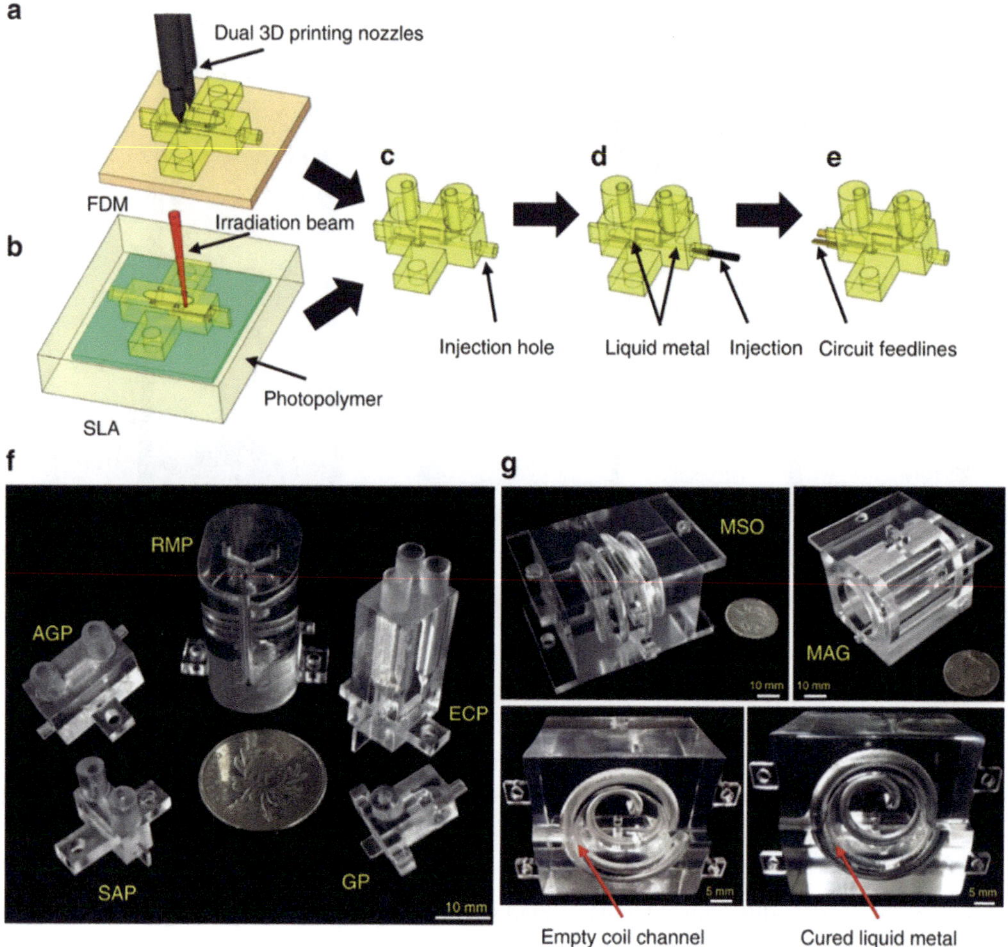

FIGURE 3.5 Both (a) fused deposition modeling (FDM) and (b) stereo lithography appearance (SLA) techniques are used to fabricate a complete probehead (c) layer-by-layer according to the simulation design. (d) Liquid metal is perfused into the model through the injection hole to form an RF coil. (e) The RF coil is connected to the matching circuit by two copper strips to form a complete probe. The entrance and exit of the liquid metal channel are completely sealed with silver paste. Various 3D-printed probeheads suitable for MR applications can be fabricated and utilized, including (f) U-tube saddle probehead (SAP), U-tube Alderman-Grant probehead (AGP), reaction monitoring probehead (RMP), electrochemical reaction monitoring probehead (ECP), gradient probehead (GP) for MR, and (g) modified solenoid imaging probehead (MSO), modified Alderman-Grant imaging probehead (MAG) for MRI. The coil channel of MSO probehead, before and after the liquid metal perfusion, are also shown. Figure 3.5 is adapted with permission [27]. Copyright: The Authors, some rights reserved; exclusive licensee Nature Publishing Group. Distributed under a Creative Commons Attribution License 4.0 (CC BY).

with tight tolerances. As of now, most commercial 3D printers for use in labs still offer tolerances in the range of one hundred microns, limiting the type of equipment that can be manufactured for NMR. Additionally, many NMR applications, particularly in MAS NMR, require airtight parts. This can often be difficult to achieve with certain printing processes such as FDM. Another challenge to consider is the chemical composition of the printing material. Most 3D-printing materials are made of organic compounds abundant in carbon and hydrogen, so care must be taken to ensure that these materials do not produce background signals in NMR spectra.

3.6 CONCLUSION AND OUTLOOK

We have herein reviewed several notable applications of 3D-printing technologies that have enabled advances in molecular and atomic spectroscopies. From applications in UV/VIS, FTIR, Raman, MS, and NMR, 3D printing has allowed researchers to perform innovative experiments that would be otherwise expensive or not readily accessible due to limitations with existing commercial technology. Across spectroscopic methods, the wide range of chemically compatible resins and material properties supporting a variety of experimental conditions illustrates the robust performance and capabilities of 3D printing. Moreover, in many applications reviewed, sub-micron tolerances were achieved using 3D printing. Continued miniaturization of spectroscopic devices is of primary interest to researchers in academia and industry alike. In the future, we anticipate additive manufacturing will continue to become a primary fabrication method in both research and commercially available spectroscopic devices. As the resolution and material properties available under 3D printing continue to evolve and advance so too will the complexity of the components fabricated.

REFERENCES

1 J. Pisaruka and M. K. Dymond, "A low volume 3D-printed temperature-controllable cuvette for UV visible spectroscopy," *Analytical Biochemistry*, vol. 510, pp. 52–55, Oct. 2016, doi: 10.1016/J.AB.2016.07.019.

2 H. D. Whitehead, J. V. Waldman, D. M. Wirth, and G. LeBlanc, "3D printed UV-visible cuvette adapter for low-cost and versatile spectroscopic experiments," *ACS Omega*, vol. 2, no. 9, pp. 6118–6122, Sep. 2017, doi: 10.1021/ACSOMEGA.7B01310.

3 M. Michalec and Ł. Tymecki, "3D printed flow-through cuvette insert for UV–Vis spectrophotometric and fluorescence measurements," *Talanta*, vol. 190, pp. 423–428, Dec. 2018, doi: 10.1016/J.TALANTA.2018.08.026.

4 E. Sirjani, M. Migas, P. J. Cragg, and M. K. Dymond, "3D printed UV/VIS detection systems constructed from transparent filaments and immobilised enzymes," *Additive Manufacturing*, vol. 33, p. 101094, 2020. doi: 10.1016/j.addma.2020.101094 2020.

5 E. Vaněčková et al., "UV/VIS spectroelectrochemistry with 3D printed electrodes," *Journal of Electroanalytical Chemistry*, vol. 857, p. 113760, Jan. 2020, doi: 10.1016/J.JELECHEM.2019.113760.

6 E. K. Grasse, M. H. Torcasio, and A. W. Smith, "Teaching UV-Vis Spectroscopy with a 3D-Printable Smartphone Spectrophotometer," *Journal of Chemical Education*, vol. 93, no. 1, pp. 146–151, Jan. 2016, doi: 10.1021/ACS.JCHEMED.5B00654

7 A. Toulouse et al., "3D-printed miniature spectrometer for the visible range with a 100 × 100 μm^2 footprint," *Light: Advanced Manufacturing*, vol. 2, no. 1, pp. 20–30, Mar. 2021, doi: 10.37188/LAM.2021.002.

8 B. Baumgartner, S. Freitag, and B. Lendl, "3D Printing for low-cost and versatile attenuated total reflection infrared spectroscopy," *Analytical Chemistry*, vol. 92, no. 7, pp. 4736–4741, Apr. 2020, doi: 10.1021/ACS.ANALCHEM.9B04043.

9 O. Aydogan and E. Tasal, "Designing and building a 3D printed low cost modular Raman spectrometer," *CERN IdeaSquare Journal of Experimental Innovation*, vol. 2, no. 2, pp. 3–14, Dec. 2018, doi: 10.23726/CIJ.2017.799.

10 J. Zhou et al., "A novel 3D printed negative pressure small sampling system for bubble-free liquid core waveguide enhanced Raman spectroscopy," *Talanta*, vol. 216, Aug. 2020, doi: 10.1016/J.TALANTA.2020.120942.

11 B. I. Karawdeniya, R. B. Chevalier, Y. M. N. D. Y. Bandara, and J. R. Dwyer, "Targeting improved reproducibility in surface-enhanced Raman spectroscopy with planar substrates using 3D printed alignment holders," *Review of Scientific Instruments*, vol. 92, no. 4, p. 043102, Apr. 2021, doi: 10.1063/5.0039946.

12 S. Almohammed, M. Alruwaili, E. G. Reynaud, G. Redmond, J. H. Rice, and B. J. Rodriguez, "3D-printed peptide-hydrogel nanoparticle composites for surface-enhanced raman spectroscopy sensing," *ACS Applied Nano Materials*, vol. 2, no. 8, pp. 5029–5034, Aug. 2019, doi: 10.1021/ACSANM.9B00940.

13 G. D. da Silveira, R. F. Quero, L. P. Bressan, J. A. Bonacin, D. P. de Jesus, and J. A. F. da Silva, "Ready-to-use 3D-printed electrochemical cell for in situ voltammetry of immobilized microparticles and Raman spectroscopy," *Analytica Chimica Acta*, vol. 1141, pp. 57–62, Jan. 2021, doi: 10.1016/J.ACA.2020.10.023.

14 M. F. dos Santos, V. Katic, P. L. dos Santos, B. M. Pires, A. L. B. Formiga, and J. A. Bonacin, "3D-Printed low-cost spectroelectrochemical cell for in situ raman measurements," *Analytical Chemistry*, vol. 91, no. 16, pp. 10386–10389, Aug. 2019, doi: 10.1021/ACS.ANALCHEM.9B01518.

15 M. Grajewski, M. Hermann, R. D. Oleschuk, E. Verpoorte, and G. I. Salentijn, "Leveraging 3D printing to enhance mass spectrometry: A review," *Analytica Chimica Acta*, vol. 1166, p. 338332, Jun. 2021, doi: 10.1016/J.ACA.2021.338332.

16 J. S. Mathieson, M. H. Rosnes, V. Sans, P. J. Kitson, and L. Cronin, "Continuous parallel ESI-MS analysis of reactions carried out in a bespoke 3D printed device," *Beilstein Journal of Nanotechnology 4:31*, vol. 4, no. 1, pp. 285–291, Apr. 2013, doi: 10.3762/BJNANO.4.31.

17 G. Scotti et al., "A miniaturised 3D printed polypropylene reactor for online reaction analysis by mass spectrometry," *Reaction Chemistry & Engineering*, vol. 2, no. 3, pp. 299–303, Jun. 2017, doi: 10.1039/C7RE00015D.

18 G. I. Salentijn, H. P. Permentier, and E. Verpoorte, "3D-printed paper spray ionization cartridge with fast wetting and continuous solvent supply features," *Analytical Chemistry*, vol. 86, no. 23, pp. 11657–11665, Dec. 2014, doi: 10.1021/AC502785J.

19 K. Iyer, B. M. Marsh, G. O. Capek, R. L. Schrader, S. Tichy, and R. G. Cooks, "Ion manipulation in open air using 3D-printed electrodes," *Journal of The American Society for Mass Spectrometry*, vol. 30, no. 12, pp. 2584–2593, 2019, doi: 10.1007/s13361-019-02307-2.

20 S. Martínez-Jarquín, A. Moreno-Pedraza, H. Guillén-Alonso, and R. Winkler, "Template for 3D printing a low-temperature plasma probe," *Analytical Chemistry*, vol. 88, no. 14, pp. 6976–6980, Jul. 2016, doi: 10.1021/ACS.ANALCHEM.6B01019.

21 J. I. Kelz, J. L. Uribe, and R. W. Martin, "Reimagining magnetic resonance instrumentation using open maker tools and hardware as protocol," *Journal of Magnetic Resonance Open*, vol. 6–7, p. 100011, 2021, doi: 10.1016/j.jmro.2021.100011.

22 A. B. Barnes et al., "Cryogenic sample exchange NMR probe for magic angle spinning dynamic nuclear polarization," *Journal of Magnetic Resonance*, vol. 198, pp. 261–270, 2009, doi: 10.1016/j.jmr.2009.03.003.

23 J. I. Kelz, J. E. Kelly, and R. W. Martin, "3D-printed dissolvable inserts for efficient and customizable fabrication of NMR transceiver coils," *Journal of Magnetic Resonance*, vol. 305, pp. 89–92, Aug. 2019, doi: 10.1016/J.JMR.2019.06.008.

24 D. Banks, B. Michael, N. Golota, and R. G. Griffin, "3D-printed stators & drive caps for magic-angle spinning NMR," *Journal of Magnetic Resonance*, vol. 335, p. 107126, Feb. 2022, doi: 10.1016/J.JMR.2021.107126.

25 P. Chen et al., "Magic angle spinning spheres," *Science Advances*, vol. 4, no. 9, Sep. 2018, doi: 10.1126/SCIADV.AAU1540

26 P.-H. Chen, C. Gao, L. E. Price, M. A. Urban, T. M. O. Popp, and A. B. Barnes, "Two millimeter diameter spherical rotors spinning at 68 kHz for MAS NMR," *Journal of Magnetic Resonance Open*, vol. 8–9, p. 100015, 2021, doi10.1016/j.jmro.2021.100015.

27 J. Xie et al., "3D-printed integrative probeheads for magnetic resonance," *Nature Communications 2020 11:1*, vol. 11, no. 1, pp. 1–11, Nov. 2020, doi: 10.1038/S41467-020-19711-Y.

28 M. Dyga, C. Oppel, and L. J. Gooßen, "RotoMate: An open-source, 3D printed autosampler for use with benchtop nuclear magnetic resonance spectrometers," *HardwareX*, vol. 10, p. e00211, Oct. 2021, doi: 10.1016/J.OHX.2021.E00211.

29 E. Montinaro et al., "3D printed microchannels for sub-nL NMR spectroscopy," *PLoS One*, vol. 13, no. 5, p. e0192780, May 2018, doi: 10.1371/JOURNAL.PONE.0192780.

4 Energy Materials for 3D Printing

Rasmita Barik[*], *Vaishali Tanwar, Paras Kalra, and Pravin P. Ingole*
Department of Chemistry, Indian Institute of Technology Delhi, New Delhi, India, 110016
*Corresponding author: rasmitabarikimmt@gmail.com

CONTENTS

4.1 INTRODUCTION

The energy and climate crisis forced researchers to find alternative energy sources. The excess exhaustion of natural sources and emission of greenhouse gases creates a dangerous threat to human beings. To avoid future disasters, alternative energy sources are very much required. Worldwide research has been trying for some breakthroughs. Although several new materials and devices, such as batteries, supercapacitors, solar cells, and fuel cells are being developed to overcome the current energy crisis issue. A reliable development is still an ongoing search. In the past decade, intensive research has been engrossed in the development of new highly conducting porous structures with higher surface area and excellent cycling stability for energy storage applications. In contrast, recently the research has intensified to the fabrication of smart, small, highly efficient devices and assembly of electrodes and electrolytes [1]–[3].

Meanwhile, a new technique, 3D printing (Additive Manufacturing (AM)), has been developed to answer most of the questions. It is the technique that can produce physical 3D objects from the geometrical representation by continuous addition of material [2]. This phenomenal technique

DOI: 10.1201/9781003296676-4

was developed by Charles Hull in 1980 and experienced remarkable growth afterward. The artificial heart pump, cornea, jewelry designing, rocket engine, Netherland steel bridge, and many more applications in the aviation industry, food industry, agriculture, locomotive industry, etc. proves this technique for futuristic applications. The layer-by-layer deposition develops the 3D structure from computer-aided design (CAD) drawings, or prototypes is the basic concept of 3D printing. This is truly a noble innovation with versatile applications. The stream of ink from the nozzle during the printing process in a vertical direction generates different preferred architectures for advanced applications. In addition, a wide range of materials such as polymers, ceramics, composites, metal alloys, food, textile, etc., are used in extrusion-type 3D printing. One of the most innovative approaches to building 3D objects is 3D printing [4], [5].

Furthermore, 3D printing is one of the best powerful techniques that help in the fabrication of devices from the macro scale to the nanoscale. It can build 3D structures with enhanced energy and power density devices by controlling the device's spatial geometries and architectures in a fast and cost-effective way. The high standards of printing inks, with anticipated viscosity, high yield stress under shear and compression, and meticulous viscoelasticity are the most important factors required to perform 3D printing. Along with that, active material properties affect the performance of devices. The 3D printing technique is subcategorized into several types such as (a) material extrusion (i.e., direct ink writing, fused deposition modelling), (b) material jetting, (c) powder bed fusion, (d) vat photopolymerization, (e) binder jetting, (f) sheet lamination (i.e., laminated object manufacturing and (g) directed energy deposition [6]–[7].

In recent years, 3D printing has been an emergent fabrication method for the energy domain, specifically energy storage owing to its ease of designing, comprehensive stages of fabrication, cost-effectiveness, and eco-friendliness. This technique further explored a broad range of applications in various advanced fields such as biotechnology, engineering, energy, electronics, etc. With the help of this emerging technique, electronic, biomedical, and energy storage fields are more benefited. The suitable inks with high viscosities are developed to print ideal architectures, and the shear-thinning rheological properties are being well studied to avoid their collapse. 3D-printed electrodes have faster ion/electron transport properties by tuneable composition and controlling the physical properties of the printed devices. Also, the thickness and active material loading in the printed electrode layers facilitates ion/electron conductivity, further enhancing the energy and power density. By following the 3D printing technique, the manufacturing costs have been reduced with the lowered material waste. It provides an electrode with a larger specific surface area, which supports the shorter ion transport distance and a higher mass loading of active materials to develop an efficient energy storage system. The enormous advantages and compatibility with curved and flexible substrates have compelled 3D printing as the most utilized and facile technique for future generations.

3D printing technology has been used to build different electrochemical devices such as rechargeable batteries [8], lithium-sulfur batteries [9], lithium-ion micro-batteries [10], supercapacitors [11], fuel cells, [12] and solar cells [13] along with various electrodes are developed for electrochemical sensing applications [14]. Furthermore, other accessories for energy storage and conversion devices such as electrodes, electrolytes, and collectors are also developed by this advanced technique. Figure 4.1 represents the energy materials used in 3D printing and their use in energy conversion and storage devices. Solid oxide fuel and electrolysis cells have also emerged as promising energy conversion candidates by using the 3D printing technique by overcoming the shape limitations.

The available different energy materials such as different polymers, different graphene-based materials, carbon-based materials, fibers, textile materials, metal-based materials (MXenes, metal-organic frameworks (MOFs), nanocomposites, etc.), and aerosol, have reinforced the modern energy world to develop energy devices with feeble limitations. Generally, all the energy materials used for the different energy storage and conversion applications can be used in the 3D technique. They are known to show promising results as compared to traditional synthesis techniques. For instance,

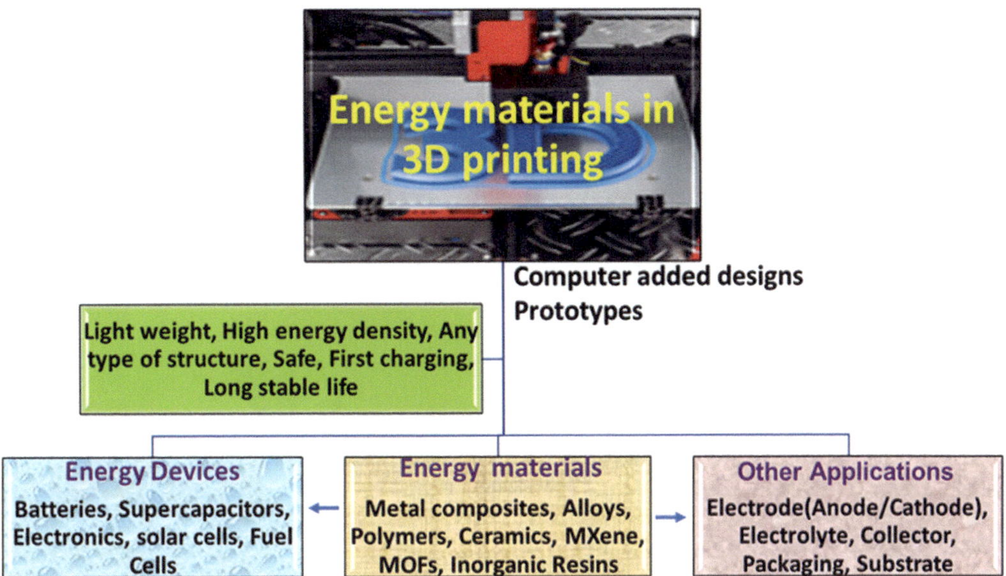

FIGURE 4.1 A flow chart on elaborating energy materials in 3D printing: Different energy materials, device fabrication, and other applications.

carbon materials with chemical stability and multipurpose nanostructures are the most used material for the 3D technique. Other explored carbon forms include graphene oxide (GO), carbon aerogel, carbon nanotubes (CNT), carbon blacks, etc. Moreover, 3D printing techniques enable us to rationalize the tailored architectural design for the graphene electrodes over multiple nanoscales. This has also been an extensive study since these as-printed devices can provide adequate electrochemical properties and could potentially be applied in printed devices and surpass even the already established technology/devices available from the conventional fabrication methods [15]–[16].

Likewise, the most challenging and futuristic energy material, MOFs and MXene, shows promising results in the field of supercapacitor and battery fabrication. The growth in the use of different energy materials in 3D printing is increasing with the passing of years can be observed in Figure 4.2. One can blindly believe that the 3D printing technique will solve the unanswered questions in the energy sector and will cover all the lacuna over the different synthesis techniques for the energy storage /conversion device fabrication. In this chapter, the state-of-the-art different energy materials used for the energy storage/ conversion device fabrication have been briefly discussed. Further, the advantages and disadvantages of energy material in 3D printing have been discussed. To close, a conclusion and future outlook are provided, with current challenges on how to make the 3D printing technique feasible, for energy device fabrication, which will further direct future research.

4.2 ENERGY MATERIALS FOR 3D PRINTING

The 3D printing processes are rational and more tailored than the traditional fabrication methods and facilitate desirable architecture or pore dimensions. In principle, designing and manufacturing the electrodes are critically the most important. However, the active materials and microscopical structures collectively influence the overall performance of the 3D-printed electrodes. This section critically discusses and summarizes the research progression in recent years, based on the different categories of material for energy storage, in printing technology.

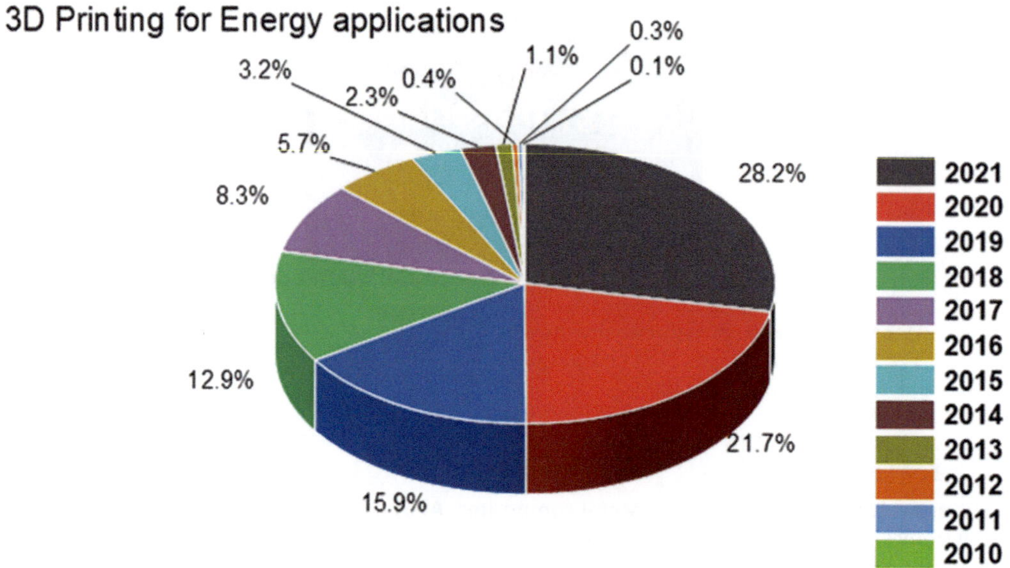

FIGURE 4.2 The progress in the use of energy materials for 3D printing applications. (Source: Scopus).

4.2.1 Carbon-Based Materials

Carbon-based materials, including graphene, carbon aerogels (CAs), carbon nanotubes (CNTs), carbon nanofibers, and activated carbons (ACs), exhibit preferable advantages such as adequate chemical stability, enormous specific surface area, and controllable porosity for the electrolyte ions [17], [18]. Therefore, various carbon-based materials are realized vividly as electrode materials with unique characteristics such as electronic conductivity and high specific surface area are used. for energy storage devices such as LIBs, SCs, and energy conversion systems.. Therefore carbon, aerogels and graphene are being extensively employed as active electrode materials in the 3D-printing for various energy applications [19], [20].

4.2.1.1 Graphene-Based Materials

Among the varied active carbon-based materials, graphene has been established with superiority in comparison to other carbon materials in terms of its applicability. It possesses adequate intrinsic capacity and appreciable mechanical strength and is endowed with flexibility and excellent electrical conductance. In recent times, significant progress has been made in the 3D-printing technique, specifically the inkjet technique for graphene-based materials, as it is widely recognized for 3D printing due to its superlative rheological properties. First, 3D-printed graphene-based SC was fabricated by Li and the group, in which GO dispersed inks were used for printing the 3D graphene electrodes [21]. Here in, a symmetric device-based supercapacitor fabricated from the 3D-printed graphene-based electrodes, which demonstrated appreciable capacity, was assembled. The as-printed electrodes have shown an ordered porous structural organization and adequate electrical conductance. Similarly, Jiang and the group designed 3D GO electrodes with hierarchical architectures. The printed GO gel ink displayed an enhanced super-capacitive and cyclic performance ascribed to perfectly designed macro-and mesopore structures and interconnected conductive skeletons [22].

However, it has not been extensively adopted by many industries, largely due to practical complications involving poor efficiency, high cost, and minimum practicability of mass production, which need to be critically addressed. As such, Tagliaferri et al. developed a cost-effective approach

FIGURE 4.3 The use of graphene as energy material for 3D printing by Fudan University. Adapted from reference [24]. Copyright (2015), Springer Nature. Distributed under a Creative Commons Attribution License 4.0 (CC BY).

based on the 3D printable graphene electrode. They also demonstrated the preparation of inks for the pristine graphene without requiring high process temperature and chemical additives. The electrodes could surpass the performance of already reported 3D-printed carbon-based materials [23]. Various kinds of printing materials have been developed with the dimensional expansion of graphene, allowing the printing method to be used for electrochemical systems. For instance., Wei et al. fruitfully fused the graphene with a thermoplastic polymer, such as acrylonitrile-butadiene-styrene, to create a 3D-printed filament (with a 5.6 wt% loading amount of graphene). This graphene filament could be manufactured into computer-designed models using a low-cost 3D printer with an electrical conductivity of 1.05×10^{-3} S/m. A functional electrode for electrochemical devices could be made using these 3D printing electrodes (Figure 4.3) [24].

Despite the well-established procedures for the 3D-printed graphene materials and successful diversity in the printed architectures such as nanowires, aerogel-based micro lattices, and complex networks. The fabrication processes of 3D objects with high precision and accuracy need to have better control over the designing and rationalizing of miniaturized devices. This new emerging field requires vigorous efforts to develop standard protocols for successful 3D fabrication of electrodes for energy storage and conversion technology. Furthermore, there is an acute need to stabilize post-treatment such as pyrolysis or post-reduction procedures for the as-printed architectures to increase the phase purity and degree of reduction of graphene. The graphene and polymer composite (Figure 4.4) are used for the 3D printing to fabricate the device, and it shows that future research must realize the importance of combining the current standardized printing procedures with other processing methods to obtain highly resolved well-structured electrode materials [24]–[26].

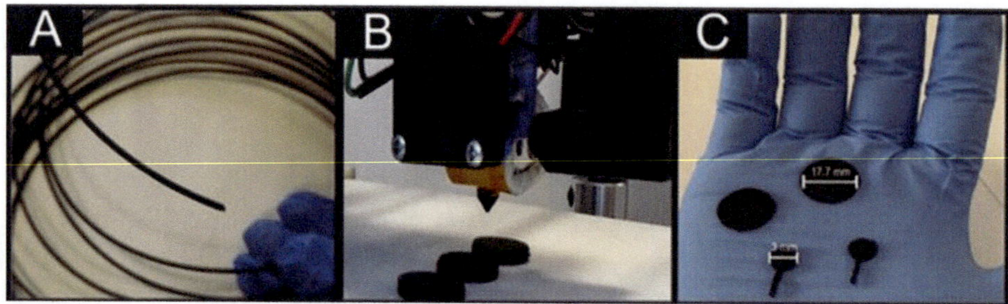

FIGURE 4.4 Optical images of the 3D printable graphene/PLA (A), the 3D printing process (B), and a variety of printed 3DEs (C). Adapted from reference [26]. Copyright (2017), Springer Nature. Distributed under a Creative Commons Attribution License 4.0 (CC BY).

4.2.1.2 Carbon Aerogel

As mentioned, the electrode materials having hierarchical macro and micropore architectures are efficient in ion diffusion, enhancing the rate performance for the SC when subjected to faster charging-discharging cycles. 3D printing generates a lattice structure with hierarchical and porous channels to be formed. For instance, Yao et al. fabricated a 3D-printing-based multidimensional carbon aerogel by combing chemical and direct ink writing techniques. The as-synthesized electrode material having high porosity and surface area of ~1750 m^2/g was also operational at low temperatures. The fabricated symmetric assembly achieved an appreciable capacitance of 148.6 F/g at 5 mV/s with capacitance retention up to 71.4 F/g at a scan rate of 200 mV/s. The study is crucial to understanding the open porous structures' influence on preserving capacitive performance at ultralow temperatures [27].

4.2.1.3 CNT-Based Materials

Considering the materials perspective, CNTs also have an immense perspective as electrode materials owing to their feeble toxicity and cost-effectiveness. Moreover, their remarkable mechanical strength, appreciable conductivity, and larger surface area can be enhanced for better performance of the as-fabricated device via attaching surface functional groups. 3D printing or additive manufacturing offers a propitious solution to reduce the cost associated with fabrication. It tends to provide less chaotic steps to generate complex 3D structures. CNTs-based 3D-printed micro-supercapacitors (MSCs) developed by Prof. Ding and his group shows a novel technique, fabricated self-standing electrodes using varied ink concentrations of CNTs. They also demonstrated exceptional durability and reliability in terms of capacity, suggesting a futuristic CNT-based device with high power and promising results in the energy storage field [28].

4.2.2 Conductive Polymer

Conductive polymers, another class of electrically conducting materials, have outstanding mechanical flexibility and appreciable printing ability. The commonly explored conducting polymers, especially for energy storage and conversion, are polypyrrole (PPy), polyaniline (PANI), and polythiophene (PTh),) and poly(3,4-ethylenedioxythiophene) PEDOT). Among them, PEDOT is majorly the potential material, because of its highly conductive nature and remarkable stability under ambient conditions, with a wide range of applications. The enhanced performance of PEDOT-based 3D-printed electrodes could be attributed to their exceptional viscoelastic nature. Considering these attributes, Wei et al. reported multidimensional structures based on the poly(3,4-ethylenedioxythiophene): polystyrene sulfonate (PEDOT: PSS) with tunable structures and structural patterns. This work provides a simple, scalable approach to synthesizing electrode materials, which could be a probable solution for developing flexible electronic devices [29], [30]. Furthermore, various conducting polymer-carbon composites can be used in 3D printing and can be directly employed as electrodes. For instance, a PEDOT: PSS-based electrode 3D printed on a cellulose template provides new horizons for the large-scale development of flexible SCs. The practicality of cellulose was to provide the adequate mechanical strength to fabricate thick-electrode SCs following a stencil-based printing procedure. Moreover, this facile amalgamation of PEDOT: PSS and carbon, having a minimum quantity of PEDOT: PSS, was reported to have enhanced performance as compared to standard commercialized SCs of activated carbon and carbon black in terms of the optimal device performance and cost-effectiveness.

Considering these attributes, a high-performance 3D printable ink based on one of the most extensively used conducting polymers PEDOT: PSS was developed. To achieve rheological qualities suitable for 3D printing, as shown in Figure 4.5. They presented a PEDOT-based high-performance 3D printable conducting polymer ink: PSS capable of fabricating highly conductive microscale structures and devices with inherent flexibility. Moreover, the work showed how to make a high-density flexible electronic circuit and a soft neural probe using 3D printing, which is facile, rapid, and greatly simplified. This research not only solves current issues in conducting polymer 3D printing but also proposes a viable fabrication technique for flexible electronics, wearable devices, and bioelectronics using conducting polymers.

In addition to the novel carbon/conducting polymer composites used for 3D printing in energy applications, a currently evolving class is conductive polymer hydrogels with the functionality of hydrogel and the electrical conductivity of carbon-based materials, conducting polymers, metal oxide, electrolytic ions, etc. The conductive polymer-based hydrogels help in inheriting the profitability of both the phases, viz., liquid, and solid, making them propitious to be used as sensors and wearable/portable electronic devices (Figure 4.6), and most importantly, flexible energy-storage devices. In the case of flexible electrochemical SCs, these distinctive conductive polymer-based

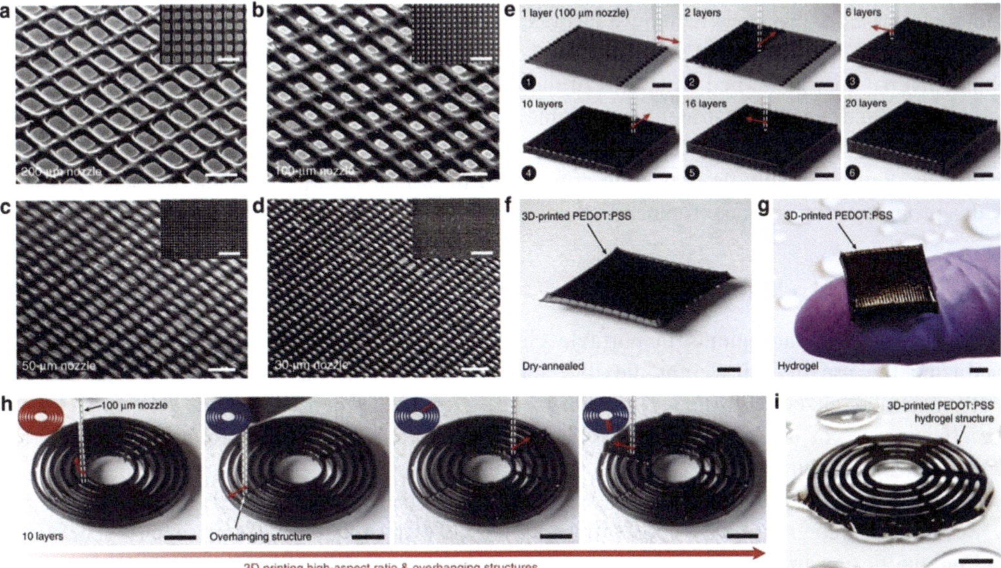

FIGURE 4.5 SEM images of 3D-printed different mesh size energy material fabricated by conducting polymers. Adapted from reference [31], Copyright: The Authors (2020), Springer Nature. Distributed under a Creative Commons Attribution License 4.0 (CC BY).

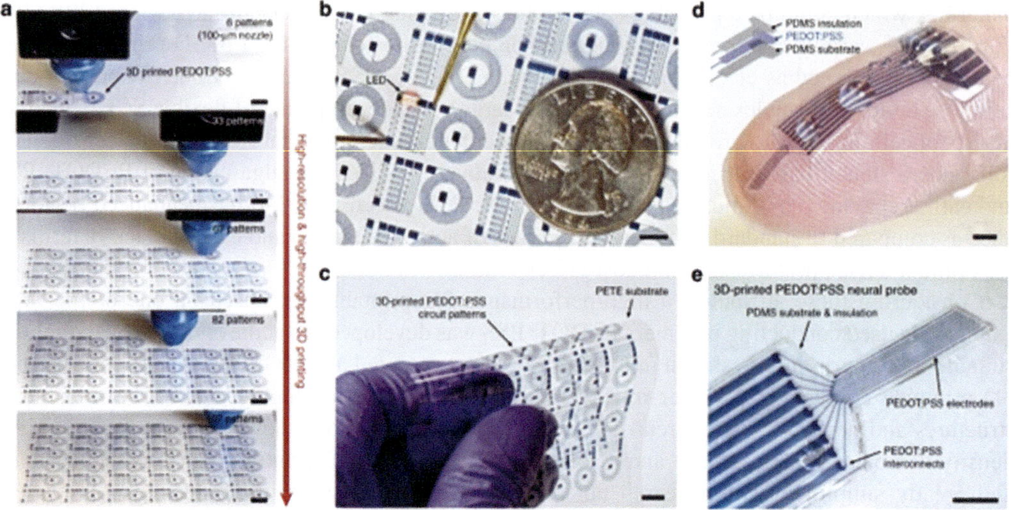

FIGURE 4.6 Sequential snapshots for 3D printing of high-density flexible electronic circuit patterns by the conducting polymer ink for lighting up an LED on the 3D-printed conducting polymer circuit. Adapted from reference [31], Copyright (2020), Springer Nature. Distributed under a Creative Commons Attribution License 4.0 (CC BY).

hydrogels can serve as practical contenders as conventional electrode and electrolyte materials due to their appropriate electrical, mechanical, and chemical properties [32].

Therefore, 3D printable technology is a facile and reliable approach for preparing conducting polymer hydrogel-based active material into a fully integrated energy device. Developing these electrically conductive polymer-based hydrogels, which are consistent in the printing, could be another futuristic approach, which must be done via acute precision in assembling the SCs or energy storage devices. For instance, Pan et al. reported a unique and approachable synthetic procedure to obtain diverse micropatterned structures of the polyaniline hydrogel using ink-jet printing. The as-assembled electrode demonstrated appreciable surface area with 3D porous microstructures with enhanced performance as SC electrodes. Considering this work, these conducting polymer-based hydrogels could probably be appealing contenders in bio-medical/implantable electronic devices with a propitious future as next-generation energy electrodes [33].

4.2.3 Fiber-Based Materials

In recent times, advancements in portable electrochemical energy storage (EES) devices have facilitated the use of lightweight, flexible and wearable electronic devices. However, the currently commercialized EES devices cannot meet the required energy demands. The hindrance to the effective use of the commercialized devices is the expenses associated with fabrication, rate performance, and durability [34]. To deal with this, numerous prospective have been followed to enhance the material's properties like flexibility and stretchability using 3D printing.

Several research groups have constantly explored novel active materials with inherited flexibility and unique design to improve their usability in energy applications. For instance, Peng and the group demonstrated flexible and wearable wire-shaped electroactive materials with excellent rate performance as the super capacitive electrode. Chou and his co-workers also showed the assembling of wire-shaped ASCs based on CNTs-based fibers and compared its rate performance with MnO_2-decorated CNT fibers. The as-designed electrode showed ideal tensile strength. Many efforts have been directed towards one-step dimensional materials, including fibers, wire, and rods, to generate extremely compressed and flexible energy storage devices which can effortlessly be woven into textile-based portable/ implantable electronics. Further, it can be used safely for a wide range of applicability in energy harvesting [35]. In the past few years, 3D printing has drawn significant attention owing to the appreciable electrochemical performance associated with the unique construction of EES devices compared to traditional fabrication methods. Zhao et al. used 3D- direct ink writing to assemble the ASCs composed of a single-walled carbon nanotube (SWCNT)/V_2O_5 fibrous cathode and an SWCNT/VN as an anode. The group reported exceptional values of specific capacitance (C_s) for the 3D printed fibers asymmetric supercapacitors, which is much higher than the reported 1D SCs [36]. A novel type of MSCs comprising hybrid multi-dimensional Fe_2O_3/graphene/Ag ink was reported by Hou and the group through 3D direct ink writing. The all-solid-state MSC device exhibited optimal performance with the areal capacitance of 412.3 mF/cm^2 at a current density of 2 mA/cm^2 and capacity retention of 89% up to 5000 charging and discharging cycles. This remarkable super capacitive performance could be attributed to adequate electrical transportation between the graphene nanosheets and 1D Ag nanowires and pseudocapacitive contributions from metal oxide nanoparticles. The printed devices exhibited reliability, flexibility and demonstrated a proficient route for advanced miniaturized EES devices [37].

4.2.4 Nanocomposites Using 3D Printing Technology

The use of nanocomposites in a wide range of applications, from electronics to electromagnetic interference shielding, energy storage to batteries, sensors, and even lightning strike protection in airplanes, is becoming increasingly popular. The production of conductive nanocomposites can be carried out using various traditional methods. Compression molding, solvent-casting, and injection molding are some of the techniques available. Even though these approaches are effective, the requirement for creating molds in these methods prompts researchers to consider any other methods that do not necessitate the creation of molds for each building [38].

3D printing technology is a highly effective technology that offers an alternative to conventional ways while also providing the advantage of not having to create molds for each structure. Various kinds of 3D printing technology techniques have been used to synthesize nanocomposites. This section of the chapter briefly summarizes different reported nanocomposites (synthesized using 3D printing technology) and used in various applications over the years. 3D printing is an adequate technique that ensures the homogenous dispersion of nanoparticles, prevent them from aggregating, and minimizes defects such as microcracks or voids in the matrix, improving their dielectric properties. Considering these facts, Kim et al. studied the effect of adding $BaTiO_3$ and MWCNT into PVDF polymer for improved dielectric properties. It was observed that 3D printed polymer exhibits a higher dielectric constant value than solvent cast polymer, which can be attributed to the extrusion process of the filament. Along with these benefits, the addition of $BaTiO_3$, has significantly enhanced the energy storage capability. With the addition of MWCNT, the size of defects such as voids and microcracks has reduced significantly [39]. When the benefits of 3D printing are merged with electrically conductive materials, obtained 3D-printed nanocomposites can be used for various applications. Electronic Sensing is one such great application. Farahani et al. fabricated 3D free-standing patterned strain sensors comprising single-walled carbon nanotubes/Epoxy nanocomposites with very high electromechanical sensitivity, i.e., prepared composites is highly sensitive to even very small mechanical disturbances (with a gauge factor of about 22) [40].

4.2.5 MOF-Based Structures Using 3D Printing Technology

Metal-Organic Frameworks (MOFs) are promising nano porous functional materials that have piqued the interest of researchers for a variety of essential applications, ranging from electrocatalysis to photocatalysis, energy storage to batteries, sensing to gas storage and separation. This class of materials was distinguished by their excellent stability, high porosity, large surface area, and excellent structural and compositional flexibility [41]. While a wide variety of MOFs has been synthesized on a smaller laboratory scale, only a few have transitioned to large-scale industrial applications and commercialization. It is imperative that after processing, the MOFs should maintain their unique properties of high surface area and high porosity to fulfill the requirements for large-scale applications. Because MOFs are typically synthesized as bulk crystalline powders that are ineffective for large-scale industrial applications, forming and consolidating MOF structures into monoliths is a critical but often disregarded element of the manufacturing process [42]–[43].

Incorporating MOFs into polymer matrix before molding them into valuable structures (such as filaments, spheres, films, and granules) is necessary to consolidate MOF structures into monoliths [2]. As of now, there are three main ways by which this goal can be achieved. The most common method is palettization, i.e., forming pellets and granules by hydraulic pressing MOF powders. Although this method is rapid and straightforward, at the same time, it blocks the accessible porous sites of MOFs and causes mass transfer limitations. The second method of achieving the goal involves growing MOF on functionalized support like carbon fibers. Apart from the various advantages of this technique, its more extended synthesis time feature limits its usage for industrial applications. Therefore, when MOFs are transformed into monoliths by the above two routes, their porosity is restricted by the matrix materials used in the conversion process, contamination from equipment can occur, and crystallinity gets altered by the mechanical forces used in the conversion process. So, the third route of 3D printing has gained much attention because it permits the fabrication of multiple complex parts using the same machine, with significantly less waste than traditional manufacturing methods. Besides this technique works without the need for expensive retooling [41].

3D printing technology has been used to manufacture various MOF-based products up to this point. These MOF-based materials, which were created with 3D printing technology, have been used in various applications, including catalysis, water remediation, sensing, biological applications, and so on. This section of the chapter briefly summarizes different reported MOF-based materials (which have been synthesized using 3D printing technology) used in various applications over the years.

4.2.5.1 MOF-Based Structures Using 3D Printing Technology for Electrocatalytic Applications

Electrically conductive materials such as glassy carbon (GC), fluorine-doped tin oxide (FTO), indium tin oxide (ITO), and conductive foams are routinely used in electrochemical energy conversion research. Still, their high cost and smaller geometric surface area limit their application to a larger scale. For the fabrication of conductive electrodes for electrocatalytic applications, with customizable shape and geometric surface area, newly developed 3D printing technology, such as selective laser melting (SLM) and fused deposition modelling (FDM), are simple and low-cost alternatives to traditional methods. By use of 3D-printed technology, Ying et al. reported ZIF-67/Ti-E electrode for electrocatalytic oxygen evolution reaction (OER). Compared to other electrodes created using the widely utilized calcination process, the ZIF-67/Ti-E electrode has shown superior OER performance with outstanding durability and a low overpotential of 360 mV at a current density of 0.01 A/cm^2 [44].

4.2.6 MXene-Based Structures Using 3D Printing Technology

Recently, the most prevalent, versatile with a wide application range, two-dimensional carbides, nitrides, and carbonitrides (of early transition metals), commonly referred to as MXenes, have drawn attention. MXenes are synthesized from MAX phases composed of layered ternary carbides with the general formula $A_{n+1}BX_n$, where A is generally a transition metal (Zr, Ta, Sc, Ti, V, Cr, Nb,

Zr, Mo), B is a group 12–16 element (Cd, Al, Pb, Ge, Si, P, Pb, S, Ga, As, In, Sn, Tl), and X is carbon or nitrogen. MXene possesses unique properties such as hydrophilicity, high surface area, high electrical conductivity, larger interlayer spacing, excellent structural flexibility, and good thermal stability, which is a promising material for energy application [45].

Despite their superior characteristics, aggregation, and self-restacking of single or multiple layers of nanosheets are typically unavoidable during the electrode fabrication process because of the strong van der Waals interactions and hydrogen bonding (existing between neighboring nanosheets). Due to the limitations in electrolyte ion accessibility caused by these processes, complete usage of the functional surfaces of MXenes is hindered, resulting in a reduced electrochemical activity. To overcome these challenges, 3D architectures using 3-D printing technology are emerging as one promising solution. By using 3-D printing technology, 3D architectures can be constructed from 2-Dimensional MXenes, with several advantages. The abundance of channels in 3D MXene frameworks favors faster electrolyte diffusion. Their excellent electrical conductivity makes them ideal active materials for the rapid transport of charge carriers in 3D MXene architectures.

Moreover, in the 3D MXene structures, the large specific surface area of MXene nanosheets can be maintained because each nanosheet is effectively prevented from restacking. As a result, many electrochemically active sites can be exposed to the electrolyte, allowing for enough electrochemical reactions to occur. There are many functional groups (OH, F) present in 3D MXene, which causes enhancement in the active sites of 3D MXene. Due to the abundance of these functional groups, 3D MXene structures can be easily constructed with varying degrees of complexity and functionality. Such 3D structures enable effective charge transport across electrodes to maximize the use of active materials [46].

Supercapacitors are regarded as promising candidates for power devices in the future. They have a greater capacity for energy storage than traditional capacitors and can deliver it at a higher power density than batteries. Combined with their excellent cyclability and stability, these characteristics make Supercapacitors an excellent energy storage solution [47]. Traditionally, materials like Carbon, Metal oxide, Conducting Polymers, and Metal sulfides are used as a material for electrodes. Although having various advantages of these materials, the electrode of these material suffers from a shorter life cycle and limited electrical conductivity. MXene is emerging as a promising solution to overcome these issues, which provides faster electron and ion transport and higher volumetric and areal capacitance [46], [48]. So, by using 3D printing technology, 2D MXenes incorporated into the 3D framework provide advantages of higher surface area, higher electrical conductivity, and faster electron and ion transport. In this regard, Redondo et al. Ti_3C_2@3DnCEs (i.e., Ti_3C_2 functionalized 3D-printed nanocarbon electrodes) with about three-fold enhanced capacitance and excellent capacitance retention ability [49]. 2D titanium carbide ($Ti_3C_2T_x$) MXene inks are developed without any additive or binary solvents, by Valeria Nicolosi and her group for 3D printing, which shows excellent electrochemical behavior with a volumetric capacitance of 562 F/cm^3 and energy density of 0.32 μWh/cm^2 [11]. The use of $Ti_3C_2T_x$ in the development of micro-supercapacitors by 3D printing is summarized by the Nicolosi group and presented in Figure 4.7.

Energy storage devices such as lithium-ion batteries and sodium-ion batteries are considered one of the most significant discoveries. In recent times, 2D MXenes are emerging as promising LIB electrode materials due to their excellent structural flexibility, high electrical conductivity, and high theoretical capacity. By using 3D printing technology, the activity of MXene can also be improved. For instance, Zhao et al. reported 3D MXene foam (synthesized using sulfur template method). The MXene foam's three-dimensional porous design provides large active sites that significantly increase the storage capacity of lithium. Furthermore, this foam structure allows the rapid lithium-ion transfer. Due to these reasons, this flexible three-dimensional porous MXene foam has a significantly increased capacity of 455.5 mAh/g at the current density of 0.05 A/g a good rate performance of 101 mAh/g at the current density of 18 A/g, and excellent capacitance retention ability even after 3500 cycles [50].

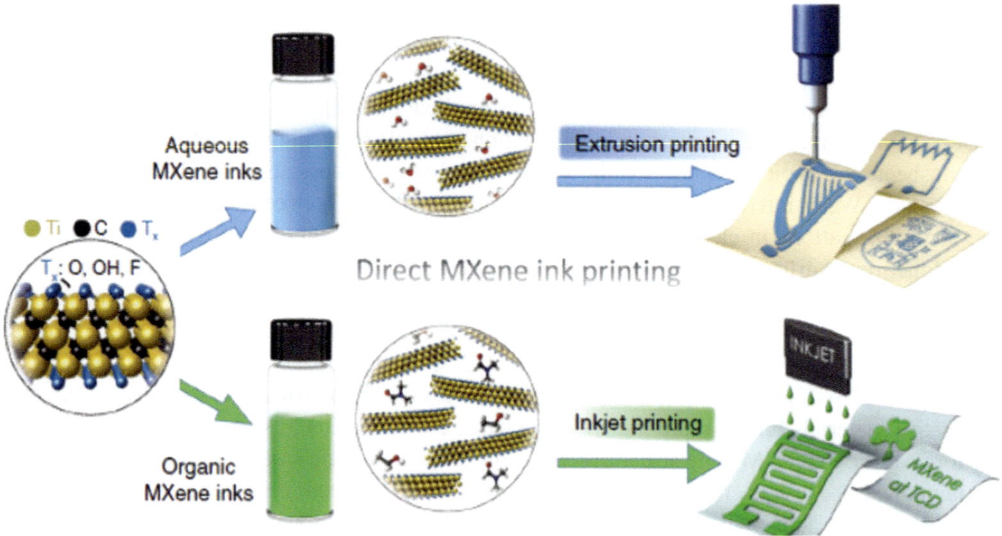

FIGURE 4.7 Schematic illustration of direct MXene ink printing. The $Ti_3C_2T_x$ organic inks, i.e., $Ti_3C_2T_x$-ethanol (molecules shown in the bottom panel) are used for inkjet printing of various patterns, such as MSCs, MXene letters, ohmic resistors, etc. Adapted from reference [11], Copyright (2015), Springer Nature. Distributed under a Creative Commons Attribution License 4.0 (CC BY).

4.3 ADVANTAGES AND DISADVANTAGES OF ENERGY MATERIAL FOR 3D PRINTING

There are several advantages to the 3D printing techniques by using energy material as it is a cost-effective technique, with ease of accessibility, and structural flexibility. In addition, this technique comes with zero waste production and rapid prototyping which makes it an environmentally viable technique. From the device fabrication point of view, specifically energy storage or conversion, the available energy materials used for them through various technique makes life easier. The commonly used energy materials such as plastics, metals, ceramics, MXenes, MOFs, composites sol-gels, metals, alloys, thermoplastic, conductive polymers, and carbon-based materials with the following advantages for 3D printing application over other synthesis techniques. It is an affordable, and easy operational procedure with wide material diversity, multi-material printing capability, and large area printing ability. it is a rapid technique with large size capability, high resolution, high surface finish, and efficiency along with a lack of chemical post-processing requirement.

Nevertheless, the 3D printing technique loses its significance with several disadvantages while using energy material for device fabrication. The 3D technique will reduce the use of manufacturing labor which will further affect the economy of countries, and the skilled people will become jobless. Simultaneously several harmful objects such as guns, knives, etc. can be manufactured, which will become a threat to society. Apart from this the key parameters such as conductivity (electronic and ionic), surface area (porosity), and mechanical stability, which are required for the functional materials in energy device fabrication, may not be achieved as the functional polymer-based composites may reduce the conductivity, sometimes reducing their stability. While the current strategy of using graphite-loaded PLA to boost electrical conductivity will make the material becomes very brittle. Along with this, there are several other disadvantages associated with the energy material utility, as listed below.

- Especially for supercapacitors, a limited exposed surface area provided by the 3D printing technique as the slurry cast over the print may reduce the performance.

- The surface area provided by the printed structures is electrically inactive.
- The electronic conductivity is low, and the voltage should be fused with the inks of the electrodes.
- Ionic diffusivity is limited.
- With the increase in surface area value, the performance of the device may be undesirably low as the expanded contact between electrode materials with electrolyte as the electrochemical reaction depends on the nature of porosity and surface area of active energy material.
- Poros of the active material may be blocked using polymers which may not support the electrochemical reaction.
- Carbon-based energy materials like CNTs. may enhance conductivity, but due to their brittleness, it affects the property.
- Thermal degradation can also offer comparable benefits to solvent etching, while the complex structures are sustained after mild thermal action or surface-selective melting.
- Still, much attention has not been paid to the current substrate, electrolyte, separators, and packaging materials, which limits the application of energy material in 3D printing for electrochemical energy storage/conversion device fabrication.
- The accessibility of printable energy materials in 3D printing is still limited as only a few selected materials are explored for this technique.

4.4 FUTURE PERSPECTIVES

Despite all the disadvantages 3D printing technique emerged as one of the most promising techniques for energy devices. The above-said limitations are the key considerations that need to be solved to achieve the required performance with valuable and durable structural characteristics. In energy devices, the most important characteristics of the 3D printing method completely depend on the printed energy material performance with acute fabrication, superior resolution, and material porosity with the desired composition, along with multilateral printing capability. Only a few approaches have been discovered to take advantage of a straightforward growth of the performance of the cells by alteration of its geometry, likely due to the strict limitations in manufacturing complex shapes. The different shapes of energy material may lead to the solve the geometry issue. Carbon fibers will be a good replacement for the carbon-based energy material to avoid brittleness. The use of thermoset printed PLA will enhance the voltage with better gravimetric energy density. The use of polymer and different carbon-based energy materials will help in increasing conductivity which will be one of the much need research topics. In energy or chemical sectors, the conversion of CO_2 into valuable products will be an exciting field in 3D printing. In the Li-ion batteries, polymeric material will help in restricting the volume expansion while the cell enclosing and limited permeability and porosity, which will be beneficial for other issues such as softening or chemical stability during solvent intake. To avoid any mechanical property loss due to the layered printing process, a rationally designed 3D structure will be helpful for electrochemical applications in energy storage and conversion. Interconnected nanoarchitectures and hierarchical porous structures will provide a particular path for electron conduction. More attention is required for reproducibility and bulk production. The impending research on electrochemical devices for any application based on 3D printing should focus on how to reduce the fabrication cost, use a wide range of materials and control the overall volume and weight. The designing of some extraordinary devices with numerous micron structures apart from the submicron or micron size is much needed for energy harvesting systems. The state of art materials from 3D printing should be diversified with 1D or 2D periodic structures for different used materials in energy devices such as current collectors, anode, cathode, separators, and electrolytes. Further research should be focused on developing advanced devices such as integrated electronics, sensors, wearable devices, wireless pharmacology, and mobile technologies.

4.5 CONCLUSIONS

Above all, the new and exciting advanced 3D printing technology enabled several solutions in the electrochemical energy device fabrication with different energy materials. The traditional energy materials used by different synthesis methods can act as futuristic materials in 3D printing. One can expect that the energy material used in 3D printing will give new direction for the future development of anodes, cathodes, separators, current collectors, etc. The very well-known carbon-based materials, MOFs, and MXenes will help fabricate advanced high-energy density energy storage devices. After all, this new advanced technique provides enormous opportunities and has the potential to cross the border of printable materials. The future of printable materials shows some new challenging opportunities in materials design, as well as broadens the scope of applications for 3D printing technologies in various advanced fields such as optics, electronics, biomedical engineering, sensors, catalysts, and environmental protection.

ACKNOWLEDGMENTS

The authors are thankful to IIT Delhi for their permission to write the book chapter. RB is grateful to Institute Postdoctoral Fellowship provided by IIT Delhi. VT is thankful to MHRD for PMRF Fellowship. PK is grateful to CSIR, India for providing the NET-JRF fellowship.

REFERENCES

1 M. P. Browne, E. Redondo, M. Pumera, 3D Printing for Electrochemical Energy Applications, *Chem. Rev*. 2020, 120, 2783–2810.

2 V. Egorov, U. Gulzar, Y. Zhang, S. Breen, C. O'Dwyer, Evolution of 3D Printing Methods and Materials for Electrochemical Energy Storage, *Adv. Mater*. 2020, 32(29), 2020, 2000556.

3 A. Pesce, A. Hornés, M. Núñez, A. Morata, M. Torrell, A. Tarancón, 3D Printing the Next Generation of Enhanced Solid Oxide Fuel and Electrolysis Cells, *J. Mater. Chem. A* 2020, 8, 16926–16932.

4 X. Tian, J. Jin, S. Yuan, C. K. Chua, S. B. Tor, K. Zhou, Emerging 3D-Printed Electrochemical Energy Storage Devices: A Critical Review, *Adv. Energy Mater*. 2017, 7(17), 1700127.

5 P. Chang, H. Mei, S. Zhou, K. G. Dassios, L. Cheng, 3D Printed Electrochemical Energy Storage Devices, *J. Mater. Chem. A* 2019, 7, 4230–4258.

6 A. Jandyal, I. Chaturvedi, I. Wazir, A. Raina, M. I. U. Haq, 3D Printing – A Review of Processes, Materials and Applications in Industry 4.0, *Sustain. Oper. Comput*. 2022, 3, 33–42.

7 N. Shahrubudin, T. C. Lee, R. Ramlan, An Overview on 3D Printing Technology: Technological, Materials, and Applications, *Procedia Manuf*. 2019, 35, 1286–1296.

8 M. Zhang, H. Mei, P. Chang, L. Cheng, 3D Printing of Structured Electrodes for Rechargeable Batteries, *J. Mater. Chem. A* 2020, 8, 10670–10694.

9 X. Gao, Q. Sun, X. Yang, J. Liang, A. Koo, W. Li, J. Liang, J. Wang, R. Li, F. B. Holness, A. D. Price, S. Yang, T.-K. Sham, X. Sun, Toward a Remarkable Li-S Battery via 3D Printing, *Nano Energy* 2019, 56, 595–603.

10 K. Sun, T.-S. Wei, B. Y. Ahn, J. Y. Seo, S. J. Dillon, J. A. Lewis, 3D Printing of Interdigitated Li-Ion Microbattery Architectures, *Adv. Mater*. 2013, 25(33), 4539–4543.

11 C. (John) Zhang, L. McKeon, M. P. Kremer, S.-H. Park, O. Ronan, A. Seral-Ascaso, S. Barwich, C. Ó Coileáin, N. McEvoy, H. C. Nerl, B. Anasori, J. N. Coleman, Y. Gogotsi, V. Nicolosi, Additive-free MXene Inks and Direct Printing of Micro-Supercapacitors, *Nat. Commun*. 2019, 10, 1795.

12 F. Calignano, T. Tommasi, D. Manfredi, A. Chiolerio, Additive Manufacturing of a Microbial Fuel Cell-A detailed study, *Sci. Rep*. 2015, 5, 17373.

13 T. E. Mogy, D. Rabea, An Overview of 3D Printing Technology Effect on Improving solar photovoltaic systems efficiency of renewable energy, *Proc. Int. Acad. Ecol. Environ. Sci*. 2021, 11(2), 52–67.

14 V. Katseli, A. Economou, C. Kokkinos, Single-Step Fabrication of an Integrated 3D-Printed Device for Electrochemical Sensing Applications, *Electrochem. Commun*. 2019, 103, 100–103.

15 K. Fu, Y. Yao, J. Dai, L. Hu, K. Fu, Y. Yao, J. Dai, L. Hu, Progress in 3D Printing of Carbon Materials for Energy-Related Applications, *Adv. Mater*. 2016, 29, 1603486.

16 Y. Wu, J. Zhu, L. Huang, A Review of Three-Dimensional Graphene-Based Materials: Synthesis and Applications to Energy Conversion/Storage and Environment, *Carbon* 2019, 143, 610–640.

17 X. Wang, Q. Li, J. Xie, Z. Jin, J. Wang, Y. Li, K. Jiang, S. Fan, Fabrication of Ultralong and Electrically Uniform Single-Walled Carbon Nanotubes on Clean Substrates, *Nano Lett.* 2009, 9(9), 3137–3141.

18 H. Wei, H. Wang, A. Li, H. Li, D. Cui, M. Dong, J. Lin, J. Fan, J. Zhang, H. Hou, Y. Shi, D. Zhou, Z. Guo, Advanced Porous Hierarchical Activated Carbon Derived from Agricultural Wastes toward High Performance Supercapacitors, *J. Alloys Compd.* 2020, 820, 153111.

19 M. R. Berber, I. H. Hafez (Eds.)., *Carbon Nanotubes: Current Progress of their Polymer Composites – Condensed Matter Physics,* 2016, London, United Kingdom, IntechOpen.

20 S. Hamid, R. Sanei, D. Popescu, 3D-Printed Carbon Fiber Reinforced Polymer Composites: A Systematic Review, *J. Compos. Sci.* 2020, 4, 98.

21 C. Zhu, T. Liu, F. Qian, T. Y. J. Han, E. B. Duoss, J. D. Kuntz, C. M. Spadaccini, M. A. Worsley, Y. Li, Supercapacitors Based on Three-Dimensional Hierarchical Graphene Aerogels with Periodic Macropores, *Nano Lett.* 2016, 16(6), 3448–3456.

22 Y. Jiang, Z. Xu, T. Huang, Y. Liu, F. Guo, J. Xi, W. Gao, C. Gao, Direct 3D Printing of Ultralight Graphene Oxide Aerogel Microlattices, *Adv. Funct. Mater.* 2018, 28(16), 1707024.

23 S. Tagliaferri, G. Nagaraju, A. Panagiotopoulos, M. Och, G. Cheng, F. Iacoviello, C. Mattevi, Aqueous Inks of Pristine Graphene for 3D Printed Microsupercapacitors with High Capacitance, *ACS Nano.* 2021, 15(9), 15342–15353.

24 X. Wei, D. Li, W. Jiang, Z. Gu, X. Wang, Z. Zhang, Z. Sun, 3D Printable Graphene Composite, *Sci. Rep.* 2015, 5, 11181.

25 J. H. Kim, S. Chang, D. Kim, J. R. Yang, T. Han, G.-W. Lee, J. T. Kim, S. K. Seol, J. H. Kim, W. S. Chang, D. Kim, J. T. Han, G.-W. Lee, S. K. Seol, J. R. Yang, J. T. Kim, 3D Printing of Reduced Graphene Oxide Nanowires, *Adv. Mater.* 2014, 27, 157–161.

26 C. W. Foster, M. P. Down, Y. Zhang, X. Ji, S. J. Rowley-Neale, G. C. Smith, P. J. Kelly, C. E. Banks, 3D Printed Graphene Based Energy Storage Devices, *Sci. Rep.* 2017, 7, 42233.

27 B. Yao, H. Peng, H. Zhang, J. Kang, C. Zhu, G. Delgado, D. Byrne, S. Faulkner, M. Freyman, X. Lu, M. A. Worsley, J. Q. Lu, Y. Li, Printing Porous Carbon Aerogels for Low Temperature Supercapacitors, *Nano Lett.* 2021, 21(9), 3731–3737.

28 W. Yu, H. Zhou, B. Q. Li, S. Ding, 3D Printing of Carbon Nanotubes-Based Microsupercapacitors, *ACS Appl. Mater. Interfaces*, 2017, 9(5), 4597–4604.

29 J. Yang, Q. Cao, X.Tang, J. Du, T. Yu, X. Xu, D. Cai, C. Guan, W. Huang, 3D-Printed Highly Stretchable Conducting Polymer Electrodes for Flexible Supercapacitors, *J. Mater. Chem. A* 2021, 9, 19649–19658.

30 D. Belaineh, R. Brooke, N. Sani, M. G. Say, K. M. O. Håkansson, I. Engquist, M. Berggren, J. Edberg, Printable Carbon-Based Supercapacitors Reinforced with Cellulose and Conductive Polymers, *J. Energy Storage* 2022, 50, 104224.

31 H. Yuk, B. Lu, S. Lin, K. Qu, J. Xu, J. Luo, X. Zhao, 3D Printing of Conducting Polymers, *Nat. Commun.* 2020, 11, 1604.

32 T. Cheng, Y.-Z. Zhang, S. Wang, Y.-L. Chen, S.-Y. Gao, F. Wang, W.-Y. Lai, W. Huang, T. Cheng, Y. Zhang, S. Wang, Y. Chen, S. Gao, F. Wang, W. Lai, W. Huang, Conductive Hydrogel-Based Electrodes and Electrolytes for Stretchable and Self-Healable Supercapacitors, *Adv. Funct. Mater.* 2021, 31(24), 2101303.

33 G. P. Hao, F. Hippauf, M. Oschatz, F. M. Wisser, A. Leifert, W. Nickel, N. Mohamed-Noriega, Z. Zheng, S. Kaskel, Stretchable and Semitransparent Conductive Hybrid Hydrogels for Flexible Supercapacitors, *ACS Nano.* 2014, 8(7), 7138–7146.

34 Y.-Q. Xiao, C.- W. Kan, Review on Development and Application of 3D-Printing Technology in Textile and Fashion Design, *Coatings* 2022, 12, 267.

35 L. Kou, T. Huang, B. Zheng, Y. Han, X. Zhao, K. Gopalsamy, H. Sun, C. Gao, Coaxial Wet-Spun Yarn Supercapacitors for High-Energy Density and Safe Wearable Electronics, *Nat Commun.* 2014, 5, 3754.

36 J. Zhao, Y. Zhang, Y. Huang, J. Xie, X. Zhao, C. Li, J. Qu, Q. Zhang, J. Sun, B. He, Q. Li, C. Lu, X. Xu, W. Lu, L. Li, Y. Yao, 3D Printing Fiber Electrodes for an All-Fiber Integrated Electronic Device via Hybridization of an Asymmetric Supercapacitor and a Temperature Sensor, *Adv. Sci.* 2018, 5(11), 1801114.

37 K. Tang, Y. Z. Tian, H. Jin, S. Hou, K. Zhou, X. Tian, 3D printed hybrid-dimensional electrodes for flexible micro-supercapacitors with superior electrochemical behaviours, *Virtual Phys. Prototyp.* 2020, 15, 511–519.

38 K. Chizari, M. A. Daoud, A. R. Ravindran, D. Therriault, 3D Printing of Highly Conductive Nanocomposites for the Functional Optimization of Liquid Sensors, *Small* 2016, 12(44), 6076–6082.

39 H. Kim, B. R. Wilburn, E. Castro, C. A. Garcia Rosales, L. A. Chavez, T. L. B. Tseng, Y. Lin, Multifunctional SENSING Using 3D Printed CNTs/$BaTiO_3$/PVDF Nanocomposites, *J. Compos. Mater.* 2019, 53(10), 1319–1328.

40 X. Wang, M. Jiang, Z. Zhou, J. Gou, D. Hui, 3D Printing of Polymer Matrix Composites: A Review and Prospective, *Compos. Part B Eng.* 2017, 110, 442–458.

41 S. Lawson, A.-A. Alwakwak, A. A. Rownaghi, F. Rezaei, Gel-Print-Grow: A New Way of 3D Printing Metal-Organic Frameworks, *ACS Appl. Mater. Interfaces*, 2020, 12(50), 56108–56117.

42 S. Mallakpour, E. Azadi, C. M. Hussain, MOF/COF-Based Materials Using 3D Printing Technology: Applications in Water Treatment, Gas Removal, Biomedical, and Electronic Industries, *New J. Chem.* 2021, 45(30), 13247–13257.

43 G. J. H. Lim, Y. Wu, B. B. Shah, J. J. Koh, C. K. Liu, D. Zhao, A. K. Cheetham, J. Wang, J. Ding, 3D-Printing of Pure Metal–Organic Framework Monoliths, *ACS Materials Lett.* 2019, 1(1), 147–153.

44 Y. Ying, M. P. Browne, M. Pumera, Metal-Organic-Frameworks on 3D-Printed Electrodes: In Situ Electrochemical Transformation towards the Oxygen Evolution Reaction, *Sustain. Energy Fuels* 2020, 4(7), 3732–3738.

45 N. K. Chaudhari, H. Jin, B. Kim, D. San Baek, S. H. Joo, K. Lee, MXene: An Emerging Two-Dimensional Material for Future Energy Conversion and Storage Applications, *J. Mater. Chem. A*, 2017, 5(47), 24564–24579.

46 K. Li, M. Liang, H.Wang, X. Wang,Y. Huang, J. Coelho, S. Pinilla, Y. Zhang, F. Qi, V. Nicolosi, 3D MXene Architectures for Efficient Energy Storage and Conversion, *Adv. Funct. Mater.* 2020, 30(47), 1–22.

47 J. Castro-Gutiérrez, A. Celzard, V. Fierro, Energy Storage in Supercapacitors: Focus on Tannin-Derived Carbon Electrodes, *Front. Mater.* 2020, 7, 217.

48 M. R. Lukatskaya, S. Kota, Z. Lin, M. Q. Zhao, N. Shpige, M. D. Levi, J. Halim, P. L. Taberna, M. W. Barsoum, P. Simon, Y. Gogotsi, Ultra-High-Rate Pseudocapacitive Energy Storage in Two-Dimensional Transition Metal Carbides, *Nat. Energy* 2017, 6, 17105.

49 E. Redondo, M. Pumera, MXene-Functionalised 3D-Printed Electrodes for Electrochemical Capacitors, *Electrochem. Commun.* 2021, 124 , 106920.

50 Q. Zhao, Q. Zhu, J. Miao, P. Zhang, P. Wan, L. He, B. Xu, Flexible 3D Porous MXene Foam for High-Performance Lithium-Ion Batteries, *Small* 2019, 15(51), 1–9.

5 Nano-Inks for 3D Printing

Mohammed Shariq[1], *Somnath Chattopadhyaya*[2], *Amit Rai Dixit*[2], *and Andrews Nirmala Grace*[1]*
[1]Centre for Nanotechnology Research, VIT University, Vellore, Tamil Nadu 632014, India
[2]Department of Mechanical Engineering, IIT(ISM) Dhanbad, Jharkhand 826004, India
*Corresponding author: anirmalagrace@vit.ac.in; anirmalagladys@gmail.com

CONTENTS

5.1 INTRODUCTION

Nanomaterials are a group of materials that consists of structural components that have at least one dimension on the nanometer scale. These are considered to be the best-suited additive building units for various applications in printed electronics, sensors (chemical, gas, and bio-sensors), bio-medical field, and energy storage devices. These nanomaterials allow the synthesis with control in different material properties that result in the enhancement of emerging properties for the evaluation of the device performance. These are classified based on their dimensions as shown in Figure 5.1. Nanomaterials with zero dimensions are called NPs such as metal and carbon-based, one dimension such as nanofibers, nanorods, two dimensions such as thin films, and three dimensions such as nanocomposites. Over the last two decades, metal and carbon-based NPs, conducting polymers, two-dimensional (2D) transitional metal dichalcogenides/carbides/carbonitrides/nitrides (MXenes), and metal oxides were introduced and developed for different deposition-based applications. The formulation of concentrated nano inks from these materials has drawn significant attention in recent years due to their high functionality These functional properties depend on the type of the synthesis method used to have the optimal size range distribution, shapes, and use of stabilizer to prevent the agglomeration and coagulation in the liquid form to increase its storage and shell life, concentration and density, and rheological properties of NPs for depositing on different substrates.

The most preferential noble metals for synthesizing nano inks are Cu, Ag, and Au. Both Cu and Ag are the most commonly and readily used metal NPs for deposition-based applications

DOI: 10.1201/9781003296676-5

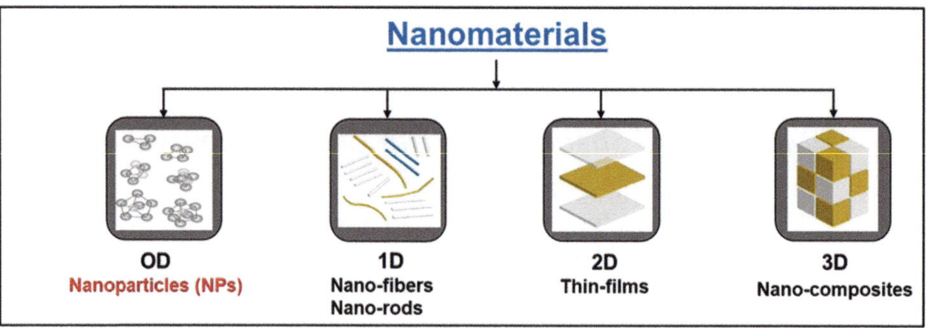

FIGURE 5.1 Schematic diagram mentioning the classification of nanomaterials based on their dimension.

due to their high conductivity and low cost as compared to AuNPs. AgNPs exhibit lower melting point temperatures making them superior for their usage in flexible electronics. But still, there are many challenges to replacing the costlier Au nano inks. Cu followed by Ag most likely degrades through the process of oxidation. Their high reactivity and proneness to oxidation limit their use. Choosing a relevant conducting polymer with a chemically less aggressive nature is an important task in these inks. This conducting polymer can also help to stabilize the NPs from aggregation. Au had shown the maximum resistance to degradation and corrosion and is highly chemically inert. AuNPs have emerged as an excellent material for applications in chemical and biological sensing renewable energy (solar cells, catalysis, and biomedical applications (drug delivery, imaging, and therapeutic agents) due to their specific unique properties such as surface plasmon resonance (SPR) and high biocompatibility [1]. SPR causes the oscillation of conduction electrons on the surface of the NPs, stimulated by incident light, and thus, AuNPs have good physical, chemical, and optical properties [2]. As Au is known as the most corrosive-resistant metal, its NPs also have the same nature. Moreover, NPs are biologically unreactive and due to their high surface area to volume ratio, they can be conjugated and functionalized with proteins, peptides, and medical drugs.

Carbon-based nano inks were widely explored for different applications due to their superior mechanical properties such as higher strength-to-weight ratio, enhanced thermal and electrical conductivities, chemical stability, ease of formulation, and more economical and bio-compatibility. Some of the developed carbon nano inks are based on graphene, graphene oxide, CNTs, and their hybrid forms. The scalable production of these carbon nano inks with well-defined functional properties is still facing a lot of challenges. One of the biggest challenges in these carbon inks is protecting the particles from agglomeration. In the case of graphene-based inks, restacking of the sheets changes the materials' properties. The second important challenge is to maintain the shell life of these inks in their original form. As the storage time increases, the shell life decreases, and the functional properties of the carbon suspensions also degrade. Some other challenges are its solubility and dispersibility in different solvents, and surface functionalization by using suitable solvents. Thereby, to meet the global demand for these nano inks, the different synthesis methods have been thoroughly summarized in the next section.

5.2 SYNTHESIS OF NPS

The different synthesis methods for NPs can be divided into bottom-up and top-down approaches, as shown in Figure 5.2. Bottom-up approaches include sol-gel, chemical vapor deposition, flame spray synthesis, various pyrolysis, and atomic or molecular condensation. It is the physiochemical preparation involving the formation of solute nucleate, nucleation, growth, and control. Top-down

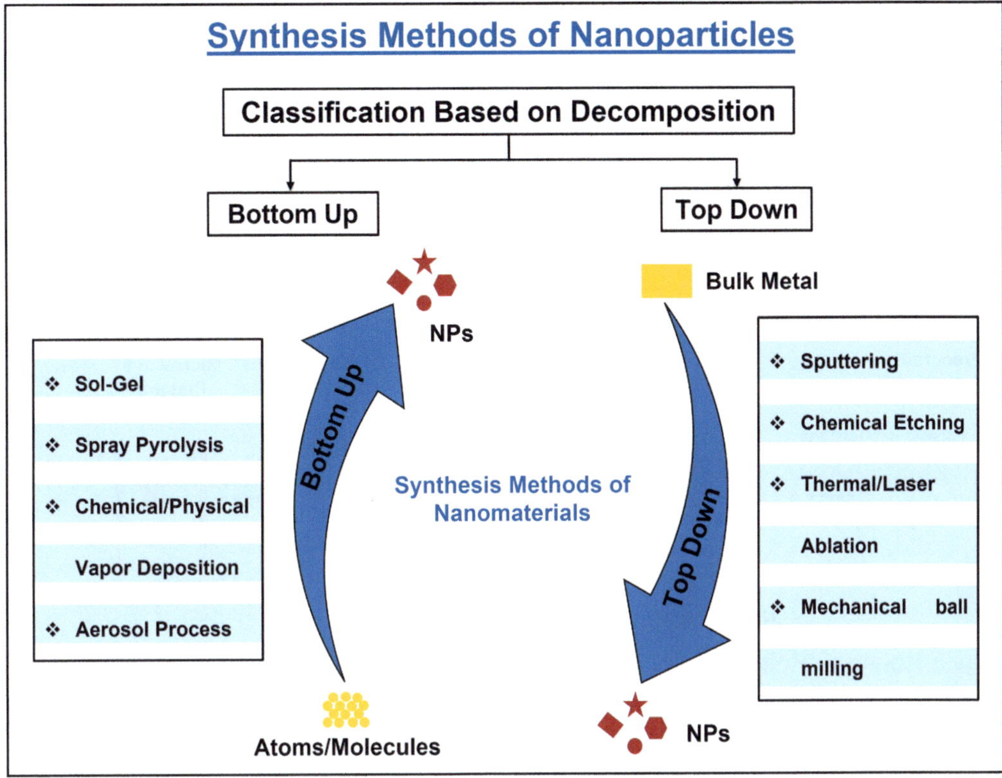

FIGURE 5.2 Classification for the synthesis of NPs based on the decomposition.

approaches include sputtering, thermal/laser ablation, nanolithography, and high-energy mechanical ball milling. It involves breaking initial powders of bulk metal to small dimensions on a nanometer-scale through crushing, milling, or grinding.

Another classification is based on the phase in which NPs are occurring: (i) Gas phase, (ii) Liquid phase, and (iii) Solid phase. In the gas phase methods (Figure 5.3), the aerosol droplets are formed from the liquid solution, which undergoes evaporation and thermal reduction in the gas phase. It results in the formation of NPs in different morphologies. These aerosols are any solid or liquid particles suspended in a gas. In the liquid phase methods, mostly, the processes will undergo in the liquid phase. The co-precipitation, sol-gel, micro-emulsions, hydrothermal and sonochemical process are some of the examples of liquid phase methods. Each method has its advantages with challenges and results in the formation of NPs with different chemical and physical properties. The solid and liquid phase methods are suitable for producing small quantities of NPs with significant variations in the shapes and sizes resulting from the production in different batches. A comparison of gas, liquid, and solid phase methods is summarized in Table 5.1.

The spray pyrolysis in the gas phase method allows easy control of process parameters with the continuous production of NPs. This continuous synthesis will result in a high yield with lower production costs and the flexibility of a large variety of nanomaterials. It involves the formation of the aerosols through different nebulization or atomization techniques of the starting precursor solution. The selection of the appropriate nebulization technique is essential as it directly impacts the aerosol droplet size and consequently the final NPs size distribution.

The production of these NPs into concentrated ink form and its deposition in micron-sized patterns involves metallization by using processes such as laser-induced deposition, physical and

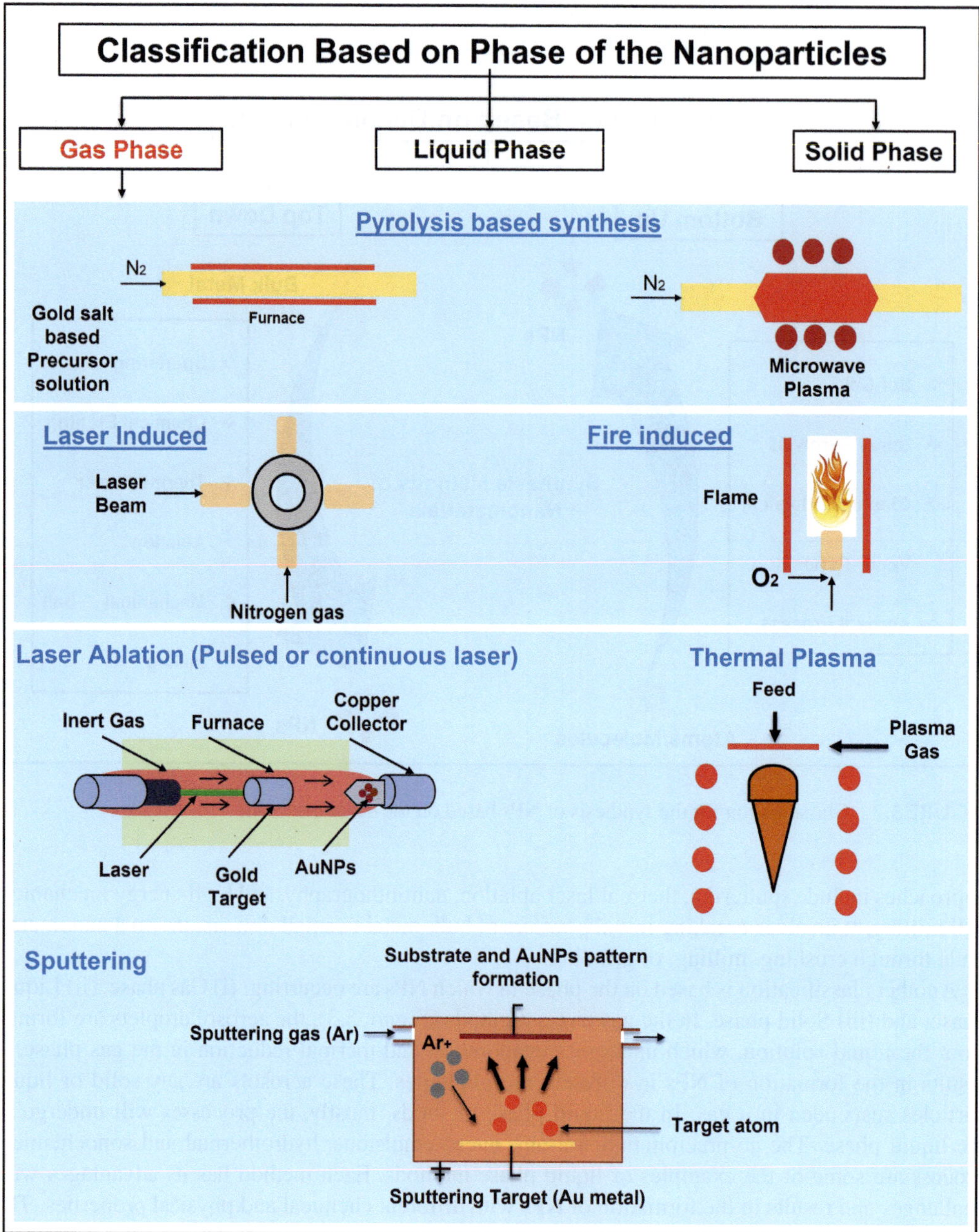

FIGURE 5.3 Schematic diagram of the synthesis method of the NPs in the gas phase.

chemical vapor deposition (PVD & CVD); dip coating (DC), etc. as shown in Figure 5.4. These methods are well developed and tested for different metal combinations over the few years in electronic, photonic, and sensing devices but lack integration into the deposition workflow. Thus, it adds a significant amount of overhead work and increases the device cost. Laser-induced and PVD/CVD-based methods offer thin-film coatings but require high-power lasers and expensive equipment. These methods operate at very high temperatures and under high vacuum pressures in a controlled environment, limiting the large-scale production of low-cost sensing devices. Due to the use of high

TABLE 5.1
Shows the Comparison of the Different Methods Based on the Phase of the NPs

Method	Production Costs/Rate	Advantages	Challenges
Gas Phase (Spray Pyrolysis)	**Low/High**	i) Easy to control process parameters ii) Possible continuous production iii) High yield with homogenous shapes iv) Material flexibility is high v) High Purity and concentration	i) Large poly-dispersity in size due to agglomeration ii) High temperature is required iii) Collection of the NPs in a continuous process is difficult
Liquid Phase	**Low/Middle**	i) Possible to produce different kinds of NPs such as nanorods, nanocomposites ii) Material flexibility is high iii) Possible to synthesize highly mono-dispersed sized particles	i) Lot of harmful chemical and gases involved ii) Number of process steps are large and need to be reduced iii) Suitable for batch production and difficult to scale up
Solid Phase	**High/Low**	i) Process is simple ii) Suitable for highly abrasive materials also iii) Reliability and ease of operation (iv) Reproducibility of the results	i) Possibility of contamination ii) High Energy Demand iii) Poly-dispersity in the size distribution with irregular shapes iv) Imperfection of the surface and crystallographic structure and properties

energy sources or high temperatures involved with these methods, it also limits the use of low-cost, flexible substrates such as photo and filter-based papers. Another method, DC, is relatively economical and utilizes concentrated nano inks. It results in non-uniform deposition and degrades the substrate properties. It consumes a large amount of time for deposition and lacks the up scaling of the process. There is an emerging group of technologies called Additive Manufacturing (AM) or 3D printing that overcome these challenges. It creates objects from the bottom-up by adding material one cross-sectional layer at a time and offers design freedom for product development, reduces waste, and minimal use of harmful chemicals. The key benefits of using 3D printing as a deposition technique: faster and reduces an excessive number of steps as compared to conventional subtractive manufacturing techniques, allows precise demand on printing for motion synchronized droplet deposition, fabricates desirable traits for developing micro-fabricated devices, flexibility and massive potential for cost-effective commercialization. Therefore, different 3D printing techniques are discussed in detail for depositing the formulated nano inks on different substrates.

5.3 AM OR 3D PRINTING

3D printing techniques are much more dominant than conventional patterning techniques as they involve several steps to fabricate the product completely. A sacrificial layer is a deposit on the substrate and the masking material is patterned through the lithography process onto it in a subtractive conventional manufacturing process. The initial sacrificial layer is etched to remove the uncovered material by the mask. After etching, the mask is removed, and the process is repeated for depositing another layer in the second device. In the 3D printing process, the material is initially deposited on the targeted substrate and subsequently processed to solidify the deposited material to achieve the desirable properties. The other layers can also be deposited in the same manner, followed by the post-treatment processes. The procedural steps are reduced significantly. In this process, the

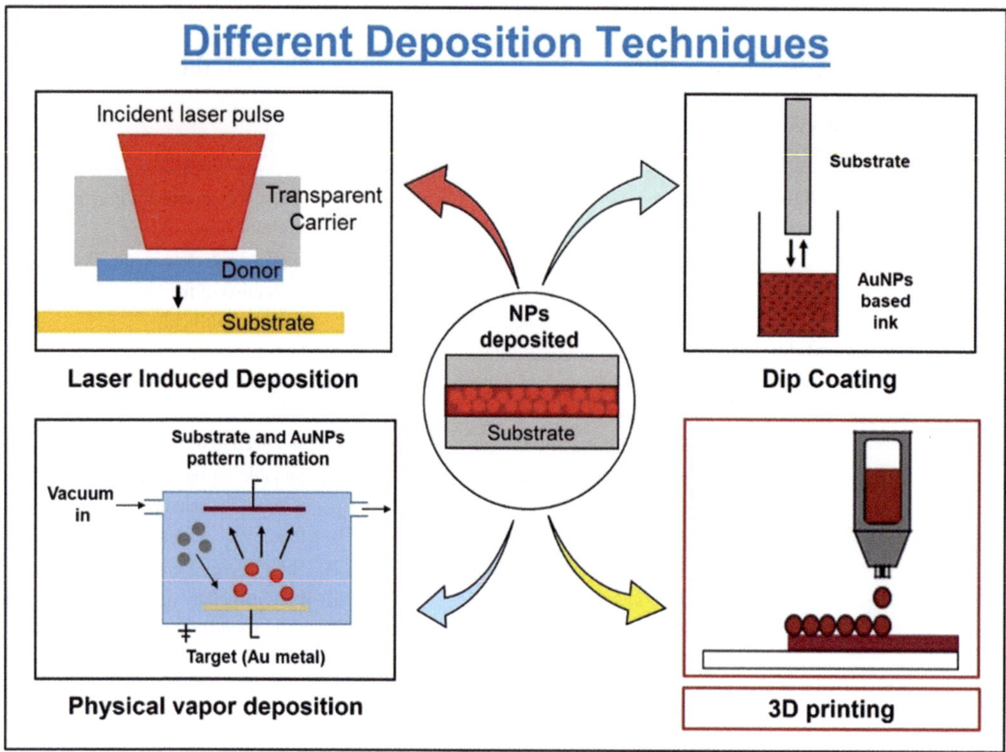

FIGURE 5.4 Shows the schematic diagram of different deposition techniques.

masking material is not required. Different material deposition techniques such as fused deposition modeling, direct writing methods including material jetting, selective laser sintering, stereolithography, and 3D printer transfer molding under the 3D printing umbrella attracted much in achieving the deposition of different materials. These techniques are compared in terms of working principle, material compatibility, and advantages with challenges in Table 5.2.

In comparing different 3D printing techniques for the patterning of nano inks, direct writing, including material jetting, allows the deposition of custom-made concentrated inks with polymer, carbon, and hybrid-based inks, wax, and metal NPs with photopolymers based liquid suspensions. It provides flexibility in the deposition of multi-materials with high resolution in a single step, high printing speed with lower production costs, and easy industrial scalability to high production quantities. Therefore, this technique has widely gained acceptance due to the mentioned capabilities and will be most suited for depositing customized nano inks. It is discussed in depth with different types, advantages, and challenges in the next section.

5.4 MATERIAL JETTING – INKJET PRINTING (IJP)

Nowadays, printing techniques are among the most attractive choices for material deposition because of their simple, economic, and high-production capabilities. It comprises the deposition on different substrates through the formulated functional nano inks. There is no limitation in the type of pattern and material printed with suitable rheological behavior of these inks. This can be easily modulated by adjusting the solvent and solute concentration. Generally, mass production printing technologies are broadly classified into two categories: Contact and non-contact printing. Contact printing involves the patterning of structures through the physical properties of structures through

TABLE 5.2
Comparison of Different 3D Printing Techniques

Technique	Working Principle	Material Compatibility	Advantages	Challenges
3D printer transfer molding	Molding	Plastics	• Flexible process • Microscopic precision was achieved and layer resolution up to 30 microns • Printing contour and complex geometrical shapes • Economical process	• Limited material flexibility • Poor product development • Low machine throughput and not suitable for mass production • Low tolerance and resolution with poor surface finish
Fused Deposition Modeling	Extrusion and deposition	Thermoplastic polymers such as PLA ABS, PC, and nylon; Polymer nano-composites – metal with ABS, wood with PLA, aluminum with nylon, glassfibers with ABS, PP, carbon fibers with ABS, resin, PLA, graphene+ABS	• Simple process • Low cost per product • More material flexibility • High strength of the printed parts • High process capability	• Results in porosity and deformed surfaces • Lower surface finish • Slower print speed • Lower dimensional accuracy • Higher processing temperatures
Direct Writing including material jetting	Pneumatic based, screw & piston driven, inkjet-based printing	Customized metal NPs with polymer and hybrid-based inks, wax, Alumina with polyurethane acrylate, short length carbon fibers with SiC whiskers, epoxy resins, metal NPs with photopolymers, CNT + PLA	• Non-contact and mask less deposition • Multi material flexibility • High resolution • Lower production cost • High speed	• Lower mechanical strength • Requires a precise control of rheological properties • Nozzle plate flooding, clogging and irregular droplet ejection
Selective Laser Sintering	Powder Bed Fusion	Thermoplastics, metals, Polymer nanocomposites – Glass with nylon 11, carbon fibers with polyamide, CNT with epoxy, Grapheme oxide with Photopolymer, Alumina powder with polystyrene, Graphene oxide with iron	• Acceptable strength • Easy removal of unwanted material such as supports • Requirement of support structures is low	• Operational cost is high • Final obtained surface consists of Powdery impurities • Limited material flexibility • Degradation of photosensitive material leading to poor performance under load
Stereo lithography	Polymeriza-tion	Photopolymers, Polymer nanocomposites – Alumina with UV cured	• Simple process • Fabrication of any self-supporting geometry • High resolutions	• Prints single material at a time • Quality is limited • Slow printing process

the substrates' physical contact with the inked surfaces. While in non-contact printing, material in the form of ink solutions is deposited on substrates through orifice openings or nozzles through the movement of the stages in a pre-programmed designed pattern.

Among the different categories of 3D printing techniques, Material Jetting (also called IJP) is defined by the American Society of Testing and Materials (ASTM) as an AM process in which the droplets of the build materials are selectively deposited. IJP is a non-contact printing technology that has emerged as the fastest-growing technology in the recent decade. In this technique, the concentrated ink is formulated in colloidal suspension form and ejected through the nozzle openings in ink droplets. The ejected droplets are deposited on the substrate under the effect of gravitational force to make the desired feature. The key benefits of using IJP in customized nano inks are – it provides good adhesion for efficient bonding to a paper, plastic, or glass substrate, offers deposition at ambient pressure and temperature, allows for precise alignment of drops, and easy handling has made this technique a dominant platform for rapid prototyping of sensors and detectors.

5.5 NANO-INKS FOR 3D PRINTING: FORMULATION, RHEOLOGY, AND CHALLENGES

Nano inks are prominently making an important place and position in the development of low-cost detection products in an industrially scaled process. These concentrated inks are highly promising due to their bio-compatibility and gained a lot of attention in the areas of sensing and diagnostics. One of the widely explored applications for nano inks is the fabrication of nano-metallic based sensors on different substrates such as colorimetric, SPR, bio-sensors, etc., as well as gas, electrical, electro-chemical sensing applications, printed and flexible electronics, and energy storage devices. Nano inks from Au-based printed patterns are used as the colorimetric sensors to provide an alternative detection system of impurities of heavy metal ions of Hg^{+}, Co^{2+}, Mn^{2+}, Pb^{2+}, Ca^{2+}, and Cd^{2+} in the drinking groundwater. This is because of the excellent bio-compatibility and unique optical and electronic properties. As the binding of these NPs takes place with the surrounding molecules to reduce the distance between them, the ZP approaches neutralization. It results in the surface plasmon coupling that causes a red shift in the absorbance wavelength of AuNPs and a color change from red to blue. This significant degree of color change can be a quantitative indicator for determining mercury concentrations. Utilizing the photothermal properties of the printed patterns from these nano-inks on the paper substrate through IJP showed the way for the fabrication of low-cost novel bio-medical devices such as the detection of glucose [3] and other biomarkers. In this case, NPs are used because of their high surface-to-volume ratio. It helps in the high-density biochemical interaction of these NPs and increases the high sensitivity analysis of the printed sensors [4]. The ability for patients to test themselves for a range of conditions within the comfort of their own homes is highly attractive. The point of care (PoC) diagnostic device market is poised to reach \$27.5 million by 2018 with a wide range of PoC technologies covering many different diseases and conditions. To ensure the commercial viability of these technologies, there is a requirement for low-cost, high-yield fabrication of such devices [5]. The use of nano inks through 3D printing is an obvious step towards the mass production of these devices at a relatively low cost when compared to the use of semi-conductor cleanroom techniques, which involve multiple processing steps using complex and expensive facilities. But, the formulation with optimum rheological properties of these inks for making different patterns on low-cost substrates is the key factor. The 3D printing of spherically shaped NPs having a size diameter of less than 50 nm on the paper and plastic substrates are in much focus for the fabrication of these diagnostic tools [6]. Their production requires inexpensive, easily scalable, and one-time usable devices that are most ideally suited to the IJP process.

The performance of the formulation of nano inks can be evaluated by the jetting behavior onto the target location. The successful jetting of these inks through the nozzle of the printer results in the formation of single droplets. Successful jettable inks are formulated by optimizing the different

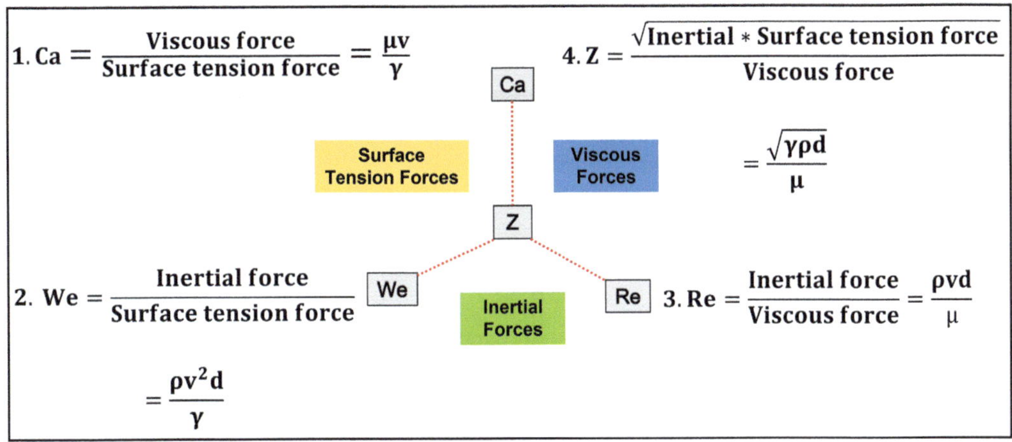

FIGURE 5.5 Pie chart diagram for defining the following dimensionless numbers: Ca, We, Re, and Z.

ink's physical and rheological properties such as size distribution, shape, stability, concentration, viscosity, and surface tension (ST). These are the different properties that need to be considered while designing the ink. Recent studies have demonstrated and explained the jettable behavior with the help of four dimensionless numbers [7]. Capillary (Ca), Weber (We), Reynold (Re) and Fromm (Z) numbers are explained in Figure 5.5.

In the above figure, the terms ρ, μ, and γ are the density, viscosity, and ST of the ink fluid respectively, d is the diameter of the printer nozzle and v is the velocity of the ink droplet. Among these four dimensionless numbers, the Fromm number (Z, reciprocal of the Ohnesorge number) is used for defining the jettability of the nano inks. Z correlates all the three forces of the fluid ink: viscous, inertial, and ST forces. While Ca, We, and Re balance only two forces. Recent studies have shown the calculated and consistent jettable range of Z value for their customized formulations of the NPs inks as $1 < Z < 10$ or $4 \leq Z \leq 14$. This Z range is called as the jettability window. Reis and Derby [8] defined the Z value ranges for the stable drop formation as $1<Z<10$ in an inkjet printer for the ink fluid. Below the lower limit of Z, the viscous forces will dominate the other two inertial and ST forces and restrict the droplet formation by stagnating/immobilizing the ink fluid flow. While the ink composition having a Z value greater than the upper limit range ($Z>10$) will behave as a low viscous fluid resulting in the formation of satellite droplets. Few works of literature have also reported the stable droplet formation in the Z value range for $Z>10$. Jang et al. [9] formulated different fluids from ethanol, ethylene glycol, and water as a solvent and studied fluid dynamic properties. The results have defined the printability range of Z value as $4 \leq Z \leq 14$ resulting in stable droplet generation. For the lower values of $Z<4$, the ink fluid had shown the formation of a long filament tail while $Z>14$ results in the formation of multiple drop formation. Nallan et al. [10] defined the Ca-We jettability window for different NPs-based inks and obtained the optimal rheological and jetting conditions (Figure 5.6). It had shown that alone Z factor is not enough to define the jettability window because the inconsistencies of different reported Z ranges. Similarly, Liu et al. [11] investigated the jettability of the ink and defined the window with the help of the We and Z number. This work has summarized the We range of printability as $2 < We < 25$ and $2 < Z < 20$ for the stable jettability of the ink. Further, it had been demonstrated that at $We = 2$, the inertial forces of the fluid will just overcome the capillary forces resulting in droplet formation whereas, at $We > 25$, the fluid will behave as unstable resulting in poor printability.

In a material jetting system, the printability of the ink is also another important factor. It includes the understanding of the wetting behavior, spreading on the substrate after its ejection from the nozzle, and the final characteristics of the printed patterns after drying [12]. The concentration of

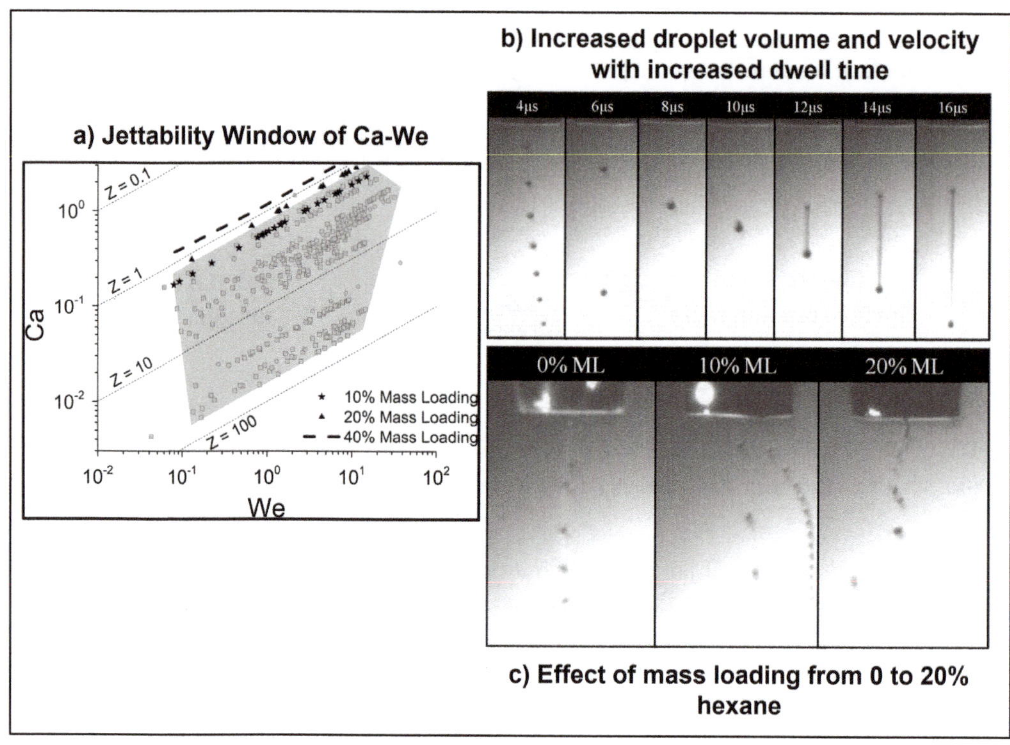

FIGURE 5.6 (a) Jettability window defined by Ca-We on different mass loading of 10, 20 and 40% solute; (b) Effect on the increased droplet volume and velocity with respect to increased dwell time; (c) Ink behavior on increasing the mass loading from 0, 10 and 20% in hexane solvent. Adapted with permission [10], Copyright (2014), American Chemical Society.

solute, viscosity and ST is the most basic properties of the ink that influences its flow and wetting behavior [13]. The wetting behavior of the ink droplet decides the resolution of the printed feature. Young's equation for determining the contact angle is mentioned in Equation (5.1) –

$$\theta_c = \cos^{-1}\left(\frac{\gamma_{SV} - \gamma_{SL}}{\gamma_{LV}}\right) \tag{5.1}$$

where θ_c is the ink droplet contact angle and γ_{SV}, γ_{SL} and γ_{LV} are the solid-vapor, solid-liquid, and liquid-vapor interface energies, respectively (Figure 5.7a). The rate of evaporation of the solvent used in the ink formulation is directly proportional to the ink droplet radius (r_i) and can be calculated as mentioned in Equation (5.2). [14]

$$R_{evap} = b_1 + b_2\left(r_i\right) \tag{5.2}$$

where b_1 and b_2' are the constants. The larger the value of the radius of the ink droplet, the higher will be the rate of evaporation. Thus, the contact angle and shape of the ink droplet influence the rate of evaporation of the solvent present in the ink (Figure 5.7b).

The formulated nano inks generally show the shear thickening followed by the shear thinning flow behavior over the mass loading of the solute concentrations. Initially, the viscosities of the ink

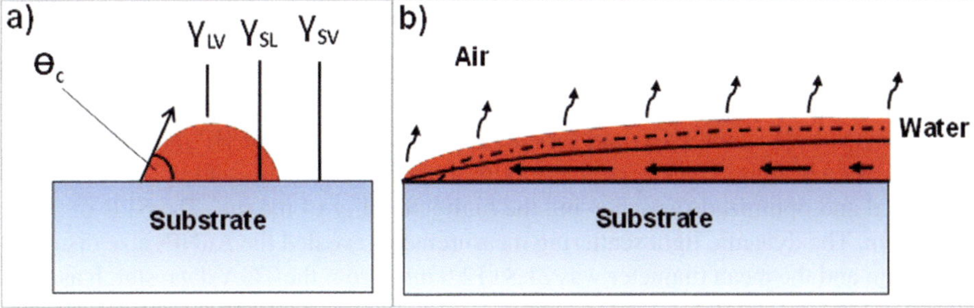

FIGURE 5.7 Schematic diagram of the (a) contact angle with interfacial energies of air, water and solid substrate, (b) the process of evaporation of the solvent present in the ink. Adapted with permission [15], Copyright (2011), Nature Publishing Group.

will increase as the shear rate increases and decreases after reaching a maximum point. The van der Waals forces are mainly involved in the NPs aggregation in these inks. Further increase in the applied shear force results in the breaking of the NPs ink network slowing down the resistance to flow. This flow behavior has always been recommended for stable drop formation in piezoelectric-based inkjet printers. Bing et al. [16] have proposed a theoretical numerical model of the ink droplet formation for the piezoelectric-based IJP by a lattice Boltzmann method. The study analyses the ink droplet formation through the force intensity, break-up time, and droplet velocity with respect to the wetting behavior and ST. By increasing the contact angle from 30°C to 150°C, the breakup time is also increased from 37.15 μs to 61.73 μs. The ink droplet velocity goes up by 56% by modifying the ST from 20 to 85 mN/m. The minimum velocity needed for the ink droplet in the nozzle needed to overcome the barrier of ST forces can be theoretically calculated as stated in Equation (5.3) [17] –

$$v_{min} = \left(\frac{4\gamma}{\rho d}\right)^{1/2} \tag{5.3}$$

The different processes such as the evaporation of the liquid solvent, sub-cooling, and absorption in the substrate occur resulting in the phase change from liquid to the solid state of the ink. The wetting behavior of the ink fluid depends on its physical properties as well as on the substrate. To have a good spreading of the ink, the droplet velocity, droplet volume, contact angle, surface, and capillary forces can be optimized.

5.5.1 Metal NPs Based Inks

Metal NPs-based inks are the colloidal suspensions of metal NPs in a suitable solvent system. These NPs are stabilized with the use of surfactants to prevent their coagulation. These surfactants are usually insulating in nature and require a sintering process to remove them. The functional properties of these inks depend on the NPs morphology (size and shape), surface charge, NPs concentration, stabilizers, or additives used in the formulation. The selection of the NPs in these inks depends on their properties such as electrical conductivity, magnetic properties, stability to oxidization, cost, and their target application. Following these parameters, Au, Ag, and Cu are the most explored metals for synthesizing and formulating them in concentrated nano ink form. Au nano inks are the most stable NPs in the ink form and have a longer shelf life [18]. Recently, Au nano inks were formulated through green and more environmentally friendly synthesis processes by using biodegradable sugar-based polyurethanes stabilizer. In these inks, the Au concentration was varied in-between 3 and

1.5% mass percent. The range of the shear viscosity and ST values were $1.87<\mu<2.09$ mPa.s and $41.79<\gamma<44.62$ mN/m, respectively. The Z values were calculated and falls in the range of $14<Z<17$ which was a little bit higher value range. The higher Z value corresponds to the more viscous nature of the ink. The AuNPs particle size was reported below 10 nm [19]. In another approach, the Au nano inks in water were synthesized and formulated by using starch for IJP through the method of microwave-assisted heating. The size of the AuNPs, yield of the process, and stability of the ink had been controlled and optimized for achieving the high jettability of the ink. The SPR of the AuNPs ink was 530 nm. The dynamic light scattering measurement revealed the AuNPs size distribution of $15 \leq 2r \leq 30$ nm and the mean diameter was 21.8±12.0 nm. From the TGA data, starch as the stabilizer helps in reducing the sintering temperature to about 300°C. The Z value of this ink formulation was 54.6 which was much higher than the limiting range of $1 \leq Z \leq 10$. Additional surfactant Triton X-100 was added to reduce the Z number value to 37.2 to improve the jettability by decreasing the ST value. This ink formulation was successfully jetted to single drop formation by using a large-sized diameter of 60 µm. It may lack jettability with a lower diameter nozzle in 20 µm range [19]. Au nano ink stabilized with polyethylene glycol (PEG) hygroscopic polymer was successfully printed on Polyethylene terephthalate (PET) films for resistive humidity sensors through a 50 µm diameter of Jetlab 4 printer. With the chemical synthesis of AuNPs ink, the reported size of the AuNPs was below 30 nm having an SPR peak of 521 nm [20].

Au nano inks stabilized with PVP were formulated in the mixed solution of water and ethylene glycol in [Au] of 50 mg/ml. UV-vis measurement showed a peak signifying the narrow AuNPs size distribution of 20 nm. The viscosity and ST values reported were 4.81±0.02 mPa.s and 33.10±0.03 mN/m respectively which showed the fluent jettability of the ink's droplets. The flat patterns of the following dimensions 29 × 21 cm onto the PET film were printed successfully with a Dimatix DMP 2831 inkjet printer with a 10 pl cartridge. These printed patterns were used for electrochemical sensing applications [21].

Ag-based nano inks offer to fabricate products at a much lower and economical cost for flexible and printed electronics. The process of sintering in Ag inks is an important step to achieving lower resistance and higher conductivity. AgNPs can be easily suspended in water-based solvents, ethanol, polyvinylpyrrolidone, PEG, polyvinyl alcohol, polyaniline, and other organic solvents. The water-based Ag nano inks are considered to be more environmentally friendly as compared to organic solvents. The printed patterns have achieved the resistivity of 3.3×10^{0} to 5.6×10^{-06} Ωcm [22]. Several studies have suggested that Ag nano inks used in PANI had reduced the sintering temperature range from 150–250°C to 110–120°C. Electromigration is the property of conductive inks that defines its capability to transfer current for a longer period. It refers to the mass transport of atoms diffusing in a unilateral direction in response to the current flow in the printed conductive lines [23]. Pure Ag nano inks lack electromigration properties. To enhance, Ag hybrid nano inks had been introduced and gained attention with Graphene. These hybrid inks with carbon materials will also reduce the cost of the manufacturer. By incorporating Graphene into the AgNPs surface, the mechanical properties will be significantly enhanced. These AgNPs/Graphene ink had achieved the sheet resistance of 9.29 kΩ/sq [24]. In another case, Ag nanowires had been used with Graphene for formulating hybrid inks. This carbon doping significantly improves the better bending resistance [25]. AgNPs also with Graphene quantum dots (GD) used for the paper-based electrochemical biosensor for the detection of respiratory diseases as PoC device. AgNPs/GD with polyvinylpyrrolidone (PVP) was formulated and printed using pen writing technology [26]. AgNPs decorated with MoS_2-based hybrid ink were used for the fabrication of the Surface Enhanced Raman Spectroscopy (SERS) array sensor. MoS_2 combined with AgNPs improves the better optical, conductivity, and electron transfer rate of these inks. This MoS_2 also helps in stabilizing the AgNPs and shows great potential for wearable sensing [27].

Cu nano inks showed great potential for replacing the Au/Ag inks for the fabrication of printed conductive patterns on different substrates. However, the biggest challenge with Cu nano inks is their tendency to get oxidized under ambient conditions. The oxidized CuNPs result in reduced electrical conductivity with the higher temperature ranges for sintering. The Cu nano inks were synthesized by using the Cu metal salts such as $CuCl_2$, $CuSO_4.5H_2O$, $Cu(NO_3)_2.3H_2O$, $Cu(CH_3COO)_2$ by using different reducing agents such as $NaBH_4$, N_2H_4, Glucose, Ascorbic acid, and Benzyl alcohol. The common stabilizers used are PVP, cetyl trimethyl ammonium bromide (CTAB), oleic acid, oleylamine, PEG, sodium oleate, and Triton X-100 [28]. Recent progress had demonstrated the new techniques for the production of anti-oxidative Cu nano inks by using $Cu(OH)_2$ as the precursor salt. Ascorbic acid with PVP was used as a reducing agent and stabilizing agent respectively in di-ethylene glycol solvent. The fabricated inks had achieved the lowest conductivities values till 5.7 μΩcm. Apart from the wet reducing chemical methods for the synthesis of these inks, the discharge method, laser ablation, and electron beam irradiation are other techniques for controlling the morphology and aggregation state of the NPs. Most of the reported works with Cu nano inks lack the issue of stability, shell life, and lower production rate. The synthesis of CuNPs from the electrolysis method directly from the solid copper plate has shown the prospect of scalable production with better control of final properties. 15 nm was the diameter of the synthesized CuNPs while achieving the lowest resistivity in the range of less than 6μΩ-cm with the above-mentioned technique [29]. The selection of the precursor in the case of CuNPs had also affected the final morphology of the CuNPs. Deng et al. synthesized the CuNPs in an aqueous medium by using the precursor solution of copper acetate salt and hydrazine as the reducing agent. Lactic acid had been used as the stabilizing agent and prevented the CuNPs from oxidation. The CuNPs in the size range of less than 10 nm were sintered at a lower temperature range of 150–200°C and achieved resistivities to 9.1±2.0 μΩ-cm [30].

5.5.2 Carbon-Based Inks: Graphene/GO/rGO, CNTs

In the last decade, Graphene has established itself as the most promising candidate for carbon-based inks due to its following properties: Quantum Hall Effect, Ultrahigh carrier mobility at room temperature, Young's modulus, Intrinsic strength, and high thermal conductivity (above 3000 WmK^{-1}), impermeable to gases, ability to sustain high densities of the electric current and high light transmittance. The different synthesis methods for Graphene nano inks are: Mechanical shear exfoliation, Oxidation-reduction method, Epitaxial Growth, CVD, Liquid Phase exfoliation, Electrochemical exfoliation, Probe Sonication, Ultrasonic assisted liquid phase exfoliation, Vortex Mixing, and Micro-fluidization of graphite [30].

Despite gaining the name of "Wonder Material", it still faces the challenge of scalable production with the desired properties. Graphene oxide (GO) and rGO are the two basic precursors for producing graphene-based nano inks. Since its solubility in water helps in the formulation of aqueous-based ink formulations. Different stabilizers and solvents used in the formulations are N-methyl-2-pyrrolidone (NMP) with a small amount of ethyl cellulose in the ethanol, NMP with vinyl acetate in isopropanol, PVP in IPA solution, cyclohexanone with small amounts of terpinol and additional ethyl cellulose stabilizer. Dimethylformamide (DMF), 1-2 dichlorobenzene, cyclo-hexanol, toluene, acetone, and ethanol, are the most commonly used solvents for graphene-based nano inks [31]. Graphene nano inks formulated and synthesized with more environment-friendly routes are in focus by using green polymers such as sodium carboxymethyl cellulose. The formulated inks had shown excellent piezoresistive and thermo-resistive responses making them suitable for the development and fabrication of printable sensors. Several reports had also been reported for the formulation of highly viscous Graphene nano inks for using it in extrusion-based 3D printing for the fabrication of sensing devices. These viscous inks have been formulated in isopropanol and Cyrene mixtures in different mass ratios for the detection of volatile organic compounds with the viscosity range of 50–60 Pa.s [32]. MoS_2/Graphene-based nano inks were developed for the fabrication of electrodes for a 2D interdigital supercapacitor. These inks have achieved a specific capacitance of

392 F/g. The loading of MoS_2 material on Graphene inks has improved the diffusion rate of the ions by facilitating better contact in-between the active MoS_2/Graphene material and electrolyte. The composite of MoS_2/Graphene materials also allows the improved inter-connection and penetration leading to the increased value of conductivity [33]. CNT/Graphene inks had been developed for the fabrication of flexible supercapacitors. CNT has poor dispersion property in different solvents. Therefore, to improve its dispersibility, Graphene was used as a surfactant. This will also enhance the performance of supercapacitors. The thermal conductivity also increases by bridging the CNT/Graphene in different solvents. CNT functions as the bridging material between the graphene flakes and particles. Aqueous rGO/MWCNT hybrid inks were also formulated for depositing it into thin films aimed for good electro-optical performance. 0.1 wt.% of MWCNT with 0.2 wt.% of the reduced GO has improved the dispersibility as well as the rheological property of the ink. These inks have shown a sheet resistance of 380 Ωsq^{-1}. By utilizing the rGO with Triton-X 100, the ST is also reduced to 30–40 mN/m and increases the viscosity from 6 mPa.s to 2 Pa.s [34]. Graphene with carbon nanoplatelets (Gr/CNP) have been used for formulating the carbon hybrid inks for fabricating micro-supercapacitors (MSC's) achieving the areal capacitance of 12.5 mF/cm^2. These MSC had shown excellent mechanical flexibility strength and water resistance properties. Incorporating nanoplatelets into the graphene structures helped to fabricate an efficient energy storage structure without diminishing its important properties of it [34].

5.5.3 MXene-Based Nano Inks

MXenes with the structural formula of $M_{n+1}X_nT_x$ where M represents the transition metal, X as carbon or nitrogen, and T as the terminal group are the 2D nanomaterials consisting of transition metal nitrides, carbides, or carbonitrides. This nanomaterial possesses high electronic and thermal conductivity, a high surface area that can be functionalized, and a high zeta potential for making them into stable colloidal suspensions. Due to these properties, MXenes had been exploited in energy storage devices, sensing applications such as gas sensors, heavy metal ions, photocatalysis, batteries, PoC diagnostic devices, CO_2 capture, drug detection, field effect transistors, printed and flexible electronics such as radio frequency identification (RFID) tags, flexible displays, and light-emitting diodes. Yury and the group demonstrated the formulation of MXene-based nano inks in water and other organic solvents such as NMP, DMF, and ethanol. These inks have shown a higher areal and volumetric capacitance when printed on different substrates. The key challenge in utilizing MXene in nano ink form is in its formulation as well as the kinetics of the drying of the solvent. The calculated Z number for these inks in organic solvents ranges in-between 2–3 and falls in the stable jettability range of $1<Z<14$. Different structures have been additively fabricated (extrusion printing and IJP) by formulating MXenes ($Ti_3C_2T_x$) inks in four organic solvents NMP, dimethyl sulfoxide (DMSO), DMF, and ethanol without using surfactants. The printed structures for micro-supercapacitors (MSCs) had shown the highest electrochemical performance, including volumetric capacitance and energy density (ED) as 562 F/cm^3 and 0.32 $\mu Wh/cm^2$, respectively [35]. Discarded sediments consisting of unetched MAX precursor and multi-layered MXenes were utilized in formulating aqueous inks for fabricating conductive patterns on paper substrate. The screen-printed structures achieved the areal capacitance and ED as 158 mF/cm^2 and 1.64 $\mu Wh/cm^2$ and high-resolution printing with uniform spacing [35]. 2D $Ti_3C_2T_x$ flakes were utilized for formulating the inks and deposited on different substrates through direct painting. IPA and toluene in a volume ratio of 2.25:1 were mixed with the synthesized flakes to optimize the rheological properties of the ink. The 40 X 40 mm^2 painted pattern having a thickness of 13.6 μm shows the excellent performance for electromagnetic interference shielding with absorption and reflection of 58.5% and 41.5% with a shielding effectiveness of up to 46.3 dB [36]. Conductive MXene inks were prepared from Ti_3AlC_2 precursor by using lithium fluoride (LiF) and HCl etchants in NMP solvent in a volume ratio of 2:1. The ink has been concentrated up to 20 mg/ml to get the desired uniformity in the printed patterns.

The patterns were printed using an HP inkjet printer on PET adhesive tape and attained the sheet resistance of 1.66±0.16 MΩ/sq to 1.47±0.16 kΩ/sq at the transmissivity of 87–24% (λ = 550 nm) [37]. Zheng et al. prepared aqueous MXene inks based on Lithium Titanate (LTO) and Lithium Iron Phosphate (LFP) from Ti_3AlC_2 precursor via LiF and HCl etching for the fabrication of flexible solar cells, MXene conductive circuits, and flexible self-powered system in a single flexible substrate through screen printing. MXene-based MSCs achieved 1.1 F/cm^2 of areal capacitance and 13.8 μWh/cm^2 of ED while MXene-based lithium-ion micro-batteries achieved an areal density of 154 μWh/cm^2 [38]. Tang et al. proposed a new chemical route for the formulation of stable MXene inks in ethanol solvent with Trifluoroacetic acid treatment with improved printable properties through electrohydrodynamic printing (EHD) printing. The printed patterns achieved high conductivity values of about 8900 S cm^{-1} and width in the range of 3–4 μm with an interplanar spacing of 0.31 nm. The obtained results showed the capabilities of MXene inks in thin film transistor (TFT) applications [38]. Zhang et al. formulated highly viscous aqueous-based MXene inks from Titanium carbide (Ti_3AlC_2) and carbonitride (Ti_3CNT_x) for the fabrication of Si/MXene-based electrodes up to the concentration value of 25 mg/mL. These electrodes achieved thickness values in the range of 450 μm and a high areal capacity of 23.3 mAh/cm^2 [39].

5.5.4 Metal Oxide-Based Nano Inks

Metal oxide-based nano inks are the colloidal suspensions of oxide species. They can be formulated into two types: NPs-based ink and precursor-based ink. The NPs-based ink can be synthesized from Al_2O_3, ZrO_2, TiO_2, HfO_2, and ZnO. Precursor-based ink can be synthesized from Al_2O_3, ZrO_2, $Sc_1Zr_1O_X$, Ta-Al-Si-alkoxide, Sr-doped Al_2O_3, HfO_2, IGZO, and ZnO. These materials are having good optical and electronic properties. The metal oxide NPs inks can be synthesized from hydrothermal, microwave, or ultrasonication. The precursor-based inks were synthesized from the sol-gel process from metal salts such as metal chloride, nitrate, sulfate, acetate, and alkoxides. The metal oxide nano inks (M-O-M) had been formed by hydrolysis and polycondensation reactions of precursor molecules. Thermal annealing is required to eliminate the organic residual or solvent in the ink after its printing. It also increases the density of the metal oxide particles. There are a few challenges in the M-O-M nano inks that need to be addressed. (a) Multi-phase inks lack uniform printing because of the uniform evaporation and drying of the ink solvent. The non-uniformity in the printing can be eliminated by adopting the different strategies in formulating the ink by selecting an appropriate and optimal morphology of the solute, using the binary mixing of solvents in suitable volume ratios. Different additives can be used for tuning the rheological and flow properties of the ink. The acidic content in the ink should be controlled by monitoring the pH value of the ink. (b) The lower temperature range for annealing the printed pattern. It can be done through microwave heating, high-pressure annealing, and UV curing. Recent studies had suggested the better suitability of the lower annealing temperatures for the fabrication of high-performance printed devices. (c) Multi-phase nano inks lack high-resolution printing. It can be improved by controlling the drop spreading and wetting behavior by selecting the appropriate substrate-solvent system for the printing. By optimizing the weight ratio of active material loading to solvent [40]. Indium and aluminum oxide-based inks were formulated for the fabrication of humidity sensors by using the screen-printing technique. These sensors were tested exhaustively for the wider range of 5–95% of the relative humidity (RH) and measured a high sensitivity of 0.85–7.76 pF/RH%. The average response and recovery time are 21.4 and 4.8 s, respectively [41]. The InO_x metal oxide nano ink with polyacrylamide in ethylene glycol in different concentrations was formulated for the fabrication of TFT's by using IJP. It resulted in high-resolution patterns with good electrical functionality of 4.2 cm^{2}V^{-1}s^{-1} with 0.7 V of the threshold voltage. These TFT's showed an on/off ratio of 10^6 and 0.30 V/dec of the sub-threshold slope [42]. In another approach, several types of metal oxide-based inks of In_2O_3, In-Ga-ZnO, Sn-doped In_2O_3, and Al_2O_3 were formulated for EHD printing of TFT devices achieving the electron transfer characteristics of up to 117 $cm^2/V.s$. These TFT's had been annealed at the lower temperature range of less than 350°C [43]. Inkjet-printed tin oxide films on polyimide substrates were fabricated for gas sensing applications. Initially, SnO_2 nano inks were synthesized through the sol-gel method and formulated with ethylene glycol and glycerol. Ethanol and 2-iso-propoxyethanol were used as the rheology modifiers for attaining the required jettability of the inks. The formulated inks have achieved a viscosity of 10 mPa.s, ST of 32 mN/m and Z value of 2.7 [44].

5.6 CONCLUSIONS AND FUTURE PROSPECTIVE

The chapter covers an in-depth summary of different production methods of the nano-inks enlisting their advantages and limitations. Different formulation techniques involved with the need of solving the challenges associated with them have been addressed. The various deposition techniques are compared to opt for the best suitable choice as 3D printing for fabricating the contour patterns. The use of metallic, carbon-based nano inks and their hybrid counterparts were shown with the extrusion and IJP. The detailed characterization framework with mass concentration, viscosity, and other rheological properties plays an important part in defining the jettability window to quantify the ink's performance in these printing techniques.

As the global market demand for these nano inks is growing, the current synthesis methods have technological as well as economic constraints for large scalability. This increasing demand cannot be met with the present restricted production rates in batch size production. These methods lack the repeatable results for the controlled synthesis of NPs with the desired morphology (size and shape), minimum impurities, and higher final concentration with the continuous production process. The literature on the use of the different precursors for making the initial precursor solutions for the scalable, robust, and reliable synthesis of nano inks is still very limited. It highlights a significant market gap for the methods that lend themselves to the mass production of nano inks and thus rendering it commercially viable. This can be a great opportunity for researchers in the coming time to develop viable solutions. As the present conventional deposition techniques also lack cost-effective commercial prospects and faster, and more precise deposition. The inkjet and extrusion printing of 3D printing techniques have shown considerable potential in meeting these challenges for the deposition of nano inks on different substrates.

REFERENCES

1 R. Raliya, D. Saha, T. S. Chadha, B. Raman, and P. Biswas, Non-invasive aerosol delivery and transport of gold nanoparticles to the brain, *Scientific Reports,* 7:44718 (2017) 1–8.

2 X. Huang, I. H. El-sayed, and M. A. El-sayed, Gold nanoparticles: Interesting optical properties and recent applications in cancer diagnostics and therapy, *Nanomedicine (London)*, 2 (2007) 681–693.

3 C. E. Krause, B. A. Otieno, A. Latus, R. C. Faria, V. Patel, J. S. Gutkind, and J. F. Rusling, Rapid microfluidic immunoassays of cancer biomarker proteins using disposable inkjet-printed gold nanoparticle arrays, *Chemistry Open*, 2 (2013) 141–145.

4 E. Skotadis, J. Tang, V. Tsouti, and D. Tsoukalas, Chemiresistive sensor fabricated by the sequential ink-jet printing deposition of a gold nanoparticle and polymer layer, *Microelectronic Engineering*, 87 (2010) 2258–2263.

5 V. Gubala, L. F. Harris, A. J. Ricco, M. X. Tan, and D. E. Williams, Point of care diagnostics: Status and future, *Analytical Chemistry*, 84 (2012) 487–515.

6 A. L. M. Marsico, B. Creran, B. Duncan, S. G. Elci, Y. Jiang, V. M. Rotello and R. W. Vachet., Inkjet-printed gold nanoparticle surfaces for the detection of low molecular weight biomolecules by laser desorption/ionization mass spectrometry, *Journal of the American Society for Mass Spectrometry*, 26 (2015) 1931–1937.

7 L. Nayak, S. Mohanty, S. K. Nayak, and A. Ramadoss, A review on inkjet printing of nanoparticle inks for flexible electronics, *Journal of Materials Chemistry C,* 7 (2019) 8771–8795.

8 N. Reis and B. Derby, Ink jet deposition of ceramic suspensions: Modeling and experiments of droplet formation, *MRS Proceedings*, 625 (2000) 117.
9 D. Jang, D. Kim, and J. Moon, Influence of fluid physical properties on ink-jet printability, *Langmuir,* 25 (2009) 2629–2635.
10 H. C. Nallan, J. A. Sadie, R. Kitsomboonloha, S. K. Volkman, and V. Subramanian, Systematic design of jettable nanoparticle-based inkjet inks: Rheology, acoustics, and jettability, *Langmuir,* 30 (2014) 13470–13477.
11 Y. Liu and B. Derby, Experimental study of the parameters for stable drop-on-demand inkjet performance, *Physics of Fluids*, 31 (2019) 32004.
12 J. Tai, H. Y. Gan, Y. N. Liang, and B. K. Lok, Control of droplet formation in inkjet printing using Ohnesorge number category: Materials and processes, in *2008 10th Electronics Packaging Technology Conference* (2008) 761–766.
13 J. R. Castrejón-Pita, G. D. Martin, S. D. Hoath, and I. M. Hutchings, A simple large-scale droplet generator for studies of inkjet printing, *Review of Scientific Instruments*, 79 (2008) 75108.
14 K. S. Birdi, D. T. Vu, and A. Winter, A study of the evaporation rates of small water drops placed on a solid surface, *Journal of Physical Chemistry*, 93 (1989) 3702–3703.
15 P. J. Yunker, T. Still, M. A. Lohr, and A. G. Yodh, Suppression of the coffee-ring effect by shape-dependent capillary interactions, *Nature*, 476 (2011) 308–311.
16 B. He, S. Yang, Z. Qin, B. Wen, and C. Zhang, The roles of wettability and surface tension in droplet formation during inkjet printing, *Scientific Reports*, 7 (2017) 11841.
17 L. Lan, J. Zou, C. Jiang, B. Liu, L. Wang, and J. Peng, Inkjet printing for electroluminescent devices: emissive materials, film formation, and display prototypes, *Frontiers of Optoelectronics,* 10 (2017) 329–352.
18 H. W. Tan, J. An, C. K. Chua, and T. Tran, Metallic nanoparticle inks for 3D printing of electronics, *Advanced Electronic Materials*, 5 (2019) 1800831.
19 B. Begines, A. Alcudia, R. A. Velazquez, G. Martinez, Y. He, G. F. Trindade, R. Wildman, M. J. Sayagues, A. J. Ruiz, and R. P. Gotor, Design of highly stabilized nanocomposite inks based on biodegradable polymer-matrix and gold nanoparticles for inkjet printing, *Scientific Reports,* 9 (2019) 1–12.
20 C.-H. Su, H.-L. Chiu, Y.-C. Chen, M. Yesilmen, F. Schulz, B. Ketelsen, T. Vossmeyer, and Y.- C. Liao, Highly responsive PEG/Gold nanoparticle thin-film humidity sensor via inkjet printing technology, *Langmuir,* 35 (2019) 3256–3264.
21 M. Deng, X. Zhang, Z. Zhang, Z. Xin, and Y. Song, A gold nanoparticle ink suitable for the fabrication of electrochemical electrode by inkjet printing, *Journal of Nanoscience and Nanotechnology,* 14 (2014) 5114–5119.
22 I. J. Fernandes, A. F. Aroche, A. Schuck, P. Lamberty, C. R. Peter, W. Hasenkamp, and T. L. A. C. Rocha, Silver nanoparticle conductive inks: synthesis, characterization, and fabrication of inkjet-printed flexible electrodes, *Scientific Reports,* 10 (2020) 8878.
23 İ. A. Kariper, Conductive ink next generation materials: Silver nanoparticle/polyvinyl alcohol/polyaniline, *Journal of Inorganic and Organometallic Polymers and Materials*, 32 (2022) 1277–1286.
24 Y. Z. N. Htwe, M. K. Abdullah, and M. Mariatti, Water-based graphene/AgNPs hybrid conductive inks for flexible electronic applications, *Journal of Materials Research and Technology*, 16 (2022) 59–73.
25 T. Liu, J. Zhao, D. Luo, Z. Xu, X. Liu, H. Ning, J. Chen, J. Zhong, R. Yao, and J. Peng, Inkjet printing high performance flexible electrodes via a graphene decorated Ag ink, *Surfaces and Interfaces*, 28 (2022) 101609.
26 H. K. Kordasht, A. Saadati, and M. Hasanzadeh, A flexible paper based electrochemical portable biosensor towards recognition of ractopamine as animal feed additive: Low cost diagnostic tool towards food analysis using aptasensor technology, *Food Chemistry*, 373 (2022) 131411.
27 X.-J. Li, Y.-T. Li, H.-X. Gu, P.-F. Xue, L.-X. Qin, and S. Han, A wearable screen-printed SERS array sensor on fire-retardant fibre gloves for on-site environmental emergency monitoring, *Analytical Methods*, 14 (2022) 781–788.
28 C. Cheng, J. Li, T. Shi, X. Yu, J. Fan, G. Liao, S. Cheng, Y. Zhong, and Z. Tang, A novel method of synthesizing antioxidative copper nanoparticles for high performance conductive ink, *Journal of Materials Science: Materials in Electronics*, 28 (2017) 13556–13564.

29 J. Cheon, J. Lee, and J. Kim, Inkjet printing using copper nanoparticles synthesized by electrolysis, *Thin Solid Films*, 520 (2012) 2639–2643.

30 J. Wang, Y. Liu, Z. Fan, W. Wang, B. Wang, and Z. Guo, Ink-based 3D printing technologies for graphene-based materials: A review, *Advanced Composites and Hybrid Materials*, 2 (2019) 1–33.

31 T. S. Tran, N. K. Dutta, and N. R. Choudhury, Graphene inks for printed flexible electronics: Graphene dispersions, ink formulations, printing techniques and applications, *Advances in Colloid and Interface Science*, 261 (2018) 41–61.

32 K. Hassan, T. T. Tung, N. Stanley, P. L. Yap, F. Farivar, H. Rastin, M. J. Nine, and D. Losic, Graphene ink for 3D extrusion micro printing of chemo-resistive sensing devices for volatile organic compound detection, *Nanoscale*, 13 (2021) 5356–5368.

33 H. Wang, D. Tran, M. Moussa, N. Stanley, T. T. Tung, L. Yu, P. L. Yap, F. Ding, J. Qian, and D. Losic, Improved preparation of MoS_2/graphene composites and their inks for supercapacitors applications, *Materials Science and Engineering: B*, 262 (2020) 114700.

34 H. Chen, S. Chen, Y. Zhang, H. Ren, X. Hu, and Y. Bai, Sand-milling fabrication of screen-printable graphene composite inks for high-performance planar micro-supercapacitors, *ACS Applied Materials & Interfaces,* 12 (2020) 56319–56329.

35 S. Abdolhosseinzadeh, R. Schneider, A. Verma, J. Heier, F. Nüesch, and C. (J.) Zhang, Turning trash into treasure: Additive free MXene sediment inks for screen-printed micro-supercapacitors, *Advanced Materials,* 32 (2020) 2000716.

36 D. Wen, X. Wang, L. Liu, C. Hu, C. Sun, Y. Wu, Y. Zhao, J. Zhang, X. Liu, and G. Ying, Inkjet printing transparent and conductive MXene (Ti_3C_2Tx) films: A strategy for flexible energy storage devices, *ACS Applied Materials & Interfaces,* 13 (2021) 17766–17780.

37 S. Zheng, H. Wang, P. Das, Y. Zhang, Y. Cao, J. Ma, S. (F.) Liu, and Z.-S. Wu, Multitasking MXene inks enable high-performance printable microelectrochemical energy storage devices for all-flexible self-powered integrated systems, *Advanced Materials*, 33 (2021) 2005449.

38 X. Tang, G. Murali, H. Lee, S. Park, S. Lee, S. M. Oh, J. Lee, T. Y. Ko, C. M. Koo, Y. J. Jeong, and T. K. An, I. In, S. H. Kim, Engineering aggregation-resistant MXene nanosheets as highly conductive and stable inks for all-printed electronics, *Advanced Functional Materials*, 31 (2021) 2010897.

39 C. (John) Zhang, S.-H. Park, A. S.-Ascaso, S. Barwich, N. McEvoy, C. S. Boland, J. N. Coleman, Y. Gogotsi, and V. Nicolosi, High capacity silicon anodes enabled by MXene viscous aqueous ink, *Nature Communications,* 10 (2019) 849.

40 Z. Zhu, J. Zhang, D. Guo, H. Ning, S. Zhou, Z. Liang, R. Yao, Y. Wang, X. Lu, and J. Peng, Functional metal oxide ink systems for drop-on-demand printed thin-film transistors, *Langmuir*, 36 (2020) 30, 8655–8667.

41 J. R. McGhee, J. S. Sagu, D. J. Southee, Peter. S. A. Evans, and K. G. U. Wijayantha, Printed, fully metal oxide, capacitive humidity sensors using conductive indium tin oxide inks, *ACS Applied Electronic Materials*, 2 (2020) 3593–3600.

42 Z. Zhu, J. Zhang, Y. Wang, H. Ning, D. Guo, W. Cai, S. Zhou, Z. Liang, R. Yao, and J. Peng, Polymer-doped ink system for threshold voltage modulation in printed metal oxide thin film transistors, *The Journal of Physical Chemistry Letters,* 10 (2019) 3415–3419.

43 Y. Liang, J. Yong, Y. Yu, A. Nirmalathas, K. Ganesan, R. Evans, B. Nasr, and E. Skafidas, Direct Electrohydrodynamic Patterning of high-performance all metal oxide thin-film electronics, *ACS Nano,* 13 (2019) 13957–13964.

44 O. Kassem, M. Saadaoui, M. Rieu, S. Sao Joao, and J.- P. Viricelle, Synthesis and inkjet printing of sol–gel derived tin oxide ink for flexible gas sensing application, *Journal of Materials Science,* 53 (2018) 12750–12761.

6 Additives in 3D Printing: From the Fabrication of Thermoplastics and Photoresin to Applications

Habdias A. Silva-Neto[1], *Lucas C. Duarte*[1], *and Wendell K. T. Coltro*[1,2*]
[1]Instituto de Química, Universidade Federal de Goiás, Goiânia, GO, 74690-900, Brazil
[2]Instituto Nacional de Ciência e Tecnologia de Bioanalítica, Campinas, SP, 13084-971, Brazil
*Corresponding author: wendell@ufg.br

CONTENTS

6.1 INTRODUCTION

Additive manufacturing (AM), popularly known as 3D printing, is the fastest-growing emergent technology in the industry sectors. The technology can allow for the creation of solid prototypes of a range of sizes through a layer-by-layer (LbL) arrangement, consuming minimal inputs with high performance in terms of reproducibility and fidelity. Through the AM approach, it is possible to model digitally the target geometry on a software program, avoiding some pertinent problems such as pre-modeling time and experimental errors. In addition, AM can improve the manufacturing cost and final quality of the customized objects [1]–[4].

The working concept of AM method can be realized in four steps: (i) the creation of computer-aided design (CAD); (ii) slicing it in modeling software, observing if support pillars are needed; (iii) their conversion to g-code format (per example); and (iv) printing the target geometry. However, some disadvantages observed in AM included anisotropic behavior, few available additives, and difficulty in obtaining smooth surfaces. For this reason, some post-manufacturing steps have been

DOI: 10.1201/9781003296676-6

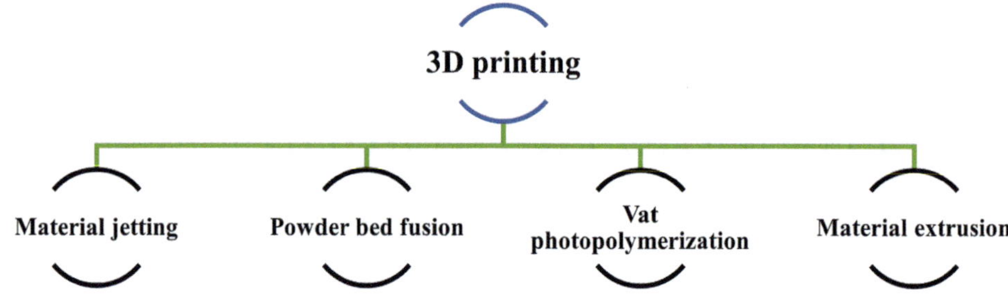

FIGURE 6.1 Main examples of 3D printing techniques.

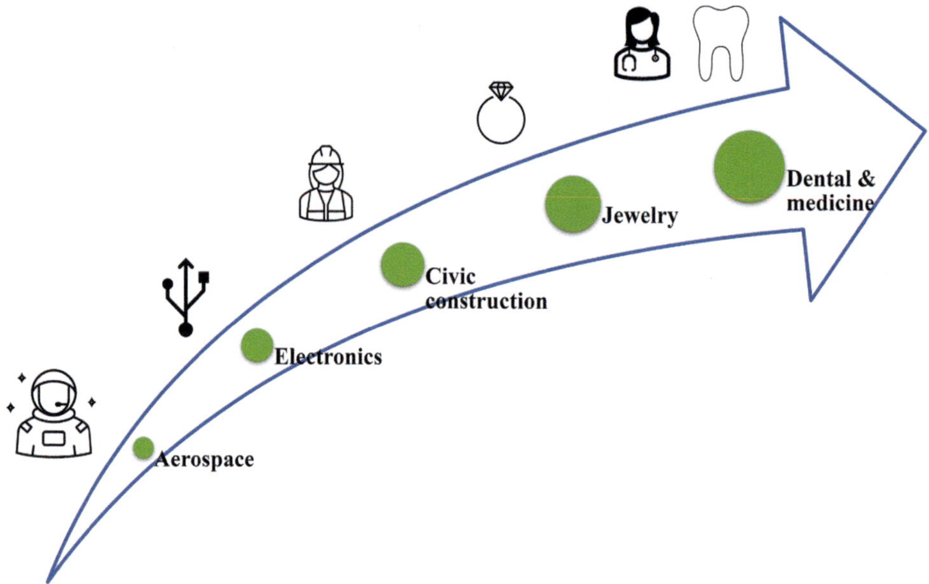

FIGURE 6.2 Most usual applications of additive materials.

developed with the aim of improving the printable geometry [4]. According to the resolution ISO/ASTM 52900:2015 [5], the 3D printing principles include material jetting, powder bed fusion, vat photopolymerization, and material extrusion (Figure 6.1).

Thanks to the recent development of the above-mentioned techniques, it is possible to create 3D segments for numerous applications, such as for the aerospace, electronics, civic construction, jewelry, and dental & medicine industries (Figure 6.2). The AM approach has demonstrated greater potentiality during the outbreak of COVID-19, in which 3D printed apparatus included face shields, plastic holder for assembling an emergency dwelling , swab sampling, testing devices, and alternative breathers [6]. The literature indicates that the manufacturing methods stereolithography (SLA) and fused deposition modeling (FDM) are the most ideal methods due to high-resolution and cost-effectiveness, respectively. The printers can be obtained commercially and exhibit the capacity to load single and/or multiple additive materials aiming to create solid structures with different colors and/or distinct physical-chemical properties [1].

The general components of a 3D printer include (i) printer head; (ii) working bed; (iii) an extruder motor or laser beam; (iv) temperature sensor; (v) and finally the additives that will be discussed further below. There are thermoset materials and thermoplastic filament as the main additive materials employed on 3D printers [2], [7]. Normally, the additive material can be manufactured by using three stages: pre-selection of the ideal properties; synthesis of the target material; a combination of polymeric composites.

Thermoset resins are dependent on light treatment for realizing the polymerization process, and they offer the possibility to construct high-quality objects. The transition between the liquid and solid phases occurs by the exposure of each layer to photoreactive UV/visible light laser at 300 to 405 nm, forming a solid layer with thicknesses ranging from 25 to 200 μm. The ideal conditions for realizing the manufacture of photopolymers depend on the planned utilization, ranging from pure elastomers and rigid material to functionalized structures. The main challenges presented in additive resin included slow polymerization, viscosity control, environmental interaction, material cost, and additional steps for creating high-defined objects [2].

Thermoplastic filaments are created by the combination of amorphous and/or semicrystalline polymeric structures so that there is a well-definite temperature of solid to semi-liquid transition, that is the melting point [8], [9]. The most popular examples of filaments include insulating, flexible, and biopolymers. Also, there are conductive composites that can be manufactured based on a mixture of polymeric binders and metals or carbon allotropes. The melting temperature depends on the loaded plastic filament which can range from 190 to 290 °C. The major drawbacks observed in the above materials are the possibilities for creating high resolution on nano-micro segments, durability materials, translucid layers, and the absence of local roughness [4].

This chapter is dedicated to demonstrating some examples of available additive materials that can be employed in the AM field. Discussion, challenges, and perspectives involving the fabrication and utilization steps of photoresin, and thermoplastic materials will be also carefully presented.

6.2 FABRICATION OF THERMOPLASTIC ADDITIVE

The approach FDM also known as fused filament fabrication (FFF), is the most popular method of 3D printing today [9]–[13]. This method consists of the controlled deposition of a thermoplastic filament through a moving nozzle or orifice heated above the glass transition temperature of the polymer. The deposited polymer solidifies on the printing table to form the desired LbL object. There are a wide variety of commercially available thermoplastic filaments in various colors, and different mechanical and electrical properties that can be used for FDM. The most commonly used are polylactic acid (PLA) and acrylonitrile butadiene styrene copolymer (ABS). Other polymeric filaments for FDM include polyvinyl alcohol (PVA), polyethylene terephthalate (PET), nylon, thermoplastic elastomeric (TPE), polycarbonate (PC), polycaprolactone (PCL), and high-impact strength polyethylene (HIPS).

The thermoplastic filaments applied in FDM are manufactured using the extrusion process. The schematic representation of a screw extruder is shown in Figure 6.3. The single-screw extruder consists of five main components, including a feeding system, drive system, screw system, barrel and heaters, head and die assembly [14]. The feed system includes the feed hopper and screw feed section. The drive system includes the motor, main gear, and thrust bearing assembly. The motor together with the gears is responsible for rotating the screw to which it is attached. On extruders that have two screws, the motor rotates them in opposite directions to promote greater material shearing efficiency. The main function of the screw system is to push the solid material along the barrel into regions of high temperature and pressure, where it is heated and compacted until it passes through a mold or dies in the shape of the desired object. The control system allows you to regulate the operating parameters including, motor speed per minute (rpm) and barrel temperature [12].

For the manufacturing of filaments, the polymer received in the form of granules or pellets is deposited directly into a hopper over the feed throats. Gravity and screw rotation ensures the polymer feed to the extrusion. Before starting the extrusion process, some polymers must go through a drying procedure to avoid their degradation due to moisture. Nylon, polyester, PET, and polycarbonate for

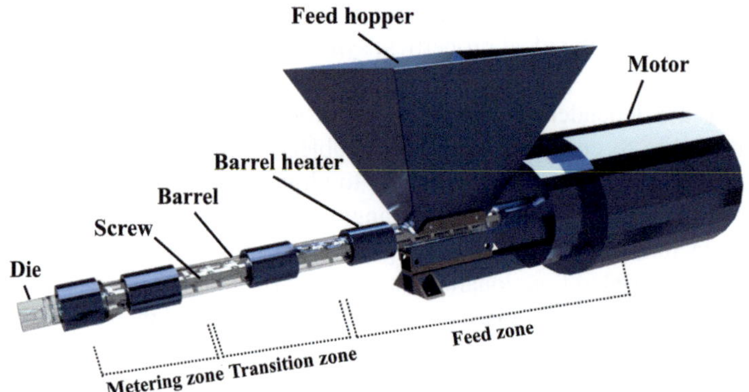

FIGURE 6.3 Schematic representation of single screw extruder setup utilized for manufacturing thermoplastic filament.

example are very hygroscopic and commonly go through a drying process before extrusion. Other materials that have not been stored in temperature and humidity-controlled environments should also be dried beforehand. This pre-treatment is important because any moisture present in the polymer will be evaporated in the extruder and, depending on the amount, may result in surface flaws such as bumps and holes in the product. In general, most polymers can be dried in dehumidifiers with dew points of −40 °C (−40 °F). In addition, additives can also be mixed with the polymer before extrusion [9], [13].

Various types of additives can be used to enable different properties for the desired product including, heat stabilizers, oxidative stability, UV stability, color pigments, flame retardants, and reinforcements, among others. To ensure uniform distribution of the additive in the polymeric material each component of the polymer/additive blend can either be added gravimetrically directly into the feed throat or be pre-blended. The feeding method will depend on the components to be mixed and the available machinery. Feeding additives directly into the extrusion throat is most efficient when you have a feeder with dimensions suitable for each component of the mix. However, in many situations, multiple feeder systems are not available, so a pre-blending process is necessary. Each component is weighed separately and added to a mixing system.

The most common mixing systems include drum mixers, V_{cone} mixers, drum rollers, and others. When mixing components of different sizes, adequate control is required. Mixing between pellets and powder, for example, may not be efficient since powder tends to flow between the pellets. As a result, the product may have more of the powder component at the beginning of the extrusion, while at the end of the process there will be more of the pellet product in the product. Coating the pellets with small amounts of mineral oil to create an adherent surface for the powder is one strategy to minimize this problem.

The properly treated and moisture-free pure polymer or polymer blend is then fed into the extruder barrel. The barrel consists of the feed zone, transition zone, and metering zone. In the feed zone, the polymer is melted and transported to the transition zone where the melting of the polymer occurs gradually [15]. The molten material fills the gaps between the screw and the barrel. In the metering zone, the fully melted material is pumped toward the die to be finally extruded. Based on the desired filament diameters, the cross-sectional dimension of the die is chosen. For filaments with a diameter of 1.75 mm, a die with a diameter of 2.5 to 3.5 mm is generally used. After extrusion, the material goes through a cooling process, which is usually done using water, air, or contact with a cold surface. The cooled filament is pulled away from the extruder using a puller. The speed of extraction is crucial to producing the correct dimensions of the product. Therefore, it is necessary to have control of the tension and speed of extraction and adjust them to the production rate of the extruder.

6.3 SYNTHESIS OF POLYMERIC PHOTORESIN

The traditional manufacturing process for constructing additive thermosets can be realized by the based combination of semi-viscous liquid photoresin and photoinitiator agents [3], [16]–[18]. Additionally, other materials such as dyes, conductive, or biomaterials can be utilized according to their final utilization [19]. Additive resin there is excellent performance in terms of mechanical properties, electrical insulation, resistance to corrosion, and chemical resistance, for example. Examples of additive resin included the ACPR-48, vinyl ester, DL260®, RTM370®, and Irix White®. Normally, the aforementioned materials are obtained from petroleum feedstock to natural resources and the final resin is composed of meth(acrylate)-based and/or epoxy-based components. Monomeric or oligomeric structures are well-known as additive options that can be manufactured by using cycloaliphatic, epoxidized, and glycated epoxy resins [1], [20].

So far, the working principle of the manufacturing method is associated with chemicals synthesized by direct oxidation of the corresponding olefin and/or through glycidation process that occurs when blending bisphenol, A, and epichlorohydrin reagents, for example. For improving the cure time, the resin is prepared in the presence of photoinitiator compounds, also nominated as (un)cleavable agents and up to now, some fabrication approaches are protected by patents reducing their accessibility. These structures can be subdivided into types I and II that correspond to cleavable and uncleavable compounds, respectively [20]. There are photoinitiators compounds such as chalcone, chromone, perylene, diketopyrrolopyrrole, acridine, naphthalimide, iodonium salts, squaraines, benzophenone, bodipy, porphyrin, coumarine, pyridinium derivatives, carbazoles, cyclohexanones, and so on [20].

Thermoset resin can be polymerized by radicalization process utilizing a combination of laser source and scanner system on builder platform also known as vat compartment. The photosensitive resin is converted from liquid to solid phase when scanned by a laser beam and the final thickness of the solid layer depends on the quality of the laser tool [2]. Traditionally, cationic or hybrid photopolymerization are preferred technique for curing the liquid resin. After finishing the printing step, the desired objects are submitted to a post-treatment process such as heating or photo-curing aim to improve the mechanical properties of the final printed layers. However, it could be noticed that manufacturing liquid resin with high viscosity is a challenge for 3D-printing research groups [3].

6.4 ADDITIVES IN 3D PRINTING

Additive materials are developed to find distinct physics–chemical properties, including, insulating, flexible, conductive, and bio-based compatible, that can be applied for numerous prototyping situations directly at home and workplaces such as labs, offices, industries, and schools. The sections below are dedicated to introducing the most popular examples of additives in the 3D-printing field, from utilized additives to applications. Figure 6.4 schematically indicates the additives materials.

6.4.1 Reinforcement on Filaments

The thermoplastic filaments can have their physical-chemical properties improved, according to the desired application, by adding other components to the polymeric matrix. Several research groups have been developing the combination of different materials. Zhong et al. [21] modified ABS thermoplastic with short glass fibers and linear low-density polyethylene (LLDPE). The authors reported that the combination of ABS and glass fiber increased the stiffness and softening temperature of the polymer. However, the polymer product became brittle at room temperature preventing its use for FDM. The addition of LLDPE coupled with compatibilizer (hydrogenated Buna-N) to ABS allowed to obtain a tough filament with adequate flexibility for handling. Singh et al. evaluated the influence of aluminum oxide particle size in combination with nylon [22].

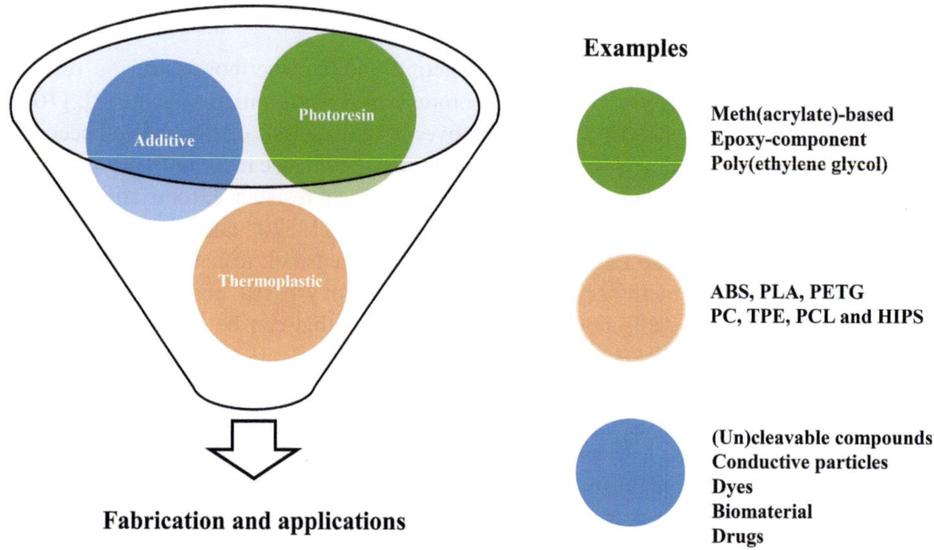

FIGURE 6.4 Some examples of materials employed for fabricating additives for 3D printing technology.

The mechanical properties of the material modified with single particle size (SPS), double particle size (DPS), and triple particle size (TPS) were studied. The experiments indicated that SPS provides better percentage elongation of the material. The best values for properties such as tensile strength, yield strength, and elastic modulus (or Young's modulus) were obtained when DPS was used. The addition of carbon nanotubes and short carbon fiber in ABS was investigated by Zhang et al. [23]. Shear tests were performed to evaluate the interfacial bonding properties between pure ABS, ABS with carbon nanotubes (CNTABS), and ABS with glass fiber (CFABS) threads. CFABBS showed the highest porosity indices; in 0° and ± 45° raster orientations, and higher tensile strength, compared to ABS and CNTABS. On the other hand, at the 90° angle the tensile strength of CFABS presents lower values than ABS and CNTABS due to the reduction of interfacial bonding from the increase of voids in the material.

The combination of ABS and short carbon fiber was also reported by Tekinalp et al. [24]. The authors also found that there was an increase in the distribution of voids in ABS with increasing fiber composition due to weak interfacial interaction between the materials. On the other hand, up to 20% of the amount of carbon fiber in the material was a significant increase in tensile strength. In addition, the fiber orientation of the FDM printed parts has the same direction as the deposition, ensuring the increase of mechanical properties of the final product. Short glass fibers have also been combined with polypropylene [25]. Filaments of polypropylene (PP) and glass fiber were evaluated in the absence and presence of polyophefin maleic anhydride (POE-g-MA) in the composition. The parts printed with PP and glass fiber showed higher modulus and strength of the composite, but with reduced flexibility. However, the addition of POE-g-MA to the components favored increased product flexibility.

The addition of fibers to photoresins can also increase the mechanical properties of SLA printed parts. Cheah et al. evaluated the influence of mixing short glass fibers with acrylic-based resin [26]. As a result, parts produced with 20% by volume glass fibers showed an increase in the modulus of elasticity and tensile strength. In addition, the shrinkage of the reinforced parts was lower than that of the pure parts. Llewellyn-Jones et al. have developed a method to manipulate the orientation of glass fibers within a photocurable resin system [27]. The fiber orientation is controlled by changing the ultrasonic stationary wave profile during printing. Various orientation angles have been achieved demonstrating the versatility of the process. This approach opens the possibility of creating complex fiber geometries within a three-dimensional printed system. Carbon fiber has been used as a reinforcement for photoresins. However, the addition of this composite in resins can be problematic, as carbon fiber is opaque and can inhibit UV light incidence making it impossible to fully cure the part. Gupta and Ogale combined UV radiation and heat treatments to cure high-volume carbon fiber reinforced resins [28]. About a quarter of the polymerization of the part is impaired, mainly in the inner regions of the carbon fibers and the resin layer. However, the tensile strength was increased by 95% after 1 h of the heat treatment process.

6.4.2 Flexible Filaments

Flexible filaments fall into the group of thermoplastic elastomers (TPE) which involve copolymers or blends of rubber with thermoplastic materials. Elastomer materials are generally characterized by tensile tests, hardness measurements, compression tests, tear tests, and shear and flexural measurements [29]. Flexible filaments used in FDM can be found on the market under trade names including Poro-Lay Series®, NinjaFlex®, UniFlex®, and FlexiFil®. TPE has been widely exploited to produce medical devices and for the manufacturing of deformable sensors. Thus, in recent years significant research efforts have been directed toward the development of elastomeric composites for 3D printing. The development of a composite ink of CNT and PDMS was reported by Wei et al. [30] for the production of a corrugated flexible electrode (Figure 6.5A and 6.5B). The connection angles of the corrugated electrode were optimized to maximize the elasticity and electrical conductivity of the device. 8% CNT in the polymeric base composition was sufficient to make the PDMS suitable for printing. The stretchable electrodes with a 45° connection angle showed high elasticity with a strain rate greater than 300% and excellent electrical stability with minimal variations upon stretching.

The combination of a highly flexible polymer with some type of highly conductive filler can also be used for the fabrication of strain sensors. The addition of multi-walled carbon nanotubes (MWCNTs) to TPU was reported by Christ et al. for printing strain sensors with high elasticity and piezoresistive properties [31]. The addition of MWCNTs improved the stiffness, modulus of elasticity, and electrical conductivity of TPU. In addition, it was observed that MWCNT composition had a significant impact on the sensitivity of the system. Kim et al. combined CNT and TPU for printing multi-axial force sensors [32]. The structural region of the sensor was fabricated using pure TPU filament, while CNT/TPU filament was employed to fabricate the sensing region. The sensor was able to measure forces from three orthogonal axes (x, y, and z) with high sensitivity allowing for detection of a submillimeter scale of deflection. Xiang et al. evaluated the CNT/TPU combination for flexible deflection sensors [33]. 1-Pyrenecarboxylic acid (PCA) was added to the polymer blend to improve the dispersion of CNT in the TPU matrix. As a result of more uniform CNT dispersion, the mechanical and electrical properties of the modified composite were increased. In addition, the sensor exhibited high sensitivity and a large detectable strain.

Photosensitive resins with elastomeric properties have also been developed by several authors. Jukka et al. synthesized a polyurethane resin for the production of elastomeric objects by combining PDMS, poly(tetramethylene) oxide (PTMO), isophorone diisocyanate (IPDI), (hydroxyethyl)methacrylate (HEMA) and 2-methyl-1propanol alcohol [34]. The light-cured objects showed Young's modulus of 2.5 MPa, a tensile strength of 3.7 MPa, and elongation at a break of 195%. Chen et al. combined commercial components including hydroxyethyl acrylate (HEA), methyltrimethoxysilane (MTMS) to obtain an ultra-low viscosity (about 10 cps) photo-curable resin suitable for printing elastomeric objects [35]. Figure 6.5C displays some objects printed with the flexible resin. The printed elastomers showed good mechanical properties with maximum yield stress and tensile strength of 3.75 MPa and 205%, respectively. Magdassi et al. [29] reported the mixing of aliphatic urethane diacrylate (AUD, Ebercryl 8413, Allnex) and epoxy aliphatic acrylate (EAA, Ebecryl 113) to obtain a highly elastomeric photocurable resin (Figure 6.5D). Different ratios between AUD

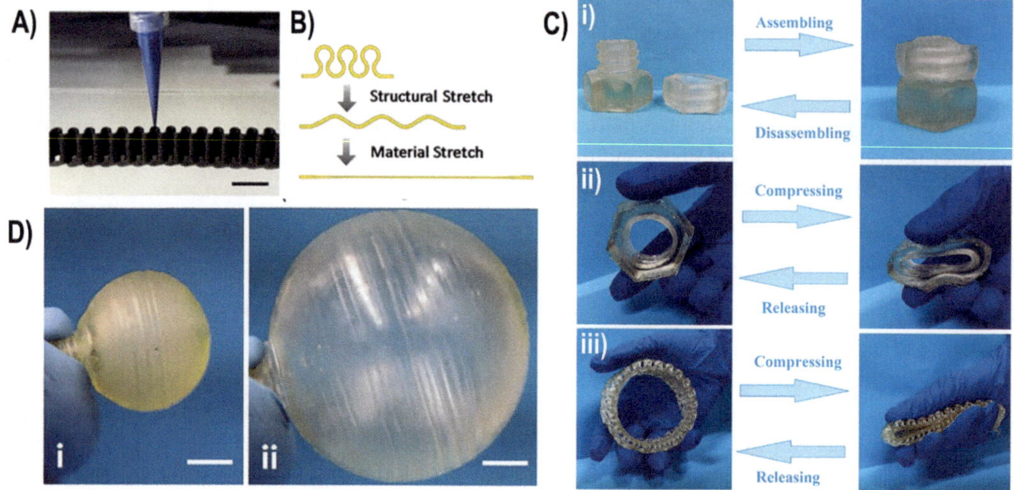

FIGURE 6.5 Devices printed with elastomeric materials. (A) Flexible CNT and PDMS sensors printed on serpentine and (B) representation showing their stretching process. Scale bar of 10 mm. (C) Objects printed with HEA and MTMS-based photoresin. Assembly and disassembly of (i) screw and nut; compression and release of (ii) nut and (iii) bracelet. (D) Highly deformable balloon printed by UV-curable resin based on AUD and EAA. (i) undeformed state; (ii) inflated state by three times the original size. Scale bar of 10 mm. ((A) and (B) Adapted with permission [30], Copyright (2017), Wiley. (C) Adapted with permission [35], Copyright (2021), Wiley. (D) Adapted with permission [29], Copyright (2017), Wiley.

and EAA generated products with Young's modulus ranging from 0.5 to 4.21 MPa. In addition, the elastomer produced exhibited elongation at break up to 1100%, about five times more than the elongation of commercial UV-cured elastomers.

6.4.3 Conductive Materials

Conductive additives are traditionally composed of synthetic or natural binder dopped with metallic particles, carbon allotropes, or conductive polymers. The most ideal metal groups dopped on binders are composed of copper, nickel, silver, aluminum, silica, and iron, for example. From the micro point of view, the metallic particles are chosen for manufacturing additives as additional components because of their short gap between the valence and conduction bond that assures excellent electrical properties to the final filament composition. The ideal mass percentage of metals that can be dopped in additive composites ranged from 5 to 30%. Higher values to a percentage of 30% can be prejudiced against the mechanical behavior of the desired additive materials.

On the other hand, another example of additional material doped on polymeric resin or thermoplastic is based on carbon particles that can be selected to fabricate additive composites due to their advantages in terms of low-cost, dispersion on solvent, functionalize, high contact area surface, and most important the capacity of electron-transfer. The conductive properties of carbon allotropes can be associated with the mobility of electrons to the π orbital localized between the lamellae from carbon structures.

Examples of allotropes carbon successfully dopped in additive materials included graphite, nanotube, graphene, carbon fiber, and carbon black. There are examples of commercially available filaments such as Proto-Pasta®, and Magic-Black® that are composed of PLA doped with carbon black and graphene, respectively. It is important to mention that the electrical resistivity of the additive material can be strongly altered depending on some information such as the choice

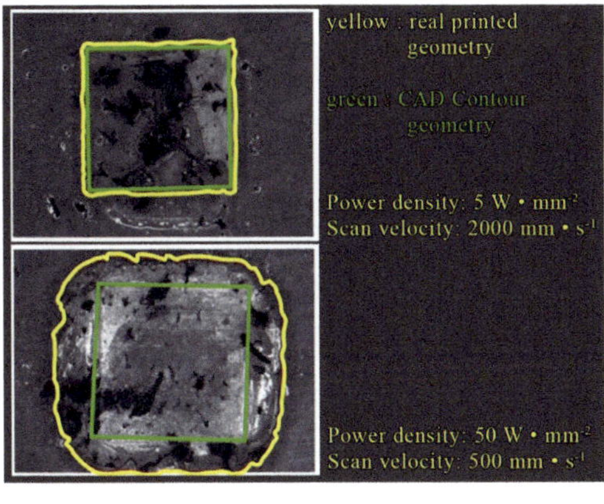

PEGDA:PEDOT (10:1)
$(2.88 \cdot 10^{-6} \pm 0.56 \cdot 10^{-6})$ $S \cdot cm^{-1}$

PEGDA:PEDOT (5:1)
$(50.24 \cdot 10^{-3} \pm 4.76 \cdot 10^{-3})$ $S \cdot cm^{-1}$

PEGDA:PEDOT (1:1)
(2.67 ± 0.20) $S \cdot cm^{-1}$

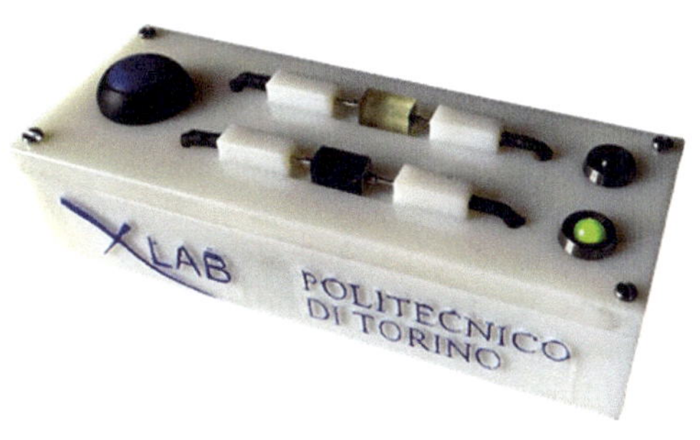

FIGURE 6.6 Example of manufactured conductive resin and final electronic device. Adapted with permission [19], Copyright (2019), Elsevier.

of proportions between the binders/conductive structures and the selected printing parameters. Considering the FDM, the parameters examples as infill density, layer height, printed speed, infill pattern, and nozzle/bed temperature must be carefully observed. Other important information must be observed when loading the additive agent on the 3D-printer tool there are printing direction and thickness of the desired objects. The printing process based on vertical direction and pre-selected thickness with a value over 10 mm are most recommended for conductive structures.

Scordo et al. demonstrated the combination of insulating and conductive polymers for creating resin with electrical and mechanical properties. The materials poly(ethylene glycol) diacrylate (PEGDA), poly(3,4-ethylenedioxythiophene) (PEDOT), and Irgacure 819® ma(photoinitiator structure) were successful mixed and exhibited two important functions, high conductivity, and biocompatibility (Figure 6.6). The authors calculated the value of conductivity to the reported resin and obtained values ~ 0.05 S/cm, emerging then as a powerful conductive additive for 3D-printing applications.

Tan et al. [36] indicated that it is possible to create conductive resin through simple integration between silver nanowires/polyvinylpyrrolidone (PVP) and the reported materials exhibited values of conductivity ~ 13.7 kΩ (0.014 S/cm) (Figure 6.7). The metal Ag doped in the additive resin does not compromise the performance of the final solid geometry and the electrical characterization showed that the reported conductive material can be explored in the most varied field of 3D-printing technology.

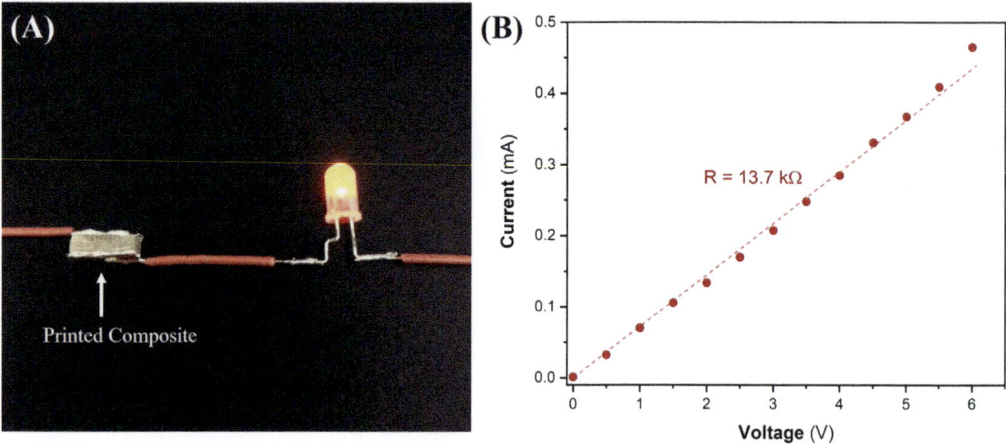

FIGURE 6.7 (A) Electronic circuit of the 3D printed conductive composite and (B) reported dependence between applied potential and current values. Adapted with permission [36], Copyright (2021), Elsevier.

Foster et al. [7] manufactured conductive filament combining native PLA pellets/nanographite and the reported filament (Ø = 1.75 mm) was loaded on an FDM printer to create carbon-based sensing electrodes. The authors' utilized carbon structures less than ~500 nm (particle size) as a conductive agent and the reported electrochemical devices revealed heterogeneous electron transfer rate constant (k_0) ~ 8.12 × 10^{-3} cm/s and successful detection of heavy metals (Cd and Pb). Kaynan et al. [14] reported that the materials carbon nanotube (CNT) and polyetherimide can be explored as additive filaments in the electronic field. The reported filament was fabricated by using CNT ~ 7 wt% and exhibited a conductivity value of ~ 0.26 S/cm. Through the reported study was possible to see that the particle size of CNT was drastically reduced after the manufacturing process (reduction ~ 40%) probably associated with the local high shear forces and mechanical stirring observed during the melting step. However, the mentioned above characteristics do not compromise the electrical properties of the target filament.

6.4.4 Pharmaceutical and Medical Applications

The introduction of drugs in a polymeric matrix is being explored for application in the pharmaceutical industry that one of the most important markets in the medical & dental fields. In general, printable raw materials require in-depth characterization of their properties. Excipients suitable for FDM have glass transition temperatures lower than the decomposition temperature of the drug and suitable rheological properties. In addition, the polymer product must have mechanical properties suitable for the printing process, since very brittle or soft filaments can be easily broken or squeezed by the feed gears making extrusion impossible. Various polymeric substrates, plasticizers, and lubricants are used to improve printability [37], [38].

Numerous studies have shown the use of these polymeric materials in the creation of oral pharmaceutical forms. Drugs can be incorporated directly into the commercial filaments by swelling the filament in API solution in a volatile solvent followed by drying or by HME. Prednisolone was added to PVA for the production of extended-release tablets by FDM [39]. Figure 6.8A depicts tablets manufactured in an ellipse shape and dissolved in a mixture of methanol and water. High-Performance Liquid Chromatography (HPLC) and dissolution test were used to validate the accuracy of the dose and drug release patterns. The printed tablets exhibited a dose accuracy range of 88.7 to 107%, while the release of prednisolone from a 3D-printed tablet was extended for 24 h.

Goyanes et al. demonstrated great efforts in modifying PVA for the preparation of oral pharmaceutical forms for FDM by both methodologies. Various drugs including, fluorescein [40], budesonide [41], paracetamol [42], and caffeine [42] were incorporated into the polymeric material to evaluate the release profile of 3D printed drugs. The method of re-extrusion of ground filaments with API by hot-melt extrusion (HME) takes advantage by demonstrating higher accuracy in the content of printed drugs. Moreover, this approach allows the production of filaments with higher drug content and mechanical properties suitable for 3D printing [40], [41].

Muwaffak et al. [43] developed antimicrobial dressings made of polycaprolactone (PCL) loaded with zinc, copper, and silver. Filaments with different concentrations of metals were used to manufacture dressings of various shapes. All dressings showed prolonged release of the different metals and bactericidal properties. However, it was concluded that the silver and copper dressings had greater bactericidal capacity. In addition, compared to conventional flat dressings, the anatomically conformable dressings exhibited greater comfort and adhesion in complex anatomical areas.

Biocompatible resin has been developed to facilitate the application of AM in the medicine industry. Grigoryan et al. [44] used PEG-diacrylate (PEGDA) to create a resin to produce a non-cellular vascularized alveolar unit that was capable of breathing (Figure 6.8B). The authors noted that the printed PEGDA-based device with a molecular mass of 6 kDa withstood more than 10,000 cycles of oxygen or nitrogen ventilation at 24 kPa and a frequency of 0.5 Hz for a period of 6h. Lim et al. [45] reported the development of a bio-compatible resin with a combination of PVA methacrylate (PVA-MA) and gelatin methacrylate (GelMA) to produce biological tissues. The ratio of 10% by weight of PVA-MA/1% by weight of GelMA allowed the construction of complex cell-laden structures suitable for bone and cartilage tissue formation.

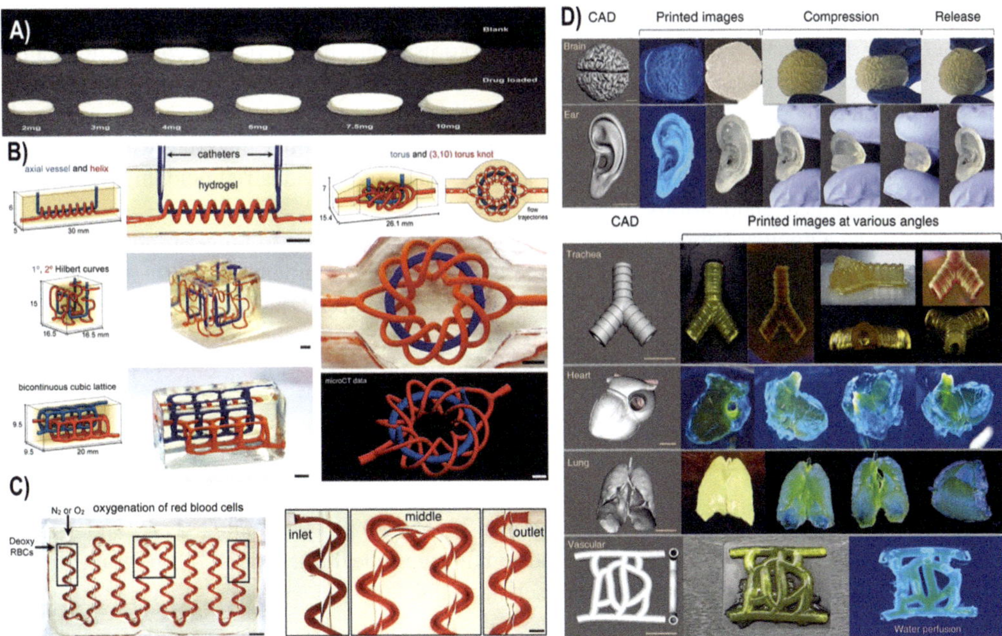

FIGURE 6.8 Devices printed with biocompatible materials. (A) PVA and prednisolone thermoplastic filament printed tablets. (B) PEGDA-based devices containing entangled vascular networks (3 mm scale bar). (C) Printed device during the process of oxygenation of red blood cells (scale bar of 3mm). (D) Cerebrum, ear, trachea, heart, lung, and vascular network printed with Sil-MA photosensitive resin. (A) Adapted with permission [39], Copyright (2015), Elsevier. (B) and (C) Adapted with permission [44], Copyright (2019), Science.

Akopova et al. [46] used chitosan prepared from crab chitin to produce a photocurable bioresin. The photo-reticulable chitosan was synthesized by introducing allyl substituents to the molecule. The allyl substituted chitosan-based photoresin was suitable for SLA allowing printing of well-defined polymeric scaffolds. Kim et al. [47] described the use of silk fibroin (SF) extracted from the silkworm *Bombyx mori* to produce a bio-ink. Glycidyl methacrylate (GMA) was used to add methacrylate groups to the side groups of SF-containing amines producing the photoresist silk methacrylate-based resin (Sil-MA). Sil-MA was successfully employed for DLP prints allowing the production of complex, biocompatible structures including the heart, vessel, brain, trachea, and ear as presented in Figure 6.8C and 6.8D.

6.5 CONCLUSION AND PERSPECTIVES

This chapter revised the main examples of additive materials and their most usual synthetic routes. Also introduced the drawbacks and advantages of the utilization of thermoplastic and photoresin on 3D printers and some grateful examples of applications. Through the fabrication process of thermoset and thermoplastic additives was possible to see additives with different colors and distinct physical-chemical properties such as flexibility, conductivity, insulating, translucid, and biocompatible. In addition, the revised manufactured protocols indicated the most usual proportions of additives agents and photoresin or thermoplastic binders.

The review benefits herein discussed included the examples of chemical synthesis for obtaining the photoresin and the ideal melting point for manufacturing the filament. There are drawbacks of the above materials, such as manufacture time, anisotropic tendency, few available additives, smooth surfaces, and high cost of photoresin. If the pure thermoplastic filaments demonstrated poor mechanical and conductive performance, the literature has successfully indicated some efforts for improving the physical-chemical properties by reinforcing the materials with glass and carbon fibers.

In terms of the manufacturing methods, the photoresin and thermoplastic composites can be obtained through the market, labs, and/or by 'do it yourself' approaches. However, the purity of additive materials can be influenced by the final quality of the desired filament, mainly in the cases involving photoresin fabrication. Considering their utilization, the additive materials when associated with 3D-printing technology can be useful for some sections of the industrial field, highlighting the electronic and medical applications.

Despite being a popular approach for manufacturing, from compact to bigger products and most modern applications, the additive materials need to be available commercially, and look forward to their finalization of patent time aiming to decrease the cost of materials and equipment. Nonetheless, the trend of the additive materials devoted to 3D-printing applications has been extremely positive, emerging as a present and future manufacturing option for the most complex objects for supporting the community and industrial needs.

ACKNOWLEDGMENTS

The authors would like to thank to CNPq (grants 307554/2020-1, 405620/2021-7, and 142412/2020-1), CAPES (grant 88887.192880/2018-00 and finance code 001), and INCTBio (grant 465389/2014-7) for the financial support and granted scholarships.

REFERENCES

1 M.S. Thompson, Current status and future roles of additives in 3D printing—A perspective, *Journal of Vinyl and Additive Technology*. 28 (2022) 3–16.

2 F. Wang, F. Wang, Liquid resins-based additive manufacturing, *Journal of Molecular and Engineering Materials*. 05 (2017) 1740004.

3 M. Peerzada, S. Abbasi, K.T. Lau, N. Hameed, Additive manufacturing of epoxy resins: Materials, methods, and latest trends, *Industrial and Engineering Chemistry Research*. 59 (2020) 6375–6390.
4 R.V. Pazhamannil, V.N. Jishnu Namboodiri, P. Govindan, A. Edacherian, Property enhancement approaches of fused filament fabrication technology: A review, *Polymer Engineering and Science*. 62 (2022) 1356–1376.
5 ISO – ISO/ASTM 52900:2015 – Additive manufacturing—General principles—Terminology, www.iso.org/standard/69669.html (accessed July 26, 2022).
6 Y.Y.C. Choong, H.W. Tan, D.C. Patel, W.T.N. Choong, C.H. Chen, H.Y. Low, M.J. Tan, C.D. Patel, C.K. Chua, The global rise of 3D printing during the COVID-19 pandemic, *Nature Reviews Materials*. 5 (2020) 637–639.
7 C.W. Foster, H.M. Elbardisy, M.P. Down, E.M. Keefe, G.C. Smith, C.E. Banks, Additively manufactured graphitic electrochemical sensing platforms, *Chemical Engineering Journal*. 381 (2020) 122343.
8 D.D. Phan, J.S. Horner, Z.R. Swain, A.N. Beris, M.E. Mackay, Computational fluid dynamics simulation of the melting process in the fused filament fabrication additive manufacturing technique, *Additive Manufacturing*. 33 (2020) 101161.
9 A.D. Valino, J.R.C. Dizon, A.H. Espera, Q. Chen, J. Messman, R.C. Advincula, Advances in 3D printing of thermoplastic polymer composites and nanocomposites, *Progress in Polymer Science*. 98 (2019) 101162.
10 X. Wang, M. Jiang, Z. Zhou, J. Gou, D. Hui, 3D printing of polymer matrix composites: A review and prospective, *Composites Part B: Engineering*. 110 (2017) 442–458.
11 P.F. Flowers, C. Reyes, S. Ye, M.J. Kim, B.J. Wiley, 3D printing electronic components and circuits with conductive thermoplastic filament, *Additive Manufacturing*. 18 (2017) 156–163.
12 M.A. Gibson, N.M. Mykulowycz, J. Shim, R. Fontana, P. Schmitt, A. Roberts, J. Ketkaew, L. Shao, W. Chen, P. Bordeenithikasem, J.S. Myerberg, R. Fulop, M.D. Verminski, E.M. Sachs, Y.M. Chiang, C.A. Schuh, A. John Hart, J. Schroers, 3D printing metals like thermoplastics: Fused filament fabrication of metallic glasses, *Materials Today*. 21 (2018) 697–702.
13 E. Gkartzou, E.P. Koumoulos, C.A. Charitidis, Production and 3D printing processing of bio-based thermoplastic filament, *Manufacturing Review*. 4 (2017) 1.
14 O. Kaynan, A. Yıldız, Y.E. Bozkurt, E. Ozden Yenigun, H. Cebeci, Electrically conductive high-performance thermoplastic filaments for fused filament fabrication, *Composite Structures*. 237 (2020) 111930.
15 E. Chaturvedi, N.S. Rajput, S. Upadhyaya, P.K. Pandey, Experimental study and mathematical modeling for extrusion using high density polyethylene, *Materials Today: Proceedings*. 4 (2017) 1670–1676.
16 Y.R. Jun Yue, Pei Zhao, Jennifer Y. Gerasimov, Marieke van de Lagemaat, Arjen Grotenhuis, Minie Rustema-Abbing, Henny C. van der Mei, Henk J. Busscher, Andreas Herrmann, *3D-printable antimicrobial composite resins, Advanced Functional Materials*. 25 (2015) 6756–6767.
17 C. Warr, J.C. Valdoz, B.P. Bickham, C.J. Knight, N.A. Franks, N. Chartrand, P.M. van Ry, K.A. Christensen, G.P. Nordin, A.D. Cook, Biocompatible PEGDA resin for 3D printing, *ACS Applied Bio Materials*. 3 (2020) 2239–2244.
18 A.W. Bassett, A.E. Honnig, C.M. Breyta, I.C. Dunn, J.J. la Scala, J.F. Stanzione, Vanillin-based resin for additive manufacturing, *ACS Sustainable Chemistry and Engineering*. 8 (2020) 5626–5635.
19 G. Scordo, V. Bertana, L. Scaltrito, S. Ferrero, M. Cocuzza, S.L. Marasso, S. Romano, R. Sesana, F. Catania, C.F. Pirri, A novel highly electrically conductive composite resin for stereolithography, *Materials Today Communications*. 19 (2019) 12–17.
20 F. Dumur, Recent advances on pyrene-based photoinitiators of polymerization, *European Polymer Journal*. 126 (2020) 109564.
21 W. Zhong, F. Li, Z. Zhang, L. Song, Z. Li, Short fiber reinforced composites for fused deposition modeling, *Materials Science and Engineering A*. 301 (2001) 125–130.
22 R. Singh, P. Bedi, F. Fraternali, I.P.S. Ahuja, Effect of single particle size, double particle size and triple particle size Al_2O_3 in Nylon-6 matrix on mechanical properties of feed stock filament for FDM, *Composites Part B: Engineering*. 106 (2016) 20–27.

23 W. Zhang, C. Cotton, J. Sun, D. Heider, B. Gu, B. Sun, T.W. Chou, Interfacial bonding strength of short carbon fiber/acrylonitrile-butadiene-styrene composites fabricated by fused deposition modeling, *Composites Part B: Engineering*. 137 (2018) 51–59.

24 H.L. Tekinalp, V. Kunc, G.M. Velez-Garcia, C.E. Duty, L.J. Love, A.K. Naskar, C.A. Blue, S. Ozcan, Highly oriented carbon fiber-polymer composites via additive manufacturing, *Composites Science and Technology*. 105 (2014) 144–150.

25 G. Sodeifian, S. Ghaseminejad, A.A. Yousefi, Preparation of polypropylene/short glass fiber composite as Fused Deposition Modeling (FDM) filament, Results in Physics. 12 (2019) 205–222.

26 C.M. Cheah, J.Y.H. Fuh, A.Y.C. Nee, L. Lu, Mechanical characteristics of fiber-filled photo-polymer used in stereolithography, *Rapid Prototyping Journal*. 5 (1999) 112–119.

27 T.M. Llewellyn-Jones, B.W. Drinkwater, R.S. Trask, 3D printed components with ultrasonically arranged microscale structure, *Smart Materials and Structures*. 25 (2016) 02LT01.

28 A. Gupta, A.A. Ogale, Dual curing of carbon fiber reinforced photoresins for rapid prototyping, *Polymer Composites*. 23 (2002) 1162–1170.

29 D.K. Patel, A.H. Sakhaei, M. Layani, B. Zhang, Q. Ge, S. Magdassi, Highly stretchable and UV curable elastomers for digital light processing based 3D printing, *Advanced Materials*. 29 (2017) 1–7.

30 C.Y.Y. Hong Wei, Kai Li, Wen Guang Liu, Hong Meng, Pei Xin Zhang, 3D printing of free-standing stretchable electrodes with tunable structure and stretchability, *Advanced Engineering Materials*. 19 (2017) 1700341.

31 J.F. Christ, N. Aliheidari, A. Ameli, P. Pötschke, 3D printed highly elastic strain sensors of multiwalled carbon nanotube/thermoplastic polyurethane nanocomposites, *Materials and Design*. 131 (2017) 394–401.

32 K. Kim, J. Park, J. hoon Suh, M. Kim, Y. Jeong, I. Park, 3D printing of multiaxial force sensors using carbon nanotube (CNT)/thermoplastic polyurethane (TPU) filaments, *Sensors and Actuators, A: Physical*. 263 (2017) 493–500.

33 D. Xiang, X. Zhang, Y. Li, E. Harkin-Jones, Y. Zheng, L. Wang, C. Zhao, P. Wang, Enhanced performance of 3D printed highly elastic strain sensors of carbon nanotube/thermoplastic polyurethane nanocomposites via non-covalent interactions, *Composites Part B: Engineering*. 176 (2019) 107250.

34 L.H. Sinh, K. Harri, L. Marjo, M. Minna, N.D. Luong, W. Jürgen, W. Torsten, S. Matthias, S. Jukka, Novel photo-curable polyurethane resin for stereolithography, *RSC Advances*. 6 (2016) 50706–50709.

35 X.Z. Zhu, Fuping Chen, Ruiqi Li, Jingxian Sun, Guoqiang Lu, Jin Wang, Bing Wu, Jingfang Li, Jun Nie, Photo-curing 3D printing robust elastomers with ultralow viscosity resin, *Journal of Applied Polymer Science*. 138 (2020) 1–12.

36 K.Y. Tan, Z.X. Hoy, P.L. Show, N.M. Huang, H.N. Lim, C.Y. Foo, Single droplet 3D printing of electrically conductive resin using high aspect ratio silver nanowires, *Additive Manufacturing*. 48 (2021) 102473.

37 T.C. Okwuosa, B.C. Pereira, B. Arafat, M. Cieszynska, A. Isreb, M.A. Alhnan, Fabricating a shell-core delayed release tablet using dual FDM 3D printing for patient-centred therapy, *Pharmaceutical Research*. 34 (2017) 427–437.

38 W. Jamróz, J. Szafraniec, M. Kurek, R. Jachowicz, 3D printing in pharmaceutical and medical applications, *Pharmaceutical Research*. 35 (2018) 1–22.

39 J. Skowyra, K. Pietrzak, M.A. Alhnan, Fabrication of extended-release patient-tailored prednisolone tablets via fused deposition modelling (FDM) 3D printing, *European Journal of Pharmaceutical Sciences*. 68 (2015) 11–17.

40 A. Goyanes, A.B.M. Buanz, A.W. Basit, S. Gaisford, Fused-filament 3D printing (3DP) for fabrication of tablets, *International Journal of Pharmaceutics*. 476 (2014) 88–92.

41 A. Goyanes, H. Chang, D. Sedough, G.B. Hatton, J. Wang, A. Buanz, S. Gaisford, A.W. Basit, Fabrication of controlled-release budesonide tablets via desktop (FDM) 3D printing, *International Journal of Pharmaceutics*. 496 (2015) 414–420.

42 A. Goyanes, M. Kobayashi, R. Martínez-Pacheco, S. Gaisford, A.W. Basit, Fused-filament 3D printing of drug products: Microstructure analysis and drug release characteristics of PVA-based caplets, *International Journal of Pharmaceutics*. 514 (2016) 290–295.

43 Z. Muwaffak, A. Goyanes, V. Clark, A.W. Basit, S.T. Hilton, S. Gaisford, Patient-specific 3D scanned and 3D printed antimicrobial polycaprolactone wound dressings, *International Journal of Pharmaceutics*. 527 (2017) 161–170.

44 B. Grigoryan, S.J. Paulsen, D.C. Corbett, D.W. Sazer, C.L. Fortin, A.J. Zaita, P.T. Greenfield, N.J. Calafat, J.P. Gounley, A.H. Ta, F. Johansson, A. Randles, J.E. Rosenkrantz, J.D. Louis-Rosenberg, P.A. Galie, K.R. Stevens, J.S. Miller, Multivascular networks and functional intravascular topologies within biocompatible hydrogels, *Science*. 364 (2019) 458–464.

45 K.S. Lim, R. Levato, P.F. Costa, M.D. Castilho, C.R. Alcala-Orozco, K.M.A. van Dorenmalen, F.P.W. Melchels, D. Gawlitta, G.J. Hooper, J. Malda, T.B.F. Woodfield, Bio-resin for high resolution lithography-based biofabrication of complex cell-laden constructs, *Biofabrication*. 10 (2018) 034101.

46 K.N. Bardakova, T.A. Akopova, A.V. Kurkov, G.P. Goncharuk, D. v. Butnaru, V.F. Burdukovskii, A.A. Antoshin, I.A. Farion, T.M. Zharikova, A.B. Shekhter, V.I. Yusupov, P.S. Timashev, Y.A. Rochev, From aggregates to porous three-dimensional scaffolds through a mechanochemical approach to design photosensitive chitosan derivatives, *Marine Drugs*. 17 (2019) 48.

47 S.H. Kim, Y.K. Yeon, J.M. Lee, J.R. Chao, Y.J. Lee, Y.B. Seo, M.T. Sultan, O.J. Lee, J.S. Lee, S.-il Yoon, I.S. Hong, G. Khang, S.J. Lee, J.J. Yoo, C.H. Park, Precisely printable and biocompatible silk fibroin bioink for digital light processing 3D printing, *Nature Communications*. 9 (2018) 1–14.

7 3D Printing for Electrochemical Water Splitting

Chong-Yong Lee[*] *and Gordon G. Wallace*
ARC Centre of Excellence of Electromaterials Science, Intelligent Polymer Research Institute, AIIM, Innovation Campus, University of Wollongong, Wollongong, NSW 2500, Australia
*Corresponding author: cylee@uow.edu.au

CONTENTS

7.1 INTRODUCTION

The increasing impacts and consequences of climate change that affect livelihood have led to global efforts in reducing CO_2 emissions [1], [2]. This momentum has accelerated the search for clean energy as part of the transition to a more sustainable society. The electrochemical energy conversion processes offer an attractive green route to produce clean energy. When driven by renewable energies such as solar, wind, and hydro, this technology formed an integral component of efforts in creating sustainable energy as an alternative to fossil fuel-based economies. Further development of electrochemical technologies is needed in advancing more efficient, and low-cost production of clean fuels to promote its widespread applications.

A prominent example of such technology is the production of green hydrogen through the electrochemical splitting of water via water electrolyzers [3], [4]. This process using electricity to split water offers a clean route compared to 'brown hydrogen' generated via a gasification process that released significant CO_2. The high energy density of hydrogen is attractive in hard-to-decarbonize and energy-intensive industries such as shipping, iron, steel, chemicals, and aviation, and hence could play a critical role in the urgent goal of limiting CO_2 emissions [5]. The emergence of additive manufacturing or 3D printing technologies has found applications in a wide range of areas,

DOI: 10.1201/9781003296676-7

including energy applications [6], [7]. 3D printing allows for the fabrication of a three-dimensional structure via a layer-by-layer deposition using a computer-created design. There is increasing evidence that 3D printing could play an important role in catalyst assemblies and in designing electrochemical water splitting cells [6], [8].

In this chapter, we discuss the applications of 3D printing technology for electrochemical water splitting. Firstly, we discuss the principle and fundamentals of electrochemical water splitting, including the light-driven photoelectrochemical process. We review the 3D printing techniques that have been employed for the electrochemical water splitting application, in both conductive (e.g., electrodes) and non-conductive (e.g., vessel, container) components. The emphasis is on the electrode fabrications which are the centers for the electrochemical reactions. This includes strategies of catalyst assemblies during electrode preparations, post-processing, and water splitting device fabrication. Finally, we discuss the aspects of ongoing challenges and the prospect of 3D printing for electrochemical water splitting.

7.2 FUNDAMENTALS OF ELECTROCHEMICAL WATER SPLITTING

Water splitting is a process of breaking down water into hydrogen and oxygen. This can be achieved by applying an electric current through water in a process called electrolysis of water. The electrochemical water splitting cell involves two half-cell reactions. The oxidation reaction generates O_2, which is known as the oxygen evolution reaction (OER). The reduction reaction that generates H_2 is known as the hydrogen evolution reaction (HER). The overall reaction is described by Equation 7.1.

$$2H_2O \rightarrow 2H_2 + O_2 \tag{7.1}$$

To drive the reaction, a minimal theoretical voltage of 1.23 V under standard conditions is required. However, an additional voltage called overpotentials must be applied to overcome the kinetic barriers, hence obtaining desirable current density to produce gases. Electrocatalysts play a role in lowering the overpotentials, which reduces the energy inputs (e.g., at lower cost) to generate gases at a given reaction rate [9], [10]. OER and HER reactions produce oxygen and hydrogen, respectively, with the stoichiometric of 1:2 as shown in Equation 7.1. The mechanism of the two half-reactions involved in water electrolysis is both multi-step processes, as described below.
OER reactions:

$$2H_2O \rightarrow O_2 + 4H^+ + 4e^- \tag{7.2}$$

$$4OH^- \rightarrow O_2 + 2H_2O + 4e^- \tag{7.3}$$

OER is a four electrons multistep transfer process, which proceeds via a kinetically slow reaction. Hence, it is considered a thermodynamically uphill and energetically demanding process. In acidic media, water is oxidized into oxygen and protons (Equation 7.2), whereas in alkaline media, the hydroxide ion is oxidized into water and oxygen (Equation 7.3).
HER reactions:

$$2H^+ + 2e^- \rightarrow H_2 \tag{7.4}$$

$$2H_2O + 2e^- \rightarrow H_2 + 2OH^- \tag{7.5}$$

HER is a two-electron process. In acidic media, it involves the reduction of two protons to generate hydrogen (Equation 7.4). In alkaline media, the HER occurs via a similar reaction pathway as in an acidic solution, except it involves the reduction of water instead of protons (Equation 7.5). The requirement to break the strong O-H bond in water resulting in HER is energetically more demanding in alkaline media, hence it is typically having slower reaction rates than that in acidic media.

Photoelectrochemical water-splitting harvests the light to split water molecules into O_2 and H_2, with the help of a suitable photocatalyst [11], [12]. In the photoelectrochemical water splitting cell, a semiconductor with an appropriate bandgap that is capable of absorbing desired wavelengths of light is deposited on a conductive substrate for charge transport in the presence of an electrolyte. Upon illumination, photoelectrodes induce intrinsic ionization of the electrodes to produce electrical charge carriers, whereby electrons are transported to the conduction band and holes to the valence bands. Photogenerated holes oxidize and split water into O_2 and H^+ ions, and O_2 is released at the anode. Electrons generated at the anode are transferred to the cathode through an external circuit, where H^+ ions are reduced to produce H_2 gas. Catalysts are employed to enhance the reaction kinetics and minimize surface and bulk charge recombination at the photoelectrodes [12].

7.3 3D PRINTING METHODS FOR ELECTROCHEMICAL WATER SPLITTING APPLICATIONS

In recent years, there are increasing research activities in 3D printing for energy conversion and storage-related applications [6], [8]. 3D printing allows rapid prototyping which is beneficial as a proof-of-concept, accelerating the progress of fundamental understanding and applied research. This technology also allows the fabrication of internal structures and complex geometries that is difficult to be attained via traditional manufacturing method. An increasingly prominent and attractive aspect of this technology is its environmental and sustainability manufacturing criteria, which decreases material waste, therefore, reducing the CO_2 footprint. For example, printing technology only used the required materials with minimal waste, whereby excessive waste materials are often produced from the manual removal or computer-controlled machining processes in traditional subtractive manufacturing.

There are wide-ranging 3D-printing methods [13], [14], and we highlight here the commonly employed 3D-printing methods for electrochemical water splitting devices. There are two distinctly different components for this application, the electrodes and the associated electrochemical reactors, which require the use of conductive and non-conductive materials, respectively. As described earlier, the technology performance is largely dependent on the electrodes consisting of electrocatalysts where the water splitting reaction takes place. Therefore, the 3D printing fabrication approach ideally should incorporate the beneficial features that optimized the water splitting reaction.

7.3.1 Fabrication of Conductive Components

The conductive components of electrochemical water splitting devices refers to the electrodes employed in both anodic and cathodic compartments. The water splitting reaction occurs at anode and cathode in the presence of catalysts, which is therefore arguably the most critical 3D printing components for this application. In general, the fabrication of conductive electrodes commonly employed printing techniques such as selective laser melting (SLM), fused deposition modeling (FDM), and direct ink writing (DIW).

Among all the above-mentioned methods, metal printing via SLM offers the most direct route to fabricating solid electrodes with high conductivity. SLM is designed to use a high power-density laser to form a 3D structure by melting and fusing metallic powders layer-by-layer [15]. The employed metal powders include titanium, copper, nickel, stainless steel, and aluminum. Titanium and its alloys are among the most common metals in 3D printing with parts produced having widespread applications across aerospace and biomedical industries [16]. Therefore, it is not surprising that a greater understanding of the printing parameters of titanium leads to it serving as electrodes or as a platform for water splitting. Note that titanium itself is known to be inefficient for water splitting, hence post-modifications of the electrode surface are required to activate the electrodes. Further details will be discussed in Section 4. Copper and aluminum offer high conductivity but are known to be unstable in both acidic and alkaline electrolytes; therefore, they are rarely printed for water splitting application. However, nickel and stainless steel are stable as metallic substrates for alkaline-based water splitting applications. Nickel-based nanomaterials particularly are known to be one of the best catalysts for alkaline water splitting [17].

FDM is a material extrusion method in which materials are extruded through a nozzle and joined together to form 3D structures [18]. The FDM process typically uses thermoplastic as feedstock materials, usually in the forms of filaments or pellets. The thermoplastic materials include polylactic acid (PLA), acrylonitrile butadiene styrene (ABS), polyurethane (PU), and polycaprolactone (PCL). In comparison to SLM, FDM offers a low-cost printing approach and hence is one of the most widely accessible printing facilities in the laboratories. However, due to the incorporation of some portion of polymer which is less conductive, electrodes for OER and HER produced from the FDM method are typically not highly conductive. Some pre-treatments of the fabricated electrodes are needed to activate the electrodes.

Another extrusion-based 3D printing by using that ink is DIW. Inks consist of rheological complex fluids, in which the electroactive materials are mixed with additives and solvents to form an extrudable paste. To obtain mechanically stable and functional electrodes, inks must be formulated to meet stringent criteria for printability with desirable rheology to meet targeted functional applications [19]. DIW is more widely employed in electrochemical storage applications such as batteries and supercapacitors, than in catalytic systems [20]. This may be due to more stringent requirements in terms of electrode mechanical and chemical stabilities for electrocatalytic application. Anh et al. fabricated NiFe layered double hydroxide (LDH) pyramid electrodes for OER (Figure 7.1a) [21]. Firstly, direct-ink-writing of a graphene pyramid array was performed, followed by electrodeposition of a copper conductive layer and NiFe-LDH electrocatalyst layer on printed pyramids. The 3D pyramid structures with NiFe-LDH electrocatalyst layers were found to increase the surface area and the active sites of the electrode and improved the OER activity. The NiFe-LDH pyramid electrode was found to have lower overpotential and higher exchange current density compared to that of the NiFe-LDH deposited Cu (NiFe-LDH/Cu) foil electrode under the same base area. The 3D-printed NiFe-LDH electrode also exhibited excellent durability with limited decay for 60 h. Therefore, by employing optimized ink rheology, DIW could be an effective approach to fabricating active, stable, and low-cost OER electrodes.

It is worth mentioning another less common electrode fabrication approach by stereolithography (SLA) [22]. This method however only serves to produce electrode support rather than a standalone electrode, which requires an additional conductive coating to serve as an electrode (Figure 7.1b(i)). For example, the rectangular-shaped mesh has been printed based on commercially available PlasClear resin that acts as insulating support (Figure 7.1b(ii)) [23]. Activation of the electrode surface was initially performed by dipping the support scaffold in commercially available electroless copper solution C (5 wt % HCl and 5 wt % $SnCl_2$), and subsequently in electroless copper solution D (5 wt % HCl and 5 wt % $PdCl_2$) for 5 min each, with oven drying at 60 °C after each step. A NiP conductive catalyst support was then achieved by immersing in a commercial electroless solution containing $NiCl_2$, $NaPO_2H_2$, $C_4H_4Na_2O_4$, and HF. The fabricated NiP/Plas Clear 3D-printed electrode exhibited excellent OER performance of 1.53 V vs reversible hydrogen electrode (RHE) at 10 mA/cm^2. The NiP coated electrodes exhibited excellent stability for both OER and HER (Figure 7.1b(ii)).

7.3.2 Fabrication of Non-conductive Components

Non-conductive components for water splitting devices are referred to as the polymeric part that is suitable to serve as a cell vessel or reactor. Practically, this component should be inert from the

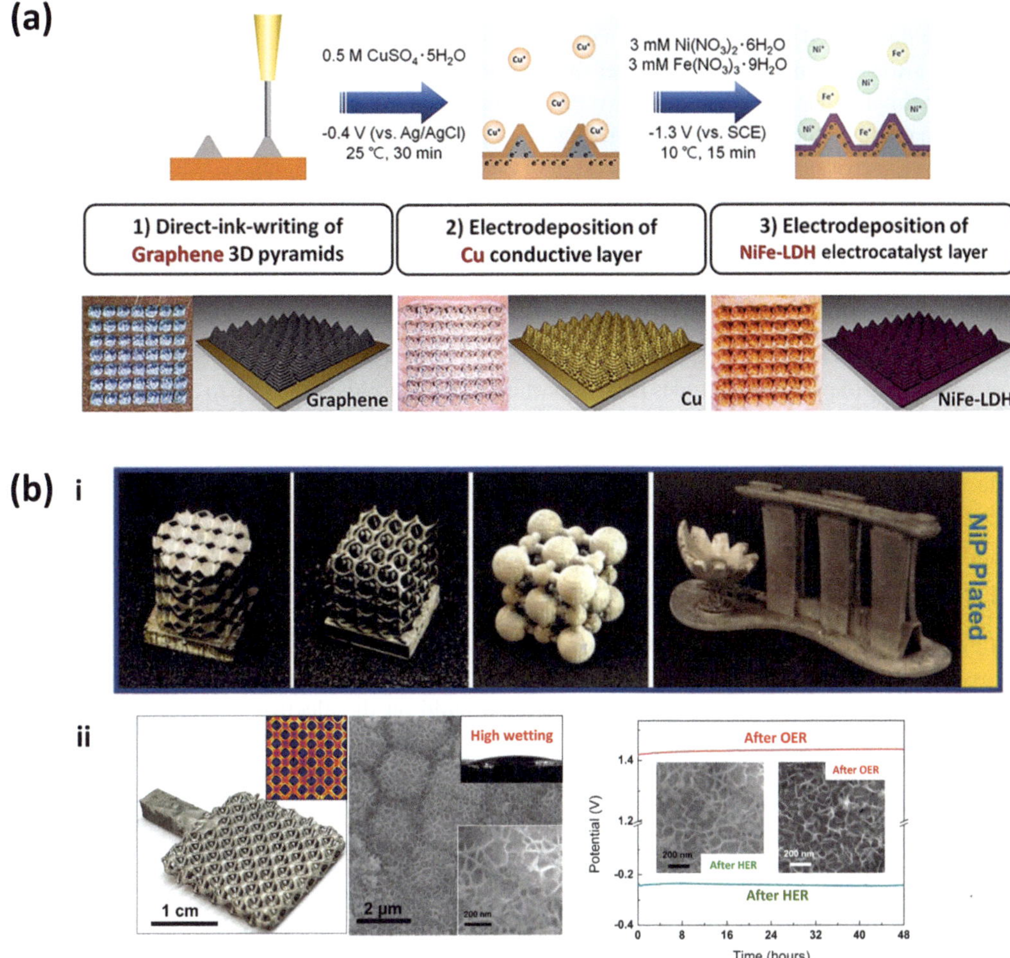

FIGURE 7.1 (a) The three steps procedure in fabricating NiFe-LDH involving direct-ink-writing. Adapted with permission [21]. Copyright: The Authors, some rights reserved; exclusive licensee Springer Nature. Distributed under a Creative Commons Attribution License 4.0 (CC BY). (b) The strategy of plated conductive catalyst (NiP) on 3D-printed polymers of different geometries (i), with optical and SEM images of the electrode and its electrochemical stability test (ii). Adapted with permission [23]. Copyright: The Authors, some rights reserved; exclusive licensee John Wiley & Sons. Distributed under a Creative Commons Attribution License 4.0 (CC BY).

electrochemical reaction, therefore highly chemical resistive and stable. For example, electrolytes employed in an optimal water splitting system that provide high current densities at low overpotentials are either highly acidic or alkaline aqueous electrolyte. This is evident from the commercial proton-exchange membrane (PEM) and alkaline-exchange membrane (AEM) electrolyzers technologies [3], [24].

There is a wide range of polymer printing technologies such as techniques based on the melting of thermoplastic filaments (FDM) or powders (selective laser sintering [SLS]), as well as reactive monomers and resins (SLA). The printable thermoplastic materials include polypropylene (PP), polystyrene (PS), polycarbonate (PC), polylactic acid (PLA), and acrylonitrile butadiene styrene (ABS). ABS and PLA filaments which are widely used in FDM printing have been employed to fabricate water splitting electrochemical reactors [25]–[27]. However, the drawback of FDM printing is the printed parts potentially have porosity and stability issues due to the inter-filament voids and artifacts from the layer-by-layer fabrication. The post-processing strategy of printed parts such as thermal annealing might improve the mechanical structure. A more commonly employed strategy to enhance the mechanical properties is by printing composite materials, which consist of the additives such as graphene, carbon fibers, and carbon nanotubes. Among all materials, polytetrafluoroethylene (PTFE) is known to be highly chemical resistive and stable. The emergence of commercially available 3M™ Dyneon™ PTFE offers an attractive route in fabricating electrochemical reactors for water splitting applications.

7.4 POST-PROCESSING OF 3D-PRINTED ELECTRODES

Post-processing of printed electrodes is an important surface modification step to achieve their desirable functional properties. These steps could involve the introduction of an additional conductive layer, incorporating catalyst layers, or modifying the surface and chemical properties of the printed surfaces. Both printed metallic and polymeric electrodes have been subject to post-processing as discussed below.

7.4.1 Metallic Electrodes

Optimization of electrode geometries which allow efficient mass transports with respect to reactant accessibility to catalyst active sites, and gas removals from the electrode surface is important in electrolyzer performance [28]. If the gases trap and build on the electrode surface, the electrode passivation could occur, which in turn will decrease the electrolysis activity. 3D printing offers opportunities in create electrode supports with such functionality in optimization of mass transport of reactant and product. Some complex geometries such as ribbons, baskets, meshes, conical arrays, and meshes have been printed for OER and/or HER reactions. In comparison to 316 L stainless steel printed in the ribbons, baskets, and meshes that were all coated with IrO_2, it was found that ribbons structure with wide open access of electrolyte showed the best OER performance [29].

Since metallic 3D-printed electrodes are readily conductive, electrodeposition is one of the most effective and common approaches to incorporating catalysts on the surface of the electrode. The low cost and simplicity of electrodeposition allow homogeneous coverage of catalyst on any type and shape of printed electrodes. The ease of transferred electrolyte bath formulas results in a large amount of OER and HER catalysts, such as Pt, IrO_2, Ni, NiP, NiFe-LDH, MoS_2, and Ni-MoS_2 electrodeposited to the printed structures [25], [30]. As expected, enhanced OER and HER performances were achieved in the presence of active catalysts (Figure 7.2a). The effect of geometric is shown, for example, higher OER performance of IrO_2 electrodeposited on ribbon shape 3D-printed stainless steel, compared to as-printed flat stainless steel [29].

Another low-cost electrochemical modification approach is anodization. Similar to electrodeposition, a vast knowledge has been developed on anodization strategies on metallic surfaces in developing porous structures that lead to catalyst activation and enhanced surface area [31], [32]. One example of combining 3D printing and anodization method is for photoelectrochemical water splitting application. Lee et al. employed SLM to produce unique microstructure electrodes to enhance reactive surface area and optimize light absorption characteristics [33]. The conical microarray structure was compounded with the growth of nanotubes via anodization that provides a distinctive blend of nano and microstructures that is beneficial for photoelectrochemical water splitting (Figure 7.3a). One of the major limitations of metal-based SLM for nanotechnology-driven applications lies in the capability in creating features in micro-instead of nanodomain. This work addressed to some extent this challenge by introducing nanofeatures onto the hierarchical 3D microstructures.

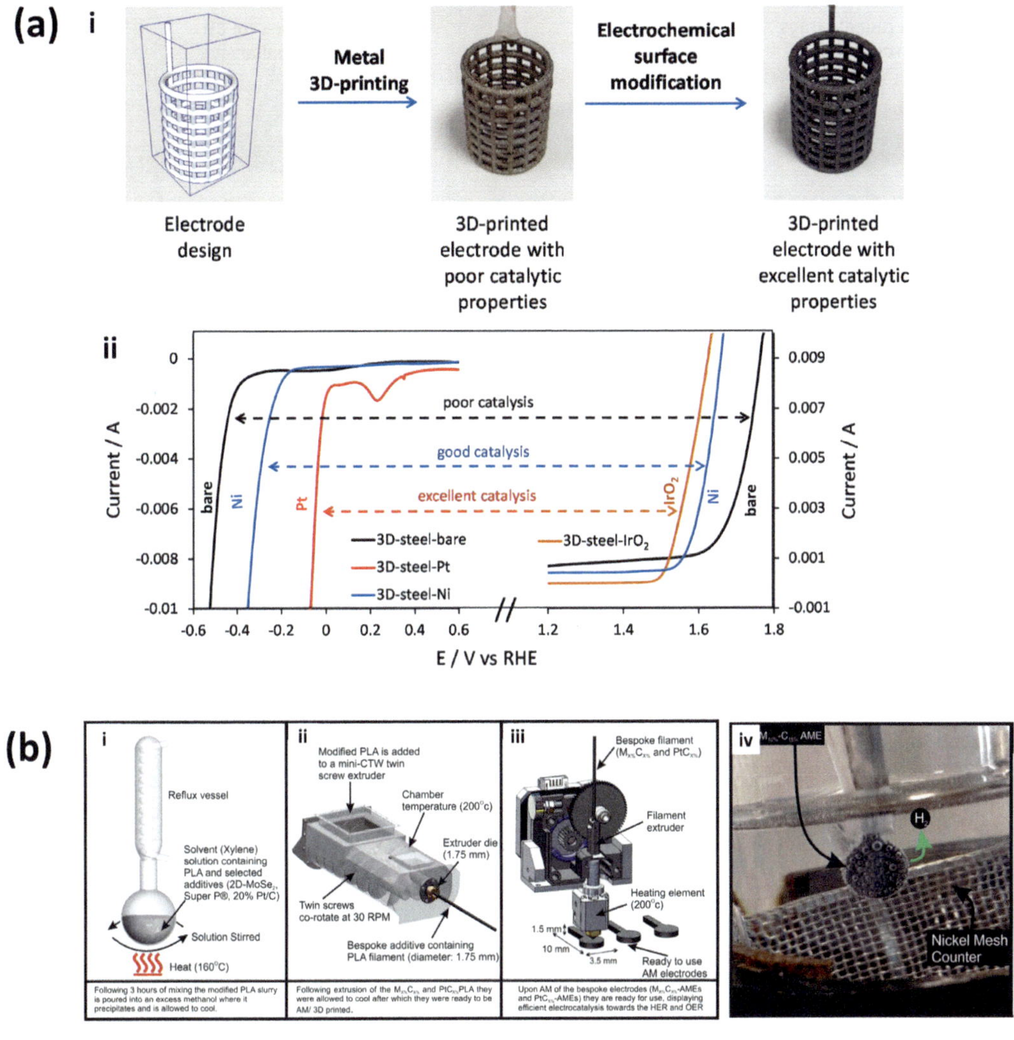

FIGURE 7.2 (a) The electrochemical surface modification on 3D-printed metallic electrodes, and the enhanced OER and HER performance with catalyst deposited on the stainless-steel electrodes. Adapted with permission [30]. Copyright (2018) John Wiley & Sons. (b) The steps to produce in-house catalyst-PLA filament containing $MoSe_2$, 20 % Pt/C and carbon black (i–iii), and the optical image of bubble formation from $MoSe_2$-PLA electrode for HER (iv). Adapted with permission [37]. Copyright: The Authors, some rights reserved; exclusive licensee The Royal Society of Chemistry. Distributed under a Creative Commons Attribution License 3.0 (CC BY).

To create an electroactive layer on the printed surface, the atomic layer deposition (ALD) strategy was employed. Browne et al. deposited TiO_2 by ALD onto stainless steel 3D-printed electrodes to serve as photoanodes for photoelectrochemical water oxidation [34]. ALD offers the advantage of precise tuning and control of the TiO_2 thicknesses. It was found that the water oxidation performance increases with photoanodes with a thicker TiO_2 layer (Figure 7.3b). ALD strategy allows precise control of the deposited thin layer of catalyst, but the challenge is the difficulty in obtaining a homogeneous deposit layer when involving complex geometries of structures. This method also is more expensive and has limited flexibility on the materials to be deposited.

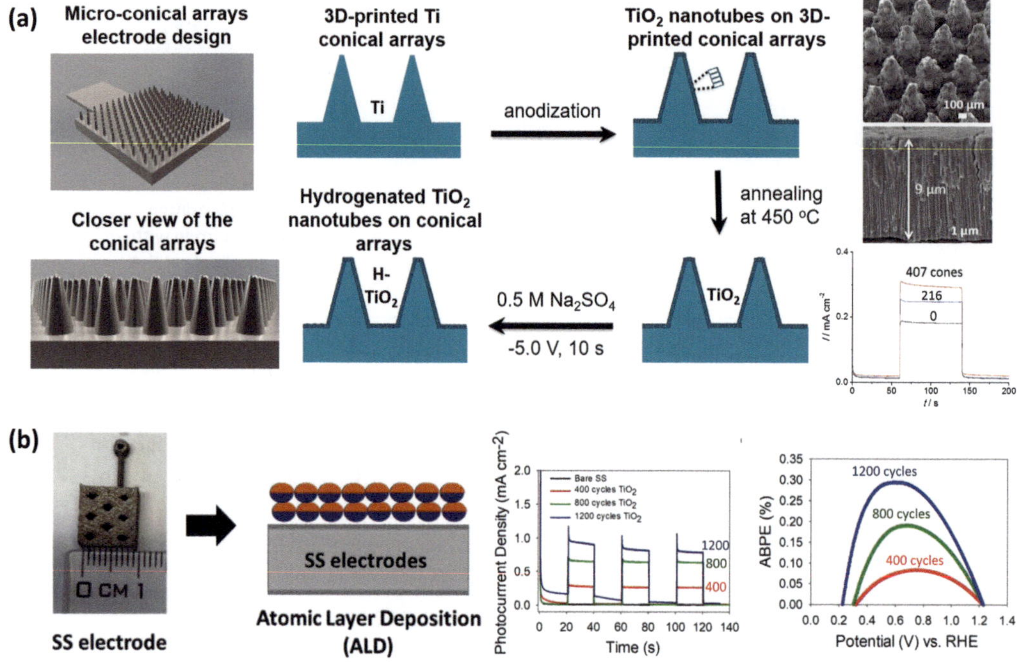

FIGURE 7.3 (a) The graphical illustration of electrochemical anodization post-processing strategy on 3D-printed conical array electrodes to form 1D nanotubular TiO_2 structures, followed by annealing and electrochemical hydrogenation. Adapted with permission [33]. Copyright (2017) John Wiley & Sons. (b) The atomic layer deposition of TiO_2 on the 3D-printed stainless electrodes shows enhanced photoelectrochemical performance with the increase in the deposition layers. Adapted with permission [34]. Copyright (2019) John Wiley & Sons.

7.4.2 Polymer-Composite Electrodes

The printed polymer electrodes consist of a composite of carbon materials and thermoplastic polymer that are fabricated by FDM printers. The incorporation of carbon materials such as graphene offers the required conductivity to be served as electrodes. For example, an electrode fabricated from commercially available 'Black Magic' filament consisting of graphene and PLA has been examined for HER activity [35]. Foster et al. reported the importance of electrode activation by continuous cycling in acidic media between 0 to -1.5 V vs. SCE [35]. The printed electrode exhibited improved HER performance after increasing the cycling from 100 to 1000 cycles. They proposed increase in HER activity was due to the hydrolyzation of the PLA chain that revealed the reactive material suspected to be titanium-derived contaminants evident from X-ray photoelectron spectroscopy analysis. This impurity was potentially created during graphene-PLA filament manufacturing. Further investigation by Browne and Pumera reveals the impurity may not be limited to Ti, but also Fe and Al [36]. Hence, it is important to have a thorough check on the impurity of polymer electrodes before being employed for electrolysis studies. Otherwise, the origin of electrocatalytic activity may be misinterpreted.

Instead of using commercial thermoplastic filaments with unknown content and impurities, making an in-house filament with a well-control procedure would be an attractive option. Hughes et al. fabricated PLA filaments containing electro-conductive carbon with electrocatalytic materials being either 2D-$MoSe_2$ or 20% Pt on carbon [37]. The $MoSe_2$ – PLA filaments were prepared by firstly refluxing xylene with $MoSe_2$ powder and conductive carbon black powder, followed by the addition of PLA with a desirable ratio and further reflux to produce a homogeneous mixture (Figure 7.2b). They investigated optimal rheology for printing by varying ratios of $MoSe_2$ to carbon black. To make filaments for 3D printing, the dried power from a mixture of carbon black and $MoSe_2$ were crystalline at 50 °C and loaded into an extruder at 200 °C. The PLA-$MoSe_2$ filaments were then printed into the electrodes for HER and OER studies. It was found a 25% mass incorporation of $MoSe_2$-carbon black represents an ideal compromise between electroactivity and printability; 10% $MoSe_2$ was found to offer highly active HER catalysis by exposing Se atoms at its active edge sites, whilst the incorporation of 15% conductive carbon offered to optimize electrical pathways through the printed electrodes.

7.5 3D-PRINTED PROTOTYPE ELECTROLYZER DEVICES

The previous section discussed the use of 3D printing to fabricate structural components related to electrochemical water splitting devices. This section focuses on the design and manufacture of various types of integrated electrolyzer cells. The electrolyzers are either partly or almost entirely made by 3D printing. The prototype cells involve both acidic and alkaline-based water electrolyzers, that are either membrane or membrane-less systems.

7.5.1 Membrane Electrolyzers

The printing of water electrolysis cells from various materials and designs is of significant interest. 3D printing facilitates the testing of new concept and design that is beneficial in improving electrochemical water splitting performance. Figure 7.4a shows a water-splitting cell that consists of metallic 3D-printed surface-patterned electrodes integrated with a custom-built reaction vessel produced via polymer-based 3D printing [26]. The electrode's conical surface structures were printed from Ti and Ni, including a curved base printed by the SLM method. The transformation of inactive Ti electrodes was achieved by the electrodeposition of an active nickel catalyst. The printed compartmented electrochemical cell was designed with both anode and cathode facing outward and separated by a Nafion membrane. This design was proposed and could also be used for photoelectrochemical water splitting experiments, particularly the curvy electrode that may improve light absorption.

Figure 7.4b shows the assembly of the alkaline electrolyzer with the metallic electrodes by SLM and the liquid/gas cell plastic (PLA) components by FDM reported by Ambrosi and Pumera [25].

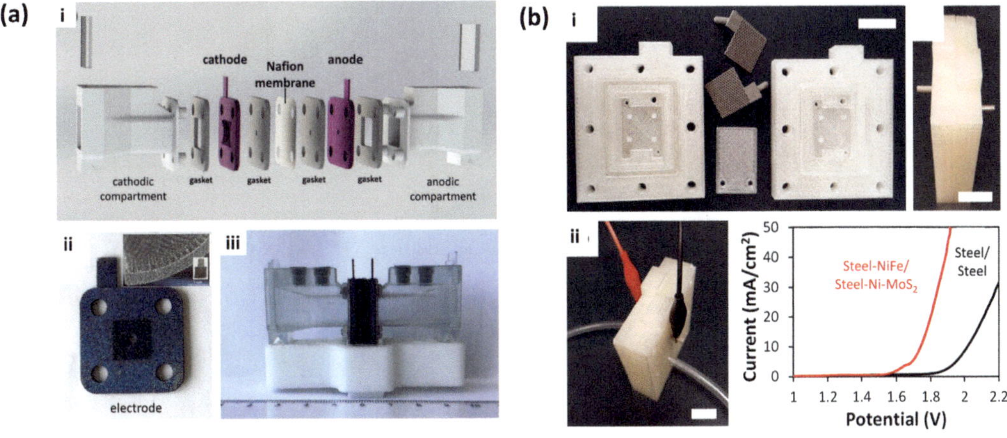

FIGURE 7.4 (a) The assembly of 3D-printed electrochemical water splitting cell consists of tailored made titanium electrodes, a translucent proprietary photopolymer vessel, and ABS base support. Adapted with permission [26]. Copyright (2019) John Wiley & Sons. (b) The electrochemical flow cell consists of 3D-printed stainless steel electrodes, and PLA polymeric vessel. The comparison of electrochemical water splitting performance was made between electrodes with and without catalysts modifications. Adapted with permission [25]. Copyright (2018) John Wiley & Sons.

The electrodes were produced in stainless steel which can be directly used for both HER and OER in alkaline electrolytes. They electrodeposited earth-abundant catalysts, and NiFe double layer hydroxide on the steel electrodes to enhance the overall water splitting performance. A thin film of Ni-MoS_2 composite was deposited on the stainless-steel cathode. The modification by these earth-abundant electrocatalysts shows a decrease in overpotential of more than 250 mV compared to the unmodified 3D-printed steel electrodes.

In an acidic water splitting system, PEM water splitting electrolyzers would require materials having mechanical and corrosion resistance properties which make them expensive to be fabricated. This high capital cost could hamper the prospect of large-scale deployment of electrolyzers. The rapid prototyping method could assist in reducing the materials fabrication and assembly cost by printing the complex flow plate architectures with lower-cost materials. Chrishom et al. first employed 3D-printing methods to fabricate a lightweight and low-cost flow plate for a PEM electrolyzer [38]. FDM was used to create a flow field from polypropylene, which was subsequently coated with a silver layer to provide conductivity. The fabricated plates were used in a commercially available membrane electrode assembly.

Another example of a 3D-printed PEM electrolyzer was reported by Yang and co-workers [39]. They demonstrated simplification in the PEM electrolyzer, where four conventional parts, namely liquid/gas diffusion layer, bipolar plate, gasket, and current distributor were integrated into one multifunctional 3D-printed plate (Figure 7.5a). This strategy was found to reduce the interfacial contact resistances between those components, resulting in excellent energy efficiency of up to 86.5% at 2 A/cm^2 and 80°C. The hydrogen generation rate was increased by 62% compared to the conventional cell. Their work highlighted the merits of fabricating highly complex inner structures

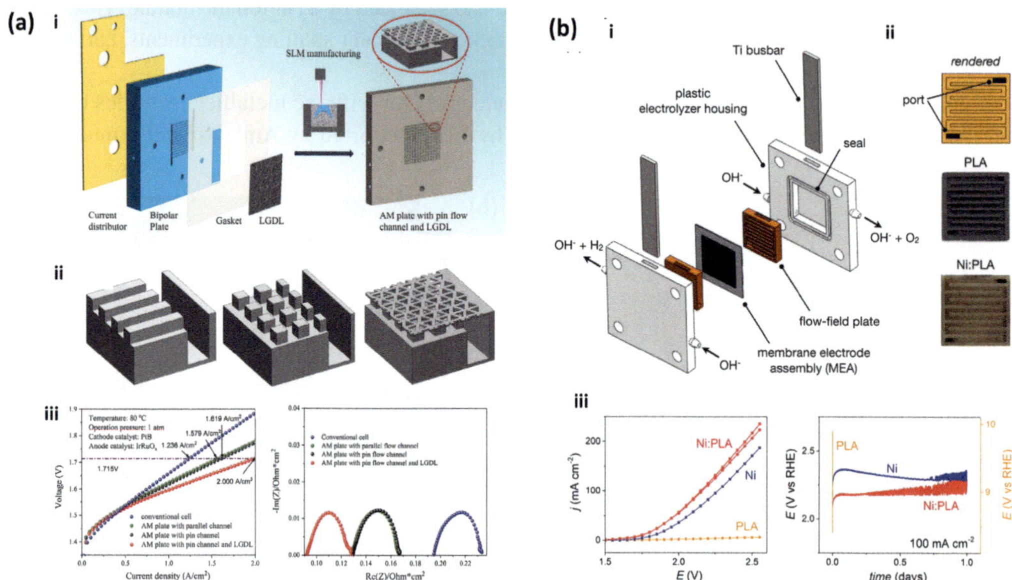

FIGURE 7.5 (a) The electrolyzer cell assembled by integrating the current distributor, bipolar plate, gasket, and LGDL into a single 3D-printed plate. The proton exchange membrane electrolysis performance test was tested on different configurations of cathode plates: parallel flow channel, pin flow channel, and pin flow channel. Adapted with permission [39]. Copyright (2018) Elsevier. (b) The 3D-printed flow-field plates of PLA electroplated with nickel that were separated by the membrane electrode assembly, with a housing made from ABS plastic (i–ii). The electrochemical performance evaluation in alkaline media (iii). Adapted with permission [40]. Copyright (2016) The Royal Society of Chemistry.

of multifunctional plates by 3D printing, which offers an advantage in compared to conventional electrolyzer manufacture methods.

The development of flow plate architectures by 3D printing is also extended to alkaline electrolyzer systems. Hudkins et al. electroplating nickel onto 3D-printed PLA flow fields (Figure 7.5b) [40]. In combination with titanium current collectors and a commercial MEA, an efficient water splitting electrolyzer was achieved. A comparison of the performance of the 3D-printed PLA-Ni plated plate to a Ni alloy plate typically used in commercial setup shows the printed plate was able to reach 100 mA/cm^2 current density at 2.2 V, whilst a higher voltage of 2.3 V is needed for the one with the Ni alloy plate. The cell was found to be stable in short term (Figure 7.5a), but it is to be examined the stability over a longer period for commercial applications. The techno-economic advantage of this approach can be justified by the raw materials costs of the bipolar plate could be reduced by half if the nickel were replaced by nickel-coated polylactic acid. It was proposed further cost savings are achievable by replacing other expensive metals such as stainless steel and titanium found in electrolyzers.

7.5.2 Membraneless Electrolyzers

Membraneless electrolyzers utilize fluidic forces, instead of solid barriers such as membranes to separate gas products from the electrolytic reaction [41], [42]. The advantages of these electrolyzers include having low ionic resistance, a simpler design, lower maintenance and the ability to operate with electrolytes at different pH. Hashemi et al. reported a membrane-less water electrolyzer by 3D printing [43]. Components such as the flow plates and the fluidic ports have been integrated into a monolithic part, which provides tight tolerances and smooth surfaces for precise flow conditioning (Figure 7.6a). They demonstrated that inertial fluidic forces are effective in milli-fluidic regimes, enabling control of the two-phase flows inside the device and preventing cross-contamination of the products. The performance of the device was examined for water splitting and the chlor-alkali reactions, and product purities were more than 99% with Faradaic efficiencies of more than 90%. They claimed the membrane-less reactor provides close to 40 times higher throughput than its microfluidic counterpart.

Bui et al. reported a membraneless design with three pairs of 3D-printed electrodes that were clicked into a printed monolithic lid containing manifolding readily for gas collection (Figure 7.6b) [44]. These electrodes were 3D printed using FDM from a composite of conductive carbon and PLA, followed by electrodeposition of nickel electrocatalyst. This membraneless architecture was operated in a stagnant KOH electrolyte through the buoyancy-driven separation of gases. The clickable nature of the electrode also offers ease of replenishment by electrodeposition. This 3D-printed electrolyzer reported an electrolysis efficiency of 48% at 50 mA/cm^2. Although such efficiency and cross-over rates did not meet the commercial application requirements, this work demonstrated the use of 3D printing as a rapid prototyping method to test new concepts and designs.

7.6 OUTLOOK AND CONCLUSIONS

Additive manufacturing for energy applications is an emerging research area that is expected to be continuously evolved along with further advancements in printing techniques and capabilities. The application of electrochemical water splitting has been widely investigated and is more advanced compared to other electrochemical energy systems. This is motivated by the significant scientific and economic interests in advancing electrolyzer technology for green hydrogen production. In addition, the simplicity of water splitting reaction in terms of the required reactant, and generated product promotes widespread studies. The water splitting performance analyses are readily evaluated through the current-voltage curves, and the products only consist of hydrogen and oxygen. Comparatively, other prominent electrochemical processes such as CO_2 electrolysis require

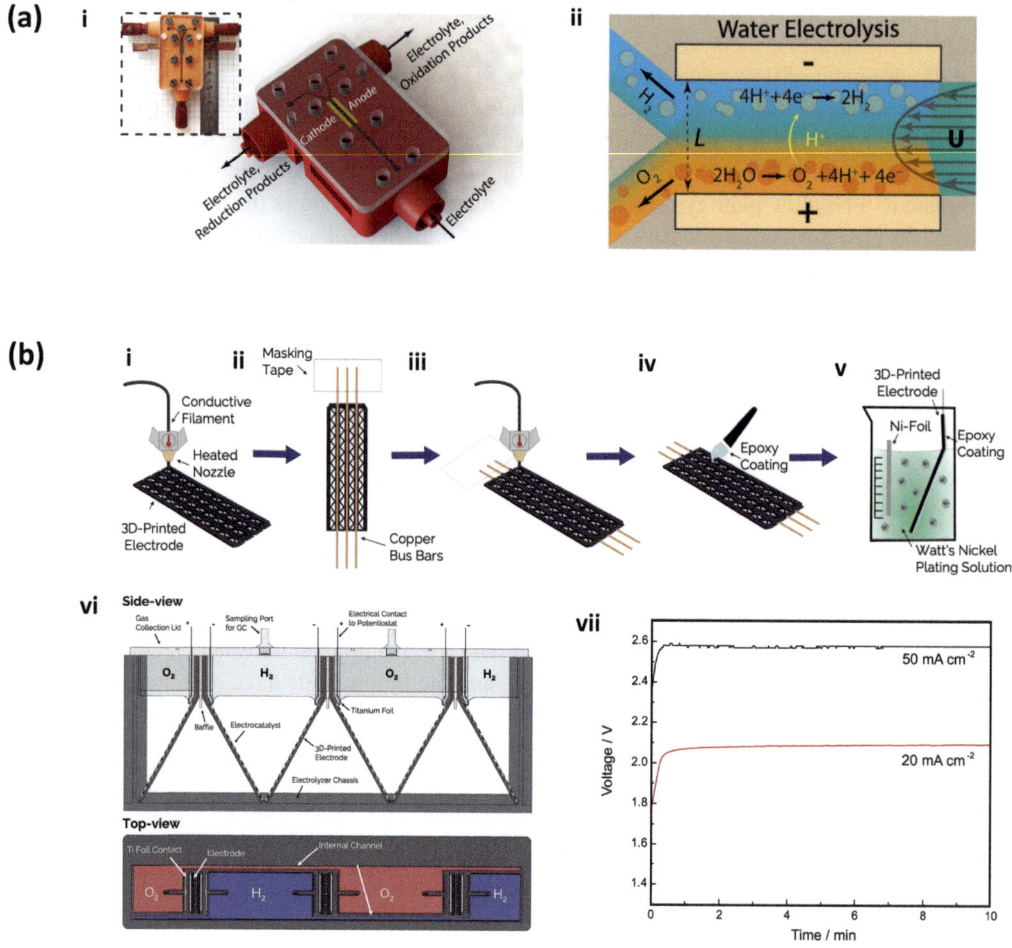

FIGURE 7.6 (a) The membraneless electrolyzer with the reactor is 3D printed by SLA with the electrodes (yellow) between the devised slots (i). The schematic diagram illustrated the essential of a fast flow rate to minimize gas crossover (ii). Adapted with permission [43]. Copyright: The Authors, some rights reserved; exclusive licensee The Royal Society of Chemistry. Distributed under a Creative Commons Attribution License 3.0 (CC BY). (b) The schematic illustration of the procedure to fabricate 'clickable' 3D-printed electrode fabrication (i–v) used in the membraneless electrolyzer (vi), and (vii) shows a six-electrode configuration voltage-time performance at 20 and 50 mA/cm^2, respectively. Adapted with permission [44]. Copyright (2020) The Royal Society of Chemistry.

the design of a specific cell to optimize the supply of CO_2 and more complex quantitative analytical measurement of multiple products.

The printing techniques for electrochemical water splitting are covering both conductive and non-conductive materials. To serve as an electrode, the basic requirement is for the electrode to be conductive and stable. To obtain high-performance catalytic reactions, the catalysts need to be in the nanoscale regions and have a high surface area. SLM method, for example, offers capability in the printing of metallic substrates such as titanium, nickel, and stainless steel, but typically at the microscale regions. This illustrated the current limitations of metallic printings which do not have the resolution required to serve directly as an electrode. Therefore, the post-processing of 3D-printed electrodes served as an essential integrated strategy to activate the electrode. The availability of materials is also another major limitation in metallic printing where the printed materials are restricted to several common metals, which metal such as Ti is inactive for water splitting reactions. The catalyst's performance also could be enhanced by a synergetic bimetallic strategy. Therefore, further development of printing strategy by allowing incorporation of two or more metals would be advantageous for electrochemical water splitting reaction. This is particularly the case since bimetallic metals such as alloy consisting of Ni and Fe served best for OER water splitting reaction.

Another common printing strategy is the printing of composite carbon-based electrode materials. This method is more widely used in the laboratory as it involves low-cost printers and materials. However, the addition of binders such as thermoplastic results in poorer conductivity that cannot support the high currents required for efficient electrochemical water splitting. A more resistive electrode also results in excess voltage needed to drive the reaction that makes such an electrode not practically feasible for commercial application. Improvement can be obtained by post-processing such as deposition of a conductive layer for catalyst support, but the question remains of the robustness of such electrodes compared to metallic-based electrodes for long-term operation. Hence, investigations of 3D-printed carbon-based electrodes is likely to stay as proof-of-concept and for fundamental studies. Extrusion methods such as DIW have been attempted to print inks consisting of nanocatalysts, but the printing resolution remained low hampering the necessity of high catalytic surface area for efficient catalytic reactions. The robustness of the printed inks to operate under commercially relevant electrolysis conditions (e.g., temperature, electrolyte, and duration) also remains to be tested.

In comparison to electrodes, the printing of non-conductive polymeric components to serve as electrochemical reactors is more accomplished. Reactors served as an essential integrated part of an electrolyzer and could lead to innovation in cell designs that would improve cell performance and efficiency. An example of these advances is the printing of membraneless water splitting reactor with tailored microfluidic channels. Innovation in flow reactor designs has highlighted the distinct advantages of 3D printing in terms of design freedom and rapid prototyping compared to traditional manufacturing. Generally, 3D printing of non-conductive components requires reduced demand on printing resolution and complexity. The key consideration lies in the selection of polymeric materials that must be resistive to the electrolytes employed in the water electrolysis conditions that either at high or low pH. Unsuitable materials may result in the leaching of dissolve polymers and other associated components, and those impurities could participate in the electrochemical oxidation or reduction reactions. This interference would severely affect the performance analysis as well as the product's quantification. Another potential non-conductive component is the printing of membranes, but the development of a suitable membrane to serve for water splitting application remains to be seen [45].

In conclusion, 3D printing technology is expected to be continuously evolved, stimulating innovations in electrodes and reactor designs for electrochemical water splitting devices. The advantages of 3D printing are its eco-friendly and rapid prototyping approach, which contributes to the fabrication of electrochemical devices. To further advance the application of 3D printing technologies in electrode fabrication, breakthroughs are needed to overcome current limitations in printing resolutions, material availability, and printing sizes. The design of metallic printers capable of achieving feature resolution closer to the nanoscale region potentially allows direct fabrication of active and high surface area electrodes. Future expansion of a wider range of printing metallic materials, including alloys, would offer the flexibility to directly print electroactive components. The work may also be extended to biological water splitting systems [46]. Other advances such as the incorporation of machine learning may assist in the design and optimization of electrodes and reactors for electrochemical water splitting [47]. These advances would promote 3D printing technology in translating fundamental scientific advances in water splitting reaction to device fabrication, bringing us closer to a breakthrough in water electrolyzer technology.

ACKNOWLEDGMENTS

Fundings from the Australian National Fabrication Facility and the Australian Research Council Centres of Excellence (CE140100012) are gratefully acknowledged.

REFERENCES

1 S. Solomon, G.-K. Plattner, R. Knutti, P. Friedlingstein, Irreversible climate change due to carbon dioxide emissions, *Proceedings of the National Academy of Sciences*, 106 (2009) 1704–1709.
2 H.S. Baker, R.J. Millar, D.J. Karoly, U. Beyerle, B.P. Guillod, D. Mitchell, H. Shiogama, S. Sparrow, T. Woollings, M.R. Allen, Higher CO_2 concentrations increase extreme event risk in a 1.5 °C world, *Nature Climate Change*, 8 (2018) 604–608.
3 O. Schmidt, A. Gambhir, I. Staffell, A. Hawkes, J. Nelson, S. Few, Future cost and performance of water electrolysis: An expert elicitation study, *International Journal of Hydrogen Energy*, 42 (2017) 30470–30492.
4 C. Coutanceau, S. Baranton, T. Audichon, Chapter 3 – Hydrogen Production From Water Electrolysis, in: C. Coutanceau, S. Baranton, T. Audichon (Eds.) *Hydrogen Electrochemical Production*, Academic Press, 2018, pp. 17–62.
5 S. van Renssen, The hydrogen solution?, *Nature Climate Change*, 10 (2020) 799–801.
6 A. Ambrosi, M. Pumera, 3D-printing technologies for electrochemical applications, *Chemical Society Reviews*, 45 (2016) 2740–2755.
7 N. Shahrubudin, T.C. Lee, R. Ramlan, An overview on 3D printing technology: Technological, materials, and applications, *Procedia Manufacturing,* 35 (2019) 1286–1296.
8 C.-Y. Lee, A.C. Taylor, A. Nattestad, S. Beirne, G.G. Wallace, 3D printing for electrocatalytic applications, *Joule*, 3 (2019) 1835–1849.
9 C.-Y. Lee, J.P. Bullock, G.G. Wallace, Chapter 4 – Earth-abundant electrocatalysts for sustainable energy conversion, in: K.Y. Cheong, A. Apblett (Eds.) *Sustainable Materials and Green Processing for Energy Conversion*, Elsevier, 2022, pp. 131–168.
10 Z.W. Seh, J. Kibsgaard, C.F. Dickens, I. Chorkendorff, J.K. Nørskov, T.F. Jaramillo, Combining theory and experiment in electrocatalysis: Insights into materials design, *Science*, 355 (2017) eaad4998.
11 A. Fujishima, K. Honda, Electrochemical photolysis of water at a semiconductor electrode, *Nature*, 238 (1972) 37–38.
12 R. Abe, Recent progress on photocatalytic and photoelectrochemical water splitting under visible light irradiation, *Journal of Photochemistry and Photobiology C: Photochemistry Reviews*, 11 (2010) 179–209.
13 C. Barnatt, *3D Printing Third Edition,* CreateSpace Independent Publishing Platform, 2016.
14 T. Duda, L.V. Raghavan, 3D metal printing technology, *IFAC-PapersOnLine*, 49 (2016) 103–110.
15 E. Yasa, Chapter 3 – Selective laser melting: principles and surface quality, in: J. Pou, A. Riveiro, J.P. Davim (Eds.) *Additive Manufacturing*, Elsevier, 2021, pp. 77–120.
16 B. Dutta, F.H. Froes, The Additive Manufacturing (AM) of titanium alloys, *Metal Powder Report*, 72 (2017) 96–106.
17 F. Ganci, S. Lombardo, C. Sunseri, R. Inguanta, Nanostructured electrodes for hydrogen production in alkaline electrolyzer, *Renewable Energy*, 123 (2018) 117–124.
18 R.B. Kristiawan, F. Imaduddin, D. Ariawan, U. Sabino, Z. Arifin, A review on the fused deposition modeling (FDM) 3D printing: Filament processing, materials, and printing parameters, *Open Engineering*, 11 (2021) 639–649.
19 J.A. Lewis, Direct ink writing of 3D functional materials, *Advanced Functional Materials,* 16 (2006) 2193–2204.
20 S. Tagliaferri, A. Panagiotopoulos, C. Mattevi, Direct ink writing of energy materials, *Materials Advances*, 2 (2021) 540–563.
21 J. Ahn, Y.S. Park, S. Lee, J. Yang, J. Pyo, J. Lee, G.H. Kim, S.M. Choi, S.K. Seol, 3D-printed NiFe-layered double hydroxide pyramid electrodes for enhanced electrocatalytic oxygen evolution reaction, *Scientific Reports,* 12 (2022) 346.

22 J. Huang, Q. Qin, J. Wang, A review of stereolithography: Processes and Systems, *Processes*, 8 (2020) 1138.
23 X. Su, X. Li, C.Y.A. Ong, T.S. Herng, Y. Wang, E. Peng, J. Ding, Metallization of 3D printed polymers and their application as a fully functional water-splitting system, *Advanced Science*, 6 (2019) 1801670.
24 I. Cerri, F. Lefebvre-Joud, P. Holtappels, K. Honegger, T. Stubos, P. Millet, A. Pfrang, M. Bielewshi, E. Tzimas, *Scientific Assessment in Support of the Materials Roadmap Enabling Low Carbon Energy Technologies: Hydrogen and Fuel Cells, Publications Office of the European Union* (2012).
25 A. Ambrosi, M. Pumera, Multimaterial 3D-printed water electrolyzer with earth-abundant electrodeposited catalysts, *ACS Sustainable Chemistry & Engineering*, 6 (2018) 16968–16975.
26 C.-Y. Lee, A.C. Taylor, S. Beirne, G.G. Wallace, A 3D-printed electrochemical water splitting cell, *Advanced Materials Technologies*, 4 (2019) 1900433.
27 C. Ponce de Leon, W. Hussey, F. Frazao, D. Jones, E. Ruggeri, S. Tzortzatos, R. Mckerracher, R. Wills, S. Yang, F. Walsh, The 3D printing of a polymeric electrochemical cell body and its characterisation, *Chemical Engineering Transactions*, 41 (2014) 1–6.
28 K.B. Oldham, C.G. Zoski, Chapter 2 – Mass transport to electrodes, in: C.H. Bamford, R.G. Compton (Eds.) *Comprehensive Chemical Kinetics*, Elsevier, 1986, pp. 79–143.
29 A. Ambrosi, J.G.S. Moo, M. Pumera, Helical 3D-printed metal electrodes as custom-shaped 3D platform for electrochemical devices, *Advanced Functional Materials*, 26 (2016) 698–703.
30 A. Ambrosi, M. Pumera, Self-contained polymer/metal 3D printed electrochemical platform for tailored water splitting, *Advanced Functional Materials*, 28 (2018) 1700655.
31 C.-Y. Lee, K. Lee, P. Schmuki, Anodic formation of self-organized cobalt oxide nanoporous layers, *Angewandte Chemie International Edition,* 52 (2013) 2077–2081.
32 C.-Y. Lee, P. Schmuki, Engineering of self-organizing electrochemistry: Porous alumina and titania nanotubes, in: *Electrochemical Engineering across Scales: From Molecules to Processes*; Alkire, R. C., Bartlett, P. N., Lipkowski, J., Eds.; John Wiley & Sons: Weinheim, chapter 5 (2015) 145–192.
33 C.-Y. Lee, A.C. Taylor, S. Beirne, G.G. Wallace, 3D-printed conical arrays of TiO_2 electrodes for enhanced photoelectrochemical water splitting, *Advanced Energy Materials*, 7 (2017) 1701060.
34 M.P. Browne, J. Plutnar, A.M. Pourrahimi, Z. Sofer, M. Pumera, Atomic layer deposition as a general method turns any 3D-printed electrode into a desired catalyst: Case study in photoelectrochemisty, *Advance Energy Materials,* 9 (2019) 1900994.
35 C.W. Foster, M.P. Down, Y. Zhang, X. Ji, S.J. Rowley-Neale, G.C. Smith, P.J. Kelly, C.E. Banks, 3D printed graphene based energy storage devices, *Scientific Reports*, 7 (2017) 42233.
36 M.P. Browne, M. Pumera, Impurities in graphene/PLA 3D-printing filaments dramatically influence the electrochemical properties of the devices, *Chemical Communications*, 55 (2019) 8374–8377.
37 J.P. Hughes, P.L. dos Santos, M.P. Down, C.W. Foster, J.A. Bonacin, E.M. Keefe, S.J. Rowley-Neale, C.E. Banks, Single step additive manufacturing (3D printing) of electrocatalytic anodes and cathodes for efficient water splitting, *Sustainable Energy & Fuels*, 4 (2020) 302–311.
38 G. Chisholm, P.J. Kitson, N.D. Kirkaldy, L.G. Bloor, L. Cronin, 3D printed flow plates for the electrolysis of water: An economic and adaptable approach to device manufacture, *Energy & Environmental Science*, 7 (2014) 3026–3032.
39 G. Yang, J. Mo, Z. Kang, Y. Dohrmann, F.A. List, J.B. Green, S.S. Babu, F.-Y. Zhang, Fully printed and integrated electrolyzer cells with additive manufacturing for high-efficiency water splitting, *Applied Energy*, 215 (2018) 202–210.
40 J.R. Hudkins, D.G. Wheeler, B. Peña, C.P. Berlinguette, Rapid prototyping of electrolyzer flow field plates, *Energy & Environmental Science*, 9 (2016) 3417–3423.
41 D.V. Esposito, Membraneless electrolyzers for low-cost hydrogen production in a renewable energy future, *Joule*, 1 (2017) 651–658.
42 G.F. Swiegers, A.L. Hoang, A. Hodges, G. Tsekouras, C.-Y. Lee, K. Wagner, G. Wallace, Current status of membraneless water electrolysis cells, *Current Opinion in Electrochemistry*, 32 (2022) 100881.

43 S.M.H. Hashemi, P. Karnakov, P. Hadikhani, E. Chinello, S. Litvinov, C. Moser, P. Koumoutsakos, D. Psaltis, A versatile and membrane-less electrochemical reactor for the electrolysis of water and brine, *Energy & Environmental Science*, 12 (2019) 1592–1604.
44 J.C. Bui, J.T. Davis, D.V. Esposito, 3D-Printed electrodes for membraneless water electrolysis, *Sustainable Energy & Fuels*, 4 (2020) 213–225.
45 B.G. Thiam, A. El Magri, H.R. Vanaei, S. Vaudreuil, 3D printed and conventional membranes – A review, *Polymers*, 14 (2022) 1023.
46 C.-Y. Lee, J. Zou, J. Bullock, G.G. Wallace, Emerging approach in semiconductor photocatalysis: Towards 3D architectures for efficient solar fuels generation in semi-artificial photosynthetic systems, *Journal of Photochemistry and Photobiology C: Photochemistry Reviews*, 39 (2019) 142–160.
47 C. Wang, X.P. Tan, S.B. Tor, C.S. Lim, Machine learning in additive manufacturing: State-of-the-art and perspectives, *Additive Manufacturing*, 36 (2020) 101538.

8 Materials and Applications of 3D Print for Solid Oxide Fuel Cells

*Jinjin Zhang**, *Xuening Pang, and Naitao Yang**
College of Chemistry and Chemical Engineering, Shandong University of Technology, 266 Xincun West Road, Zhangdian District, Zibo City, Shandong Province, 255000, P.R.China
*Corresponding author: jjzhang@sdut.edu.cn; naitaoyang@126.com

CONTENTS

8.1 INTRODUCTION

Solid oxide fuel cell (SOFC) is an energy conversion device that can directly convert the chemical energy of fuel into electrical energy. Because of its advantages of high energy conversion efficiency, environment friendly, fuel flexibility, and low catalyst cost, SOFC is considered as a potential next-generation power system. Recently, 3D-printing technology has been widely applied to SOFC field because it can flexibly prepare highly complex and accurate structures and components. For a single cell, 3D-printing technology improves the performance of the cell by designing the electrolyte surface pattern [1]–[3], printing thin and dense electrolytes [4]–[5], and improving the distribution of catalyst and cocatalyst in the porous electrode [6]. In practical applications, the power of single cells is limited, and multiple single cells are often assembled into a stack through connectors in parallel or in series to meet the power demand. As shown in Figure 8.1(a–b), the manufacturing process of the planar-cell stack [7] and the tubular-cell stack is complicated [8], and there are lots of connectors and seals in the stack. The thermal expansion coefficients of different materials are quite different,

DOI: 10.1201/9781003296676-8

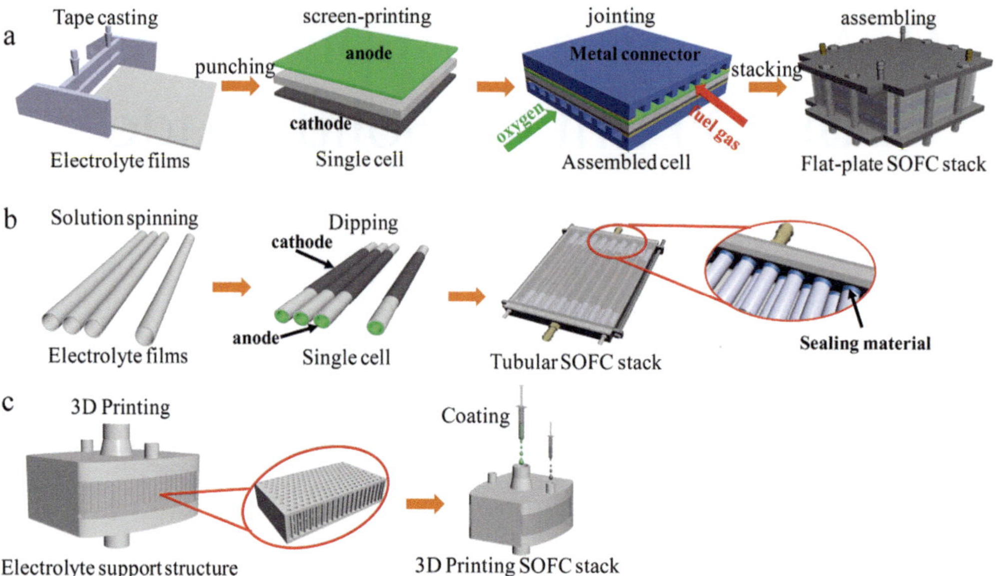

FIGURE 8.1 (a) Planar and (b) Tubular SOFC stack prepared by traditional method (c) Fabrication of SOFC stack by 3D printing. Adapted with permission [7]. Copyright (2019), Elsevier.

resulting in thermal stress during the working process of the cell, which greatly affects the reliability and durability of the SOFC system, restricting its commercialization process.

The emergence of 3D printing technology provides a new method for the preparation of the SOFC stack, as shown in Figure 8.1(c). Firstly, the electrolyte support of the stack is prepared integrally, and then the electrode layer is deposited on it. The integrated SOFC stack module can be obtained without the steps of single-cell preparation, connection, and manual assembly. 3D-printing technology can improve manufacturing accuracy, flexibly adjust manufacturing dimensions, design and customize complex architecture, and ensure standardization and repeatability of the manufacturing process [9]. In addition, 3D printing is expected to improve transport performance and realize its efficient and long-term stable operation through scientific design and optimization of the structure of the stack and auxiliary system. This chapter briefly introduces several commonly used 3D-printing technologies, and mainly summarizes many attempts to prepare SOFC cathode [10]–[11], anode [6], [12]–[13], electrolyte [4]–[5], stack components, stack adaptive functional devices, and auxiliary systems by 3D-printing technology [7], [14]–[16]. The challenges faced by 3D printing technology in the SOFC field are analyzed, and the development direction of 3D-printing technology in the SOFC field in the future is put forward.

8.2 MATERIALS OF 3D PRINT IN SOFC

The main components of SOFC include the electrolyte, anode, and cathode. Generally, 3 mol%/ 8 mol% yttria-stabilized zirconia (3YSZ/8YSZ), gadolinia-doped ceria (GDC), scandia-stabilized zirconia (SSZ) are commonly used electrolytes, and NiO is widely used as the anode in SOFC. $La_{0.6}Sr_{0.4}Co_{0.2}Fe_{0.8}O_{3-\delta}$(LSCF), $Ba_{0.5}Sr_{0.5}Co_{0.8}Fe_{0.2}O_{3-\delta}$(BSCF), $La_{0.8}Sr_{0.2}MnO_{3-\delta}$ (LSM), $La_{0.6}Sr_{0.4}CoO_{3-\delta}$ (LSC), and $Sm_{0.2}Ce_{0.8}O_{2-\delta}$-$Sm_{0.5}Sr_{0.5}CoO_{3-\delta}$ (SDC-SSC) and $La_{0.6}Sr_{0.4}Co_{0.2}Fe_{0.8}O_{3-\delta}$-$Ce_{0.9}Gd_{0.1}O_{1.95}$ (LSCF-CGO) composite cathodes are the frequently used cathode. 3D-printing technology uses digital slices in a 3D computer-aided design (CAD) model to manufacture devices by adding materials through the steps of point-by-point, line-by-line, or layer-by-layer superposition [17].

TABLE 8.1
Common 3D Printing Technologies and Features in SOFC

Printing Technologies	Materials Status	Prototyping Principle	Limit of Material	Merits	Demerits	Application in SOFC
IJP [18]	Suspension	Binder bonding	Powder needs physical or chemical combination	Simple equipment, low cost, high precision and precise control	Poor ink stability, difficult to print high aspect ratio devices	Porous electrodes and dense electrolyte films, e.g. 8YSZ [4], [5], [18], [19], SDC/SSC [10], LSCF [11], NiO-8YSZ [6], [18], BSCF [19], GDC [20]
SLA [2], [3]	Suspension or paste	Photopolymerization reaction	Photocuring materials	Printable for complex structure, fine size and large devices	Relatively low shaping speed	Electrolyte or electrolyte support, e.g. 3YSZ [2], 8YSZ [3]
DLP [21]	Suspension	Photopolymerization reaction	Photocuring materials	Precise printing ability	Limited molding size	Electrolyte sheet or electrolyte support, e.g. 8YSZ [1], [7], SSZ [22]
FFF [23]/ FDM [24]	Solid wire	Melting and cooling	Thermoplastic polymer	Low cost, suitable for lots of materials, and simple post-treatment process	Low resolution of printing devices and low strength in thickness direction	Porous electrode, e.g. NiO-8YSZ [23]
DIW [25]	Paste	Extrusion and rheology	Slurry with good rheological property	Low cost, high speed, flexible printing structure	Low surface quality and resolution of printing devices	Porous electrode, e.g. NiO-8YSZ [26]

At present, the commonly used 3D-printing technologies in SOFC mainly include inkjet printing (IJP), fused deposition modeling (FDM)/fused filament fabrication (FFF), and direct ink writing (DIW). Photocuring printing can be divided into stereolithography apparatus (SLA) and digital light processing (DLP) according to the difference in light source and forming mode. Each printing technology is used for different purposes due to their unique forming principle and printing characteristics. Table 8.1 gives the main characteristics of different 3D-printing methods and the specific materials in SOFC.

8.3 APPLICATION OF 3D-PRINTING TECHNOLOGY IN SOFC

3D-printing technology is applied in the SOFC field through three strategies including improving single cell performance, flexibly designing stack structure, and simplifying the stack component preparation process. Specifically, the microstructure, composition, and thickness of functional films are precisely tailored by adjusting the composition and printing parameters of printing materials. The exposure of effective surfaces and reactive active sites can be increased by designing complex shapes and optimizing layered geometry [27]. Through the integrated design of flexible and complex SOFC structure, the heat and mass transfer effects can be improved, the assembly process avoided, the preparation cycle shortened, and the cost saved [28]. Moreover, 3D-printing technology is expected to overcome the limitations and reliability problems caused by traditional SOFC manufacturing methods and improve its durability and specific power per unit mass or volume [26].

8.3.1 3D Printing Cathode

The cell microstructure and the distribution of cathode active components are the main factors affecting the electrochemical performance of SOFC. The traditional spraying technology will lead to an uneven distribution and structure of active substances on the cathode surface [29]. The precise positioning function of inkjet 3D printing can realize the design and preparation of the expected structure, ensure the uniform distribution of active components, and then improve cell performance. For instance, after LSC nanoparticles were inkjet printed onto the LSCF cathode surface, the surface catalysis was significantly enhanced, the polarization impedance of cathode film was significantly reduced, and the power was increased by five times [11], which indicated the effectiveness of inkjet printing on cathode surface modification. In addition, a porous SDC/SSC composite cathode layer was also prepared by inkjet printing technology [10], and the peak power density (PPD) of the assembled cell at 750°C was up to 940 mW/cm^2. Inkjet printing technology can also accurately control the thickness and composition of cathode porous layers. Sukeshini et al. [30] inkjet printed LSM cathode layer by adjusting the rheological properties (solid content) and printing parameters. Han et al. [31] controlled the inkjet amount by adjusting the gray level of the printed image and inkjet printing the LSCF cathode. The results showed that the electrochemical performance was comparable to that of the cathode prepared by the conventional method. Moreover, the LSCF-CGO porous layer was inkjet printed between the traditionally coated cathode and electrolyte, which expanded the three-phase interface length and improved the cell performance by more than 30% [32]. This showed that inkjet printing technology has a broad application prospect in cathode layer manufacturing.

8.3.2 3D Printing Anode

3D printing can realize the addition of anode functional layer (AFL), uniform distribution of pores in the electrode, appropriate grain size control of electrode material, and structural design of electrode surface. The printing of the NiO-8YSZ functional layer was realized by inkjet printing, and the maximum power density of 500 mW/cm^2 was obtained at 850°C [18]. Besides, the co-catalyst $BaZr_{0.9}Y_{0.1}O_{3-\delta}$ (BZY) was well injected into the NiO-8YSZ anode by the inkjet printing technology, and the distribution of BZY in the electrochemically active region of NiO-8YSZ anode was accurately controlled. In the humidified H_2 and dried CH_4 fuel, the BYZ impregnated NiO-8YSZ anode obtained by inkjet printing technology showed better performance than the traditional NiO-8YSZ anode [6], which proved that inkjet 3D-printing technology played an important role in the manufacturing of SOFC microstructure.

In addition to controllable preparation of electrode morphology, 3D-printing technology can also improve electrochemical performance by increasing the interface area between SOFC electrode and electrolyte. Direct writing printing can prepare anode-supported SOFC with different interface area increase ratio by changing the coverage area of extruded slurry [26]. Melt spinning 3D-printing technology was also used to prepare NiO-8YSZ anode support, and the maximum power density at 850°C achieved 400 mW/cm^2 [33].

3D-printing technology does not only prepare the planar SOFC modules as mentioned above but also microtubular cells. The first microtubular solid oxide fuel cell (MT-SOFC) with high power output, long-term stability, and thermal cycle stability prepared by 3D-printing technology achieved long-term operation of more than 4000 hours at a constant current of 18.5 A and performed more than 1000 rapid thermal cycles without cell failure [34]. This can be attributed to the high thermal shock resistance of the microtubular cell and the close adhesion between the electrolyte and the electrode provided by inkjet printing. The above research shows that 3D-printing technology can realize controllable anode preparation and obtain excellent performance.

8.3.3 3D Printing Electrolyte

8.3.3.1 3D Printing Electrolyte Film

8YSZ electrolyte films have been successfully prepared by a variety of technologies, including RF magnetron sputtering [15], chemical vapor deposition (CVD) [35], and electron beam deposition [36], but it is difficult to scale up. Tape casting [37] and screen-printing methods are low-cost and easy to commercialize, but it is difficult to prepare a thickness less than 10 μm. Liquid phase deposition technologies, such as spray, electrostatic spray deposition (ESP), dip coating, or spin coating, are both economical and suitable for preparing thin coatings but require multi-layer coatings and complex sintering processes. Compared with these traditional manufacturing methods, direct ceramic inkjet printing can easily produce a thickness of less than 10 μm by controlling and optimizing printing parameters such as pressure, nozzle opening time, and droplet overlap [38].

The dense 8YSZ with the thickness of about 6 μm prepared on the casting porous NiO-8YSZ anode carrier by inkjet printing was assembled to the single cell of NiO-8YSZ/8YSZ(~6 μm)/LSM-8YSZ/LSM structure, the open circuit voltage of which was close to the theoretical value [4], which successfully proved that inkjet printing was a simple and economical technology for manufacturing 8YSZ electrolyte film. However, the surface energy of nanoparticles in the inkjet ink is commonly high, causing them easy to agglomerate and resulting in poor ink stability.

Two strategies can be adopted to solve the above problems. On the one hand, the thermal inkjet technology can print high solid content ink, and the organics in the ink can be evaporated through the heating plate. The cell with a 7.5 μm thick 8YSZ electrolyte film and 2 μm SDC functional layer had a maximum power density of 1050 mW/cm^2 at 750°C [19]. On the other hand, reducing the concentration of nanoparticle ink can also improve printability and stability. Esposito et al. [5] successfully prepared the dense 8YSZ electrolyte of 1.2 μm thickness using a low-cost HP Deskjet 1000 DOD inkjet printer and optimized 3.7 vol% 8YSZ colloidal water-based ink by inkjet printing technology. With the dense 8YSZ electrolyte, the prepared SOFC can reach the theoretical open circuit voltage and power density of 1.5 W/cm^2 at 800°C, which further proved the feasibility of inkjet printing technology for SOFC electrolyte preparation.

8.3.3.2 3D Printing to Increase the Three-Phase Boundary and Specific Surface Area

The performance of SOFC is mainly controlled by microstructure characteristics such as three-phase boundary length and interface-specific surface area [29]. The cathode polarization resistance can be reduced by approximately 30% by enlarging the electrode-electrolyte interface through a laser micro pattern [39]. When the specific surface area of electrolyte prepared by the engraving method was increased by 73%, the cell performance was improved by 59% [40]. The electrolyte with a cubic pattern was prepared by the cold stamping method, and the cell performance was improved by 64% [41].

However, compared with 3D-printing technology, the above strategies for preparing patterned electrolytes were complex and expensive, which is not conducive to the industrial preparation of electrolytes. The photocuring 3D-printing technology can simplify the preparation process of patterned electrolytes, easily and quickly realizing the preparation of electrolyte-support SOFC [2] with planar and honeycomb-patterned electrolytes. Under the same thickness, compared with the standard planar cell, the honeycomb structure and corrugated electrolyte prepared by 3D-printing technology can increase the electrode/electrolyte interface, and the cell performance was improved by 32% [1] and 15% [2], respectively. When the specific surface area of corrugated electrolyte continues to increase, the cell performance can be improved by 60% [3], and the PPD can reach 410 mW/cm^2 at 900°C, which proves that the improvement of corrugated cell performance is positively related to the increase in the effective contact area between electrode and electrolyte caused by the structural design of electrolyte surface. The 3D-printing method can improve the design freedom of high-end complex equipment, effectively improve the SOFC performance, and is an important step to reform the energy industry.

8.3.4 3D Printing Components of the Cell Stack

3D printing integrated preparation of cell stack components cannot only save materials and time but also improve the reliability and durability of the SOFC system. Figure 8.2 shows the 8YSZ cell module prepared by DLP 3D printing [14]. Although the module presented porous microstructure after sintering, which made it difficult to be used directly for cell performance tests, the appearance of the complete component has provided a good theoretical basis and technical support for the use of 3D-printing technology in SOFC modules. In addition to the 8YSZ cell module, the independent three-dimensional channel cell stack support was prepared by DLP 3D technology [15]. This structure provides a smooth gas flow channel for the operation of the SOFC stack and greatly improves the heat and mass transfer effect, but also poses greater challenges to the preparation process. It is necessary to repeatedly iterate and optimize the cell structure preparation process and process parameters. The microstructure can be improved by increasing the solid content of the slurry, adjusting the printing process parameters, and optimizing the debinding and sintering processes [7]. For instance, Jia et al. [42] successfully prepared the electrolyte support of 3D printing independent tubular cell stack using 43 vol.% photocuring ceramic suspension and optimized the heat treatment procedure of vacuum debinding and air sintering, and obtained a peak power density of 230 mW/cm^2 at 850°C, which shows that the application of 3D-printing technology in the integrated preparation of cell stack is feasible. However, the complete integrated preparation of porous electrode, dense electrolyte, and connector still needs to improve from the aspects of multi-material 3D-printing technology research and development, slurry preparation process optimization, and debinding and sintering process.

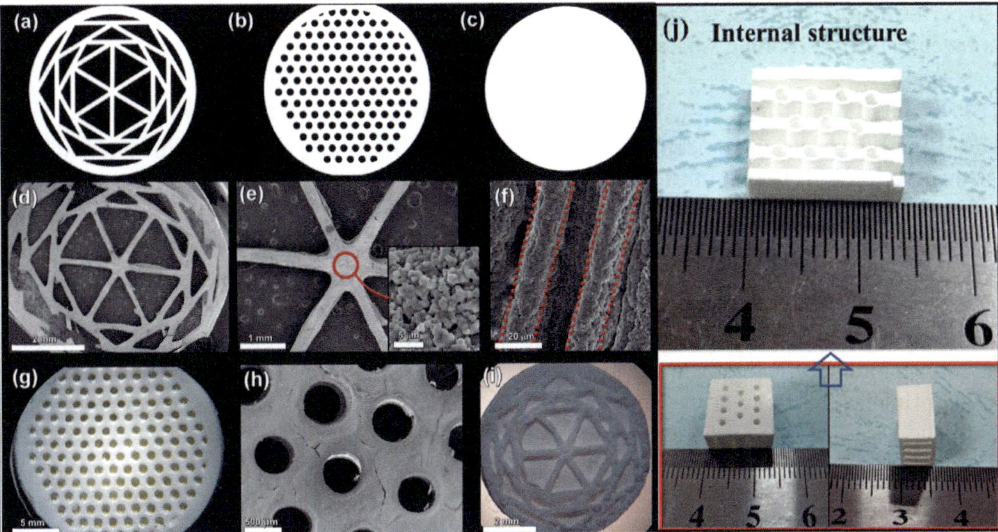

FIGURE 8.2 (a–i) SOFC assembly with good shape and appearance prepared by 8YSZ. Adapted with permission [14]. Copyright (2014), Elsevier, and (j) independent tubular SOFC stack electrolyte support. Adapted with permission [15]. Copyright (2020), Elsevier.

8.3.5 3D Printing Stack Auxiliary Device

Compared with large-diameter tubular cells, the micro-tube solid oxide fuel cell (MT-SOFC) has the advantages of fast start-up and high power density. To solve the rapid prototyping of the fuel intake manifold and economically optimize the fuel path, 3D-printing technology plays an important role in optimizing the design of the fuel distribution inlet by its advantages of easy prototyping of complex designs and visualization of concepts [16]. However, the plastic manifold would deform at high temperatures, which need to be overcome by using 3D-printed ceramic materials to prepare manifolds.

In addition to manifolds, 3D printing can also be used to prepare gas injectors. For anode gas recirculation, a blower or gas ejector was applied to return part of the gas at the outlet of the SOFC anode channel back to the reformer [43], increasing the efficiency by 5–16% [44]. By changing the channel shape between the suction chamber wall and the nozzle outer wall, stretching the mixing chamber and diffuser and eliminating the surface roughness, the total ejector efficiency increased by 30% [43]. This showed that 3D printing can prepare not only the electrolyte and electrodes but also accessories of SOFC. It has broad development prospects in the overall preparation of SOFC.

8.3.6 Challenges of 3D Printing in SOFC

The application of 3D-printing technology in the SOFC field has achieved initial success. In 2015, Northwestern University successfully printed some independent components of solid oxide fuel cells using 3D printers for the first time [45]. Cell3ditor prepared all ceramic, jointless SOFC modules with embedded fluid and current integration. Compared with the traditional process, 3D printing effectively reduced energy consumption and material waste, reduced assembly costs, simplified the manufacturing process, and shortened the manufacturing cycle of products [46]. Relevant patents [47], [48] also provide specific methods, but the integrated preparation of SOFC stacks with high performance, long life, and stable operation by using 3D-printing technology requires further improvement in 3D-printing software, materials, equipment, etc.

8.3.6.1 High Resolution and High Precision Ceramic 3D-Printing Technology

SOFC usually expects area-specific resistance (ASR) < 0.15 Ωcm^2 [28] and electrolyte thickness less than 50 μm [49] to obtain excellent cell performance. Inkjet printing is appropriate for preparing the electrolyte film, but the printing speed is relatively slow and the particles in the ink are easy to agglomerate and precipitate. For high-precision photocuring printing technology, the layer thickness is usually larger than 20 μm. Generally, electrolyte with multiple layer thickness is desirable to get a defect-free and densified body because it is difficult to achieve sintering densification for the green body with only a single layer. However, thick electrolyte often brings in bad cell performance. Moreover, the repeated positioning accuracy and forming accuracy of the present printer are usually at the micron level. Therefore, it is necessary to develop 3D-printing technology with high control accuracy to realize direct real-time controllable adjustment of microstructure.

8.3.6.2 Manufacturing of Multi-Material and Hybrid 3D Printer

SOFC contains cathode, anode, electrolyte, and other materials. Direct inkjet printing is suitable to prepare high-performance functional films, but it is difficult to prepare SOFC stacks which have a certain high aspect ratio. Photocuring 3D-printing technology can realize the preparation of high aspect ratio stacks. However, due to the curing principle limitations of SLA and DLP printing, it is difficult to print SOFC anode and cathode materials directly. Therefore, it is necessary to further develop multi-material and hybrid 3D printers, combined with the unique functions of different printing technologies to build a high aspect ratio multi-material SOFC with the required micro- and nano-level high precision, thus realizing the integrated preparation of SOFC stack. However, to realize the integrated preparation of porous electrodes, dense electrolytes, and connectors, cross-scale preparation is required, and multi-disciplinary cooperation from equipment, control, materials, chemical engineering, and design is also needed. The relationships between the slurry preparation process, printing process, heat treatment process, structure control, and cell performance need to be explored in depth, and the technical problems such as slurry universality, multi-material synchronous printing, and accurate equipment regulation need to be solved.

8.4 SUMMARY AND PROSPECT

3D-printing technology is a breakthrough technology for manufacturing devices with any complex geometry, which greatly simplifies the design and manufacturing process. In the field of SOFC, 3D-printing technology provides a new way to simplify the preparation process of SOFC stack and improve the electrochemical performance, provides the possibility for the preparation of integrated SOFC stack, excavates potential in maximizing the power density per unit mass and unit volume, and also provides a strong guarantee for the stable operation and durability of the stack. 3D printing can flexibly design the complex three-dimensional channel of gas flow in the cell stack, which is conducive to heat and mass transport and increase the effective electrode area. It can greatly improve the electrochemical performance of SOFC by simply and accurately controlling the microstructure, composition, and thickness of the functional film. In addition, the manufacturing of functional accessories of SOFC, such as manifold and gas ejector, is also crucial. In the future, 3D printing will be comprehensively and systematically applied in the SOFC field. Inkjet printing is suitable for the preparation of SOFC thin films. The micro shape can be printed to increase and expose effective surfaces to improve the performance of SOFC. However, it is difficult to prepare the stack with a certain aspect ratio. Photocurable 3D-printing technology has unique advantages in stack preparation, but the integrated preparation of multi-materials still needs further research and development. Due to the limited development of 3D-printing materials at present, the development of functional materials or pastes in the SOFC field will be a hot spot in the future. In addition, multi-material 3D-printing systems and high-resolution and high-precision 3D-printing technology may also be important research directions in the future.

REFERENCES

1 B. Xing, Y. Yao, X. Meng, W. Zhao, M. Shen, S. Gao, Z. Zhao, Self-supported yttria-stabilized zirconia ripple-shaped electrolyte for solid oxide fuel cells application by digital light processing three-dimension printing, *Scripta Mater.* 181 (2020) 62–65.

2 S. Masciandaro, M. Torrell, P. Leone, A. Tarancón, Three-dimensional printed yttria-stabilized zirconia self-supported electrolytes for solid oxide fuel cell applications, *J. Eur. Ceram. Soc.* 39 (2020) 9–16.

3 A. Pesce, A. Hornés, M. Núñez, A. Morata, M. Torrella, A. Tarancón, 3D printing the next generation of enhanced solid oxide fuel and electrolysis cells, *J. Mater. Chem. A.* 8 (2020) 16926–16932.

4 R. I. Tomov, M. Krauz, J. Jewulski, S. C. Hopkins, J. R. Kluczowski, D. M. Glowacka, B. A. Glowacki, Direct ceramic inkjet printing of yttria-stabilized zirconia electrolyte layers for anode-supported solid oxide fuel cells, *J. Power Sources* 195 (2010) 7160–7167.

5 V. Esposito, C. Gadea, J. Hjelm, D. Marani, Q. Hu, K. Agersted, S. Ramousse, S. H. Jensen, Fabrication of thin yttria-stabilized-zirconia dense electrolyte layers by inkjet printing for high performing solid oxide fuel cells, *J. Power Sources* 273 (2015) 89–95.

6 H. Shimada, F. Ohba, X. Li, A. Hagiwar, M. Ihara, Electrochemical behaviors of nickel/yttria-stabilized zirconia anodes with distribution controlled yttrium-doped barium zirconate by ink-jet technique, *J. Electrochem. Soc.* 159 (2012) F360–F367.

7 L. Wei, J. Zhang, F. Yu, W. Zhang, X. Meng, N. Yang, S. Liu, A novel fabrication of yttria-stabilized-zirconia dense electrolyte for solid oxide fuel cells by 3D printing technique, *Int. J. Hydrogen Energ.* 44 (2019) 6182–6191.

8 Q. Chen, Q. Qiu, X. Yan, M. Zhou, Y. Zhang, Z. Liu, W. Cai, W. Wang, J. Liu, A compact and seal-less direct carbon solid oxide fuel cell stack stepping into practical application, *Appl. Energ.* 278 (2020) 115657.

9 G. Jin, Y. Zeng, J. Chen, Fine lattice structural titanium dioxide ceramic produced by DLP 3D printing, *Ceram. Int.* 45 (2019) 23007–23012.

10 C. Li, H. Chen, H. Shi, O. M. Tade, Z. Shao, Green fabrication of composite cathode with attractive performance for solid oxide fuel cells through facile inkjet printing, *J. Power Sources* 273 (2015) 465–471.

11 M. Kim, D. H. Kim, G. D. Han, H. J. Choi, J. H. Shim, Lanthanum strontium cobaltite-infiltrated lanthanum strontium cobalt ferrite cathodes fabricated by inkjet printing for high-performance solid oxide fuel cells, *J. Alloy. Compd.* 843 (2020) 155806.

12 T. Mahata, S. R. Nair, R. K. Lenka, P. K. Sinha, Fabrication of Ni-YSZ anode supported tubular SOFC through iso-pressing and co-firing route, *Int. J. Hydrogen Energ.* 37 (2012) 3874–3882.

13 S. Toshio, F. Yoshihiro, Y. Toshiaki, Y. Fujishiro, M. Awano, Development of cube-type SOFC stacks using anode-supported tubular cells, *J. Power Sources* 175 (2008) 68–74.

14 E. M. Hernández-Rodríguez, P. Acosta-Mora, J. Méndez-Ramos, E. Borges Chinea, P. Esparza Ferrera, J. Canales-Vázquez, P. Núñez, J. C. Ruiz-Morales, Prospective use of the 3D printing technology for the microstructural engineering of solid oxide fuel cell components, *Boletín de la Sociedad Española de Cerámica y Vidrio* 53 (2014) 213–216.

15 J. Zhang, L. Wei, X. Meng, F. Yu, N. Yang, S. Liu, Digital light processing-stereolithography three-dimensional printing of yttria-stabilized zirconia, *Ceram. Int.* 46 (2020) 8745–8753.

16 A. D. Meadowcroft, K. Gulia, K. Kendall, 3D printing and prototyping manifolds for microtubular solid oxide fuel cells (mSOFCs), *Int. J. Sci. Res.* 4 (2015) 709–712.

17 S. A. M. Tofail, E. P. Koumoulos, A. Bandyopadhyayet, S. Bose, L. O. Donoghue, C. Charitidis, Additive manufacturing: Scientific and technological challenges, market uptake and opportunities, *Mater. Today* 21 (2018) 22–37.

18 M. A. Sukeshini, R. Cummins, T. L. Reitz, R. M. Miller, Ink-jet printing: A versatile method for multilayer solid oxide fuel cells fabrication, *J. Am. Ceram. Soc.* 92 (2009) 2913–2919.

19 C. Li, H. Shi, R. Ran, C. Su, Z. Shao, Thermal inkjet printing of thin-film electrolytes and buffering layers for solid oxide fuel cells with improved performance, *Int. J. Hydrogen Energ.* 38 (2013) 9310–9319.

20 P. Qu, D. Xiong, Z. Zhu, Z. Gong, Y. Li, L. Fan, Z. Liu, P. Wang, C. Liu, Z. Chen, Inkjet printing additively manufactured multilayer SOFCs using high quality ceramic inks for performance enhancement, *Addit. Manuf.* 48 (2021) 102394.

21 H. Quan, T. Zhang, H. Xu, S. Luo, J. Nie, X. Zhu, Photo-curing 3D printing technique and its challenges, *Bioact. Mater.* 5 (2020) 110–115.

22 D. A. Komissarenko, P. S. Sokolov, A. D. Evstigneeva, I. V. Slyusar, A. S. Nartov, P. A.Volkov, N. V. Lyskov, P. V. Evdokimov, V. I. Putlayev, A. E. Dosovitsky, DLP 3D printing of scandia-stabilized zirconia ceramics, *J. Eur. Ceram. Soc.* 41 (2021) 684–690.

23 C. Berges, A.Wain, R. Andujar, J. A. Naranjo, A. Gallego, E. Nieto, G. Herranz, R. Campana, Fused filament fabrication for anode supported SOFC development: Towards advanced, scalable and cost-competitive energetic systems, *Int. J. Hydrogen Energ.* 46 (2021) 26174–26184.

24 B. Wittbrodta, J. M. Pearce, The effects of PLA color on material properties of 3-D printed components, *Addit. Manuf.* 8 (2015) 110–116.

25 H. Elsayed, P. Colombo, E. Bernardo, Direct ink writing of wollastonite-diopside glass-ceramic scaffolds from a silicone resin and engineered fillers, *J. Eur. Ceram. Soc.* 37 (2017) 4187–4195.

26 H. Seo, H. Iwai, M. Kishimoto, C. Ding, M. Saito, H. Yoshida, Microextrusion printing for increasing electrode–electrolyte interface in anode-supported solid oxide fuel cells, *J. Power Sources* 450 (2020) 227682.

27 M. Dudek, R. I. Tomov, C. Wang, B. A. Glowacki, P. Tomczyka, R. P. Socha, M. Mosialek, Feasibility of direct carbon solid oxide fuels cell (DC-SOFC) fabrication by inkjet printing technology, *Electrochim. Acta* 105 (2013) 412–418.

28 J. C. Ruiz-Morales, A. Tarancón, J. Canales-Vázquez, J. Mendez-Ramos, L. Hernández-Afonso, P. Acosta-Mora, J. R. Marin Ruedac, R. Fernández-Gonzáleza, Three dimensional printing of components and functional devices for energy and environmental applications, *Energ. Environ. Sci.* 10 (2017) 846–859.

29 H. Moussaoui, R. K. Sharma, J. Debayle, Y. Gavetb, G. Delettea, J. Laurencin, Microstructural correlations for specific surface area and triple phase boundary length for composite electrodes of solid oxide cells, *J. Power Sources* 412 (2019) 736–748.

30 A. Sukeshini, R. Cummins, T. Reitz, R. Miller, Inkjet printing of anode supported SOFC: Comparison of slurry pasted cathode and printed cathode, *ECS Solid State Lett.* 12 (2009) B176–B179.

31 G. Han, H. Choi, K. Bae, H. Choi, D. Jang, J. Shim, Fabrication of lanthanum strontium cobalt ferrite-gadolinium doped ceria composite cathodes using a low-price inkjet printer, *ACS Appl. Mater. Inter.* 9 (2017) 39347–39356.

32 N. Yashiro, T. Usui, K. Kikuta, Application of a thin intermediate cathode layer prepared by inkjet printing for SOFCs, *J. Eur. Ceram. Soc.* 30 (2010) 2093–2098.

33 C. Berges, A. Wain, R. Andújar, J. Naranjo, A. Gallego, E. Nieto, G. Herranz, R. Campana, Fused filament fabrication for anode supported SOFC development: Towards advanced, scalable and cost-competitive energetic systems, *Int. J. Hydrogen Energ.* 46 (2021) 26174–26184.

34 W. Huang, C. Finnerty, R. Sharp, K. Wang, B. Balili, High-performance 3D printed microtubular solid oxide fuel cells, *Adv. Mater. Technol.* 2 (2017) 1600258.

35 K. Choy, Chemical vapour deposition of coatings, *Prog. Mater. Sci.* 48 (2003) 57–170.

36 B. Meng, X. He, Y. Sun, M. Li, Preparation of YSZ electrolyte coatings for SOFC by electron beam physical vapor deposition combined with a sol infiltration treatment, *Mat. Sci. Eng. B-Adv.* 150 (2008) 83–88.

37 I. Polishko, S. Ivanchenko, R. Horda, N. Lysunenkoa, L. Kovalenko, Tape casted SOFC based on Ukrainian 8YSZ powder, *Mater. Today: Proceedings* 6 (2019) 237–241.

38 D. Young, A. Sukeshini, R. Cummins, H. Xiao, M. Rottmayer, T. Reitz, Ink-jet printing of electrolyte and anode functional layer for solid oxide fuel cells, *J. Power Sources* 184 (2008) 191–196.

39 J. Cebollero, R. Lahoz, M. Laguna-Bercero, A. Larrea, Tailoring the electrode-electrolyte interface of solid oxide fuel cells (SOFC) by laser micro-patterning to improve their electrochemical performance, *J. Power Sources* 360 (2017) 336–344.

40 A. Konno, H. Iwai, K. Inuyama, A. Konnoa, H. Iwai, K. Inuyama, A. Kuroyanagi, M. Saitoa, H. Yoshida, K. Kodani, K. Yoshikata, Mesoscale-structure control at anode/electrolyte interface in solid oxide fuel cell, *J. Power Sources* 196 (2011) 98–109.

41 A. Chesnaud, F. Delloro, M. Geagea, A. Abellard, J. Ouyang, D. Li, T. Shi, B. Chi, R. Ihringer, M. Cassir, A. Thorel, Corrugated electrode/electrolyte interfaces in SOFC: Theoretical and experimental development, *ECS Trans.* 78 (2017) 1851–1863.

42 K. Jia, L. Zheng, W. Liu, J. Zhang, F. Yu, X. Meng, C. Li, J. Sunarso, N. Yang, A new and simple way to prepare monolithic solid oxide fuel cell stack by stereolithography 3D printing technology using 8 mol% yttria stabilized zirconia photocurable slurry, *J. Eur. Ceram. Soc.* 42 (2022) 4275–4285.

43 V. A. Munts, Y. V. Volkova, M. I. Ershov, Development and testing of the 3D printed plastic ejector prototype for the SOFC anode gas recirculation (2018) *International Multi-Conference on Industrial Engineering and Modern Technologies*.

44 T. I. Tsai, S. Du, A. Dhir, A. Williams, R. Steinberger-Wilckens, Modelling a methane fed solid oxide fuel cell with anode recirculation system, *ECS Trans.* 57 (2013) 2831–2839.

45 Northwest university uses 3D printing technology to manufacture fuel cells, *Powder Technol.* 39 (2015) 435.

46 Cell3Ditor project on 3D printing tech for SOFC stacks, *Fuel Cells Bull.* 2016 (2016) 12.

47 N. Yang, X. Meng, J. Zhu, B. Meng, X. Tan, Method for three-dimensional printing honeycomb solid oxide fuel cell with three-dimensional channel, CN201510833332.2, 2015-11-25.

48 J. Zhang, F. Yu, N. Yang, L. Wei, X. Meng, B. Meng, S. Liu, A method for fabricating solid oxide fuel cell stack without connector anode support by 3D printing, CN201810364487.X., 2018-04-23.

49 R. T. Nishid, S. B. Beale, J. G. Pharoah, Comprehensive computational fluid dynamics model of solid oxide fuel cell stacks, *Int. J. Hydrogen Energ.* 41 (2016) 20592–20605.

9 3D-Printed Integrated Energy Storage

Additive Manufacturing of Carbon-based Nanomaterials for Batteries

Joshua B. Tyler[1,2,3], *Gabriel L. Smith*[2], *Arthur V. Cresce*[2], *and Nathan Lazarus*[2*]
[1]Oak Ridge Associated Universities, 2800 Powder Mill Road, Adelphi, MD 20783, USA
[2]US Army Research Laboratory, 2800 Powder Mill Road, Adelphi, MD 20783, USA
[3]Department of Material Science and Engineering, University of Maryland, College Park, MD 20742, USA
*Corresponding author: nathan.s.lazarus.civ@army.mil

CONTENTS

9.1 INTRODUCTION

3D printing is a means to rapidly prototype and design complex devices difficult to manufacture using traditional processing. Design and manufacturing of batteries in three dimensions, particularly the electrodes, has significant advantages over traditional 2D manufacturing due to higher areal loading density and shorter ion-diffusion distances. Electrode designs can also be tailored for specific applications and form factors.

Carbon materials are easily added to 3D-printing techniques by incorporating a variety of carbon allotropes into printable filaments and inks. The carbon composite printable materials have shown

DOI: 10.1201/9781003296676-9

success in extrusion-based 3D printing of battery electrodes. Though Fused Filament Fabrication (FFF) has shown the ability to print a full coin cell using multilateral printing for each component of the cell, the poor performance of such a battery limits the applications of the technology. Direct Ink Writing (DIW) has emerged as the most widely used 3D printing technology for battery development, especially with the printing of carbon electrodes. Many examples of DIW carbon-composite electrodes have been demonstrated and the technology shows vast potential in developing next-generation batteries. The addition of carbon materials into lithography-based 3D-printing technologies has shown promise for increasing mechanical properties but is less widely used for electrical conductivity. However, through thermal degradation of the thermoset materials used in stereolithography (SLA), digital light processing (DLP), and two-photon polymerization (2PP), amorphous carbon materials similar to those used in conventional battery electrodes can be made. By using these 3D-printed and pyrolyzed resins multiple carbon electrochemical cells have been demonstrated to show great promise for the development of customized batteries with high energy and/or power density.

This chapter will discuss the current state of 3D printing of carbon materials for use as battery electrode materials and will reflect on how this technology will be influential in the future development of 3D-printed electronics. We will first introduce battery chemistry and what makes carbon materials beneficial as electrode materials. We will then examine the multitude of ways in which carbon electrodes for batteries can be 3D printed followed by a discussion of what future directions in 3D printing carbon-based batteries may look like.

9.2 BATTERY CHEMISTRY

9.2.1 Battery Chemistry Introduction

To better understand how 3D printing can be beneficial in designing and manufacturing batteries, it is imperative to understand the current battery landscape as well as the chemical processes happening inside lithium-ion batteries. Lithium-ion batteries provide the highest energy density of any electrochemical cell-based device and therefore have become commoditized as an enabling technology for nearly all of the mobile electronic and electrified transportation markets. As of 2022, individual cylindrical 18650 and 21700 lithium-ion battery *cells* can provide off-the-shelf energy densities of 250 watt-hours per kilogram (Wh/kg), far eclipsing what is typically delivered by nickel-metal hydride (NiMH) at 110 Wh/kg, alkaline Zn-MnO_2 at 80 Wh/kg, or lead-acid at 50 Wh/kg. Figure 9.1a depicts a typical lithium-ion battery with a graphite anode and a $LiCoO_2$ cathode. Lithium-ion battery *packs*, composed of many cells working together, typically deliver between 100 and 200 Wh/kg, the greater energy density sufficient to push the range of electric vehicles to a plateau of about 400 miles (~640 km) per charge. Lithium-ion batteries suffer however from safety issues stemming from the combined possibility of rapid self-heating and the use of flammable organic electrolytes. Serious safety failures have been partially mitigated using advanced materials and engineering controls, such as non-flammable electrolytes [1], non-shrinking separator materials [2], battery health monitoring [3], and pack control systems.

Lithium-ion batteries derive their high energy density from two primary factors: the volume utilization of lithium storage in electrode materials, and the relatively high voltage existing between electrodes in each electrochemical stack. Graphite, as an anode material, has a capacity of about 380 mAh/g compared to 250 mAh/g typically extracted from the $Ni^{2+/3+}$ reaction at the $Ni(OH)_2$ cathode of Ni-Cd or NiMH batteries. Lithiated (charged) graphite, composed of the intercalation compound LiC_6, has an electrode potential of about -3.0V vs. the standard hydrogen electrode potential, putting it very close to the plating potential of Li^+ on Li metal. The structure of LiC_6 is depicted in Figure 9.1b. The high electronegativity of LiC_6 allows typical lithium-ion battery electrode pairings

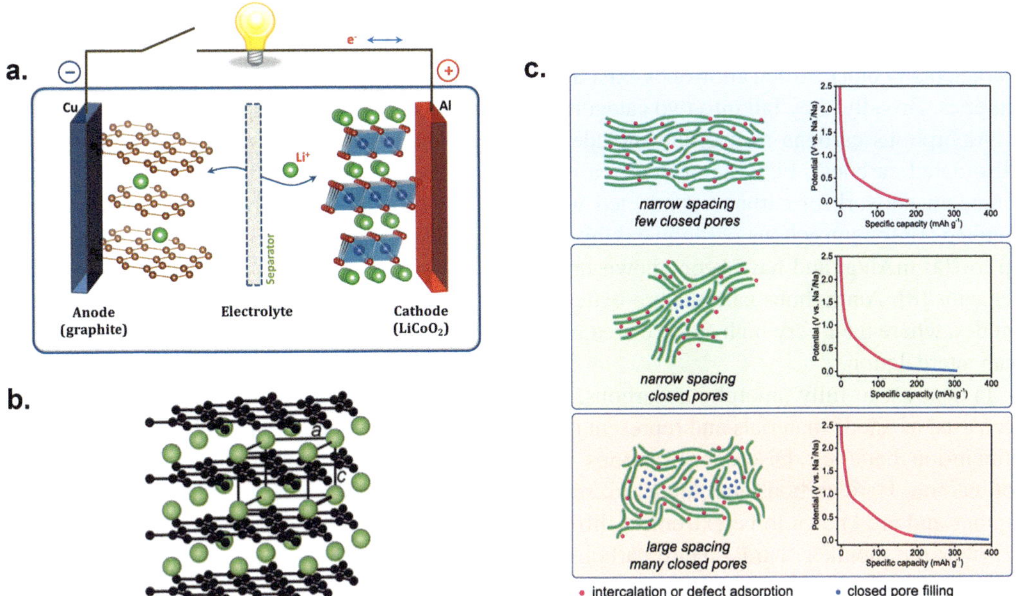

FIGURE 9.1 (a) Basic lithium-ion battery structure showing Li^+ (unconnected spheres) intercalation into anode (graphite) and movement through electrolyte/separator to intercalation in cathode (here, $LiCoO_2$). Adapted with permission [5], Copyright (2013), American Chemical Society. (b) Storage of Li^+ (unconnected spheres) in graphite (connected spheres) as LiC_6, showing the accumulation of Li^+ in the spaces between individual graphene sheets. Adapted with permission [6], Copyright (2022), American Chemical Society. (c) A depiction of the arrangement of graphene sheets and the accompanying capacity as a function of voltage in soft carbon (top), intermediate hard & soft carbon (center), and hard carbon (bottom). Adapted with permission [7], Copyright (2021), Elsevier.

to routinely exceed 3.0V as a single-cell potential. Energy in a battery can be expressed as the product of capacity in amp-hours and cell potential in volts:

$$Energy = capacity \times cell\ potential \tag{9.1}$$

It is the combination of the extraordinary cell potential of lithium-ion batteries with high-capacity electrode materials that give lithium-ion batteries their energy density advantage. A further component however is needed to ensure cell operation at potentials greater than 3.0V: the organic electrolyte. Water-based electrolytes, such as those used in NiMH, typically undergo water electrolysis at cell potentials between 1.5V-2.0V and are operated at no more than 1.5V to avoid cell gassing issues. The organic electrolyte of lithium-ion batteries based on ethylene carbonate possesses the ability to sustain high cell potentials through electrode passivation by direct redox decomposition of the electrolyte components [4]. This insoluble passivation layer termed the 'solid electrolyte interphase', or SEI, (a) prevents further redox decomposition of the electrolyte, (b) allows transport of Li^+ to the electrode active material, and (c) can be replenished by the electrolyte as defects occur in the passivation layer due to chemical or physical processes.

9.2.2 Carbon for Batteries

Graphite is the only carbon allotrope that sees widespread use in the current lithium-ion battery manufacturing industry. The availability and variety of both natural and synthetic graphite have led to significant fine-tuning of the graphite particle structure to suit the needs of batteries that sustain high charging rates or long-term cycling stability. Though graphite is dominant in the existing market, many other carbon allotropes exist that have potential applications in 3D-printed lithium-ion batteries. Broadly, they fall into two categories: amorphous carbons and carbon nanotubes (CNTs).

Amorphous carbons as a group include fully amorphous carbons as well as what are called 'disordered carbons'. Fully amorphous carbons, like carbon black and activated charcoal, contain a random network of carbons connected with varying ratios of sp^2- and sp^3-type carbon-carbon bonding. Fully amorphous carbons exhibit some capacity for intercalation of Li^+, in some cases up to 200 mAh/g, and have been shown in research-scale cells to have a stable charge-discharge behavior [8]. Amorphous carbons are better known for their ability to function as supercapacitor anodes, where their very high surface area is used for Li^+ non-Faradic (surface layer) storage rather than intercalation.

In contrast to fully amorphous carbons, disordered or partially graphitic carbons have already seen used as anode materials and represent a fertile area for development in the arena of 3D-printed lithium-ion batteries. Disordered carbons include hard carbon or non-graphitizing carbon, and soft carbon. Hard carbons are characterized by graphitic domains separated by amorphous carbon regions and are known to be extremely difficult to convert fully into graphite using heat treatment [7] (thus the moniker 'hard'). Hard carbons are often derived from heat treatment of biomass [9] and have been shown to have the capacity for both Li^+ or Na^+ above 300 mAh/g [10]. The goal of hard carbons as anode materials is not to attempt graphitization, but to take advantage of the storage capacity inherent in the house-of-cards structure. Two areas of concern for hard carbon anodes are (a) low first-cycle Coulombic efficiency and (b) high slope of capacity as a function of electrode potential. First-cycle efficiency problems have been partially addressed through materials research [11], but the sloping capacity profile of hard carbons typically results in lower average cell potential of individual cells as well as increasing the potential range across which energy is delivered, complicating the use of such a cell in electronic devices.

Soft carbons have a historical place as one of the first carbons utilized as an anode in a production lithium-ion battery. Nobel laureate Akira Yoshino and his group used petroleum coke in the mid-1980s to produce lithium-ion battery cells delivering about 80 Wh/kg [12], so there is a solid precedent for the use of soft carbons in lithium-ion batteries. Soft carbons have graphitic domains embedded in a matrix of amorphous carbon, and soft carbons differ from hard carbons in that soft carbons have stronger graphite (002) diffraction peak and a generally more overlapping and interlocked graphitic structure interspersed with amorphous carbon. They are called 'soft carbons' because they are graphitizable at temperatures lower than 3000 °C in contrast to hard carbons. Soft carbons have a Li^+ specific capacity of ~ 190 mAh/g, and the capacity is delivered with a sloping profile as a function of electrode potential. While soft carbons have lower Li^+ capacity than hard carbons, soft carbons have a much higher initial Coulombic efficiency than hard carbons, and they are typically synthesized from precursors at lower temperatures than hard carbons [13]. The lower processing temperatures of soft carbons are particularly suited to the process of 3D printing and pyrolyzing carbon-based structures that could be used as lithium-ion battery anodes. A graphical summary of the storage of Li^+ in soft and hard carbons can be seen in Figure 9.1c.

CNTs have also been explored as anode materials for lithium-ion batteries, though no production cell uses CNTs as the ion-storage active material. CNTs are typically valued for their participation in composite binders for lithium-ion batteries that exploit the high tensile strength, electrical conductivity, and enhanced percolation ability of single and multiwall carbon nanotubes. Li^+ has a high energy barrier of entry into the interior of a single CNT that is short and open-ended without significant mechanical or chemical modification to the walls of the nanotube, regardless of nanotube chirality. This issue makes single-walled carbon nanotubes (SWCNTs) an unlikely candidate for 3D-printed carbon anodes and explains why they are mostly ignored as a lithium-ion battery anode material [14]. Multiwalled carbon nanotubes (MWCNTs) intercalate Li^+ in the gallery space between successive nanotubes, giving them an effective capacity of about 350 mAh/g, close to that of graphite [15]. Pyrolysis of polymeric precursors to form MWCNT anode materials has been reported to occur at temperatures of 800 °C or less, putting them in a processing range suitable for 3D-printed precursors [16]. However, the real value of SWCNTs and MWCNTs appears to be as high tensile strength and conductive scaffolds upon which other kinds of high-capacity active anode materials can be grown, such as SnO_2 and Si/SiO_2 [15]. In this way, CNTs are a promising carbon allotrope to consider for 3D-printed battery systems, both as active materials and scaffolding for the directed growth of other kinds of anode materials.

3D-printed lithium-ion batteries represent an opportunity to explore the fabrication of two kinds of batteries difficult to make by other means: batteries that are integrated into an electronic board to provide small-scale energy storage with custom dimensions and minimal packaging, and device-integrated batteries that are permanently embedded into the structure of the device. Such small form factors are not easily manufactured using current techniques. As the cell size decreases, packaging material generally becomes a greater proportion of both cell mass and volume, resulting in gross inefficiencies of gravimetric and volumetric energy density with decreasing size. To summarize, the variety of carbon allotropes from soft and hard carbons to CNTs that have been studied in the lithium-ion battery research community brings a host of possibilities, advantages, and disadvantages to 3D-printed batteries. Graphite structures, while difficult to create through pyrolysis, might be part of a printed anode using a curable binder. Other carbons, including amorphous carbons and CNTs, could be created through the pyrolysis or processing of 3D-printed precursor structures, combining battery materials with the ability to design specific geometries through 3D-printed structures. We will now examine the multitude of ways in which 3D printing has shown the ability to print carbon materials suitable for 3D-printed lithium-ion batteries.

9.3 3D PRINTING OF CARBON

9.3.1 FFF

FFF is a 3D-printing technology based on building up layers of thermoplastic extruded from a heated nozzle to form a completed part. FFF, alternatively known as fused deposition modeling (FDM), is one of the oldest and most common AM approaches and can print a wide variety of thermoplastics and thermoplastic composites. The technique has the advantage of being low cost, widely available, and capable of printing usable electrodes with no or minimal post-processing [17], although typically at a lower resolution than alternatives like SLA. FFF is also capable of multi-material printing through the use of multiple extruder heads, enabling the ability to print a complete battery in a single print [18].

Adding nano-dimensional carbons to thermoplastic is a well-established approach for creating conductive filaments within FFF. High aspect ratio 1D/2D carbon fillers such as CNTs and graphene have been demonstrated to have both higher conductivity and larger surface areas due to a low percolation threshold [19]. The nanocarbon is first composited with a 3D-printable thermoplastic such as PLA to form a filament (Figure 9.2a); this filament is then extruded through a heated extrusion nozzle to form a conductive part that can be used for a battery electrode (Figure 9.2b) [17]. The conductive carbon filaments can also be used in combination with active lithium-based filaments to print a complete lithium-ion battery. In one notable example [20], several filaments based on different carbon fillers (graphene, MWCNT, and carbon black) were tested as anodes and cathodes of a coin cell in conjunction with polylactic acid (PLA) infused with Li salts to serve as an electrolyte (Figure 9.2c–d). Lithium manganese oxide and lithium titanate particles were also added to the cathode and anode materials, respectively, and a copper-loaded filament was used to print the current collectors due to its higher conductivity. The approach was used to print an integrated battery directly into 3D-printed glasses and a wrist band, demonstrating an important advantage of FFF-based 3D printing of batteries.

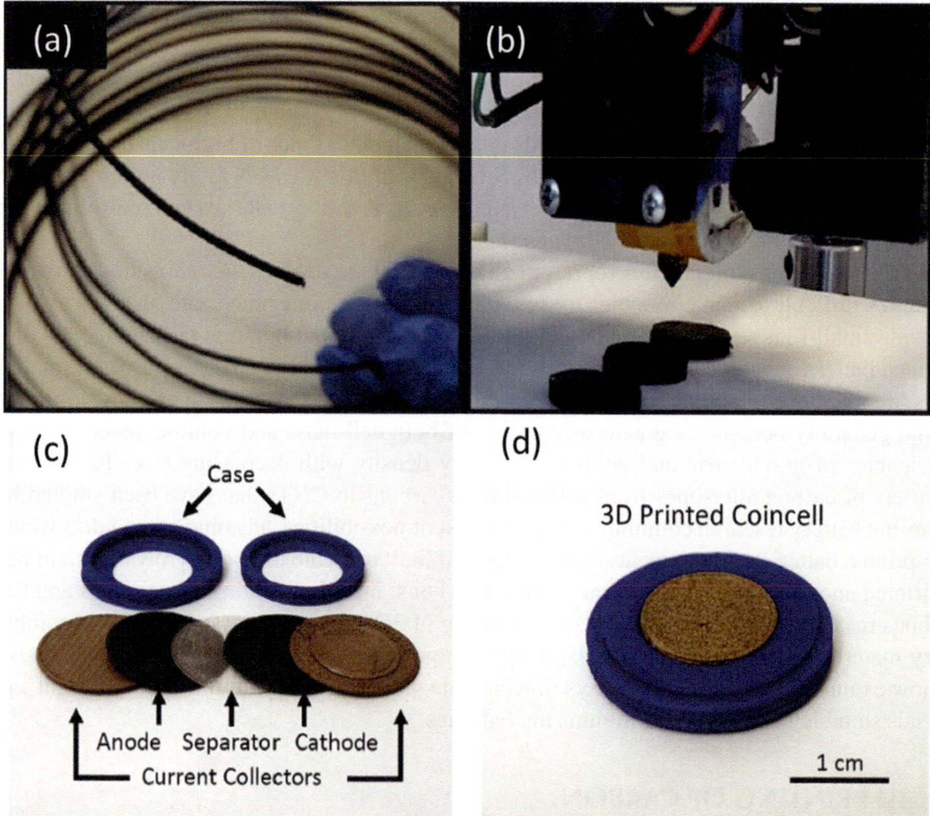

FIGURE 9.2 (a) Graphene/PLA filament and (b) printing of disc electrode. (a,b) adapted with permission [17]. Copyright: The Authors, some rights reserved; exclusive licensee Springer Nature. Distributed under a Creative Commons Attribution License 4.0 (CC BY). (c,d) Coin cell printed using FFF. Adapted with permission [20], Copyright (2018), American Chemical Society.

A major limitation of FFF-based batteries has been that performance has remained relatively poor compared to more traditional manufacturing approaches. The battery manufactured by Reyes et al., for instance, has a capacity roughly two orders of magnitude lower than a conventional lithium-ion battery, resulting largely from the lower concentration of conductive filler possible without degrading the printability of the filament. There has been recent work on addressing this problem through the addition of a plasticizer to bring the active material in the filament up to above 60% by weight, with substantially higher resulting capacity (up to 200 mAh/g of graphite) [18]. While FFF-based nanomaterial battery printing remains less common than other 3D-printing approaches such as DIW, the improving performance along with low cost has generated increasing interest in recent years.

9.3.2 Direct Write

DIW has emerged as one of the most widely used AM techniques for developing batteries due to its affordability, ease of use, and vast material library. During DIW, the printed ink material is patterned by extruding it through a nozzle with compressed air and allowing it to solidify. The versatility of layer-by-layer deposition of custom inks and slurries with high precision and resolution has been instrumental in the 3D printing of electrode materials. The freedom in material selection with DIW also makes it appealing for electroactive devices. DIW is a highly customizable process, albeit with certain limitations resulting from ink fabrication and nozzle size. Thus, precise rheology of fabricated inks as well as appropriate nozzle size is required to prevent structural deformations and defects.

The first reported account of a fully 3D-printed battery was performed using DIW by Sun et al. [21]. They demonstrated that by using DIW of Li-based inks it is possible to manufacture an interdigitated Li-ion battery (Figure 9.3a). Printing the anodes and cathodes directly onto the current collector allows for ease of integration for the device. They developed aqueous inks for the cathode and anode containing $LiFePO_4$ (LFP) and $Li_4Ti_5O_{12}$ (LTO). Following printing and processing, the packaged battery had a capacity of 1.2 mAh/cm^2. Their demonstration has led to further research into developing composite electrodes manufactured with DIW.

Carbon composite inks for DIW have been investigated for electrode materials due to their high electrical conductivity. Graphene oxide (GO) and reduced GO (rGO) based inks have come to the forefront of carbon 3D printing with DIW due to their printability and unique viscoelastic properties which allow high concentrations of GO to be incorporated in the ink. During printing GO flakes tend to align along the direction of extrusion due to the shear stress brought on by the nozzle further enhancing the electrical properties of the printed inks. This alignment assists in improving properties like conductivity, making GO-based inks a promising material candidate for DIW printed electrodes.

Successful printing of a DIW GO electrode has been demonstrated and shown to be useful in the manufacturing of 3D printed Li-ion batteries [22]. Fu et al. developed aqueous GO-based inks consisting of high concentrations of GO sheets and electroactive Li materials (LTO and LFP) that was easily processible, safe, and low-cost. Thermal treatment of the inks was done to reduce the GO of the printed electrodes. The electrical conductivities of the annealed electrode materials are reported as 3,160 and 610 S/m for LFP/GO and LTO/GO respectively, improving by approximately seven orders of magnitude from the untreated material. These inks were used to fabricate one of the electrodes in an interdigitated battery (Figure 9.3b–c). Charge and discharge testing of the individual LFP/GO and LTO/GO electrodes demonstrate stable capacitance after cycling up to 20 cycles, with specific capacities of ~160 mAh/g and ~170 mAh/g, respectively. Using the LFP/GO and LTO/GO as the cathode and anode and by printing a solid electrolyte (poly(vinylidene fluoride)-co-hexafluoropropylene (PVDF-co-HFP) and Al_2O_3 nanoparticles) into the channel between the electrodes demonstrates the ability to DIW a complete battery cell. This full cell was able to achieve charge and discharge capacities of 117 and 91 mAh/g demonstrating the ability to fully 3D print a battery using GO composite inks using DIW. The demonstration of a fully 3D-printed battery using GO composite inks using DIW has influenced further developments of batteries using DIW.

The development of a GO composite ink with an already established high-capacity active anode material is an avenue in which higher capacity batteries can be developed through 3D printing. A GO composite ink was developed containing SnO_2 quantum dots (QD) [23]. Figure 9.3d shows the ability to print complex structures with the composite ink including mosquito coils, spiral rectangles, zigzag lines, as well as high surface area periodic lattices which can be used as electrodes in batteries. The 3D-printed QD/GO electrodes demonstrated an ultrahigh charge capacity of 991.6 mAh g^{-1} at 50 mA g^{-1} during the first cycle and the reversible capacity was increased to 1004.9mAh g^{-1} after 50 cycles, showing fantastic cycling stability and increased performance over pure SnO_2 QDs. This order of magnitude increase in specific reversible capacity of the electrodes over the previous work discussed by Hu et al. may enable the development of higher capacity faster response 3D-printed full cell Li-ion batteries.

DIW has been examined as a means to prototype and manufacture advanced energy storage devices for next-generation batteries. Research efforts have been directed towards the development of Li-S batteries due to their high specific capacity (2600 Wh/kg). The development of a sulfur graphene composite ink for DIW is useful in the 3D printing of electrodes for Li-S batteries [24].

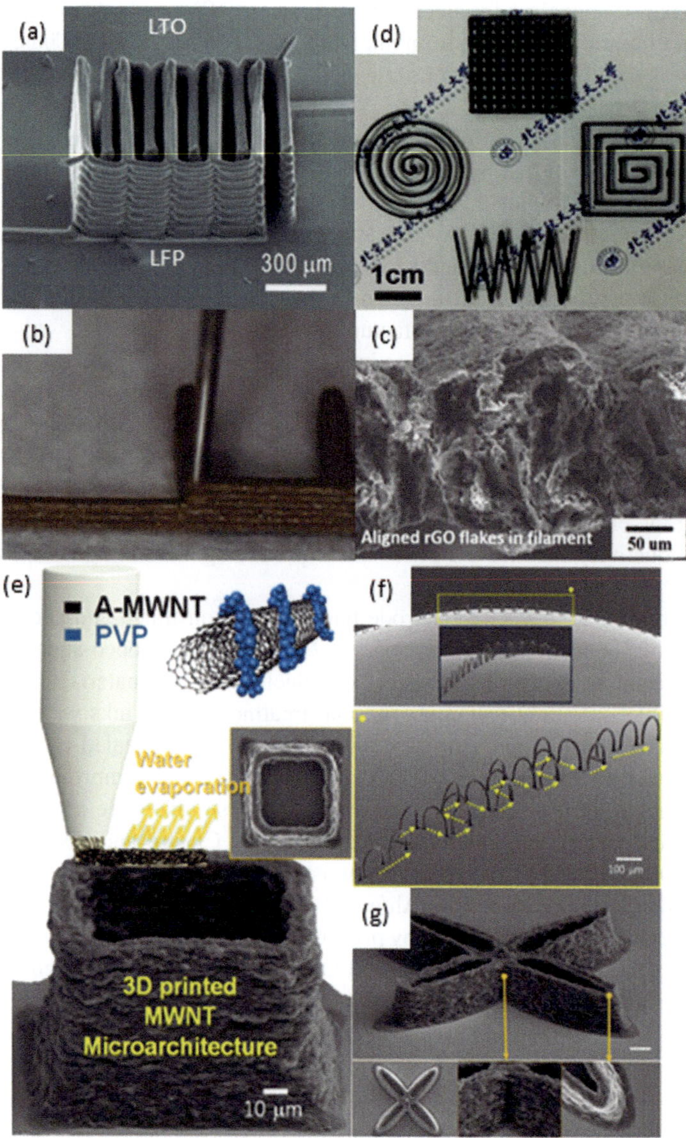

FIGURE 9.3 (a) SEM micrographs of the first 3D-printed battery made up of 16-layer interdigitated LTO-LFP electrode architectures. Adapted with permission [21], Copyright (2013), John Wiley and Sons. (b) Digital image of the printing process for GO-based electrodes for 3D-printed batteries. (c) SEM micrograph of electrodes cross-section showing GO alignment in the filament. Adapted with permission [22], Copyright (2016), John Wiley and Sons. (d) Optical image of complex patterns printed with SnO_2 QD/GO ink. Adapted with permission [23], Copyright (2018), Royal Society of Chemistry. (e) Schematic illustration of 3D printing CNT-based ink. (f) SEM micrograph of complex MWCNT bridges deposited on a curved glass substrate, showing flexibility in the printing process. (g) SEM micrographs of hollow curved architectures printed with MWCNT ink (scale bar 20 μm). Adapted with permission [25], Copyright (2016), American Chemical Society.

The printable ink uses S particles, 1,3-diisopropenylbenzene (DIB), and dispersed condensed GO. The formulation for the ink allowed for ease of print using DIW due to the high viscosity and shear thinning properties of the ink. 3D printing of well-designed periodic micro-latices with the ink should allow for ease of access of the electrolyte of a battery developed with this material to the active electrode. Electrodes developed with the S/DIB/GO composite show a high reversible capacity of 812.8 mAh/g and stable cycling performance when integrated as the cathode of a Li-S battery.

CNTs are widely used in the manufacturing of electronics and energy storage devices, owing to CNTs' high mechanical strength, chemical stability, large surface area, and excellent electrical and thermal properties. 3D-printable MWCNT materials have been shown for high resolution complex highly conductive materials using a DIW method (Figure 9.3e–g) [25]. By DIW of MWCNT, microelectrodes for a Li-S battery have also been developed [26]. The MWCNT electrode demonstrated high conductivity and ultrahigh porosity, which as previously discussed benefit the electron and ion transport of the electrode. The printed micro-battery demonstrated high areal capacities of more than 5 mAh/cm^2 with high cycling stability. This result shows potential for the development of on-chip energy storage using 3D printing.

DIW has developed as one of the most commonly used techniques in the fabrication of 3D-printed batteries. The optimization of the inks used in DIW with carbon materials has allowed for the development of higher-performing electrodes for batteries. While there remain areas of active research to address limitations such as the weak mechanical strength between the layers of DIW that have caused some concern, the development of complex 3D printed electrodes with DIW shows significant potential in the prototyping and manufacturing of next-generation batteries including on-chip energy storage for electronics.

9.3.3 SLA

SLA is a layer-by-layer 3D printing process in which a photochemical resin is exposed to light causing the chemical monomers and oligomers to cross-link forming a solid polymer structure (Figure 9.4a). Since being introduced as the first commercially available 3D-printing technology in 1986 by Charles Hull [27], SLA has advanced to a point where cheap commercial printers are widely available with an immense catalog of printable materials. The use of photocurable resins as the printable material allows for the ease of incorporation of additives into the thermoset precursors allowing tunability of the material properties of as-printed parts. Carbonizing of SLA-printed parts through either additives or post-processing allows SLA 3D printing to be a useful tool in developing carbon electrode materials for batteries. These carbon materials can also be used as template materials for the deposition of more electrochemically active electrode materials such as Cu and Ni.

Carbon-based nanomaterials can be used as additives in thermoset 3D-printable resins to enhance the properties of the as-printed material. However, the reduction in printability caused by reduced transparency to UV light and the decreased ductility caused by the addition of carbon nanomaterials such as graphene have limited use of carbon nanomaterials as an additive in the SLA process. Carbon nanomaterial additives can block UV light, and therefore limit photopolymerization during the 3D printing process, and as a result, only small quantities of carbon additive are usually incorporated into SLA resins. Graphene as an additive has been studied, but it has been reported that to maintain the appearance and the mechanical strength of the parts, the concentration in the SLA polymer must be less than 5 wt% [28]. This result limits the ability to take full advantage of the properties that make graphene desirable as a battery electrode. More often, GO fillers are used in SLA composite printing due to the strong interfacial bonds generated between GO fillers and polymer matrices allowing for more reliable prints than pure graphene additive. Many of the current studies on GO reinforced photopolymers focus on the enhancement of mechanical and thermal properties [29], [30]. There has been a continued interest in GO as a material candidate for battery and energy storage applications due to the unique surface properties, large surface area, and layered structure [31] and there is likely to be a related investigation of GO/resin composites for battery electrode manufacturing.

CNTs have similarly been studied as an additive to increase the properties of photopolymerizable materials. Sandoval et al. were one of the first studies to demonstrate the ability to enhance the SLA printed material properties by dispersing MWCNTs into an SLA resin [32]. Incorporating an MWCNT concentration of 0.05% (w/v) showed an increase in the ultimate tensile stress and fracture stress by an average of 17% and 37%, respectively. An increase of MWCNT concentration to 0.5% (w/v) showed resiliency in parts at higher operating temperatures. These small quantities of CNT are insufficient for use in electrical components but may open future work toward use as a template for further processing towards an electrochemical cell.

Amorphous carbons developed from pyrolysis of polymeric precursors in an inert atmosphere have been widely used in electrochemical applications due to the wide potential stability window, chemical inertness, low cost, low background currents, and electrocatalytic activity for redox reactions [33], [34]. Pyrolysis of SLA printed parts has proven successful in the development of complex 3D carbon electrodes [35]. Pyrolysis of the as-printed polymer structure results in an amorphous material that has shrunk by about 2.5 X (Figure 9.4b). This pyrolysis process requires specific design and fabrication constraints as large solid parts deform due to outgassing, leading to parts needing to be latticed. This latticing also increases the surface area of the 3D carbon electrode, which potentially provides a larger electrode/electrolyte interface. Cyclic volumetry and electrochemical impedance spectroscopy experimentation performed on an electrochemical cell manufactured using pyrolysis of an SLA printed part showed excellent electrochemical stability, repeatability, and reversibility of the electrode. The designed electrode also exhibited similar kinetics of electron transfer behavior to glassy carbon electrodes, which is already known as a stable working electrode [36]. This process can conceivably be extended to the development of 3D-architected mesoscopic battery electrodes, whether for individual electrodes or complex interdigitated designs for dual carbon electrode batteries.

9.3.4 DLP

Another lithography-based 3D-printing technique that has shown promise in the development of batteries is DLP. The DLP 3D-printing process consists of a photocurable resin being patterned layer-by-layer by a projected UV light. Since DLP is based on the same photo-chemical process as SLA, except with a projected image rather than laser-induced, there has been similar interest in the development of carbon electrodes using DLP. Similar pyrolysis processing as was shown using SLA has been successfully used to develop 3D-archited periodic carbon structures with high surface area for use as an electrode (Figure 9.4c–d) [37]. These glassy carbon complex electrodes were able to withstand compressive stresses of 27 MPa and show an areal capacity of 4 mAh/cm^2 at 0.38 mA/cm^2 over 100 cycles and 3.2 mAh/cm^2 at 2.4 mA/cm^2 with a gradual decrease up to around 1 mAh/cm^2 over 500 cycles. Pyrolysis of DLP parts proves to be a means by which large-scale processing of periodic microscale high surface area carbon electrodes can be manufactured.

9.3.5 2PP

2PP is a method of AM that uses a non-linear optical process to photopolymerize a thermoset resin. A femtosecond pulsed laser is focused into a volume of a photosensitive resin at which point the light pulses initiate polymerization when two photons are absorbed in the resin, thus two-photon polymerization. Because of the nonlinear nature of the process, a resolution beyond the diffraction limit can be obtained, resulting in submicron resolution. As such, 2PP is a promising technique to reliably print complex 3D micro-devices. The high spatial resolution as well as the true 3D fabrication of

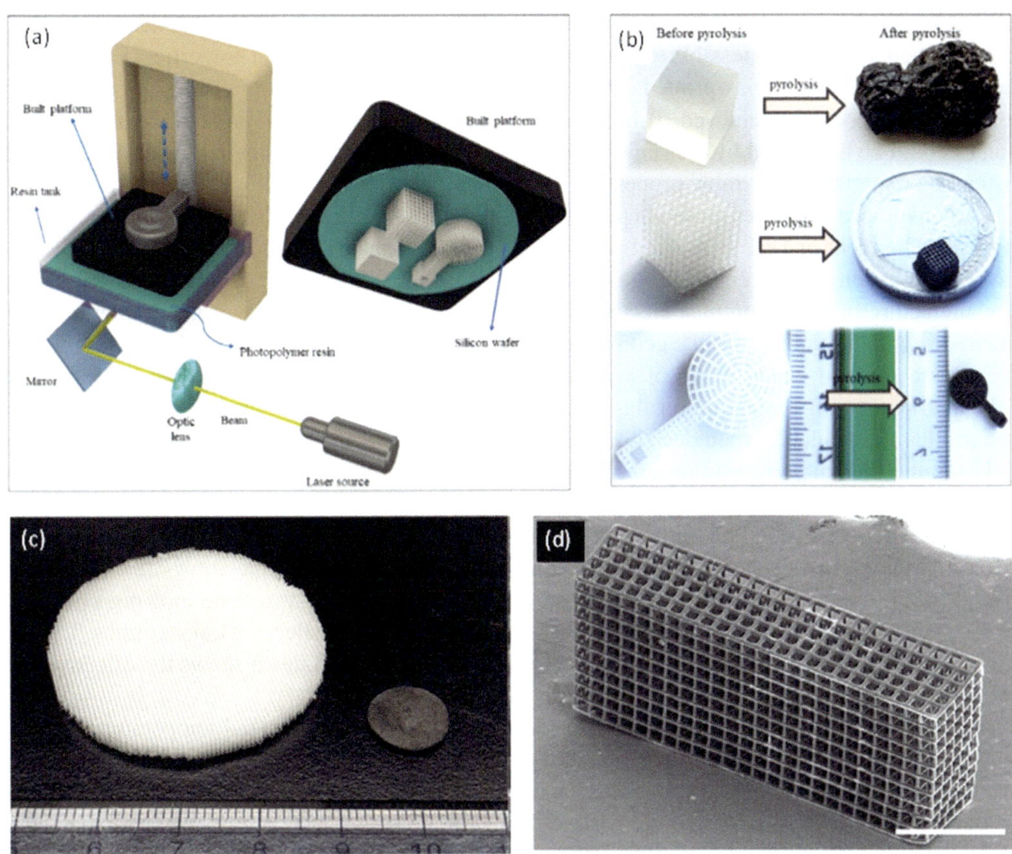

FIGURE 9.4 (a) SLA printing of polymer precursor electrode, (b) before and after pyrolysis of as-printed SLA polymer. Adapted with permission [35], Copyright (2020), Elsevier. (c–d) DLP printed periodic structures pre and post-pyrolysis (scale bar 1 mm). Adapted with permission [37], Copyright (2020), John Wiley and Sons.

parts allows for innovation in the fabrication of micro/nanoscale 3D-printed devices. The ease with which 3D micro-devices can be manufactured using 2PP has allowed for extreme design freedom in a size scale that was previously limited to mostly 2–2.5D processing. Printing microscale high surface area lattice devices using 2PP has attracted much interest in the fields of microfluidics, biological growth, chemical sensing, and energy storage. There has also been interesting in using additives in the 2PP resin to develop composite polymer/metal matrices followed by burnout to leave a strictly metal structure for electrical device design [38]. Unfortunately, the rough surface finish and long processing time for metal 2PP devices limit their possible use for integration with electrical systems. Though these processes may be suitable for developing battery electrodes via 2PP, this chapter is focused on how 2PP of carbon materials can be used in the development of batteries.

One of the most common means by which to manufacture carbon devices using 2PP is through pyrolysis of the as-printed polymer, similarly to the pyrolysis of SLA and DLP discussed above. Multiple reports on the pyrolysis of 2PP photoresists such as the Nanoscribe IP resist series have shown shape retention in lattice structures upon pyrolysis as well as increased resolution [39]. These studies focus on the ability to create some of the strongest known devices, approaching a theoretical strength limit [40]. There have also been developments in the creation of carbon MEMS devices out of the 2PP photoresist pyrolyzed carbon material [41], [42], taking advantage of the electrical conductivity of glassy carbon.

Characterization of the pyrolyzed glassy carbon 2PP material has shown that by varying the final pyrolysis temperature it is possible to tune the electrical and microstructural properties (Figure 9.5a–c) [43]. By pyrolyzing the as-printed 2PP Nanoscribe IP-Dip resin at 1400 °C a highly graphitized conductive material was achieved. This characterization study showed that predictable passive electronics could be developed by altering the final pyrolysis temperature, showing the ability to tailor the carbon material for desirable properties.

Carbon latticed structures made via pyrolysis of 2PP parts, similar to what has been manufactured in the development of DLP pyrolyzed carbon electrodes for batteries, have been widely studied in the search for ultralight-weight high-strength microstructures (Figure 9.5d) [40]. With the ability to manufacture fully 3D complex electrodes the development of completely isolated interdigitated electrodes with optimized surface geometries for increased battery performance is foreseeable. As 2PP uses a rastered laser instead of a projected image as in DLP, 2PP may prove to be a more influential processing technique for the research and development of unconventional and complex designs that are available when designing electrodes layer by layer on the submicron level.

Using the incredible resolution of 2PP as well as the conductivity achieved through pyrolysis of the parts, carbon electrodes for neural sensing [42] have been successfully manufactured (Figure 9.5e–f). Free-standing carbon electrodes were fabricated by printing photopolymer resin using 2PP on the tips of metal wires followed by pyrolysis. This technique allows for large batch processing of normally difficult-to-manufacture electrodes. CV testing was done showing a peak-to-peak separation similar to other carbon electrodes demonstrating fast electron transfer rates. Though this study was done for manufacturing neurochemical sensing electrodes it shows the effectiveness of carbon electrodes manufactured by pyrolysis of 2PP resists.

The carbon 2PP material has shown improved conductivities at elevated temperatures caused by graphitization. The conductivity is reported to be ~ 10^4 S m^{-1} at 1400 °C which is still orders of magnitude below that of bulk metals. This, as well as the brittle nature of carbon materials, creates an issue when trying to integrate these structures into electrical systems. An idea that has been studied to help increase the conductivity and mechanical properties of the 2PP pyrolyzed material is through electroplating of the conductive carbon [44]. Simple two-electrode electroplating using the carbon 2PP material as the cathode and Cu or Ni as the anode shows a greater than 50X decrease in the electrical resistance. Compressive loading of Ni-plated carbon lattice structures (Figure 9.5g–h) demonstrates a 4X increase in the maximum compressive load over purely pyrolyzed 2PP lattices. This first example of electroplating 2PP pyrolyzed carbon devices opens up avenues for manufacturing microscale Ni electrode batteries, taking the ease of manufacturing complex 3D electrodes with the fast response and high energy density of Ni electrode batteries. Another possibility in using electroplating of 2PP pyrolyzed devices for batteries is in the development of two material electrodes for batteries directly on a device making dedicated power supplies on-chip at the microscale.

9.4 FUTURE APPLICATIONS

Most of the demonstrated work in the field of printing carbon-based batteries relies on minor modifications to conventional AM like SLA, FFF, DLP, and DIW approaches, previously described above. To realize the true potential of these new materials and electrode form factors, other techniques of integration need to be investigated. Across industry, defense, and academia the field of AM is evolving to a concept of convergent manufacturing (CM) for greater complexity of features and materials. The US National Academies describes CM as a unified manufacturing system platform that converges heterogeneous interfaces in design, materials, processes (additive, subtractive, and transformative), and diagnostics with physical and digital signatures as inputs producing functional devices components and complete systems as an output at the point of use. We can imagine combining AM techniques, mentioned previously with subtractive approaches like conventional

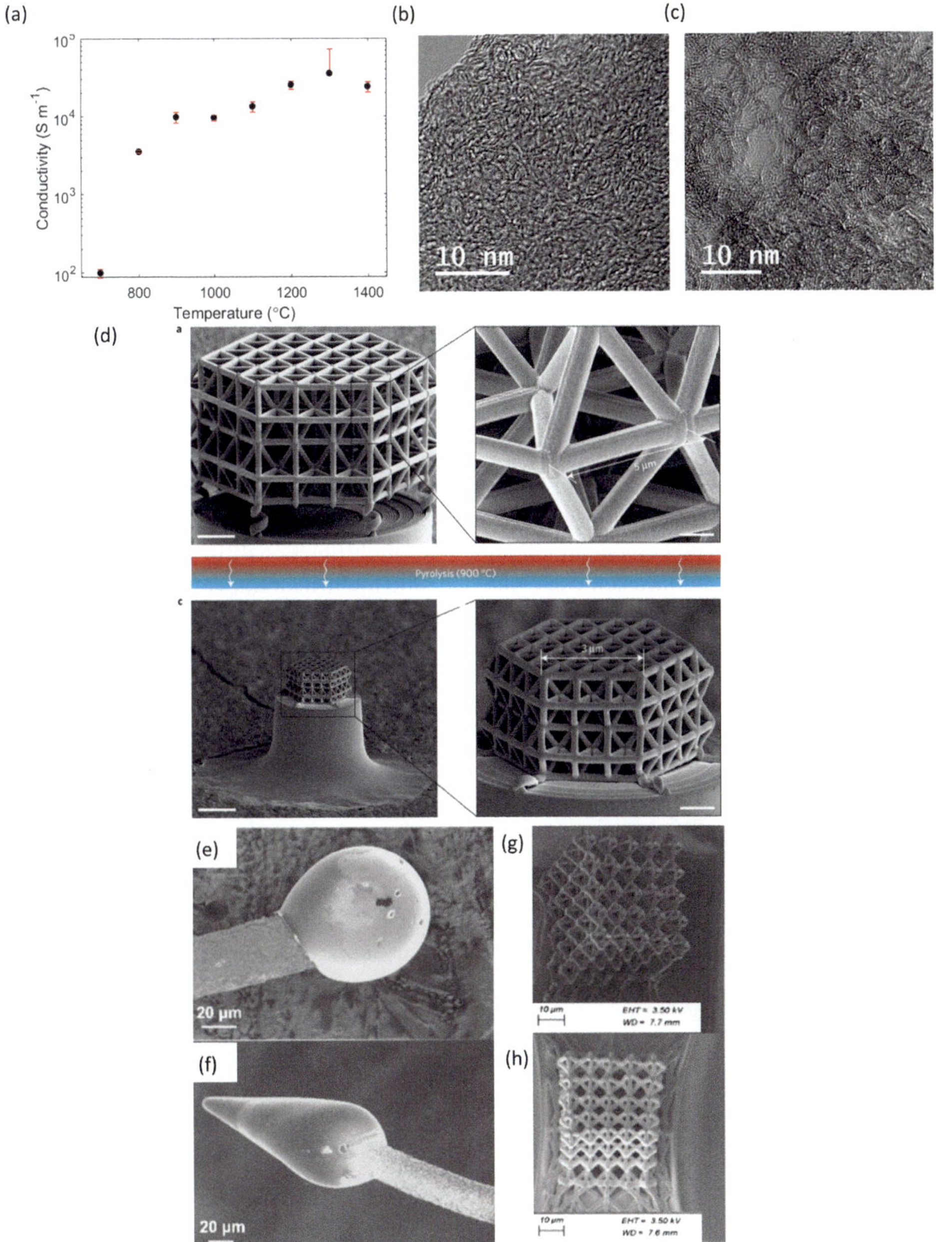

FIGURE 9.5 (a) Conductivity of 2PP pyrolyzed carbon material at differing final pyrolysis temperatures. TEM micrographs of 2PP pyrolyzed carbon material (b) at 900 °C and (c) at 1400 °C. Adapted with permission [43], Copyright (2020), John Wiley and Sons. (d) As-printed and pyrolyzed 2PP micro-lattice structure. Adapted with permission [40], Copyright (2016), Springer Nature. (e–f) Pyrolyzed 2PP carbon neural sensors. Adapted with permission [42], Copyright (2018), John Wiley and Sons. (g) Pyrolyzed 2PP carbon lattice and (h) the same lattice electroplated in Ni. Adapted with permission [44], Copyright (2022), IEEE.

machining, laser ablation, ion milling, and photo etching; along with cleanroom/MEMS techniques like electrodeposition, evaporation, sputtering, ALD, plasma ashing, chemical etching; and with thermal processes like laser forming, inert pyrolysis, annealing. Clever combinations of these approaches will allow battery electrode designers to improve the producibility and volumetric efficiency of printed carbon electrode batteries.

Moreover, CM will allow for optimal battery shapes that conform to the application, leading to ideal placement, thermal control, predictable stress profiles, overall weight reduction, and integration of multifunctional materials with carbon electrodes. In 2021, Asp et al. reported on a structural battery composite with an energy density of 24 Wh/kg and an elastic modulus of 25 GPa, and tensile strength exceeding 300 MPa (Figure 9.6a). The battery was made from multifunctional reinforcing carbon fibers acting as an electrode and current collector. A structural electrolyte is used for load transfer and ion transport and a glass fiber fabric separates the carbon fiber electrode from an aluminum foil-supported lithium–iron–phosphate positive electrode [45]. While this battery was not printed via AM or CM processes, the idea of printed carbon lattices as structural multifunctional geometries is a natural extension.

These optimal shape approaches could integrate micro carbon lattice electrodes that would also minimize the effect of localized impact on the battery shown by Portelea et al. [46] (Figure 9.6b). Expanded material sets via computational designed multifunctionality with vat polymerized hydrogel which with directed infusion can create metallic multi-material composites, and compositional freedom to enable the integration of multicomponent alloys in high surface area geometries into 3D batteries. This approach has been demonstrated by Yee et al. [47] by printing lithium cobalt oxide from hydrogels which could be combined with a glassy carbon lattice anode in a CM process.

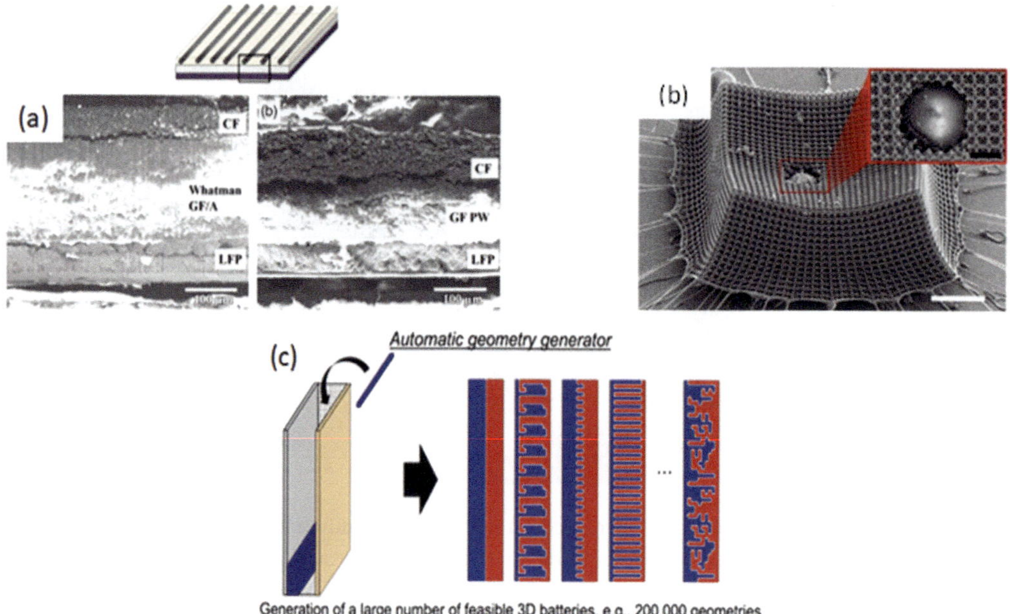

FIGURE 9.6 (a) Structural battery where the multifunctional carbon fiber serves both electrochemical and mechanical functions. Adapted with permission [45]. Copyright: The Authors, some rights reserved; exclusive licensee John Wiley and Sons. Distributed under a Creative Commons Attribution License 4.0 (CC BY). (b) Results of an impact test on carbon lattices (scale bar 20 μm). (Adapted with permission [46], Copyright 2021, Springer Nature.) (c) Automatic geometry generator creating optimal 3D microbattery architectures from over 200,000 configurations. Adapted with permission [49], Copyright (2020), Elsevier.

Another CM approach to metalizing components by combining AM and vacuum evaporation or sputtering of metal allows for the single-step creation of independent electrodes through printed shadow masking features like overhangs, similar to Kim et al. [48].

Advanced computational modeling design of batteries with CM is likely critical for the optimal performance of printed carbon-based Li-Ion batteries. Another example of a CM concept proposed by Gu et al. is the Material-structure-performance integrated AM (MSPI-AM). They posit, that to meet unique performance demands, novel multi-material structures. The diversification of approaches is necessary for further enhancing MSPI-AM. Numerical simulations based on 'Virtual manufacturing' or 'digital-twin' integrated with real production can provide multiscale modeling and accurate prediction and optimization of the CM . Applied to batteries this could result in multifunctional performance including unprecedented electrochemistry, mechanical, electrical, and thermal control. The advancement of computational algorithms and codes in other technical processes also translates to process control methods needed to achieve a complete understanding of physics in the optimal printed battery manufacture via CM (Figure 9.6c) [49].

While many of the individual requisite AM processes have greatly matured in the last decade, for CM to be truly realized in batteries, advancements in material enhancement, diversity, and interfacial performance must be better understood. Moreover, computer-aided design for topology optimized geometry and process feedback control will need to be advanced as well, combined with the development of the manufacturing hardware and software for a wide range of complementary technologies to collaborate with 3D printers to fabricate fully multifunctional batteries [50].

9.5 SUMMARY

We have examined the many ways in which carbon materials can be 3D printed for developing battery technology. Using FFF it was shown that a relatively low-performance full battery coin cell can be manufactured through multi-material printing followed by assembly. DIW has proven itself as the current frontrunner of 3D-printing technologies for batteries with many examples of successful high capacity, reliable batteries using DIW of carbon composite electrodes. Pyrolysis of lithography-based 3D-printed parts has shown the ability to make highly complex amorphous carbon electrodes from the macroscale down to the micro/nanoscale. Though limited studies of using 3D-printed pyrolyzed parts for batteries are currently available, the increased resolution and implementation of higher conductivity amorphous carbons have great potential for high-capacity high strength complex battery electrodes. It is foreseeable that pyrolyzed 3D-printed SLA, DLP, or 2PP carbon parts are going to be more influential in the future development of advanced battery technology than DIW electrodes. Furthermore, by implementing CM techniques for higher capacity, more optimized batteries can be designed in the future, with possible interdigitated multi-component batteries being developed for on-chip integrated microscale batteries. The ability to 3D print batteries is ever evolving and the development of complex carbon electrodes using 3D printing is helping to lead the way.

REFERENCES

1 H. Q. Pham, H. Y. Lee, E. H. Hwang, Y. G. Kwon, and S. W. Song, "Non-flammable organic liquid electrolyte for high-safety and high-energy density Li-ion batteries," *J. Power Sources,* vol. 404, pp. 13–19, Nov. 2018.

2 S. Kalnaus, Y. Wang, and J. A. Turner, "Mechanical behavior and failure mechanisms of Li-ion battery separators," *J. Power Sources,* vol. 348, pp. 255–263, Apr. 2017.

3 R. Xiong, L. Li, and J. Tian, "Towards a smarter battery management system: A critical review on battery state of health monitoring methods," *J. Power Sources,* vol. 405, pp. 18–29, Nov. 2018.

4 K. Xu, "Electrolytes and interphases in Li-ion batteries and beyond," *Chem. Rev.,* vol. 114, no. 23, pp. 11503–11618, Dec. 2014.

5 J. B. Goodenough and K. S. Park, "The Li-ion rechargeable battery: A perspective," *J. Am. Chem. Soc.,* vol. 135, no. 4, pp. 1167–1176, Jan. 2013.

6 C. Ertural, R. P. Stoffel, P. C. Müller, C. A. Vogt, and R. Dronskowski, "First-principles plane-wave-based exploration of cathode and anode materials for Li- and Na-ion batteries involving complex nitrogen-based anions," *Chem. Mater.,* vol. 34, no. 2, pp. 652–668, Jan. 2022.

7 D. Cheng, X. Xhou, H. Hu, Z. Li, J. Chen, L. Miao, X. Ye, and H. Zhang, "Electrochemical storage mechanism of sodium in carbon materials: A study from soft carbon to hard carbon," *Carbon N. Y.,* vol. 182, pp. 758–769, Sep. 2021.

8 S. Villagómez-Salas, P. Manikandan, S. F. Acuña Guzmán, and V. G. Pol, "Amorphous carbon chips Li-ion battery anodes produced through polyethylene waste upcycling," *ACS Omega,* vol. 3, no. 12, pp. 17520–17527, Dec. 2018.

9 M. Thompson, Q. Xia, Z. Hu, and X. S. Zhao, "A review on biomass-derived hard carbon materials for sodium-ion batteries," *Mater. Adv.,* vol. 2, no. 18, pp. 5881–5905, Sep. 2021.

10 D. A. Stevens and J. R. Dahn, "High capacity anode materials for rechargeable sodium-ion batteries," *J. Electrochem. Soc.,* vol. 147, no. 4, p. 1271, Apr. 2000.

11 Y. Wan, Y. Liu, D. Chao, W. Li, and D. Zhao, "Recent advances in hard carbon anodes with high initial Coulombic efficiency for sodium-ion batteries," *Nano Mater. Sci.*, Feb. 2022.

12 "US4668595A – Secondary battery – Google Patents." [Online]. Available: https://patents.google.com/patent/US4668595A/en. [Accessed: 13-Jun-2022].

13 L. Xie, C. Tang, Z. Bi, M. Song, Y. Fan, C. Yan, X. Li, F. Su, Q. Zhang, and C. Chen, "Hard carbon anodes for next-generation Li-ion batteries: Review and perspective," *Adv. Energy Mater.,* vol. 11, no. 38, p. 2101650, Oct. 2021.

14 C. De Las Casas and W. Li, "A review of application of carbon nanotubes for lithium ion battery anode material," *J. Power Sources,* vol. 208, pp. 74–85, Jun. 2012.

15 D. Di Lecce, P. Andreotti, M. Boni, G. Gasparro, G. Rizzati, J. Hwang, Y. Sun, and J Hassoun, "Multiwalled carbon nanotubes anode in lithium-ion battery with LiCoO2, Li[Ni1/3Co1/3Mn1/3]O_2, and LiFe1/4Mn1/2Co1/4PO_4 cathodes," *ACS Sustain. Chem. Eng.,* vol. 6, no. 3, pp. 3225–3232, Mar. 2018.

16 N. Mishra, G. Das, A. Ansaldo, A. Genovese, M. Malerba, M. Povia, D. Ricci, E. Di Fabrizio, E. Di Zitti, M. Sharon, and M. Sharon, "Pyrolysis of waste polypropylene for the synthesis of carbon nanotubes," *J. Anal. Appl. Pyrolysis,* vol. 94, pp. 91–98, Mar. 2012.

17 C. W. Foster, M. P. Down, Y. Zhang, X. Ji, S. J. Rowley-Neale, G. C. Smith, P. J. Kelly, and C. E Banks, "3D printed graphene based energy storage devices," *Sci. Reports* 2017 71, vol. 7, no. 1, pp. 1–11, Mar. 2017.

18 A. Maurel, S. Grugeon, M. Armand, B. Fleutot, M. Courty, K. Prashantha, C. Davoisne, H. Tortajada, S. Panier, and L. Dupont, "Overview on lithium-ion battery 3D-printing by means of material extrusion," *ECS Trans.,* vol. 98, no. 13, pp. 3–21, Sep. 2020.

19 W. Gao and M. Pumera, "3D printed nanocarbon frameworks for Li-ion battery cathodes," *Adv. Funct. Mater.,* vol. 31, no. 11, p. 2007285, Mar. 2021.

20 C. Reyes, R. Somogyi, S. Niu, M. A. Cruz, F. Yang, M. J. Catenacci, C. P. Rhodes, and B. J. Wiley, "Three-dimensional printing of a complete lithium ion battery with fused filament fabrication," *ACS Appl. Energy Mater.,* vol. 1, no. 10, pp. 5268–5279, Oct. 2018.

21 K. Sun, T. S. Wei, B. Y. Ahn, J. Y. Seo, S. J. Dillon, and J. A. Lewis, "3D printing of interdigitated Li-ion microbattery architectures," *Adv. Mater.,* vol. 25, no. 33, pp. 4539–4543, Sep. 2013.

22 K. Fu, Y. Wang, C. Yan, Y. Yao, Y. Chen, J. Dai, S. Lacey, Y. Wang, J. Wan, T. Li, Z. Wang, Y. Xu, and L. Hu, "Graphene oxide-based electrode inks for 3D-printed lithium-ion batteries," *Adv. Mater.,* vol. 28, no. 13, pp. 2587–2594, Apr. 2016.

23 C. Zhang, K. Shen, B. Li, S. Li, and S. Yang, "Continuously 3D printed quantum dot-based electrodes for lithium storage with ultrahigh capacities," *J. Mater. Chem. A,* vol. 6, no. 41, pp. 19960–19966, Oct. 2018.

24 K. Shen, H. Mei, B. Li, J. Ding, and S. Yang, "3D printing sulfur copolymer-graphene architectures for Li-S batteries," *Adv. Energy Mater.,* vol. 8, no. 4, p. 1701527, Feb. 2018.

25 J. H. Kim, S. Lee, M. Wajahat, H. Jeong, W. S. Chang, H. J. Jeong, J. Yang, J. T. Kim, and S. K. Seol, "Three-dimensional printing of highly conductive carbon nanotube microarchitectures with fluid ink," *ACS Nano,* vol. 10, no. 9, pp. 8879–8887, Sep. 2016.

26 C. Milroy and A. Manthiram, "Printed microelectrodes for scalable, high-areal-capacity lithium-sulfur batteries," *Chem. Commun.,* vol. 52, no. 23, pp. 4282–4285, Mar. 2016.
27 C. W. Hull, "The birth of 3D printing," *Res. Manag.*, vol. 58, no. 6, pp. 25–30, 2015.
28 D. Wang, X. Huang, J. Li, B. He, Q. Liu, L. Hu, and G. Jiang, "3D printing of graphene-doped target for 'matrix-free' laser desorption/ionization mass spectrometry," *Chem. Commun.,* vol. 54, no. 22, pp. 2723–2726, Mar. 2018.
29 D. Lin, S. Jin, F. Zhang, C. Wang, Y. Wang, C. Zhou, and G. J. Cheng, "3D stereolithography printing of graphene oxide reinforced complex architectures," *Nanotechnology,* vol. 26, no. 43, p. 434003, Oct. 2015.
30 J. Z. Manapat, J. D. Mangadlao, B. D. B. Tiu, G. C. Tritchler, and R. C. Advincula, "High-strength stereolithographic 3D printed nanocomposites: Graphene oxide metastability," *ACS Appl. Mater. Interfaces,* vol. 9, no. 11, pp. 10085–10093, Mar. 2017.
31 D. R. Dreyer, S. Park, C. W. Bielawski, and R. S. Ruoff, "The chemistry of graphene oxide," *Chem. Soc. Rev.,* vol. 39, no. 1, pp. 228–240, Dec. 2009.
32 J. H. Sandoval, K. F. Soto, L. E. Murr, and R. B. Wicker, "Nanotailoring photocrosslinkable epoxy resins with multi-walled carbon nanotubes for stereolithography layered manufacturing," *J. Mater. Sci.,* vol. 42, no. 1, pp. 156–165, Jan. 2007.
33 Y. Lim, J. Il Heo, M. Madou, and H. Shin, "Monolithic carbon structures including suspended single nanowires and nanomeshes as a sensor platform," *Nanoscale Res. Lett.,* vol. 8, no. 1, pp. 1–9, Nov. 2013.
34 B. Hsia, M. S. Kim, M. Vincent, C. Carraro, and R. Maboudian, "Photoresist-derived porous carbon for on-chip micro-supercapacitors," *Carbon N. Y.,* vol. 57, pp. 395–400, Jun. 2013.
35 B. Rezaei, J. Y. Pan, C. Gundlach, and S. S. Keller, "Highly structured 3D pyrolytic carbon electrodes derived from additive manufacturing technology," *Mater. Des.,* vol. 193, p. 108834, Aug. 2020.
36 J. Kim, X. Song, K. Kinoshita, M. Madou, and R. White, "Electrochemical studies of carbon films from pyrolyzed photoresist," *J. Electrochem. Soc.,* vol. 145, no. 7, pp. 2314–2319, Jul. 1998.
37 K. Narita, M. A. Citrin, H. Yang, X. Xia, and J. R. Greer, "3D architected carbon electrodes for energy storage," *Adv. Energy Mater.,* vol. 11, no. 5, p. 2002637, Feb. 2021.
38 A. Vyatskikh, S. Delalande, A. Kudo, X. Zhang, C. M. Portela, and J. R. Greer, "Additive manufacturing of 3D nano-architected metals," *Nat. Commun.*, vol. 9, p. 593, 2018.
39 G. Seniutinas, A. Weber, C. Padeste, I. Sakellari, M. Farsari, and C. David, "Beyond 100 nm resolution in 3D laser lithography—Post processing solutions," *Microelectron. Eng.,* vol. 191, pp. 25–31, May 2018.
40 J. Bauer, A. Schroer, R. Schwaiger, and O. Kraft, "Approaching theoretical strength in glassy carbon nanolattices," *Nat. Mater.*, vol. 15, pp. 438–443, 2016.
41 A. Zakhurdaeva, P. Dietrich, H. Hölscher, C. Koos, J. Korvink, and S. Sharma, "Custom-designed glassy carbon tips for atomic force microscopy," *Micromachines,* vol. 8, no. 9, p. 285, Sep. 2017.
42 C. Yang, Q. Cao, P. Puthongkham, S. T. Lee, M. Ganesana, N. V. Lavrik, and B. J. Venton, "3D-printed carbon electrodes for neurotransmitter detection," *Angew. Chemie – Int. Ed.,* vol. 57, no. 43, pp. 14255–14259, Oct. 2018.
43 J. B. Tyler, G. L. Smith, A. C. Leff, P. M. Wilson, J. Cumings, and N. Lazarus, "Understanding the electrical behavior of pyrolyzed three-dimensional-printed microdevices," *Adv. Eng. Mater.,* vol. 23, no. 1, p. 2001027, Jan. 2021.
44 J. B. Tyler, G. L. Smith, J. Cumings, and N. Lazarus, "3D printing metals at the microscale: Electroplating pyrolyzed carbon mems," *IEEE Symp. Mass Storage Syst. Technol.*, vol. 2022-January, pp. 531–534, 2022.
45 L. E. Asp, K. Bouton, D. Carlstedt, S. Duan, R. Harnden, W. Johannisson, M. Johansen, M. K. G. Johansson, G. Lindbergh, F. Liu, K. Peuvot, L. M. Schneider, J. Xu, D. Zenkert, "A structural battery and its multifunctional performance," *Adv. Energy Sustain. Res.,* vol. 2, no. 3, p. 2000093, Mar. 2021.
46 C. M. Portela, B. W. Edwards, D. Veysset, Y. Sun, K. A. Nelson, D. M. Kochmann, J. R. Greer, "Supersonic impact resilience of nanoarchitected carbon," *Nat. Mater.,* vol. 20, no. 11, pp. 1491–1497, Jun. 2021.
47 D. W. Yee, M. A. Citrin, Z. W. Taylor, M. A. Saccone, V. L. Tovmasyan, and J. R. Greer, "Hydrogel-based additive manufacturing of lithium cobalt oxide," *Adv. Mater. Technol.,* vol. 6, no. 2, p. 2000791, Feb. 2021.

48 S. Kim, C. Velez, R. S. Pierre, G. L. Smith, and S. Bergbreiter, "A two-step fabrication method for 3D printed microactuators: Characterization and actuated mechanisms," *J. Microelectromechanical Syst.,* vol. 29, no. 4, pp. 544–552, Aug. 2020.
49 K. Miyamoto, T. Sasaki, T. Nishi, Y. Itou, and K. Takechi, "3D-microbattery architectural design optimization using automatic geometry generator and transmission-line model," *iScience,* vol. 23, no. 7, p. 101317, Jul. 2020.
50 E. MacDonald and R. Wicker, "Multiprocess 3D printing for increasing component functionality," *Science.* vol. 353, no. 6307, Sep. 2016.

10 3D-Printed Graphene-Based Electrodes for Batteries

Yaroslav Kobzar[1], *Igor Tkachenko*[2], *and Kateryna Fatyeyeva*[1*]
[1]Univ Rouen Normandie, INSA Rouen Normandie, Polymerès Biopolymères Surfaces (PBS) UMR 6270 CNRS, 76000 Rouen, France
[2]Institute of Macromolecular Chemistry, National Academy of Sciences of Ukraine, Department of Chemistry of Oligomers and Netted Polymers, 02160, Kyiv, Ukraine
*Corresponding author: kateryna.fatyeyeva@univ-rouen.fr

CONTENTS

10.1 INTRODUCTION

Today, industrialization progress requires not only significant energy resources but also the development of energy storage technologies. Electrical energy storage (EES) systems are an important element in the development of sustainable energy technologies [1]–[6]. They are designed to store electrical energy from an external source in special storage devices for some time. Energy storage devices have a similar process mechanism but differ in many other aspects, such as the lifespan, number, and duration of charge/discharge cycles, architecture, type of stored energy (electrochemical or electrical), operating voltage, etc., and, therefore, they have different applications.

Based on specific elaboration technology and performance characteristics energy storage devices can be divided into batteries and supercapacitors. One of the main differences between a battery and a supercapacitor is that the battery has a much higher energy density. Consequently, the battery is appropriate for higher energy density applications, e.g. an application where the system must be able to work for long periods on a single charge. Fabrication of electrolytes and electrodes as well as their assembly play an important role in obtaining high-performance batteries. The existing fabrication technology has some significant limitations, which directly affect the battery performance. These limitations are connected with the form and architecture of electrodes and solid-state electrolytes [7].

Three-dimensional (3D) printing that allows us to create a physical object from a digital design plays a special role in the field of battery fabrication due to free-form construction and controllable structure [5], [8]–[14]. A layer-by-layer deposition technology is of great interest and its application is constantly growing year after year. Nowadays, 3D printing can be successfully used for battery printing (mainly for electrode printing) or individual battery tools. For example, 3D-printed resin templates were used for the fabrication of aligned microchannels in the graphene oxide/Li electrode manufacturing process [15].

DOI: 10.1201/9781003296676-10

Graphene-based materials attract special attention in the production of batteries due to their unique conductivity and mechanical flexibility [16]–[19]. There are two most common techniques of 3D printing for preparing graphene-based batteries: fused deposition modeling (FDM) and direct ink writing (DIW). The FDM technique is based on layer-by-layer depositing of the melted filament material over a platform. The filament is a polymer composite that consists of a polymer matrix (mainly poly(lactic acid) (PLA)) and various additives (graphene, Si, etc.) [20]–[22]. The filament can be prepared by the joint thermal extrusion of all components. Note that the extruded components can be mixed either before the extrusion process or directly in the extruder during the extrusion [20]–[22]. The advantages of FDM printing are its simplicity and speed of the process and also the fact that there are a lot of 3D printers adapted for such a technique (i.e., there no need to modify existing devices). However, difficulties in the preparation of a printable filament by extrusion, the impossibility to obtain a filament with a high mass fraction of filler, very low electrical conductivity of polymer matrix, heat treatment during extrusion and printing, the impurities presence, and the absence of a porous structure or rather low porosity of the prepared electrodes limit the FDM industrial application in battery filed [20]–[23].

The DIW technique is another powerful technique of 3D printing [24]–[27]. This technique is similar to FDM; however, in this case, the 3D layer-by-layer structure is obtained through the deposition of colloid-based ink. The DIW technique allows us to adapt effectively and purposefully the multiscaled graphene structure – from nano-level (nanometer graphene oxide (GO)) to macro-level (GO macroscopic monoliths) [28]. Highly concentrated GO water dispersions form suitable inks for 3D printing. They are viscous gels at rest with shear-thinning under extrusion [29]–[31]. The composition modification of GO suspensions by adding polymers and other additives (i.e., binders, viscosifiers, etc.) allows us to control and improve the ink viscoelasticity and physical properties of final electrodes [31], [32]. Moreover, the 3D-printed graphene-based elements could strengthen battery performance and facilitate their application in energy storage due to the high porosity. It should be noted that the DIW printing technique greatly simplifies and speeds up the process of obtaining such 3D graphene porous architecture [33]. Among the disadvantages of DIW printing, there are difficulties in the preparation of stable and printable ink, a lack of 3D printers adapted for such a technique (i.e., some modifications are required), and a required post-treatment procedure for printed materials. Generally, such a procedure includes freeze-drying (at -40°C or -50°C for ~24h) and thermal annealing (at 600–800°C in Ar or Ar/H_2 mixed atmosphere for 1–3h) or chemical reduction (in presence of hydrazine hydrate) to form microporous structure and to obtain reduced graphene oxide (rGO), respectively. However, other additional post-treatment steps are also possible to enhance the efficiency of printed electrodes (e.g., deposition of Na or Ni on electrode surfaces, etc.) [28], [34], [35].

In the present review, the trends of applying the 3D-printing technique for electrode elaboration for battery application are discussed. The review mainly focuses on 3D-printing techniques that are most widely used for graphene-based electrode fabrication. The performances of fabricated electrodes are discussed in relationship with the used 3D-printing method, electrode composition, structure, and morphology.

10.2 FDM 3D-PRINTED ELECTRODES

The first example of FDM-printed electrodes is anode electrodes for a Li-ion battery based on 3D-printed solid filaments [20]. Such filaments consist of graphite-PLA composites designed to be used in conventional FDM 3D printers (Table 10.1, electrode 1, and Figure 10.1).

3D technology was used for the elaboration of Li-ion anode type electrode based on PLA/graphene filament (Table 10.1, electrode 2) [21]. The proposed filament was fabricated in several steps: (i) the dispersion of graphene in xylene with subsequent PLA solubilization and recrystallization in methanol to obtain the PLA/graphene powder; (ii) the extrusion of as-prepared powder to obtain the filament (1.75 mm diameter).

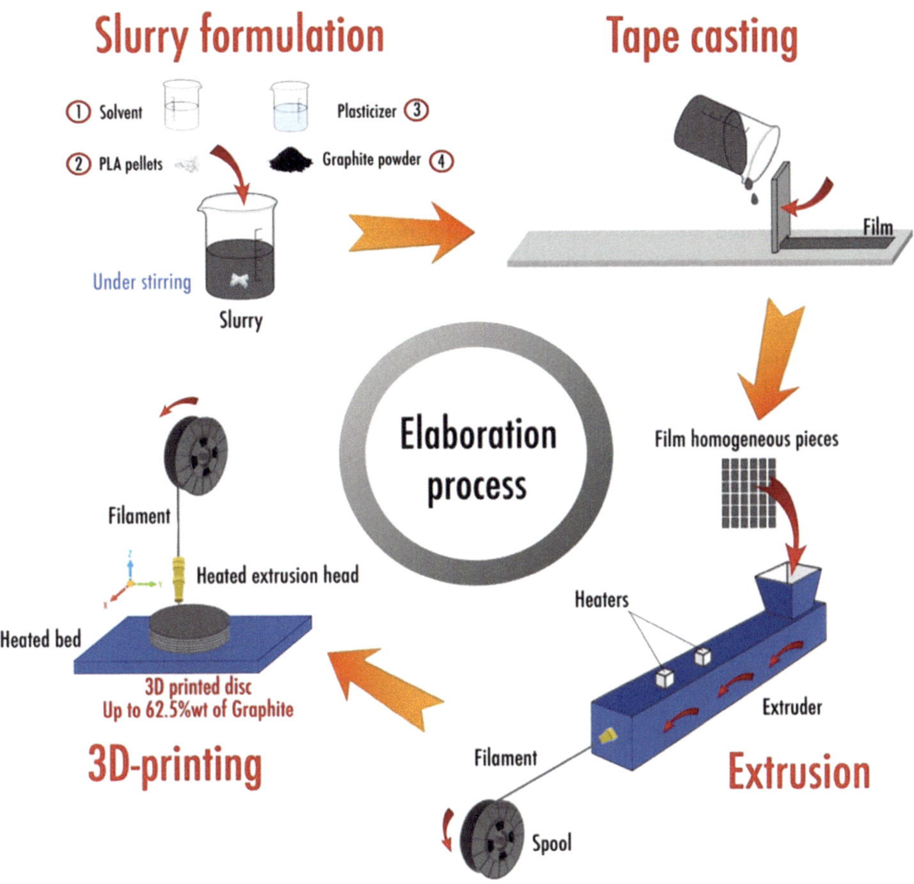

FIGURE 10.1 Elaboration process of graphite-PLA composite and 3D-printed negative electrode disc. Adapted with permission [20]. Copyright (2022) American Chemical Society. The elaboration process includes several steps. The first one is the slurry preparation, the second is tape casting with film fabrication, the third one is the extrusion of the as-prepared film to transform it into filament and the fourth is electrode 3D printing.

It is found that the optimal graphene concentration is 15–20 wt.%. At the same time, the filament with 20 wt.% of graphene showed the highest conductivity (resistivity of ~100 Ohm/cm) and electrochemical activity due to the best graphene dispersion in PLA. The graphene concentration above 20 wt.% leads to brittle and fragile material, while the material obtained with graphene concentration below 10 wt.% does not offer sufficient percolation, as the resistivity value is higher than 100 kOhm/cm. Therefore, the PLA/graphene filament with 20 wt.% of graphene was used for printing the test anode with the geometry of a CR2016 coin cell (i.e., 17.75 mm and 1 mm in diameter and thickness, respectively).

The Li as both counter and reference electrode, polypropylene film separator, and 1 M solutions of $LiPF_6$ in ethylene carbonate (EC)/dimethyl carbonate (DEC) (1/1) was used for battery tests. It is found that such a battery shows a low specific capacity. The initial battery capacity at the current density of 40 mA/g is ~20 mA·h/g. The battery shows substantial capacity loss over 200 scans. The low electrochemical behavior can be explained by the low porosity of the printed electrode according to topography data. However, a pre-treatment of such electrodes by NaOH is found to significantly improve the material's electrochemical behavior. Such treatment induces porosity increase and, consequently, leads to an increase in the specific capacity (~500 mA·h/g at a current density of 40 mA/g) in a 200-fold [21].

Another strategy for the anode preparation is based on using wet-jet milling-exfoliated few-layers graphene (WJM-FLG) (Table 10.1, electrode 3) [22]. The PLA/carbon black doped polypyrrole (PPy) blend with the inclusion of Si nanoparticles was prepared for such a goal. The general procedure of the filament fabrication and 3D-printing process is shown in Figure 10.2.

The filament with 45 wt.% of PLA (i.e., weight ratio of PLA/PPy/Si/WJM-FLG is 45/20/29.5/5.5) is still printable and the electrode-specific capacity equals 334 mA·h/g at the current density of 20 mA/g due to high content of Si nanoparticles. In addition, the uniform distribution of Si nanoparticles inside the matrix ensures high stability of cyclic performance for all 3D-printed electrodes. The specific capacity measured at different values of the current density of 20, 30, 40, and 50 mA/g reveals the decrease from 334 (at 20 mA/g) to 308, 288, and 275 mA·h/g at 30, 40, and 50 mA/g, respectively, for material with 45 wt.% of PLA. Nevertheless, such materials show 95% of their initial specific capacity at the current density of 20 mA/g and 82.3% of their initial specific capacity at a current density of 50 mA/g. Moreover, the electrodes with 45 wt.% of PLA show capacity retention of 95% after 350 cycles confirming their high-performance stability (specific capacity is 327 mA·h/g) [22].

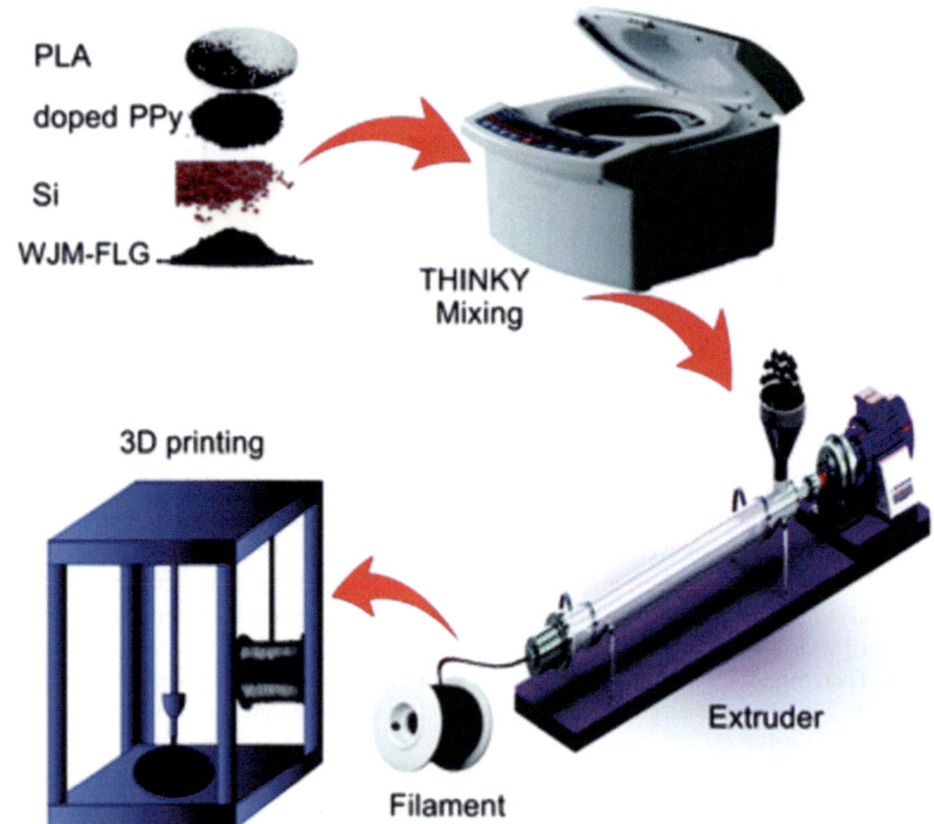

FIGURE 10.2 The procedure of the filament fabrication and 3D printing process [22]. This article is licensed under Creative Commons Attribution 3.0. The elaboration process of PLA/PPy/Si/WJM-FLG 3D-printed electrode includes several steps. The first one is the mixing of components by using a planetary mixer, the second is the extrusion of the as-prepared mixture to prepare filament and the fourth is electrode 3D printing.

TABLE 10.1
Batteries Composition and Performances

#	Composition	Conductivity (S/cm)	Cell Components			Cell Performance		Ref.
			Type of Battery	Electrodes (Anode @ Cathode)	Electrolyte	Initial Discharge Capacity	Cycling Performance	
1	PLA/graphite/poly(ethylene glycol) dimethyl ether (Mn~500)/carbon black Timcal Super-P	~0.56	Li-ion	PLA/graphite/poly(ethylene glycol) dimethyl ether (Mn~500)/carbon black Timcal Super-P @ Li	1M $LiPF_6$ in EC/DEC (1:1 weight ratio)	85 mA·h/g at 18.6 mA/g	-	[20]
2	PLA/graphene	-	Li-ion	PLA/graphene @ Li	1M $LiPF_6$ in EC/DEC (1:1 weight ratio)	~20 mA·h/g at 40 mA/g (before NaOH treatment) ~590 mA·h/g at 40 mA/g (after NaOH treatment)	-	[21]
3	PLA/PPy/Si/WJM-FLG	5.19	Li-ion	PLA/PPy/Si/WJM-FLG @ Li	1M $LiPF_6$ in EC/DEC (1:1 volume ratio)	327 mA·h/g at 20 mA/g	Capacity retention: 96% at 20 mA/g after 350 cycles	[22]

Note: EC – ethylene carbonate; DEC – diethyl carbonate.

10.3 DIW 3D-PRINTING TECHNIQUE

The DIW technique was applied to fabricate the porous 'O_2 breathable' cathode for Na-O_2 batteries (Table 10.2, electrode 4) [36]. The aqueous-based GO ink (with a GO concentration of ~15.0 mg/mL) without any binders was used for printing such an electrode. The standard treatment procedure, namely freeze-drying (for 24h) with subsequent thermal treating (at 800 °C in an H_2/Ar atmosphere for 2h), was applied to manufacture the rGO-based electrode. The fabricated electrode with mesh structure possesses an architecture that provides continuous conduction of Na^+ ions, O_2, and electrons due to interconnected macro- and micropores. The micropores (with diameters between 2 and <50 nm) formed by rGO sheets provide a dual function: electrolyte reservoir and for NaO_2 accommodation, while the macropores (with diameter >50 nm) provide access for O_2 across the whole electrode.

Three electrode types with different Brunauer–Emmett–Teller (BET) surface area values, namely electrodes with no open pores (NOP) (BET=342.8 m^2/g), with small open pores (SOP) (BET=362.6 m^2/g) and large open pores (LOP) (BET=369.6 m^2/g), were printed by varying the space between the filaments. Swagelok-type cells were used to evaluate the electrochemical performance of electrodes. The tested cell was assembled by using a metallic Na anode with carbon paper as a protector. The material that was obtained by impregnation of glass fiber with a solution of 1 M $NaSO_3CF_3$ in diethylene glycol/dimethyl ether was used as a separator and an electrolyte. It is found that the electrode porosity has a strong effect on the performance of such batteries, namely discharge capacity. The highest discharge capacity value was found for SOP electrodes, while the lowest value was noted for NOP electrodes (Table 10.2, electrode 4). Moreover, the SOP electrode has an areal capacity value of 9.1 mA·h/cm^2 at 0.2 A/g, whereas NOP and LOP electrodes showed areal capacity values equal to 5.3 and 5.8 mA·h/cm^2, respectively. In addition, the battery based on SOP showed high performance and a remarkable capacity value equal to 13484.6 mA·h/g (9.1 mA·h/cm^2) at 0.2 A/g as well as stable productivity in cycling mode (final capacity of 500 mA·h/g at 0.5 A/g after~ 120 cycles) [36].

Another type of GO-based porous cathode was printed by using highly concentrated (≈100 mg/mL) aqueous-based ink which was fabricated based on a porous 2D nanomaterial, namely holey graphene oxide (hGO) (Table 10.2, electrode 5) [31]. The freeze-drying was applied to the material after printing to form a microporous structure of 3D-printed meshes. A feature of such 3D hGO-based material is the presence of different porosity: small (4–25 nm), medium (~10 μm), and large (<500 μm). Such porosity provides interfacial reactions to ensure full utilization of the active sites which in turn is advantageous for the performance of energy systems, especially for energy storage devices. The printed hGO frame is a three-layered mesh structure (0.6 cm × 0.6 cm) obtained by layer-by-layer deposition of six zig zag constructions (0.8 mm spacing between lines). Traditionally 3D hGO mesh was converted into reduced hGO (rhGO) by thermal treatment (in an inert atmosphere for 2h at 1000 °C). The rhGO mesh was used as Li-O_2 cathode in CR2032 cells with a Li as a counter and reference electrode. The deep discharge areal capacity tested at 0.1 mA/cm^2 was 13.3 mA·h/cm^2 (~4.8 mA·h in total capacity or ~3879 mA·h/g).

The same coin cell with rhGO vacuum-filtrated film cathode (with identical areal dimensions) instead of 3D-printed rhGO-based mesh showed discharge areal capacity only around 0.21 mA·h/cm^2 (<0.03 mA·h in total capacity or ~92 mA·h/g). The difference in discharge areal capacity can be connected with reduced porosity of rhGO vacuum-filtrated film cathode as compared with rhGO 3D-printed meshes (owing to reduced interfacial reactions as well as to low availability of active sites). The porous architecture of the 3D-printed electrodes confirms its advantages in cycle mode performance tests under charge/discharge at 1 mA·h/cm^2 and shows enhancing Li-O_2 device productivity. Thus, the rhGO mesh supports ~13 cycles at 0.1 mA/cm^2 until the voltage of 4.5 V is reached, while even one full cycle under the same testing conditions cannot be reached in the case of rhGO vacuum-filtrated film electrode [31]. In addition, the importance of 3D-printed electrode porosity was demonstrated by testing electrodes from various GO-based substances, such as GO-based on natural graphite and Vor-X, which are generally used to create holey graphene. Traditionally, after printing, GO-based structures were thermally treated before testing as Li-O_2 battery cathodes. The lowest value of capacity under full discharge conditions was found for rGO from natural graphite mesh – 3.1 mA·h/cm^2, ~600 mA·h/g, or 0.77 mA·h in total capacity. At the same time, rGO based on Vor-X graphene frame reveals a discharge value of 10.5 mA·h/cm^2 (2.63 mA·h in total capacity or ~2471 mA·h/g). Note that the full areal capacity value of the 3D meshes is underestimated, as the inherent macroporosity is not extracted from the dimension of the electrode before the test [31].

Micro-lattice aerogels were fabricated as Na-ion host material for the Na-ion batteries of the future with reduced local current density and dendritic formation (Table 10.2, electrodes 6 and 7) [35]. Two types of electrodes were prepared, i.e., rGO/carbon nanotube (CNT) (as the anode) (Table 10.2, electrode 6) and rGO/$Na_3V_2(PO_4)_3$(NVP)/C (as the cathode) (Table 10.2, electrode 7). Both electrodes were printed using GO aqueous-based ink. The ink for the anode was fabricated by mixing GO (200 mg) and CNT (500 mg) in 5 mL of deionized water, while the ink for the cathode was prepared by mixing GO (80 mg) and NVP/C (160 mg) in 2 mL of deionized water. The printed GO frameworks (1×1 cm) were treated by freeze-drying (at -40°C) with a subsequent thermally annealed step (at 600°C under the Ar/H_2 (95/5%) mixed gas). The BET value of rGO/CNT is found to be 416.6 m^2/g. Unfortunately, the BET of rGO/NVP/C has not been determined yet. Finally, electrochemical deposition of Na onto printed rGO/CNT is performed to fabricate Na/rGO/CNT micro-lattice anode aerogel. The battery based on prepared electrodes and fiberglass used as a separator is tested. The charge/discharge capacity value of the cathode is equal to 98.8/91.1 mA·h/g, while the charge/discharge capacity value of the full battery is around 101.3/93.1 mA·h/g (at the seventh cycle) at a density of the current equal 100 mA/g. A capacity value of 67.6 mA·h/g was found in the long-term cycling performance test at 100 mA/g after 100 cycles confirming battery durability [35].

The GO-based cathode for Li-CO_2 battery with a high density of energy fabricated by 3D printing and thermal shock treatment is also proposed (Table 10.2, electrode 8) [34]. For this purpose, an aqueous-based ink is made by mixing 200 mg of GO in 0.5 mL of deionized water. The standard treating procedure is applied for the post-printed GO framework with some modifications, namely after freeze-drying (at -50 °C) and reduction at high temperature (in Ar atmosphere at 300 °C for 1 h) the obtained rGO framework is dipped into $NiCl_2$ solution for 54 ms at 1900 K (1626.85 °C). In this case, a GO framework with anchored Ni nanoparticles (rGO/Ni) is obtained. The rGO/Ni framework is electrochemically examined as a cathode in CR 2032-type battery with a glass fiber separator. The good distribution of the catalytic Ni nanoparticles, as well as the high thickness of the electrode, resulted in a high areal capacity value of 14.6 mA·h/cm^2 (at a discharge density of an electric current of 0.1 mA/cm^2) and the specific capacity value of 8991.0 mA·h/g, and a low overpotential value of 1.05 V at 100 mA/g. Furthermore, high and stable performance of up to 1000 mA/g was found for the battery at the cycle mode test during 100 cycles. Note that such performance is higher than the performance of Ni cathode for the Li-CO_2 energy storage system reported previously [34], as the maximum rate and overpotential are only 200 mA/g and 1.9 V, respectively.

The elaboration strategy of electrodes with improved performance for Na-ion battery using the GO/NVP inks was proposed (Table 10.2, electrodes 9–11) [37]. The mass ratio of NVP to GO is fixed at 9:1 (GO/NVP_9), 7:3 (GO/$NVP_{7/3}$), and 1:1 (GO/NVP_1), and the concentration values of solid material in ink are 200, 67, and 40 mg/mL, respectively. The GO concentration in the used aqueous-based solution is 20 mg/mL while the loading concentration of NVP is 180, 47, and 20 mg/mL in H_2O for GO/NVP_9, GO/$NVP_{7/3}$, and GO/NVP_1, respectively. The usual hierarchical porous rGO/NVP framework is obtained by freeze-drying and thermal treatment of as-printed 3D structures. The corresponding 3D porous framework is labeled as rGO/$NVP_{6.8}$ (electrode 9), rGO/$NVP_{4.2}$ (electrode 10), and rGO/$NVP_{2.8}$ (electrode 11) (according to thermogravimetric analysis) (Table 10.2).

rGO/NVP$_{6.8}$ electrode showed a higher conductivity value as compared with other rGO/NVP compositions. The electrochemical performance is determined by galvanostatic discharge/charge measurements of 3D-printed rGO/NVP sodium storage frameworks used as cathode from 0.2 to 20C (1C=2.1 mA·h/cm^2) [37]. The capacity value of 1.26 mA·h/cm^2 is achieved at 0.2C for the rGO/NVP$_{6.8}$ electrode while the capacity value of ~0.67 mA·h/cm^2 is found in the case of rGO/NVP$_{4.2}$ and rGO/NVP$_{2.8}$ [37]. rGO/NVP$_{6.8}$ possesses still high capacity at 5C and 20C (0.89 mA·h/cm^2 and 0.65 mA·h/cm^2, respectively) that is higher as compared with the C/NVP composites [37]–[40]. Note that rGO/NVP$_{6.8}$ showed a high value of capacity retention up to 90.1% after 900 cycles at 1C [37]. Furthermore, the electrochemical capacity of the rGO/NVP$_{6.8}$ anode used for Na-ion storage is 0.32 mA·h/cm^2 at 2C. Such performance is much higher than that obtained for the other rGO/NVP frameworks (0.08–0.23 mA·h/cm^2). Moreover, asymmetric full 2032 coin cell based on rGO/NVP$_{6.8}$ framework used as the cathode and anode, and glass fiber film used as the separator reveals a great capacity value of 0.39 mA·h/cm^2 at 1C [37]. Note that the capacity value is higher than that of other graphite-NVP full cells [41].

Fu et al. proposed the strategy of the fabrication of the water-based GO ink to prepare 3D-printed electrodes (both anode and cathode) for Li-ion batteries (Table 10.2, electrodes 12 and 13) [30]. Such a strategy is based on the mixing of GO sheets and different active substrates. Lithium ferrophosphate (LFP) or lithium-titanate (lithium titanium oxide (LTO)) were used in ink for the fabrication of cathode (electrode 13) and anode (electrode 12), respectively. The concentration value of GO in water was equal to 80 mg/mL and the weight ratio of LFP or LTO to GO was 7:3. The solid material concentration was 285 mg/mL. It is shown that such a concentration provides optimum viscoelastic properties as well as sufficient material viscosity and capacity binding of electrode materials, which enables 3D printing of porous electrodes. The electrodes were fabricated by deposition 6, 12, and 18 layers one by one. The layer thickness was equal to ~0.18 mm. Freeze-drying and thermal treatment (at 600 °C in mix atmosphere of argon and hydrogen for 2 h) were used after 3D printing to obtain the final porous rGO electrodes [30].

The poly(vinylidene fluoride-*co*-hexafluoropropylene) (PVDF-*co*-HFP)/Al_2O_3 nanocomposite film and composite polydimethylsiloxane membrane impregned by the solution of $LiPF_6$ dissolved in a mixture of ethylene carbonate/diethyl carbonate (1:1) serve as a separator and an electrolyte, respectively, for full cell fabrication. As-prepared rGO/LFP cathode and rGO/LTO anode showed stable performance in cycle mode test with values of specific capacity ~160 mA·h/g and ~170 mA/g, respectively. The full cell possesses an electrode mass loading ~of 18 mg/cm^2 (normalized to the overall battery area). Such full cell showed the primary charge and discharge capacity of 117 and 91 mA·h/g with great stability at cycle mode tests (over 20 cycles) [30].

3D freeze printing accompanied by freeze casting and ink-jet printing is proposed and successfully applied for the elaboration of the materials for sodium-ion battery anodes (Table 10.2, electrode 14) [42]. Note that such a method is very similar to DIW but freezing and printing are performed simultaneously. This approach allows using the process of nucleation and growth of ice microcrystals during printing as useful tools for controlling the electrode porosity on micro and macro levels. The ink was prepared by mixing GO and ammonium thiomolybdate (as the MoS_2 precursor) in water. The rGO macroporous electrode with MoS_2 nanoparticles was fabricated by using the post-treatment process, namely freeze-drying with subsequent reductive thermal treatment. The resulting anode – rGO/MoS_2 aerogel is used for the 2032 coin cell (Table 10.2, electrode 14). The metallic Na disk was used as a counter electrode (cathode), while the glass fiber disk impregnated by 1 M solution of sodium perchlorate monohydrate in propylene carbonate was used as a separator and electrolyte, respectively. The interconnected network in such aerogels allows for enhancing the electron conductivity and system mechanical stability. In addition, the pores (~3–5 μm) promote fast ion transport. Such electrode aerogel showed a high initial specific capacity over 429 mA·h/g at a C/3.3 rate in the potential range of 2.5–0.10 V (vs Na^+/Na) during the first 10 cycles. It was found that the two parallel processes occurred in the system: (i) reversible connected with 2Na^+ insertions

TABLE 10.2
Batteries Composition and Performances

#	Composition	Conductivity (S/cm)	Cell Components				Cell Performance		Ref.
			Type of Battery	Electrodes (Anode @ Cathode)	Mass Loading (mg/cm²)	Electrolyte	Initial Discharge Capacity	Cycling Performance	
4	rGO	–	Na–O_2	Na @ rGO	–	1 M $NaSO_3CF_3$ in DEGDME	3425.8 mA·h/g (NOP) 11935.3 mA·h/g (SOP) 10900.0 mA·h/g (LOP) at 0.5 A/g	–	[36]
5	rhGO	–	Li–O_2	Li @ rhGO	5.42	1 M LiTFSI in DMSO	–	–	[31]
6	Na/rGO/CNT	$8.0 \cdot 10^{-2}$	Na-ion	Na/rGO/CNT @ rGO/NVP/C	–	1 M $NaPF_6$ in diglyme	–	–	[35]
7	rGO/NVP/C	–			15				
8	rGO/Ni	–	Li-CO_2	Li @ rGO/Ni	–	1 M LiTFSI in TEGDME/DMSO	–	–	[34]
9	rGO/$NVP_{6.8}$	$6.3 \cdot 10^{-3}$	Na-ion	$NVP_{6.8}$/rGO @ $NVP_{6.8}$/rGO	–	1 M $NaClO_4$ in EC/PC (1/1 w/w) with 5 wt.% FEC	–	Capacity retention: 90.1% at 1C after 900 cycles	[37]
10	rGO/$NVP_{4.2}$	$2.3 \cdot 10^{-3}$			–	–	–	–	
11	rGO/$NVP_{2.8}$	$1.2 \cdot 10^{-3}$			–	–	–	–	
12	rGO/LFP	31.6	Li-ion	rGO/LTO @ rGO/LFP	18	$LiPF_6$ in EC/DEC (1:1)/polydim ethylsiloxane film	185 mA·h/g at 10 mA/g (rGO/LTO, half cell) 91 mA·h/g at 50 mA/g (full cell)	–	[30]
13	rGO/LTO	6.1							
14	rGO/MoS_2	–	Na-ion	rGO/MoS_2 @ Na	–	1 M $NaClO_4$ in PC	~111 mA·h/g at 100 mA/g	–	[42]
15	GO/V_8C_7–VO_2/Super P carbon	–	Li–S	GO/V_8C_7–VO_2/Super P carbon @ Li	~0.8	0.5 M LiTFSI/0.5 M $LiNO_3$ in DME/DO (1:1 by volume)	–	–	[43]
16	GO/V_8C_7–VO_2/S/Super P carbon			Li @ GO/V_8C_7–VO_2/S/Super P carbon	~1.5 (sulfur)			Capacity retention: 89.5% at 0.5C after 100 cycles	

(continued)

TABLE 10.2 (Continued)
Batteries Composition and Performances

#	Composition	Conductivity (S/cm)	Cell Components				Cell Performance		Ref.
			Type of Battery	Electrodes (Anode @ Cathode)	Mass Loading (mg/cm^2)	Electrolyte	Initial Discharge Capacity	Cycling Performance	
17	AgNW/rGO/LTO	1740	Li-ion	AgNW/rGO/LTO @ Li	~30	1 M $LiPF_6$ in EC/DEC/DMC (by volume 1:1:1)	–	–	[44]
18	rGO/CNTs/α-Fe_2O_3	–	Ni-Fe	rGO/CNTs@α-Fe_2O_3 @ rGO/CNTs@$Ni(OH)_2$	130	6 M PVA/KOH	–	Capacity retention: ~91.3% at 300 mA/cm^2 after 900 cycles	[45]
19	rGO/CNTs/$Ni(OH)_2$	–							
20	GO/SCP	–	Li–S	Li @ GO/SCP	–	1 M LiTFSI in DME/DO (1:1 by volume) with 1% $LiNO_3$	~490 mA·h/g at 50 mA/g	Capacity retention: 43.4% after 50 cycles	[46]
21	Na/rGO		Na	Na/rGO @ Na	–	1 M $NaPF_6$ in diglyme	–	–	[28]
22	rGO/LFP/PVDF	–	Li-ion	rGO/LTO/PVDF @ rGO/LFP/PVDF	–	1 M $LiPF_6$	–	Capacity retention: 98% after 500 cycles	[47]
23	rGO/LTO/PVDF								

Note: EC – ethylene carbonate; DEC – diethyl carbonate; DMC – dimethyl carbonate; DME – dimethoxyethane; DO – 1,3-dioxolane; PC – propylene carbonate; FEC – fluoroethylene carbonate; TEGDME – tetraethylene glycol dimethyl ether; DMSO – dimethyl sulfoxide; DEGDME – diethylene glycol dimethyl ether.

and (ii) irreversible connected with the conversion of MoS_2 to metallic form. Note that the Na^+ intercalation process is faster and with good stability, whereas the conversion process is slower due to suppression [42].

Attempts to improve the characteristics of Li-S batteries have led to the creation of 3D-printed GO electrodes based on V_8C_7-VO_2 (Table 10.2, electrodes 15 and 16). The V_8C_7-VO_2 was acted as a Li stabilizer and a site for the confinement of polysulfide [43]. The inks for lithium anode (lithium stabilizer) and sulfur cathode (polysulfide immobilizer) were prepared by mixing of one part of V_8C_7-VO_2 or V_8C_7-VO_2/S with one part of Super P carbon and eight parts of GO. The aqueous solution of 25 mg/mL of GO was used for this goal. Before assembling the Li-S battery, a freeze-drying process was applied to the all electrodes. CR2032 coin cell was assembled from V_8C_7-VO_2/S-based cathode, Li anode, and polypropylene monolayer membrane separator (Celgard 2400). The cell for the V_8C_7-VO_2-based anode was fabricated in the same manner. The solution of 0.5 M LiTFSI/$LiNO_3$ in dimethoxyethane/1,3-dioxolane (1:1, v/v) acts as electrolyte. The cathode sulfur loading was ~1.5 mg/cm^2 and the anode mass loading was ~0.8 mg/cm^2. It is shown that the lipophilic V_8C_7-VO_2 plays important role in preventing dendrite formation and stabilizing the lithium anode, whereas V_8C_7-VO_2/S is responsible for polysulfide catching and transforming. The cathode rate capacity (643.5 mA·h/g at 6.0 C) and cycling stability (a capacity decay of 0.061% per cycle at 4.0 C after 900 cycles is noted) are quite correct. Moreover, the battery with a V_8C_7-VO_2-based anode and V_8C_7-VO_2/S-based cathode at a sulfur loading of 9.2 mg/cm^2 shows a high areal capacity equal to 7.36 mA·h/m^2 [43].

Electrode 17 was fabricated to improve battery performance, namely ion transport and decreased internal stress at the charge and discharge (Table 10.2, electrode 17) [44]. A such electrode was printed by using 3D-printable mixed functional aqueous-based ink consisting of Ag nanowires (AgNWs) ((AgNWs)/GO/LTO/H_2O with a mass ratio of 8:1:21:50, respectively). Also, pure GO/LTO/H_2O (3/7/65) ink was prepared for comparison. All inks were fabricated by standard mixing procedure. It was shown that the optimal ink viscosity and ideal shear-thinning for printing were achieved at a GO concentration of 46 mg/mL and ~20 mg/mL for GO/LTO and AgNWs/GO/LTO inks, respectively. The freeze-drying with subsequently treating by hydrazine monohydrate vapor (at 80 °C for 12 h) to obtain rGO was applied for all electrodes. As-prepared inks were labeled as AgNWs/rGO/LTO and rGO/LTO. The ratio between the total surface of an electrode and its volume (specific surface area) of AgNWs/rGO/LTO and rGO/LTO was 336.6 m^2/g and 101 m^2/g. The pore volume values of the mentioned electrodes were 0.14 cm^3/g and 0.04 cm^3/g. The electrode thickness was in the range of 150 to 1500 μm. As it is expected, the AgNWs/rGO/LTO electrode conductivity value is higher as compared with the conductivity of rGO/LTO electrode (~1.4 S/cm) due to the AgNWs presence (Table 10.2, electrode 17). The electrochemical performance of both electrodes was investigated by using a Li-based counter electrode and membrane separator (Celgard 2300) with 1 M electrolyte solution of $LiPF_6$ in ethylene carbonate/diethyl carbonate/dimethyl carbonate with a volume ratio of 1:1:1. The specific capacity value of the rGO/LTO electrode (192.2 mA·h/g) is a little higher as compared with the AgNWs/rGO/LTO electrode (177.5 mA·h/g) at 0.5 C due to a higher GO concentration. However, the specific capacity value of rGO/LTO is significantly lower at 10 C (i.e., 19.4 mA·h/g for rGO/LTO and 121 mA·h/g for AgNWs/rGO/LTO). The rGO/LTO electrode exhibits a significantly lower efficiency than gNWs/rGO/LTO-based electrode. Besides, the productivity of the AgNWs/rGO/LTO electrode as a function of the thickness decreases at 5 C with the thickness increasing from 300 to 900 μm. The AgNWs/rGO/LTO electrode with 900 μm thickness still delivers a specific capacity of 132.8 mA·h/g at 2 C, which is higher as compared with 3D-printed LTO/carbon electrodes [44]. Moreover, the AgNWs/rGO/LTO electrodes show high areal capacity equal to 4.74 mA·h/cm^2 and maintain an areal capacity of 3.55 mA·h/cm^2 (capacity retention of ~95.5%) after 100 cycles at the current density of 0.76 mA/cm^2. Note that such cycling performance as well as areal capacity values are comparable or even higher than the values obtained for Si and metal oxide-based batteries [44].

The performance of Ni-Fe batteries was improved by the fabrication and application of the self-supporting rGO/CNTs aerogel micro lattices [45]. An aqueous solution of GO (100 mL, 20 mg/

mL) was mixed with a sample of CNTs (500, 1000, and 2000 mg) to prepare ink. Such ink was used to prepare different frameworks with subsequent immersion into liquid N_2, freeze-drying (two days), and annealing (at 650 °C for 3 h under an inert atmosphere). Finally, rGO/CNTs hybrid aerogel micro lattices were used for obtaining the rGO/CNTs/Ni$(OH)_2$-based cathode (Table 10.2, electrode 19) and rGO/CNTs/α-Fe_2O_3 anode (Table 10.2, electrode 18) by a simple water bath or hydrothermal growth processing with future annealing, respectively. A quasi-solid-state battery was assembled from fabricated micro-lattices electrodes using 6M composite of poly(vinyl alcohol) (PVA)/KOH as both electrolyte and separator. 3D-printed batteries with thicknesses from 2 mm to 8 mm were fabricated and their performance was evaluated at various current densities – in the range of 10 - 300 mA/cm^2. The batteries keep a stable performance level during 1500 cycles even at mechanical stress. In addition, the battery with a thickness of 8 mm shows an impressive areal capacity value equal to 13.2 mA·h/cm^2 with a specific capacity of 206.4 mA·h/g. Moreover, this battery reveals a high value of volumetric energy density equal to 28.1 mWh/cm^3 at various power densities (10.6 mW/cm^3, 13.9 mW·h/cm^3) and an ultrahigh power density value equal to 318.8 mW/cm^3. Such performance is higher as compared with the already reported batteries [45]. Moreover, super-high energy density and superior long-term cycling durability (~91.3% capacity retention after 10,000 cycles) were found for Ni-Fe batteries with active material loading over 130 mg/cm^2. Interestingly, the capacity retention value of ~85.6% after 15,000 cycles was found for the same battery with a 3M solution of KOH used as an electrolyte [45].

3D-printed sulfur copolymer graphene cathode with well-designed periodic micro-lattices was elaborated for Li-S batteries (Table 10.2, electrode 20) [46]. In this case, the synthesis of copolymer occurs inside the electrode after printing during treatment. The ink for 3D-printing was prepared in a few steps: (i) mixing GO suspension (4 mg/mL of GO in H_2O) and sublimed sulfur (mass ratio of 4:1); (ii) formation of gel-like ink (50 mg/mL of GO) by its concentration; (iii) combination of gel-like ink with 1,3-diisopropenylbenzene. After printing, the electrodes were subjected to several stages of processing: (i) freeze-dried (formation of porous structure) and (ii) formation of sulfur copolymer on the graphene surface (by the thermal treatment at 200 °C) (Table 10.2, electrode 20). The tests of the electrode at 50 mA/g show that the combination of graphene and sulfur copolymer ensures a great reversible capacity of 812.8 mA·h/g and cycle performance. However, the declared value is at the lower limit of the reversible capacity values range previously reported for sulfur copolymer Li-S batteries (800–1000 mA·h/g). Such difference can be explained by a thicker wall (600 μm) and material resistance [46]. Although the electrode shows low-capacity retention (43.4% after 50 cycles) it still supports a high level of reversible capacity (186 mA·h/g) even after increasing the current density from 50 to 800 mA·h/g [46].

3D-printable ink with binders is also successfully used for electrode elaboration by 3D printing. The rGO lattice for the advanced Na metal anode was fabricated by 3D printing (Table 10.2, electrode 21) [28]. The ink with suitable viscoelasticity and rheological properties was prepared by adding 2 vol % of ethylenediamine (with a concentration of ~1%) to 10 mg/mL of GO aqueous suspension. Printed material was treated in several steps: (i) hydrothermal treatment; (ii) dialysis; (iii) freeze-drying under vacuum; and (iv) thermal annealing. The rGO lattice with a BET value equal to 52.8 m^2/g was obtained. The final stage of the anode preparation was the deposition of metallic Na (4 mA·h/cm^2 at 1 mA/cm^2). The Na with a smooth and compact morphology is formed in empty voids of rGO lattice. These results can be linked with the fact that the high current density holes can serve for Na deposition. The reversible and stable Coulomb efficiency of the formation of Na plating/stripping on rGO lattice surfaces was equal to 99.84% after 500 cycles (nearly 1000 h) at 1 mA/cm^2 and measured using Na foil as the counter electrode. It validates the high reversibility of Na plating/stripping behavior obtained for rGO lattice due to its multiscale hierarchical ordered structure. The cycling stability of the prepared anode was evaluated by using a symmetrical CR2032 coin cell at a cycling capacity of 1 mA·h/cm^2 and a current density of 1 or 2 mA/cm^2. The rGO lattice revealed a stable cycling behavior at a current density of 1 mA/cm^2 for a period of ~600 h with a small voltage hysteresis value of ~40 mV. The material shows high stability of life (over 400 h) at 2 mA/cm^2 and a slight increase in voltage hysteresis. The above-mentioned lattice cycling lifetime is comparable with that of previously reported Na metal anodes [28]. Moreover, rGO lattice has a higher specific capacity than that of materials based on a hard carbon support. These results confirm that the printed rGO lattice can serve as an anode in Na batteries [28].

The formation of electrode 22 (Table 10.2) was performed similarly as in the case of electrodes 12 and 13 (Table 10.2). N,N-methylpyrrolidone (NMP), and polyvinylidene fluoride (PVDF) were used as the solvent and the binder, respectively [47]. Thus, 3D printable inks consist of active materials (LTO or LFP), rGO, and PVDF in a weight ratio of 7:2:1 dispersed in NMP. The ink with the concentration of 3 mL/g showed the most printable characteristics, namely a high viscosity state without back flowing. Consequently, such ink is used for the 3D-printing. The electrodes were washed with water followed by freeze-drying after printing. The electrodes were tested in half battery cells with 1 M solution of $LiPF_6$ as an electrolyte. The discharge capacity value was up to 168 mA·h/g at 0.5 C for the battery based on a three-layer LFP-based cathode with a 16.3 mg/cm^2 mass loading (calculated from the actual area) Even at the rate of 3 C the discharge capacity value was still high and equal to 125 mA·h/g [47]. Moreover, the three-layer LFP-based cathode showed stable performance with high-capacity retention equal to 99%. Thus, electrode 23 demonstrates a discharge capacity equal to 116 mA·h/g after 500 charge/discharge cycles at 3 C. The specific discharge capacity of the three-layer LTO-based anode (Table 10.2, electrode 22) was equal to 180 mA·h/g which is higher than the calculated theoretical value (175 mA·h/g) [47]. Besides, the increasing test rates from 0.5 C to 3 C for the anode (Table 10.2, electrode 23) did not show strong capacity changes. The specific capacity of the three-layer anode was equal to 171, 166, 162, and 155 mA·h/g at 1, 1.5, 2, and 3 C, respectively. Moreover, such anode demonstrates long-term cycling stability after 500 charge/discharge cycles and capacity retention of 99.9%. A full-cell coin (CR2032 type) assembled based on two layers of LFP cathode and LTO anode showed a reversible capacity equal to 155, 140, 119, 107, and 91 mA·h/g registered at 0.5, 1, 1.5, 2 and 3 C, respectively. It is interesting that the cell showed high-performance stability and capacity retention of 98% after almost 500 cycles [47].

10.4 CONCLUSION AND FUTURE TRENDS

Graphene-based materials attract special attention for the production of EES devices, especially batteries, due to their unique conductivity and mechanical flexibility. 3D printing is a powerful technique that allows converting a digital design into a physical object. Thus, the combination of both graphene-based materials and 3D printing opens a huge perspective in the fabrication of next-generation batteries. Nowadays, FDM and DIW are the two most commonly used techniques of 3D printing for preparing graphene-based electrodes for batteries. Based on the reported results, one can predict that future development in the field of 3D-printed batteries will be focused on the design of new high-performance electrodes. Most probably, an advancement of DIW printed electrodes will continue. However, new trends and directions can also be foreseen. New 3D-printing technologies (e.g., selective laser sintering) can have a promising prospects. Also, the combination of 3D printing with artificial intelligence can significantly improve the performance of printed electrodes by improving their design and architecture. The reported recent achievements in 3D-printed graphene-based batteries demonstrate that further studies and developments in this field would be very beneficial for industrial application.

REFERENCES

1 S. Zhang, Y.Q. Liu, J.N. Hao, G.G. Wallace, S. Beirne, and J. Chen, 3D-printed wearable electrochemical energy devices. *Advanced Functional Materials,* 2022. **32**(3): p. 2103092.

2 D. Moldovan, J. Choi, Y. Choo, W.S. Kim, and Y. Hwa, Laser-based three-dimensional manufacturing technologies for rechargeable batteries. *Nano Convergence,* 2021. **8**(1): pp. 1–16.

3 V. Krishnadoss, B. Kanjilal, A. Hesketh, C. Miller, A. Mugweru, M. Akbard, A. Khademhosseini, J. Leijten, and I. Noshadi, In situ 3D printing of implantable energy storage devices. *Chemical Engineering Journal,* 2021. **409**: p. 128213.

4 W.L. Gao and M. Pumera, 3D printed nanocarbon frameworks for Li-ion battery cathodes. *Advanced Functional Materials,* 2021. **31**(11): p. 2007285.

5 H. Ma, X.C. Tian, T. Wang, K. Tang, Z.X. Liu, S.E. Hou, H.Y. Jin, and G.Z. Cao, Tailoring pore structures of 3D printed cellular high-loading cathodes for advanced rechargeable zinc-ion batteries. *Small,* 2021. **17**(29): p. 2100746.

6 R.P. Chen, Y.M. Chen, L. Xu, Y. Cheng, X. Zhou, Y.Y. Cai, and L.Q. Mai, 3D-printed interdigital electrodes for electrochemical energy storage devices. *Journal of Materials Research,* 2021. **36**(22): pp. 4489–4507.

7 F. Zhang, M. Wei, V.V. Viswanathan, B. Swart, Y.Y. Shao, G. Wu, and C. Zhou, 3D printing technologies for electrochemical energy storage. *Nano Energy,* 2017. **40**: pp. 418–431.

8 Z. Qi, J.C. Ye, W. Chen, J. Biener, E.B. Duoss, C.M. Spadaccini, M.A. Worsley, and C. Zhu, 3D-printed, superelastic polypyrrole-graphene electrodes with ultrahigh areal capacitance for electrochemical energy storage. *Advanced Materials Technologies,* 2018. **3**(7): p. 1800053.

9 X.C. Tian, J. Jin, S.Q. Yuan, C.K. Chua, S.B. Tor, and K. Zhou, Emerging 3D-printed electrochemical energy storage devices: A critical review. *Advanced Energy Materials,* 2017. **7**(17): p. 1700127.

10 W. Zhang, H.Z. Liu, X.A. Zhang, X.J. Li, G.H. Zhang, and P. Cao, 3D printed micro-electrochemical energy storage devices: From design to integration. *Advanced Functional Materials,* 2021. **31**(40): p. 2104909.

11 C. Zhu, T.Y. Liu, F. Qian, W. Chen, S. Chandrasekaran, B. Yao, Y. Song, E.B. Duoss, J.D. Kuntz, C.M. Spadaccini, M.A. Worsley, and Y. Li, 3D printed functional nanomaterials for electrochemical energy storage. *Nano Today*, 2017. **15**: pp. 107–120.

12 P. Chang, H. Mei, S.X. Zhou, K.G. Dassios, and L.F. Cheng, 3D printed electrochemical energy storage devices. *Journal of Materials Chemistry A*, 2019. 7(9): pp. 4230–4258.

13 T.T. Gao, Z. Zhou, J.Y. Yu, J. Zhao, G.L. Wang, D.X. Cao, B. Ding, and Y.J. Li, 3D printing of tunable energy storage devices with both high areal and volumetric energy densities. *Advanced Energy Materials,* 2019. **9**(8): p. 1802578.

14 M.P. Browne, E. Redondo, and M. Pumera, 3D printing for electrochemical energy applications. *Chemical Reviews,* 2020. **120**(5): pp. 2783–2810.

15 S.Y. Ni, J.Z. Sheng, C. Zhang, X. Wu, C. Yang, S.F. Pei, R.H. Gao, W. Liu, L. Qiu, and G.M. Zhou, Dendrite-free lithium deposition and stripping regulated by aligned microchannels for stable lithium metal batteries. *Advanced Functional Materials,* 2022. **32**(21): p. 2200682.

16 B.B. Guo, G.J. Liang, S.X. Yu, Y. Wang, C.Y. Zhi, and J.M. Bai, 3D printing of reduced graphene oxide aerogels for energy storage devices: A paradigm from materials and technologies to applications. *Energy Storage Materials,* 2021. **39**: pp. 146–165.

17 N. Wang, S.J. Yan, S.K. Peng, X. Chen, and S.L. Dai, Research progress on 3D printed graphene materials synthesis technology and its application in energy storage field. *Cailiao Gongcheng – Journal of Materials Engineering*, 2017. **45**(12): pp. 112–125.

18 C.W. Foster, M.P. Down, Y. Zhang, X.B. Ji, S.J. Rowley-Neale, G.C. Smith, P.J. Kelly, and C.E. Banks, 3D printed graphene based energy storage devices. *Scientific Reports*, 2017. **7**: pp. 1–11.

19 V.G. Rocha, E. Garcia-Tunon, C. Botas, F. Markoulidis, E. Feilden, E. D'Elia, N. Ni, M. Shaffer, and E. Saiz, Multimaterial 3D printing of graphene-based electrodes for electrochemical energy storage using thermoresponsive inks. *ACS Applied Materials & Interfaces,* 2017. **9**(42): pp. 37136–37145.

20 A. Maurel, M. Courty, B. Fleutot, H. Tortajada, K. Prashantha, M. Armand, S. Grugeon, S. Panier, and L. Dupont, Highly loaded graphite-polylactic acid composite-based filaments for lithium-ion battery three-dimensional printing. *Chemistry of Materials*, 2018. **30**(21): pp. 7484–7493.

21 C.W. Foster, G.Q. Zou, Y.L. Jiang, M.P. Down, C.M. Liauw, A.G.M. Ferrari, X.B. Ji, G.C. Smith, P.J. Kelly, and C.E. Banks, Next-generation additive manufacturing: Tailorable graphene/polylactic(acid) filaments allow the fabrication of 3D printable porous anodes for utilisation within lithium-ion batteries. *Batteries & Supercaps*, 2019. **2**(5): pp. 448–453.

22 H. Beydaghi, S. Abouali, S.B. Thorat, A.E.D. Castillo, S. Bellani, S. Lauciello, S. Gentiluomo, V. Pellegrini, and F. Bonaccorso, 3D printed silicon-few layer graphene anode for advanced Li-ion batteries. *RSC Advances*, 2021. **11**(56): pp. 35051–35060.

23 K. Ghosh, S. Ng, C. Iffelsberger, and M. Pumera, Inherent impurities in graphene/polylactic acid filament strongly influence on the capacitive performance of 3D-printed electrode. *Chemistry – A European Journal,* 2020. **26**(67): pp. 15746–15753.

24 J.X. Zhao, H.Y. Lu, X.X. Zhao, O.I. Malyi, J.H. Peng, C.H. Lu, X.F. Li, Y.Y. Zhang, Z.Y. Zeng, G.C. Xing, and Y.X. Tang, Printable ink design towards customizable miniaturized energy storage devices. *ACS Materials Letters,* 2020. **2**(9): pp. 1041–1056.

25 M. Wei, F. Zhang, W. Wang, P. Alexandridis, C. Zhou, and G. Wu, 3D direct writing fabrication of electrodes for electrochemical storage devices. *Journal of Power Sources*, 2017. **354**: pp. 134–147.

26 Y.B. Zhang, G. Shi, J.D. Qin, S.E. Lowe, S.Q. Zhang, H.J. Zhao, and Y.L. Zhong, Recent progress of direct ink writing of electronic components for advanced wearable devices. *ACS Applied Electronic Materials*, 2019. **1**(9): pp. 1718–1734.

27 J.X. Zhao, Y. Zhang, X.X. Zhao, R.T. Wang, J.X. Xie, C.F. Yang, J.J. Wang, Q.C. Zhang, L.L. Li, C.H. Lu, and Y.G. Yao, Direct ink writing of adjustable electrochemical energy storage device with high gravimetric energy densities. *Advanced Functional Materials,* 2019. **29**(26): pp. 1900809.

28 Y.K. Yu, Z.Y. Wang, Z. Hou, W.R. Ta, W.H. Wang, X.X. Zhao, Q. Li, Y.S. Zhao, Q.Q. Zhang, and Z.W. Quan, 3D printing of hierarchical graphene lattice for advanced Na metal anodes. *ACS Applied Energy Materials*, 2019. **2**(5): pp. 3869–3877.

29 S. Naficy, R. Jalili, S.H. Aboutalebi, R.A. Gorkin III, K. Konstantinov, P.C. Innis, G.M. Spinks, P. Poulin, and G.G. Wallace, Graphene oxide dispersions: tuning rheology to enable fabrication. *Materials Horizons,* 2014. **1**(3): pp. 326–331.

30 K. Fu, Y.B. Wang, C.Y. Yan, Y.G. Yao, Y.A. Chen, J.Q. Dai, S. Lacey, Y.B. Wang, J.Y. Wan, T. Li, Z.Y. Wang, Y. Xu, and L.B. Hu, Graphene oxide-based electrode inks for 3D-printed lithium-ion batteries. *Advanced Materials,* 2016. **28**(13): pp. 2587–2594.

31 S.D. Lacey, D.J. Kirsch, Y.J. Li, J.T. Morgenstern, B.C. Zarket, Y.G. Yao, J.Q. Dai, L.Q. Garcia, B.Y. Liu, T.T. Gao, S.M. Xu, S.R. Raghavan, J.W. Connell, Y. Lin, and L.B. Hu, Extrusion-based 3D printing of hierarchically porous advanced battery electrodes. *Advanced Materials,* 2018. **30**(12): p. 1705651.

32 C. Zhu, T. Han, E.B. Duoss, A.M. Golobic, J.D. Kuntz, C.M. Spadaccini, and M.A. Worsley, Highly compressible 3D periodic graphene aerogel microlattices. *Nature Communications,* 2015. **6**(1): pp. 1–8.

33 X. Wu, F. Mu, and Z. Lin, Three-dimensional printing of graphene-based materials and the application in energy storage. *Materials Today Advances,* 2021. **11**: p. 100157.

34 Y. Qiao, Y. Liu, C.J. Chen, H. Xie, Y.G. Yao, S.M. He, W.W. Ping, B.Y. Liu, and L.B. Hu, 3D-printed graphene oxide framework with thermal shock synthesized nanoparticles for Li-CO_2 Batteries. *Advanced Functional Materials,* 2018. **28**(51): p. 1805899.

35 J. Yan, G. Zhi, D.Z. Kong, H. Wang, T.T. Xu, J.H. Zang, W.X. Shen, J.M. Xu, Y.M. Shi, S.G. Dai, X.J. Li, and Y. Wang, 3D printed rGO/CNT microlattice aerogel for a dendrite-free sodium metal anode. *Journal of Materials Chemistry A,* 2020. **8**(38): pp. 19843–19854.

36 X.T. Lin, J.W. Wang, X.J. Gao, S.Z. Wang, Q. Sun, J. Luo, C.T. Zhao, Y. Zhao, X.F. Yang, C.H. Wang, R.Y. Li, and X.L. Sun, 3D printing of free-standing "O-2 breathable" air electrodes for high-capacity and long-life Na-O-2 batteries. *Chemistry of Materials*, 2020. **32**(7): pp. 3018–3027.

37 J.W. Ding, K. Shen, Z.G. Du, B. Li, and S.B. Yang, 3D-printed hierarchical porous frameworks for sodium storage. *ACS Applied Materials & Interfaces,* 2017. **9**(48): pp. 41871–41877.

38 Z. Jian, L. Zhao, H. Pan, Y.- S. Hu, H. Li, W. Chen, and L. Chen, Carbon coated Na3V2 (PO4) 3 as novel electrode material for sodium ion batteries. *Electrochemistry Communications,* 2012. **14**(1): pp. 86–89.

39 J. Liu, K. Tang, K. Song, P.A. van Aken, Y. Yu, and J. Maier, Electrospun $Na_3V_2(PO_4)3$/C nanofibers as stable cathode materials for sodium-ion batteries. *Nanoscale,* 2014. **6**(10): pp. 5081–5086.

40 Z. Jian, W. Han, X. Lu, H. Yang, Y.S. Hu, J. Zhou, Z. Zhou, J. Li, W. Chen, and D. Chen, Superior electrochemical performance and storage mechanism of $Na_3V_2(PO_4)3$ cathode for room-temperature sodium-ion batteries. *Advanced Energy Materials,* 2013. **3**(2): pp. 156–160.

41 S. Li, Y. Dong, L. Xu, X. Xu, L. He, and L. Mai, Effect of carbon matrix dimensions on the electrochemical properties of Na_3V_2 (PO_4) 3 nanograins for high-performance symmetric sodium-ion batteries. *Advanced Materials,* 2014. **26**(21): pp. 3545–3553.

42 E. Brown, P.L. Yan, H. Tekik, A. Elangovan, J. Wang, D. Lin, and J. Li, 3D printing of hybrid MoS2-graphene aerogels as highly porous electrode materials for sodium ion battery anodes. *Materials & Design*, 2019. **170**: p. 107689.

43 J.S. Cai, J. Jin, Z.D. Fan, C. Li, Z.X. Shi, J.Y. Sun, and Z.F. Liu, 3D printing of a V_8C_7-VO_2 bifunctional scaffold as an effective polysulfide immobilizer and lithium stabilizer for Li-S batteries. *Advanced Materials,* 2020. **32**(50): p. 2005967.

44 C. Sun, S.R. Liu, X.L. Shi, C. Lai, J.J. Liang, and Y.S. Chen, 3D printing nanocomposite gel-based thick electrode enabling both high areal capacity and rate performance for lithium-ion battery. *Chemical Engineering Journal*, 2020. **381**: p. 122641.

45 D.Z. Kong, Y. Wang, S.Z. Huang, B. Zhang, Y.V. Lim, G.J. Sim, P.V.Y. Alvarado, Q. Ge, and H.Y. Yang, 3D printed compressible quasi-solid-state nickel-iron battery. *ACS Nano*, 2020. **14**(8): pp. 9675–9686.

46 K. Shen, H.L. Mei, B. Li, J.W. Ding, and S.B. Yang, 3D printing sulfur copolymer-graphene architectures for Li-S batteries. *Advanced Energy Materials,* 2018. **8**(4): p. 1701527.

47 X.C. Tian, T. Wang, H. Ma, K. Tang, S.E. Hou, H.Y. Jin, and G.Z. Cao, A universal strategy towards 3D printable nanomaterial inks for superior cellular high-loading battery electrodes. *Journal of Materials Chemistry A,* 2021. **9**(29): pp. 16086–16092.

11 3D-Printed Metal Oxides for Batteries

Anupam Patel and Rajendra K. Singh**
Ionic Liquid and Solid-State Ionics Lab, Department of Physics, Institute of Science, Uttar Pradesh, India
*Corresponding author: anupam3184@gmail.com, rajendrasingh.bhu@gmail.com

CONTENTS

11.1 INTRODUCTION

At present, the energy and climate crises are affecting the entire world. The rate at which humans are depleting the world's supply of fossil fuels, such as gas, oil, and coal, and creating greenhouse gas emissions is frightening. The availability of clean, sustainable energy is an important pressing scientific challenge in front of people in the twenty-first century as it is predicted that global energy consumption will rise. Due to this, researchers have concentrated on developing renewable energy sources, such as wind, solar, and hydroelectricity, to replace fossil fuels, which have caused significant environmental concerns. Currently, due to energy demands and discourse on growing

DOI: 10.1201/9781003296676-11

environmental concerns, batteries are expected to play a significant role in the future usage of electrical energy in technologies such as autonomous sensors, digital phones, laptops, healthcare portable devices, smart cards, electrical vehicles, etc. [1], [2]. In the meantime, conventional batteries are made-up by using anodes, cathodes, and separator membranes with fixed shapes and sizes and then packing these components into the pouch or cylindrical shapes after liquid-electrolyte injection. Particularly, with liquid electrolytes, cell construction raises major issues because they need precise packaging materials to prevent leakage issues as well as separator membranes to prevent electrical contact between electrodes [3]. Therefore, the batteries produced using standard battery materials and assembly techniques have constrained form factors and mechanical flexibility, making it extremely challenging to integrate them into intricately structured electronic devices. To overcome the above-mentioned challenges with design diversity and flexibility, new battery systems such as printed power sources have lately become a new battery solution. Simple, affordable, and scalable printing techniques are used to create printed batteries. Printed batteries with various advantages like thinner, lighter, mechanically flexible, small size, compact, customizable, integration, and assembly of large-area devices can be easily made using many low-cost printing techniques [4]. At present, due to its easy processing, reproducibility, and diversity, printing skill is extensively employed in the range of application domains (particularly printed electronics), where rationally prepared inks (containing pastes) are printed in pre-made forms. The development of battery shapes and sequence form, the synthesis of electrolytes, electrodes, and separator membranes by using appropriate printing methods and the tunable rheological properties of inks are currently being pursued as the directions for future research on printed batteries [3], [5].

Recently, a large amount of research focused on three-dimensional (3D) printing technology, which provided an excellent opportunity to produce electrodes for energy storage devices like supercapacitors and batteries. The term '3D printing' denotes an additive manufacturing process that builds 3D structure layer by layer while being guided by computer-aided design (CAD) software. In recent years, a variety of 3D-printing processes and innovative frameworks have emerged. 3D-printing techniques are divided by the American Society for testing and materials (ASTM) into the following seven categories [6]–[9].

1. Material extrusion (e.g., direct ink writing (DIW) and fused deposition modeling (FDM))
2. Material jetting (e.g., inkjet printing)
3. Binder jetting (BJ)
4. Powder bed fusion (PBF) (e.g., selective laser sintering and selective laser melting)
5. Directed energy deposition (DED)
6. Vat photopolymerization (stereolithography (SLA))
7. Sheet lamination

Charles Hull invented 3D printing in 1986 and used an ultraviolet (UV) laser to cause resin materials to polymerize. The name of this 3D printing method is stereolithography (SLA). Other 3D-printing techniques have advanced quickly since the invention of SLA. The schematic representation of seven 3D-printing techniques is shown in Figure 11.1.

The Li-ion, Li-S, Li-O_2, Zn-air, Na-ion, etc., and other types of batteries have different structures and compositions of electrode materials (negative or positive). The positive electrode (cathode) materials printed by 3D techniques for batteries such as Li-ion, metal-air, Na-ion, etc., has been described for even whole cells [6]. In this chapter, we have focused on the 3D-printable cathode and anode materials for Li-ion batteries. Mostly, the negative anode of Li-ion batteries consists of active material, a polymer binder, and a conductive additive, that can experience reversible lithiation and delithiation during the charging and discharging process. Lithium titanate (e.g., $Li_4Ti_5O_{12}$, LTO), graphite, and graphene are the most utilized anode-active materials in 3D-printed Li-ion batteries [7], [10], [11]. Also, in this chapter, we have discussed the various 3D-printing methods in detail for printed 3D-electrode materials as well as applications of 3D-printing batteries.

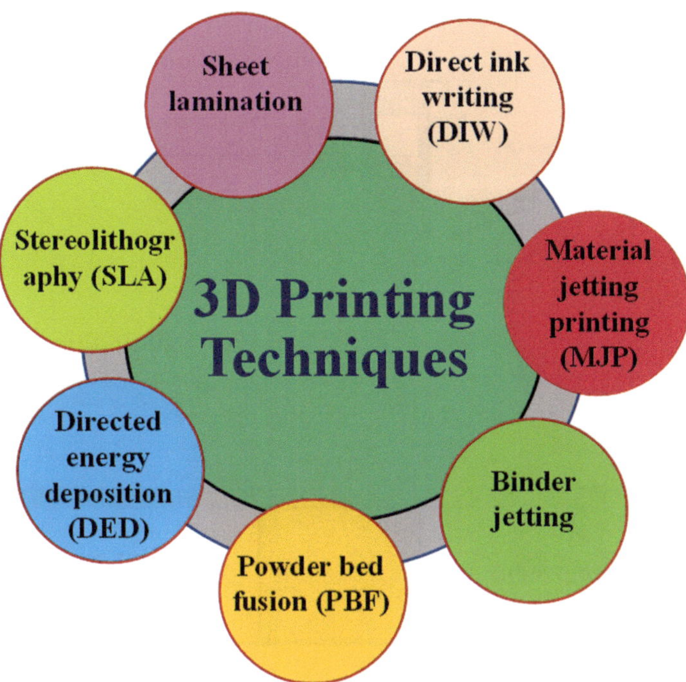

FIGURE 11.1 Schematic of the major 3D printing technologies.

11.2 3D-PRINTED TECHNIQUES

11.2.1 Material Extrusion (Direct Ink Writing (DIW))

The DIW technique is the most popular 3D-printing technique for making batteries, which has the advantages of being inexpensive, simple to use, versatile in terms of materials, and maskless. The nozzle diameter of this technique is based on the ejection of ink ingredients at room temperature and determines its clarity. In this method, a 3D layer-by-layer structure is created by depositing liquid-phase ink along a digitally determined route from a tiny nozzle at a regulated flow rate. With the help of a nozzle, the heated material is deposited layer by layer on the platform. The substrate and nozzle can be moved vertically up and down and horizontally after each new layer has been applied [12], [13]. The schematic illustration of DIW method is shown in Figure 11.2a.

In recent work, the Lewis group used the DIW technology to create printed Li-ion batteries having thick electrodes. The packaging and battery parts including the cathode, anode, and separator, were all printed using the DIW technique. SnO_2 quantum dot inks were used by Zhang et al. to print 3D electrodes using DIW. The production of LIBs micro-batteries by direct writing of anode $Li_4Ti_5O_{12}$ (LTO), cathode LFP, and ink was demonstrated by Wei et al. [10]. The yield stress and storage modulus for DIW's gel-based viscoelastic inks must be sufficiently high. Furthermore, there is a desperate need for a solution to the weak mechanical strength between the layers. Therefore, significant effort must be made to enhance the DIW technique's use in battery manufacture.

11.2.2 Material Jetting

Material jetting printing (MJP) is the common droplet-based deposition technique that uses nozzles to directly deposit ingredients onto platforms like paper, plastic, or other materials to produce high-resolution complex patterns and customizable thickness according to the number of droplets settled. The materials are released on the constructed platform through the horizontally moving nozzle.

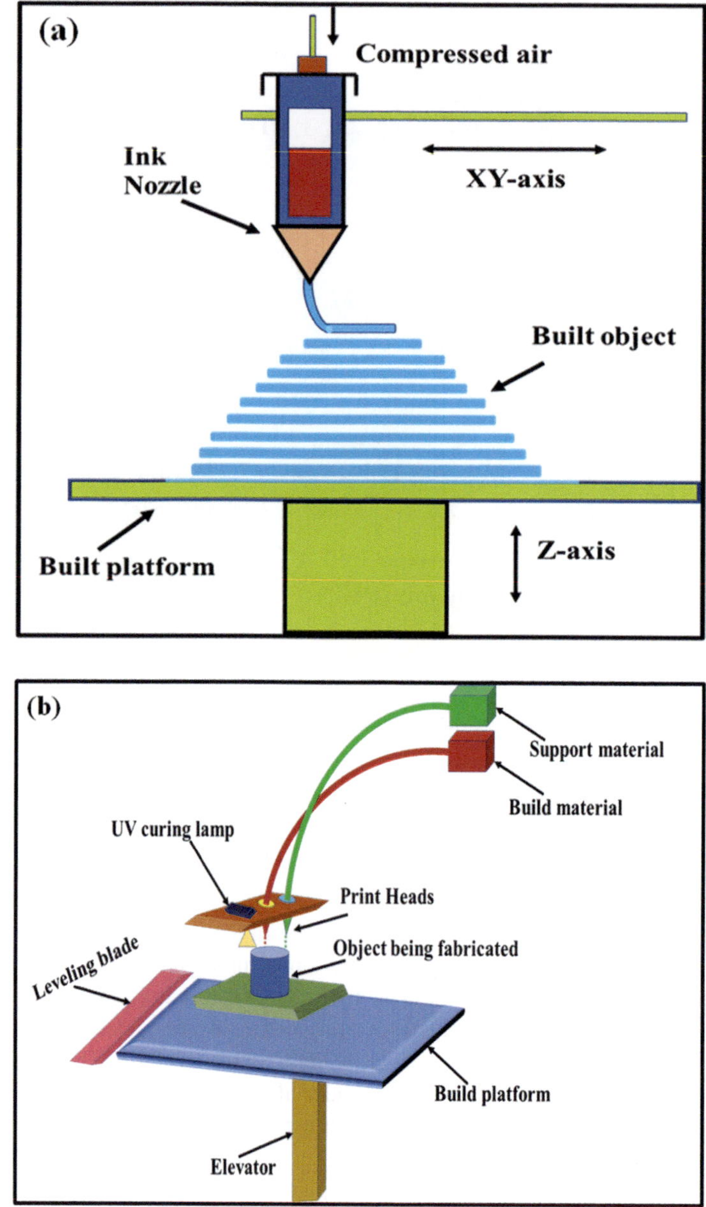

FIGURE 11.2 Schematic representation of (a) direct ink writing and (b) material jetting 3D printing techniques.

The material layer is cured by using ultraviolet (UV) light. Due to their viscosity and propensity to form drops, polymers and waxes are excellent materials for the MJP technique. The schematic illustration of the MJP method is shown in Figure 11.2b. The MJP technique has been used to make energy storage devices (batteries and supercapacitors). For example, Lawes et al. developed thin-film silicon anodes for Li-ion batteries using a desktop inkjet printer method [14]. In addition, Hu et al. developed the 3D inkjet printing of Li-ion [15]. MJP can print different designed patterns with excellent resolution, high material consumption, and good multi-material capability, which helps enhance battery performance. The slow printing speed and demanding ink formulation needs are the major drawbacks of the MJP.

11.2.3 Binder Jetting

Binder jetting (BJ) technique is used for the fabrication of complex structures using a powder-based and a binder. Generally, liquid from glue acts as a binder between the layers of dust. In this technique, layers of binding and building material are alternately deposited by moving the print head horizontally along the *x*- and *y*-axes of the machine. A leveling roller is used during production to evenly disperse a micron layer of powder across the bed. Following the creation of the bed powder, the printer head's nozzle jets the binder onto the powder, enabling the joining of the powder to the subsequent layers. Generally speaking, there are two basic subgroups of the BJ: both ink-jet and aerosol printing. The schematic illustration of the binder jetting method is shown in Figure 11.3a.

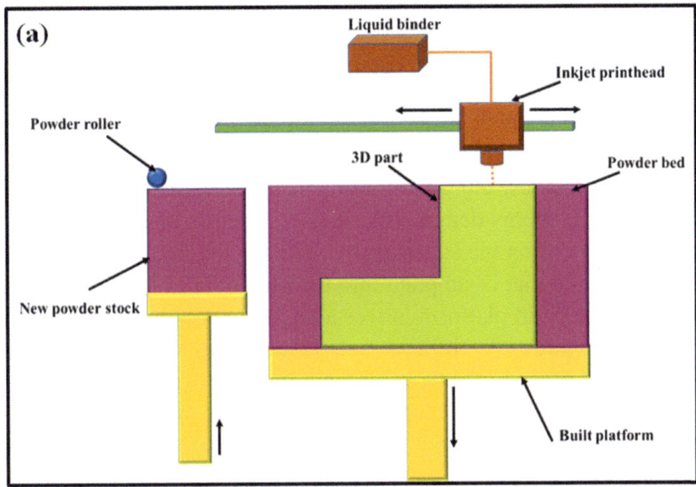

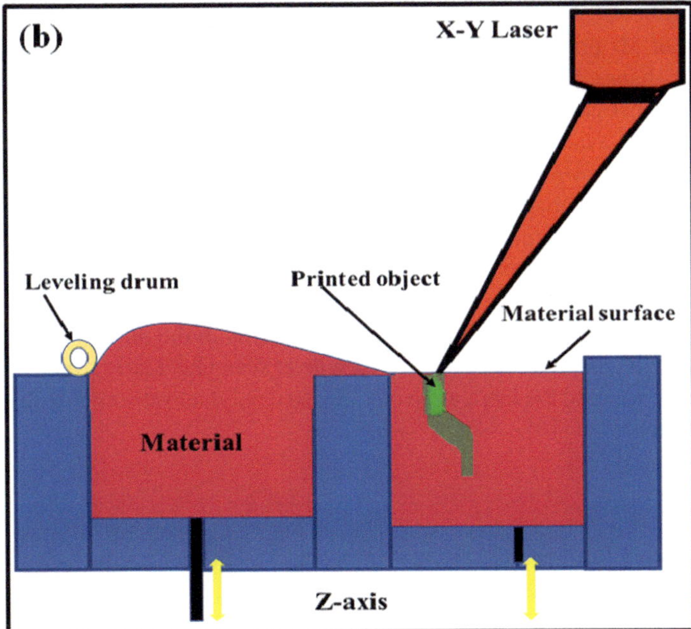

FIGURE 11.3 Schematic illustration of (a) binder jetting and (b) powder bed fusion 3D printing techniques.

11.2.4 Powder Bed Fusion (PBF)

Some parts of the powder bed fusion (PBF) technique have similar to the BJ printing technique. BJ and PBF both have a powder bed and a power supply, but PBF uses light sources (lasers or electron beams) to melt and fuse the powder in particular places, enabling the construction of complicated structures. Along the *x*- and *y*-axes, a mirror directs the light source towards the bed powder, while the build platform travels along the *z*-axis. The newly added powder is then dispersed over the previously created layer using a leveling roller. With the help of the light source, the fresh power layer melts and is simultaneously fused with the earlier layer to create a three-dimensional structure. Selective laser sintering (SLS), Direct metal laser sintering (DMLS), selective laser melting (SLM), electron beam melting (EBM), selective heat sintering (SHS), and are several subtypes of PBF. The schematic illustration of the powder bed fusion (or SLA) method is shown in Figure 11.3b.

11.2.5 Directed Energy Deposition (DED)

The DED technique is used for the deposition of high-performance materials, including ceramics, titanium-based alloys, composites, cobalt-based alloys, high-entropy alloys, nickel-based alloys, and functionally graded materials, aluminum alloys, and steels with these compositions. In this technique, heat sources (high energy density) like a laser, electron beam, or plasma/electric arc) are used to make a small melt pool on the substrate by focusing on the substrate and at the same time melting the feedstock material that is supplied into the melt pool in the form of wire or powder. In addition, in this process, the build platform moves along the *x*-, *y*-, and *z*-axes while the print head is stationary. When adjacent paths of solidified material are created by the print head, they combine to form printed layers. This process can be commonly used for high-performance alloy formation. Laser net shape engineering and directed light fabrication are the two subcategories of DED. The schematic illustration of the directed energy deposition method is shown in Figure 11.4a.

11.2.6 Vat Photopolymerization Stereolithography (SLA)

SLA is a method of 3D printing that selectively solidifies photocurable resin using light to form items with various shapes. SLA uses liquid monomers to photopolymerize with UV assistance. To cure the liquid monomer in specific locations as determined by the tool paths, a UV laser is scanned over the monomer layer. One coating of resin is finished, and then another is applied over the dried layer. The procedure is known as re-coating. During the printing process, the build platform moves along the *z*-axis while the light source is directed towards the vat along the *x*- and *y*-axes, permitting the 3D layer-by-layer creation of functional components. The schematic illustration of the SLA method is shown in Figure 11.4b. Nowadays, photopolymers (introduced in the late 1960s) are used in many different industries and are mainly used in the microelectronics sector. Charles (Chuck) Hull developed the first additive manufacturing SLA process in the mid-1980s. Hull created solid items out of liquid photopolymer resins by layer-by-layer scanning a laser over a photopolymer vat. In 2018, Cohen et al. reported the fabrication of perforated polymer substrates with high surface area by using the SLA technique [16].

11.2.7 Sheet Lamination

Laminated object manufacturing (LOM) and ultrasonic additive manufacturing (UAM) are two methods for sheet lamination. In the sheet lamination process, the light source is directed towards the sheet along with the *x*- and *y*-axes with the help of a mirror. In addition, the build platform moves along the *z*-axis to receive the stack of sheets with the cut area. Metal plates or metal strips

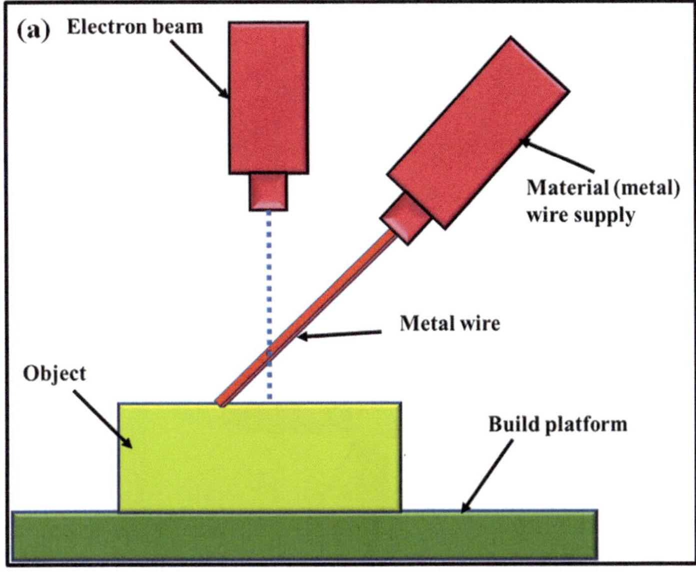

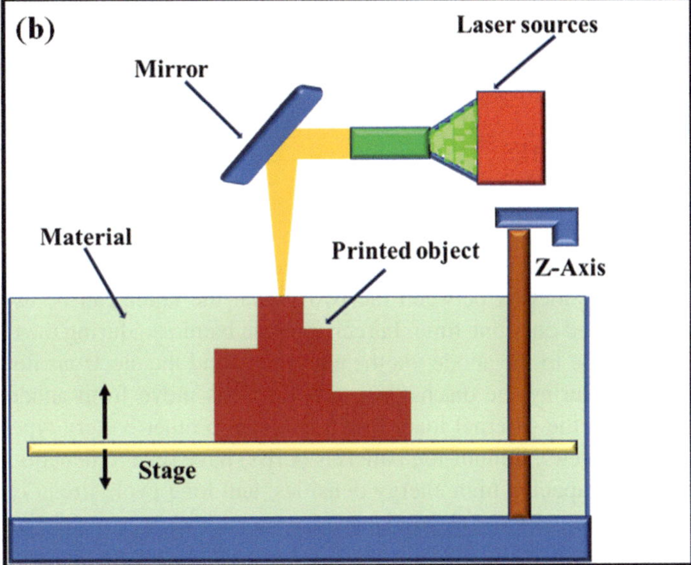

FIGURE 11.4 Schematic illustration of (a) directed energy deposition and (b) vat photopolymerization stereolithography (SLA) 3D printing techniques.

are welded together using ultrasound during the manufacturing of ultrasonic additives. The welding procedure is frequently accompanied by further CNC machining that eliminates loose metals. LOM uses a similar method of building up layers, except they use glue and paper as their medium instead of welding. Cross-shadowing is used by the LOM process during printing to make extraction following extraction straightforward. Aluminum, titanium, stainless, copper, steel, and other metals are used by UAM. Internal geometries can be created by the cryogenic process. Because the metal does not melt, this technique can combine various elements and uses comparatively little energy. The schematic illustration of the sheet lamination method is shown in Figure 11.5.

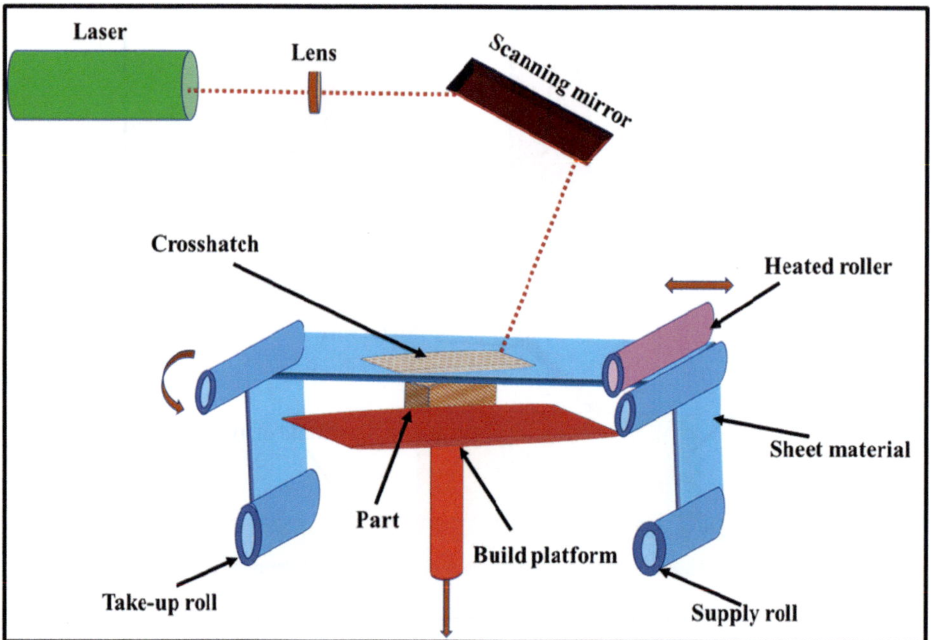

FIGURE 11.5 Schematic representation of sheet lamination 3D printing technique.

11.3 PRINTING BATTERIES

Today batteries (created by Alessandro Volta, 1800) are the primary electrochemical power source that revolutionized the idea of road mobility (electric and hybrid vehicles) as well as the consumer electronics market. Batteries are typically divided into primary and secondary (rechargeable, several uses); the distinction between the two is that the latter can be used several times while the former can be used only one time. In rechargeable batteries, during the charging process, ions move from the cathode to the anode via the electrolyte and the electrons flow through to the external circuit. Further, during the discharging process, ions move from anode to cathode and the electrons flow through the external load. In comparison to other battery types like Zn, NiCd, Ni-metal hydride (NiMH), etc., lithium-ion batteries (LIBs) have several benefits like lightweight, compact, high discharge capacity, high energy densities, and long cycle life [17], [18]. The most popular battery types in printed batteries are Li and Zn based, and only the electrodes are printed. The separator is typically not printed because it is made of a porous membrane that contains an electrolyte solution (salts dissolved into a polar solvent). By uptake (soaking) or injection, the electrolyte solution is delivered into the porous membrane. Cathodes, anodes, electrolytes, and separators are the components of batteries, whose design and manufacture are crucial to improving the electrochemical performance of batteries such as power and volume energy density, cycling life, and safety. It is challenging to find an ink that satisfies the functional requirements of every component in a 3D-printed battery. The majority of current battery research is concentrated on electrode and electrolyte materials. Batteries have a much higher degree of complexity because the electroactive materials they use are reactive and because anodes and cathodes have physically intricate structures. It is exceedingly challenging to create versions of these materials for 3D printing and even once produced, they must maintain their electrical connections, control chemical reactions among the parts, and permit the battery to be charged and discharged again [4], [19]. As a result, we will evaluate existing printable battery components in the next section from an architectural standpoint.

11.4 ELECTRODE MATERIALS FOR 3D-PRINTED BATTERIES

Several cathode materials like lithium cobalt oxides ($LiCoO_2$), lithium manganese oxides ($LiMn_2O_4$), lithium iron phosphate ($LiFePO_4$), lithium transition-metal phosphates ($LiMn_{1-x}Fe_xPO_4$), and $LiNi_{0.8}$-$Co_{0.15}Al_{0.05}O_2$ (NCA) have been successfully printed for lithium-ion batteries [15], [20]–[22]. Among them, the $LiFePO_4$ is the most developed cathode, which has been printed using a variety of methods, including DIW, FDM, and IJP. Na-ion batteries, a less expensive alternative to LIBs, have also their printed cathodes made, for example, $NaMnO_2$ with a cylinder shape by FDM and $Na_3V_2(PO_4)_3$ with a mesh structure via DIW [23]. In addition, varieties of cathode materials are printed for other batteries system for example sulfur and carbon composite for Li-S batteries, reduced graphene oxide (RGO) with carbon nanotube for Zn-air batteries, RGO with a hierarchical porous structure for Na-O_2 batteries, etc. [24]. Some electrodes for 3D printing batteries are discussed with their properties in the next subsection.

11.4.1 Carbon Materials-Based Electrodes

Superior ink production capabilities, distinctive viscoelastic characteristics, and functional qualities appropriate for 3D printing have been demonstrated by graphene oxide (GO). Reduced graphene oxide, or RGO, is a promising material for battery electrodes because it has strong electrical conductivity and is easily converted from graphene oxide (GO) by a thermal annealing process. Most 3D-printed structures, including aerogel micro lattices, nanowires, periodic scaffolds, and complicated networks, have been produced using graphene oxide [25]. A printed highly conductive CNT microarchitecture was described by Kim et al. using a dispenser printing process [26]. Milroy and Manthiram created an MWCNT-based microelectrode for a Li-S battery [27]. High conductivity and extremely high porosity of the MWCNT electrodes are advantages for ionic transport and electronic as well as electrolyte transit within a battery.

11.4.2 Cellulose Nanofiber-Based Electrodes

Because there are many hydroxyl groups on the cellulose molecules, cellulose nanofibers (CNF) are highly soluble in water. Since CNF has a negative zeta potential of up to 60 mV, it can be used as a surfactant to improve the uniform distribution of additional materials in an aqueous solution [28]. The carbonized CNFs combined with Li metal and LFP after an oxidation treatment procedure can be used to print 3D interdigitated cathode and anode, respectively. All of the ink containing CNF had a high viscosity and stuck to the bottom of the vial when it was turned upside down. In 3D-printed lithium-ion batteries, Kohlmeyer et al. used CNF as a porous scaffold, charge collector, and conductive additive [21].

11.4.3 $Li_4Ti_5O_{12}/LiFePO_4$ Based Electrodes

During the charging and discharging process, the cathode (positive electrode) and anode (negative electrode) act in a somewhat similar sense, except for the direction of lithium-ion. LCO, LMO, LFP, and LMFP cathode materials are used in the manufacturing of 3D printing for Li-ion batteries. Among them, LFP is the most extensively studied for 3D printing lithium-ion batteries as it possesses high stability, less volumetric expansion, batter rate capability, and excellent processability properties. Several LTO/LFP-based electrodes have been reported for 3D printing batteries using different 3D-printing techniques. For example, Delannoy et al. studied the ink-jet printed LFP-based composite porous electrodes formulated with carbon super p (C_{SP}) and poly-acrylic-co-maleic acid (PAMA) for micro battery application [29]. This printed electrode has an active mass loading of 0.35 mg/cm^2 and is made up of 85 weight percent $LiFePO_4$, 10 weight percent CSP, and 5 wt.% PAMA after the water solvent has been evaporated. The electrochemical performance of printed

TABLE 11.1
Comparison of Metal or Metal Oxide 3D-printed Electrodes for Li-Ion Batteries

Active Materials	Main Additive	3D Techniques (mAh/g)	Specific Capacity	Ref. No.
LCO	Graphite, carbon black, PVDF–HFP Laser direct	Laser direct writing (LDW)	120 at 100 μA/cm^2	[30]
LTO	Acetylene black, Poly, PVDF, NMP	DIW	160 at 0.2 C	[31]
LTO	Acetylene black, CMC, 1,4 dioxane	LTDW	155 at 0.2 C 115 at 2 C	[32]
LFP	Carbon black, SCMC, carbon	IJP	151.3 at 15 mA/g	[33]
LFP	C65, NaCMC, T X-100	IJP	140 at 1C	[34]
LTO/LFP	HPC, HEC	DIW	131/160 at 1 C	[20]
LFP/LTO/LCO	CNF, graphite, carbon black, PVDF-HFP, NMP	DIW	156/150/137 at 0.2 C 106/89/80 at 5	[21]
LMO	Carbon black, PVDF, NMP	DIW	117 at 0.1 C	[35]
LFP/LTO	CNT, CMC	DIW	156.3 at 0.2 C 140 at 1 C	[36]
SnO_2	Acetylene black, carboxymethyl cellulose	IJP	812.7 at 33 μA/cm^2	[37]

LFP electrodes has been tested by the Swagelok cell using lithium foil as a negative electrode electrolyte (LP30 organic solvent or Li-TFSI in PYR13-TFSI ionic liquid). Delannoy found 80 mAh/g to 70 mAh/g capacity at corresponding cycling rates of 9 C and 90 C. Additionally, after 100 cycles, a low polarization of just 100 mV is obtained at 90 C with no loss capacity at 9 C. As result, printed LFP electrodes show good performance. In addition, Sun et al. demonstrated the 3D interdigitated micro battery architectures (3D-IMA) with a high areal energy density of 9.7 J/cm^2 at a power density of 2.7 mW/cm^2 using LTO/LFP materials [20]. The comparison of several 3D-printed metal oxide-based batteries is summarized in Table 11.1. Overall, the metal and metal oxide 3D printing electrode with high surface area, high porosity and microporous structure ensure the enhancement of 3D-printed batteries' performance.

11.4.4 Anode Materials for 3D Printing Batteries

Various types of anode materials such as LTO, carbon (graphite and graphene), silicon (Si) $SnO_{2,}$ quantum dots, etc. have been printed successfully with higher specific capacities. Among them, LTO is the most common printed anode, which can be demonstrated by DIW and FDM techniques [22]. For instance, Foster reported a 3D-printable porous anode by adjusting the weight ratios of graphite/PLA/plasticizer when processing the filaments, and a printable graphite anode with a high graphite loading of 49.2 wt% was successfully constructed utilizing the FDM approach [38]. The analogous Na-ion anodes, made of graphene, TiO_2, and MoS_2-graphene composite, have also been manufactured, much like the printed cathodes of Na-ion batteries. Furthermore, current Joule research seems to concentrate on the 3D-printed structure for the next-generation Li metal anodes, which have a very low redox potential of 3.04 V compared to ordinary hydrogen electrodes and an incredibly high theoretical capacity of 3,860 mAh/g [39]. For example, the current collectors for the Li metal anode have been built using 3D-printed porous frameworks made of carbon, 2D MXenes ($Ti_3C_2T_x$), and copper metal. The Li metal anode is associated with some issues like uncontrollable Li dendrite formation and significant volume interface changes, so to solve the problems, 3D-printed porous architecture is highly used [40], [41]. The literature survey shows that the direct ink writing (DIW) technique is mostly used for the fabrication of anodes.

11.5 ELECTROLYTES FOR 3D PRINTING BATTERIES

The electrolyte is a major component of lithium-ion batteries, which plays an important role in electrochemical performance, cycle life, and protection of the batteries. The development of better electrolytes with advantages like low activation energy, low electronic conductivity, and high ionic conductivity, has received more and more attention for 3D-printed batteries. Recently, the electrolyte of batteries might be directly printed to advancements in 3D-printing technology, which simplifies the fabrication processes and lowers production costs. The properties like low mechanical flexibility and poor electrochemical stability, limit the inorganic electrolytes in the practical application of 3D-printed batteries [42]. Kim and coworkers created an innovative, all-solid-state Li-ion battery using printed gel composite electrolytes (GCE) to address this problem [43].

In 2017, Blake et al. demonstrated a unique technique for 3D printing flexible, high-performance ceramic-polymer electrolytes (CPEs) [44]. In another method, without the use of any additional processing steps, Cheng et al. demonstrated a novel technique for fabricating hybrid solid-state electrolytes using elevated-temperature direct ink writing [45]. In which, Cheng to enhance the Li-ion diffusion and mechanical stability, poly(vinylidene fluoride-hexafluoropropylene) (PVDF-HFP) matrices were used, while Li-ionic was used as electrolyte. Also, other materials, such as garnet-type $Li_7La_3Zr_2O_{12}$ (LLZO), are utilized as electrolytes for solid-state lithium batteries. Traditional die-pressing and tape-casting techniques are restricted to random porosities and flat geometries, whereas 3D printing of solid electrolytes enables the creation of specifically shaped electrodes and electrolytes [46].

11.6 APPLICATION OF 3D PRINTING BATTERIES

Manufacturers must specify a specific size and shape for the battery when building devices like cell phones, which can prevent waste space and restrict shape design. It is theoretically possible to create entire devices using 3D-printing technology, including batteries, structures, and electronic components. 3D-printed batteries are widely used in several application fields (Healthcare and Cosmetics, the internet of things and people, RFID, air vehicles, automobiles, smart packaging wearable devices, and electronic equipment), due to their high power density, long-term durability, low price, and favorable safety [5], [35]. The schematic representation of 3D-printed batteries-based field applications is shown in Figure 11.6.

11.6.1 3D-Printed Batteries Are Edible, with Many Medical Device Applications

The researchers of Carnegie Mellon University are targeting to resolve the issues associated with traditional batteries by demonstrating biodegradable batteries composed of natural materials. A prototype battery that they have created can deliver 5 mW of power for up to 18 hours. This energy is sufficient to detect the development of dangerous microorganisms inside the body or to gently distribute drugs over several hours. The developed battery is created by the use of Melanin pigment, which is found in skin, hair, and nails. By absorbing harmful free radicals and UV rays, the pigment shields the body from harm. Additionally, it can bind to metallic ions, making it the ideal material for batteries. The anodic/cathodic terminal of the battery can make from Melanin, at the second terminal magnesium oxide is used and GI fluid serves as the electrolyte. The materials are contained in a 3D-printed capsule which is made using polylactic acid, or PLA.

11.6.2 In Automobiles

Recently, Sakuu Corporation developed an electric vehicle by using a solid-state battery 3D-printing system. To complete the task, the 3D-printing system comprises two forms of printing. It features two heads: one for jet deposition, which effectively squirts material out to precise specifications,

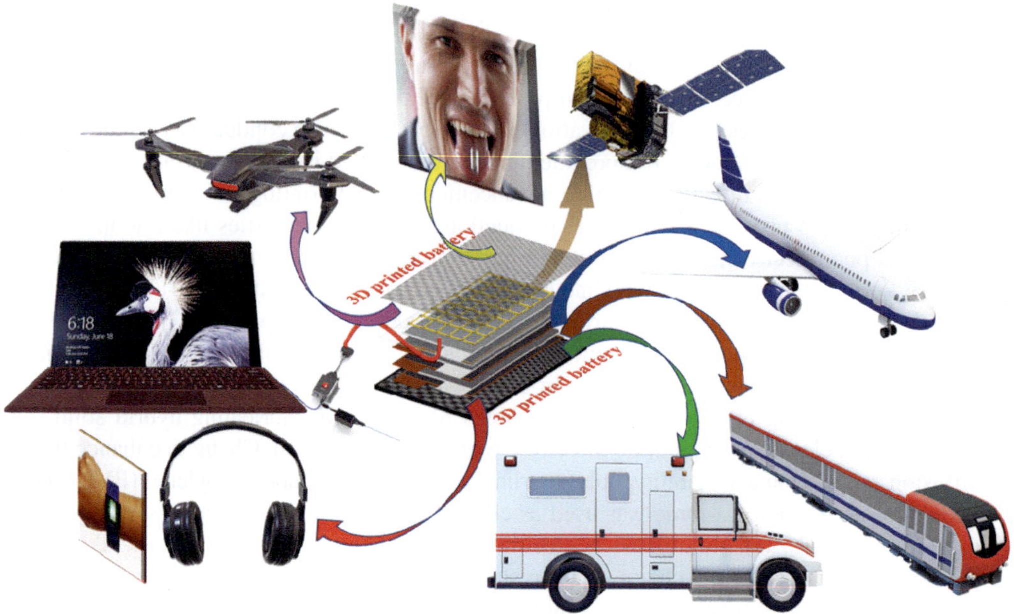

FIGURE 11.6 Schematic representation of 3D printed batteries-based applications.

and one for powder bed system, which sinters material into a solid form. These first-generation batteries use lithium metal, a customized printed ceramic separator, and 30 sub-cells. The battery has been built to support even higher voltage cathodes in the future, which might produce up to 25% more energy than the current industry standard cathode materials. Given its benefits in safety and energy density, this makes the new battery ideal for consumer, aerospace, and mobility applications. Scooters and other electric vehicles are among the products that now use 3D-printed batteries.

11.6.3 In Aerospace Vehicles Applications

Nowadays, 3D-printed technology is the most promising system to minimize the weight and shape of vehicles. The 3D-printed batteries with lightweight, small size, high power density, long cycle life, and safety properties are used in aerospace, light drones, and other light electronic device applications. KULR Technologies, based in San Diego, California, creates cutting-edge carbon fiber heat management solutions for batteries and electrical systems. Most industries, including those producing electric vehicles, electronics, and aerospace, use KULR solutions. The KULR technology group created a PPR (passive propagation resistant) battery design for use in space applications in June 2020. Additionally, KULR recently broke the major news that NASA wants them to construct 3D-printed batteries in space. The NASA Artemis project is a lunar exploration initiative that will deliver a man to the Moon's surface in the coming years. Furthermore, longer missions would be possible if batteries could be printed in space using 3D technology.

11.6.4 In Electronic Equipment

3D-printed batteries could have a significant impact on both consumer electronics and electric vehicles. Recently, 3D-printed batteries can be developed in any desired shape and small size along with lightweight by using various 3D techniques (DIW, SLA, DED, sheet lamination, etc.), which can be used in laptops, smartphones and wearable devices, etc. Li-ion batteries have been formulated with 3D-printing technology and are easily available in various shapes and sizes, due to the increasing demand for powerful batteries for energy storage. For example, S. Praveen demonstrated the shape of glass frames, wrist bands, and hearing aids by using 3D-printed Li-ion batteries. [22] Various 3D-printed battery-based electronic devices are shown in Figure 11.6.

11.7 CHALLENGES AND PROSPECT

1. Nowadays, only a few active materials are printable, which are used as inks for 3D-printing batteries. Therefore, it is necessary to create new electrochemically active materials to achieve maximum electrochemical performance. Additionally, it is expected that the electrodes fabricated through 3D-printing techniques will show a larger specific surface area as compared to electrodes developed by traditional methods. The porous structures of the 3D-printed electrodes have enhanced the power and energy density.
2. Since, 3D-printed things are created layer by layer, the interfaces between layers often have weak bonds. Therefore, more study is required to enhance the mechanical characteristics of 3D-printed batteries, particularly in some applications like wearable and flexible electronics.
3. The performance of electrolyte and active electrode materials is significantly impacted by the amount of oxygen and moisture in the air. For 3D-printed Li-ion batteries to avoid being harmed by oxygen/water molecules and other substances, effective encapsulation is necessary. Therefore, it is also important to create new materials and methods for encapsulation or packaging in the future.
4. Currently, research focused on electrode and electrolytes materials that are used for completely 3D-printed batteries. The compatibility of inks with components to be printed successfully for 3D printing batteries is a major challenge. With the continuous growth of printing technology and materials, it is expected that 3D-printed batteries with mechanical stability, high energy, power density, long life, favorable safety, and low price will be demonstrated for large-scale applications.

11.8 CONCLUSION

A lot of attention has been directed to the 3D printing of improved batteries as a cutting-edge technology that has the potential to fundamentally alter the design and architecture of materials, modules, and devices. We studied recent developments in 3D-printed batteries in this chapter from the viewpoints of printable materials, printing techniques, and design considerations at both the module and full cell device levels. Further, we have discussed the different 3D-printing technologies in detail such as DIW, IJP, SLA, etc. which are used for the fabrication of 3D-printed batteries. Furthermore, several types of electrodes (metal or metal oxide) or electrolytes have been discussed with better performance for 3D printing batteries. We have found that 3D-printed batteries have many advantages such as complex architecture, lightweight design, ease of operation, control of the thickness and the shape of the electrodes, materials flexibility, structural stability, low cost, and environmental friendliness. For future generations, 3D-printed batteries will be used extensively in applications like wearables, medical equipment, micro-electronic devices, electric vehicles, aerospace vehicles, and smart grids.

ACKNOWLEDGMENTS

R.K. Singh thankfully acknowledges the financial assistance from SERB, New Delhi, and IOE, BHU to carry out the research work. A. Patel is thankful to UGC, New Delhi, for the award of SRF.

REFERENCES

1 T.M. Gür, Review of electrical energy storage technologies, materials and systems: Challenges and prospects for large-scale grid storage, *Energy Environ. Sci.* 11 (2018) 2696–2767.

2 C.W. Foster, G.Q. Zou, Y. Jiang, M.P. Down, C.M. Liauw, A. Garcia-Miranda Ferrari, X. Ji, G.C. Smith, P.J. Kelly, C.E. Banks, Next-generation additive manufacturing: Tailorable graphene/polylactic(acid) filaments allow the fabrication of 3D printable porous anodes for utilisation within lithium-ion batteries, *Batter. Supercaps.* 2 (2019) 448–453.

3 K.-H. Choi, S.-Y. Lee, Design of printed batteries, *Print. Batter.* (2018) 112–143.

4 C.M. Costa, R. Gonçalves, S. Lanceros-Méndez, Recent advances and future challenges in printed batteries, *Energy Storage Mater.* 28 (2020) 216–234.

5 K.H. Choi, D.B. Ahn, S.Y. Lee, Current status and challenges in printed batteries: Toward form factor-free, monolithic integrated power sources, *ACS Energy Lett.* 3 (2018) 220–236.

6 Z. Lyu, G.J.H. Lim, J.J. Koh, Y. Li, Y. Ma, J. Ding, J. Wang, Z. Hu, J. Wang, W. Chen, Y. Chen, Design and manufacture of 3D-printed batteries, *Joule.* 5 (2021) 89–114.

7 S. Zhou, I. Usman, Y. Wang, A. Pan, 3D printing for rechargeable lithium metal batteries, *Energy Storage Mater.* 38 (2021) 141–156.

8 M. Cheng, R. Deivanayagam, R. Shahbazian-Yassar, 3D printing of electrochemical energy storage devices: A review of printing techniques and electrode/electrolyte architectures, *Batter. Supercaps.* 3 (2020) 130–146.

9 S.C. Ligon, R. Liska, J. Stampfl, M. Gurr, R. Mülhaupt, Polymers for 3D printing and customized additive manufacturing, *Chem. Rev.* 117 (2017) 10212–10290.

10 T.S. Wei, B.Y. Ahn, J. Grotto, J.A. Lewis, 3D printing of customized Li-ion batteries with thick electrodes, *Adv. Mater.* 30 (2018) 1–7.

11 A. Maurel, S. Grugeon, B. Fleutot, M. Courty, K. Prashantha, H. Tortajada, M. Armand, S. Panier, L. Dupont, Three-dimensional printing of a $LiFePO_4$/graphite battery cell via fused deposition modeling, *Sci. Rep.* 9 (2019) 1–14.

12 N.W.S. Pinargote, A. Smirnov, N. Peretyagin, A. Seleznev, P. Peretyagin, Direct ink writing technology (3d printing) of graphene-based ceramic nanocomposites: A review, *Nanomaterials.* 10 (2020) 1–48.

13 Y. Xu, X. Wu, X. Guo, B. Kong, M. Zhang, X. Qian, S. Mi, W. Sun, *The Boom in 3D-Printed Sensor Technology* 5 (2017) 1166.

14 S. Lawes, Q. Sun, A. Lushington, B. Xiao, Y. Liu, X. Sun, Inkjet-printed silicon as high performance anodes for Li-ion batteries, *Nano Energy.* 36 (2017) 313–321.

15 J. Hu, Y. Jiang, S. Cui, Y. Duan, T. Liu, H. Guo, L. Lin, Y. Lin, J. Zheng, K. Amine, F. Pan, 3D-printed cathodes of $LiMn_{1-x}Fe_xPO_4$ nanocrystals achieve both ultrahigh rate and high capacity for advanced lithium-ion battery, *Adv. Energy Mater.* 6 (2016) 1–8.

16 E. Cohen, S. Menkin, M. Lifshits, Y. Kamir, A. Gladkich, G. Kosa, D. Golodnitsky, Novel rechargeable 3D-Microbatteries on 3D-printed-polymer substrates: Feasibility study, *Electrochim. Acta.* 265 (2018) 690–701.

17 J.B. Goodenough, K. Park, J.B. Goodenough and K.-S. Park, J. Amer. The Li-Ion Rechargeable Battery: A Perspective, *J. Am. Chem. Soc.* 135 (2012) 1167.

18 C.M. Costa, Y.H. Lee, J.H. Kim, S.Y. Lee, S. Lanceros-Méndez, Recent advances on separator membranes for lithium-ion battery applications: From porous membranes to solid electrolytes, *Energy Storage Mater.* 22 (2019) 346–375.

19 R.E. Sousa, C.M. Costa, S. Lanceros-Méndez, Advances and Future Challenges in Printed Batteries, *ChemSusChem.* 8 (2015) 3539–3555.

20 K. Sun, T.S. Wei, B.Y. Ahn, J.Y. Seo, S.J. Dillon, J.A. Lewis, 3D printing of interdigitated Li-ion microbattery architectures, *Adv. Mater.* 25 (2013) 4539–4543.

21 R.R. Kohlmeyer, A.J. Blake, J.O. Hardin, E.A. Carmona, J. Carpena-Núñez, B. Maruyama, J. Daniel Berrigan, H. Huang, M.F. Durstock, Composite batteries: A simple yet universal approach to 3D printable lithium-ion battery electrodes, *J. Mater. Chem. A.* 4 (2016) 16856–16864.

22 S. Praveen, P. Santhoshkumar, Y.C. Joe, C. Senthil, C.W. Lee, 3D-printed architecture of Li-ion batteries and its applications to smart wearable electronic devices, *Appl. Mater. Today.* 20 (2020) 100688.

23 J. Ding, K. Shen, Z. Du, B. Li, S. Yang, 3D-printed hierarchical porous frameworks for sodium storage, *ACS Appl. Mater. Interfaces.* 9 (2017) 41871–41877.

24 X. Lin, J. Wang, X. Gao, S. Wang, Q. Sun, J. Luo, C. Zhao, Y. Zhao, X. Yang, C. Wang, R. Li, X. Sun, 3D printing of free-standing "o2Breathable" air electrodes for high-capacity and long-life Na-O_2 batteries, *Chem. Mater.* 32 (2020) 3018–3027.

25 K. Fu, Y. Wang, C. Yan, Y. Yao, Y. Chen, J. Dai, S. Lacey, Y. Wang, J. Wan, T. Li, Z. Wang, Y. Xu, L. Hu, Graphene oxide-based electrode inks for 3D-printed lithium-ion batteries, *Adv. Mater.* 28 (2016) 2587–2594.

26 J.H. Kim, S. Lee, M. Wajahat, H. Jeong, W.S. Chang, H.J. Jeong, J.R. Yang, J.T. Kim, S.K. Seol, Three-dimensional printing of highly conductive carbon nanotube microarchitectures with fluid ink, *ACS Nano.* 10 (2016) 8879–8887.

27 C. Milroy, A. Manthiram, Printed microelectrodes for scalable, high-areal-capacity lithium-sulfur batteries, *Chem. Commun.* 52 (2016) 4282–4285.

28 H. Zhu, S. Zhu, Z. Jia, S. Parvinian, Y. Li, O. Vaaland, L. Hu, T. Li, Anomalous scaling law of strength and toughness of cellulose nanopaper, *Proc. Natl. Acad. Sci. U. S. A.* 112 (2015) 8971–8976.

29 P.E. Delannoy, B. Riou, T. Brousse, J. Le Bideau, D. Guyomard, B. Lestriez, Ink-jet printed porous composite $LiFePO_4$ electrode from aqueous suspension for microbatteries, *J. Power Sources.* 287 (2015) 261–268.

30 J. Pröll, H. Kim, A. Piqué, H.J. Seifert, W. Pfleging, Laser-printing and femtosecond-laser structuring of $LiMn_2O_4$ composite cathodes for Li-ion microbatteries, *J. Power Sources.* 255 (2014) 116–124.

31 A. Izumi, M. Sanada, K. Furuichi, K. Teraki, T. Matsuda, K. Hiramatsu, H. Munakata, K. Kanamura, Rapid charge and discharge property of high capacity lithium ion battery applying three-dimensionally patterned electrode, *J. Power Sources.* 256 (2014) 244–249.

32 Z. Chen, T. Xiang, Z. Feng, X. Li, J. Huang, X. Shen, A NiO/NiS_2 nanosheet integrated electrode for high area specific capacity alkaline metal battery, *Mater. Lett.* 283 (2021) 128771.

33 Y. Gu, A. Wu, H. Sohn, C. Nicoletti, Z. Iqbal, J.F. Federici, Fabrication of rechargeable lithium ion batteries using water-based inkjet printed cathodes, *J. Manuf. Process.* 20 (2015) 198–205.

34 I. Ben-Barak, Y. Kamir, S. Menkin, M. Goor, I. Shekhtman, T. Ripenbein, E. Galun, D. Golodnitsky, E. Peled, Drop-on-demand 3D printing of lithium iron phosphate cathodes, *J. Electrochem. Soc.* 166 (2019) A5059–A5064.

35 J. Li, M.C. Leu, R. Panat, J. Park, A hybrid three-dimensionally structured electrode for lithium-ion batteries via 3D printing, *Mater. Des.* 119 (2017) 417–424.

36 L. Zhou, W. Ning, C. Wu, D. Zhang, W. Wei, J. Ma, C. Li, L. Chen, 3D-printed microelectrodes with a developed conductive network and hierarchical pores toward high areal capacity for microbatteries, *Adv. Mater. Technol.* 4 (2019) 1–7.

37 C. Wei, G. Zhang, Y. Bai, D. Yan, C. Yu, N. Wan, W. Zhang, Al-doped SnO_2 hollow sphere as a novel anode material for lithium ion battery, *Solid State Ionics.* 272 (2015) 133–137.

38 A. Maurel, M. Courty, B. Fleutot, H. Tortajada, K. Prashantha, M. Armand, S. Grugeon, S. Panier, L. Dupont, Highly loaded graphite-polylactic acid composite-based filaments for lithium-ion battery three-dimensional printing, *Chem. Mater.* 30 (2018) 7484–7493.

39 Z. Lyu, G.J.H. Lim, R. Guo, Z. Pan, X. Zhang, H. Zhang, Z. He, S. Adams, W. Chen, J. Ding, J. Wang, 3D-printed electrodes for lithium metal batteries with high areal capacity and high-rate capability, *Elsevier B.V.* 24 (2020) 336–342.

40 D. Cao, Y. Xing, K. Tantratian, X. Wang, Y. Ma, A. Mukhopadhyay, Z. Cheng, Q. Zhang, Y. Jiao, L. Chen, H. Zhu, 3D printed high-performance lithium metal microbatteries enabled by nanocellulose, *Adv. Mater.* 31 (2019) 1–24.

41 K. Shen, B. Li, S. Yang, 3D printing dendrite-free lithium anodes based on the nucleated MXene arrays, *Energy Storage Mater.* 24 (2020) 670–675.

42 M. Marcinek, J. Syzdek, M. Marczewski, M. Piszcz, L. Niedzicki, M. Kalita, A. Plewa-Marczewska, A. Bitner, P. Wieczorek, T. Trzeciak, M. Kasprzyk, P. Łężak, Z. Zukowska, A. Zalewska, W. Wieczorek, Electrolytes for Li-ion transport – Review, *Solid State Ionics.* 276 (2015) 107–126.

43 S.H. Kim, K.H. Choi, S.J. Cho, J. Yoo, S.S. Lee, S.Y. Lee, Flexible/shape-versatile, bipolar all-solid-state lithium-ion batteries prepared by multistage printing, *Energy Environ. Sci.* 11 (2018) 321–330.

44 A.J. Blake, R.R. Kohlmeyer, J.O. Hardin, E.A. Carmona, B. Maruyama, J.D. Berrigan, H. Huang, M.F. Durstock, 3D printable ceramic–polymer electrolytes for flexible high-performance Li-ion batteries with enhanced thermal stability, *Adv. Energy Mater.* 7 (2017) 1602920.

45 M. Cheng, Y. Jiang, W. Yao, Y. Yuan, R. Deivanayagam, T. Foroozan, Z. Huang, B. Song, R. Rojaee, T. Shokuhfar, Y. Pan, J. Lu, R. Shahbazian-Yassar, Elevated-temperature 3D printing of hybrid solid-state electrolyte for Li-ion batteries, *Adv. Mater.* 30 (2018) 1800615.

46 D.W. McOwen, S. Xu, Y. Gong, Y. Wen, G.L. Godbey, J.E. Gritton, T.R. Hamann, J. Dai, G.T. Hitz, L. Hu, E.D. Wachsman, 3D-printing electrolytes for solid-state batteries, *Adv. Mater.* 30 (2018) 1707132.

12 3D-Printed MXene Composites for Batteries

V Raja[1], J Vigneshwaran[1], Kenan Song[2], A M Kannan[3], and Sujin P Jose[1]*

[1]Advanced Materials Laboratory, School of Physics, Madurai Kamaraj University, Madurai 625021, Tamil Nadu, India

[2]Advanced Materials Advanced Manufacturing Laboratory, The Polytechnic School (TPS) & School of Manufacturing Systems and Networks (MSN) & School of Engineering of Matter, Transport and Energy (SEMTE), Ira A. Fulton Schools of Engineering, Arizona State University, Mesa, AZ 85212, USA

[3]Fuel Cell Laboratory, The Polytechnic School, Ira A. Fulton Schools of Engineering, Arizona State University, Mesa, AZ 85212, USA

[*]Corresponding author: sujamystica@yahoo.com

CONTENTS

DOI: 10.1201/9781003296676-12

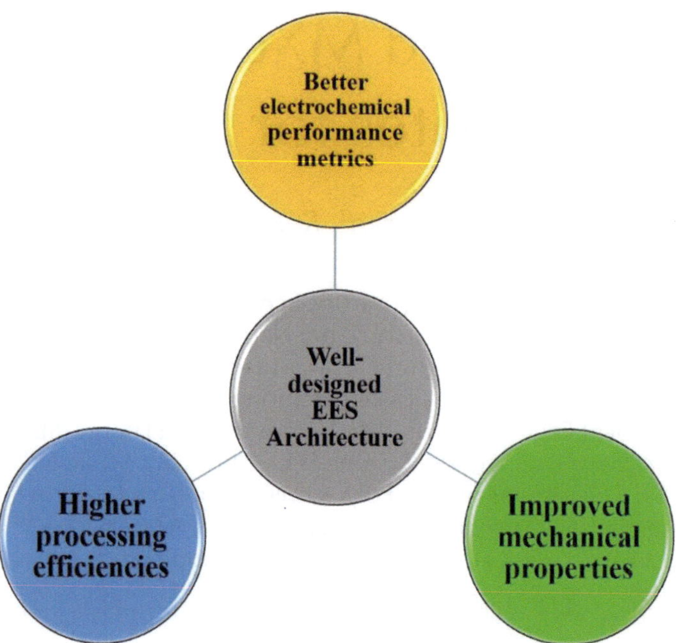

FIGURE 12.1 Schematic of Electrochemical Energy Storage Devices (EES devices) with well-designed architecture for better electrochemical performance, higher processing efficiencies, and improved mechanical properties.

12.1 INTRODUCTION

12.1.1 Electrochemical Energy Storage Devices

In the past couple of decades, both the energy industry and academic institutions around the world have become progressively interested in the development of energy storage and conversion devices [1]. However, most of the academic research has only focused on the discovery of components such as electrode and electrolyte materials, with inadequate devotion to the design, development, and demonstration of devices. Scientific discoveries and material innovation are two significant entities whereas process engineering is highly essential in integrating these building blocks composed and frequently defining their role in real-time applications. High energy and power density along with environmentally benign procedures are the prerequisites for electrochemical energy storage devices (EESDs) usually identified with (i) electrochemical capacitors and (ii) rechargeable batteries which constitute potential energy storage solutions for various applications. An electrochemical capacitor typically consists of two electrodes, an electrolyte, current collectors, and separators [2]. As illustrated in Figure 12.1, three crucial processes such as electrode fabrication, electrolyte alteration, and device assembly could significantly influence the electrochemical performance, effective processing, durability, and mechanical stability of the energy storage devices.

Supercapacitors are widely employed in electronic circuits that need quick and regular charging due to their high power density, faster discharge rate, and extended cycle life [3]. Nevertheless, the relatively lower energy density (< 10 Wh kg^{-1})is the issue to be addressed. The fairly low power density and higher energy density (50 to 250 Wh kg^{-1}) result in a substantially longer discharging duration. For example, Li-ion batteries with the highest energy density compared to all other systems (lead-acid, Ni-Cd, and Ni-MH) are being employed in portable, stationary, and automotive applications [4].

12.1.2 Lithium-Ion Batteries (LIBs) and Beyond

Rechargeable lithium-ion batteries (LIBs) in cylindrical, prismatic, and pouch configurations with low-cobalt cathodes are the most common EESDs in daily life for personal devices and electric vehicles [5]. In general, LIBs operate with Li intercalation/deintercalation back and forth into the anode and cathode reversibly at higher redox potentials. In the recent years, researchers have been focusing on lithium-air (Li-air) and lithium-sulfur (Li-S) batteries due to their superior energy density compared to LIBs. Li-S batteries work by cutting down the amount of sulfur at the cathode followed by the combination of the generated polysulfides with the opposite Li electrodes during the discharge process [6]. Given the scarcity of the resources for the LIBs on the planet, new rechargeable battery solutions like sodium, aluminum, magnesium, and calcium-based batteries are also gaining prominence due to their abundance of related raw materials [5]. The working principles of the above-mentioned new energy storage systems are on par with those of lithium-based batteries, but in most cases, they behave differently towards their electrochemical activity [7].

Electrochemical capacitors popularly acclaimed as supercapacitors belong to EESDs and their importance is attributed to the higher power density than that of the secondary batteries. The electrochemical capacitors are mainly divided into two types, electrical double-layer capacitors (EDLCs) and pseudo capacitors that store energy by using fast charge storage mechanisms. When high power density is the aim of the delivery, the usage of electrolytic ion absorption on the surface of the electrode along with the redox reactions is envisaged in electrochemical capacitors for supplementing or replacing the batteries [8].

12.1.3 Conventional and State-of-Art 3D Printing

Additive manufacturing through the 3D printing process enables precise layer-by-layer construction of the structured functional materials, and the device architecture [9], [10]. However, for some uses of functional materials, including electrochemical energy storage, the printing techniques might be constrained by their exclusive but crucial device design and development [11,12]. Hence, to consider the 3D printing strategies, one has to understand the device structures along with several other parameters that describe the performance.

Additive manufacturing (AM) is rapidly becoming an influential technique for engineering efficient 3D structures for automotive, aerospace, and other consumer products [12]. It became so popular that it has grown up among hobbyists to engineering professionals due to its outstanding control in manipulating innovative architectures right from the computer-aided design (CAD) software. To go for highly efficient electrochemical energy storage applications, 3D printing has emerged with unique rewards related to the conservative manufacturing methods such as:

1. The desired architectures can be directly written down with controlled chemistry, designed shapes, and interconnected porosity;
2. Simple fabrication procedure and ensuring customized configurations during the prototyping of electrodes and other components;
3. Ability in realizing full device manufacturing;
4. On-chip EESDs can be obtained with optimized thicknesses ranging from hundreds of nanometers to millimeters, resulting in increased efficiency. Meanwhile, printing offers an expedient procedure for microfabricating asymmetric electrodes as well as encapsulation of the miniaturized EESDs, and
5. Promotion of direct integration of EESDs or co-fabrication towards external electronics, evading device assembly and packaging steps.

The exceptional device production mode, where fundamental device architectures and numerous performance factors play vital roles, is the usual benefit of 3D printing procedures for the creation

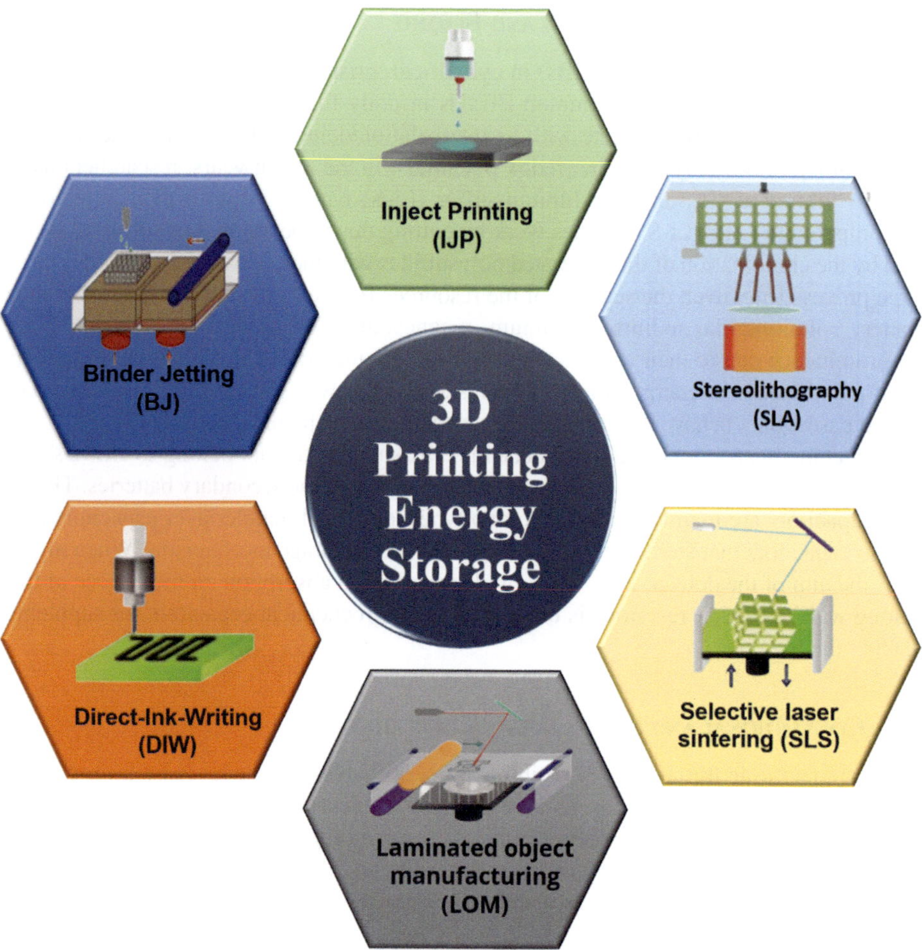

FIGURE 12.2 Representation of the various 3D printing strategies for energy storage applications.

of EESDs, with optimized ink preparation and printing processes. Similarly, both the segments are likewise regarded as critical components for electrochemical energy storage [12]. Various printing technologies for developing EESDs are given in Figure 12.2. The prepared ink/raw material is very important for various printing methods, particularly for the direct ink writing of functional materials. Each printed EESD component layer's surface tension and viscosity are strongly influenced by the ink characteristics, particularly their rheological behaviors [13]. The adaptable viscoelastic properties under external stimuli like pressure, temperature, and electromagnetic field project colloidal gels, hydrogels, and liquid polymer composites as good candidates for unceasing filament writing or droplet jetting of multifaceted architectures. For creating the inks, polymeric, colloidal, or polyelectrolyte building blocks are often suspended or dissolved in a liquid or heated to provide satisfactory flow characteristics [14].

To formulate the printable inks, two significant conditions are addressed. To begin with, in extrusion/vibration deposition, the viscoelastic properties must be controllable for them to stream through the printing nozzles [15]. Liquid inks having low viscosity are used as they can easily form droplets and deposit them on the substrate. In contrast, extremely viscous inks typically form a continuous rodlike filament. The 3D extrusion-based printing process benefits from highly shear thinning behavior. Second, inks must have enough mechanical stiffness and strength to assist the whole structure during the period of ink deposition followed by the rapid solidification processes. These measures necessitate suitable ink formulations and rheological properties to produce a steady dispersion, which encourages the fluid-to-gel transition, ensuring the synchronized holding of printed forms/shapes and blending with formerly deposited films [16], [17].

Other 3D printing techniques like direct ink writing processes are crucial methods for creating useful materials. Although it is now easier to access the raw materials used for printing, the powder characteristics are still crucial for SLS/SLM processing. The mechanical and electrochemical performance of functional 3D-printed structures can be significantly influenced by particle sizes, forms, and chemical composition of the powder. Similarly, control of the printing parameters also has an impact on the characteristics of the final device.

When choosing appropriate 3D printing technologies, EESD designers should take into account both the materials preparation and printing conditions. Although design factors have been extensively developed and optimized in some areas, understanding of electrochemical energy storage is still in its early stage, necessitating rigorous investigation to achieve desired EESDs [18].

12.2 MATERIALS AND SYNTHESIS STRATEGY – MXENES TOWARDS 3D PRINTING

12.2.1 Materials – Definition of MXenes

MXenes are created by exfoliating their corresponding MAX phases, which are three-dimensional and are specifically etched to remove the Group IIIA or IVA elements denoted by A in the MAX phase and corrosion-resistant d-block metals denoted by M in the MAX and C or N (X in MAX) [19]. Overall, different three-dimensional MAX phases are preferentially etched to gather the corresponding MXenes as illustrated in Figure 12.3a. Two important methods are used in synthesizing MXenes as illustrated in Figure 12.3b. They are (i) formation of layered ternary metal

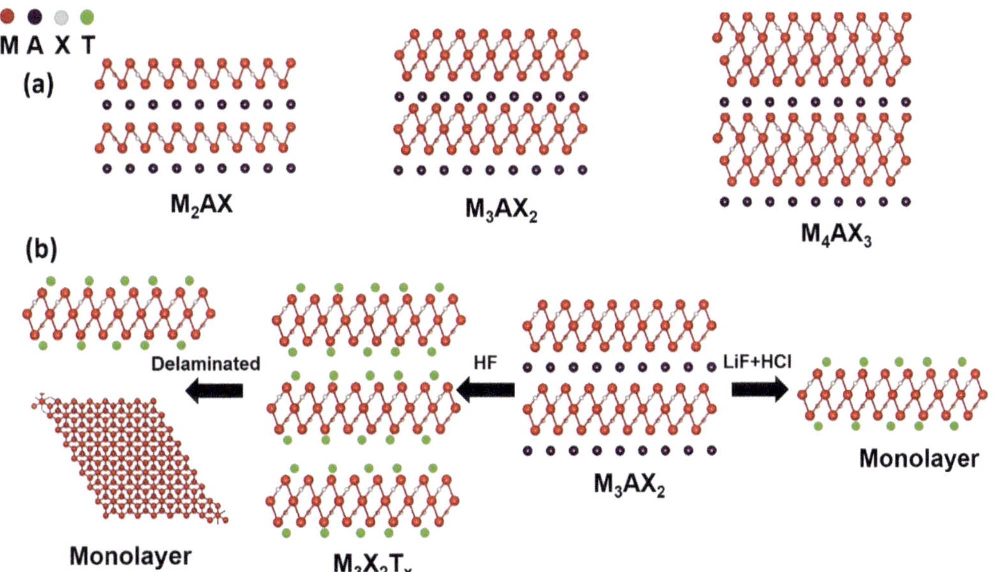

FIGURE 12.3 (a) Schematic of crystal structures of M_2AX, M_3AX_2, and M_4AX_3 phases, and (b) Synthesis of multilayer MXenes by HF etching or monolayer MXenes by LiF + HCl etching. Green, light blue, brown, and red balls represent M (transition metal), A (usually Al or Si), X (C or N), and T (O, F, or OH) respectively.

carbides and nitrides resulting from the selectively erodible elements, by HF or LiF/HCl and (ii) appropriate etching/exfoliation environments that exercised an imperative impact on products and phase conversion of MXenes. The widely explored MXene, Ti_3C_2 can also be exfoliated ultrasonically by intercalation with urea, dimethyl sulfoxide, potassium cations, and ammonia [20]. Mild sonication in water is employed to delaminate MXenes with controllable/changeable morphology. Because of its ceramic nature, MXene has an apparent micro-sized firm layered morphology; two-dimensional MXene has an ultra-large layer spacing compared to the graphite layer (0.337 nm). The synthesis of lamellar 2D Ti_3C_2 establishes an invigorate direction for MXenes. Annealed MXenes in Ar, N_2, and N_2/H_2 atmospheres kept their original chemical structure which may be destroyed if unveil to a high-temperature oxidation atmosphere [21].

In summary, MXenes have seen gradual technological progress, from ultrasonic exfoliation to intercalation of nanoparticles between the MXene layers. Numerous spin-off techniques resulted in an abundance of functional groups and steric configuration, as well as the unique chemical-physical properties of MXene. Significantly, MXenes offer surface functional groups to adsorb the cations. These findings showed that all MXene surfaces are endowed with exceptional properties. This in turn confirms that the fully functionalized MXene surfaces exhibit more thermodynamical stability. In summary, MXene exfoliation and delamination procedures have been progressively established in one direction [22]. To provide some significant theoretical insights into the physical property data like structural stability and conductivity which are key factors for energy storage applications DFT studies were carried out [12,20,23]. MXenes display several electronic properties extending from metallicity and semiconductivity to topological insulative, which can be attributed to the compositional diversity, numerous possibilities for surface functionalization, and the option to go for flexible thickness controllability. It is interesting to note that most of the bare MXenes and their surface functionalized counterparts are metallic. Fascinatingly, the work functions (WFs), of metallic MXenes were calculated and they span across a wide range, that is, from 1.8 to 8 eV. Also, it is identified that the work functions of MXenes depend on their surface chemistry [22]. Let us consider an MXene and related to its bare surface, the presence of OH (O) decreases its WF. The F decoration demonstrates the tendency to count on the specific material. Further, it is noted that the WFs of a few O-terminated MXenes are on the higher side than that of Pt which shows the highest WF of all elemental metals [24]. The F-terminated MXenes come up with WFs that lie between OH- and O-terminated counterparts. On surface functionalization, a change in the WF is obtained which can be ascribed to the change in surface dipole moment brought by functionalization [25].

12.2.2 Synthesis Strategy – MXenes Towards 3D Printing

MXenes can be realized by both top-down and bottom-up methods of synthesis analogous to other 2D materials; however, top-down methods, which have a higher production capacity and are less expensive, are of greater interest [26]. The stack of $M_{n+1}X_n$ layers in MXenes is interweaved by added layers with robust metallic connections which differ from other 2D materials like graphene in which the layers are stacked by weak van der Waals attractions. The parent crystal is known as the MAX phase when the extra layer is a monoatomic single layer (marked as "A"; an element from groups 13 and 14 of the periodic table) (e.g., Ti_3AlC_2). The 'A' layers need to be taken out of the MAX phase to prepare individual sheets of MXene [27].

The elimination of diatomic multilayers, that is, Al_3C_3 from $Zr_3Al_3C_5$, is a step in the creation of MXenes from non-MAX phases. Because of the high chemical activity of the aforementioned layers in both situations, elimination with a particular etching agent is an easy approach. The quality of MXenes and the exotic properties from mechanical to chemical that the MXenes can provide profoundly be subjected to their synthesis protocols. Researchers all over the world nowadays working on developing diverse etching protocols for achieving phase pure MXenes [28]. Also, the types of the functional groups and their ratio together with the structural defect density are dictated by the etching techniques and their relevant surroundings. Because of their high selectivity, an aqueous solution of hydrofluoric (HF) acid has been widely used to remove the 'A' element. The environmentally hazardous nature and the difficulties in the handling of HF acid instigated us to look for alternate techniques like combinations of fluoride salts and strong acids accomplished to form HF (in-situ, e.g., LiF + HCl) [29], [30]. The mild etching strategies employing ammonium hydrogen difluoride, ammonium fluoride, and strong alkali solutions for the HF-free approach were established to be effective etchants. Significantly individual optimization is advocated as the various parameters like concentration, time, temperature, and other etching conditions vary to a greater extent for various MAX phases. Overall, the experimental protocols are favoured by increasing the atomic number of the transition metal and/or 'n' (in $M_{n+1}X_n$) along with a longer duration of the etching process and the usage of sturdier etchants [31].

12.3 PROPERTIES OF MXENES TOWARDS 3D-PRINTED ELECTRODES

12.3.1 Interfacial Chemistry and Properties

MXene as a 2D electrode candidate renders high surface-active sites, has higher conductivity, supports intercalation, and is receptive to a wide range of composites (Figure 12.4a). MXenes are typically synthesized by an etching process using HF or fluoride-based salt solution. This in turn leaves MXenes impulsively terminated with ample functional groups such as O, -OH, and -F [32]–[36]. The commonly used techniques to explore the surface chemical conformations and the chemical states of MXene are XPS and nuclear magnetic resonance (NMR). These are identified as effective tools for the qualitative and quantitative analysis of surface terminations on MXene. The electrochemical properties characterized by electronic/ionic conductivity and hydrophily and energy storage efficiency are ascribed to the nature of surface functional groups. That is the surface groups of -F and -OH may hinder the ions' transport causing a decline in their electrochemical capacity. Hence, fine-tuning the content and distribution of the surface functional groups are of remarkable implication for the performance optimization of MXenes electrodes. Wang and co-workers synthesized two kinds of $Ti_3C_2T_x$ MXenes by etching Ti_3AlC_2 precursors using aqueous HF solution at low (6 M) and high (15 M) concentrations. The low-concentration HF-etching endowed an unusually higher capacitance that may be ascribed to the abundance of O groups on the surface of MXene offer more pseudocapacitance active sites which were confirmed by in-situ Raman spectroscopy and X-ray photoelectron spectroscopy (XPS). However, it is to be noted that the F groups cannot be removed by reducing the HF concentration [37]–[39]. To evade the undesirable effect of F groups, another protocol of fluoride-free synthesis of MXenes has been utilized for the effective synthesis of bare MXenes and MXenes with O, NH, S, Cl, Se, Br, and Te surface terminations. The results in MXenes demonstrate characteristic structural and electronic properties like tunable interatomic distances, extended in-plane lattice, and functional group-dependent superconductivity [37]–[40].

12.3.2 Chemical Stability and Storage of MXenes Inks

MXene suspensions have been shown to disperse very well in a variety of solvents, particularly water. Previous research has revealed that oxygen dissolved in water takes up an important part in the chemical degradation of delaminated MXene dispersed in aqueous solutions as shown in Figure 12.4b–d. This degradation procedure is typically accelerated in colloidal suspensions of single or a few-layered MXenes determined by a higher surface-area-to-volume ratio that makes them extremely susceptible to oxidation [41]–[43]. As an outcome of the unimpeded development of transition metal oxides on MXenes, such as TiO_2 on the surface of $Ti_3C_2T_x$, the structural breakdown and shelf life of the articulated MXene inks would occur. Such structural deformation and phase transformation are exacerbated when MXene dispersions are exposed to light or high temperatures.

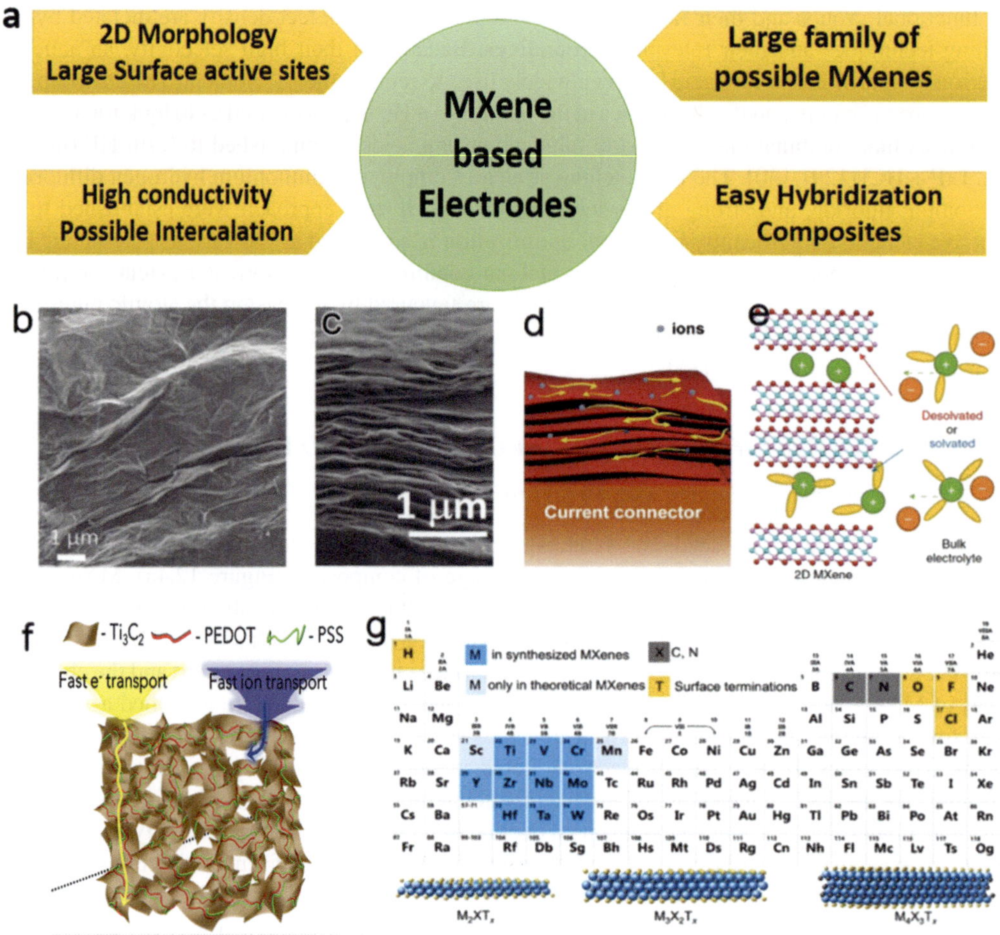

FIGURE 12.4 (a) Properties of MXene enabling them as active materials for battery electrodes, (b) SEM image of MXene film displaying the 2D morphology and wrinkled surface [54], (c) Cross-section of SEM image of MXene film presenting the multiple layers, (d) Schematic of distinctive in-plane ion transport channels in MXene film, (e) Mechanism of MXene-based battery electrodes [55], (f) Schematic displaying the synergistic effect of MXene, (g) Periodic table representing the family of theoretically predicted and/or experimentally synthesized. MXenes and MAX phases. Adapted with permission [54] [55], Copyright (2018, 2019), Wiley, Cell Press.

In general, oxidation starts along the edges of the MXene flakes and spreads to the entire basal plane. Hence, smaller flakes characterized by more active sites are more susceptible to oxidation. Oxidation arises invariably in all types of media, including air and liquid and solid media, with the liquid media oxidizing at the fastest rate and solid media oxidizing at the slowest. A general desolvation mechanism has been illustrated in Figure 12.4e. Thus, removing dissolved oxygen from water by saturating it with nitrogen or argon is an efficient way to inhibit oxidation. Nonetheless, in addition to dissolved oxygen, water itself is a mild oxidant of MXenes. In this regard, refrigeration was found to significantly slow down the degradation rate caused by H_2O-induced oxidation. For example, MXene suspensions stored under Ar at low temperatures only degraded by percent after 25 days, with the possible oxidation explained. As a result, it is now the standard procedure to store stable MXene inks in Ar-sealed vials, away from light and below the refrigeration temperature (4 °C) [44]–[46].

12.3.3 Rheological Properties of MXene Inks

The rheological properties of MXene inks are defined by the physical behavior of MXenes under applied stresses and these properties have a significant impact on existing printing techniques, reproducibility of the printing methods, and pattern quality. Shear-thinning liquids with viscoelastic and thixotropic properties are looked for in several printing/coating processes such as screen printing, spray coating, and extrusion printing. When a force is applied to a liquid, it can flow through the printer nozzle and instantly recuperate its high viscosity after leaving the nozzle. This ensures the preservation of the desired patterns after the printing process. However, the rheological necessities for various printing/coating methods differ significantly, for example, from viscous paste for extrusion 3D printing and screen printing to low-concentration dispersion for inkjet printing [47]. Shear experiments can be conducted for single-layer MXene flake suspension as well as for multi-layer MXene dispersions with various concentrations to investigate the rheological properties of MXenes. Both suspensions exhibited shear-thinning behaviors, with increased viscosity as concentrations increased. Another important parameter for determining dispersion in rheology is the elastic modulus to viscous modulus ratio (G′/G″). Spray coating, for example, frequently necessitates superior processing rates with large viscous modules, whereas extrusion printing necessitates high elastic modules to maintain the shape of the designs. Single layer and multilayer MXenes exhibit distinct rheological behaviors, and the extensive variety of rheological behavior of dispersions with varying concentrations provides a variety of printing options for MXene inks. The rheological properties of single-layer MXene dispersion at a low concentration of less than 10 mg have also been carried out [48].

Fascinatingly, MXene flakes of high aspect ratio can be tuned effectively rheologically and Yang et al. illustrated that the viscosity parameters of single-layer MXene dispersions, that is up to 15 mg mL^{-1} were on a comparable magnitude as extremely concentrated (2.33 g mL^{-1} 70 wt%) multilayer MXene dispersions [49]. They have also engaged the vacuum evaporation method to prepare MXene dispersion of a high concentration of 50 mg mL^{-1}. When the yield stress (critical shear stress) is not reached, G′ is greater than G″, suggesting a normal elastic solid. G′ begins to be higher than G″ as shear stress rises above yield stress, which causes the transition of the ink from solid-like to liquid-like behavior [50]–[52]. Additionally, Fan et al. [53] generated nitrogen-doped Ti_3C_2 dispersion of varying concentrations up to 300 mg mL^{-1} and observed a reduction in shear stress to 170 Pa. A scheme of MXene with polymer composite has been illustrated in Figure 12.4f. MXene a receptive structure towards hybridization shall improve the desired properties, Figure 12.4g illustrates the periodic table for the possible/attempted feasible compounds as MXene elements.

12.4 3D PRINTING DESIGNS AND MODULES – ELECTRODE PREPARATION TECHNOLOGY

12.4.1 3D Printing in Energy Storage

3D Printing designates appending one layer of material upon another for deliverables with intended 3D geometries and necessary compositions [ASTM F42 and TC261]. Charles W. Hull was the first to coin the name stereolithography (the first reported type of 3D printing) and to file a patent ("US4575330A") in 1984. The patent describes the system to generate a cross-sectional pattern upon stimulation to form a stepwise laminar build-up of desired three–dimensional objects [56]. The major advantages of 3D printing over traditional methods are (i) flexible and customized geometrical, structural, and compositional design, (ii) getting rid of time-consuming tool engineering, (iii) accelerated transformation from design to manufacturing, (iv) controllable waste materials, and (v) feasibility of local production [23, 25, 31]. The 3D printing technique's compelling advantage over traditional cathode fabrication is high customizability in forming an ordered structure including patterning dots, thin-width lines, systematic lattices, cellular scaffolds, and functionally graded structures. Notably, the customizable morphology of powdered material lends support to improving product properties like resolution, interlayer composition, physical-chemical properties, and manufacturing speed [57].

12.4.2 Types of 3D Printing Technologies

12.4.2.1 Inkjet Printing (IJP)

Inkjet printing is a direct droplet-based material deposition technology that is non-contact. IJP deposits a dilute suspension or solution of polymer/nanoparticle over the substrate, which is then solidified via photon irradiation, heat transfer, or chemical reaction. IJP is an excellent technique to produce complex patterns with a wider scope of printing material with high resolution. Electro-hydrodynamic (EHD) jetting involves the application of an electrical voltage between the jetting nozzle and the substrate. By overcoming capillary forces, EHD produces a high-resolution end product. Polymer jetting and binder jetting are the two types of jetting-based printing. The InkJet print head deposits photosensitive polymer over the substrate in the polymer jetting method, which is then cured with a UV light source [58]. Binder jetting is a dry-material-based technique, that selectively ejects small molecules over powder substrate gluing it to formulate a cross-section of the desired pattern. Binder jetting requires subsequent heat treatment to achieve a defect-free high close packing product.

12.4.2.2 Stereolithography (SLA)

Stereolithography (SLA) is a vat polymerization of a photocurable material that is printed on a substrate. Curing printed polymers with visible, UV, and laser sources is required. SLA is classified into three types based on its curing method and dimensions: micro-SLA (SLA), Continuous Liquid Interface Processing (CLIP), and Digital Light Processing (DLP). CLIP technology shall be an intermediate step along with other 3D print production techniques. DLP has been realized with projection-based printing in which selective polymerization of large surfaces leads to the potential for industrial-scale production. Since wet input-based techniques are widely used in the fabrication of cathode materials. Extrusion-based printing technologies are used in dry input techniques. Powders, plaster, laminate, or filament are used in dry input techniques [59]–[64].

12.4.2.3 Extrusion and Direct Ink Writing (DIW)

In this technique, the material is extruded through a die/nozzle to form layer-by-layer structures in 3D printing. Extrusion-based techniques will be expanded to include multi-material and multi-nozzle capabilities, giving an advantage in industrial-scale production. Direct Ink Writing (DIW) uses shear thinning to extrude paste-like ink. The ink under consideration should have a high stress and storage modulus to tolerate shape retention and distortion-free filament spanning. DIW and IJP are versatile techniques for creating energy storage devices. The primary advantage of DIW as a fabrication technique is its large mass loading that enhances areal capacitance and energy density [60].

12.4.2.4 Freeze Nano Printing (FNP)

Freeze nano printing (FNP) is a process of direct writing aqueous ink onto a freezing substrate. The ability of FNP to form complex-shaped structures such as hierarchical pore structures/aerogels with flexibility in mass transportation balancing, ion diffusion, and diffusion length support higher capacities and rate capability. Freezing conditions, ink/slurry concentration, and additives all contribute to changes in pore size and distribution [61]. Increased material loading and a hierarchical porous structure in FNP improve the high specific energy and power. FNP holds the record of forming the least dense 3D-printed structure.

12.5 3D-PRINTED MXENE AND MXENE COMPOSITE FOR BATTERIES

12.5.1 MXene Electrodes for Batteries

By reducing dependence on a fixed power source, rechargeable LIBs empowered the much-needed but the unforeseen rise of electric vehicles and portable electronics. The high energy density LIBs led to the miniaturization of electronics with a smaller footprint along with faster charging. MXene has unexpectedly emerged as a strong competitor, not only because it improves the performance of LIBs but also ensures the feasible development of batteries other than Li-ion. Naguib et al. [65] demonstrated MXene as an anode material of LIB for the first time in 2011 and found it to be promising because of its large specific surface area, low interlayer forces, open structure, and functional groups on the surface (Figure 12.5a and b). Because MXene is made up of M (early transition metals), X (Carbon and/or Nitrogen), and T (functional groups -O, -OH, and -F), there are a plethora of variations that could be employed to tune the characteristics. $Ti_3C_2F_2$ and $Ti_3C_2(OH)_2$ with their near-Fermi level band structure and the corresponding Li to C distance in the case of optimized lithiation demonstrate the impact of functional groups on the nature of lithium intake as illustrated in Figure 12.5c and d.

Chlorides, non-native functional groups of MXenes can offer outstanding performance by playing with increasing the interlayer distance and mitigating the negative effects of OH and F functional

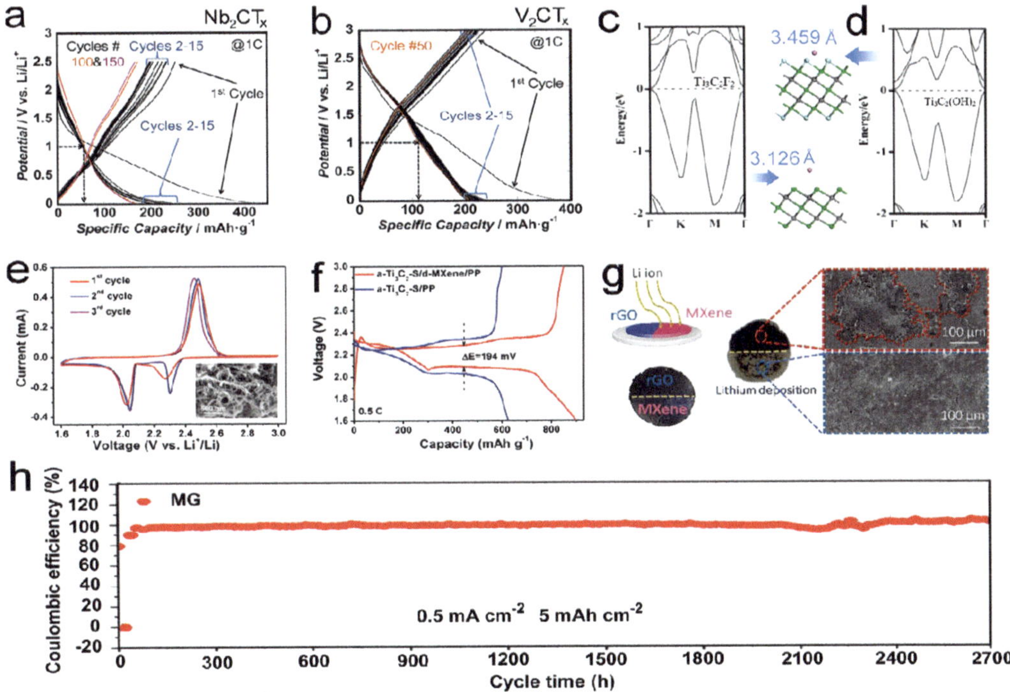

FIGURE 12.5 (a), (b) Voltage profiles of Li/Li^+ vs. Nb_2CT_x and V_2CT_x at various voltage ranges [66], (c) Illustrations of the effect of functional groups on lithium uptake behavior, (d) Near-Fermi level band structures of $Ti_3C_2F_2$ and $Ti_3C_2(OH)_2$, respectively, and the corresponding Li to C distance in the case of optimal lithiation [67], (e) CV curves, (f) GCD profiles of Li-S batteries with sulphur electrodes hosted in Ti_3C_2 nanoribbons (inset-the SEM image of Ti_3C_2 nanoribbon and sulphur hybrid [68], (g) Photograph and SEM image of lithium deposition on graphene and MXene demonstrating the benefit of homogeneous lithium distribution on MXene film. (h) Coulombic efficiency of MXene/graphene-Li anode measured over 2700 h at a current density of 0.5 mA cm^{-2} [69]. Adapted with permission reference [66], [67], [68], [69] Copyright (2013, 2012, 2018, 2019), American Chemical Society.

groups. In fact, because of varying the degree of surface properties, even the capacity of LIBs can be indirectly impacted by stoichiometric ratios of "M" from the same species generated using various manufacturing techniques. Ti_2CT_x with higher surface energy than $Ti_3C_2T_x$ typically has a lower c-lattice parameter, preventing Li-ion intercalation. Moreover, Figure 12.5e and f depict the CV and GCD plots of Li-S batteries with sulfur electrodes hosted in Ti_3C_2 nanoribbons. Figure 12.5g represents the SEM images of lithium deposited on graphene and MXene whereas Figure 12.5h shows the anode made up of MXene/graphene-Li with its coulombic efficiency (~99%) for 2700 hours at a current density of 0.5 mA cm^{-2}.

12.5.2 3D-Printed MXene Electrodes for Batteries

To effectively address the limitations of the theoretical capacity of MXene and MXene-based composites, there is yet another way of implementing MXene-based composites as anodes for LIBs in the form of 3D printing. Also, the rheological characteristics MXene inks, which determine their adaptability and processability depend on their concentrations. Other factors like the dispersion of the particle size, the amount of functionalization, and the aspect ratio can affect the rheological characteristics and these factors should be taken into account when developing the inks [64].

In general, three unique regimes in the rheological behavior of the MXene inks can be observed by altering the concentration of the inks (low, medium, and high). For example, a sol-gel transition (viscous at low frequencies and gel at high frequencies) can be seen for an m-MXene dispersion with a medium particle size of 5 μm at medium concentrations (e.g. 30 wt. percent), whereas at low concentrations (e. g. 10 wt. percent), particles are randomly dispersed with little interaction with each other. A colloidal gel with significant yield strength is created by the percolated network of nanosheets in high concentrations (>40 weight percent). Each of these behavioral regimes can be employed to formulate inks for a variety of techniques in the liquid phase, ranging from spray and spin coating, as well as screen and extrusion printing (by simply adjusting the solid content). The suitable printing or coating procedure, and thus the appropriate ink, are chosen for the preparation of the battery based on its application. While thick film deposition is preferred in most applications, thin film formation is required in some cases. Some samples of recently developed MXene inks for various battery types and primary printing/coating methods, rheological requirements, and electrochemical characteristics are explained in the following sections [70].

Chaohui Wei et al. [71], demonstrated a general method to create in situ MO_x-MXene (M: Ti, V, and Nb) heterostructures (Figure 12.6a and b) as heavy and multifunctional hosts with synchronous polysulfide immobilization and conversion to manifest good battery performance. According to theoretical estimation, the in situ grafted oxides increase the kinetics of the polysulfide transition without changing the MXene intrinsic conductivity as evident from figures, Figure 12.6c and d. Hence, the VO_X-V_2C/S electrode can achieve excellent volumetric capacity (1645 mAh cm^{-3} at 0.2 C) and cycling stability (retaining 631 mAh cm^{-3} across 1500 cycles at 2.0 C with a capacity deterioration of 0.03 percent per cycle) as shown in Figure 12.6e and f. More encouragingly, 3D printed enhanced VO_x-V_2C / sulfur electrodes exhibit an aerial capacity of 9.74 mAh cm^{-2} at 0.05 C at an elevated sulfur loading of 10.78 mg cm^{-2} with a potential to become practically viable Li-S batteries (Figure 12.6g and h).

Significantly, the electrode structure created through 3D printing is eventually in favor of improving sulfur redox reactions and ion diffusivity, as well as reducing volume change during discharge/charge. This generic and practical approach should enable the development of useful LSBs for futuristic energy storage applications.

Kai Shen et al. [72], demonstrated a dendrite-free lithium electrode, that is an anode with a low overpotential of 10 mV, long cyclic stability up to 1200 h, and enhanced areal capacities fabricated using extrusion-type 3D printing (Figure 12.7a). MXene arrays with large interspaces assist lithium nucleation and regulated both lithium-ion flux and electric field, substantially inhibiting

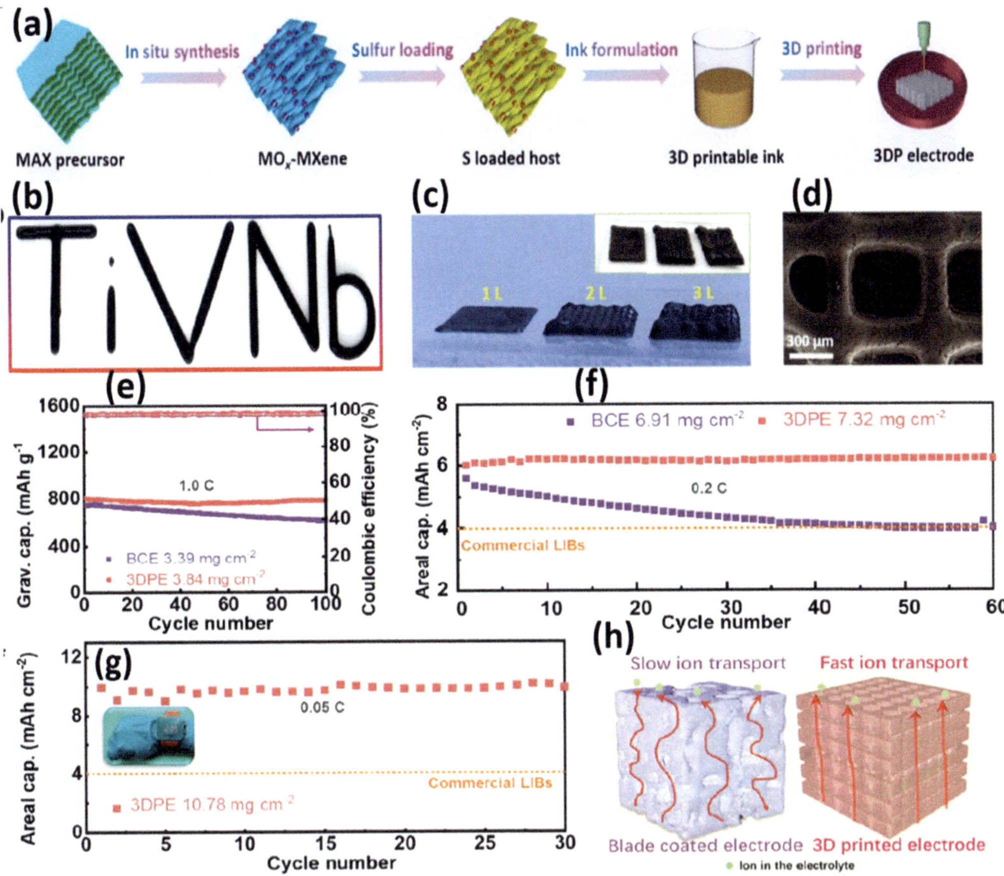

FIGURE 12.6 (a) Diagram illustrating the fabrication procedure of 3DP MO_x-MXene/S electrode, (b) Digital image displaying printing patterns on a PET substrate with the designed dimension 6 cm × 2 cm in size, (c) Optical images showing the top- and side-view of 3DP sulfur electrodes (2 cm × 2 cm) harnessing different layer numbers, (d) Top-view SEM image of a representative 3DP sulfur electrode, (e) BCE and 3DPE cycling results for VO_x-V_2C/S at 1.0 C, (f) BCE and 3DPE cycling results at 0.2 C with sulfur loadings of 6.91 and 7.32 mg cm^{-2}, respectively, (g) Performance of 3DPE while cycling at 0.05 C and a sulfur loading of 10.78 mg cm^{-2}, and (h) Diagram illustrating the pathway of ion transporters in the BCE and 3DPE models (shown by red arrows) [71]. Adapted with permission [71], Copyright (2022), American Chemical Society.

the growth of lithium dendrites. Also, provide sufficient space for the enormous growth of large cobblestone-like lithium crystals. As depicted in Figure 12.7b, a full cell is comprehended by integrating a 3D-printed anode and a $LiFePO_4$ cathode and possesses high-rate capabilities up to 30 °C and long cycle life (more than 300 cycles). This study illuminated the possibility of 3D-printed metal-based electrodes and energy storage devices of high areal capacities. Another interesting work done by Zixuan Wang et al. [73], established a 3D hierarchical porous sodiophilic V_2CT_x/rGO-CNT microgrid aerogel synthesized by using direct-ink writing technique and used as the Na metal matrix to create a Na@V_2CT_x/rGO-CNT sodium metal anode (Figure 12.7c). The V_2CT_x/rGO-CNT electrode has demonstrated a cycling life of >3000 h (2 mA/cm^{-2}, 10 mAh cm^{-2}) and average Coulombic efficiency of >99 %. In this context, it was capable of functioning uninterruptedly for over 900 hours at 5 mA cm^{-2} with an exceptional areal capacity of 50 mAh cm^{-2}. V_2CT_x with abundant sodiophilic functional groups can efficiently promote sodium metal

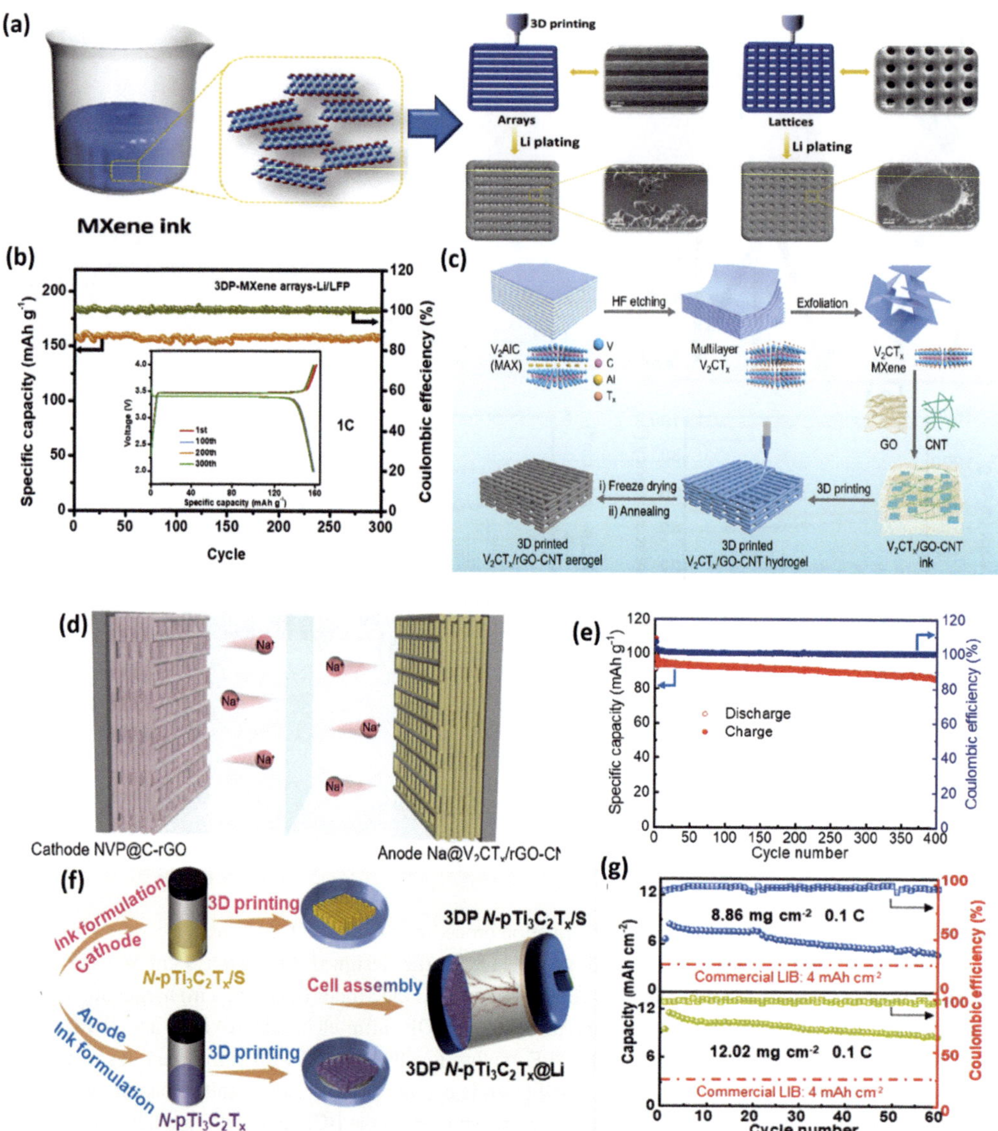

FIGURE 12.7 (a) MXene ink with a high concentration of 300 mg mL^{-1} that is used to 3D print the MXene arrays and lattices to guide the nucleation and the growth of lithium. Lithium that resembled cobblestones was grown in the arrays along the direction of the interspaces between the printed filaments. Lithium was developed extensively in the lattices around the hole walls and moving towards the hole centers, (b) Cycling stability of 3DP-MXene arrays-Li/LFP (inset figure is the typical voltage profile) [72], (c) Schematic representation of the creation of a 3D-printed V_2CTx/rGO-CNT microgrid aerogel electrode, (d) Diagram of Na@$V_{2CT}{}^{x}$/rGO-CNT||NVP@C-rGO full battery electrochemical performance evaluation, (e) Cycling performance and CE of the entire cell at 100 mA g^{-1} [73], (f) A schematic illustration depicting the process of producing LSB dual electrodes using the 3DP framework of Np$Ti_3C_2T_x$, (g) Cyclic studies of 3DP N-pTi3$C^{2T}{}_{x}$/S||3DP N-pTi3$C^{2T}{}_{x}$@Li LSB full cells with sulfur loadings of 8.86 and 12.02 mg cm^{-2} at 0.1 C (E/S ratio: 5 μL mg^{-1}) [70]. Adapted with permission [72], [73], [70] Copyright (2020, 2022, 2021), Elsevier, American Chemical Society, and Elsevier.

nucleation and uniform deposition, offering a dendrite-free morphology, according to in situ and ex situ characterizations and density functional theory simulation studies. Moreover, (Figure 12.7d) a full cell pairing a Na@V2CT$_x$/rGO-CNT anode with a Na3V$^{2(PO}$$_{4)}$^{3}@C-rGO cathode can deliver a high reversible capacity of 86.27 mAh g^{-1} after 400 cycles at 100 mA g^{-1} (Figure 12.7e). Cutting-edge Na metal anodes could be fabricated via 3D printing and this approach explains an improved Na deposition chemistry on the sodiophilic V_2CT_x/rGO-CNT microgrid aerogel electrode. Chaohui Wei et al. [70] demonstrated the uncontrolled anode dendritic growth followed by poor cathode high-loading performance. This in turn necessitated an innovative method for continuously modulating both electrodes. An adaptable 3D-printed (3DP) framework of nitrogen-doped porous Ti_3C_2 MXene (N-p$Ti_3C_2T_x$) capable of regulating the dual electrodes of Li-S batteries (Figure 12.7f) was fabricated. The fabricated 3DP scaffold was attributed to rich nitrogen sites, high conductivity, and ordered porosity to assimilate lipophilic and sulfiphilic properties. The 3DP N-p$Ti_3C_2T_x$ interlayer was able to dissipate the local current and homogenize Li deposition, performing as a dendritic inhibitor and enabling the dendrite-free anode to sustain an extremely extended lifespan of up to 800 h at 5.0 mA cm^{-2} / 5.0 mAh cm^{-2}. In the meantime, the 3DP N-p$Ti_3C_2T_x$ host repressed the polysulfide shuttle and enhanced sulfur electrochemistry, particularly under raised sulfur loadings. Figure 12.7g depicts, the 3D-printed Li-S full cells (3DPbN-p$Ti_3C_2T_x$/S||3DP N-p$Ti_3C_2T_x$@Li) that have a capacity decline of 0.06% each cycle and can run continuously for 250 cycles with a sulfur loading of 7.56 mg cm^{-2}. More significantly, after 60 cycles at 12.02 mg cm^{-2}, a maximum capacity of 8.47 mAh cm^{-2} was obtained.

To meet the need for flexible and wearable electronics, high-capacity, durable batteries are urgently needed. To increase capacity, a lot of work has gone into developing new electrode materials and electrolytes. On the other hand, the electrochemical performance of batteries can potentially be enhanced by using revolutionary manufacturing processes like smart printing. Zhang et. al. illustrated that the wonderful 2D nanosheets MXene could be a conductive binder for a simple and scalable slurry-casting method of fabrication of silicon electrodes without the interference of any other additives. In this procedure, nanoscale Si powders (nSi) and graphene-wrapped Si (Gr-Si) were added to MXene dispersion to produce their ink, which is homogeneous, viscous, and endowed with proper rheological properties. This ink was then coated on Cu foil with no other binders or conductive additive (Figure 12.8a). For nSi and Gr-Si slurries, respectively, the coated electrodes could approach a thickness 650 and 2100 μm, representing the much desirable high mass-loading of the electrode. Figure 12.8b showcases the evaporation process. After the coating, MXene sheets were randomly deposited to incorporate multiple nanosheet networks with wrapped Si nanoparticles. This nanoscale network turned out to be a highly durable composite electrode with outstanding mechanical properties during its evolution as depicted in Figure 12.8c. The concentration (or viscosity) of the MXene ink could be adjusted to modify the electrode shape, size, and thickness. One of the important properties look for high-performance batteries is their electrical conductivity and MXene nanosheets are earmarked by excellent electrical conductivity. Hence the addition of 30% MXene into Si powder would boost the conductivity of the composites nSi/MX-C (3448 S m^{-1}), nSi/MX-N (336 S m^{-1}), and Gr-Si/MX-C (5333 S m^{-1}), respectively. This may be attributed to the synergistic interaction between the MX-C sheets and the Si particles, which is supported by the strong network of the composites (Figure 12.8d and e). Furthermore, after repetitive bending, the composite electrodes uphold their outstanding electrical conductivity and admirable mechanical elasticity, leading to the possibility of flexible and wearable electronics. As shown in Figure 12.8f, the electrochemical performance was investigated and the nSi/MX-C anode shows rational stability of 280 charge/discharge cycles, Besides the nSi/MX-C electrode displayed remarkable Coulombic efficiency. Nevertheless, the nSi/MX-N electrode exhibited less cycling stability related to nSi/MX-C. The low fracture energy, and lower thermodynamic stability along with the low side conductivity of MX-N sheets decide the poor cycling stability of the networks. To enhance the mass loading and capacity at a larger level, the Gr-Si/MX-C anode was employed. Raising the MX-C mass fraction in the Gr-Si/MX-C electrode (Figure 12.8g and h) provided enhanced anode capacity that reached the theoretical values with the add-ons like mechanical properties. Furthermore, the mass loading of Gr-Si anode had an impact on its capability to cycle. The cell-level specific capacity of the composite anodes to the previously reported Si/carbon binder systems was compared and the performance of composite electrodes was evaluated based on the mass of the Cu foil, electrode, and half of the separator. On the cell level, specific capacity, nSi/MX-C and Gr-Si/MX-C electrodes demonstrated higher capacity than Si/C-binder systems as displayed in Figure 12.8i. The capacities of the nSi/MX-C and Gr-Si/MX-C electrodes were 790 mAh/g at 3.8 mg cm^{-2} areal mass loading and 815 mAh g^{-1} at 13 mg cm^{-2} areal mass loading, demonstrating that practically all Si active materials were used.

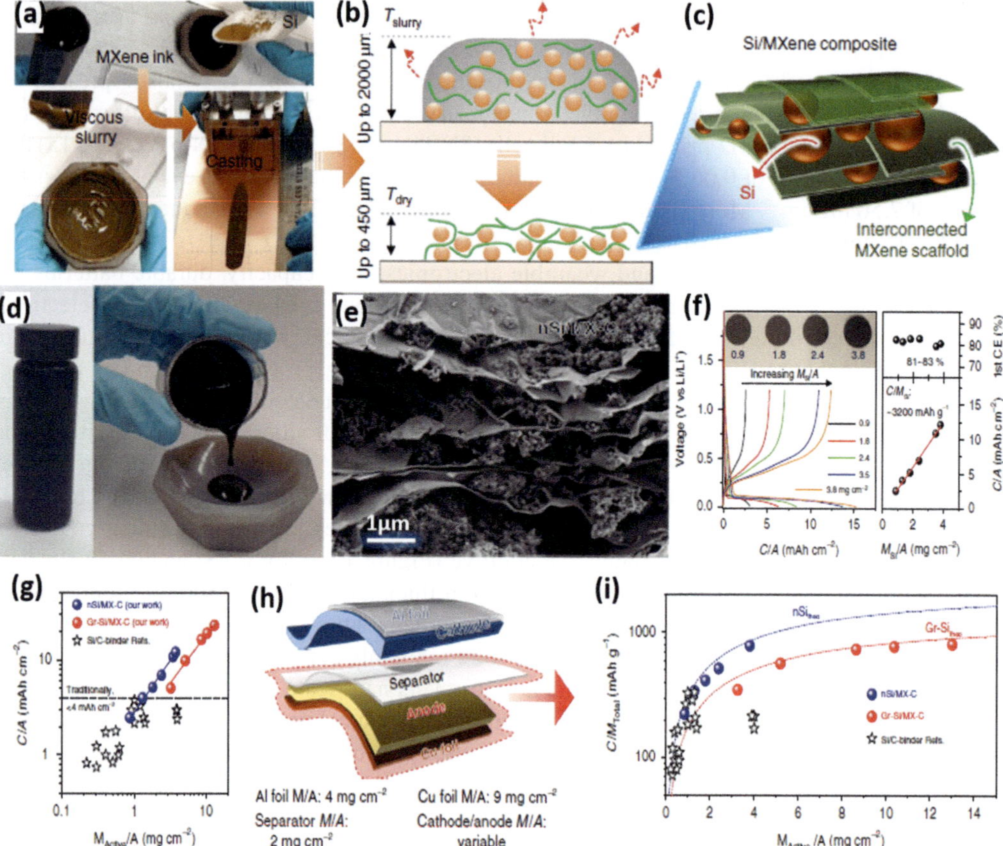

FIGURE 12.8 (a) Fabrication of composite electrode and composite electrode preparation from Si/MXene ink-based slurry, (b) the slurry drying process, (c) scheme displaying the resulting Si/MXene composite, (d) optical images of the MX-C ink, showing its viscous nature, (e) cross-sectional scanning electron microscopy (SEM) images of nSi/MX-C electrode, (f) first cycle charge-discharge curves at a 0.15 A g^{-1} (~1/20 C-rate) of the nSi/MX-C electrodes (MX-C Mf = 30 wt%) with the MSi/A ranging from 0.9 to 3.8 mg cm^{-2}. Insets are the optical images of these electrodes. Right: Coulombic efficiency, (g) areal capacity comparison of this work to other Si/conductive-binder (Si/C-binder) systems, (h) scheme of components inside the cell, highlighting the importance of utilizing high MActive/A electrodes in reducing the contribution from the inactive components, (i) cell-level specific capacity (C/MTotal) on the anode side plotted as a function of MActive/A, and compared to the reported Si/C-binder systems. Dashed lines indicate the theoretical performance of the nSi (blue line) and Gr-Si (red line) particles [74]. Adapted with permission [74], Copyright (2019), Nature.

12.5.3 Merits and Limitations of 3D-Printed MXene Electrodes for Battery

3D printing has a high degree of process flexibility and geometry control that allows the move from prototyping to production faster. 3D printing uses an eclectic range of materials, including powder, liquid solution, filament, and laminate. These adaptable methods could greatly aid in the development of advantageous nanoarchitectures ranging from zero-dimensional (0D) carbon-based nanoparticles to one-dimensional (1D) carbon nanotubes to two-dimensional (2D) graphene. 3D printing can easily realize very complex shaped and precise devices. The complexity could be in both planar and 3D space. Favorable interdigital designs that are difficult to realize using traditional techniques have easily been comprehended using different 3D printing techniques. Fast ion transport is enabled by specific space constructions with periodic or aligned pores draw advantageous to speed up the charge/discharge and improved mass loading. Figure 12.9 schematic of the chronicle condition of the printed/coated MXene process including the synthesis of MXene, optimization of electrodes in printing/coating processes, and device fabrications which have to be collectively optimized.

3D printing also known as additive manufacturing is a layer-by-layer deposition process with the help of computer-created design. This process enables us to precisely manage the thickness of electrodes. For flexible and wearable devices, thin-film energy devices could be created simply by depositing material over a few layers to build a 3D pattern. Additionally, thick electrodes, which are responsible for increased energy density per foot area can also be printed. In general, 3D printing layer-wise assembly characteristic enables good control over electrode thickness outside of the plane and thereby of high precision.

3D-printed energy devices characteristically outperform their corresponding bulk counterparts in terms of performance. The areal energy and power densities of the 3D micro battery are very high because of the use of a direct writing technique that enables the fabrication of high aspect ratio structures within a small areal footprint. [63]. Compared to the other reported carbon-based electrodes, the printed 3D and hierarchical graphene aerogel-based supercapacitor electrode

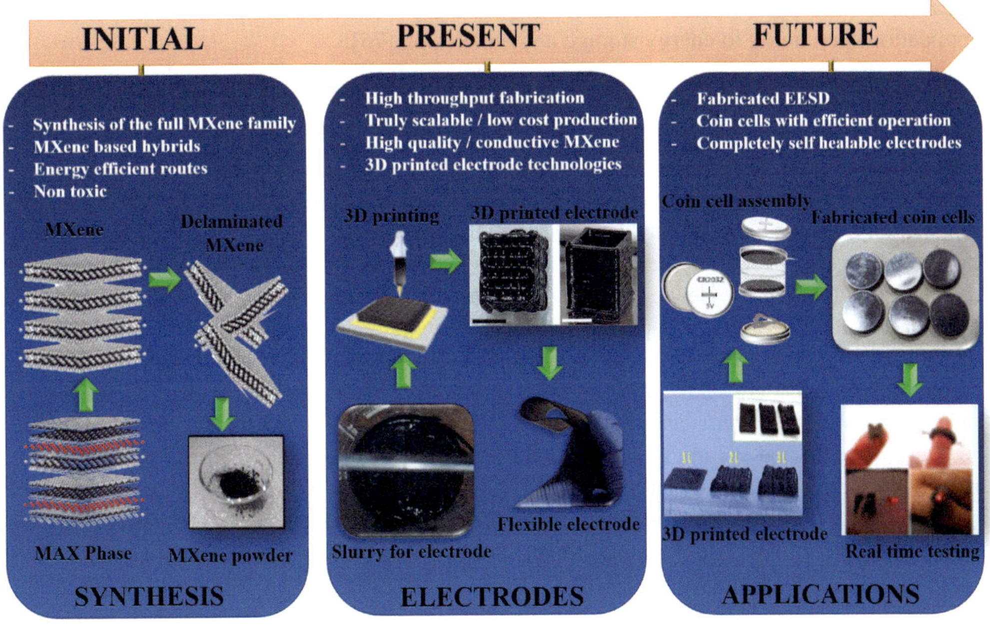

FIGURE 12.9 Scheme representing chronological order of progress of 3D printed MXene.

demonstrated the highest rate capability. [64]. Thanks to the 3D-printed macro-architecture that enables fast ion diffusion through the thick electrode.

3D printing is a low-cost and environmentally friendly approach. This is due to the greatly simplified process that allows a facile one-step method of fabrication. The additive manufacturing technique is used in 3D printing, which means that the material can be deposited on demand. Furthermore, this characteristic minimizes material waste, making it far more energy-efficient and eco-friendly than the traditional techniques.

12.6 CONCLUSIONS AND FUTURE PERSPECTIVE

3D printing is promising in terms of cracking the evolution of energy materials and altering the design and implementation of future energy systems. Despite significant progress in recent years, there are a few challenges that must be overcome before 3D printing can be used directly as a manufacturing method for the large-scale production of energy storage devices. (i) At first, most of the studies on 3D-printed EES devices have not paid attention to the packaging materials, current collectors, as well as separators compared to electrodes and electrolytes [69]. Even if DIW and IJP enable multi-material printing by preparing inks for each component that is from electrodes to packaging materials, the assembly of all-printed integrated EES devices in a single printing step is required for fully integrated manufacturing and industrial applications remains a significant challenge for 3D printing techniques. (ii) Furthermore, only a limited number of materials can currently be used in 3D printing techniques. Most of the reported studies on energy storage applications have focused on conventional inorganic materials. Currently, there are few printed materials available, especially ones that are electrochemically active for energy storage devices. To enhance the electrochemical behavior of 3D-printed energy storage devices, it is envisaged to add novel electrochemically active materials with high specific capacity and voltages to the electrode inks. From a different angle, there is a tone of room to grow the sector to include a wider range of printable materials. Since there are so many different materials that can be printed, there will be more chances for materials design and a wider range of uses for 3D printing technologies, including biomedical engineering, optics, sensors, and catalysts. (iii) Although further refinement is required before commercialization, 3D-printed architectures characterized by distinct morphologies and diverse features are especially beneficial for applications pertaining to energy storage and conversion [75].

Device performance is always governed by the electron and ion transports, and the electrode or device fabrication by using a network of nano architectures or ordered porous structures is important to accomplish superior charge storage capacity and appreciable rate ability. The fully interconnected particle networks can offer well-developed and preferentially aligned conduction channels for electrons and ions in the printed electrodes that will improve the accessible surface area for the electrolyte. These constraints can be overcome by the higher resolution 3D printing methods which are beneficial to engineering new and relevant energy storage systems with amazing size and shape diversity and micro/nanoscale dimensions. Hence empowering advanced structural and functional breakthroughs are indispensable to progress the 3D-printed energy storage systems like batteries and supercapacitors.

The 3D printing process must focus on the integration of printed energy storage devices with electronic devices and energy-harvesting systems. Integrating the 3D-printed energy storage devices with electronic circuits, systems and sensors offers the advantage of a direct and effective reduction in the overall weight and volume, lower fabrication costs, and diversity of exotic designs formulating power-source integrated electronic systems. Careful integration of energy storage devices into energy-harvesting systems like triboelectric and piezoelectric nanogenerators can easily deal with the concerns of energy-harvesting systems as well as the energy density problems of energy storage devices ensuring the reliability and stability [70]. It is strongly believed that with the unceasing growth, development, modification, and amendment of the corresponding technologies, high-build-speed 3D printing with an ultra-fine spatial resolution (high-resolution 3D printing emanates with a tradeoff) and an extensive material choice from the MXene family will ultimately emerge. This will perform a critical role in offering unprecedented freedom in commercially viable manufacturing techniques for the perfect trifecta of improved performance, complex geometries, and simplified fabrication for the next-generation energy storage devices and paving way for more value-added activity.

ACKNOWLEDGEMENT

SJ and AMK thank SPARC (SPARC/2018–2019/P392/SL) for nurturing the work; VR and SJ acknowledge financial support by UGC – Dr D. S. Kothari PDF (OT/20-21/2008/0008); SJ and JV thank financial assistance from SERB-ASEAN, India (CRD/2021/000482).

REFERENCES

1 K.J. Stevenson, Electrochemical Energy Storage, *Acc. Chem. Res.* 46 (2013) 1051–1052.

2 Y. Zhai, Y. Dou, D. Zhao, P.F. Fulvio, R.T. Mayes, S. Dai, Carbon Materials for Chemical Capacitive Energy Storage, *Adv. Mater.* 23 (2011) 4828–4850.

3 D.-W. Wang, F. Li, M. Liu, G.Q. Lu, H.-M. Cheng, 3D Aperiodic Hierarchical Porous Graphitic Carbon Material for High-Rate Electrochemical Capacitive Energy Storage, *Angew. Chemie Int. Ed.* 47 (2008) 373–376.

4 J. Xu, K. Wang, S.-Z. Zu, B.-H. Han, Z. Wei, Hierarchical Nanocomposites of Polyaniline Nanowire Arrays on Graphene Oxide Sheets with Synergistic Effect for Energy Storage, *ACS Nano.* 4 (2010) 5019–5026.

5 Z. Yang, J. Zhang, M.C.W. Kintner-Meyer, X. Lu, D. Choi, J.P. Lemmon, J. Liu, Electrochemical Energy Storage for Green Grid, *Chem. Rev.* 111 (2011) 3577–3613.

6 H.-S. Kim, J.B. Cook, H. Lin, J.S. Ko, S.H. Tolbert, V. Ozolins, B. Dunn, Oxygen Vacancies Enhance Pseudocapacitive Charge Storage Properties of MoO_{3-x}, *Nat. Mater.* 16 (2017) 454—460.

7 S. Chu, A. Majumdar, Opportunities and Challenges for a Sustainable Energy Future, *Nature.* 488 (2012) 294–303.

8 B. Mendoza-Sánchez, Y. Gogotsi, Synthesis of Two-Dimensional Materials for Capacitive Energy Storage, *Adv. Mater.* 28 (2016) 6104–6135.

9 M. Pumera, Graphene-Based Nanomaterials for Energy Storage, *Energy Environ. Sci.* 4 (2011) 668–674.

10 Z.-S. Wu, G. Zhou, L.-C. Yin, W. Ren, F. Li, H.-M. Cheng, Graphene/Metal Oxide Composite Electrode Materials for Energy Storage, *Nano Energy.* 1 (2012) 107–131.

11 J.-G. Wang, F. Kang, B. Wei, Engineering of MnO_2-Based Nanocomposites for High-Performance Supercapacitors, *Prog. Mater. Sci.* 74 (2015) 51–124.

12 X. Tian, J. Jin, S. Yuan, C.K. Chua, S.B. Tor, K. Zhou, Emerging 3D-Printed Electrochemical Energy Storage Devices: A Critical Review, *Adv. Energy Mater.* 7 (2017) 1700127.

13 Y. Zhu, S. Murali, M.D. Stoller, K.J. Ganesh, W. Cai, P.J. Ferreira, A. Pirkle, R.M. Wallace, K.A. Cychosz, M. Thommes, D. Su, E.A. Stach, R.S. Ruoff, Carbon-Based Supercapacitors Produced by Activation of Graphene, *Science.* 332 (2011) 1537–1541.

14 X.Y. Yu, X.W. (David) Lou, Mixed Metal Sulfides for Electrochemical Energy Storage and Conversion, *Adv. Energy Mater.* 8 (2018) 1701592.

15 J.-G. Wang, H. Liu, H. Sun, W. Hua, H. Wang, X. Liu, B. Wei, One-Pot Synthesis of Nitrogen-Doped Ordered Mesoporous Carbon Spheres for High-Rate and Long-Cycle Life Supercapacitors, *Carbon N. Y.* 127 (2018) 85–92.

16 X. Cai, Y. Luo, B. Liu, H.- M. Cheng, Preparation of 2D Material Dispersions and Their Applications, *Chem. Soc. Rev.* 47 (2018) 6224–6266.

17 M. Cakici, R.R. Kakarla, F. Alonso-Marroquin, Advanced Electrochemical Energy Storage Supercapacitors Based on the Flexible Carbon Fiber Fabric-Coated with Uniform Coral-Like MnO_2 Structured Electrodes, *Chem. Eng. J.* 309 (2017) 151–158.

18 N. Cirigliano, G. Sun, D. Membreno, P. Malati, C.J. Kim, B. Dunn, 3D Architectured Anodes for Lithium-Ion Microbatteries with Large Areal Capacity, *Energy Technol.* 2 (2014) 362–369.

19 K. Chen, D. Xue, Materials Chemistry Toward Electrochemical Energy Storage, *J. Mater. Chem. A.* 4 (2016) 7522–7537.

20 B. Scrosati, J. Garche, Lithium Batteries: Status, Prospects and Future, *J. Power Sources.* 195 (2010) 2419–2430.

21 X. Li, B. Wei, Supercapacitors Based on Nanostructured Carbon, *Nano Energy.* 2 (2013) 159–173.

22 J. Gao, L. Li, J. Tan, H. Sun, B. Li, J.C. Idrobo, C.V. Singh, T.- M. Lu, N. Koratkar, Vertically Oriented Arrays of ReS_2 Nanosheets for Electrochemical Energy Storage and Electrocatalysis, *Nano Lett.* 16 (2016) 3780–3787.

23 W. Xu, S. Jambhulkar, D. Ravichandran, Y. Zu, M. Kakarla, Q. Nian, B. Azeredo, X. Chen, K. Jin, B. Vernon, D G. Lott, J L. Cornella, O. Shefi, G M. Garner, Y. Yang, K. Song, 3D printing-enabled nanoparticle alignment: a review of mechanisms and applications, *Small, 17* (2021), 2100817.

24 D. Xiong, X. Li, Z. Bai, S. Lu, Recent Advances in Layered $Ti_3C_2T_x$ MXene for Electrochemical Energy Storage, *Small.* 14 (2018) 1703419.

25 W. Xu, Y. Zu, D. Ravichandran, S. Jambhulkar, M. Kakarla, M. Bawareth, S. Lanke, K. Song, Review of fiber-based three-dimensional printing for applications ranging from nanoscale nanoparticle alignment to macroscale patterning. *ACS Applied Nano Materials, 4* (2021), 7538–7562.

26 M. Beidaghi, C. Wang, Supercapacitors: Micro-Supercapacitors Based on Interdigital Electrodes of Reduced Graphene Oxide and Carbon Nanotube Composites with Ultrahigh Power Handling Performance, *Adv. Funct. Mater.* 22 (2012) 4500.

27 K. Fu, Y. Yao, J. Dai, L. Hu, Progress in 3D Printing of Carbon Materials for Energy-Related Applications, *Adv. Mater.* 29 (2017) 1603486.

28 A.D. Valentine, T.A. Busbee, J.W. Boley, J.R. Raney, A. Chortos, A. Kotikian, J.D. Berrigan, M.F. Durstock, J.A. Lewis, Hybrid 3D Printing of Soft Electronics, *Adv. Mater.* 29 (2017) 1703817.

29 B.G. Compton, J.A. Lewis, 3D Printing: 3D-Printing of Lightweight Cellular Composites, *Adv. Mater.* 26 (2014) 6043.

30 X. Liu, R. Jervis, R.C. Maher, I.J. Villar-Garcia, M. Naylor-Marlow, P.R. Shearing, M. Ouyang, L. Cohen, N.P. Brandon, B. Wu, 3D-Printed Structural Pseudocapacitors, *Adv. Mater. Technol.* 1 (2016) 1600167.

31 D. Ravichandran, W. Xu, S. Jambhulkar, Y. Zu, M. Kakarla, M. Bawareth, K. Song, Intrinsic Field-Induced Nanoparticle Assembly in Three-Dimensional (3D) Printing Polymeric Composites. *ACS Applied Materials & Interfaces, 13* (2021), 52274–52294.

32 M. Wei, F. Zhang, W. Wang, P. Alexandridis, C. Zhou, G. Wu, 3D Direct Writing Fabrication of Electrodes for Electrochemical Storage Devices, *J. Power Sources.* 354 (2017) 134–147.

33 B. Hsia, J. Marschewski, S. Wang, J. Bin In, C. Carraro, D. Poulikakos, C.P. Grigoropoulos, R. Maboudian, Highly Flexible, All Solid-State Micro-Supercapacitors from Vertically Aligned Carbon Nanotubes, *Nanotechnology.* 25 (2014) 55401.

34 C. Ladd, J.-H. So, J. Muth, M.D. Dickey, 3D Printing of Free Standing Liquid Metal Microstructures, *Adv. Mater.* 25 (2013) 5081–5085.

35 D. Pech, M. Brunet, P.-L. Taberna, P. Simon, N. Fabre, F. Mesnilgrente, V. Conédéra, H. Durou, Elaboration of a Microstructured Inkjet-Printed Carbon Electrochemical Capacitor, *J. Power Sources.* 195 (2010) 1266–1269.

36 S.S. Hoseini, M.G. Sauer, Molecular Cloning Using Polymerase Chain Reaction, an Educational Guide for Cellular Engineering, *J. Biol. Eng.* 9 (2015) 2.

37 J. Kim, R. Kumar, A.J. Bandodkar, J. Wang, Advanced Materials for Printed Wearable Electrochemical Devices: A Review, *Adv. Electron. Mater.* 3 (2017) 1600260.

38 S.C. Ligon, R. Liska, J. Stampfl, M. Gurr, R. Mülhaupt, Polymers for 3D Printing and Customized Additive Manufacturing, *Chem. Rev.* 117 (2017) 10212–10290.

39 M. Zhu, F. Zhu, O.G. Schmidt, Nano Energy for Miniaturized Systems, *Nano Mater. Sci.* 3 (2021) 107–112.

40 J.C. Ruiz-Morales, A. Tarancón, J. Canales-Vázquez, J. Méndez-Ramos, L. Hernández-Afonso, P. Acosta-Mora, J.R. Marín Rueda, R. Fernández-González, Three Dimensional Printing of Components and Functional Devices for Energy and Environmental Applications, *Energy Environ. Sci.* 10 (2017) 846–859.

41 S. Lawes, A. Riese, Q. Sun, N. Cheng, X. Sun, Printing Nanostructured Carbon for Energy Storage and Conversion Applications, *Carbon N. Y.* 92 (2015) 150–176.

42 A. Zhakeyev, P. Wang, L. Zhang, W. Shu, H. Wang, J. Xuan, Additive Manufacturing: Unlocking the Evolution of Energy Materials, *Adv. Sci.* 4 (2017) 1700187.

43 C. Zhu, T. Liu, F. Qian, W. Chen, S. Chandrasekaran, B. Yao, Y. Song, E.B. Duoss, J.D. Kuntz, C.M. Spadaccini, M.A. Worsley, Y. Li, 3D Printed Functional Nanomaterials for Electrochemical Energy Storage, *Nano Today.* 15 (2017) 107–120.

44 P.-Y. Chen, M. Liu, Z. Wang, R.H. Hurt, I.Y. Wong, From Flatland to Spaceland: Higher Dimensional Patterning with Two-Dimensional Materials, *Adv. Mater.* 29 (2017) 1605096.

45 M. Mao, J. He, X. Li, B. Zhang, Q. Lei, Y. Liu, D. Li, The Emerging Frontiers and Applications of High-Resolution 3D Printing, *Micromachines*, 8 (2017) 113.

46 J.A. Lewis, J.E. Smay, J. Stuecker, J. Cesarano, Direct Ink Writing of Three-Dimensional Ceramic Structures, *J. Am. Ceram. Soc.* 89 (2006) 3599–3609.

47 S.V Murphy, A. Atala, 3D Bioprinting of Tissues and Organs, *Nat. Biotechnol.* 32 (2014) 773–785.

48 M.A. Skylar-Scott, S. Gunasekaran, J.A. Lewis, Laser-Assisted Direct Ink Writing of Planar and 3D Metal Architectures, *Proc. Natl. Acad. Sci.* 113 (2016) 6137–6142.

49 J. Li, X. Liang, R. Panat, J. Park, Enhanced Battery Performance through Three-Dimensional Structured Electrodes: Experimental and Modeling Study, *J. Electrochem. Soc.* 165 (2018) A3566–A3573.

50 A. Ambrosi, M. Pumera, 3D-Printing Technologies for Electrochemical Applications, *Chem. Soc. Rev.* 45 (2016) 2740–2755.

51 A. Izumi, M. Sanada, K. Furuichi, K. Teraki, T. Matsuda, K. Hiramatsu, H. Munakata, K. Kanamura, Development of High Capacity Lithium-Ion Battery Applying Three-Dimensionally Patterned Electrode, *Electrochim. Acta.* 79 (2012) 218–222.

52 K. Sun, T.-S. Wei, B.Y. Ahn, J.Y. Seo, S.J. Dillon, J.A. Lewis, 3D Printing of Interdigitated Li-Ion Microbattery Architectures, *Adv. Mater.* 25 (2013) 4539–4543.

53 X. Fan, Doping and Design of Flexible Transparent Electrodes for High-Performance Flexible Organic Solar Cells: Recent Advances and Perspectives, *Adv. Funct. Mater.* 31 (2021) 2009399.

54 C. (John) Zhang, M.P. Kremer, A. Seral-Ascaso, S.-H. Park, N. McEvoy, B. Anasori, Y. Gogotsi, V. Nicolosi, Stamping of Flexible, Coplanar Micro-Supercapacitors Using MXene Inks, *Adv. Funct. Mater.* 28 (2018) 1705506.

55 G.S. Gund, J.H. Park, R. Harpalsinh, M. Kota, J.H. Shin, T. Kim, Y. Gogotsi, H.S. Park, MXene/Polymer Hybrid Materials for Flexible AC-Filtering Electrochemical Capacitors, *Joule.* 3 (2019) 164–176.

56 T.-S. Wei, B.Y. Ahn, J. Grotto, J.A. Lewis, 3D Printing of Customized Li-Ion Batteries with Thick Electrodes, *Adv. Mater.* 30 (2018) 1703027.

57 K. Fu, Y. Wang, C. Yan, Y. Yao, Y. Chen, J. Dai, S. Lacey, Y. Wang, J. Wan, T. Li, Z. Wang, Y. Xu, L. Hu, Graphene Oxide-Based Electrode Inks for 3D-Printed Lithium-Ion Batteries, *Adv. Mater.* 28 (2016) 2587–2594.

58 R.R. Kohlmeyer, A.J. Blake, J.O. Hardin, E.A. Carmona, J. Carpena-Núñez, B. Maruyama, J. Daniel Berrigan, H. Huang, M.F. Durstock, Composite Batteries: A Simple yet Universal Approach to 3D Printable Lithium-Ion Battery Electrodes, *J. Mater. Chem. A.* 4 (2016) 16856–16864.

59 Y. Wang, C. Chen, H. Xie, T. Gao, Y. Yao, G. Pastel, X. Han, Y. Li, J. Zhao, K. (Kelvin) Fu, L. Hu, 3D-Printed All-Fiber Li-Ion Battery Toward Wearable Energy Storage, *Adv. Funct. Mater.* 27 (2017) 1703140.

60 C. Liu, X. Cheng, B. Li, Z. Chen, S. Mi, C. Lao, Fabrication and Characterization of 3D-Printed Highly-Porous 3D $LiFePO_4$ Electrodes by Low Temperature Direct Writing Process, *Materials* 10 (2017) 934.

61 J. Hu, Y. Jiang, S. Cui, Y. Duan, T. Liu, H. Guo, L. Lin, Y. Lin, J. Zheng, K. Amine, F. Pan, 3D-Printed Cathodes of $LiMn_{1-x}Fe_xPO_4$ Nanocrystals Achieve Both Ultrahigh Rate and High Capacity for Advanced Lithium-Ion Battery, *Adv. Energy Mater.* 6 (2016) 1600856.

62 S. Naficy, R. Jalili, S.H. Aboutalebi, R.A. Gorkin III, K. Konstantinov, P.C. Innis, G.M. Spinks, P. Poulin, G.G. Wallace, Graphene Oxide Dispersions: Tuning Rheology to Enable Fabrication, *Mater. Horiz.* 1 (2014) 326–331.

63 J.H. Kim, W.S. Chang, D. Kim, J.R. Yang, J.T. Han, G.-W. Lee, J.T. Kim, S.K. Seol, 3D Printing of Reduced Graphene Oxide Nanowires, *Adv. Mater.* 27 (2015) 157–161.

64 G. Sun, J. An, C.K. Chua, H. Pang, J. Zhang, P. Chen, Layer-by-Layer Printing of Laminated Graphene-Based Interdigitated Microelectrodes for Flexible Planar Micro-Supercapacitors, *Electrochem. Commun.* 51 (2015) 33–36.

65 M. Naguib, M. Kurtoglu, V. Presser, J. Lu, J. Niu, M. Heon, L. Hultman, Y. Gogotsi, M.W. Barsoum, Two-dimensional nanocrystals produced by exfoliation of Ti_3AlC_2, *Adv. Mater.* 23 (2011) 4248–4253.

66 M. Naguib, J. Halim, J. Lu, K.M. Cook, L. Hultman, Y. Gogotsi, M.W. Barsoum, K.M. Cook, L. Hultman, Y. Gogotsi, M.W. Barsoum, New Two-Dimensional Niobium and Vanadium Carbides as Promising Materials for Li-Ion Batteries, *J. Am. Chem. Soc.* 135 (2013) 15966–15969.

67 Q. Tang, Z. Zhou, P. Shen, Are MXenes Promising Anode Materials for Li Ion Batteries? Computational Studies on Electronic Properties and Li Storage Capability of Ti_3C_2 and $Ti_3C_2X_2$ (X = F, OH) Monolayer, *J. Am. Chem. Soc.* 134 (2012) 16909–16916.

68 Y. Dong, S. Zheng, J. Qin, X. Zhao, H. Shi, X. Wang, J. Chen, Z.-S. Wu, All-MXene-Based Integrated Electrode Constructed by Ti_3C_2 Nanoribbon Framework Host and Nanosheet Interlayer for High-Energy-Density Li–S Batteries, *ACS Nano.* 12 (2018) 2381–2388.

69 H. Shi, C.J. Zhang, P. Lu, Y. Dong, P. Wen, Z.-S. Wu, Conducting and Lithiophilic MXene/Graphene Framework for High-Capacity, Dendrite-Free Lithium–Metal Anodes, *ACS Nano.* 13 (2019) 14308–14318.

70 C. Wei, M. Tian, Z. Fan, L. Yu, Y. Song, X. Yang, Z. Shi, M. Wang, R. Yang, J. Sun, Concurrent Realization of Dendrite-Free Anode and High-Loading Cathode via 3D Printed N-Ti_3C_2 MXene Framework Toward Advanced Li–S Full Batteries, *Energy Storage Mater.* 41 (2021) 141–151.

71 C. Wei, M. Tian, M. Wang, Z. Shi, L. Yu, S. Li, Z. Fan, R. Yang, J. Sun, Universal in situ Crafted MOx-MXene Heterostructures as Heavy and Multifunctional Hosts for 3D-Printed Li-S Batteries, *ACS Nano.* 14 (2020) 16073–16084.

72 K. Shen, B. Li, S. Yang, 3D Printing Dendrite-Free Lithium Anodes Based on the Nucleated MXene Arrays, *Energy Storage Mater.* 24 (2020) 670–675.

73 Z. Wang, Z. Huang, H. Wang, W. Li, B. Wang, J. Xu, T. Xu, J. Zang, D. Kong, X. Li, H.Y. Yang, Y. Wang, 3D-Printed Sodiophilic V_2CTx/rGO-CNT MXene Microgrid Aerogel for Stable Na Metal Anode with High Areal Capacity, *ACS Nano.* 16 (2022) 9105–9116.

74 C. (John) Zhang, S.-H. Park, A. Seral-Ascaso, S. Barwich, N. McEvoy, C.S. Boland, J.N. Coleman, Y. Gogotsi, V. Nicolosi, High Capacity Silicon Anodes Enabled by MXene Viscous Aqueous Ink, *Nat. Commun.* 10 (2019) 849.

75 Z. Wang, Q. Zhang, S. Long, Y. Luo, P. Yu, Z. Tan, J. Bai, B. Qu, Y. Yang, J. Shi, H. Zhou, Z.-Y. Xiao, W. Hong, H. Bai, Three-Dimensional Printing of Polyaniline/Reduced Graphene Oxide Composite for High-Performance Planar Supercapacitor, *ACS Appl. Mater. \& Interfaces.* 10 (2018) 10437–10444.

13 3D-Printed Nanocomposites for Batteries

Yash Desai[1,2], *Kwadwo Mensah Darkwa*[3], *and Ram K. Gupta*[1,2*]
[1]Department of Chemistry, Pittsburg State University, Pittsburg, Kansas 66762, USA
[2]National Institute for Material Advancement, Pittsburg, Kansas 66762, USA
[3]Department of Materials Engineering, College of Engineering, Kwame Nkrumah University of Science and Technology, Kumasi, Ghana
*Corresponding author: ramguptamsu@gmail.com

CONTENTS

13.1 INTRODUCTION

Energy in the world is directly linked with the well-being and prosperity of mankind. It comes in many forms and sources. Subsequently, energy is required to run everything in the world and outer space. The surge in energy makes it a very important aspect; without energy, life is impossible. The Sun is the largest source of energy for the Earth. However, in the present day, we are most reliant on fossil fuels for generating and supplying energy, these are non-renewable sources of energy. However, due to the overexploitation of these resources, severe adverse effects have been observed in the atmosphere. Solar and tidal energy are good sources but they aren't continuous in supplying power, which forces us to develop new and better energy storage systems. However, with the growing population and increasing consumption of energy, the development of new technologies becomes vital to support the functioning of the planet. The development of new technologies to generate energy will need efficient storage methods. The world is progressing toward decarbonization

DOI: 10.1201/9781003296676-13

which will escalate the growth of the energy storage industry. Decarbonization is the term used to denote the process of reducing the dependence on the energy sources such as fossil fuels which give out carbon dioxide gas while supplying energy and have adverse effects on the atmosphere such as global warming. These issues enable new opportunities and approaches for building different types of storage energy devices such as mechanical, thermal, electrochemical, and chemical. Among the techniques, electrical and electrochemical storage devices have been widespread due to their easy operation and user-friendly operation.

From the above categories, electrochemical and electrical comprise several devices, such as batteries, flow batteries, supercapacitors, and superconducting magnetic energy storage, which have gained popularity due to their high performance. These devices display good performance due to their high energy density and long lifespan. The performance of batteries and supercapacitors depends on electrochemical processes that determine the relative energy and power density. Many recent studies have prevailed that several materials demonstrate the high power density of batteries as well as the long cycle life and short charging times of supercapacitors which will be a breakthrough for manufacturing energy storage devices (ESD) [1]. Furthermore, materials used to manufacture energy storage devices play a very important role as they make up the largest proportion of the system cost, performance, and properties of energy storage devices. Polymeric materials such as polyvinylidene fluoride (PVDF) have been a point of research for their applications in energy storage devices such as an electrolyte. PVDF and its copolymers are also being used as a host for gel polymer electrolytes [2]. For the last two centuries, batteries have been mainly fabricated using liquid/semiliquid electrolytes due to their high ionic conductivity and excellent wetting properties of the electrodes, but they were vulnerable to corrosion and leakage. These issues are even more concerning due to the increasing demand for energy storage devices for handheld devices, so replacing liquid electrolytes with inorganic material-based solid-state electrolytes (SSE) is necessary. The batteries with SSE display exceptionally high power density, durability, and better safety [3].

Apart from conductive polymers and inorganic metals, nanocomposites and nanomaterials are the materials which been a great area of research for electrochemical energy storage devices. Nanomaterials have enabled electrically pseudocapacitive materials like transition metal oxides to be used more efficiently in batteries and capacitors [4]. Nanomaterials in form of natural biopolymers such as nano-cellulose are used in several ESDs due to their unique and attractive mechanical and electrochemical properties that make nano-cellulose highly suitable for manufacturing several energy storage components such as electrolytes, separators, binders, and substrates [5]. Apart from nano-cellulose, researchers have also focused on building ESDs through several eco-friendly bionanomaterials such as lignin, cotton, and protein (silk) by incorporating electroactive nanomaterials for example carbon nanotubes (CNT), graphene, and metallic nanomaterials to make them electronically conductive [6]. Once we select the appropriate materials for manufacturing energy storage devices, there is a challenge for the process selection, here 3D printing comes into play. 3D printing has been a popular method in manufacturing quick prototypes of the product and parts which can be further tested to know if they comprise the desired properties or need improvement to enhance the overall performance. Upon producing the prototypes 3D printing also enable us to produce innovative state-of-the-art products which are a leap ahead of the energy storage technologies [7].

13.2 CHARACTERISTICS AND TYPES OF BATTERIES

To understand the working principle of batteries, it's necessary to study the fundamental characteristics and functioning of a battery. This section describes the basic terminologies used in batteries, such as anode, cathode, theoretical potential of battery, theoretical capacity, specific capacity, theoretical energy, specific energy, energy density, coulombic efficiency, and C-rate.

13.2.1 Anode and Cathode

The anode is even noted as a negative electrode that undergoes an oxidation reaction during the discharging process of an electrochemical process. The cathode is known as the positive electrode where a reduction reaction occurs during the discharge process.

13.2.2 Theoretical Voltage

Theoretical voltage is described based on the electrode redox potentials. It is also denoted as standard cell voltage calculated as the difference between cathode standard potential and anode standard potential.

13.2.3 Theoretical and Specific Capacity

The theoretical capacity of the cell is the maximum capability to store electric charge by using the overall potential of the electrode materials. The unit of capacity is the coulomb. However, the specific capacity is the capability of the electrode material to store charge per unit mass of area.

13.2.4 Theoretical and Specific Energy

Theoretical energy is the highest energy output by the cell with a theoretical capacitance and voltage. Its unit is Watt-hour. While the specific energy is the highest energy produced by a cell from the unit mass of the active material and has a unit of Wh/g.

13.2.5 Coulombic Efficiency, C-Rate, and Current Density

The coulombic efficiency is the ratio of the overall charge output of the battery to the overall charge total charge in the battery during the complete cycle or can be simplified as the ratio of discharge and charge capacities of the cell. While the C-rate is defined as the ratio of the current required for charging to the current obtained on discharging of the cell. The current density of the cell can be explained as the current flow per cross-sectional area of the electrode.

13.2.6 Types of Batteries

Every battery runs on almost the same principle of converting chemical energy into electrical energy. Redox reaction occurs during the transformation of energy into the cell. Here redox reaction is the reduction and oxidation of the different electrode materials dipped into the electrolyte solution and connected to an external circuit. This mechanism remains nearly the same on different cells and batteries. The batteries are categorized based on charge and discharge mechanism, the cells in which the redox reaction cannot be reversed are known as primary cells and in contrast, the secondary cells, hold the potential for redox reversibility. So, the secondary batteries can be recharged and discharged repeatedly. Batteries are also differentiated based on the materials they are made from, and they are mostly secondary batteries. Metal-based secondary cells consist of electrodes of two different metals such as copper, silver, etc. These electrodes are dipped into electrolytes and separated with a permeable membrane which allows the flow of charge but restricts electrolytes to mix. Li-ion batteries are the most popular type nowadays due to their performance. While they follow the same configuration of the components as discussed above but they possess high energy density which makes them applicable for various applications for example in mobile phones, laptops, and electric vehicles as well.

They follow the most common layered structure which helps in quick charge transfer and good retention of electrolytes. However, the same layered structure has some drawbacks such as they tend to overheat and get sluggish after some time. Other than the metal-based batteries there are other types such as organic batteries which display a better redox reaction than Li-ion batteries and have an enhanced reaction rate. The organic batteries also have different subcategories such as p-type, n-type, and b-type. Further differentiating, p-type materials undergo oxidation producing cations that are processed on the anode. On the other hand, n-type cell configuration undergoes reduction which produces anions and a processed on the cathode. Both oxidation and reduction are carried out on the b-type cell which is used on the anode and cathode respectively. The metal-organic type of cells is the other type of configuration for the cells which is considered as the anodic material and organic cathode of p-type or n-type.

Batteries and cells even have different housing/casings which comprise coin cells, these coin cells are mostly known for testing the electrodes. They have a porous membrane separating the electrodes, due to their minimal size fabrication and testing of the cell becomes fast and easy. Another cell housing types are pouch cell which has a sandwich type of structure where several layers of cathode and anode are packed together with foils in between which inhabit the direct contact of the electrodes. Current is passed through the current collector present on each terminal and the final housing is made using an aluminum plastic box. The commercially used cells are the cylindrical housing type due to their design which makes them mechanically stable. They share the same sequence of component placement as the other cell where the cathode and anode electrodes are separated with a porous membrane.

13.3 3D-PRINTED NANOCOMPOSITES FOR BATTERIES

13.3.1 Layered Materials-Based Nanocomposites

3D printing has been in popularity for the last 20 years with several applications which could be only possible due to several methods and materials. The same versatility of additive manufacturing technology and the combination of nanomaterials has been useful in manufacturing energy storage devices and has helped enhance the performance of the devices quite drastically. Among all the techniques of additive manufacturing, 3D printing with layered structure nanomaterials comes with the most cost-effective process and has the potential for mass manufacturing. Following this, some researchers developed high-performance Li-ion batteries with microarchitecture materials such as poly-lactic acid, lithium iron phosphate, and carbon nanotubes [8]. Here in this additive manufacturing process, they prepared the mixture of materials through solvent casting, where the electrode materials were dissolved into dichloromethane for processing. Once the material composition was prefixed in the solvent, this slurry was then processed on a twin-screw extruder which gave out filaments from the same. The nanomaterial is then processed onto a commercial 3D printer and with controlled process parameters, micro-architected cell lattices were formed (Figure 13.1). The 3D-printed electrodes attained porous frameworks and a better conducting network is achieved through CNT which enhances the ion and electron transfer reactions [8].

Other than the liquid phase exfoliation, several printing techniques have been used for manufacturing the 3D-printed micro and nanoarchitecture structures. Some researchers worked on the facile technique which fabricated interdigitated micro battery architectures composed of $Li_2Ti_5O_{12}$ (LTO) and $LiFePO_4$ (LFP) which were subjected as anode and cathode materials respectively (Figure 13.2) [9]. These materials were used as ink slurries for 3D-printed Li-ion batteries. Inks were prepared as suspension of nanomaterials (LTO and LFP) in a solution composed of various solvents such as deionized water, ethylene glycol, glycerol, and cellulose-based viscofiers via a multi-step process of particle dispersion, centrifugation, and homogenization. After mixing the materials in the solution, electrodes were printed on a glass substrate through a cylindrical nozzle of 30 μm control ink. This process had a drawback of solidification of the ink. So, to prevent solidification and adhesion during the patterning process a graded volatility solvent system was used in which water evaporation during printing allowed partial solidification of the printed structures ensuring the structural integrity, other

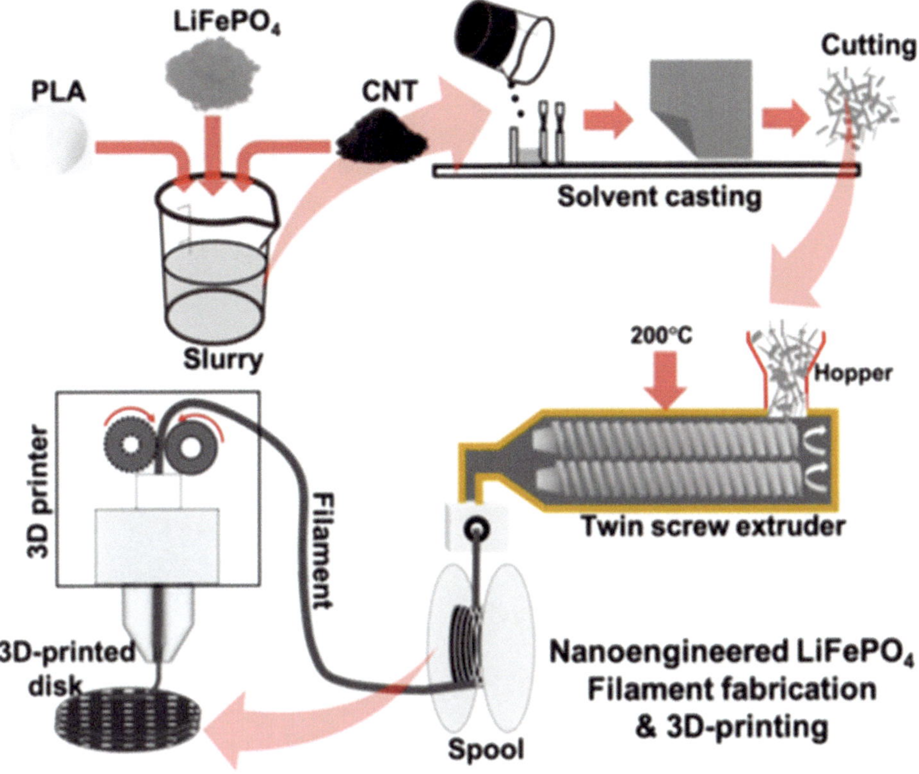

FIGURE 13.1 Representing the process of manufacturing 3D-printed Li-ion batteries with carbon nanotubes, polylactic acid, and lithium iron phosphate. Adapted with permission [8], Copyright (2021), Elsevier.

solvents such as ethylene glycol and glycerol serve as humectants that promote bonding between the individual layers. Once the layer-by-layer structure was ready the electrodes were processed at high temperatures (up to 600 °C) to remove organic additives and allow nanoparticle sintering. Particle sintering at high temperatures allowed high porosity which made the surface favorable for electrode penetration. Using the LTO-LFP system, deflected 3D-IMA with high areal energy of 9.7 J/cm^2 and a power density of 2.7 mW/cm^2. This attests to the potential application of micro/nano batteries in electronics and biomedical devices.

Moreover, with high areal energy and power density, cycle life is also a crucial characteristic that is considered while manufacturing and performing batteries. Wherefore, sodium ion-based batteries with graphene oxide-based ink used for building filament and multi-hole gridding frameworks were developed by some researchers. Such a structure enables the quick exchange of sodium ions as well as electrons. Therefore, the frameworks possessed high specific energy with an excellent rate of performance and periodic steadiness for sodium storage batteries [10]. Graphene oxide and $Na_3V_2(PO_4)_3$ (NVP) inks were made by providing different concentrations of NVP in the solutions (Figure 13.3) [10]. Furthermore, to test the electrochemical performance of the 3D-printed frameworks, 2-layer frameworks were formed with an electrode width of 200 μm and spacing of 800 μm on stainless steel sheets followed by freeze-drying and annealing at 200 °C for two hours in an inert atmosphere. The areal loading of 3D-printed NVP-reduced GO frameworks was 18 mg/cm^2. The structural framework electrode material displayed a promising capacity of 117.6 mAh/g for the above-fixed configuration and areal loading of NVP and rGO-induced 3D-printed sodium storage systems. The

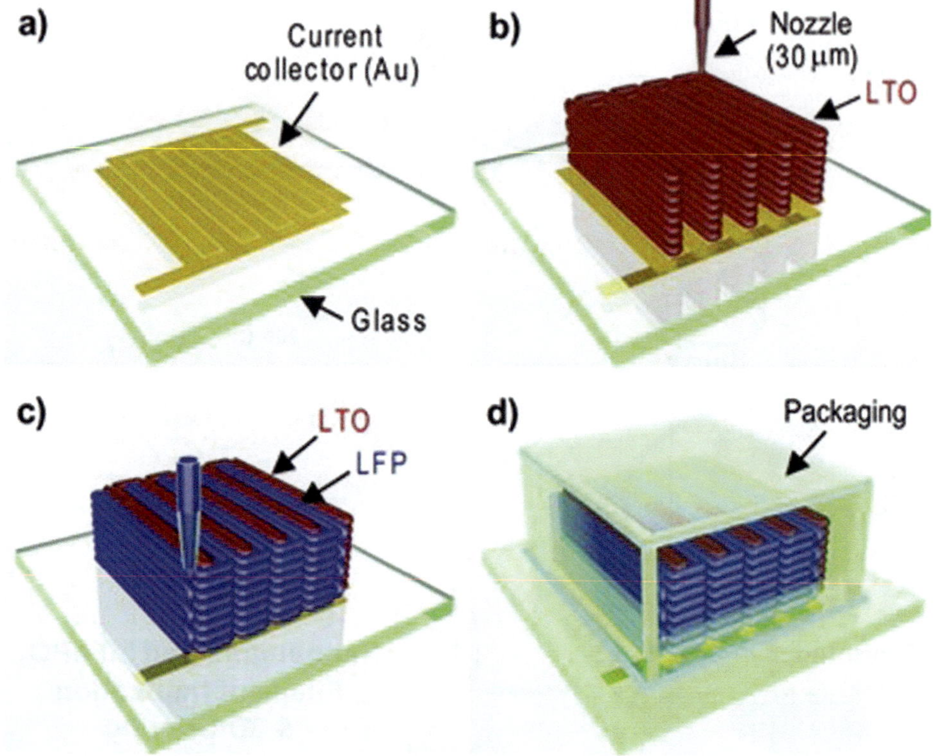

FIGURE 13.2 Schematic representation of 3D-printed microbattery. (a) gold current collector, (b) LTO, (c) LFP, and (d) packaging. Adapted with permission [9], Copyright (2013), John Wiley and Sons.

framework of the filaments is porous and crosslinked by numerous flexible nanosheets which ease the flow of sodium ions and electrons through the system.

Flexible 3D layered batteries have also been a field of interest for researchers when energy storage devices are considered. Flexible and 3D-printed energy storage devices mostly use polymers such as poly (vinylidene fluoride) (PVDF). Phase inversion and high ceramic loading were used to manufacture Li-ion battery electrolytes to demonstrate high performance. The polymeric material PVDF was first dissolved into N-methyl-2-pyrrolidone (NMP) and glycerol. These solvents imparted high porosity and phase inversion which has been a successful method even in other research to achieve highly porous structures. However, a large cellular structure is formed which is not at all safe and suitable due to its incapability to inhabit dendritic lithium growth. So, to troubleshoot this issue aluminum hydroxide (Al_2O_3) nanoparticles were introduced into the polymer matrix. The nanoparticles helped to reduce the pore size and shrinkage. Subsequently, the mixture did inhabit the dendritic lithium growth and gained thermal stability as well as electrolyte wetting has been enhanced by this mixture. Even the characterization of electrolyte ink on a scanning electron microscope (SEM) attested that interaction of the elements generated sub-micrometer porous structures. The mechanical properties of the ink had a high strain to failure which made the concentrate suitable for flexible printed battery applications. Interestingly, in the current density test of the electrolyte, it was observed that the electrolyte was functional for more than 4000 hours and deflected enhanced tortuosity and uniform current flow. The SEM also confirmed that layer-by-layer printing provided excellent interfacial adhesion throughout the interface of the electrodes and electrolytes. The device also showed uniform discharge voltage stability under a mechanical application such as stressful bending [11].

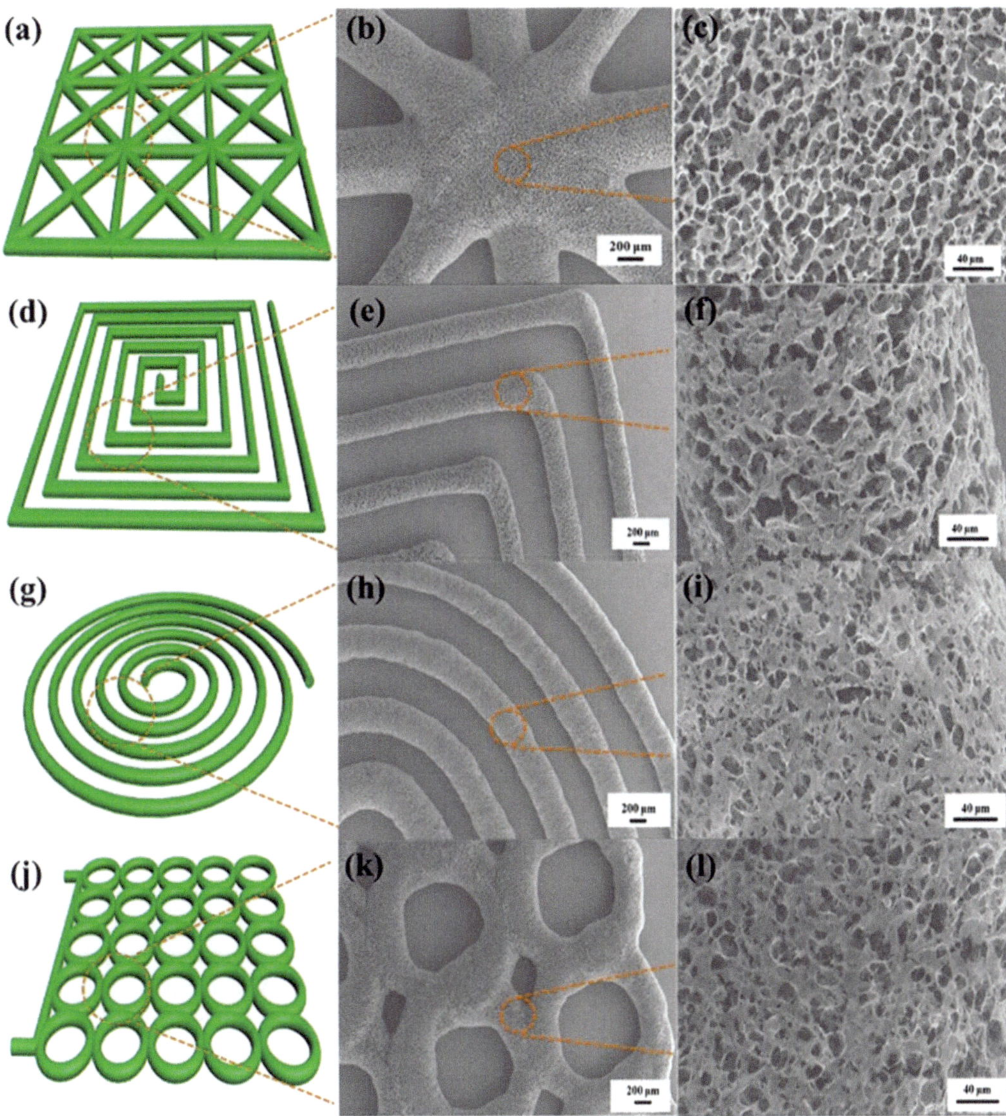

FIGURE 13.3 Schematics and SEM images of 3D-printed frameworks. (a) Schematic and (b, c) SEM images of staggered grids printed by NVP9-GO ink. (d) Schematic and (e, f) SEM images of square coils. (g) Schematics and (h, i) SEM images of mosquito coils. (j) Schematics and (k, l) SEM images of circular arrays. Adapted with permission [10], Copyright (2017), American Chemical Society.

In contrast with the 3D-printing technique which utilized the regular solvent evaporation methods, which have also been discussed above in ceramic polymer electrolyte manufacturing, some researchers derived a new and innovative method of additive manufacturing of layered 3D-printed Li-ion-based energy storage configuration that was processed and manufactured under extremely low temperature up to -40 ^{0}C. The electrodes produced showed interconnected pores and graded porosity which is the most important property of the electrodes. The temperature-based conventional extrusion process had drawbacks such as the structural integrity of the system isn't good enough and affecting the mechanical properties of the products. Such drawbacks were minimized with the subsequent technique. As well as the electrodes have improved porosity which is considered beneficial

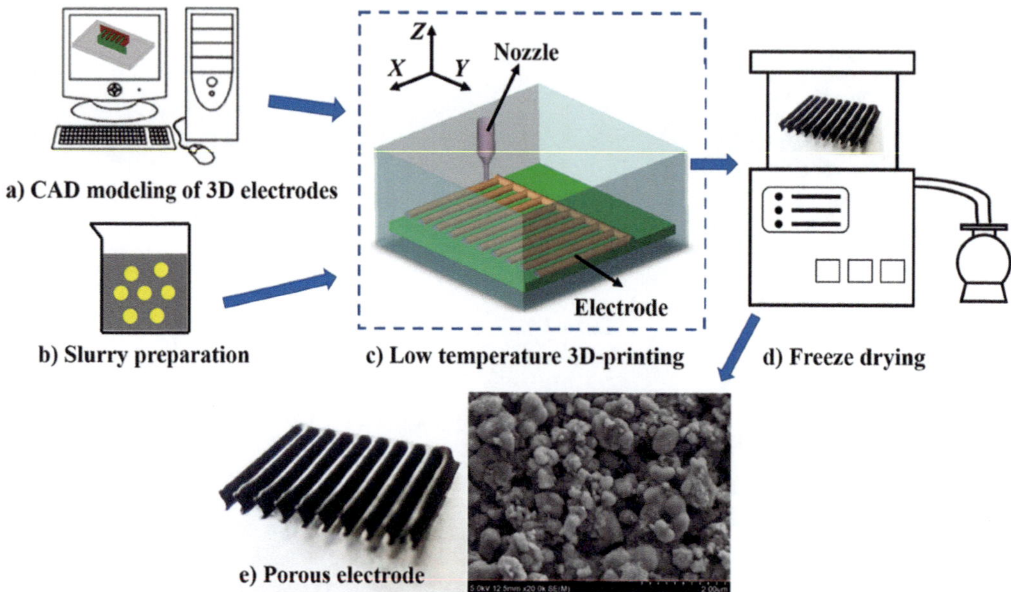

FIGURE 13.4 Schematics of the process to manufacture low-temperature direct written electrodes through freeze-drying and actuating a solventless method. (a) CAD modeling of 3D electrodes. (b) slurry preparation. (c) low-temperature direct writing process. (d) freeze-drying process. (e) porous electrodes. Adapted from reference [12], Copyright (2017) by the authors. Licensee MDPI, Basel, Switzerland. This article is an open access article distributed under the terms and conditions of the Creative Commons Attribution (CC BY) license.

for the lithium-ion diffusion and interface and contact of the electrode material and electrolyte is also enhanced. The prototype of the modeling electrode was first designed on the CAD 3D modeling software called Solidworks and on the other side ink was produced for additive manufacturing which was processed under the influence of cooling agent 1,4 dioxane. A required viscosity of the ink was obtained which was extruded on the substrate through the printer and the printed electrode was then transferred to the freeze-drying machine to get rid of the excess solvent to obtain a porous electrode [12]. The process is shown through a flow chart in Figure 13.4. The low-temperature direct writing (LTDW) technique has been beneficial in many aspects, considering the 3D printing system of the electrodes. The configuration obtained showed greater pore volume compared to the product received through the conventional additive manufacturing process through a direct extrusion system.

13.3.2 Metal Oxide-Based Nanocomposites

Transition metal oxides are prominent candidates for manufacturing electrochemical batteries, primarily because of their low processing costs, ease of production, and ability to tune their morphology and electrochemical properties. With the introduction of nanostructured materials and advancements in manufacturing techniques, there has been a huge leap in the electrochemical performance of the subsequent metal oxide materials [13]. These products also perform well from the commercialization perspective. High-performance micro batteries are of great importance for manufacturing microelectronic devices. As advancement in this field, self-supported 3D-printed metal oxide technologies have been introduced which leads to the enhancement of energy and power density of devices. If 2D and 3D electrodes are compared, 3D architectures achieve greater height by providing higher power density and shortening the diffusion length.

Earlier, while manufacturing the 3D electrodes, researchers combined the conductive materials with binders and additives to achieve fabricated electrodes. However, to eradicate the drawbacks caused by the binder and additives, researchers made self-supported electrodes comprising nano-architecture. Nanowires and nanotubes from a variety of inorganic materials are prepared by the sol-gel method [14]. A range of substrates such as titanium, platinum, or silica can be used. These substrates can be coated via very simple methods such as the dip-coating method. This produces 3D templates of $LiMn_2O_4$ with a porosity of 400 nm and an aerial capacity of 3μAh/cm^2. Later to increase the areal capacity, MnO_2 is electrodeposited to increase the areal capacity up to 76% under special conditions such as a high temperature of 185 °C. However, the cost of manganese has been flaring. To revive from such issues, Li-S batteries are introduced which are cost-effective, environmentally sustainable, and also show good performance and a high energy density of 2600 Wh/kg. However, their low cyclability and rate capacity due to the insulating nature of sulfur are major concerns. Therefore, carbon nanomaterials such as carbon nanotubes and other hybrid forms of nanomaterials have been utilized as hosts for sulfur cathodes for manufacturing Li-S batteries. These carbon-based materials increase the surface area and pore volumes which provide an elongated charge pathway. Subsequent materials when combined with extrusion-based 3D-printing technology produce adaptable customized energy storage devices. Following the 3D-printing technology sulfur hosted lithium-sulfur batteries reflect catalytic and trapping ability which enhances the ability of the electrodes to attain high gravimetric and volumetric capacity values. The sulfur/carbon electrodes formed by this method evolved a high gravimetric capacity.

The metal oxide electrodes were manufactured using a hydrothermal method in which the materials were exposed to HF vapor to improve the etching efficiency and produce heterostructures that propose the wide application of the material. Sulfur and metal oxides heterostructures were then thoroughly mixed and heated at 155 °C for 6 hours to prepare a sulfur-based composite for the cathode slurry [15]. The usage of nanotubes and sulfur for the production of batteries showed enhanced electrochemical properties due to the impressive porous structure and sufficient interaction of the electrolytes. Metal oxide batteries have excellent electrochemical performance but to attain the rheological qualities and thermal properties of the metal oxide batteries, 3D-printed hybrid polymer-based electrolytes play a crucial role. Some researchers have worked out preparing a new electrolyte comprising of silane boron-nitride additive which was initially mixed with N-methyl-2 pyrrolidone to exfoliate nanosheet via sonication. A coin cell assembly was prepared to test the electrochemical properties of the electrolyte where charging and discharging cycling were tested to show the effectiveness of the electrolyte in the battery. The battery showed a cycle time of 10 hours at 25 °C, and displayed outstanding electrochemical properties even after 150 cycles. Besides the electrochemical properties, the addition of hexagonal boron nitride filler enhanced the thermal properties of the cell by lowering the temperature as compared to the cells with liquid electrolytes.

13.3.3 Chalcogenide-Based Nanocomposites

The energy demand on the planet has been increasing exponentially and due to this energy production and storage becomes a crucial task. However, generating energy is an easy task compared to its storage and it is required to derive new materials and techniques which ease the task of storing energy. Among the energy storage systems, lithium-ion batteries (LIBs) have gained wholesome popularity but they lack electrochemical performance such as the maximum energy capacity they can deliver is 400 Wh/kg which is not sufficient [16]. Chalcogenide combined with lithium has been in good resemblance and have been possessing high energy density. The combination of selenium and sulfur has been beneficial in different ways such as more conductivity and cycle stability compared to the properties of the two individual elements respectively. Some researchers synthesized SeS_2 and carbon hollow spheres as high-performance cathode electrodes which delivered a high specific capacity of 930 mAh/g at 0.2 C and long cycle life of 89% capacity revival even after 900 cycles [17]. Furthermore to utilize these excellent properties in a better way the chalcogen materials in a powder form and processed on the 3D-printing technology to provide maximal performance for the electrochemical energy storage device. Initially, the SeS_2 powder was mixed with a host material ketjenblack (KB) in a proportion of 7:3, respectively. KB was selected due to its high micropores structure. Ethanol was used as the solvent when the sample mixture was processed into an autoclave under the influence of argon and heated at 160 °C for 12 hrs to obtain the composite matrix. Once the mixture is prepared, it is mixed with multi-walled carbon nanotubes and PVDF with a weight ratio of 7:2:1 to obtain a suitable solution to process it as dispersion ink with a concentration of 333.3 mg/mL [16]. The ink was then extruded on the 3D printer to get the final electrodes and then coagulated into the water for phase inversion for 12 h and then freeze-dried for the sake of solvent evaporation. The composite of KB/SeS_2 is thoroughly mixed with the CNTs and the conductive and porous

The structural framework provides a coherent pathway for the transportation of both ions and electrons which is beneficial for the electrochemical performance of the storage device. The printed electrode displayed a high areal capacity of 5.6 mAh/cm^2 for more than 80 cycles, which was higher compared to other electrodes made of different materials and techniques. The respective characterization and comparison of the material suggested that electrodes possessed high mass loading and enhanced electrochemical performance. The 3D-printing technology used here covered the theoretical energy density to a great extent combined with good chemical, and mechanical stability as well as outstanding electrochemical performance due to the high areal loading of the thick electrodes. This confirms the reliability of lithium chalcogenide batteries and their usefulness [16].

Apart from the lithium-ion batteries which have been in mass usage throughout several applications, there are certain issues while using lithium-ion batteries such as acute safety issues, mediocre energy density, and limited sources of lithium metal. Therefore, it is necessary to derive new alternatives. To tackle these situations, new alternatives are being searched in a way where the resources are wholesome amount as well as the electrochemical properties of the electrode should be better than its competitors. One example of such electrode materials is the metal chalcogen configuration, such as Na-S, K-S, and K-Se. K-Se is the focus of research nowadays due to its outstanding properties of Se such as efficient electronic conductivity and high theoretical capacity. However, there is no doubt that the chalcogen material has good electrochemical properties but there are many problems and challenges when they are converted to cathodic material. For example, while selenium metal is being utilized as cathodic material it was observed that the metal goes through enormous volumetric changes and results in rapid capacity damage. So to overcome this loss and issue Se is encapsulated into three-dimensional flexible carbon nanotubes and carbon nanosheets which retards the chemical mobilization of selenides and the loss of active Se materials. Ding et al. managed to fabricate a metal-organic framework of Se/$CoNiSe_2$ nanoreactor to construct 3D-printed K-Se batteries (Figure 13.5) [18]. While the configuration of nickel and cobalt was carried out by the solvothermal reaction which provided a high crystal structure framework. With Se ink to be printed as the cathodic electrode and K as a metal anode, it was observed that Se had been throughout the structure with uniform loading [18]. Se/$CoNiSe_2$ also represented an increase in size without any shape deformation and the interlayer spacing of 0.25 nm attested that the preparation of the nanocomposite was successful. During the charge and discharge process, it was also observed that the overpotential of the Se/$CoNiSe_2$- nanoreactor (NR) type configuration was surprisingly lower than the Se/$NiSe_2$-NR which subsequently reflected enhanced kinetics. As discussed before the chalcogen metal delivered a good amount of capacity of 484.5 mAh/g at the initial stages and started decaying at the rate of 0.22% per cycle after 100 cycles. But with the increase in the loading of metal layers, the material started to degrade at a higher rate percent and even at shorter cycle counts. But the lightest loading configuration of 1.7 mg/cm^2 had an open geometric architecture that readily served for rapid ion exchange and also enabled enhanced penetration of the electrolytes [18]. Conclusively, the combination of the 3D-printing technology with the chalcogen metal framework

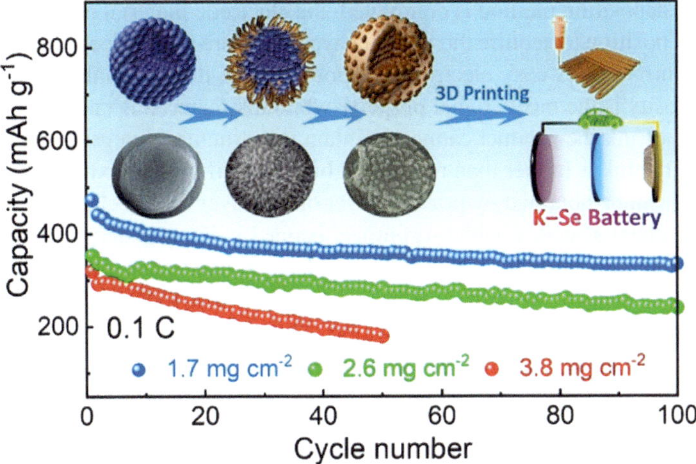

FIGURE 13.5 Schematics and electrochemical performance of K-Se battery. Adapted with permission [18], Copyright (2022), American Chemical Society.

batteries implemented high-capacity output and long lifetime cycling of the energy storage devices. Apart from this, the technology also helps in reducing the over usage of other battery configurations such as lithium-ion batteries.

13.3.4 Nanocomposites for Flexible Batteries

The growing demand for energy storage devices and the evolution in manufacturing techniques provide us with a new platform to turn out some products with advanced properties, where one such quality which is in large upsurge is highly flexible and wearable energy storage batteries. While discussing the versatility and popularity of energy storage devices lithium-based batteries are the most renowned ones. While conventional batteries provide good performance, they lack mechanical performance due to rigidity and fragile behavior. Such rigidity is largely due to the use of metallic foils which serve as the supportive layers and current collectors for the electrodes. So when the device experiences a bend the foils tend to move from their place and lose contact with the electrodes, with eventual failure of the device [19]. The challenge is to provide an energy storage device that withstands frequent mechanical stress without affecting the device's performance.

Several improvements have been made to the manufacturing techniques and new materials certainly, nanomaterials are introduced as well as many modifications have been made to the structural design of the electrodes to enhance their subsequent flexibility. Configurations for such types of batteries as lithium sulfur-based batteries have an ultra-high theoretical energy density of 2600 Wh/kg and have a lower manufacturing cost than conventional Li-ion batteries [20]. While this material configuration is utilized in additive manufacturing technology which yields cost-effective products and the wastage of material can also be minimized. The flexible Li-S battery being discussed was manufactured into a bracelet design so it can be used for the desired application of flexible battery type. Firstly, the ink was prepared with graphene/phenol formaldehyde to obtain the desired viscoelasticity. The reason for combining phenolic resins with the ink is due to their excellent mechanical strength after curing which save provides the device with better shock resistance and saves it from getting ruptured. Along with the material selection, two distinct 3D-printing techniques were merged where the graphene and silicon dioxide dispersed ink were printed through a syringe with the help of air pressure. Post-printing the electrodes were cured in an oven where the graphene oxide was reduced and silicon oxide was extracted through hydrofluoric acid.

Once the direct deposition method is completed, the electrode material is transferred to the other process of fused deposition to acquire the filament-type structure of the electrode material. Like this, the filament structure enables easy electron transport through the material. During the printing of the filaments, viscosity is the most crucial property while ink material is taken into consideration as with low viscosity of ink the product cannot maintain the structural integrity and on the other side of the viscosity of the ink is higher than required it becomes difficult to extrude through the needle. High sulfur loaded can be obtained by printing layer on the layer for the electrode to homogenously distribute in the skeletal structure which promotes better Li^+ ion transfer and improve the charge capability of 505.4 mAh/g at 0.2 C after 500 cycles of the product [20].

Apart from the chalcogen materials, some researchers also focused on building novel self-supportive electrode materials such as metal oxide. The researcher selected lithium nickel cobalt aluminum oxide as the cathode active material due to its high energy density, low manufacturing cost, and outstanding cycle life. Subsequently, vapor-grown carbon fibers (VGCF) were chosen as the anodic materials due to their high surface area, and electric conductivity and the material possesses a porous structure that enhances ion transportation. Furthermore, VGCF demonstrates a coiled structure that makes the material a competent material with excellent electrical properties as well as good flexibility and print affinity. PVDF and NMP were materials used as a binder and solvent for the preparation of the 3D-printing ink, respectively. The composition and proportion of all the materials were optimized so the ink had the required viscosity to get it printed on a 3D-printer extrusion nozzle.

Once the ink was ready it was processed on the Nordson EFD desktop 3D extrusion-type printer with a nozzle diameter of 0.8mm. With the help of compressed air which was constantly kept at 80 psi to maintain the printing speed of 50 mm/s to sustain the structural integrity of the electrodes, additionally, the electrodes were printed on a glass substrate, and once fully printed they were processed on a hot plate for 15 min at 110 °C to remove the excess solvent from the product and then kept immediately into a vacuum oven at 120 °C to dry the material [21]. To measure the electrochemical properties of the electrodes, the CR2032 configuration was used to fabricate a coin cell-type device. Electrical conductivity tests were conducted with the respective cell configuration types where the electrodes displayed a conductivity between 1.1 and 5.5 S/cm which is considered good concerning self-standing Li-ion electrodes which could be increased particularly by increasing the loading proportion of VGCF into the mixture.

Apart from the electrochemical properties, electrodes were mechanically tested to prove their flexibility and strength towards bending application. For this reason, the electrodes were subjected to tensile strain. The test showed that the increase in polymer loading into the electrode increased the mechanical properties with the highest tensile stress of 1.11 MPa at 1.6% elongation and the electrochemical performance of the electrodes was also tested while bending the electrode material, no significant difference was observed in the electrical conductivity even when the material was being flexed. This attested to the electrochemical capabilities of the 3D-printed batteries in a situation of forceful bending [21].

13.4 CONCLUSION

In this chapter, the novel technology of 3D printing was discussed to manufacture batteries with several configurations and design structures. Nanocomposites and nanomaterials were also included during the manufacturing process which enhanced the capabilities of the batteries in terms of energy density and charge-discharge capacity. Li-ion batteries are the once most used nowadays and the same were discussed in this chapter. We have moved a long way further in the field of energy storage devices from rigid and fragile energy storage devices to flexible and highly efficient devices. Batteries nowadays have been advanced to the next level, having superior qualities such as high capacitance identical combined with excellent power delivery and higher cycle times. Storage devices today are getting smaller in size but this does not affect the performance of the batteries, which becomes the interesting part. This is due to the modeling and design merged with nanomaterials which make the production of such microelectronics possible. From liquid electrolytes which had some safety issues and hazards to irradicating such issues, gel-based electrolytes were introduced. With time designs are getting smarter and more efficient than ever before. Na-ion and K-ion-based batteries have also been introduced as a convenient alternative to Li-ion batteries from the commercialization point of view, which reduces the dependence on the same. They also help in minimizing the manufacturing cost because the cost of lithium has been flaring with time.

Chalcogen batteries are also getting popular nowadays due to their long cycle life, which is around 900 cycles. So not only does the battery configuration enhance the properties, but materials have also been a helping part in providing intelligent as well as efficient energy storage services. Such materials consisted of biomaterials such as nanocellulose and other nanomaterials like carbon nanotubes, nanowires, and graphene oxide which enhance the porosity of the electrodes to enable favorable and efficient ion transfer and better penetration of the electrolytes through the system. 3D printing is also somewhere responsible for the structural perfections of the devices and the technique also enables the manufacturing of miniature as well as flexible storage devices. Excluding the manufacturing of batteries, 3D-printing or additive manufacturing technology has been used for versatile applications such as building novel types of self-supported solid-state supercapacitors and many other devices. In the concluding part, we can say this technology of 3D printing will have a bright future in terms of manufacturing and will have even more quality-oriented and high-performance-based energy storage devices to meet the energy demands of the world.

REFERENCES

1 P. Simon, Y. Gogotsi, B. Dunn, Where do batteries end and supercapacitors begin?, *Science* 343 (2014) 1210–1211.

2 S. Chen, A. Skordos, V.K. Thakur, Functional nanocomposites for energy storage: chemistry and new horizons, *Mater. Today Chem.* 17 (2020) 100304.

3 W. Xia, Y. Zhao, F. Zhao, K. Adair, R. Zhao, S. Li, R. Zou, Y. Zhao, X. Sun, Antiperovskite electrolytes for solid-state batteries, *Chem. Rev.* 122 (2022) 3763–3819.

4 Y. Gogotsi, R.M. Penner, Energy storage in nanomaterials – Capacitive, pseudocapacitive, or battery-like?, *ACS Nano.* 12 (2018) 2081–2083.

5 C. Chen, L. Hu, Nanocellulose toward advanced energy storage devices: Structure and electrochemistry, *Acc. Chem. Res.* 51 (2018) 3154–3165.

6 O. Faruk, D. Hosen, A. Ahmed, M.M. Rahman, Functional bionanomaterials—Embedded devices for sustainable energy storage, in *Biorenewable Nanocomposite Mater. Vol. 1 Electrocatal. Energy Storage* Editor(s): Deepak Pathania and Lakhveer Singh, Chapter 1, pp. 1–23.

7 U. Gulzar, C. Glynn, C. O'Dwyer, Additive manufacturing for energy storage: Methods, designs and material selection for customizable 3D printed batteries and supercapacitors, *Curr. Opin. Electrochem.* 20 (2020) 46–53.

8 V. Gupta, F. Alam, P. Verma, A.M. Kannan, S. Kumar, Additive manufacturing enabled, microarchitected, hierarchically porous polylactic-acid/lithium iron phosphate/carbon nanotube nanocomposite electrodes for high performance Li-Ion batteries, *J. Power Sources.* 494 (2021) 229625.

9 K. Sun, T.-S. Wei, B.Y. Ahn, J.Y. Seo, S.J. Dillon, J.A. Lewis, 3D printing of interdigitated Li-ion microbattery architectures, *Adv. Mater.* 25 (2013) 4539–4543.

10 J. Ding, K. Shen, Z. Du, B. Li, S. Yang, 3D-printed hierarchical porous frameworks for sodium storage, *ACS Appl. Mater. Interfaces.* 9 (2017) 41871–41877.

11 A.J. Blake, R.R. Kohlmeyer, J.O. Hardin, E.A. Carmona, B. Maruyama, J.D. Berrigan, H. Huang, M.F. Durstock, 3D printable ceramic–polymer electrolytes for flexible high-performance Li-ion batteries with enhanced thermal stability, *Adv. Energy Mater.* 7 (2017) 1–10.

12 C. Liu, X. Cheng, B. Li, Z. Chen, S. Mi, C. Lao, Fabrication and characterization of 3D-printed highly-porous 3D $LiFePO_4$ electrodes by low temperature direct writing process, *Materials (Basel).* 10 (2017) 1–13.

13 Q. Liu, Z. Hu, M. Chen, C. Zou, H. Jin, S. Wang, S.L. Chou, S.X. Dou, Recent progress of layered transition metal oxide cathodes for sodium-ion batteries, *Small.* 15 (2019) 1–24.

14 W. Xiong, Q. Xia, H. Xia, Three-dimensional self-supported metal oxides as cathodes for microbatteries, *Funct. Mater. Lett.* 7 (2014) 1–10.

15 C. Wei, M. Tian, M. Wang, Z. Shi, L. Yu, S. Li, Z. Fan, R. Yang, J. Sun, Universal in situ crafted MOx-MXene heterostructures as heavy and multifunctional hosts for 3D-printed Li-S batteries, *ACS Nano.* 14 (2020) 16073–16084.

16 C. Shen, T. Wang, X. Xu, X. Tian, 3D printed cellular cathodes with hierarchical pores and high mass loading for Li–SeS_2 battery, *Electrochim. Acta.* 349 (2020) 136331.

17 H. Zhang, L. Zhou, X. Huang, H. Song, C. Yu, Encapsulation of selenium sulfide in double-layered hollow carbon spheres as advanced electrode material for lithium storage, *Nano Res.* 9 (2016) 3725–3734.

18 Y. Ding, J. Cai, Y. Sun, Z. Shi, Y. Yi, B. Liu, J. Sun, Bimetallic selenide decorated nanoreactor synergizing confinement and electrocatalysis of Se species for 3D-printed high-loading K-Se batteries, *ACS Nano.* 16 (2022) 3373–3382.

19 K.H. Choi, S.J. Cho, S.H. Kim, Y.H. Kwon, J.Y. Kim, S.Y. Lee, Thin, deformable, and safety-reinforced plastic crystal polymer electrolytes for high-performance flexible lithium-ion batteries, *Adv. Funct. Mater.* 24 (2014) 44–52.

20 C. Chen, J. Jiang, W. He, W. Lei, Q. Hao, X. Zhang, 3D printed high-loading lithium-sulfur battery toward wearable energy storage, *Adv. Funct. Mater.* 30 (2020) 1–7.

21 S. Praveen, G.S. Sim, N. Shaji, M. Nanthagopal, C.W. Lee, 3D-printed self-standing electrodes for flexible Li-ion batteries, *Appl. Mater. Today.* 26 (2022) 100980.

14 3D-Printed Carbon-Based Nanomaterials for Supercapacitors

Digambar S. Sawant[1,2], *Onkar C. Pore*[1], *Shrinivas B. Kulkarni*[2], *and Gaurav M. Lohar*[*1]
[1]Department of Physics, Lal Bahadur Shastri College of Arts, Science and Commerce, Satara, Maharashtra 415002 India
[2]Department of Physics, Institute of Science, Dr. Homi Bhabha State University, Mumbai, Maharashtra 400032 India
*Corresponding author: gauravlohar24@gmail.com

CONTENTS

14.1 INTRODUCTION

Due to increasing energy demand, worldwide energy sources, especially fossil fuels, are on the way to exhaustion. So, society has been moving toward the utilization of sustainable and renewable energy resources [1]–[3]. To store energy generated from these unequal resources ESD like conventional capacitors, fuel cells, batteries, and SCs have attracted much attention from researchers. Among the various ESD, SC plays a significant role in energy storage because of its promising features such as low cost, higher power density, quick charge-discharge process, and long cyclic life [4]–[6]. 3D-printing technology is a computer-aided layer-by-layer Additive Manufacturing (AM) process [7]–[9]. The 3D-printing process gives phenomenal benefits in geometric form creation and quick prototyping mainly with large surface area complex 3D structure productions [10]. In the 3D-printing method, the development of a material design is in the simplest form. So, the 3D-printing process is used to develop many functional devices. Also, this technology reduces the

DOI: 10.1201/9781003296676-14

loss of valuable innovative materials. Many innovative materials including nanoparticles, aerogels, biomaterials [11], electronic device materials [12], metamaterials [13], and even concrete that dries quickly have been developed by using 3D-printing technology. In current years, due to increasing demand for portable and wearable electronics, 3D-printing technology is being used. In the case of SCs, 3D-printing technology is used to prepare materials with desired shapes along with improvement in their capacitance, energy density, power density, and cyclic stability [14].

Carbon-based materials are the most prominently used materials for 3D printing because of their properties such as good charge storage capacity, cyclic stability, and the form of different nanostructures. In this chapter, we are discussing graphene, CNTs, active carbon materials, and carbon-based complex materials all have shown their excellent power as well as energy density properties. Graphene is favorable material that is used in energy storage applications such as SCs due to its high surface area, high electrical conductivity, and high intrinsic specific capacity. Due to their skills to create links graphene and graphene oxide (GO) and reduced graphene oxide (rGO) inks are used for 3D printing on large scale [15]. In electrochemical devices, CNTs are used to improve their electrochemical performance, surface area, and current-carrying ability. CNT is classified specifically in single-walled carbon nanotubes (SWCNT) and multi-walled carbon nanotubes (MWCNT) [16]. The symmetric and asymmetric devices have been developed of carbon-based materials and their composite with other prominent materials such as SiO_2 [12], $CoNi_2S_4$ [17], MnO_2 [18], Fe_2O_3 [19], V_2O_5 [20] for improving their specific capacitance, energy as well as power density and cyclic stability.

14.2 3D-PRINTING METHODS

In coming years 3D-printing technology will be a strong production approach for the development of functional 3D structures with their applications ranging from automotive, and aerospace to a variety of consumer products [21], [22]. Depending on the printing technology and required applications, each cross-sectional printed layer has a thickness ranging from 15 to 500 m. Currently, 3D printing is only used for prototypes and custom-built small parts, However, crucial characteristics such as layer thickness of the material, build volume (the printable size of the material), and build speed (defined as the height of an object constructed in a given time or as a volumetric rate ($mm^3\ h^{-1}$) are required to represent its use as a preferred industrial manufacturing tool and the qualities of the materials need to be improved and optimized [6]. Several crucial factors have improved over the previous decade, as well as the introduction of new 3D-printing and 3D-printing-based technologies. As a result, the International Committee American Society for Testing and Materials (ASTM) has categorized 3D printing into the following distinct methods:

- Vat Photopolymerization (VAT-P)
- Direct Energy Deposition (DED)
- Binder Jetting (BJ)
- Powder Bed Fusion (PBF)
- Sheet Lamination (SL)
- Material Jetting (MJ) is known as inkjet printing (IJP) and
- Material Extrusion (ME) is known as Direct ink writing (DIW)

Based on physical conditions, such as the form of solid/liquid material used and the process used to combine the building material, we can further classify these technologies into mechanical or optical. The MJ and VAT-P methods need a liquid phase to construct 3D materials while the remaining method used solid powder or filament to fabricate 3D materials. In the 3D-printing technology MJ, ME, and BJ are used to build 3D structures based on mechanical forces on the

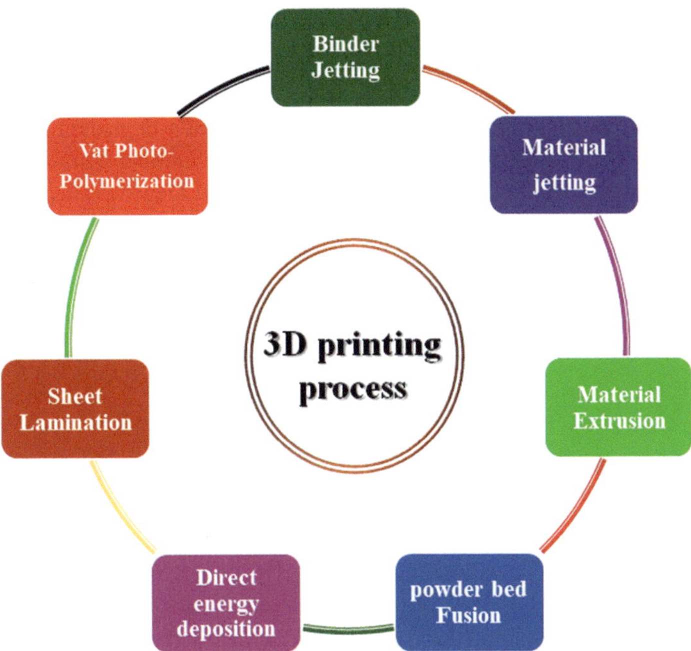

FIGURE 14.1 Classification of the 3D printing process. Adapted with permission [23]. Copyright, (2017), Elsevier.

other hand PBF, VAT-P, SL, and DED use an electromagnetic beam of high power made by light or electrons to build a 3D structure by melting, blending or inducing. As per the review, a large amount of the carbon-based SCs materials have been printed by using the MJ and ME method [21]. These methods include terms like laser-designed net shaping, focused beam fabrication, direct metal deposition, 3D laser cladding, etc. The ME method involves fused filament fabrication or fused deposition modeling (FDM). The VAT-P method is better known as stereolithographic apparatus and digital light processing printing, and MJ involves IJP and associated technologies [23]. Figure 14.1 presents different 3D printing processes.

14.2.1 Vat Photopolymerization (VAT-P)

It is a liquid-to-solids technique in which a vat of liquid resins is irradiated with light and solid objects are created using computer-spatially controlled photopolymerization [24]. To develop high-resolution smooth electrodes VAT-P method is used. High resolution, accuracy capability, high fabrication speed, and low imaging specific energy are the main advantages of this method. By using the UV range a large number of photopolymers may be available in a good manner and these are used for commercial AM methods. The connecting single monomers linked with chain-like polymers via a photochemical process is known as photopolymerization. The polymers must always be suitably cross-linked such that the polymerized molecules do not disintegrate back into the liquid phase monomers. It must also be strong enough to withstand varied loads while being structurally sound [23]. On the other hand, traditional VAT-P is a slow production process. The disadvantages of these methods are that they require a special processing unit to take out support and cure for enhanced strength. Many new and speedier approaches, such as digital optical processing and continuous liquid interfaces, have now been created [25].

14.2.2 Direct Energy Deposition (DED)

The DED is another 3D-printing method that involves directing energy into a small surface of the electrode which is used to heat and the material is deposited in the form of metal powder or wire at the same time by a melting process. The quantity of metal powder deposited for the DED process has a direct impact on the resolution of the printed item. Most melting metal powders are usually done with a laser, electron beam, or plasma/electric arc. In the DED method deposition rate and utilization of material are excellent for 3D printing [26]. In one essential sense, the DED method differs from PBF in that the use of laser for the higher density electron beam is concentrated on a coating of metal powder that is pre-deposited [23]. The DED printing technology takes a long time and needs inert conditions, making it an expensive manufacturing process. Poor dimensions accuracy and resolution and limited materials for production purposes are the main disadvantages of the DED printing process [27].

14.2.3 Binder Jetting (BJ)

Binder jetting (BJ) involves a specific coating of a liquid state bonding material on top of powder particles to bind them together, resulting in the creation of a 3D item. The platform is then lowered to distribute a second layer of powder after the print head drops a bonding material over the powder. This creates a 3D structure by feeding a layer of powder and adding binders in a two-dimensional pattern. BJ method requires low-imaging specific energy. BJ's benefits are the lack of support, material flexibility, huge build capacity, fast printing speed, and availability at minimum cost. But demerits of the BJ-printing methods are the strength of a material is very poor and additionally processing unit is required for eliminating moisture and enhancement of strength of the given material [23].

14.2.4 Powder Bed Fusion (PBF)

Another 3D-printing technique is PBF in which a high-temperature heat source like a laser or electron beam is used to generate partial or complete fusion between small particles of powder. Another powder layer is added to these small particles of powder and smoothed using a roller or blade recoated. This process is also known as laser sintering. The material fabrication speed of the PBF method is very high. In recent years direct metal laser sintering, selective laser melting, and electron beam melting are the most common powder bed melting processes that are used in the PBF process [28]. In solid-state sintering, just the surface of the particles binds together and resulting in intrinsic porosity in the component. Whereas in liquid-state melting, most particles melt and merged resulting in a fully dense part with nearly no porosity. In the PBF method, the thickness of the deposited powder layer is about 1 micrometer (μm) and the speed of the laser system is 1 m/s [29]. The main disadvantage of PBF is the 3D material build rate of the PBF method is slow as compared to other printing processes. Also, the machines which are used to make 3D materials are costly. The construction period of 3D materials might range from hours to several weeks. PBF and BJ are two types of 3D-printing technologies that employ powders as building components (metals or polymers) [8].

14.2.5 Sheet Lamination (SL)

Sheet lamination (SL) is the technique of assembling and laminating thin layers of material to create a 3D item with high speed. In the SL method, it is possible to develop 3D-printing materials with several colors. The lamination method involves bonding, and ultrasonic welding/brazing while at the last stage-specific shape is accomplished either by laser cutting or by using ultrasound. Each material sheet may be thought of as one of the solid item's cross-sectional layers. The SL process can be classified into two categories viz laminated object manufacturing (LOM) and ultrasonic additive manufacturing (UAM). The LOM is further classified into adhesive bonding, thermal bonding, and clamping. The UAM is a hybrid SL method that incorporates computer numerical controlled cutting and ultrasonication metal welding. In the SL method when the materials are in the internal cavity removing the support is very hard. Further toxic gases are produced during thermal cutting and large-scale material is wasted these are the drawbacks of SL methods [30].

14.2.6 Material Jetting (MJ) or Inkjet Printing (IJP)

Material jetting (MJ) or inkjet printing (IJP) is a drop-casting-based material deposition technology that was first utilized in 2D space for graphic or document printing. The IJP is a promising technology to develop high-resolution 3D-printing designs of multiple materials with multiple color combinations. It was developed as a direct material deposition method without touch for a variety of uses, including electrical and biological ones [31]. Polymer jetting and BJ are two methods included in IJP for their innovative technology properties in the printing process. In the polymer jetting method, the inkjet print top head deposits light-healed photosensitive polymer onto the current collector. The binder is sprayed onto a powder bed during the BJ process to form a cross-section of the surface. This is identified as 3D printing because the liquid substance used in IJP is so close to that used in traditional procedures such as the Meyer rod coating. IJP is also widely utilized for printing SCs [32] due to the high resolution of IJP technology and its great various types of printing capacity on large areas [33].

In most cases, the active ingredient is dispersed in a solvent to prepare the ink and the resulting final material for IJP is used in the form of a diluted liquid. To avoid blockage in printing, IJP ink must be well written and free of agglomerates. To obtain the requisite ink with the desired fluid behavior, the synthetic substance that is blended with ink must also be considered. For the effective work of the IJP processes ink parameters such as viscosity μ, surface tension σ, density ρ with specified ranges, and nozzle size also affects the particular constraints for a fixed nozzle diameter d [34].

14.2.6.1 Principles of IJP Technique

To develop high-resolution and homogeneous forms, the IJP technique involves jetting ink drops from a nozzle. It produces an effective coating of drops of ink on the target material surface in a quick and controlled mode. The IJP technique can be separated into two types based on the droplet jetting mechanism: continuous inkjet (CIJ) and drop-on-demand (DOD) [4]. Benefiting from surface tension forces, the CIJ process produces a nonstop stream flow of droplets that separates into a stream of droplets. After that, an electric charge is introduced to certain droplets via an electrostatic field. In this process, charged droplets are diverted towards the electrode whereas the chargeless droplets are collected for reuse. As a result, individual ink drops are created just when needed in DOD systems. DOD IJP is classified as thermal or piezoelectric based on the ink-drops-driven mechanism. In piezoelectric DOD inkjet printers, an electric field is delivered to the piezoelectric material to cause the reservoir's shape to change. The ink of a thermal DOD inkjet printer is quickly heated to create small air bubbles inside the ink chamber. These bubbles collapsed and clean ink droplets passed out from the nozzle. Droplets of smaller diameters are created in comparison to CIJ, which are reached by developing higher resolutions [35]. Besides the printing equipment, the ink composition, ink-substrate interactions, and post-treatment techniques can all have a crucial impact on the quality of printed form. But in the IJP methods variety of printable inks are limited.

14.2.7 Material Extrusion (ME) or Direct Ink Writing (DIW)

Another 3D-material deposition process is ME or DIW, used to deposit printing materials such as pastes or hydro-gels. This method has multiple benefits, including material versatility and cost savings. It is a straightforward printing method that requires no post-processing with low-cost printing machines. In this process by applying constant pressure, the material is pushed out from a nozzle. Increasing the viscosity of a printed structure can help it keep its shape [23]. To allow for form preservation of printed lines, the printing materials must be thixotropic, i.e., shear thinning. Also, it has sufficiently high yield stress and storage modulus. This technique helps with the computer-assisted stacking of successive layers into a pile-like structure. Direct writing has been used to build batteries and SCs as a suitable 3D-printing technology due to its adaptability and high material loading [36]. In comparison to IJP, this printing technology enables large mass-loading materials to be printed with a significantly lower chance of nozzle blockage [21]. Furthermore, very large mass loadings of active materials may be employed to boost areal capacitance and energy density tremendously. However, achieving high viscosity ink with shear thinning behavior requires extensive management and rheological properties of links and cannot be constructed by piercing outside corners of that materials.

In the ME printing process, after exiting the nozzle the extruded material has deposited at a consistent pace and entirely hardened on the substrate. Furthermore, the material must connect with the preceding material to build a solid component that will stay in that structure throughout the procedure. Thermoplastic materials are used in FDM. These FDM materials can provide material qualities such as UV resistance, biocompatibility, translucence, and toughness. These qualities make them ideal for use in demanding environments such as those found in the automotive, aerospace, medical, and other sectors [37]. The extrusion approach increases the material selection flexibility for SC electrodes by allowing the use of large surface area materials such as CNT, graphene, and pseudo material. Rheology testing, on the other hand, is a necessary part of the extrusion material selection process. With increasing shear rate, the viscosity of extrusion material decreases, showing shear thinning behavior. The printing material's viscosity is temporarily reduced by extrusion, but it returns to its previous condition after exiting the nozzle. This is significant because enhancing viscosity can aid in the form-keeping of a printed structure.

14.3 SUPERCAPACITOR PERFORMANCE OF 3D-PRINTED CARBON-BASED MATERIALS

The SCs materials consist of carbon-containing material, conductive polymers, metal oxides (MOs), carbides, sulfides, and nitrides. But in the actual fabrication of SC devices, carbon-based materials like AC, CNT, graphene, and its derivatives are widely used. Their cost effectiveness, outstandingly fast charge-discharge ability, high power density, and long cycle lifetime make them a preferable candidate for making SC devices [38]. In recent years most researchers fabricated symmetric and asymmetric SC devices of carbon-based materials by using IJP and ME 3D-printing processes. We said in the aforementioned concept that these devices are easy to construct, and less vulnerable to ionic conductivity boundaries of the internal bulk volume of constructing ingredients band in aqueous-based electrolytes a several polymeric or plastic materials are more stable. IJP is a popular technique used in SCs printing.

Huan et al. [39] synthesized $K_2Co_3(P_2O_7).2H_2O$ by hydrothermal process. Thereafter they fabricated a flexible solid-state micro-SC on a polyethylene terephthalate substrate used as a current collector which was coated with silver ink. Then the slurry of active material $K_2Co_3(P_2O_7).2H_2O$ nanocrystal whiskers coated on the electrode surface. The device consists of $K_2Co_3(P_2O_7).2H_2O$ positive electrode and a graphene nanosheet used as a negative electrode using a solid-state KOH/PVA gel electrolyte. The device delivered a specific capacitance of 6.0 F cm^{-3} at a current density of 10 mA/cm^3 with good cyclic stability up to 5000 cycles with 94.4 % of its initial capacitance. This device also showed good flexibility with a tilt angle ranging from 0 to 180° and a good 0.96 mWh/cm^3 energy density. Li et al. [40] developed a solid-state micro-supercapacitor (MSC) device of rGO/MoO_3 nanosheets by IJP technique on polyamide (PI) film in air atmosphere through a

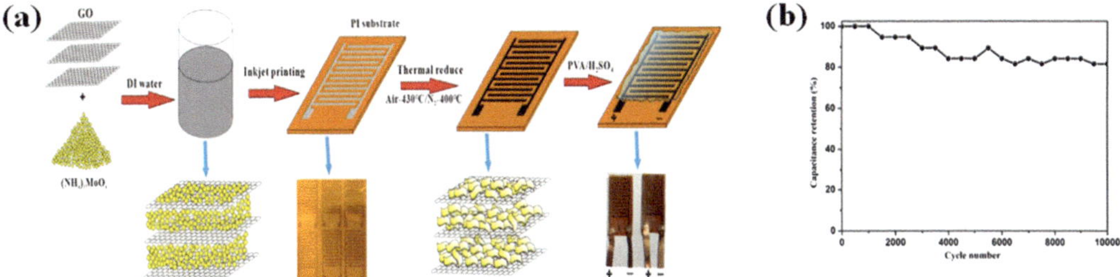

FIGURE 14.2 (a) Schematic construction of rGO/ MoO_3 nanosheets by IJP for solid-state MSCs device and (b) Cyclic stability of rGO/MoO_3 MSCs device in 10000 cycles. Adapted with permission [40]. Copyright (2019), American Chemical Society.

thermal route. Figure 14.2a represents the construction process of rGO/MoO_3 by an inkjet method. The prepared rGO/MoO_3 by inkjet symmetric MSCs using (PVA)-H_2SO_4 gel electrolyte takes good flexibility, attains volumetric capacitance of 22.5 F/cm^3 at 4.4 mA/cm^3 current density, and sustains capacity retention 82% after 10000 galvanostatic charge-discharge (GCD) cycles (Figure 14.2b). It exhibits a high energy density of 2 mWh/cm^3 concerning a power density of 0.018 W/cm^3.

Wang et al. [5] synthesized asymmetric supercapacitor (ASCs) device with MnO_2/Ag/MWNT as positive and MWNT as a negative electrode by using HP Deskjet 1010 inkjet printer. The fabricated ASC device exhibited volumetric capacitance of 5.3 F/cm^3 at a scan rate of 10mV/s Also it delivered high energy density (1.28 mWh/cm^3) at corresponding power density (96 mW/cm^3) and good cyclic stability of 96.9% after 3000 GCD cycles. The MJ-method able to attain higher mass loading of multi-materials structures. Also, it allows good contact between cathode, electrolyte, and cathode. Further, it allows the electrode to create more composite geometries and helps to eliminate the discrepancy in specific mass, print area, and size differences in SC devices. In comparison to ASCs applications, MJ-based electrodes for symmetric SCs (SSCs) have received and achieved greater development. Le et al. [41] synthesized GO nanosheets by inkjet method on Ti current collector electrode and thermally reduced at 200°C in the presence of nitrogen, which exhibits specific capacitance of 132 F/g.

Le et al. [7] prepared inkjet printer graphene-based MSCs on different current collector electrode which exhibits the maximum areal capacitance of about 0.7 mF/cm^2 and AC-based materials were developed with the IJP method on silicon current collector electrode. They connected more than 100 devices in both combinations (series and parallel). The best thing about these developed MSCs is that they have been shown to have a good performance after eight months of fabrication when charged up to 12 V. Pech et al. [42] fabricated the MSCs device by preparing an ink mixture of AC powder with polytetrafluoroethylene (PTFE) in ethylene glycol with a gold metal collector and depositing it onto a silicon substrate by IJP method. It has been reported to show a cell capacitance of 2.1 mF/cm^2 in SSC using the MJ technique.

Li et al. [43] developed that printing graphene with the dimethylformamide (DMF) used as a solvent for dispersions as inks in SSCs exhibits a high areal capacitance of 0.82 mF/cm^2 at a minimum scan rate of 10 mV/s and delivered power density of 8.8 mW/cm^2. Xu et al. [44] mentioned that MJ provides excellent control over thickness as a layer-by-layer process and SSCs exhibit a power density of 124 Wh/kg with respect to an energy density of 2.4 Wh/kg of NGP/PANI thin film when graphene inks are printed on polyaniline. Choi et al. [33] fabricated an SWNT/AC SCs device by direct IJP on high-resolution print on an A4 paper sheet. The fabricated SC device exhibited an areal capacitance of 100 mF/cm^2 with good capacity retention after 10000 GCD cycles. Chi et al. [3] synthesized a symmetric flexible nanohybrid paper electrode based on an IJP graphene hydrogel polyaniline (GH-PANI) current collector. In Figure 14.3a. shows the stepwise fabrication process of the GH–PANI/GP electrode. Firstly, graphene hydrogel (GH) is synthesized hydrothermally from

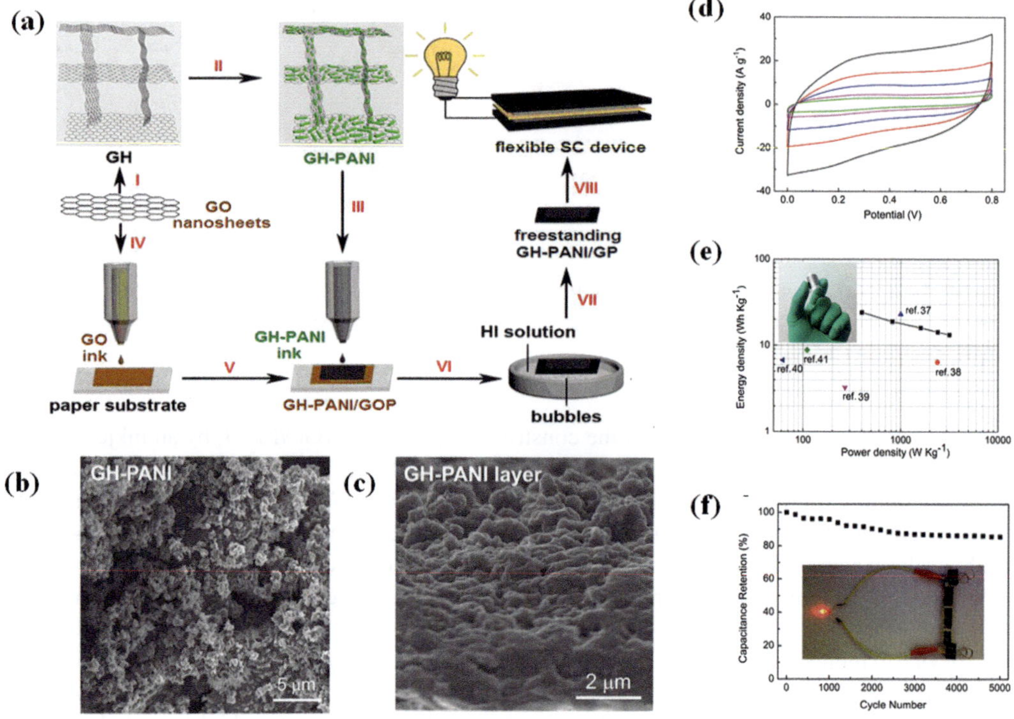

FIGURE 14.3 (a) Schematic Illustration of the Fabrication Process of GH–PANI/GP. (b) SEM image of GH–PANI. (c) SEM image of GH–PANI layer on GH–PANI/GP at high magnification. (d) CV curves of the all-solid-state symmetric GH–PANI/GP SC. (e) Ragone plots of the GH–PANI/GP SC device and (f) Cycling behavior of the GH–PANI/GP SC device at a current density of 8 A g^{-1}. Adapted with permission [3]. Copyright (2014), American Chemical Society.

GO slurry and after PANI is deposited on GH. These GH-PANI were deposited on prepared GO ink paper substrate. The GH-PANI electrode exhibited a coral-like structure (Figure 14.3b). In the next step, the GH-PANI nanocomposite was printed on GP. The micrograph of layered GH–PANI/GP at high magnification is presented in Figure 14.3c. The CV curve of the GH–PANI/GP electrode is depicted in Figure 14.3d. It attains 190.6 mF/cm^2 at 0.5 mA/cm^2. The Regone plot of GH–PANI/GP electrode shows a power density of 0.4 kW/kg and an energy density of 24.02 Wh/kg in Figure 14.3e. The inset of Figure 14.3e showed the actual photograph of the prepared device. The cyclic stability of the device is mentioned in Figure 14.3f. which exhibited 85.6% capacity retention over 5000 cycles at 8 A/g. The inset of Figure14.3 f showed the actual working device with a glowing red LED. Zang et al. [45] prepared a printing of flexible, coplanar MSCs using Mxene inks which exhibited areal capacitance of 61 mF/cm^2 at μA/cm^2 and delivered excellent GCD cyclic stability 82% at 200 mV/s over 10000 cycles, as well as the good energy density of 0.76 mWh/cm^2 with respect to a power density of 0.63 mWh/cm^2. They used a stamping system for the construction of the Mxene device. Yang et al. [46] fabricated device delivers a maximum areal capacitance of 2.1 F/cm^2 at 1.7 mA/cm^2 and also exhibits gravimetric capacitance of 242.5 F/g at 0.2 A/g with capacity retention of 90 % after 10,000 cycles.

Yao et al. [47] developed graphene aerogels with the DIW-printing of GO inks. They collected ready-to GO for using aerogel ink formulation. GO ink was then deposited by the DIW method and the resulting 3D material was frozen in liquid nitrogen to form an aerogel. Graphene aerogel was obtained by calcinating these prepared samples and fabricating process of a 3D-printed graphene

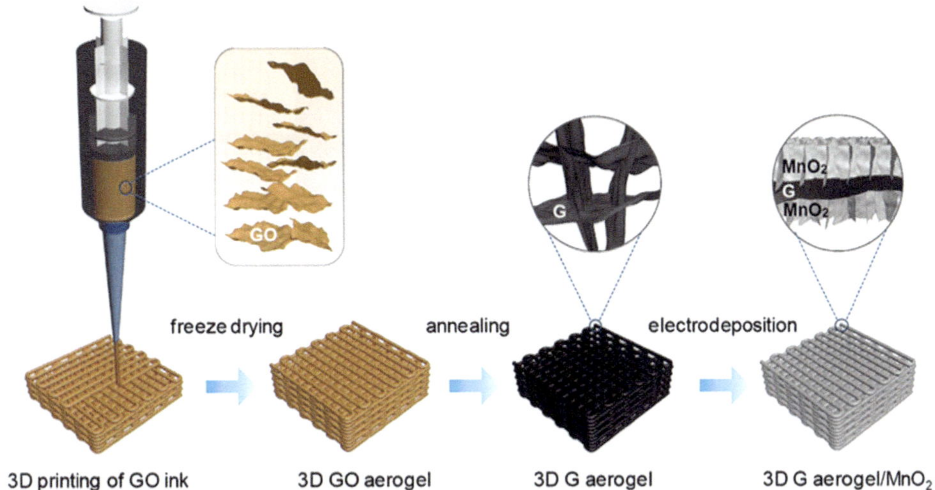

FIGURE 14.4 Schematic diagram of the fabricated 3D printed graphene aerogel/MnO_2 electrode. Adapted with permission [47]. Copyright (2014), Elsevier.

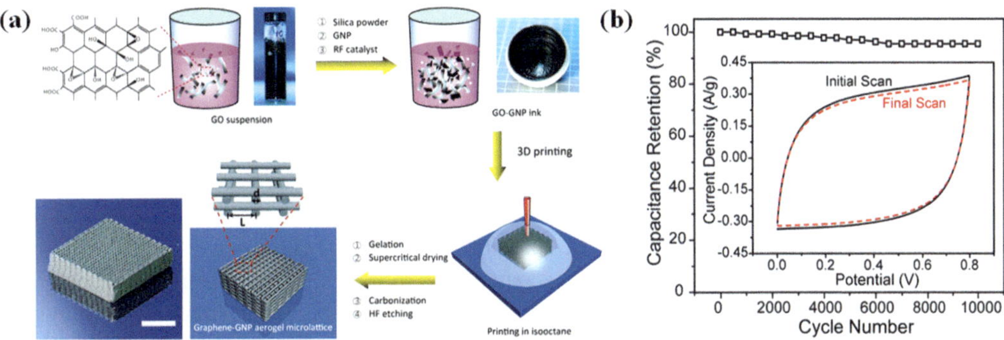

FIGURE 14.5 (a) Schematic presentation of preparation of 3D-GCA microlattices. (b) Capacity retention vs cycle number of 3D-GCA quasi solid-state SSCs device (Inset: First and last CV cycle of cyclic stability). Adapted with permission [12]. Copyright (2016), American Chemical Society.

Aerogel/MnO_2 electrode by electrodeposition is given in Figure 14.4. It exhibited areal capacitance of 44.13 F/cm^2. The most important fact of these Aerogel/MnO_2 material electrodes which attain excellent capacitance stabilized the specific area volume and gravimetry. These electrode materials have contributed to the evolution of pseudocapacitive device fabrication.

Xue et al. [48] used a stereolithographic 3D-printing method for device fabrication. The fabricated device exhibits high areal capacitance of 57.75 mF/cm^2 and rate capability with 70% retention between 2-40 mA/cm^2 with long cyclic stability of 96% after 5000 cycles. Azhari et al. [49] prepared a device using the binder jet AM method. The Synthesized device showed 260 F/g gravimetric capacitance and 700 mF/cm^2 areal capacitance at a scan rate of 5 mV/s in 1 M H_2SO_4 electrolyte. The device retained 80% stability after 1000 cycles.

Zhu et al. [12] prepared a symmetric GO/GNP/SiO_2 electrode device by ME 3D printing process. The schematic of a fabrication process of the GO/GNP/SiO_2 electrode is depicted in Figure 5a. It exhibited a specific capacitance of 4.76 F/g at 0.4 A/g. Also, it delivered excellent cyclic stability of 95.5% after testing for 10,000 cycles (Figure 14.5b). The excellent specific capacitance of the

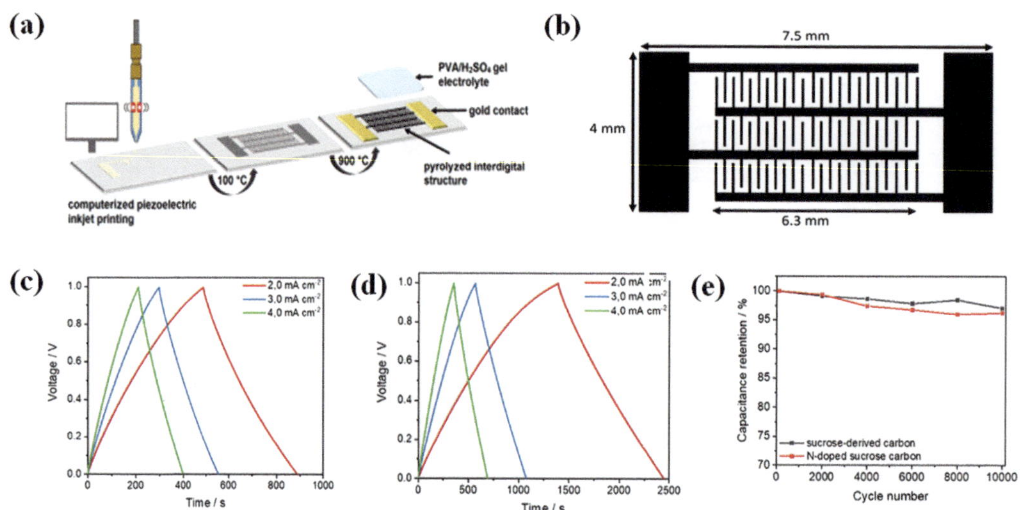

FIGURE 14.6 (a) Schematic process of MSC processing via piezo-printing. (b) Dimensions of the printed interdigital structure. (c) GCD curves of sucrose-derived carbon. (d) GCD measurements of N-doped sucrose carbon and (e) cycling stability of MSCs of sucrose-derived carbon (black) and N-doped sucrose carbon (red) over 10,000 cycles at 15 A/cm^2. Adapted with permission [51]. Copyright (2021), American Chemical Society.

electrode may be due to the sheet-like morphology. Jiang et al. [50] developed a symmetric GAMs device using the ME process, GAMs show low density, high porosity, tailored lattice patterns, strong elasticity, and adjustable electrical conductivity, with excellent promise for numerous multifunctional applications. Yao et al. [18] fabricated a 3D GA/MnO_2//SF-3D GA asymmetric SC device by ME process which attains high areal capacitance 1700.6 mF/cm^2 at 5 mA/cm^2. It shows nanosheets 3D-printed morphology and excellent cyclic stability 93% after 100,000 cycles.

Brauniger et al. [51] fabricated an N-doped sucrose carbon SSCs device using a computerized piezoelectric 3D IJP process presented in Figure 14.6a. The optimized structure presented in Figure 14.6b is based on 69 electrode fingers on an area of 4 × 7.5 mm and two outer contacts. The GCD curves of N-doped sucrose carbon showed doubled charge-discharge time as compared to sucrose-derived carbon and are depicted in Figure 14.6c and Figure 14.6d respectively. It showed stable 3D structures with specific capacitance up to 151 F/cm^3 (3.9 mF/cm^2). The fabricated SC device showed excellent cyclic stability of 96% after 10,000 cycles (Figure 14.6e). Yu et al. [52] fabricate AC/CNT/MXene-N/GO nanosheets symmetric device by ME 3D-printing processes which attains specific capacitance of 8.2 F/cm^2 at 10 mV/s and also attains good cyclic stability 96.2% after 5000 cycles.

Yao et al. [53] synthesized 3D-MCA at a low-temperature symmetric device by an ME printing process. The fabrication process of the 3D-MCA electrode with micron pores morphology is given in Figure 14.7a. The CV curves of 3D-MCA and MCA through the fast and slow kinetic processes at a scan rate of 100 mV/s are given in Figure 14.7b. The prepared electrode reported an outstanding capacitance of 148.6 F/g at a minimum scan rate of 5 mV/s, which is six to seven times greater than the non-3D printed MCA. Figure 14.7c presents the 80% capacitance contribution of 3D-MCA by the fast kinetic process while MCA electrodes contribute 57.8% by the slow kinetic process. In the case of MCA, the value of charge transfer resistance is too much greater than the 3D-MCA electrode at -70 ^{0}C as mentioned in Figure 14.7d.

Lang et al. [54] prepared EG/CNT/AgNW symmetric device by an inkjet 3D-printing process. It also showed morphology like graphene nanosheets, CNTs, and silver nanowires. It reported a specific capacitance of 21.6 mF/cm^2 at a scan rate of 0.01 mV/s and GCD cyclic stability of 93.1%

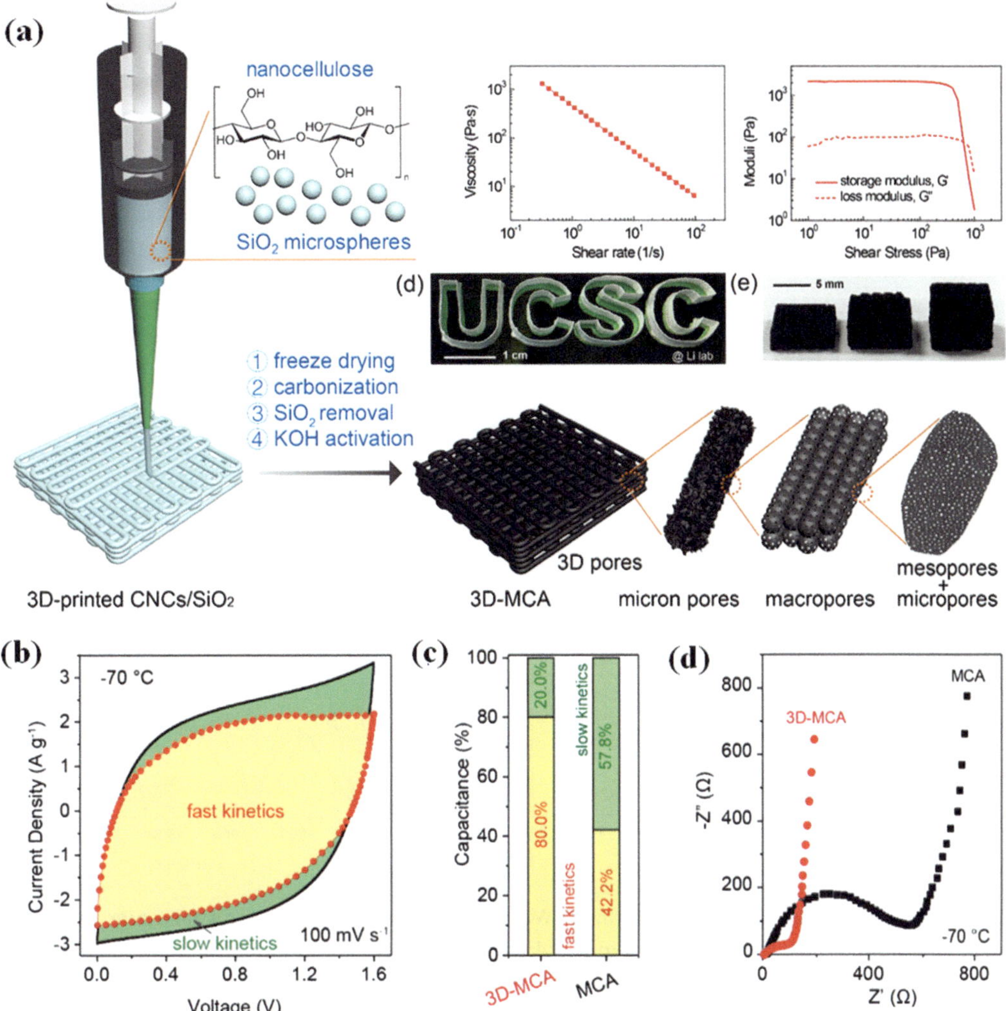

FIGURE 14.7 (a) Schematic illustration of carbon aerogels by DIW using cellulose nanocrystalline-based inks. (b) Histogram of the capacitance contribution of the 3D-MCA and MCA through the fast and slow kinetic processes. (c) CV of 3D-MCA obtained at 100 mV/s showing the capacitance contribution from the fast-kinetic processes and slow-kinetic processes. (d) Nyquist plots of the 3D-MCA and MCA. Adapted with permission [53]. Copyright (2021), American Chemical Society.

after 2000 cycles. Lai et al. [19] developed an N-STC/Fe_2O_3//N-STC ACS asymmetric SC device by ME process which shows a specific capacitance of 1.63 F/cm^2 at a scan rate of 2 mV/s. The device retained 93.2% capacity over their 10,000 cycles under various temperature ranges (-10 to 60 °C). Yang et al. [46] developed nanoflakes of 3D printed $Ti_3C_2T_x$ symmetric device by ME process which achieved high areal capacitance of 2.1 F/cm^2 at 1.7 mA/cm^2 and excellent GCD cyclic stability of 90% after 10,000 cycles. Shen et al. [55] developed nanosheets of G-VNQDs/rGO//V_2O_5/rGO asymmetric device by an inkjet method and it showed a specific capacitance of 207.9 mF/cm^2 at 0.63 mA/cm^2 and good cyclic stability 65% after 8000 cycles.

Orangi et al. [14] prepared $Ti_3C_2T_x$ MXene ink and it was printed into fine lines and predetermined shapes in a layer-by-layer way on paper, polymer films substrate. Figure 14.8a shows layer material

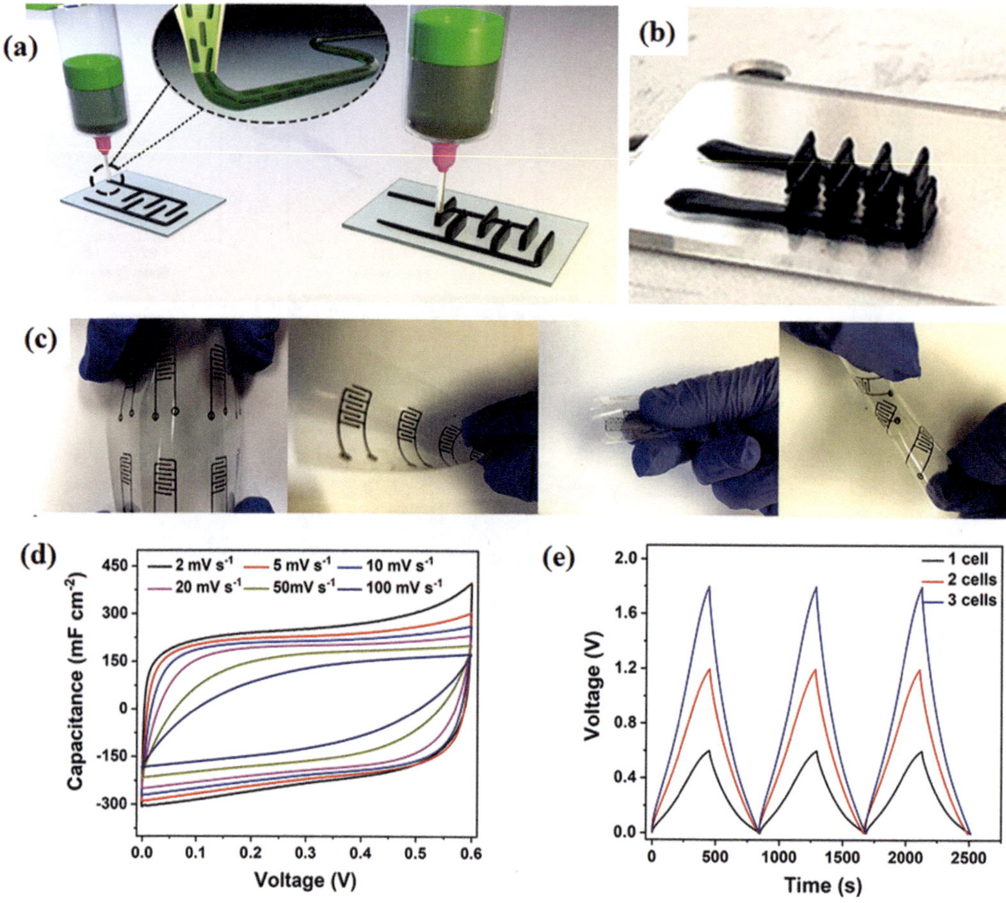

FIGURE 14.8 (a) Schematic diagram of the DIW of micro-SC with forked finger architecture. (b) Optical image of the MSC-10 device printed on a glass substrate before drying. (c) Optical images of MSC devices during repeated bending and twisting. (d) CV curves of MSCFs at various scan rates and (e) Voltage profile of fabricated MSCs at 0.2 mA cm^{-2} for a different number of cells connected in series. Adapted with permission [14]. Copyright (2020), American Chemical Society.

deposited on the current collector and the deposited layers and finger architecture obtained at specific heights by the DIW method. while Figure 14.8b presents the optical image of the different layers (2,5,10) of MSCs material deposited on a substrate before drying. MSCs printed on polymer substrates were bent and twisted in various directions and at various angles, yet they exhibited no defects or separation from the substrate following the test and they have shown in Figure 14.8c. Figure 14.8d presents the CV curves of flexible MSCs at various scan rate ranges from (2-100 mV/s). The fabricated electrode achieved high areal and volumetric energy densities of about 51.7 μWh/cm^2 and 56 mWh/cm^3 respectively. 3D interdigital electrode design at room temperature, which gained a high specific capacitance of 1035 mF/cm^2 at a minimum scan rate of 2 mV/s. The voltage profiles of MSCs devices arranged in a series combination are shown in (Figure 14.8e).

Kang et al. [56] developed GO/PVA/CNT electrodes. The fabrication process of GO/PVA/CNT electrode is presented in Figure 14.9a. The fabricated SC exhibited a good areal capacity of 2.9 F/cm^2 and delivered an excellent energy density of 0.18 mWh/cm^2. A working device with three red LEDs flashing as shown in Figure 14.9b. The prepared 18-layer electrode device delivers good cyclic stability of 98.5% after 10,000 cycles at 10 A/g (Figure 14.9c).

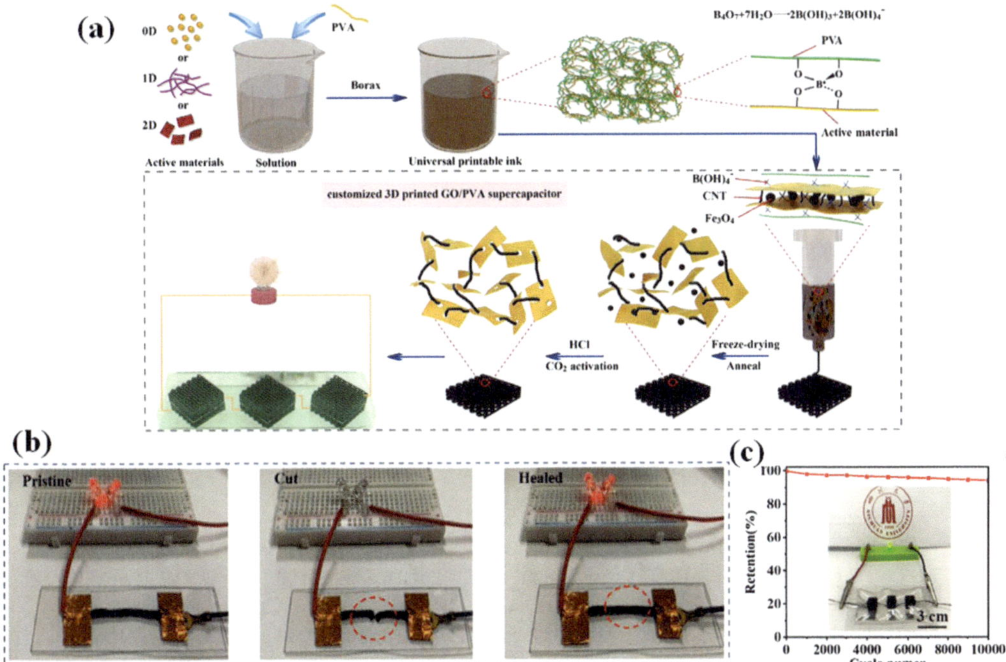

FIGURE 14.9 (a) A schematic representation of the gelation-induced ink formation and the subsequent 3D printing process. (b) A digital photo showing the self-healing function on the restoration of conductivity of a cut electrode and (c) The cycling stability of the device in 10000 cycles at 2 A g^{-1}. Adapted with permission [56]. Copyright (2020), Elsevier.

The given 3D-printed method discussed above and studied their prominent electrochemical performances. 3D printing is the future of SCs devices and is helpful for device fabrication. The fabricated SCs devices are expected to be applied to the next-generation energy storage system. We also mentioned different carbon-based materials synthesized by different 3D-printed methods with their electrochemical performance, cyclic stability and electrolyte in Table 14.1.

14.4 CONCLUSION

This chapter describes an overview of 3D-printing methods for ESD. In this overview we have explained the choice of standard 3D-printing methods such as (a) MJ (b) ME (c) BJ (d) DED (e) VAT-P (f) PBF and (g) SL. IJP and ME are mostly usable 3D-printing methods for SC applications due to their good chemical properties such as higher specific capacitance, long cyclic life, and high power, as well as the energy density of 3D-printed carbon-based materials. Lastly, we discussed 3D-printed carbon-based materials and their electrochemical performances for SCs application. Additionally, it is important to give careful thought to how to lower 3D printer maintenance expenses as well as how to increase the productivity of IJP and ME. However, as more pertinent research is published, we anticipate that these difficulties will soon be resolved.

ACKNOWLEDGMENT

Dr. G. M. Lohar is thankful to DST-SERB, Government of India, for providing funds under the ECRA scheme File No: ECR/2017/002099.

TABLE 14.1
Summary of Electrochemical SC Properties of 3D-printed Carbon-based Materials

Sr. No.	Electrode Material	3D Printing Method	Specific Capacitance	Potential Window (V)	Cyclic Stability	Electrolyte	Ref
1.	GO/GNP/SiO_2//GO/GNP/ SiO_2	ME	4.76 F·g^{-1} at 0.4 A·g^{-1}	0.0 to 0.8	95.5% after 10000 cycles	KOH	[12]
2.	3D-MCA//3D-MCA	ME	148.6 F g^{-1} at 5 mV s^{-1}	0.0 to 1.6	-	$TEABF_4$/ AN/MF	[53]
3.	$CoNi_2S_4$/Ni/OTL//CM/ Ni/OTL	VAT-P	157.6 F·g^{-1} at 1 mA cm^{-3}	0.0 to 0.8	77.3% after 1800 cycles	KOH	[17]
4.	GAMs//GAMs	ME	213 F g^{-1} at 0.5 A g^{-1}	0.0 to 0.8	90% after 50000 cycles	H_2SO_4	[50]
5.	3D GA/MnO_2//SF-3D GA	ME	1700.6 mF cm^{-2} at 5 mA cm^{-2}	0.0 to 2.0	93% after 100000 cycles	LiCl	[18]
6.	N-doped sucrose carbon// N-doped sucrose carbon	MJ	2.8 mF cm^{-2} at 5 mVs^{-1}	0.0 to 1.0	96% after 10000 cycles	PVA/H_2SO_4	[51]
7.	EG/CNT/AgNW//EG/ CNT/AgNW	MJ	21.6 mF cm^{-2} at 0.01 V s^{-1}	0.0 to 0.8	93.1% after 2000 cycles	PVA/H_3PO_4	[54]
8.	N-STC/Fe_2O_3//N-STC ACS	ME	1.63 F cm^{-2} at 2 mV s^{-1}	0.0 to 4.0	93.2 % after 10000 cycles	$EMIMBF_4$	[19]
9.	AC/CNT/MXene-N/GO// AC/CNT/MXene-N/GO	ME	8.2 F cm^{-2} at 10 mV s^{-1}	0.0 to 0.6	96.2% after 5000 cycles	PVA/H_2SO_4	[52]
10.	Ti_3C_2Tx//Ti_3C_2Tx	ME	2.1 Fcm^{-2} at 1.7 mA cm^{-2}	0.0 to 0.6	90 % after 10000 cycles	PVA/H_2SO_4	[46]
11.	MXene-MnONWs-AgNWs-C_{60}// MXene-MnONWs-AgNWs-C_{60}	ME	216.2 mF cm^{-2} at 10 mV s^{-1}	0.0 to 0.8	50% after 1000 cycles	PVA/KOH	[57]
12.	G-VNQDs/rGO//V_2O_5/ rGO	MJ	207.9 mF cm^{-2} at 0.63 mA cm^{-2}	0.0 to 2.0	65% after 8000 cycles	PVA/LiCl	[55]
13.	V_2O_5/SWCNT//VNNW/ SWCNT	ME	116.19 mF cm^{-2} at 0.6 mA cm^{-2}	0.0 to 1.6	91.9% after 8000 cycles	PVA/KOH	[20]
14.	V_2O_5/MWCNT//VNNW/ MWCNT	ME	152.7 mF cm^{-2} at 5 mV s^{-1}	0.0 to 1.6	93.1% after 12000 cycles	PVA/KOH	[58]
15.	CNT@SF//CNT@SF	ME	26.42 mF cm^{-2} at 0.42 mA cm^{-2}	0.0 to 0.6	Good cyclic stability after 15000 cycles	PVA/H_3PO_4	[59]
16.	AC//AC	ME	1.48 F cm^{-2}	0.0 to 0.8	56% after 500 cycles	PVA/H_3PO_4	[60]
17.	Ni-Co–O@3D rGO// MnO_2@3D rGO	ME	384.9 mF cm^{-2} at 2 mV s^{-1}	0.0 to 1.3	89.9% after 10000 cycles	PVA/KOH	[61]
18.	Ti_3C_2Tx//Ti_3C_2Tx	ME	1035 mF cm^{-2} at 2 mV s^{-1}	0.0 to 0.6	-	PVA/H_2SO_4	[14]
19.	3D G/MnO_2//3DG/MnO_2	ME	18.74 F cm^{-2} at 1 mA cm^{-2}	0.0 to 0.8	92.9% after 20000 cycles	LiCl	[47]
20.	GO/PVA/CNT//GO/PVA/ CNT	ME	2.9 F cm^{-2}	0.0 to 1.0	94.3% after 10000 cycles	H_2SO_4	[56]
21.	GO/ZnV_2O_6@$Co_3V_2O_8$// GO/VN	ME	149.71 F g^{-1} at 0.5 A g^{-1}	0.0 to 1.6	95.5% after 10000 cycles	KOH	[62]
22.	GO-MnO_2//AC	IJP	1.586 F cm^{-2} at 4 mA cm^{-2}	0.0 to 2.0	89.6 % after 9000 cycles	PVA-LiCl	[63]
23.	CQDs/GO	IJP	4.2 mF cm^{-2} at 1 mV s^{-1}	0.0 to 1.0	98% after 1000 cycles	PVA/H_2SO_4	[64]

TABLE 14.1 (Continued)
Summary of Electrochemical SC Properties of 3D-printed Carbon-based Materials

Sr. No.	Electrode Material	3D Printing Method	Specific Capacitance	Potential Window (V)	Cyclic Stability	Electrolyte	Ref
24.	MWCNT	IJP	235 F g^{-1} at 5 mV s^{-1}	0.0 to 3.0	90 % after 5000 cycles	$EMIMBF_4$	[16]
25.	MWCNT	IJP	268 F g^{-1} at 5 mV s^{-1}	0.0 to 0.5	95 % after 1000 cycles	KOH	[65]
26.	Graphene	IJP	0.14 mF cm^{-2} at 5 mV s^{-1}	0.0 to 1.0	90 % after 1000 cycles	PVA/H_3PO_4	[66]
27.	Graphene	IJP	9.3 F cm^{-1} at 0.25 A cm^{-3}	0.0 to 1.0	98 % after 10000 cycles	PVA/H_3PO_4	[67]
28.	$Ti_3C_2T_x$ MXene	IJP	108.1 mF cm^{-2} at 1Ag^{-1}	-	94.7% after 4000 cycles	PVA/H_2SO_4	[68]

REFERENCES

1 O. C. Pore, A. V. Fulari, R. K. Kamble, A. S. Shelake, N. B. Velhal, V. J. Fulari, G. M. Lohar, Hydrothermally synthesized Co_3O_4 microflakes for supercapacitor and non-enzymatic glucose sensor, *J. Mater. Sci. Mater. Electron.* 32 (2021) 20742–20754.

2 O. C. Pore, A. V. Fulari, N. B. Velha, V. G. Parale, H. H. Park, R. V. Shejwal, V. J. Fulari, G. M. Lohar, Hydrothermally synthesized urchinlike NiO nanostructures for supercapacitor and nonenzymatic glucose biosensing application, *Mater. Sci. Semicond. Process.* 134 (2021) 105980.

3 K. Chi, Z. Zhang, J. Xi, Y. Huang, F. Xiao, S. Wang, Y. Liu, Freestanding graphene paper supported three-dimensional porous graphene-polyaniline nanocomposite synthesized by inkjet printing and in flexible all-solid-state supercapacitor, *ACS Appl. Mater. Interfaces* 6 (2014) 16312–16319.

4 L. W. T. Ng, R. C. T. Howe, Z. Yang, T. Hasan, G. Hu, X. Zhu, C. G. Jones, *Printing of Graphene and Related 2D Materials: Technology, Formulation and Applications*, Springer International, 2018.

5 S. Wang, N. Liu, J. Tao, C. Yang, W. Liu, Y. Shi, Y. Wang, J. Su, L. Li, Y. Gao, Inkjet printing of conductive patterns and supercapacitors using a multi-walled carbon nanotube/Ag nanoparticle based ink, *J. Mater. Chem. A* 3 (2015) 2407–2413.

6 S. A. M. Tofail, E. P. Koumoulos, A. Bandyopadhyay, S. Bose, L. O'Donoghue, C. Charitidis, Additive manufacturing: Scientific and technological challenges, market uptake and opportunities, *Mater. Today* 21 (2018) 22–37.

7 J. Li, S. Sollami Delekta, P. Zhang, S. Yang, M. R. Lohe, X. Zhuang, X. Feng, M. Östling, Scalable fabrication and integration of graphene microsupercapacitors through full inkjet printing, *ACS Nano.* 11 (2017) 8249–8256.

8 R. D. Farahani, M. Dubé, D. Therriault, Three-dimensional printing of multifunctional nanocomposites: Manufacturing techniques and applications, *Adv. Mater.* 28 (2016) 5794–5821.

9 A. Zolfagharian, A. Z. Kouzani, S. Y. Khoo, A. A. A. Moghadam, I. Gibson, A. Kaynak, Evolution of 3D printed soft actuators, *Sensors Actuators A Phys.* 250 (2016) 258–272.

10 P. Chang, H. Mei, S. Zhou, K. G. Dassios, L. Cheng, 3D printed electrochemical energy storage devices, *J. Mater. Chem. A* 7 (2019) 4230–4258.

11 S. V. Murphy, A. Atala, 3D bioprinting of tissues and organs, *Nat. Biotechnol.* 32 (2014) 773–785.

12 C. Zhu, T. Liu, F. Qian, T. Y. J. Han, E. B. Duoss, J. D. Kuntz, C. M. Spadaccini, M. A. Worsley, Y. Li, Supercapacitors based on three-dimensional hierarchical graphene aerogels with periodic macropores, *Nano Lett.* 16 (2016) 3448–3456.

13 X. Zheng, H. Lee, T. H. Weisgraber, M. Shusteff, J. DeOtte, E. B. Duoss, J. D. Kuntz, M. M. Biener, Q. Ge, J. A. Jackson, S. O. Kucheyev, N. X. Fang, C. M. Spadaccini, Ultralight, ultrastiff mechanical metamaterials, *Science* 344 (2014) 1373–1377.

14 J. Orangi, F. Hamade, V. A. Davis, M. Beidaghi, 3D printing of additive-free 2D Ti_3C_2Tx (MXene) ink for fabrication of micro-supercapacitors with ultra-high energy densities, *ACS Nano* 14 (2020) 640–650.

15 C. Zhu, T. Y. J. Han, E. B. Duoss, A. M. Golobic, J. D. Kuntz, C. M. Spadaccini, M. A. Worsley, Highly compressible 3D periodic graphene aerogel microlattices, *Nat. Commun.* 6 (2015) 1–8.

16 S. K. Ujjain, P. Ahuja, R. Bhatia, P. Attri, Printable multi-walled carbon nanotubes thin film for high performance all solid state flexible supercapacitors, *Mater. Res. Bull.* 83 (2016) 167–171.

17 J. Song, Y. Chen, K. Cao, Y. Lu, J. H. Xin, X. Tao, Fully controllable design and fabrication of three-dimensional lattice supercapacitors, *ACS Appl. Mater. Interfaces* 10 (2018) 39839–39850.

18 B. Yao, S. Chandrasekaran, H. Zhang, A. Ma, J. Kang, L. Zhang, X. Lu, F. Qian, C. Zhu, E. B. Duoss, C. M. Spadaccini, M. A. Worsley, Y. Li, 3D-printed structure boosts the kinetics and intrinsic capacitance of pseudocapacitive graphene aerogels, *Adv. Mater.* 32 (2020) 1906652.

19 F. Lai, C. Yang, R. Lian, K. Chu, J. Qin, W. Zong, D. Rao, J. Hofkens, X. Lu, T. Liu, Three-phase boundary in cross-coupled micro-mesoporous networks enabling 3D-printed and ionogel-based quasi-solid-state micro-supercapacitors, *Adv. Mater.* 32 (2020) 2002474.

20 J. Zhao, Y. Zhang, Y. Huang, J. Xie, X. Zhao, C. Li, J. Qu, Q. Zhang, J. Sun, B. He, Q. Li, C. Lu, X. Xu, W. Lu, L. Li, Y. Yao, 3D printing fiber electrodes for an all-fiber integrated electronic device via hybridization of an asymmetric supercapacitor and a temperature sensor, *Adv. Sci.* 5 (2018) 1801114.

21 V. Egorov, U. Gulzar, Y. Zhang, S. Breen, C. O'Dwyer, Evolution of 3D printing methods and materials for electrochemical energy storage, *Adv. Mater.* 32 (2020) 2000556.

22 Y. Zhang, F. Zhang, Z. Yan, Q. Ma, X. Li, Y. Huang, J. A. Rogers, Printing, folding and assembly methods for forming 3D mesostructures in advanced materials, *Nat. Rev. Mater.* 2 (2017) 17019.

23 J. Y. Lee, J. An, C. K. Chua, Fundamentals and applications of 3D printing for novel materials, *Appl. Mater. Today* 7 (2017) 120–133.

24 X. Xu, A. Awad, P. Robles-Martinez, S. Gaisford, A. Goyanes, A. W. Basit, Vat photopolymerization 3D printing for advanced drug delivery and medical device applications, *J. Control. Release* 329 (2021) 743–757.

25 A. Zhakeyev, P. Wang, L. Zhang, W. Shu, H. Wang, J. Xuan, Additive manufacturing: Unlocking the evolution of energy materials, *Adv. Sci.* 4 (2017) 1700187.

26 D. Svetlizky, M. Das, B. Zheng, A. L. Vyatskikh, S. Bose, A. Bandyopadhyay, J. M. Schoenung, E. J. Lavernia, N. Eliaz, Directed energy deposition (DED) additive manufacturing: Physical characteristics, defects, challenges and applications, *Mater. Today* 49 (2021) 271–295.

27 B. Mueller, D. Kochan, Laminated object manufacturing for rapid tooling and patternmaking in foundry industry, *Comput. Ind.* 39 (1999) 47–53.

28 L. E. Murr, S. M. Gaytan, D. A. Ramirez, E. Martinez, J. Hernandez, K. N. Amato, P. W. Shindo, F. R. Medina, R. B. Wicker, Metal fabrication by additive manufacturing using laser and electron beam melting technologies, *J. Mater. Sci. Technol.* 28 (2012) 1–14.

29 W. E. King, A. T. Anderson, R. M. Ferencz, N. E. Hodge, C. Kamath, S. A. Khairallah, A. M. Rubenchik, Laser powder bed fusion additive manufacturing of metals; physics, computational, and materials challenges, *Appl. Phys. Rev.* 2 (2015) 041304.

30 I. Gibson, D. Rosen, B. Stucker, *Additive Manufacturing Technologies: 3D Printing, Rapid Prototyping, and Direct Digital Manufacturing*, second edition, Springer, New York, 2015.

31 M. Singh, H. M. Haverinen, P. Dhagat, G. E. Jabbour, Inkjet printing-process and its applications, *Adv. Mater.* 22 (2010) 673–685.

32 P. E. Delannoy, B. Riou, T. Brousse, J. Le Bideau, D. Guyomard, B. Lestriez, Ink-jet printed porous composite LiFePO4 electrode from aqueous suspension for microbatteries, J. *Power Sources* 287 (2015) 261–268.

33 K. H. Choi, J. T. Yoo, C. K. Lee, S. Y. Lee, All-inkjet-printed, solid-state flexible supercapacitors on paper, *Energy Environ. Sci.* 9 (2016) 2812–2821.

34 B. Derby, Inkjet printing of functional and structural materials: Fluid property requirements, feature stability, and resolution, *Annu. Rev. Mater. Res.* 40 (2010) 395–414.

35 N. C. Raut, K. Al-Shamery, Inkjet printing metals on flexible materials for plastic and paper electronics, *J. Mater. Chem. C* 6 (2018) 1618–1641.

36 Q. Shi, K. Yu, X. Kuang, X. Mu, C. K. Dunn, M. L. Dunn, T. Wang, H. Jerry Qi, Recyclable 3D printing of vitrimer epoxy, *Mater. Horizons* 4 (2017) 598–607.

37 M. Toyoda, T. Matsuo, Fused deposition of ceramics: A comprehensive experimental, analytical and computational study of material behavior, fabrication process and equipment design, *Fused Depos. Ceram.* 427 (1999) 375–381.

38 J. Yesuraj, O. Padmaraj, S. A. Suthanthiraraj, Synthesis, characterization, and improvement of supercapacitor properties of $NiMoO_4$ nanocrystals with polyaniline, *J. Inorg. Organomet. Polym. Mater.* 30 (2020) 310–321.

39 H. Pang, Y. Zhang, W. Y. Lai, Z. Hu, W. Huang, Lamellar $K_2Co_3(P_2O_7)_2{\cdot}2H_2O$ nanocrystal whiskers: High-performance flexible all-solid-state asymmetric micro-supercapacitors via inkjet printing, *Nano Energy* 15 (2015) 303–312.

40 B. Li, N. Hu, Y. Su, Z. Yang, F. Shao, G. Li, C. Zhang, Y. Zhang, Direct inkjet printing of aqueous inks to flexible all-solid-state graphene hybrid micro-supercapacitors, *ACS Appl. Mater. Interfaces* 11 (2019) 46044–46053.

41 L. T. Le, M. H. Ervin, H. Qiu, B. E. Fuchs, W. Y. Lee, Graphene supercapacitor electrodes fabricated by inkjet printing and thermal reduction of graphene oxide, *Electrochem. Commun.* 13 (2011) 355–358.

42 D. Pech, M. Brunet, P. L. Taberna, P. Simon, N. Fabre, F. Mesnilgrente, V. Conédéra, H. Durou, Elaboration of a microstructured inkjet-printed carbon electrochemical capacitor, *J. Power Sources* 195 (2010) 1266–1269.

43 J. Li, F. Ye, S. Vaziri, M. Muhammed, M. C. Lemme, M. Östling, Efficient inkjet printing of graphene, *Adv. Mater.* 25 (2013) 3985–3992.

44 Y. Xu, I. Hennig, D. Freyberg, A. James Strudwick, M. Georg Schwab, T. Weitz, K. Chih-Pei Cha, Inkjet-printed energy storage device using graphene/polyaniline inks, J. *Power Sources* 248 (2014) 483–488.

45 C. J. Zhang, M. P. Kremer, A. Seral-Ascaso, S. H. Park, N. McEvoy, B. Anasori, Y. Gogotsi, V. Nicolosi, Stamping of flexible, coplanar micro-supercapacitors using MXene inks, *Adv. Funct. Mater.* 28 (2018) 1705506.

46 W. Yang, J. Yang, J. J. Byun, F. P. Moissinac, J. Xu, S. J. Haigh, M. Domingos, M. A. Bissett, R. A. W. Dryfe, S. Barg, 3D printing of freestanding MXene architectures for current-collector-free supercapacitors, *Adv. Mater.* 31 (2019) 1902725.

47 B. Yao, S. Chandrasekaran, J. Zhang, W. Xiao, F. Qian, C. Zhu, E. B. Duoss, C. M. Spadaccini, M. A. Worsley, Y. Li, Efficient 3D printed pseudocapacitive electrodes with ultrahigh MnO_2 Loading, *Joule* 3 (2019) 459–470.

48 J. Xue, L. Gao, X. Hu, K. Cao, W. Zhou, W. Wang, Y. Lu, Stereolithographic 3D printing-based hierarchically cellular lattices for high-performance quasi-solid supercapacitor, *Nano-Micro Lett.* 11 (2019) 1–13.

49 A. Azhari, E. Marzbanrad, D. Yilman, E. Toyserkani, M. A. Pope, Binder-jet powder-bed additive manufacturing (3D printing) of thick graphene-based electrodes, *Carbon N. Y.* 119 (2017) 257–266.

50 Y. Jiang, Z. Xu, T. Huang, Y. Liu, F. Guo, J. Xi, W. Gao, C. Gao, Direct 3D Printing of ultralight graphene oxide aerogel microlattices, *Adv. Funct. Mater.* 28 (2018) 1707024.

51 Y. Brauniger, S. Lochmann, J. Grothe, M. Hantusch, S. Kaskel, Piezoelectric inkjet printing of nanoporous carbons for micro-supercapacitor devices, *ACS Appl. Energy Mater.* 4 (2021) 1560–1567.

52 L. Yu, Z. Fan, Y. Shao, Z. Tian, J. Sun, Z. Liu, Versatile N-doped MXene ink for printed electrochemical energy storage application, *Adv. Energy Mater.* 9 (2019) 1901839.

53 B. Yao, H. Peng, H. Zhang, J. Kang, C. Zhu, G. Delgado, D. Byrne, S. Faulkner, M. Freyman, X. Lu, M. A. Worsley, J. Q. Lu, Y. Li, Printing porous carbon aerogels for low temperature supercapacitors, *Nano Lett.* 21 (2021) 3731–3737.

54 L. Liu, J. Y. Lu, X. L. Long, R. Zhou, Y. Q. Liu, Y. T. Wu, K. W. Yan, 3D printing of high-performance micro-supercapacitors with patterned exfoliated graphene/carbon nanotube/silver nanowire electrodes, *Sci. China Technol. Sci.* 64 (2021) 1065–1073.

55 K. Shen, J. Ding, S. Yang, 3D printing quasi-solid-state asymmetric micro-supercapacitors with ultrahigh areal energy density, *Adv. Energy Mater.* 8 (2018) 1800408.

56 W. Kang, L. Zeng, S. Ling, R. Yuan, C. Zhang, Self-healable inks permitting 3D printing of diverse systems towards advanced bicontinuous supercapacitors, *Energy Storage Mater.* 35 (2021) 345–352.

57 X. Li, H. Li, X. Fan, X. Shi, J. Liang, 3D-printed stretchable micro-supercapacitor with remarkable areal performance, *Adv. Energy Mater.* 10 (2020) 1903794.

58 J. Zhao, H. Lu, Y. Zhang, S. Yu, O. I. Malyi, X. Zhao, L. Wang, H. Wang, J. Peng, X. Li, Y. Zhang, S. Chen, H. Pan, G. Xing, C. Lu, Y. Tang, X. Chen, Direct coherent multi-ink printing of fabric supercapacitors, *Sci. Adv.* 7 (2021) eabd6978.

59 M. Zhang, M. Zhao, M. Jian, C. Wang, A. Yu, Z. Yin, X. Liang, H. Wang, K. Xia, X. Liang, J. Zhai, Y. Zhang, Printable smart pattern for multifunctional energy-management e-textile, *Matter* 1 (2019) 168–179.

60 M. Areir, Y. Xu, D. Harrison, J. Fyson, 3D printing of highly flexible supercapacitor designed for wearable energy storage, *Mater. Sci. Eng. B* 226 (2017) 29–38.

61 T. Wang, X. Tian, L. Li, L. Lu, S. Hou, G. Cao, H. Jin, 3D printing-based cellular microelectrodes for high-performance asymmetric quasi-solid-state micro-pseudocapacitors, *J. Mater. Chem. A* 8 (2020) 1749–1756.

62 J. Zhao, Y. Zhang, X. Zhao, R. Wang, J. Xie, C. Yang, J. Wang, Q. Zhang, L. Li, C. Lu, Y. Yao, Direct ink writing of adjustable electrochemical energy storage device with high gravimetric energy densities, *Adv. Funct. Mater.* 29 (2019) 1900809.

63 P. Sundriyal, S. Bhattacharya, Inkjet-printed electrodes on A4 paper substrates for low-cost, disposable, and flexible asymmetric supercapacitors, *ACS Appl. Mater. Interfaces* 9 (2017) 38507–38521.

64 J. Liu, J. Ye, F. Pan, X. Wang, Y. Zhu, Solid-state yet flexible supercapacitors made by inkjet-printing hybrid ink of carbon quantum dots/graphene oxide platelets on paper, *Sci. China Mater.* 62 (2019) 545–554.

65 S. K. Ujjain, R. Bhatia, P. Ahuja, P. Attri, Highly conductive aromatic functionalized multi-walled carbon nanotube for inkjet printable high performance supercapacitor electrodes, *PLoS One* 10 (2015) e0131475.

66 J. Li, V. Mishukova, M. Östling, All-solid-state micro-supercapacitors based on inkjet printed graphene electrodes, *Appl. Phys. Lett.* 109 (2016) 123901.

67 L. Li, E. B. Secor, K. S. Chen, J. Zhu, X. Liu, T. Z. Gao, J. W. T. Seo, Y. Zhao, M. C. Hersam, High-Performance solid-state supercapacitors and microsupercapacitors derived from printable graphene inks, *Adv. Energy Mater.* 6 (2016) 1600909.

68 C. W. Wu, B. Unnikrishnan, I. W. P. Chen, S. G. Harroun, H. T. Chang, C. C. Huang, Excellent oxidation resistive MXene aqueous ink for micro-supercapacitor application, *Energy Storage Mater.* 25 (2020) 563–571.

15 Recent Progress in 3D-Printed Metal Oxides Based Materials for Supercapacitors

K.A.U. Madhushani[1,2], A.A.P.R. Perera[1,2], Felipe M. de Souza[2], and Ram K. Gupta[1,2]*

[1]Department of Chemistry, Pittsburg State University, Pittsburg, Kansas 66762, USA

[2]National Institute for Materials Advancement, Pittsburg State University, Pittsburg, Kansas 66762, USA

*Corresponding author: ramguptamsu@gmail.com

CONTENTS

15.1 INTRODUCTION

With the development of science and technology, the machinery which is operated through the combustion of fossil fuels was invented and with growing populations, the energy demand is increasing daily. Consequently, it led to the depletion of fossil resources and the extensive use of fossil resources is creating detrimental effects on the environment and living beings. One of the solutions is to use advanced technologies to convert renewable resources such as wave, air, tidal, geothermal, water, and sunlight into green energy [1]–[3]. However, the availability of these energy sources is unforeseeable with fluctuations in energy output due to factors such as time, region, and season [1], [4]. Therefore, it is important to store the energy produced from renewable resources for later use. In that sense, electrochemical energy storage (EES) devices appear viable candidates. However, the demanding requirements such as the need for rapid energy storage, high energy capacity, long device life, and cost-effectiveness are needed to fulfill their range of applications [5].

High-performance devices can be designed with a plethora of electroactive materials that can be synthesized in various ways while allowing the manufacture of smaller, thinner, flexible, and more

DOI: 10.1201/9781003296676-15

efficient energy storage devices [4]. Batteries and supercapacitors (SCs) are among the most popular EES devices [6], [7]. A supercapacitor (also known as Faraday quasi capacitor, ultracapacitor, or electrochemical capacitor) is becoming more popular as a convenient, efficient, and environmentally friendly energy storage device due to its many promising characteristics [5]. Although SCs have stupendous advantages of rapid charge-discharge, lengthy durability, and high power density, they often present an unsatisfactory energy density compared to rechargeable batteries. The low energy density of SCs can be overcome with the development of new materials and technologies. The energy density (E) of a supercapacitor is expressed in Equation 15.1.

$$E = \frac{1}{2}CV^2 \tag{15.1}$$

E is proportional to the capacitance (C) and the square of the voltage (V). Thus, the energy density can be enhanced by increasing C and V. This can be achieved via improving electrode components with high electrical capacity, using wide potential windows materials and electrolytes, and developing nanocomposites [5]. The materials for the electrodes can be synthesized by various techniques, and from those, 3D printing is one of the famous electrode fabrication methods due to its benefits of excellent process flexibility, geometry and thickness controllability, cost-effectiveness, and eco-friendliness [8]. Through this chapter, examples of the 3D-printed metal oxides (MOs) based materials for SCs are discussed.

15.2 FUNDAMENTALS OF SUPERCAPACITOR

Supercapacitors usually have four components: anode, cathode, electrolyte, and separator. The charge in SCs is stored at the interface between the electrode and electrolyte [9]. The constituents of electrodes are a decisive factor in EES devices as energy storage mainly occurs through the transference of ions at the junction of electrodes and electrolytes [1]. Therefore, with the changes in the electrochemical properties of the electrode, a supercapacitor electrode with high specific capacitance, and cyclic constancy can be obtained. Carbon (e.g., graphene, carbon nanotubes, activated carbon, porous carbon), conducting polymers (e.g., polyaniline, polypyrrole, polythiophene, poly (3,4-ethylene dioxythiophene)), metal oxides/ hydroxides/ sulfides/ phosphides, and their composites are commonly used electrode materials [1].

Carbon-based electrodes are used for commercial SCs due to their easy availability, low cost, and diversity as a natural resource [10], [11]. In a carbon-based electrode, the capacitance is attained via pure electro-sorption of the electrical double-layer [12]–[17]. Although carbon-based electrode materials have superior cyclic stability, their energy density is often low due to their energy storage mechanism. Electrodes based on conducting polymers (CPs) provide high energy density due to redox reactions, however, their electrochemical cyclic stability is low [18]. Moreover, CPs provide low power density because of their dense structure. As electrode components, transition metal oxides/hydroxides (TMOs/TMHs) deliver high power density and stability compared to carbon and CPs based electrodes [19], [20]. However, the low conductivity and processability are some of the limitations which could be overcome by controlling dimensionality and morphology.

In addition to electrode materials, electrolytes also play a vital role in the performance of the EES devices as they can affect power efficiency, resistance of the device, activity, thermal working range, durability, self-discharge ability, etc. (Figure 15.1). Large varieties of electrolytes such as aqueous, organic, ionic liquids, redox-active, and solid or semi-solid can be used to cover a wide range of SCs. An ideal electrolyte should have characteristics such as a wide operating potential, high conductance, high chemical/electrochemical stability, low volatileness, non-combustibility, inexpensiveness, and eco-friendliness [5].

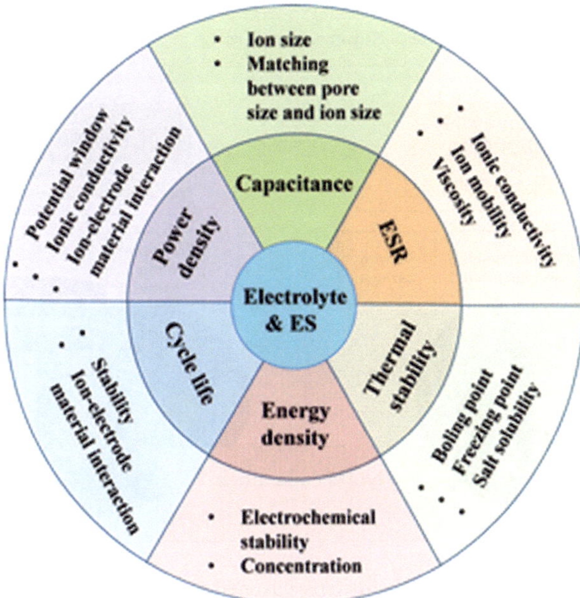

FIGURE 15.1 Characteristics of an ideal electrolyte for EES devices. Adapted with permission [5]. Copyright (2015), Royal Society of Chemistry.

15.3 TYPES OF SUPERCAPACITORS AND THEIR MECHANISMS

Supercapacitors can be classified in many ways based on the principle of charge storing mechanism, electrolytes, the components of the electrodes, and the arrangement of the cell. According to the principle of energy storage, SCs are categorized into three main groups namely electric double-layer capacitors (EDLCs), pseudocapacitors, and hybrid supercapacitors (Figure 15.2) [5].

15.3.1 Electric Double-Layer Capacitors

Electric double-layer capacitors, also known as electrostatic capacitors, store the charge based on a double-layer mechanism. Carbon-based materials which do not participate in the redox process are electrode materials for this category. The accumulation of ions at the electrode's surface is due to the applied voltage which creates electric double layers at the interface of the electrode and electrolyte. The capacitance of the dual layers is linked with the accumulated opposite electric charges at the boundary of the electrode and electrolyte (Figure 15.3). In addition, the excess or shortage electrons are directed to the exterior conducting wire of the electrode, and the ions in the solvent are balanced with the opposite ions to keep the electrolyte as a neutral solution and due to this reason, the EDLC charge storage mechanism can be defined as a surface process [21].

In the charging process (applying an exterior voltage), cations of electrolyte move towards the negatively charged electrode whereas anions migrate to the positively charged electrode, and electrons move via an external circuit from the cathode to the anode forming a double layer at the surface of the electrodes. This is reversely true for the process of discharging. When the charging process is finished, the positive ions on the electrode are bound with the negative ions in the electrolyte solution as well as the anions on the electrode are bound with the cations in the solution [11]. The theory behind the electric charge accumulation on the interior layer indicates that the amount of the anions and cations in electrolytes are maintained at a steady state throughout the process of

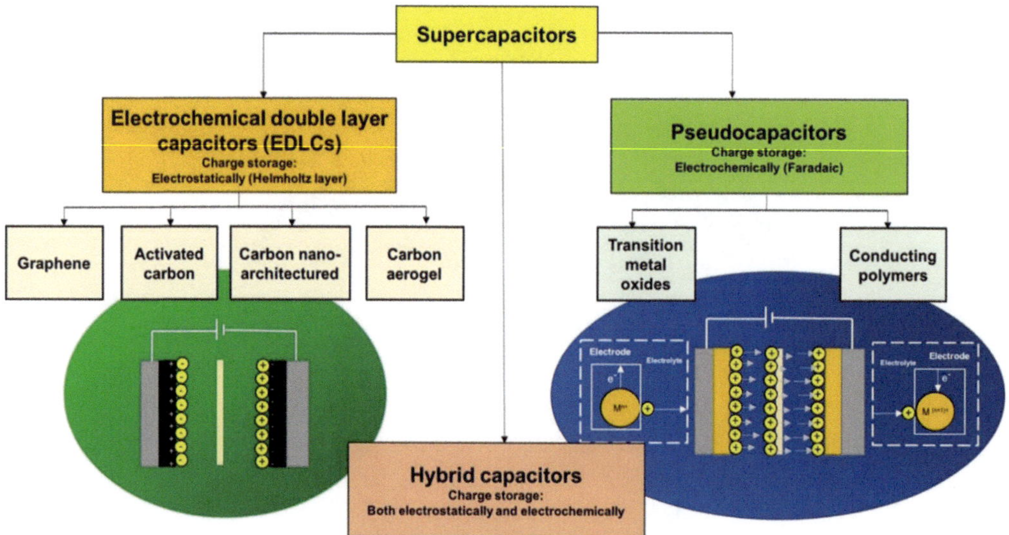

FIGURE 15.2 Classification of supercapacitors based on the charge storage mechanism. Adapted with permission [21]. Copyright (2020), Elsevier.

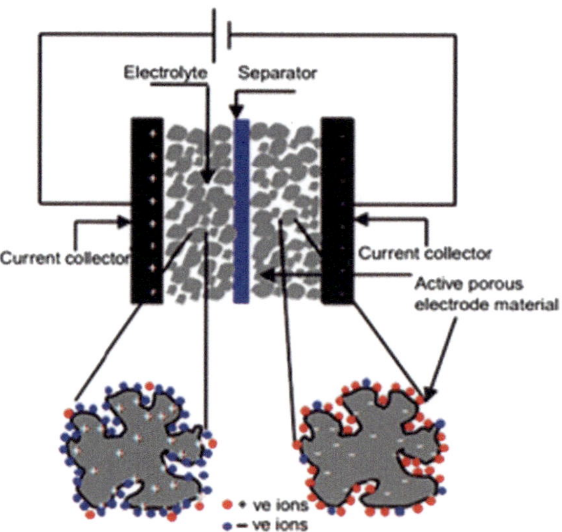

FIGURE 15.3 Graphical view of electric double-layer capacitors (EDLCs). Adapted with permission [11]. Copyright (2018), Elsevier.

charging and discharging. The characteristic of the surface can greatly influence the capacitance. The surface polarization, a large number of adsorbed ions from the electrolyte, defects in the morphological structures, and polarizable species are some factors that are related to the mechanism of surface electrode charge storage. Carbon-related materials including carbon nanotubes (CNTs), graphene, and porous carbon are used as electrodes because of their benefits such as high surface area, availability, abundance, and low cost [11].

15.3.2 Pseudocapacitors

The pseudocapacitors present a diverse mechanism of electric charge storage compared to EDLCs and it occurs due to the fast, reversible redox reactions at the surface of the electrodes [21]. When an external voltage is supplied to charge the pseudocapacitor the faradaic (redox) reaction occurs at the electrodes, usually due to the chemical incorporation of the charged species from the electrolyte within the electrode's structure. This process leads to the chemical potential energy that is stored in the electrode. It also engages the movement of ions to form a double layer likewise in the EDLC process. When the system is discharged there is the disincorporation of electrolytes that causes the release of electrons that are transferred through the external conducting circuit. As a result, a faradaic current is passed across the supercapacitor cell [22].

CPs, TMOs, TMHs, etc. are commonly used materials for pseudocapacitors due to their redox behavior. In such materials, three types of redox reactions can take place. They are reversible adsorption, e.g., bonding of hydrogen on the outer layer of Pt or Au, redox reactions of TMOs, and changeable electrochemical doping–de-doping in both directions in electrodes made of CP. An important aspect of these reactions is that they are involved with the enhancement of both the operating voltage and the specific capacitance. They usually show higher capacitance and energy density compared to EDLCs-based materials. However, since the rate of the redox reactions is slower than that of the non-faradaic processes, pseudocapacitor delivers lower power density than EDLCs. The cycling stability in pseudocapacitors is lower (like batteries) than EDLCs as redox reactions introduce chemical strain to the electrode [22].

15.3.3 Hybrid Capacitors

A hybrid capacitor (HC) shows a combination of EDLC and pseudocapacitor mechanisms for charge storage. This combination leads to some advantages such as an increase in overall energy density at a higher rate. The mechanisms of both EDLC and faradaic capacitance can take place simultaneously in hybrid SCs. Based on these concepts, large surface area, appropriate pore size, high conduction, and redox activeness are some of the decisive features of electrode materials to provide high capacitance values. HC can provide high energy density derived from redox reactions as well as high power density and stable cyclic performance provided by the non-faradaic capacitive electrode [11].

15.4 INTRODUCTION TO 3D-PRINT TECHNOLOGY

3D printing or additive manufacturing (AM) is a fast-growing, cutting-edge technology that is used for constructing structures via layer-by-layer deposition. Computer-aided design (CAD) is used to construct the structure files for print. This technique plays a vital role in manufacturing EES devices with a wide range of nano to macro sizes while controlling the geometric factors of the device, including dimension, porosity, shape, and layout with improved definite energy and power densities [8]. This technology can be applied to a wide range of electrode materials, including carbon, polymers, metals, MOs, and ceramics [23]. Developing novel 3D-print technologies in the field of EES devices can bring unique advantages in different applications such as automotive, healthcare, aerospace, and consumer products as it provides a simplistic freeform production and fast 3D structural prototyping in customized patterns [24], [25]. In addition, the 3D-printing technology allows the fabrication of flexible materials at a low production cost for the fabrication of EES devices with unique 3D structures [3].

The 3D-printing method consists of several steps starting from the model design by software to post-treatment. As a first step, the required components (active materials, binders, etc.) are prepared as a paste. Normally, the CAD with the 3D model of various shapes is used for the design in software that is later stored as a 3D-printing adapted format. The created design model is transformed into sliced layer data by using slicer programs to adjust other variables of the printing setup such as the height of the layers, the thickness of the model, printing speed, and filling ratio as pre-conditions. Finally, printing and post-manufacturing steps such as removal of supports, sanding, and filling are carried out to get the customizable 3D objects with the demand qualities including conductive probes with modern technology [23], [25]. The scheme for the 3D design process is presented in Figure 15.4. Depending on the unique characteristics and their applications, there are several categories in 3D-printing processes such as photopolymerization, extrusion, powder-based, and lamination. Among them, the appropriate printing approach is chosen based on the requirements of the EES devices.

Photopolymerization was the first developed 3D-printing technique that can be classified into several sections. Stereolithography (SLA) follows the mechanism of photopolymerization. This is one of the most popular methods for the production of 3D objects using photopolymers. The powdered polymer layers extruded are bound together using a UV light, and the power bed is dropped down to form the next layer on top after scanning each cross-section. This process is carried out until the intended model is completed [25]. This technique is largely applied for manufacturing EES devices including electrolytes, membranes, current collectors, and conductors using a photocurable resin which includes mostly ceramics [23].

Another convenient method for 3D printing is the extrusion-based method which is versatile and can include a variety of electrode materials. This method is based on two steps: (a) pretreatment of the material and (b) modeling deposition through a nozzle tip. Other methods such as fused deposition modeling (FDM), direct ink writing (DIW), and inkjet printing (IJP) follow the extrusion-based mechanism [3]. The FDM process is used for printing filament-based supplies such as acrylonitrile butadiene styrene (ABS), polycarbonate (PC), polylactic acid (PLA), polyphenyl sulphone (PPSF), polyvinyl alcohol (PVA), and polyamide (PA). In this process, thermoplastic materials are converted to a semi-molten phase by heating. Subsequently, 3D devices are produced by depositing these materials on the platform layer-wise via extrusion from the dispenser nozzle. Then, they are allowed to solidify on a printing bed [25]. One of the major challenges of this process is to maintain the firmness and superiority of the object which is influenced by increasing the filament extrusion using a high ratio of nanofillers [27].

Through the DIW techniques, 3D electronic devices and structures can be printed with the high mass filling of active electrode materials. Based on the shear-thinning rheological behavior those substances can be prepared as high viscous and shear-thinning flow to keep a smooth extrusion and to maintain the shape of the objects [23]. It is necessary to maintain the viscosity at higher levels otherwise it can cause damage to the structure of the desired objects. The exactness and resolution of

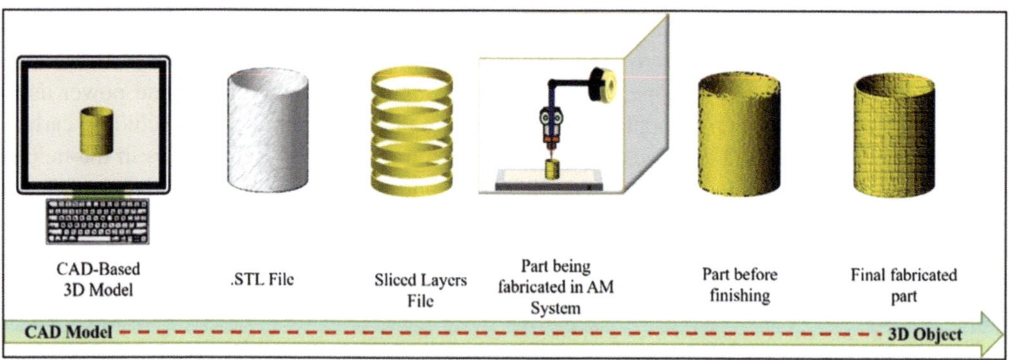

FIGURE 15.4 Basic steps of the 3D printing process. Adapted with permission [26]. Copyright (2021), Taylor & Francis. This is an open-access article distributed under the terms of the Creative Commons CC BY license.

the ink can be regulated by changing the printing parameters such as the rate of the flow, the radius of the needle, the speediness of the printed head, the thickness of the ink, and the temperature of the outlet [27]. Although the ability to store charge can be enhanced using this method, it is difficult to increase the resolution and volume of the printed device [23].

The inkjet process also follows extrusion-based ink deposition, where toner is ejected from the tip of the outlet and deposited onto a substrate. Particularly, this process can be classified into two methods: continuous inkjet (CIJ) and drop-on-demand (DOD). In the CIJ process, a steady movement of ink droplets is deposited, and excess ink is recycled while in DOD, a discrete ink droplet is dropped at the pre-designed area. In this sense, the droplet ejection depends on the viscidness, interfacial tension, and concentration of the ink. The DOD process exhibits the benefits of greater material selectivity, less leakage, low expenditure, less contamination, quick program, and scalability from tiny droplets to the wide designed areas. The ink's properties can be related to the nanoparticle size of which the ink is composed. Material accumulation, moving distance of droplets, and surface tension of the ink are some key factors that are considered for handling this process [28].

Compared to others, powder-based printing is contrary to liquid deposition methods, yet it is like SLA because of the light-based printing technique. In that sense, 3D components are manufactured by sintering the solid matters which act as the building structures from the high-powered source of a laser beam. Particularly, selective laser sintering (SLS) which followed the powder-based principle, is a productive way of manufacturing 3D EES devices. The main advantage of this process is that unreactive portions can be recycled for further usage in the next processes minimizing the wastage of the ingredients. Moreover, laminated object manufacturing (LOM) is considered another AM mechanism that is used for producing sandwich-like EES devices from several substrates such as paper, plastic, and metals. In this procedure, the sheets of material are combined with heat and pressure, then it is cut using a knife or laser to get the desired product. Additionally, it was modified by using drilling machines [3]. The basic schematics for SLA, FDM, SLS, and DIW are schematized in Figure 15.5.

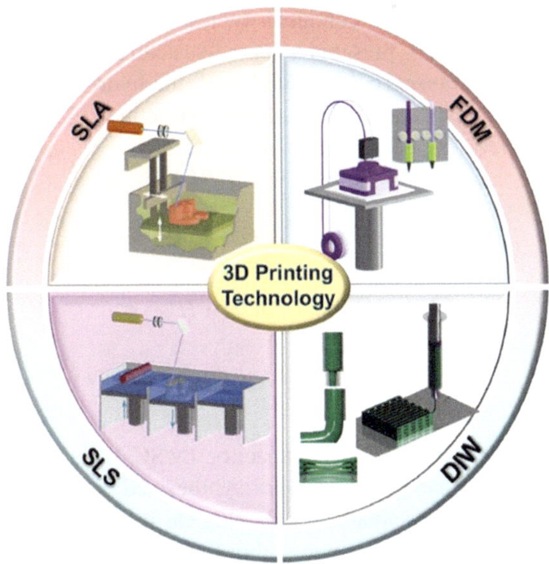

FIGURE 15.5 3D printing technologies for fabricating EES devices. Adapted with permission [4]. Copyright (2022), Elsevier. This is an open-access article distributed under the terms of the Creative Commons CC BY license.

15.5 SUPERCAPACITORS USING 3D-PRINT TECHNOLOGY

15.5.1 Metal Oxide-Based 3D-Printed Supercapacitors

Transitional metal oxides are considered ideal electrode materials for electronic devices due to their superior chemical and physical properties compared to CPs and carbon-based materials. Comparatively, TMOs have specific capacitance ranging from 50 to over 2,000 F/g, which are much higher than carbon and CPs based materials [29]. Additionally, the capacitance of TMOs can be increased by tailoring their structure with desired defects and interfaces. TMOs are becoming more desired components for the 3D printing of electronic devices due to properties such as rich availableness, eco-friendliness, diversity in morphology including high surface area, and ability to tune their properties. Although they indicate a higher increment in their energy density to a certain level, their uncontrollable volume enlargement, their poor electrical conductivity, and inactive ions diffusion in the bulk phase have interrupted the rate of capability and cycling stability of the electrodes [30].

Metal oxide thin films can be fabricated using techniques such as electron beam and vacuum evaporation, sol-gel method, chemical vapor deposition (CVD), electrostatic spray deposition, and spin coating. However, these techniques have some drawbacks, including a requirement for a post-annealing process which causes defects due to high temperature, complex operation procedures, and expensive equipment. As a convenient method for synthesizing MOs, the IJP 3D printing method is mostly used due to its profits of simplicity, low charge, facile operation along with allowing more control of parameters such as particle size, and device design. The main challenge regarding the use of MOs is the preparation of a stable colloidal dispersion of chemically active ingredients for the ink. A wet ball-milling method has been an approach to stabilize the MO nanoparticles. Through that, a smooth and uniform thin film can be printed using this 3D-print method. Also, the thickness of the active layer can be adjusted by varying the ratio of abundance moles to the volume of the MOs in the dispersion system or through the number of printing times [31].

Among MOs, nickel oxide (NiO) is considered promising electrode material for SCs due to its high theoretical capacitance (3750 F/g), reasonable price, and environmental friendliness. The redox reactions involved in NiO are shown in Equations 15.2–15.5. Despite the defined redox process for NiO, it presents a relatively low electric conductivity, low operational potential, and poor cycling stability. On the other hand, $Ni(OH)_2$ presents a crystal structure with a relatively larger interlayer spacing which facilitates ionic transport. Along with that, the hydroxyl groups improve their interaction with water, which prompts its diffusion within the structure [32].

$$NiO + OH^- \leftrightarrow NiOH + e^- \tag{15.2}$$

$$NiO + H_2O \leftrightarrow NiOH + H^+ + e^- \tag{15.3}$$

$$Ni(OH)_2 \leftrightarrow NiOOH + H^+ + e^- \tag{15.4}$$

$$Ni(OH)_2 + OH^- \leftrightarrow NiOOH + H_2O + e^- \tag{15.5}$$

Giannakou et al. fabricated a micro-supercapacitor (MSCs) via the IJP process using NiO nanoparticles [33]. 50 nm NiO nanoparticles were synthesized and dissolved in ethylene glycol and surfactant for 3D printing. The satisfactory rheological properties of the solvent allowed the formation of a stable ink that forms a continuous thin film over a substrate. The NiO morphology accompanied by the formation of a thin film yielded an appreciable conductivity of 210 S/m. The device fabrication was performed in four steps. First, the IJP of Ag nanoparticles was applied to serve as the current collector matrix. Second, NiO nanoparticles were IJP over Ag nanoparticles.

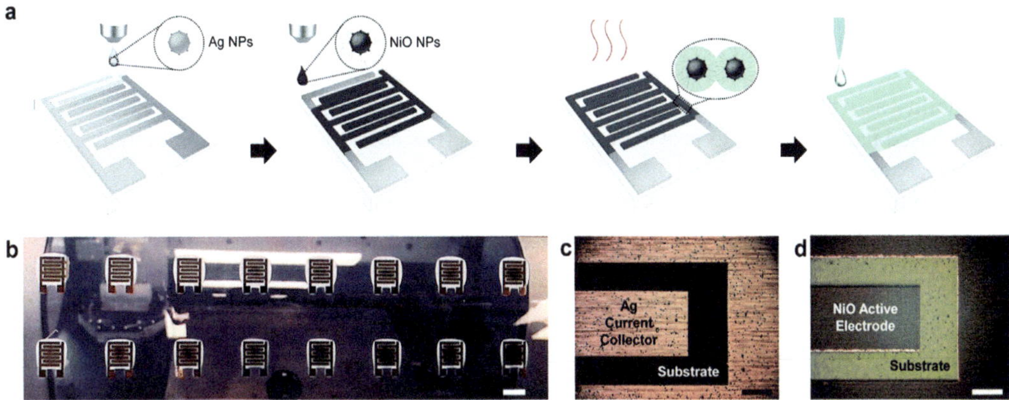

FIGURE 15.6 (a) Use of IJP technique to obtain NiO electrodes for MSCs. (b) Reproducibility of IJP for the manufacture of NiO-based MSC. Adapted with permission [33]. Copyright (2019), Royal Society of Chemistry.

Third, an annealing process was performed to properly adhere and sinter the nanoparticles at 150 °C. This was an important step to decrease the contact resistance between the nanoparticles. Fourth, a drop-casting technique was used to add the electrolyte (Figure 15.6a). This technique presents the advantage of the scalability and versatile use of substrate which can be either flexible or rigid (Figure 15.6b). The combination of techniques along with proper contact between the NiO and Ag nanoparticles yielded a device with volumetric as well as areal capacitances of 705 F/cm^3 and 155 mF/cm^2, respectively.

Among redox-active TMOs, vanadium oxides (VO_x) are promising pseudocapacitive electroactive materials due to their several stable oxidation states along with their unique layered structures that can improve the ionic transfer processes. In terms of 3D-printing approaches, it has been shown that V_2O_5 electrodes can be printed through the IJP method [28]. For the 3D printing of CuO, the FDM and sintering techniques can be employed for electrode fabrication. The synthesis can be carried out using sol-gel and solid-state calcite processes which are relatively non-costly and simple. A mixture of Cu particles and polylactic acid (PLA) can be squeezed out into the filament of FDM [34]. After depositing the Cu blend composite, the model can be annealed to form a CuO electrode. Importantly, the heating and cooling steps during the annealing process should be maintained at a slow speed otherwise, it could lead to cracks due to the thermal stress. Although homogeneous, the 3D-shaped Cu particles presented a high rate of diffusion within the sintering process. After the conversion to CuO, the surface became rough leading to the formation of pores somewhat like foams. The physical characteristics of this outcome such as stiffness, strength of the compression, and the ratio of mass over volume were lower than Cu-blend composite. Also, the usage of high-temperature sintered CuO presented moderately low resistance to electric current. Moreover, the photosensitive features of the final 3D-printed product were determined through the results of the photoluminescence and UV-visible absorption techniques.

MOs have become a promising class of electrode materials for synthesizing EES devices due to their properties while nanocomposite MOs are mostly used to overcome the limitations of the single TMOs. The atomic arrangement, nanostructures, electrical conductivity, and vacancies of oxygen are some of the factors that contribute to the improvement of the physical and chemical properties of MOs for SCs. In addition, the bimetallic metal oxides are attractive as they present two metallic centers that often display better electrochemical properties compared to the single MO due to the synergy effect. For example, $NiCo_2O_4$ has better conductivity compared to NiO and Co_3O_4. Also, the morphology of $NiCo_2O_4$ is favorable to increasing the specific capacitance by reducing the charge resistance of EES devices. Also, the porous structure in nano MOs provides a high surface area. Through that, a larger number of ionized species can permeate within the structure which aids in the EDLC as well as pseudocapacitive behavior in MOs. However, MOs may display low electrochemical stability as well as conductance. Hence, to address that, a common strategy is to incorporate carbon-based nanomaterials to provide improved properties such as electrical conductivity, specific capacitance, and rate capability [30].

Among the MOs, the composites of RuO_2 have been extensively used as an idyllic capacitive compound for printing SCs because of their high capacitance even though it is a scarce, toxic, and expensive material. Printing RuO_2-based SCs via 3D techniques can be considered a suitable process. Fabricating the composites can enhance the utilization of RuO_2 and reduce the manufacturing cost while increasing the electrochemical performance of the electrode because of the synergetic process between the additive component and RuO_2. For instance, RuO_2 can be combined with metal sulfides, MO, and carbon-based for the synthesis of electrode materials [29].

Another viable option lies in the compositing of MnO_2, which is relatively cheap and can be synthesized through several methods. Through that, the properties such as the conductivity, surface area, and cycling stability of MnO_2 can be improved, which are inherently low. Therefore, the incorporation of various elements such as Au, Ag, Cu, Ni, Co, Fe, Al, Zn, Mo, and Sn into the MnO_2 structure is an effective method to resolve these matters [32]. Carbonaceous materials such as CNTs, graphene, carbon nanowires (CNWs), and carbon nanofibers (CNFs) have excellent electrical conductivity and stability with a large surface area and can be used as a platform for depositing MnO_2 nanostructures [29]. As an example, Wenrui Cai et al. synthesized hollow N-doped carbon (HNC)@MnO_2 composite which showed a good electrochemical, and retention capacity [35]. Lei et al. synthesized MnO_2 nanosheets (MnO_2 NS) over CNT (MnO_2 NS@CNT) by CVD [36]. The nanocomposite exhibited a high specific capacitance of 574.4 F/g and a longer lifespan, as a result of the correlation between MnO_2 and conductible CNT networks. Here, charges are stored through the merging of the mechanisms of bilayer capacitance of CNTs and faradaic reactions of MnO_2. Also, the magnitude of electric current increases through the development of repetition rate, and steady distributions of the CV curves which indicated the rapid degree of the ion migration within the MnO_2@CNTs electrode.

Sunriyal and Bhattacharya fabricated a flexible asymmetric supercapacitor by using paper as substrate and GO-MnO_2 nanocomposite via the IJP method [37]. One of the major challenges with using individual MnO_2 is the limited potential window which leads to a decrease in energy density. A viable way to address this issue is to produce an asymmetric SCs by composing two dissimilar materials as anode and cathode. SC was fabricated by grouping with a conducting pattern of GO-MnO_2 nanocomposite ink on paper which acted as a positive electrode whereas the negative electrode was printed with stimulated carbon-based ink. In particular, the conducting region in flexible A4 sheets was developed with the deposition of reduced GO ink via the IJP method. Both printed electrodes were connected through a PVA-LiCl gel as the ion-rich source for making SCs (Figure 15.7). The composite showed a comparatively high energy density with high potential window flexibility, good cyclic stability, and high-rate capability.

Yao et al. designed a conductor for SCs composing graphene aerogel and MnO_2 via the DIW process [38]. A 3D platform suitable as a scaffold for pseudocapacitive materials was printed by using an ink filled with GO suspension and 5% hydroxypropyl methylcellulose. It was then freezedried to form aerogels. The conversion of GO into reduced GO (rGO) was performed through annealing under nitrogen gas. As a final step, MnO_2 nanosheets were electrodeposited on the graphene aerogel lattice (Figure 15.8). The MnO_2 was deposited uniformly on the exterior as well as interior regions of the lattice. This uniform layer provided good electro-conductivity of graphene aerogel and the efficient diffusion of electrolyte through the 3D-printed lattice. Overall, this composite exhibited excellent energy storage properties based on normalized volumetric, areal, and gravimetric capacitance. The real capacitance enhances linearly with the weight of MnO_2 and the thickness of the

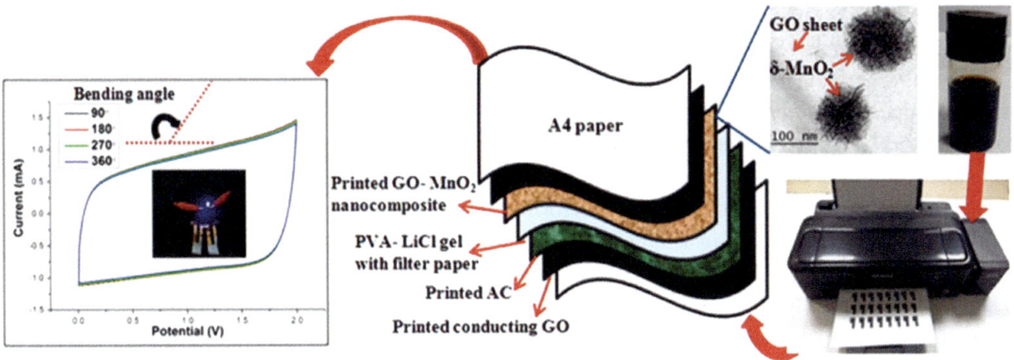

FIGURE 15.7 Inkjet-printed GO-MnO_2 composite as pseudocapacitive electrodes for SCs. Adapted with permission [37]. Copyright (2017), American Chemical Society.

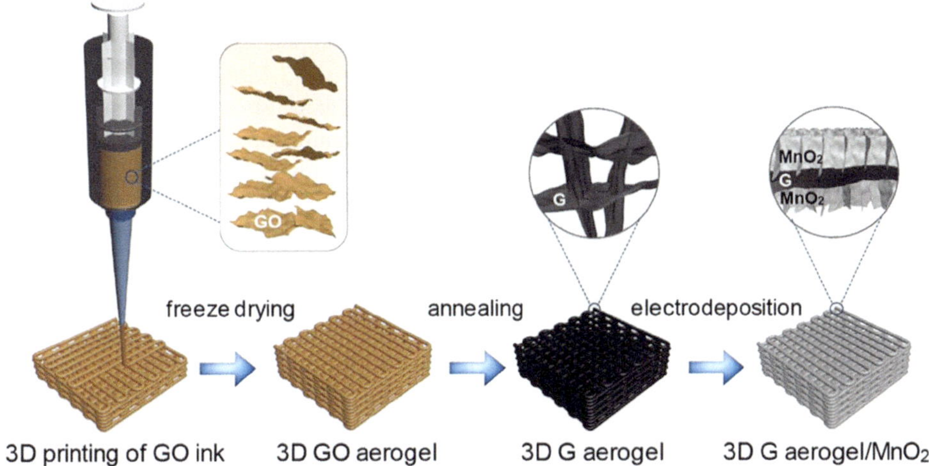

FIGURE 15.8 3D printed graphene aerogel/MnO_2 composite for SCs. Adapted with permission [38]. Copyright (2019), Elsevier.

electrode while the gravimetric capacitance was stable, illustrating that the capability of the capacitance was not controlled by ion diffusion even at high mass loading.

Currently, composing MO with CNT became more popular in the application of EES devices. Generally, CNT-based inks can be obtained via the IJP which is a conventional printing method. Although MnO_2 has a high theoretical capacitance of 1370 F/g, its electric conductivity is poor. Therefore, the incorporation of MnO_2 into flexible conductive CNT is a proper solution for fabricating SCs by overcoming these limitations. For instance, the fabrication of two different electrodes including MnO_2–Ag–MWCNTs as the anode and filtered multi-walled CNTs (MWCNTs) as the cathode via the DIW method gave a better electrochemical performance to the flexible energy storage devices [39].

Cobalt oxide-based materials such as Co_3O_4 can have their electrochemical properties improved through compositing as it possesses high theoretical capacitance but has drawbacks of low energy density and poor conductivity, which disturbs thecycling stability and rate capability. Co_3O_4 can be incorporated with carbon materials such as graphene, CNT, CNF, and amorphous carbon to increase electron and ions transportation. As another method, to improve the properties of SCs, nanostructured Co_3O_4 such as nanowires, nanospheres, nano-cubes, and nano-heterostructures can be prepared to improve the electrochemical performance by supplying the transfer routes for the ions and alternative active sites to promote the redox process [30], [40]. As an example, Tao Liu et al. introduced an active positive conductor for top-performance SCs by synthesizing a composite based on Co_3O_4 and N-doped carbon capsules named Co_3O_4/NHCSs. Due to the high porosity, enhancement in conductivity, and a larger number of defects due to the doping of N, this composite showed a better electrochemical performance compared to the single Co_3O_4 [41]. Li et al. synthesized 3D porous carbon (3DPC)/Co_3O_4 composites by pyrolysis of 3D graphene/CO-MO framework precursor for improved electrode material [40].

Furthermore, due to the low capacitance of tin oxide, it can be combined with other materials through doping or by compositing to increase the efficiency of the EES devices. Yan et al. [42] fabricated a composite by coating amorphous MnO_2 on SnO_2 nanowires (MWs). SnO_2 was prepared using the CVD technique and MnO_2 nanorods were fabricated using a solution-based method. Some of the advantages of the nanostructured composite were related to providing a shorter ion diffusion route to improve the reversible redox reaction of MnO_2 by supplying a direct way to transfer electrons with high conductivity of SnO_2 NWs while creating channels by SnO_2 for the electrolyte transportation.

Solid-state (SS) SCs which are composed of rigid form components are optimized as cutting-edge EES devices for clothing applications. For instance, quasi-SS-MSCs have properties of high electron transmission rate and satisfactory stability, allowing their use as a power supply for portable and miniaturized devices. These types of SCs can be manufactured via IJP, SLS, and silicon-based molding methods. For example, an asymmetric MSC negative electrode was fabricated using G-MnO_2 and a positive electrode was prepared using graphene-ferric oxyhydroxide (G-FeOOH) through a laser-induced process followed by electrodeposition [43]. The main challenge of this device was making this with high energy compactness, and cost-effectively. Shen et al. proposed a 3D-manufacturing strategy of asymmetric MSC which was composed of a cathode and anode made with V_2O_5 and graphene–vanadium nitride quantum dots (G–VNQDs) with highly concentrated GO dispersions, respectively [43]. The V_2O_5 is a suitable material due to its high reduction-oxidation activity, and availability. On the other hand, vanadium nitride which was used as anode material has high pseudocapacitance with negative operating potential. Consequently, the cathode and anode inks were made by separately mixing V_2O_5 and VNQDs with high concentrated GO dispersions in equal portions. It is worth noting that, high viscosity and shear-thinning rheology made them suitable for use in the extrusion-based printing process for the fabrication of MSCs devices. As a preparation procedure, the dual-charged conducting inks were separately loaded onto the plungers and printed layer-wise on the finger-shaped contemporary collectors. Subsequently, vacuum freeze-drying was used to clear away the solvent to solidify the 3D-printed structures, and hydrazine hydrate was used as a reducing agent to reduce GO. As a final step, LiCl-PVA as a gel-like electrolyte was applied into arrays (Figure 15.9). It acted as an electric separator and provided wettability as an electrolyte while enhancing the performance of the printed MSCs. This device showed an excellent structural integration with substantial open macrospores which act as numerous channels for stimulating the high amount of weight flow rate per area (3.1 mg/cm^2) and a wide electrochemical potential of 1.6 V [43].

15.5.2 Metal Oxide-Based 3D-Printed Wearable Supercapacitors

With the growing demand for EES devices, newly invented flexible and wearable SCs play an important role in smart homes, medical, and entertainment due to their properties of high power densities, flexibility, lightweight, easy integration, fast charge/discharge rate, shape compatibility, and safety features. These wearable SCs can be fabricated using different active matters such as MOs, CPs, carbon-based materials, and other emerging materials such as titanium carbide. Within this line, textile-based SCs composed of MOs are further discussed. A schematic presentation of

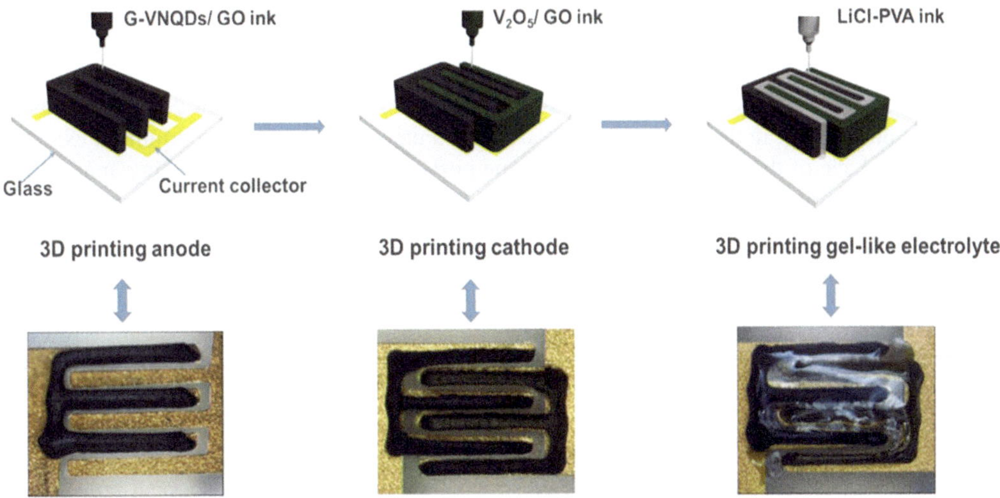

FIGURE 15.9 Schematic representation of 3D printing of electrodes in asymmetric micro-supercapacitor. Adapted with permission [43]. Copyright (2018), John Wiley and Sons.

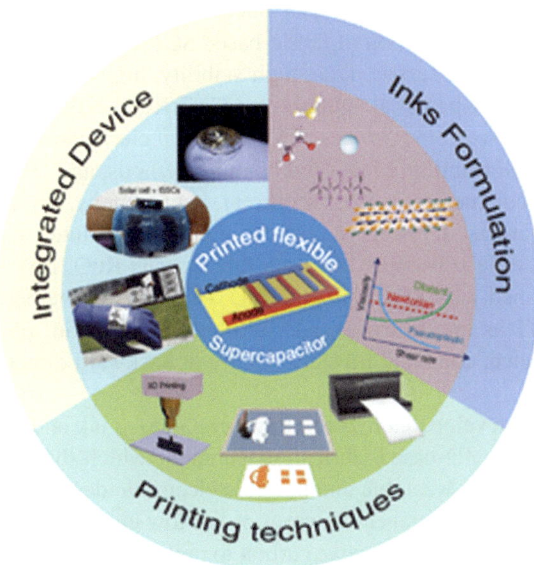

FIGURE 15.10 Schematic presentation of flexible wearable devices. Adapted with permission [44]. Copyright (2021), AIP Publishing.

flexible wearable devices based on ink formulations, printing methods, and integrated applications is shown in Figure 15.10.

There are numerous techniques for manufacturing wearable EES devices including 3D, screen, and transfer printing. Nowadays 3D printing has become more popular among other printing processes because of its advantages as mentioned above. IJP is one of the famous 3D-printing practices which are used for fabricating metal-based wearable SCs due to its sustainability, the possibility of production at desired scales, reasonable working temperature operation, low cost, and multi-purposed applications into objects with adjustable measurements at selected areas. In printing

SCs, some considerable factors such as the ability to bend in any direction without breaking, robustness, long-term stability, and high energy capacity are required for practical applications of wearable EES devices. Therefore, the toner would have a definite resistance to fluid flow and internal friction to ensure the quality of the printing process. In addition, the radius of the components which are used as ink must be in the scope, which is around 50 times smaller than the printhead extent, to avoid the blockage of the nozzles. Moreover, some post-printing processes including annealing or film compression are carried out to enhance the quality and the performance of the printed devices, but they should be well suited to the printed resources considering the facts of material degradation, flexibility, and steadiness [28].

Flexible substrates including textiles, papers, and polymers play a vital role in the printing of flexible SCs (FSCs). Some parameters such as the tendency of spreading, market price, the softness of the outward, and transparency are considered for large-scale production. These factors are changed with the mode of application. Some features such as permeability, capillary effect, direct deposition of ink deprived of any treatments, and defined thinness are important to consider. These properties can be applied or performed on paper as they are suitable as substrates for printing FSCs via IJP. On the other hand, polymers can be used as substrates for SCs due to their functional properties of uniform lucidity, aquatic resistance, and thermal stability. However, the wearing uncomfortability of polymers and papers limits their use as substrates for wearable devices. Therefore, it is important to make FSCs using textile substrate because of its benefits of luxury, lifetime, and narrowness [44]. In the next section, some innovations of textile-based SCs are discussed.

Different types of conductive constituents such as CNT, conductive polymers, and MO composites are used for the construction of fabric-based SCs. Usually, CNT is an ideal material for SCs due to its highly accessible surface area, high stability, high electrical conductivity, mechanical flexibility, and low cost. Chen et al. synthesized CNT/RuO_2 NW SCs on fabrics using the IJP process that can be used as wearable devices. CNT thin-film electrodes can be produced in various methods such as roll-to-roll printing, Meyer rod coating methods, and IJP which has the capability of controlling geometry, conduction of electric current, homogeneity, depth of the layer, and design locations while printing are used. Herein, an off-the-shelf inkjet printer was used to synthesize the SWCNT films on the flexible substrates and cloth fabrics. The RuO_2 NW synthesized through the CVD method was combined with SWCNT. The RuO_2 can improve the performance of the SCs by composing with SWCNT as an electrode material due to its relevant physical characteristics of ultrahigh pseudocapacitance, conductibility, and ability in backward directions of exterior faradaic reactions [45].

In addition, Tang et al. developed a hybrid Fe_2O_3/graphite/Ag ink and squeezed out it into MSC conductors via the DIW for a 3D-based fabricating technique which showed excellent electrochemical behavior and flexibility for wearable and portable electronic devices [46]. Typically, like other MOs, Fe_2O_3 has appreciable capacitance; however, it has low electrical conductivity. As a solution to this problem, it was mixed with conductive carbon to prepare printable toner with improved electrical conductivity. In this sense, the poor electron transportation during the charge-discharge process in Fe_2O_3 was enhanced by developing stable inks containing Fe_2O_3 nanoparticles, 2D G NSs, and 1D Ag NWs with the binder. Moreover, with the changes in ink formulations, the ideal MSCs with high areal capacitance, high energy density, and high flexibility can be obtained via the 3D-printing process.

Lin et al. synthesized planar SCs using unique hierarchical Ni@MnO_2 nanocomposites via the IJP process. For that approach, Ni@MnO_2 was deposited via electrochemical deposition on interdigitated metal finger arrays in flexible substrates which were printed through the IJP process. The fabrication method of printable planar SCs is schematically shown in Figure 15.11. Interdigitated finger electrode arrays were inkjet printed on substrate using ink of Ag nanoparticle, and consequently, three-layer depositions were carried out as a thin gold (Au) layer on Ag electrodes, Ni nano-corals on Au, and pseudocapacitive MnO_2 nano-flakes on top of the Ni layer. Finally, Ni@MnO_2 conductors with a

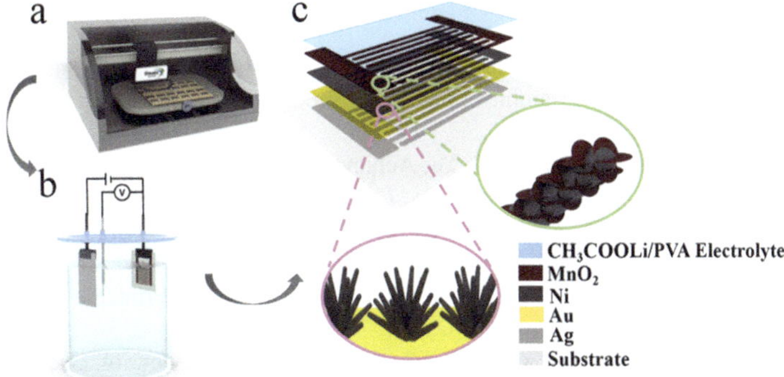

FIGURE 15.11 Schematic representation of 3D printed solid-state planar SCs with Ni@MnO_2 nano-coral nanostructures. Adapted with permission [47]. Copyright (2017), John Wiley and Sons.

larger exterior area were obtained and lithium acetate/poly(vinyl acetate) (CH_3COOLi/PVA) gel was used as an electrolyte. The simplistic design of nano-coral arranged Ni@MnO_2 on the electrodes with high active areas modified the activity of the SC excessively and this device can be easily fixed onto wearable substrates as an EES gadget. Especially, an acceptable amount of MnO_2 deposition was required for energy storage. Excessive addition of this compound may decrease ion transportation and electric conductance. In wearable applications, flexibility is a key reason for designing EES tools and IJP is a suitable and profitable way for the fabrication of SCs to achieve the preferred design [47]. Overall, the integration of 3D-printing technologies like IJP, FDM, and DIW to produce EES devices, and other electronics applications on affordable flexible substrates might meaningly satisfy the necessity of manufacturing sustainable wearable items.

15.4 CONCLUSION AND PERSPECTIVE

The main goal of supercapacitors is energy storage without applying strain on the environment. Although carbon-based materials and CPs can be used as electrode materials, TMOs are promising electrode materials due to their high energy density, excellent specific capacitance, and better stability. Their properties can be further enhanced by making nanocomposites with other materials. Nanocomposites of MOs lead to enhancements in the electrochemical performances by overcoming the drawbacks of single TMOs including poor electric current transmission, low specific capacitance, low steadiness in lifecycles, and small operating voltage. On the other hand, although the fabrication of electrodes is carried out by many traditional techniques such as spray deposition, CVD, ALD, LBL, sol-gel, etc., the 3D-printing process became popular due to better control in the geometry, thickness controllability, cost-effectiveness, along with environmental credentials. Thus, using 3D-printing technology to introduce flexibility can enhance power and energy density, range of applicability, and contribution to the development of wearable EES devices. Even though 3D-printing technology is relatively new it allows for the fabrication of SCs with novel architecture. When considering the use of 3D-printing techniques, most of the metal oxide-based SCs are fabricated via IJP, FDM, and DIW methods. Even though the printing techniques of the SCs have been evolving, some limitations remained. Some of these challenges are the displaying of poor energy storage because of the small extent and low quantity of weight usage of substances for conductors, the gradual decrease in the productivity of the electrode components without mixing of inactive additives or binders as an energetic agent, and the insufficient resolution of the printed SCs. Therefore, these issues should be resolved in the future to get the better electrochemical performance of the 3D-printed SCs. It is also essential to select suitable inks or filament electrodes, electrolytes, and current collectors for the fabrication of SCs using 3D printing.

REFERENCES

1 M. Guan, Q. Wang, X. Zhang, J. Bao, X. Gong, Y. Liu, Two-dimensional transition metal oxide and hydroxide-based hierarchical architectures for advanced supercapacitor materials, *Front. Chem.* 8 (2020) 1–14.

2 H. Li, L. Peng, Y. Zhu, X. Zhang, G. Yu, Achieving high-energy-high-power density in a flexible quasi-solid-state sodium ion capacitor, *Nano Lett.* 16 (2016) 5938–5943.

3 X. Tian, J. Jin, S. Yuan, C.K. Chua, S.B. Tor, K. Zhou, Emerging 3D-printed electrochemical energy storage devices: A critical review, *Adv. Energy Mater.* 7 (2017) 1–17.

4 Y. Gao, X. Guo, Z. Qiu, G. Zhang, R. Zhu, Y. Zhang, H. Pang, Printable electrode materials for supercapacitors, *ChemPhysMater.* 1 (2022) 17–38.

5 C. Zhong, Y. Deng, W. Hu, J. Qiao, L. Zhang, J. Zhang, A review of electrolyte materials and compositions for electrochemical supercapacitors, *Chem. Soc. Rev.* 44 (2015) 7484–7539.

6 R.K. Gupta, *Organic Electrodes: Fundamental to Advanced Emerging Applications*, Springer Nature, 2022.

7 S. Bhoyate, P.K. Kahol, R.K. Gupta, Nanostructured materials for supercapacitor applications, in: P.J. Thomas, N. Revaprasadu (Eds.), Nanosci. Vol. 5, *The Royal Society of Chemistry*, 2019: pp. 1–29.

8 F. Zhang, M. Wei, V.V. Viswanathan, B. Swart, Y. Shao, G. Wu, C. Zhou, 3D printing technologies for electrochemical energy storage, *Nano Energy*. 40 (2017) 418–431.

9 M. Cheng, R. Deivanayagam, R. Shahbazian-Yassar, 3D printing of electrochemical energy storage devices: A review of printing techniques and electrode/electrolyte architectures, *Batter. Supercaps.* 3 (2020) 130–146.

10 X. Chen, J. Zhang, B. Zhang, S. Dong, X. Guo, X. Mu, B. Fei, A novel hierarchical porous nitrogen-doped carbon derived from bamboo shoot for high performance supercapacitor, *Sci. Rep.* 7 (2017) 1–11.

11 L. Zhou, C. Li, X. Liu, Y. Zhu, Y. Wu, T. van Ree, *Metal Oxides in Supercapacitors*, Elsevier Inc., 2018.

12 J. Choi, T. Wixson, A. Worsley, S. Dhungana, S.R. Mishra, F. Perez, R.K. Gupta, Pomegranate: An eco-friendly source for energy storage devices, *Surf. Coatings Technol.* 421 (2021) 127405.

13 H.S. Using, P. Kahol, R. Gupta, Waste coffee management: Deriving nitrogen-doped coffee-derived carbon, *C.* 5 (2019) 44.

14 S. Bhoyate, C.K. Ranaweera, C. Zhang, T. Morey, M. Hyatt, P.K. Kahol, M. Ghimire, S.R. Mishra, R.K. Gupta, Eco-friendly and high performance supercapacitors for elevated temperature applications Using Recycled Tea Leaves, *Glob. Challenges*. 1 (2017) 1700063.

15 C.K. Ranaweera, P.K. Kahol, M. Ghimire, S.R. Mishra, R.K. Gupta, Orange-peel-derived carbon: Designing sustainable and high-performance supercapacitor electrodes, *C.* 3 (2017) 25.

16 C. Zequine, C.K. Ranaweera, Z. Wang, P.R. Dvornic, P.K. Kahol, S. Singh, P. Tripathi, O.N. Srivastava, S. Singh, B.K. Gupta, G. Gupta, R.K. Gupta, High-performance flexible supercapacitors obtained via recycled jute: Bio-waste to energy storage approach, *Sci. Rep.* 7 (2017) 1–12.

17 C. Zequine, C.K. Ranaweera, Z. Wang, S. Singh, P. Tripathi, O.N. Srivastava, B.K. Gupta, K. Ramasamy, P.K. Kahol, P.R. Dvornic, R.K. Gupta, High per formance and flexible supercapacitors based on carbonized bamboo fibers for wide temperature applications, *Sci. Rep.* 6 (2016) 1–10.

18 E. Mitchell, J. Candler, F. De Souza, R.K. Gupta, B.K. Gupta, L.F. Dong, High performance supercapacitor based on multilayer of polyaniline and graphene oxide, *Synth. Met.* 199 (2015) 214–218.

19 J. Choi, T. Ingsel, D. Neupane, S.R. Mishra, A. Kumar, R.K. Gupta, Metal-organic framework-derived cobalt oxide and sulfide having nanoflowers architecture for efficient energy conversion and storage, *J. Energy Storage*. 50 (2022) 104145.

20 K. Thompson, J. Choi, D. Neupane, S.R. Mishra, F. Perez, R.K. Gupta, Tuning the electrochemical properties of nanostructured $CoMoO_4$ and $NiMoO_4$ via a facile sulfurization process for overall water splitting and supercapacitors, *Surf. Coatings Technol.* 421 (2021) 127435.

21 M.A.A. Mohd Abdah, N.H.N. Azman, S. Kulandaivalu, Y. Sulaiman, Review of the use of transition-metal-oxide and conducting polymer-based fibres for high-performance supercapacitors, *Mater. Des.* 186 (2020) 108199.

22 G. Wang, L. Zhang, J. Zhang, A review of electrode materials for electrochemical supercapacitors, *Chem. Soc. Rev*. 41 (2012) 797–828.
23 T. Chu, S. Park, K. Fu, 3D printing-enabled advanced electrode architecture design, *Carbon Energy*. 3 (2021) 424–439.
24 V. Egorov, U. Gulzar, Y. Zhang, S. Breen, C. O'Dwyer, Evolution of 3D printing methods and materials for electrochemical energy storage, *Adv. Mater*. 32 (2020) 1–27.
25 T. Abudula, R.O. Qurban, S.O. Bolarinwa, A.A. Mirza, M. Pasovic, A. Memic, 3D printing of metal/metal oxide incorporated thermoplastic nanocomposites with antimicrobial properties, *Front. Bioeng. Biotechnol*. 8 (2020) 1–8.
26 A.S.K. Kiran, J.B. Veluru, S. Merum, A.V Radhamani, M. Doble, T.S.S. Kumar, S. Ramakrishna, Additive manufacturing technologies: An overview of challenges and perspective of using electrospraying, *Nanocomposites*. 4 (2018) 190–214.
27 I.J. Gómez, N. Alegret, A. Dominguez-Alfaro, M. Vázquez Sulleiro, Recent advances on 2D Materials towards 3D Printing, *Chemistry (Easton)*. 3 (2021) 1314–1343.
28 T.T. Huang, W. Wu, Scalable nanomanufacturing of inkjet-printed wearable energy storage devices, *J. Mater. Chem. A*. 7 (2019) 23280–23300.
29 R. Liang, Y. Du, P. Xiao, J. Cheng, S. Yuan, Y. Chen, J. Yuan, J. Chen, Transition metal oxide electrode materials for supercapacitors: A review of recent developments, *Nanomaterials*. 11 (2021) 1248.
30 C. An, Y. Zhang, H. Guo, Y. Wang, Metal oxide-based supercapacitors: progress and prospectives, *Nanoscale Adv*. 1 (2019) 4644–4658.
31 Y. Zhao, Q. Zhou, L. Liu, J. Xu, M. Yan, Z. Jiang, A novel and facile route of ink-jet printing to thin film SnO_2 anode for rechargeable lithium ion batteries, 51 (2006) 2639–2645.
32 D.P. Dubal, N.R. Chodankar, P. Gomez-Romero, D.-H. Kim, *Fundamentals of Binary Metal Oxide–Based Supercapacitors*, Elsevier Inc., 2017.
33 P. Giannakou, M.G. Masteghin, R.C.T. Slade, S.J. Hinder, M. Shkunov, Energy storage on demand: ultra-high-rate and high-energy-density inkjet-printed NiO micro-supercapacitors, *J. Mater. Chem. A*. 7 (2019) 21496–21506.
34 A. Salea, R. Prathumwan, J. Junpha, K. Subannajui, Metal oxide semiconductor 3D printing: Preparation of copper(II) oxide by fused deposition modelling for multi-functional semi-conducting applications, *J. Mater. Chem. C*. 5 (2017) 4614–4620.
35 W. Cai, R.K. Kankala, M. Xiao, N. Zhang, X. Zhang, Three-dimensional hollow N-doped ZIF-8-derived carbon@MnO_2 composites for supercapacitors, *Appl. Surf. Sci*. 528 (2020) 146921.
36 R. Lei, J. Gao, L. Qi, L. Ye, C. Wang, Y. Le, Y. Huang, X. Shi, H. Ni, Construction of MnO_2 nanosheets@graphenated carbon nanotube networks core-shell heterostructure on 316L stainless steel as binder-free supercapacitor electrodes, *Int. J. Hydrogen Energy*. 45 (2020) 28930–28939.
37 P. Sundriyal, S. Bhattacharya, Inkjet-printed electrodes on A4 paper substrates for low-cost, disposable, and flexible asymmetric supercapacitors, *ACS Appl. Mater. Interfaces*. 9 (2017) 38507–38521.
38 B. Yao, S. Chandrasekaran, J. Zhang, W. Xiao, F. Qian, C. Zhu, E.B. Duoss, C.M. Spadaccini, M.A. Worsley, Y. Li, Efficient 3D printed pseudocapacitive electrodes with ultrahigh MnO_2 Loading, *Joule*. 3 (2019) 459–470.
39 S. Wang, N. Liu, J. Tao, C. Yang, W. Liu, Y. Shi, Y. Wang, J. Su, L. Li, Y. Gao, Inkjet printing of conductive patterns and supercapacitors using a multi-walled carbon nanotube/Ag nanoparticle based ink, *J. Mater. Chem. A*. 3 (2015) 2407–2413.
40 S. Li, K. Yang, P. Ye, K. Ma, Z. Zhang, Q. Huang, Three-dimensional porous carbon/Co_3O_4 composites derived from graphene/Co-MOF for high performance supercapacitor electrodes, *Appl. Surf. Sci*. 503 (2020) 144090.
41 T. Liu, L. Zhang, W. You, J. Yu, Core–shell nitrogen-doped carbon hollow spheres/Co_3O_4 nanosheets as advanced electrode for high-performance supercapacitor, *Small*. 14 (2018) 1–8.
42 J. Yan, E. Khoo, A. Sumboja, P.S. Lee, Facile coating of manganese oxide on tin oxide nanowires with high-performance capacitive behavior, *ACS Nano*. 4 (2010) 4247–4255.
43 K. Shen, J. Ding, S. Yang, 3D printing quasi-solid-state asymmetric micro-supercapacitors with ultrahigh areal energy density, *Adv. Energy Mater*. 8 (2018) 1–7.
44 J. Liang, C. Jiang, W. Wu, Printed flexible supercapacitor: Ink formulation, printable electrode materials and applications, *Appl. Phys. Rev*. 8 (2021) 021319.

45 P. Chen, H. Chen, J. Qiu, C. Zhou, Inkjet printing of single-walled carbon nanotube/RuO_2 nanowire supercapacitors on cloth fabrics and flexible substrates, *Nano Res*. 3 (2010) 594–603.

46 K. Tang, H. Ma, Y. Tian, Z. Liu, H. Jin, S. Hou, K. Zhou, X. Tian, 3D printed hybrid-dimensional electrodes for flexible micro-supercapacitors with superior electrochemical behaviours, *Virtual Phys. Prototyp*. 15 (2020) 511–519.

47 Y. Lin, Y. Gao, Z. Fan, Printable fabrication of nanocoral-structured electrodes for high-performance flexible and planar supercapacitor with artistic design, *Adv. Mater*. 29 (2017) 1–8.

16 3D-Printed MXenes for Supercapacitors

Dipanwita Majumdar[1], Padma Sharma[2], and Niki Sweta Jha[2]*
[1]Department of Chemistry, Chandernagore College, Strand Road, Barabazar, Chandannagar, Hooghly, Pin-712136, West Bengal, India
[2]Department of Chemistry, National Institute of Technology, Patna – 800005, Bihar, India
*Corresponding author: wbesdmajumdar@gmail.com

CONTENTS

16.1 INTRODUCTION

For the last two decades, scientists and technologists have been continuously engaged in developing renewable, sustainable, and eco-friendly benign electronic gadgets that can efficiently store and transform energy to address the critical issues of the global electrical power crisis and environmental pollution problems [1]–[2]. Side by side, the advancements of the Internet of Things (IoT) have extensively promoted the use of flexible and wearable devices in almost every sector of today's society [3]–[4]. These very concerns have set up huge challenges in devising eco-compatible, flexible, high-performing electrochemical energy storage (EES) units in the form of supercapacitors to provide uninterrupted power support as well as revolutionize their applications in portable electronic gadgets, electric automobiles, and emergency power supplies in biomedical, astronomical and in almost all important domains of practical day to day needs [5]–[8]. This realizations have triggered proficient employment of 3D printing technology in fabricating smart and diligent electronic gadgets in the recent years.

16.1.1 3D-Printing Technique in Supercapacitor Technology

The advancement of fabrication technology has urged the manufacturing of smart, slender, and low-dimensional, highly sophisticated integrated 'on-chip' flexible microsupercapacitors with maximized

DOI: 10.1201/9781003296676-16

areal and volumetric yields within limited footprint areas, through proper choice of electrode materials and 3D device architectural designing, ensuring ultrasmooth interfaces for improved electron/ion transport properties across the electrodes as well as decent geometrical flexibility to withstand the imposed mechanical stress, with minimum maintenance demand while undergoing long and non-stop operative periods [9]–[11]. In this context, the 3D-extrusion printing technique has been widely adopted as it is an inexpensive and versatile procedure, mostly relying on three-axis motion steps to create well-distinct periodic geometries vide layer-by-layer stacking strategy. This popular method involves the deposition of inks in the colloidal form in a layer-by-layer or additive manner onto a chosen substrate by extruding them vide a nozzle to develop 3D objects or patterns, as indicated in steps 1–3 in the inset of Figure 16.1a [12]. High viscous inks are essential to maintain the 3D-printed construction without any damage or collapse. The morphology of 3D-printed systems on various substrates is needed to solidify immediately for promoting precise and rapid prototyping patterns.

It has been well perceived that the scalable fabrication of such high-performance 3D energy storage and conversion devices is a very challenging task in reality. Nonetheless, simple and cost-effective methodologies with improved 3D-printing quality boosted by component placement, and good electrical interconnections are very necessary to enjoy the smart integration of electronic components using flexible supercapacitors to yield highly sophisticated automated gadgets [13]–[15]. The additive 3D-printing technology ensures the production of various geometries with controlled tuning on lateral extension and thickness on diverse curved and flexible substrates like plastic, paper, fabrics, ceramics, etc., with a high extent of compatibility, guaranteeing modulated composition, curtailing unwanted material wastage besides improving physical features of 3D-printed devices in an optimal

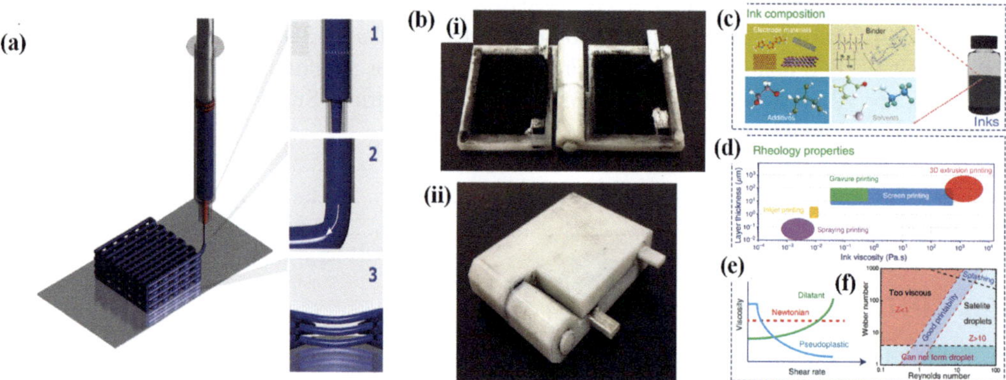

FIGURE 16.1 (a) Schematic presentation of steps involved in the 3D extrusion printing process involving ink flow through the syringe barrel and nozzle. Insets- 1, 2, and 3 show ink ejection from the nozzle, deposition onto the substrate, and formation of self-standing structure respectively. Adapted with permission [12], Copyright: The Authors, some rights reserved; exclusive licensee [Royal Society of Chemistry]. Distributed under a Creative Commons Attribution License 4.0 (CC BY). (b) (i) Optical image of electrode slurry matter printed on the top of the silver paint used to frame 3D printed microsupercapacitor and (ii) optical image of a complete 3D printed microsupercapacitor. Adapted with permission [15], Copyright (2017), Elsevier. (c) Schematically illustrated components and rheology properties of printing ink that embrace the major components like electrode materials, additives, binder, and solvents of printable inks. (d) Graphical (qualitative) representation of the general relationship between ink viscosity and thickness of the deposited layers in different printing techniques. (e) Graphical (qualitative) response of shear rate with the viscosity of different ink fluids. (f) Variation of Reynolds and Weber's Numbers required for printing performances. Adapted with permission [6], Copyright (2022), AIP Publishing.

economic approach. Such programmable designs using this technique would surely increase ion accessibility rates as well as evade undesired mechanical and morphological constraints produced as a result of structural disintegration of the electrodes during large cycling processes, thereby cumulatively upgrading the electrochemical as well as mechanical performances integrally. Hence lies the inevitability of designing strategies for fabricating printable inks composed of highly flexible, decent conducting, and electrochemically active electrode materials to satisfy the ideal foundation of smart and diligent printed 'on-chip' microsupercapacitors, as reflected through the optical images provided in Figure 16.1b respectively [15].

The popular printable functional inks required for the above procedure, as delineated in Figure 16.1c, mostly comprise of electro-active components, binding agents, and additives suspended together in dispersible solvents. The binders and additives ensure ink flowability and stability [16]. Moreover, they often also promote the formation of the conductive and porous structures to expose more electroactive sites besides augmenting ink's viscosity and printing pliability. However, such additions frequently lead to a fall in the electrical conductivity and alteration in intrinsic rheological and mechanical properties, and accordingly, binder-free inks are nowadays vastly encouraged [17].

16.1.2 Criteria of Inks Formulations in 3D-Printing Technology

Currently, researchers are targeting to explore simple and novel ink formulations, ensuing large mass loadings of electro-active materials, binder-free and non-additive compositions, to obtain high conducting, well-dispersed, eco-friendly inks with minimum post-processing requirements, to achieve high-class, large-scale printed supercapacitors with the remarkable electrochemical response and maximum energy density [6]. By general definition, a printable ink can be well extruded as a continuous filament through a particular nozzle, to figure out different structures very much resembling the digital model. Particularly, the ink should be smooth flowing, without creating bulky and solid agglomerates which may otherwise clog the nozzle. From a rheological perspective, as highlighted in Figure 16.1d–f respectively, the shear thinning behavior of the ink (i.e., decreasing viscosity with rising shear rate) is usually chosen to meet these criteria [6]. Figure 16.1d depicts the different viscosity demands for various printing techniques following the order inkjet printing (low) < gravure printing (medium) < screen printing (high).

As evident, compared to other printing approaches, in the 3D technique, the employed ink viscosity ranges around $0.03–6 \times 10^4$ Pa s. The ejecting ink must rapidly emerge as a shear-thinning fluid and then after deposition must immediately solidify to retain its shape. The printable ink should therefore possess good rheological properties, with remarkable shear-thinning and viscoelastic features, which would enable shape-retaining characteristics with each layer as well as sufficient fluidity for strong substrate and interlayer adhesion. Such factors can be effectively tuned through the appropriate choice of additives, binder materials, and solvents as well as the chemical nature and structural features of the employed active material used for the ink formulation.

The commonest rheological tests involve flow ramps that are used to determine the shear thinning gradation of inks. Besides, the oscillatory tests are employed to evaluate the yield stress and viscoelastic properties of inks at rest. Accordingly, Figure 16.1e focuses on the behavior of different kinds of fluids; signifying the fact that Newtonian fluids with small surface tension (25–50 mN m^{-1}) favor formulating fine droplets from the nozzles for efficient printing. Further, for the formation of ink droplets for effective printing, two important parameters are observed, known as Reynolds Number and Weber Number, respectively [6]:

$$\text{(Reynolds Number) } R_e = dv\rho/\eta \tag{16.1}$$

$$\text{(Weber Number) } W_e = v^2 d\rho/\sigma \tag{16.2}$$

where, the symbols indicate the density (ρ), velocity (ν), nozzle diameter (d), viscosity (η), and surface tension (σ) of the printable inks respectively. Thus, Figure 16.1f shows the dependence of the Reynolds Number with the Weber Number, thereby outlining the necessary conditions for effective printability of the formulated inks [6].

Another important parameter that is essential to determine for ensuing shape reliability of printed electrodes is the storage modulus of the ink. It is dependent on the length of the spanning element, diameter, and specific weight of the ink. The elastic modulus (G′) delivers evidence of the rigidity and nature of the particle networks that comprise the studied system. On the contrary, the viscous modulus (G") measures the degree of readiness at which an imposed stress on the system gets relaxed. Thus, the ratio of storage modulus (G') and the viscous modulus (G") is monitored as a function of applied stress to understand the degree of the brittleness of the formulated ink and can be correlated with its printability [12].

Again, for attaining fine-resolution printing, stable jetting of MXene inks is a must which is indicated by the inverse Ohnesorge number (Z). It measures the figure of merit of stable drop formation, given by: $Z = (\gamma\rho D/\eta^9)^{1/2}$, where Z is related to the surface tension (γ), viscosity (η), density (ρ), and nozzle diameter (D) parameters respectively. Accordingly, the value reported for ethyl alcohol-MXene ink (Z~2.6), DMSO-based (Z~ 2.5), and NMP (Z~ 2.2) based MXene inks afford in high-quality 3D inkjet printing [18].

16.2 MXENES IN THE FABRICATION OF 3D-PRINTED SUPERCAPACITORS

16.2.1 Why Are MXenes Special in Supercapacitor Technology?

Inspired by the compliant features of recently developed 2D-layered nanomaterials, such as graphene, MoS_2, and various similar transition metal dichalcogenides, borides, phosphides, nitrides, carbonitrides, carbides, etc., the study of MXenes is also rapidly emerging as a very interesting and prosperous class of layered materials, because of their versatile tunable physical and chemical properties, as indicated in Figure 16.2a [19].

They are transition metal carbides and nitrides having the general formula $M_{a+1}X_aT_{b,}$ where the value of a = 1, 2, or 3, M indicates an early transition metal (Ti, Cr, V, etc.), X denotes carbon and/or nitrogen, while T_b indicates nature and number of surface functional groups (-O, -OH, and -F), correspondingly, latter remaining arbitrarily spread over the surfaces of the 2D flakes. They exhibit an attractive amalgamation of physicochemical properties including ultrafast electrical charge transportation, outstanding electrochemical behavior, and superior hydrophilic character which make

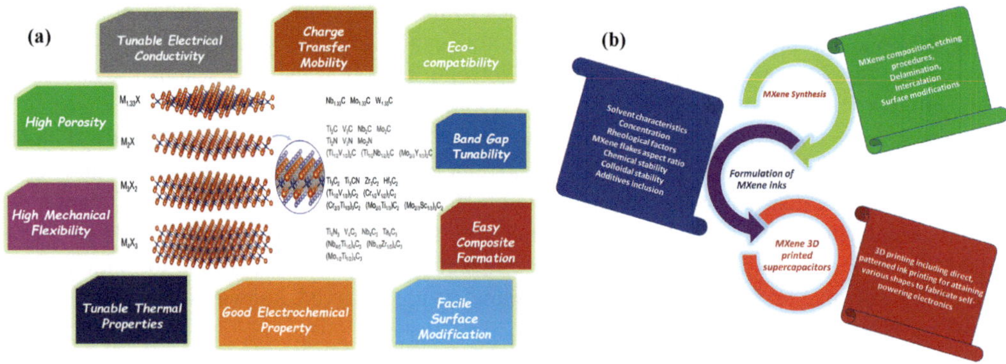

FIGURE 16.2 (a) Schematic presentation of some versatile features of MXenes. Adapted with permission [19], Copyright (2019), Elsevier. (b) General steps for fabricating 3D printed supercapacitors with MXene-based inks.

them, a versatile group of solids for framing multifunctional flexible microsupercapacitors for vivid functions [20]–[21]. The literature survey discloses that the pseudocapacitive performance of $Ti_3C_2T_x$ was initially explored using freestanding and binder-free film electrodes, demonstrating remarkable volumetric capacitances (300–400 Fcm^{-3}), utterly beyond the electrochemical responses of analogous 2D carbon nanomaterials exhibiting typical electrical double-layer capacitors [22]. Later, a higher degree of improved performances of $Ti_3C_2T_x$ electrodes through further modifications, hybridization with other nanomaterials, and controlled engineering have remarkably elevated the volumetric capacitances to a value as high as ~1500 Fcm^{-3} for pristine and ~1682 Fcm^{-3} for conductive polymers hybrid systems, respectively [23]–[24].

16.2.2 Role of MXenes in Fabricating 3D-Printed Supercapacitors

Rigorous and systematic investigations have un-curtained several versatile, exciting and unique structural, mechanical, and electronic features of MXenes that make them particularly suitable as printable inks for patterning and coating technology [25]–[29]. The synthesis tunability allows the liberty of designing MXene fragments to retain optimum hydrophilicity owing to the availability of adequate surface functional moieties that can readily form stable and homogeneous colloidal dispersions in numerous aqueous and non-aqueous solvents [30]–[31]. Consequently, in Figure 16.2b, generalized steps have been proposed for establishing MXenes-based complex 3D architectures and multifaceted utilities via printing/ patterning via an easily adaptable procedure to promote scalable, reproducible, and economic designing methodology in the true sense, and accordingly, researchers exploring MXenes should fundamentally thrust more on the synthesis particularly taking place during the etching and delamination stages, modification, and application process sequences.

The chemical nature of MXene surfaces greatly impacts the kinetic and thermodynamic stability of the ink and printing process. The hydrophilicity, delamination, and oxidation of MXenes are largely guided by the nature of surface functionalization present therein in the MXene system [32]. The proportion of the surface terminations is largely related to the etching process and reaction conditions, while the degree of MXene flakes oxidation varies with the delamination process and post-synthesis treatment procedures. These surface functional moieties contribute to negative zeta potentials and thus, dictate the hydrophilicity of MXenes layers. Such characteristics result in the formation of stable dispersions of MXene flakes in absence of external stabilizing agents and/ or rheology modifiers, which is very decisive for easy and smooth printing of MXenes inks [33].

16.2.3 Chemical Stability and Storage of MXenes Inks for 3D Printing

Recent reports reveal that the aqueous suspensions of MXenes display very good dispersion stability [32]. However, detailed investigations have exposed the fact that the dissolved oxygen in an aqueous medium often leads to chemical oxidation of delaminated single/ few-layered MXenes which triggers structural collapse and overall deterioration of the formulated MXene inks [33]–[34]. Such phase transformation and structural deformation owing to oxide formation also get accelerated at elevated temperatures or on light exposure. Thus, elimination of the dissolved oxygen in an aqueous medium by purging with inert gases such as dinitrogen or argon can effectively suppress oxidation. Even so, not only dissolved oxygen, but the water itself also oxidizes MXenes, and thus this chemical instability may be considerably controlled via refrigeration [35]. Thus, for storage of MXene inks for a longer time duration, it is currently a popular upcoming practice to keep them inside argon-sealed, non-transparent vials, devoid of light and under refrigeration which has yielded successful results considerably [36].

Nonetheless, irrespective of the above-mentioned precautions, the long-term storage of MXenes can be substantially improved through storage in a non-aqueous medium (for several months) and has been rightly encouraged in recent times [37]. While often to evade MXenes oxidation, filtration of the suspensions and subsequent storage in the form of dried and desiccated films under vacuum has been the other effective option. However, such a drying process often leads to delaminated-nanosheets restacking, analogous to various other 2D materials like graphenes, layered metallic chalcogenides, etc., especially when exposed to elevated temperatures. In another strategy as reported by Zhao et al. for retarding the restacking of MXenes in suspensions, antioxidants like sodium l-ascorbate have been used to escape the degradation of $Ti_3C_2T_x$ even in the dissolved oxygen-containing aqueous medium [38]. Hence, sodium l-ascorbate interacts with the positively charged surface terminals of the MXene flakes, thus curbing the reactive MXene edges from reacting with water or dissolved dioxygen molecules, thereby comprehensively increasing the stability to an appreciable extent [31]. Besides, the high reactivity of MXene flakes can also be restricted through the involvement of large-sized polyanions like polyphosphates, polyborates, and polysilicates, which successfully curbs away the sites and thus, suppresses the oxidation of MXene flakes, thus stabilizing aqueous dispersion for about one month. Therefore, the unrequired post-treatment advantages, as well as cost-effectiveness of polyanionic salts, have promoted their usages in the scalable production of long-term stable MXene inks admirably [39].

Even the solvent and MXene interactions are vital to determine the relative stability of the dispersed inks [40]–[41]. To illustrate, the stability of 50% HF etched $Ti_3C_2T_x$ MXene has been subjected to qualitative assessment by observing the nature of solvents that offer stable dispersions versus those which promote MXene agglomeration and precipitation over the experimental timeline as reflected in Figure 16.3a [42]. It was perceived that out of the several studied ones, the less polar, hydrophobic organic solvents offered negligible dispersions just after sonication, the trend thus following the declining polarity of organic solvents from 1,2-dichlorobenzene to hexane correspondingly when studied for one day and subsequently for four days standing duration. Even it has been observed that MXenes with large lateral dimensions promote multiple solvent–nanosheet interactions that facilitate better stability of the dispersion due to minimum loss on breaking solvent–solvent interactions [42].

16.2.4 Factors Affecting the Rheology of MXene Inks for 3D Printing

Observation of the rheological responses of the MXene suspensions is a vital aspect of evolving high-quality MXene inks. [40], [41] The rheological features of MXene inks are demarcated by the physical response of MXenes under applied stress, which guides the choice of printing technique, reproducibility, and the quality of printed patterns. It is important to note that different printing/coating techniques adopt different rheological requirements of inks. For instance, extrusion 3D printing involves viscous paste while screen printing/inkjet printing is manageable with a low/moderate concentration of ink dispersion. Thus, the fluidity of ink is defined by the variation of its viscosity (η) with shear rate. Figure 16.3b–c shows the variation of viscosity as a function of shear rate for monolayered and multi-layered MXene flakes dispersions, respectively, as a function of concentrations (mg/mL) and corresponding volume fractions (ϕ). The key difference between the above two studied systems stands to be the effect of shear on the viscosity at high shear rates ($\gamma = 100\ s^{-1}$). At low shear rates ($\gamma = 0.01\ s^{-1}$), both systems showed a greater than four orders of magnitude rise in viscosity with increasing solid loading (ϕ). But at high shear rates ($\gamma = 100\ s^{-1}$), the monolayered dispersions exhibited a variance of a factor of four only, while the multi-layered inks still maintained a difference of three orders of magnitude between the highest (70 wt %) and lowest (10 wt %) concentrations. Therefore, the above study affirms that the number of MXene layers can modify the viscosity factors of the inks significantly. The authors further disclosed that monolayered-$Ti_3C_2T_x$ inks of high concentrations possess suitable rheology for extrusion printing as also reflected by their G'/G" ratio being higher than ~2, exhibiting gel-like behavior [40].

In another systematic study, variation of the surface tension in different dispersing solvents has also been projected. Thus, lately, a combination of organic N-methyl-2-pyrrolidone and ethanol

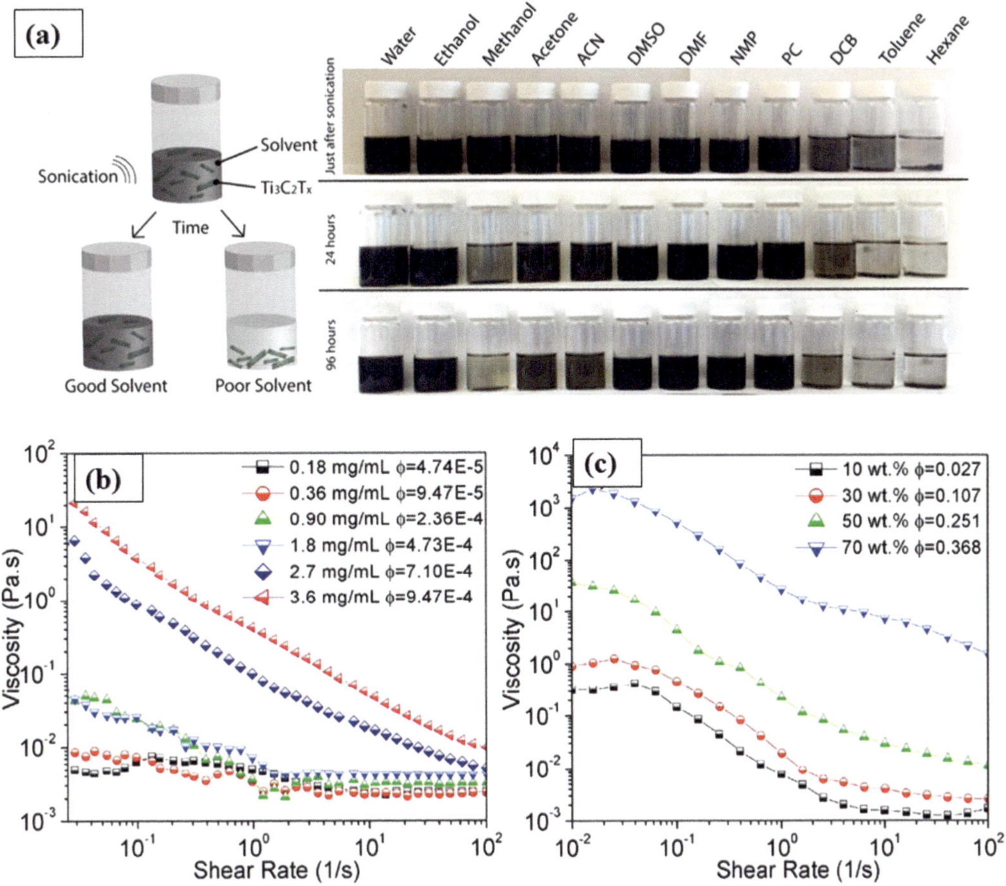

FIGURE 16.3 (a) Dispersions of 50% HF-etched $Ti_3C_2T_x$ in the twelve enlisted solvents. Time points of just after sonication (top), 1 day after sonication (middle), and 4 days after sonication (bottom) were used to monitor stability. Adapted with permission [42], Copyright (2017), American Chemical Society. (b) Variation of viscosity with the shear rate for monolayered MXenes (c) Variations of viscosity with the shear rate for multi-layered MXenes. Both (b) and (c) are adapted with permission [40], Copyright (2018), American Chemical Society.

was employed for $Ti_3C_2T_x$ MXene ink preparation to augment the surface tension and accordingly succeeded in improving the printing resolution significantly [27]. Water is the most commonly used dispersion medium for $Ti_3C_2T_x$ that exhibited a surface tension of 72 mNm^{-1} , signifying that $Ti_3C_2T_x$ may want solvents with surface tensions of at least > 40 mNm^{-1} to procure stable colloidal solution, as reflected in Figure 16.4a. This ensures decent matching of the surface tension between inks and substrates for good pattern printing.

Additionally, solvent characteristics such as boiling points, molecular weights, viscosity, surface tension, polarity, etc., are equally vital for the formulation of good quality MXene inks with excellent rheological parameters. Thus, in Figure 16.4b, the variation of the concentration of $Ti_3C_2T_x$ against the boiling point of the solvents infer that the less volatile solvents are more effective in dispersing $Ti_3C_2T_x$ in general [42]. Moreover, solvent molecular weight (Figure 16.4c) can likewise be to some extent correlated with MXene dispersion concentration, although the relation is quite poor in general but can be enhanced if some of the poor solvents such as hexane, toluene, and dichlorobenzene, water, etc., are excluded [35]. The relation between the viscosity of the solvent

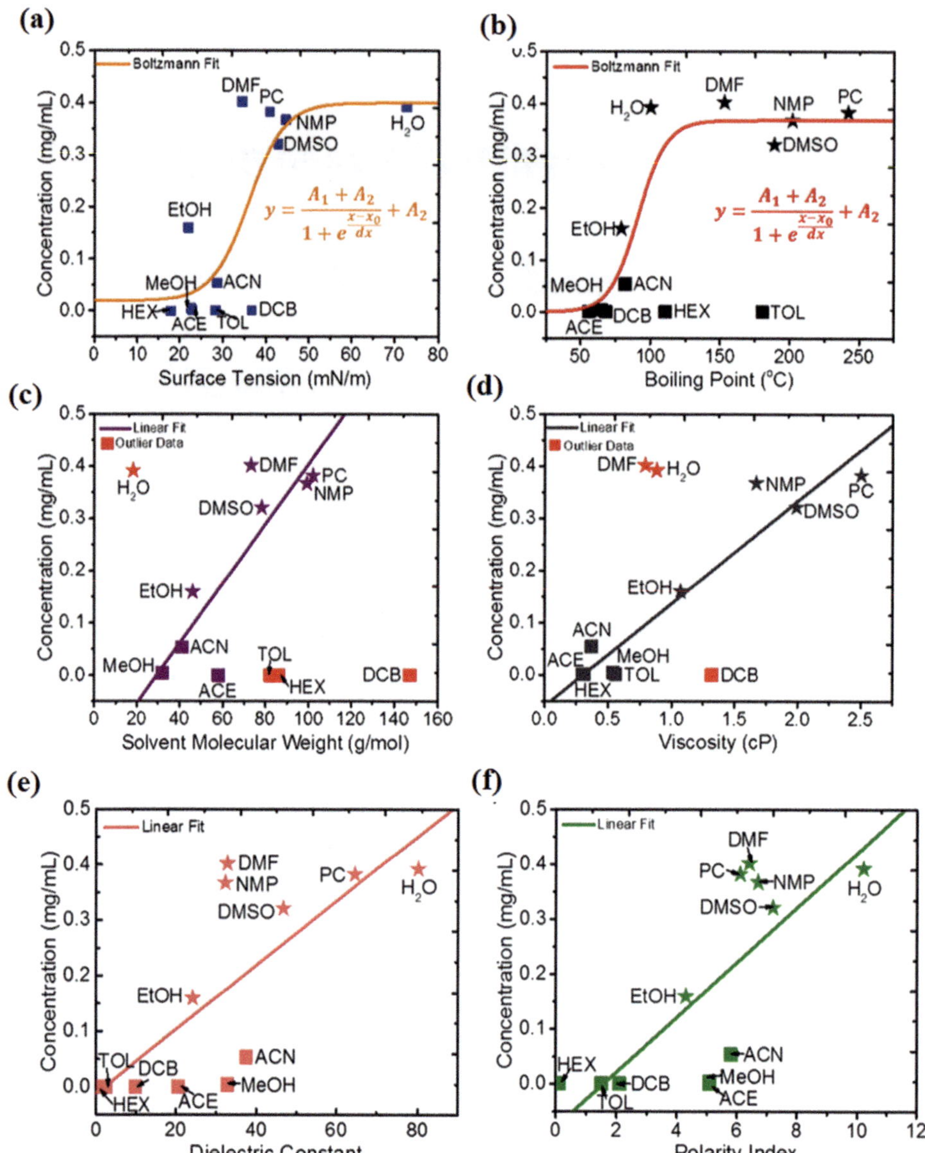

FIGURE 16.4 Variation of MXene concentrations in dispersions plotted against different solvent properties: (a) surface tension, (b) boiling point of the solvent, both graphs fitted using Boltzmann functions with the equations as shown within the insets, (c) Variation of concentration with solvent molecular weight (g/mol), graph fitted linearly not including the outlier solvents – water and nonpolar solvents (toluene, hexane, and dichlorobenzene), and (d) Variation of solvent viscosity with concentrations also fitted linearly excluding the solvents like DCB, DMF, and water. Here, the asterisk on data points signifies "good/ well dispersed" solvents. (e) Variation of $Ti_3C_2T_x$ concentration with respect to the solvent dielectric constant, with linear fitted plot and (f) Variation of $Ti_3C_2T_x$ concentration against solvent polarity index. Adapted with permission [42], Copyright (2017), American Chemical Society.

and the dispersion stability is also crucial. In general, solvents with high viscosity offer large kinetic dispersion stability, obviously different from true thermodynamic dispersion stability, but may be extremely advantageous for storage and applications which necessitate the projection of the nanomaterial through a nozzle, for instance, in printing, where the viscous nature of the solvent promotes the formation of precise droplets of the material to be printed, or in the spray coating where the homogeneous flow of the nanosheets is extremely essential. As depicted in Figure 16.4d, apart from DMF, water, and DCB solvents, in general, a linear correlation of dispersion concentration with solvent viscosity has been accomplished for many other solvents [42].

Furthermore, it has been conveyed the solvents that provide the best dispersion stability (DMF, NMP, DMSO, PC, and water) have high dielectric constants as reflected in Figure 16.4e. Similarly, high dielectric constant solvents offer a greater ability to stabilize the charged particles in solution; therefore, can better disperse MXene because it has an inherent negative surface charge evidenced by a high zeta potential (ζ). The dielectric constant correlates to the solvent polarity index ($\varepsilon > 20$). The polarity of the solvent matches with the polarity of the material based on the 'like dissolves like' concept. By plotting the concentration versus polarity index for each solvent (Figure 16.4f), a linear fit so-obtained revealed that solvents with a higher polarity index are, indeed, best for dispersing $Ti_3C_2T_x$ [42]. Thus, comparing the variation of MXene dispersion concentrations with various solvent parameters such as surface tension, boiling point, polarity, etc., recommend that polar solvents with high surface tension, elevated boiling point, and high dielectric constant, in general, offer better dispersion stability for $Ti_3C_2T_x$.

16.2.5 Formulation of Additive-Free MXene Inks for 3D Printing

Typical extrusion printing used for patterned interdigitated designs needs inks with appropriate viscosity, surface tension, and related properties, typically attained by introducing surfactants and/or polymer stabilizers, which after printing is over, necessitate absolute removal by heating to ensure maximum device signaling output. Nevertheless, such procedures often result in unwanted complexities and inefficient post-fabrication finishing processes. MXene inks, in this situation, is very much superior and can be utilized directly without the requirement for any additives. Accordingly, Zhang et al. took up the challenge to prepare additive-free organic MXene concentrated inks, to avoid the unwanted 'coffee ring effect' issues originating from additives-containing inks [18].

Again, the dimensions of MXenes platelets considerably affect the rheology of the dispersion and it can be customized to improve its printability. Thus, Yang et. al communicated an additive-free, aqueous MXene ink containing large, high aspect ratio based 2D $Ti_3C_2T_x$ sheets obtained via less aggressive etching and delaminating protocol, for extrusion-based 3D printing of freestanding, high specific surface area architectures for fabricating current collector-free energy storage devices of various morphologies [41]. They claimed that the bi-dimensionality of these MXenes containing charged surfaces and edges would promote easy intercalation of water molecules to control the rheological properties without the necessity for sacrificial additives. The resultant 3D-printed device displayed areal capacitance as high as 2.1 Fcm^{-2} for active material loading of about 8.5 $mgcm^{-2}$ with decent capacitance retention efficacy of ~ 90 % even after undergoing 10000 charging/ discharging cycles [41].

16.2.6 Printing and Patterning of MXene Inks on Various Substrates for Flexible Supercapacitors

Printing, as well as patterning of well-defined architectures with MXene inks on various substrates for designing highly flexible supercapacitors, is a very essential issue as far as formulation expertization is concerned. The preparation of inks should fundamentally ensure smooth and effortless printing and patterning on varying substrate surfaces under ambient conditions. Thus, high concentrations of large, decent quality MXene inks, obtained via the Minimally Intensive Layer Delamination synthesis approach (MILD) technique, recently ensured the production of high-density Ti_3C_2 inks of concentrations up to 30 mg mL^{-1} which could be readily loaded into roller ball pens for smooth and reliable writing without any leakage or clogging but can be printed directly on paper, plastic sheets, polymer membranes, and textiles effortlessly. Besides, well-desired patterns of various choices can be drawn manually or by using automatic drawing devices conveniently [26]. In another work by the Orangi group, ultrahigh concentrations of printable MXene aqueous inks were developed for the fabrication of 3D interdigital electrode architecture on a variety of substrates, including flexible polymer films and papers, for delivering superior areal and volumetric energy densities via room temperature direct ink writing procedure. As MXenes inks do not require post-printing high-temperature processes, they are extensively compatible with a variety of substrates, such as paper or plastic. Thus, the aqueous MXene dispersion was subjected to beads of re-usable superabsorbent polymer for water absorption from the dispersion under continuous stirring at 400 rpm to speed up the absorption process as well as precisely tune the dispersion concentration. The superabsorbent polymer concentrated the dispersion as high as 290 mg/mL, with exquisite viscoelastic properties favoring smooth and continuous coating of numerous layers of active material successfully. Further, such optimization of ink composition assisted in fine-tuning of electrode heights as well as regulation of specific active materials loading by the number of deposited layers [16]. The research group also fabricated macroscopic porous aerogels with vertically aligned $Ti_3C_2T_x$ sheets to construct unidirectional, well-aligned electrodes via freeze casting and inkjet-based 3D-printing combinations to ensure smooth electron transport across the three-dimensional printed structures. The tailored micro-/macro- architected aerogels showed excellent electromechanical performance, withstanding nearly 50% compression and recovering the original shape without losing their electrical conductivities during continuous compression cycles. The present technique can conveniently deploy water (ice) as a supporting matrix to construct true-3D constructions with overhang features, unlike extrusion-based 3D printing, and does not necessitate the usage of viscoelastic shear-thinning inks. Additionally, all-MXene micro-supercapacitor composed of the current collector and porous electrodes with horizontally as well as vertically oriented MXene sheets well lifted the overall electrical conductivity and ionic diffusion kinetics, thus demonstrating excellent electrochemical responses with thickness independent capacitive behavior [43].

For well-aligned and positional patterning of 2D MXenes, a scalable and simple hybrid 3D-printing approach involving surface topology designing and directed assembling of MXene flakes vide capillary-driven direct ink writing was proposed. Anisotropically-aligned MXene flakes on 3D-printed flexible substrates using MXene ink suspensions under leveraged microforces experienced preferential alignment vide layer-by-layer additive depositions. The so-obtained printed gadgets, as presumed, exhibited various multifunctional behavior, with a broad sensing range, elevated sensitivity, high-speed response time, and mechanical durability [44]. Additive-free, appropriate rheological properties based on 2D MXene aqueous inks consisting of essentially the sediments of un-etched precursors and unprocessedmulti-layered-MXenes during delamination procedure have been used for scalable screen printing. Notably, these inks when introduced on paper yield high resolution and spatial uniformity ideally fit for designing printed microsupercapacitors, conductive tracks, integrated circuit pathways, etc, with sustaining mechanical integrity as well as conductive interconnecting network. The resultant energy storage 3D-printed gadgets displayed superior areal capacitance (158 $mFcm^{-2}$) and energy density (1.64 $\mu Whcm^{-2}$) compared to other printed devices based on MXene or graphenes. This ink formulation scheme of "turning trash into treasure" outlines the prospect of utilizing wasted remnants of MXene sediment for large-scale and effective printing productions of next-generation wearable smart electronics [45].

Very recently, Wu group proposed a multi-scale structural engineering approach for fabricating ordered-MXene hydrogel printed supercapacitor electrodes for upgraded electrochemical outputs.

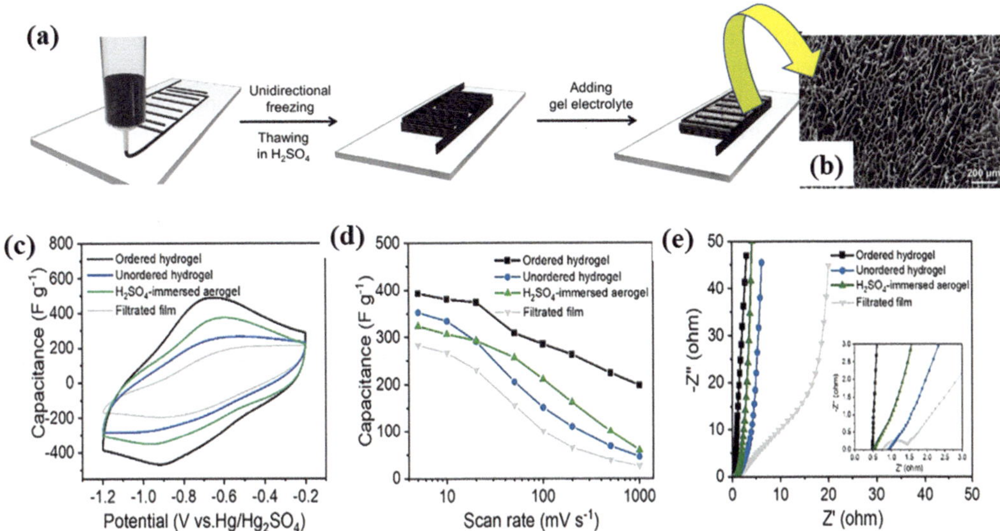

FIGURE 16.5 (a) Schematic presentation of the 3D-printing all-MXene micro-supercapacitor fabrication process via Multi-Scale Structural Engineering Strategy. (b) SEM images of (morphology) of ordered MXene hydrogel sample (top-view) (c) Comparative CV profiles of all samples indicated @ voltage scan rate of 100 mV/s, (d) Comparative rate performance (variation of capacitance with current density), and (e) Comparative EIS profiles of unordered MXene hydrogel, ordered MXene hydrogel, sulphuric acid-immersed MXene aerogel, and filtrated-MXene film samples respectively. Figure adapted with permission [46], Copyright: The Authors, some rights reserved; exclusive licensee [Wiley-VCH GmbH]. Distributed under a Creative Commons Attribution License 4.0 (CC BY).

Usage of unidirectional freezing treatment aided MXene sheets to create a hierarchical structure with vertically oriented paths (Figure 16.5a) which could be extended to construct 3D-printed MXene micro-supercapacitor electrodes via direct ink writing printing having honeycomb-type compartmental porous morphology, as confirmed from the SEM image provided in the Figure 16.5b. The so-designed energy storage device recorded superior areal capacitance as high as 2.0 Fcm^{-2}@ current density of 1.2 $mAcm^{-2}$ with high rate-capacity (capacitance of 1.2 Fcm^{-2}@ high current density of 60 $mAcm^{-2}$) and energy density of 0.1 $mWhcm^{-2}$at 0.38 $mWcm^{-2}$ (as reflected from the graphs placed in the Figure 16.5c–e respectively) compared to films obtained via other techniques such as directly freezing-dried MXene aerogel and compact-stacked MXene films. The vertically oriented architecture in these 3D-printed electrodes facilitated intercalation of protons due to enhanced interlayer spacing that promoted augmented electrolyte ions diffusion kinetics, and increased surface-active area, resulting in raised electrochemical responses [46].

Similarly, Liang's research group employed direct ink writing printing technology in combination with a unidirectional freezing method to fabricate highly stretchable microsupercapacitors that could restore stable electrochemical responses even under a strain of up to 50% owing to sliding, highly flexible honeycomb-like porous construction of MXene nanosheets [47]–[48],

16.3 CONCLUSION AND PROSPECTS

Thus, the above discussion highlights the current status of MXenes inks-based 3D-printing technology deployed for designing highly flexible and lightweight auto-powering electronic gadgets. Nonetheless, there are still a vast number of challenges in commercializing 2D MXene materials into printable inks. Currently, the primary necessity of the MXene fabrication industry is to carry out serious revision and improvement in the existing MXene synthesis strategies by avoiding the employment of destructive methods but designing economically workable procedures for devising large-scale flexible electronic devices in bulk measures to furnish the sky-high demands of consumer electronic industry. Moreover, other varieties of MXene inks apart from Ti_3C_2, with diverse compositions are needed to be systematically explored. Room temperature-based scalability, as well as stable ink preparation, is yet a deep concern.

Besides, MXenes having adequate chemical as well as colloidal stability is yet an additional task for industrially feasible applications. Study of the decay kinetics and other insights of the colloidal dispersions inks in various solvents vide various probes over time such as measuring the extinction and absorption spectra, etc., are crucial. Like many other 2D materials, MXenes are highly susceptible to sheet agglomeration, as well as high-rate oxidation, especially in aqueous conditions. Even they do often experience very tiny shelf lifespan and therefore, need critically controlled measures like the addition of external binders, surface modifications, co-solvents effects, etc., to cope with such vital issues.

In addition, many details of MXenes properties such as optical, mechanical, electrochemical, chemical, etc., are yet to be comprehensively understood and more methodical investigations on their rheological properties are urged, especially on the substrate-specific surface area and its surface energy, wetting conditions, ink viscosity, solvent volatility, polarity, solvent evaporation kinetics, etc., while proposing simplified and common inks formulation with manageable physical and chemical aspects which are in fine agreement with various requirements of prevailing printing technologies. Furthermore, while designing printed MXene patterns, care should be taken for suitable substrate selection so that they are not spoiled or deformed, and the charge conduction, as well as intended electrochemical responses, get fully recovered upon releasing the strains even after executing several cycles of deformation activities.

In addition, the simplicity, reproducibility, and reliability of 3D-printing technology and conforming MXene ink characteristics should be recognized for smooth, straight, and even printed patterns on different substrates to overcome associated issues such as the 'coffee ring effects' that tend to degrade the overall printing quality. Lately, the formulation of bio-compatible MXene inks is becoming extremely important for promoting smart bioelectronic devices and wearable electronic gadgets which can be readily implanted on the muscle, joints, or skin to capture necessary automatic signals. Therefore, it is clear that the present objective lies in overcoming the challenges that stand in scheming eco-friendly, chemically and thermodynamically stable, additive/binder-free, MXene inks via easy, scalable cost-effective routes for framing smart and flexible, high-performing auto-powering electronics shortly.

ACKNOWLEDGMENT

DM acknowledges Chandernagore College, Chandannagar, Hooghly, West Bengal, Pin-712136, India, for providing permission to do honorary research.

REFERENCES

1 D. Verma, K.R.B. Singh, A.K. Yadav, V. Nayak, J. Singh, P.R. Solanki, R.P. Singh, Internet of things (IoT) in nano-integrated wearable biosensor devices for healthcare applications, *Biosens Bioelectron X* 11 (2022) 100153.

2 D. Majumdar, S. Ghosh, Recent advancements of copper oxide-based nanomaterials for supercapacitor applications, *J Energy Storage* 34 (2021) 101995.

3 D. Majumdar, M. Mandal, S.K. Bhattacharya, Journey from supercapacitors to supercapatteries: recent advancements in electrochemical energy storage systems, *Emergent Mater* 3 (2020) 347–367.

4 D. Majumdar, Recent progress in copper sulfide based nanomaterials for high energy supercapacitor applications, *J Electroanal Chem* 880 (2021) 114825.

5 Y. Yao, K.K. Fu, C. Yan, J. Dai, Y. Chen, Y. Wang, B. Zhang, E. Hitz, L. Hu, Three-dimensional printable high-temperature and high-rate heaters, *ACS Nano* 10 (2016) 5272–5279.
6 J. Liang, C. Jiang, W. Wu, Printed flexible supercapacitor: ink formulation, printable electrode materials and applications. *Appl Phys Rev* 8 (2021) 21319.
7 V. Augustyn, P. Simon, B. Dunn, Pseudocapacitive oxide materials for high-rate electrochemical energy storage, *Energy Environ Sci* 7 (2014) 1597–1614.
8 Poonam, K. Sharma, A. Arora, S.K. Tripathi, Review of supercapacitors: materials and devices, *J Energy Storage* 21 (2019) 801–825.
9 K. Shen, J. Ding, S. Yang, 3D printing quasi-solid-state asymmetric micro-supercapacitors with ultrahigh areal energy density, *Adv Energy Mater* 8 (2018) 1800408.
10 P. Chang, H. Mei, S. Zhou, K.G. Dassios, L. Cheng, 3D printed electrochemical energy storage devices, *J Mater Chem A* 7 (2019) 4230–4258.
11 M. Areir, Y. Xu, D. Harrison, J. Fyson, R. Zhang, Development of 3D printing technology for the manufacture of flexible electric double-layer capacitors, *Mater Manuf Process* 33 (2018) 905–911.
12 S. Tagliaferri, A. Panagiotopoulos, C. Mattevi, Direct ink writing of energy materials, *Mater Adv* 2 (2021) 540–563.
13 C. (John) Zhang, B. Anasori, A. Seral-Ascaso, S.H. Park, N. McEvoy, A. Shmeliov, G.S. Duesberg, J.N. Coleman, Y. Gogotsi, V. Nicolosi, Transparent, flexible, and conductive 2D titanium carbide (MXene) films with high volumetric capacitance, *Adv Mater* 29 (2017) 1702678.
14 J.J. Yoo, K. Balakrishnan, J. Huang, V. Meunier, B.G. Sumpter, A. Srivastava, M. Conway, A.L. Mohana Reddy, J. Yu, R. Vajtai, P.M. Ajayan, Ultrathin planar graphene supercapacitors, *Nano Lett* 11 (2011) 1423–1427.
15 A. Tanwilaisiri, Y. Xu, R. Zhang, D. Harrison, J. Fyson, M. Areir, Design and fabrication of modular supercapacitors using 3D printing, *J Energy Storage* 16 (2018) 1–7.
16 J. Orangi, F. Hamade, V.A. Davis, M. Beidaghi, 3D printing of additive-free 2D Ti_3C_2Tx (MXene) ink for fabrication of micro-supercapacitors with ultra-high energy densities, *ACS Nano* 14 (2020) 640–650.
17 B. Zazoum, A. Bachri, J. Nayfeh, Functional 2D MXene Inks for Wearable Electronics, *Mater. (Basel)* 14 (2021) 6603.
18 C. (John) Zhang, L. McKeon, M.P. Kremer, S.H. Park, O. Ronan, A. Seral-Ascaso, S. Barwich, C.Ó. Coileáin, N. McEvoy, H.C. Nerl, B. Anasori, J.N. Coleman, Y. Gogotsi, V. Nicolosi, Additive-free MXene inks and direct printing of micro-supercapacitors, *Nat. Commun.* 10 (2019) 1795.
19 L. Verger, V. Natu, M. Carey, M.W. Barsoum, MXenes: an introduction of their synthesis, select properties, and applications, *Trends Chem* 7 (2019) 656–669.
20 B. Anasori, M.R. Lukatskaya, Y. Gogotsi, 2D metal carbides and nitrides (MXenes) for energy storage, *Nat Rev Mater* 2 (2017) 16098.
21 P. Das, Z.-S. Wu, MXene for energy storage: present status and future perspectives, *J Phys Energy* 2 (2020) 32004.
22 M. Ghidiu, M.R. Lukatskaya, M.-Q. Zhao, Y. Gogotsi, M.W. Barsoum, Conductive two-dimensional titanium carbide 'clay' with high volumetric capacitance, *Nature* 516 (2014) 78–81.
23 K. Das, D. Majumdar, Prospects of MXenes/graphene nanocomposites for advanced supercapacitor applications, *J Electroanal Chem* 905 (2022) 115973.
24 D. Majumdar, Role of MXenes/polyaniline nanocomposites in fabricating innovative supercapacitor technology. *Adv Energy Convers Mater* 3 (2022) 30–53.
25 J. Ma, S. Zheng, Y. Cao, Y. Zhu, P. Das, H. Wang, Y. Liu, J. Wang, L. Chi, S. Liu, Aqueous MXene/PH1000 hybrid inks for inkjet-printing micro-supercapacitors with unprecedented volumetric capacitance and modular self-powered microelectronics, *Adv Energy Mater* 11 (2021) 2100746.
26 E. Quain, T.S. Mathis, N. Kurra, K. Maleski, K.L. Van Aken, M. Alhabeb, H.N. Alshareef, Y Gogotsi, Direct writing of additive-free MXene-in-Water ink for electronics and energy storage, *Adv Mater Technol* 4 (2019) 1800256.
27 S. Uzun, M. Schelling, K. Hantanasirisakul, T.S. Mathis, R. Askeland, G. Dion, Y. Gogotsi, Additive-free aqueous MXene inks for thermal inkjet printing on textiles, *Small* 17 (2021) 2006376.
28 Y. Wang, M. Mehrali, Y.-Z. Zhang, M.A. Timmerman, B.A. Boukamp, P.-Y. Xu, J.E. ten Elshof, Tunable capacitance in all-inkjet-printed nanosheet heterostructures, *Energy Storage Mater* 36 (2021)318–325.

29 C.-W. Wu, B. Unnikrishnan, I-.W.P. Chen, S.G. Harroun, H-.T. Chang, C-.C. Huang, Excellent oxidation resistive MXene aqueous ink for micro-supercapacitor application, *Energy Storage Mater* 25 (2020) 563–571.

30 D. Singh, V. Shukla, N. Khossossi, A. Ainane, R. Ahuja, Harnessing the unique properties of MXenes for advanced rechargeable batteries, *J Phys Energy* 3 (2020) 12005.

31 A. Iqbal, J. Hong, T.Y. Ko, C.M. Koo, Improving oxidation stability of 2D MXenes: synthesis, storage media, and conditions, *Nano Converg* 8 (2021) 1–22.

32 S. Abdolhosseinzadeh, X. Jiang, H. Zhang, J. Qiu, C. (John) Zhang, Perspectives on solution processing of two-dimensional MXenes, *Mater Today* 48 (2021) 214–240.

33 S. Biswas, P.S. Alegaonkar, MXene: evolutions in chemical synthesis and recent advances in applications, *Surfaces* 5 (2021) 1–34.

34 A. Bhat, S. Anwer, K.S. Bhat, M.I.H. Mohideen, K. Liao, A. Qurashi, Prospects challenges and stability of 2D MXenes for clean energy conversion and storage applications, *npj 2D Mater. Appl.* 5 (2021) 1–21.

35 Y. Zhang, Y. Wang, Q. Jiang, J.K. El-Demellawi, H. Kim, H.N. Alshareef, MXene printing and patterned coating for device applications, *Adv Mater* 32 (2020) 1908486.

36 J. Azadmanjiri, T.N. Reddy, B. Khezri, L. Děkanovský, A.K. Parameswaran, B. Pal, S. Ashtiani, S. Wei, Z. Sofer, Prospective advances in MXene inks: screen printable sediments for flexible micro-supercapacitor applications, *J Mater Chem A* 10 (2022) 4533–4557.

37 F. Cao, Y. Zhang, H. Wang, K. Khan, A. K. Tareen, W. Qian, H. Zhang, H. Ågren, Recent advances in oxidation stable chemistry of 2D MXenes, *Adv. Mater.* 34 (2022) 2107554.

38 X. Zhao, A. Vashisth, J.W. Blivin, Z. Tan, D.E. Holta, V. Kotasthane, S.A. Shah, T. Habib, S. Liu, J.L. Lutkenhaus, pH, nanosheet concentration, and antioxidant affect the oxidation of $Ti_3C_2T_x$ and Ti_2CT_x MXene dispersions, *Adv Mater Interfaces* 7 (2020) 2000845.

39 V. Natu, J.L. Hart, M. Sokol, H. Chiang, M.L. Taheri, M.W. Barsoum, Edge capping of 2D-MXene sheets with polyanionic salts to mitigate oxidation in aqueous colloidal suspensions, *Angew Chemie* 131 (2019) 12785–12790.

40 B. Akuzum, K. Maleski, B. Anasori, P. Lelyukh, N.J. Alvarez, E.C. Kumbur, Y. Gogotsi, Rheological characteristics of 2D titanium carbide (MXene) dispersions: a guide for processing MXenes, *ACS Nano* 12 (2018) 2685–2694.

41 W. Yang, J. Yang, J.J. Byun, F.P. Moissinac, J. Xu, S.J. Haigh, M. Domingos, M.A. Bissett, R.A.W. Dryfe, S. Barg, 3D printing of freestanding MXene architectures for current-collector-free supercapacitors, *Adv Mater* 31 (2019) 190272.

42 K. Maleski, V.N. Mochalin, Y. Gogotsi, Dispersions of two-dimensional titanium carbide MXene in organic solvents, *Chem Mater* 29 (2017) 1632–1640.

43 H. Tetik, J. Orangi, G. Yang, K. Zhao, S. Bin Mujib, G. Singh, M. Beidaghi, D. Lin, 3D printed MXene aerogels with truly 3D macrostructure and highly engineered microstructure for enhanced electrical and electrochemical performance, *Adv Mater* 34 (2022) 2104980.

44 S. Jambhulkar, S. Liu, P. Vala, W. Xu, D. Ravichandran, Y. Zhu, K. Bi, Q. Nian, X. Chen, K. Song, Aligned $Ti_3C_2T_x$ MXene for 3D micropatterning via additive manufacturing, *ACS Nano* 15 (2021) 12057–12068.

45 S. Abdolhosseinzadeh, R. Schneider, A. Verma, J. Heier, F. Nüesch, C. Zhang, Turning trash into treasure: additive free MXene sediment inks for screen-printed micro-supercapacitors, *Adv Mater* 32 (2020) 2000716.

46 X. Huang, J. Huang, D. Yang, P. Wu, A multi-scale structural engineering strategy for high-performance MXene hydrogel supercapacitor electrode, *Adv Sci* 8 (2021) 2101664.

47 X. Li, H. Li, X. Fan, X. Shi, J. Liang, 3D-printed stretchable micro-supercapacitor with remarkable areal performance, *Adv Energy Mater* 10 (2020) 1903794.

48 W. Zong, Y. Ouyang, Y.-E. Miao, T. Liu, F. Lai, Recent advance and perspective of 3D printed micro-supercapacitor: from design to smart integrated devices. *Chem Commun.* 58 (2022) 2075–2095.

17 3D-Printed Nanocomposites for Supercapacitors

Shiva Bhardwaj[1,3], *Sudhakar Reddy Madhira*[1,3], *and Ram K. Gupta*[2,3*]

[1]Department of Physics, Pittsburg State University, Pittsburg, Kansas 66762, USA

[2]Department of Chemistry, Pittsburg State University, Pittsburg, Kansas 66762, USA

[3]National Institute for Material Advancement, Pittsburg, Kansas 66762, USA

*Corresponding author: ramguptamsu@gmail.com

CONTENTS

17.1 INTRODUCTION

The possible fatigue of petroleum derivatives and the natural influences related to ozone harm has prompted worldwide requests to create practical energy supplies. Scientists are looking for alternative solutions to prevent the continuous depletion of petroleum derivatives via renewable resources like solar and wind energy, but the portable energy demand tends researchers to move forward to energy storage devices (ESDs). ESDs store energy in various forms and help to decrease the dependency on petroleum derivatives by storing and delivering the energy created via renewable resources [1].

Among the variety of ESDs, batteries, and supercapacitors (SCs) addresses the two driving electrochemical energy-stockpiling advancements as shown in Figure 17.1 of energy vs. power densities known as a Ragone plot. Batteries store energy chemically and transform it electrically when required. They have high energy density and are utilized in purchaser electronics. However, because of an assortment of resistive misfortunes from drowning electron and particle transport, batteries get heated up and further develop dendrite at high power consumption, creating genuine issues [2]. This has come about in a few widely discussed disappointments, including the electric vehicle made by Tesla and the Dreamliner plane made by Boeing.

Therefore, SCs, otherwise called electrochemical capacitors, can supplement and support the batteries in specific applications since they can securely provide the high and fast power required and can easily run for very long-life cycles (>100,000 cycles). A large number of interests have

DOI: 10.1201/9781003296676-17

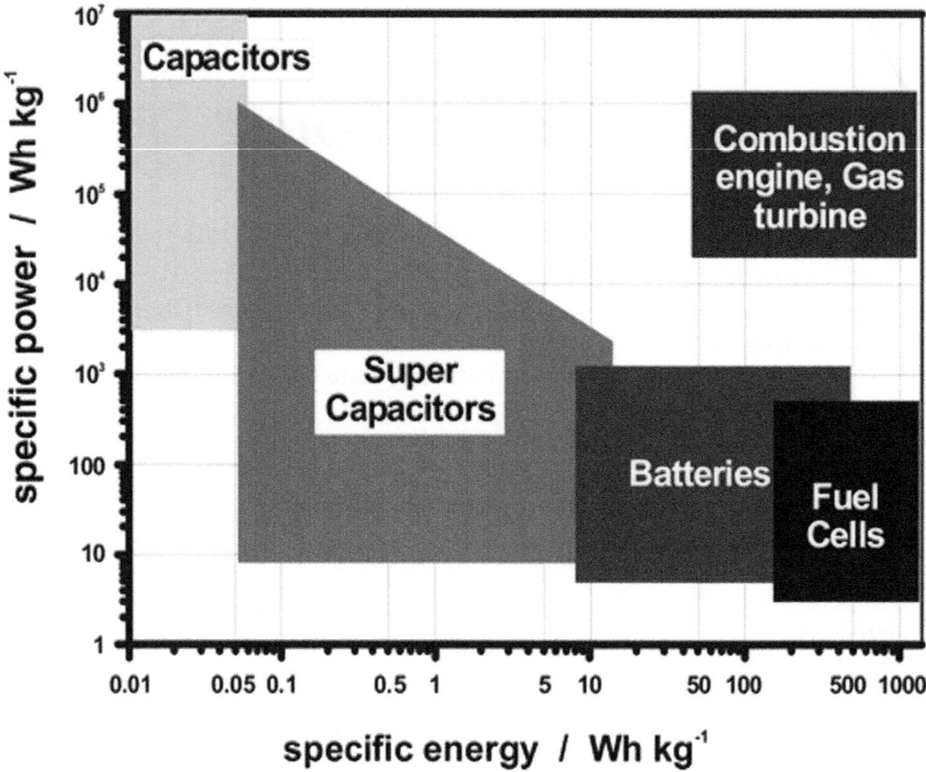

FIGURE 17.1 A schematic plot between power density *vs.* energy density of various energy storage devices. Adapted with permission [3]. Copyright (2004), American Chemical Society.

been developed for SCs, like quick charging, incredible cycling security, and high power thickness. Although economically accessible, SCs can convey a lot higher energy thickness than custom strong state electrolytic capacitors but are still lagging behind the batteries. Consequently, the boundless utilization of SCs has been developed, and broad examination endeavors are determined to accomplish the energy-stockpiling capacities [4].

SCs can be ordered into three parts based on the charge storage mechanism: electric double-layer capacitors (EDLCs), pseudocapacitors, and hybrid SCs. EDLCs store charges based on an electrostatically formed double layer on the surface of electrodes which results in coulobmic interaction. Pseudocapacitors store charges based on a faradaic reversible redox reaction, whereas the hybrid SCs refer to the combination of EDLCs and pseudocapacitors. There are number of factors on which the performance of SC depends among which energy density is an essential parameter. This energy factor is directly proportional to the applied voltage, which is given in Equation (17.1).

$$E = \frac{1}{2}CV^2 \tag{17.1}$$

Where C is the specific capacitance (C_{sp}) of SC, Equation (17.1) contains the square of voltage in the equation, which expresses the 2-fold increase in voltage results in a four -fold increase in energy for the same value of specific capacitance [5]. One method by which scientists have attempted to address the test of relatively low energy thickness is by creating hybrid SCs. In this unique situation, asymmetric SCs cover various gadget configurations. Therefore, there is a requirement for developing a variety of synthesis routes for the fabrication of materials for SCs, which could easily offer the prospect of an increase in energy density and specific capacitance for applications in real-time devices. Furthermore, given the rapid improvements in this technology, asymmetric SCs are expected to play a vital role in the energy storage industry where two different materials are used as cathode and anode in the fabrication of SCs [6]. There are different 3D-printing methods like inkjet printing, stereolithography, etc., which helps in the development of hierarchical porous 3D SCs. 3D printing enhances the performance of the devices by reducing the error in the composition of the fabricated materials. In this regard, various materials and their fabrication techniques are discussed further in the following sections.

17.2 MATERIALS FOR SUPERCAPACITORS

The demand for SCs is increasing due to their high power density and fast charge-discharge capability. Moreover, the selection of material plays a vital role in determining the properties and performance of SCs. Various materials like metal-organic framework (MOF), MXenes, conductive polymer, composite, and many other materials have been employed to develop SCs. The materials that have to be selected must be processed quickly and possesses a high active surface area with a greater number of active sites participating in the reaction [7]. Among all the materials, conducting polymer demonstrates various properties for SCs and attracted much interest in past decades. The most accessible conductive polymer that can be fabricated easily is polyaniline (PANI). PANI demonstrates doping and de-doping chemistry, which allows it to show high electrical conductivity. Moreover, PANI has moderate solubility, allowing it to be prepared using electrospinning, spin-coating, inkjet printing, electrodeposition, and 3D technique. These days scientists are focusing on 3D-printed materials, which among the polymers have become the famous technology for the processing and fast fabrication of complicated objects. A 3D PANI/reduced graphene oxide (rGO) has been prepared by Wang et al. [8] using modified Hummer's method and 3D-printing technology to fabricate electrodes later. The electrochemical measurement of printed SC shows a specific capacitance of 1329 mF/cm^2. Figure 17.2 shows the synthesis route used to develop PANI/rGO.

Nanocomposites of different materials (like metal hydroxide, metal oxide, etc.) possess high specific capacitance but suffer life cycles and sluggish kinetics, limiting their application in various SC-based devices. Therefore, scientists are looking toward novel nanocomposites for hybrid SC electrodes. Makkar et al. [9] fabricated cobalt-iron-oxide ($CoFe_2O_4$) and porous carbon (PC, derived

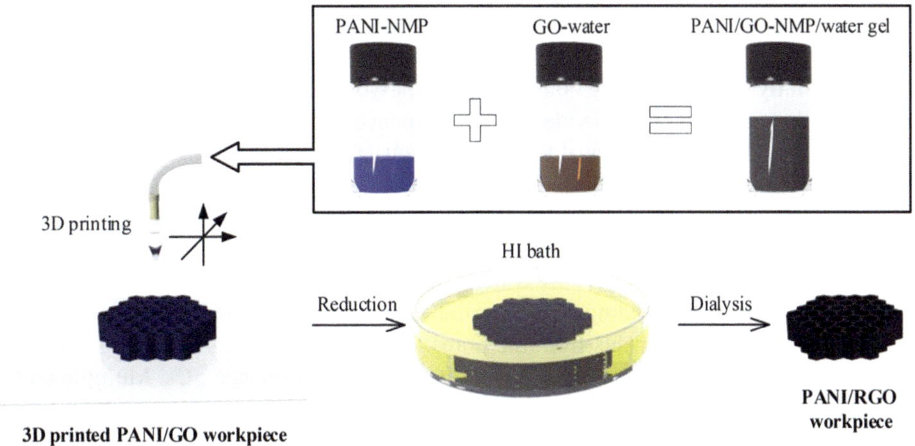

FIGURE 17.2 A schematic representation of PANI/rGO. Adapted with permission [8]. Copyright (2018), American Chemical Society.

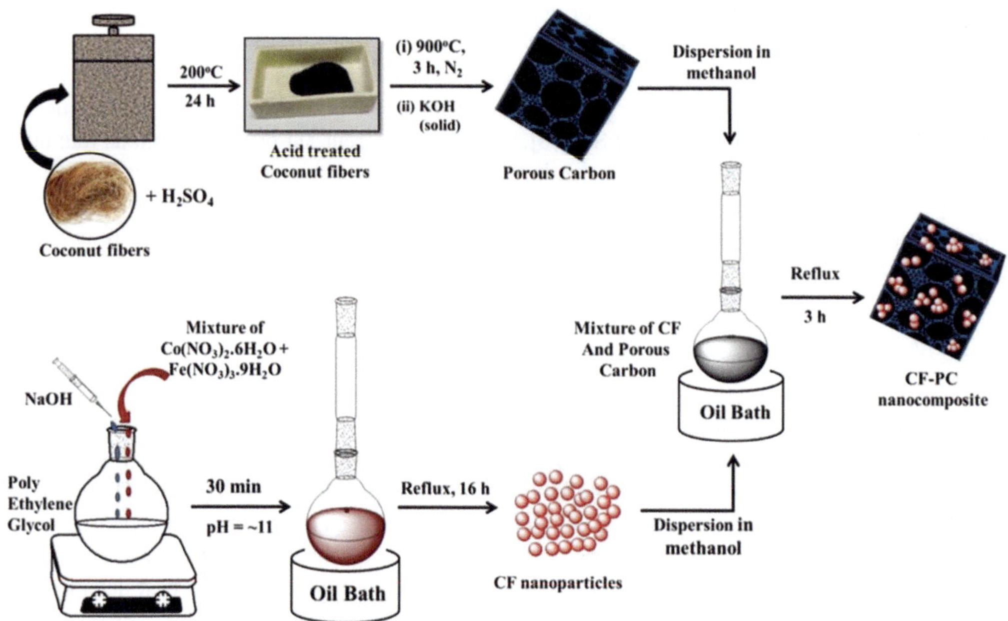

FIGURE 17.3 A methodology depicting the formation of the nanocomposite. Adapted with permission [10]. Copyright (2022), American Chemical Society.

from coconut fiber) doped nanocomposite using co-precipitation and wet-impregnation techniques (Figure 17.3). Electrochemical measurements were performed in a three-electrode system for the reality check of the developed electrode. The highest specific capacitance of 272 F/g with 93% retention in capacitance after 5000 cycles represents its high performance and stability. This retention in capacitance was due to the large surface area, heterogenous catalysis, good power performance, and high energy density of $CoFe_2O_4$, which acted as a suitable cathode material with pseudocapacitive behavior. The as-synthesized nanocomposites of $CoFe_2O_4$ with PC derived from coconut fibers are eco-friendly and sustainable materials with a good life cycle.

There is always a race to find different classes of materials to develop energy storage devices. In this race, scientists discovered a new material called MOF. MOFs are multifunctional frameworks with large surface areas, adjustable pores, and crystalline structures which make them suitable materials for SCs. Yaghi's group has played an essential role in developing MOF since 1995 [11]. MOFs are generally prepared by the combination of polydentate organic ligands and metallic ions and possess high porosity which provides a stable connection between the electrodes and the electrolyte [12]. During the earlier stages of the development of MOFs, cobalt-based MOFs (Co-MOFs) were the promising candidates for the SCs behavior. Lee et al. [13] synthesized Co-MOF using Co (II) and terephthalic acid as an electrode material which shows a specific capacitance of 206.76 F/g with cyclic stability of 98.5% after 1000 cycles.

It has been found that Co-MOFs have a high surface area and large pores depending upon the type of organic linker used, leading to the partial failure of framework structure because of the incomplete vaporization of solvents from the pores. Similarly, single ligands along with multiple-metal MOFs have been prepared to meet the needs of high-performance SCs. Multiple metals and multiple ligands have multivalent ions in the material, achieving the synergistic effect to improve the electrochemical properties. As an organic ligand, the multivalent Co-MOF using 2-methyl benzimidazole and pyromellitic dianhydride has been synthesized for SC electrodes. Sun et al. [14] replace Co^{2+} with Ni^{2+} by isomorphism, exhibiting a high specific capacitance of 2422.7 F/g with an

86% retention after 3000 cycles. The design of molecular and adjustable crystal structures can show significant improvements in the electrical conductivity of pristine MOFs on 3D substrates, further showing greater aspects for SC devices.

Continuous research is going on to discover and develop different materials for SC devices. A unique structure of transition metal nitrides and carbides is referred to as MXenes which show desirable properties that make them attractive for use in SCs. These 2D inorganic compounds constitute a few atomic layers of transition metal carbides or nitrides [15]. Generally, they are synthesized by etching the element A from their MAX phases ($M_{n+1}AX_n$, n=1,2,3), where M is the transition metal, A is the leading group element, and X is the nitrogen or carbon [16]. Lukatskaya et al. [17] construct a 13 µm thick $Ti_3C_2T_x$ film, reaching a specific capacitance of 310 F/g at 10 mV/s in 3M H_2SO_4 compared to traditional graphene SC which delivers 78 F/g. The big difference in specific capacitance was due to different intercalation types shown by the MXenes and graphene. Equation (17.1) shows the participation of the surface functional group in the charge storage mechanism, leading to the consideration of the redox process as the protonation of oxygen functional groups.

$$Ti_3C_2T_x(OH)_yF_z + \delta e^- + \delta H^+ = Ti_3C_2T_{x-}\delta(OH)_{y+}\delta F_z \quad (17.2)$$

Ghidiu et al. [18] synthesized a 5 µm thick $Ti_3C_2T_x$ film, reaching the volumetric capacitance of 910 F/cm^3 at the scan rate of 2 mV/s with capacitance retention of 81% after 10000 cycles. The loss in capacitance was due to the intercalation and de-intercalation via the charging-discharging phenomenon. However, several challenges like complex chemical compositions, structures, and lack of systematic discussion on the application of MXenes prevent them from becoming more practical SC electrodes. Different methods for fabrication, etching treatment, and intercalation of additives like hydrochloric acid (HCl), hydrofluoric acid (HF), and LiF/HCl mixture and molten salts are practiced to produce high-performance SCs. Even with the above development, MXenes are in their infancy. Yet, based on 2D structures and surface groups, MXenes are promising pseudocapacitive electrodes.

17.3 RECENT DEVELOPMENT IN 3D-PRINTED SUPERCAPACITORS USING NANOCOMPOSITES

17.3.1 Nanocomposites of 2D Materials for 3D-Printed Supercapacitors

Materials with multilayer morphology are more suitable for SCs due to their higher porosity and large electrochemically active surface area. The 2D materials used in SC electrodes are graphene, graphene oxide (GO), reduced graphene oxide (rGO), functional graphene oxides, and MXenes. These 2D nanocomposite materials generally consist of many thin layers. In addition, their high surface-to-volume ratio means more surface atom exposure during the electrochemical reactions. Among all the given materials, graphene has been explored as 2D material possessing a high surface area. Its structure can be coiled to form 1D nanotubes, as well as its honeycomb network-like structure, can be stacked later for 3D graphite formation. Graphene has many functional, mechanical, and electrical properties. Other important properties are a quantum hall effect at room temperature and tuneable bandgap. It also has high electron mobility, high surface area, stability, and intrinsic capacitance, making it an ideal 2D material for SCs. The access of ions to the surface determines their practical capacitance. Theoretically, the ideal specific capacitance of graphene is about 550 F/g, where its entire surface is electrochemically active [19].

Furthermore, the capacitance depends on how the graphene structure is arranged to achieve optimal surface access for electric double layer (EDL) formation, which is influenced by the volume and pore size of the graphene electrode. The condition under which maximum specific capacitance can be obtained is when the size of pores of electrodes and electrolytic ions matches. Ma et al. [20] fabricated the MnO_2/graphene aerogel (GA) based nanocomposite via hydrothermal growth of 2D MnO_2 nanosheets on 3D-printed graphene aerogel along with their connected micropores

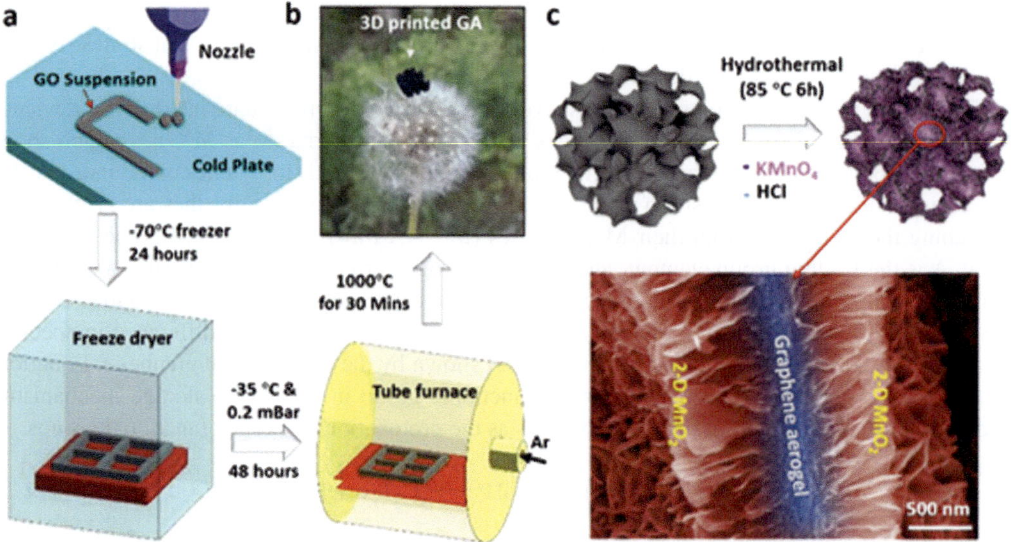

FIGURE 17.4 (a) The schematic diagram is shown along with the optical image in (b) represents the glowing bulb-like structure. (c) the SEM image of the obtained product. Adapted with permission [20]. Copyright (2019), Elsevier.

(Figure 17.4). The formation of surface voids and an insufficient amount of bond energy during the intermolecular diffusion process generates several issues which are removed by using the cold-plated freeze-drying method to fabricate unidirectional GA plates. The synthesized MnO_2/GA during electrochemical characterization shows a specific capacitance C_{sp} of 275 F/g at 1 A/g along with 99% capacitance retention after 5000 cycles at 5 A/g. The high capacitance retention is obtained due to the porous nature of GA. Wetting the electrode surface via electrolyte further helps to increase the contact area between the electrolyte and electrode. The real-time 3D printed MnO_2/GA SC shows the discharge-specific capacitance of 95 F/g at 1 A/g, presenting the realization of self-powered nanosystems.

Polymeric chains of PANI were fabricated into the defective sites of 2D materials to activate faulty electrochemical areas. However, PANI showed the thick and distorted structures that allow the migration of ions through the electrolyte, leading to poor cyclability and low charge-discharge rate. The semiconducting 2D materials, like black phosphorus (BP) and MXene, are suitable for the hybrid nanocomposites with the conducting PANI as they provide ample surface area. Vaghasiya et al. [21] synthesized a flexible SC electrode using 3D-printing polyjet technology for the BP/PANI-based nanocomposites. The prepared nanocomposite when subjected to Galvanostatic charge-discharge (GCD) and stability test showed a specific capacitance of 96 F/g at 0.5 A/g along with their flexibility check at 180 °C offering the retention in specific capacitance of 93.2 F/g at 0.5 A/g. The cyclic durability shows retention of 85.5% at 10 A/g after 4000 cycles. This decrease in retention is due to a large number of cycles which allows the slow insertion of ions across the BP/PANI clusters.

Although many composites have recently been investigated as electrodes for SCs, scientists found MXenes as an exciting class of 2D materials. They represent an emerging class for SC applications, with the general formula of $M_{n+1}X_nT_x$ where M is the transition metal, X is carbon or nitrogen, and T_x represents the surface contamination. They are generally produced by etching their different layers via HF treatment, as depicted in Figure 17.5.

Moreover, easy oxidation, restacking of the MXene layer, mechanical flexibility, and weak electrical conductivity are the drawbacks of pristine MXene. These limitations can easily be reduced via

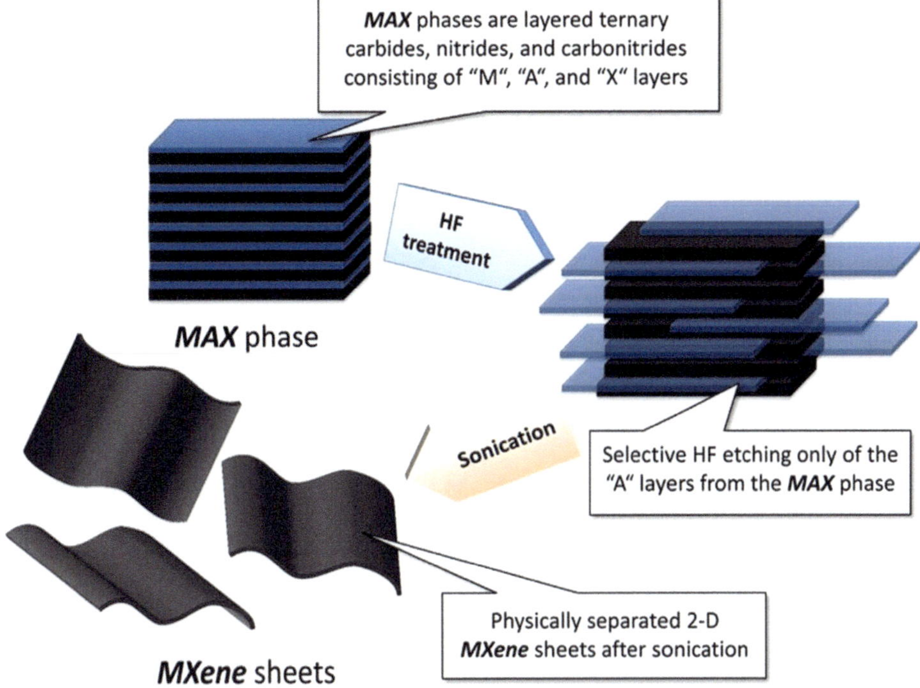

FIGURE 17.5 (a) Represents the MAX phase of MXenes and (b) HF-treated MXene sheets after sonication. Adapted with permission [16]. Copyright (2012), American Chemical Society.

fabricating nanocomposites of metals with MXenes. Their combination shows an admirable variety of properties, where additives can accommodate MXene layers and helps to suppress the restacking issue. The nanocomposites, along with MXenes, enhance the specific capacitance by five times compared to the pristine MXene. Furthermore, the exfoliation of MXene shows a stacked-layer structure that allows the intercalation of electrolytic ions, which further results in materials that can significantly store energy for various applications. In addition, MXenes have hydrophilic properties, resulting in higher electrode stability for the fabrication of SCs. Among the multiple types of MXenes, titanium carbides ($Ti_3C_2T_x$) demonstrate better cyclic stability and high volumetric capacitance. Depending on the various steps followed for the fabrication of MXene-based nanocomposites electrodes exhibit different performances and store an appreciable amount of energy. Li et al. [22] fabricated the MXene-silver nanowire-MnO_2-C-60 (MXene/AgNW/MnO_2/C-60) for supercapacitor applications. The as-prepared nanocomposites were further subjected to the 3D printing benchtop robot along with the ink controller, whose synthesis route is shown in Figure 17.6, and targeted for the electrochemical performance where GCD and stability were tested. The device showed a specific capacitance of 216.2 mF/cm^2 at 10 mV/s along with a high areal energy density of 19.2 μWh/cm^2 at a power density of 0.86 mW/cm^2. The real-time fabricated device showed capacitance retention of 85% after 10,000 cycles at 200 mV/s. The introduction of C-60 allows the proper reduction in friction between the adjacent layers of MXenes, decreasing the resistance and allowing them to slip over each other and the entire bulk material. This decrease in resistance enhances the capacitance due to a large number of ion transport behavior along with the interpenetrated hierarchical network structure in the printed electrodes.

Electrostatic interaction controls the weak gelation in MXenes, making it difficult to fabricate a variety of 3D MXene aerogels. Therefore, many scientists use an additive approach for the fabrication of MXenes. But adding additives leads to a high relative ink concentration, which further

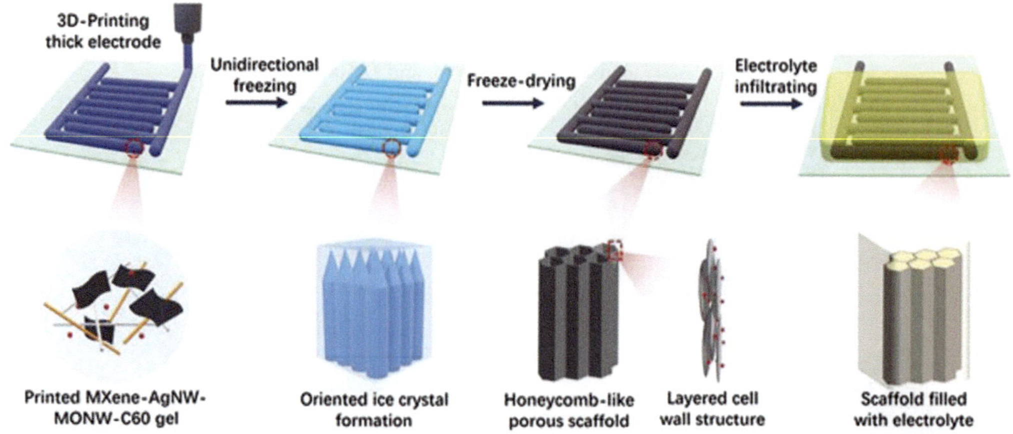

FIGURE 17.6 Synthesis route showing the fabrication of 3D printed MXene-based nanocomposites electrode. Adapted with permission [22]. Copyright (2020), John Wiley and Sons.

results in agglomeration of these nanosheets along with low diffusion kinetics and surface accessibility. To resolve the issue, Yang et al. [23] and his team fabricated an additive-free MXene structure using 3D printed template-assisted assembly. The exfoliation of the MXene sheet is followed by the freeze-drying process using a significant direct ink writing amount which is further followed by the hollow resin template via digital light processing 3D printing using stereolithography and insertion of Na-ions for the formation of the nanocomposites. The developed MXene microlattice structure was further transferred for the electrochemical testing which exhibited discharge-specific capacitance of 7.5 F/cm^2 at 0.5 mA/cm^2 with higher mass loading of 54.1 mg/cm^2 with an areal energy density of 0.38 mWh/cm^2 at a power density of 0.66 mW/cm^2. The higher mass loading effectively achieves high-performance SCs, providing a large surface-area-to-volume ratio and stopping the solid diffusion into the electrodes. However, high areal capacity might lead to the dead zone of active materials resulting in poor cyclability, which is overcome by the insertion of Na-ions into the MXene structure leading to the tortuosity-designable structure providing efficient ion-transportation and electrolyte permeation pathways.

17.3.2 Nanocomposites of Metal Oxides for 3D-Printed SCs

The porous nanostructure provides many active sites, fast electrode/electrolyte interaction, and ion transport/electron exchange, enhancing the SCs rate capability and power density. Besides all the factors, cost consideration is also an important key leading to considerable research for developing, designing, and synthesizing different metal oxide-based compounds and composites. Metal oxides generally show pseudocapacitive behavior for energy storage applications. They are widely explored due to their high theoretical specific capacitance and reversible redox reactions, leading to higher specific capacitance than carbon-based materials [24]. Combining these metal oxides with other compounds to form nanocomposites might result in a synergistic effect leading to their better performance than metallic oxide; the combination also features the enhancement of conductivity and energy density via multiple oxidation states and a conducting substrate. The noble metal oxide-based composite ruthenium oxide (RuO_2) with different combinations is preferred during the ongoing global research due to its theoretical specific capacitance ($\approx$ 2000 F/g) and long-life cycle. RuO_2 is so dense that ion/electron insertion and extraction are challenging, leading to the fall of its practical specific capacitance to 700 F/g on Ti substrate. However, its nanocomposite with CNT resulted in a specific capacitance of 1170 F/g [25]. Due to its low quantity in the Earth's crust, Ru has become an expensive material, and its commercialization has become difficult for scientists. Therefore, researchers look for different nanocomposites.

Among other metal oxides, manganese dioxide (Mn_xO_y) and its nanocomposites have attracted scientists' attention due to the high theoretical specific capacitance of manganese dioxide (~1370 F/g), low cost, and low toxicity. Its abundance in nature allows for overcoming the shortcomings issues of rare earth metals-based oxides. However, the Mn_xO_y electrode's performance and characteristics are limited due to its low conductivity. The crystal structure and crystallinity of Mn_xOy could affect the charge storage capacity. The effect of crystallinity can be seen in different levels of crystalline materials, such as a specific capacitance of 200 F/g at 1 A/g was observed for poorly crystalline Mn_xO_y which could be due to the intergrowth of a distinct tunnel leading to difficulty in cation diffusion and results in a higher resistance, whereas, the highly crystalline Mn_xO_y showed a specific capacitance of 539 F/g. The energy storage process in Mn_xO_y is mainly owned by the transition between the different oxidation states and the reversible redox reactions. The microstructure is largely responsible for cycling stability, whereas chemically hydrated conditions are primarily responsible for high specific capacitance. The combination of Mn_xO_y with various materials shows improvement in crystallinity and enhances the electrical conductivity of the $Mn_xO_{y,}$ which further enhances their specific capacitance. Rezaei et al. [26] synthesize the Mn_3O_4/pyrolytic carbon (PC) nanocomposites using the acrylated monomers to print complex 3D structures via a pyrolysis furnace. The obtained complex was further doped by the Mn_xO_y using a one-step wet chemical reduction method, as shown in Figure 17.7. The obtained structure during the SC performance showed a specific capacitance of 186 F/g at 0.5 mA/cm^2. Moreover, they leach the sample and perform the testing in an HCl electrolyte, where the sample exhibits a specific capacitance of 457 F/g at 0.5 F/cm^2 and capacitance retention of more than 92% after 5000 cycles. The prepared electrodes demonstrate high cyclability due to the substrate's low ionic and intrinsic resistance, allowing the high capacitance performance.

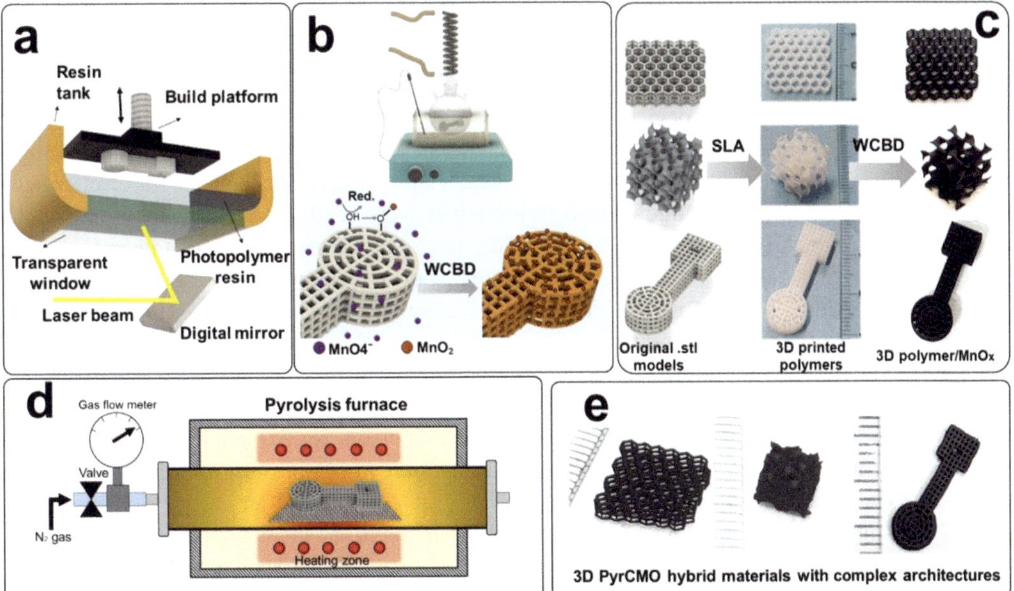

FIGURE 17.7 The schematic diagram (a) represents the free-standing procedure configuration. (b) Shows the synthesis of MnO_2 on the 3D printed polymer structure. (c) The various steps of the model and their printed structure. (d) Shows the pyrolysis and drying process if synthesized 3D model. (e) Magnified image of the as-prepared electrodes. Adapted with permission [26]. Copyright (2022), American Chemical Society.

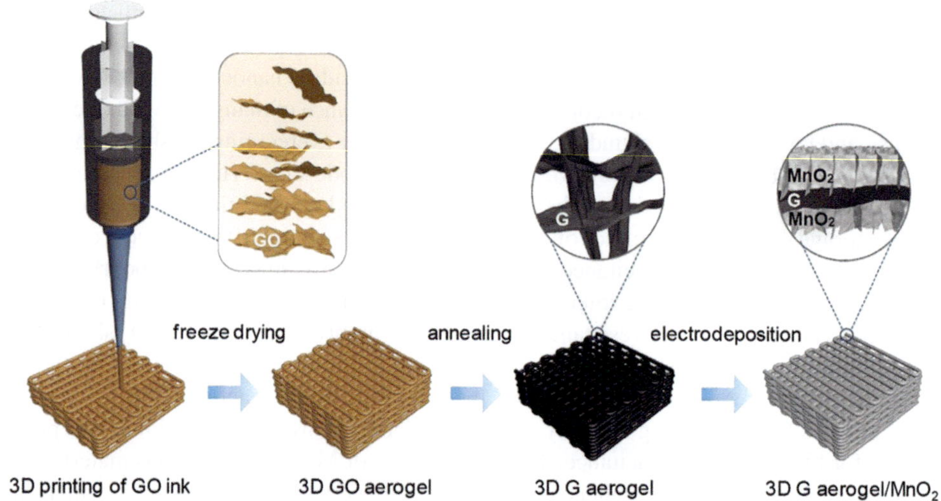

FIGURE 17.8 The schematic diagram represents the 3D G aerogel/MnO_2. Adapted with permission [27]. Copyright (2018), Elsevier.

Yao et al. [27] fabricated the MnO_2/graphene aerogel (3D-G) using single-layer GO sheets which were sonicated and transferred to the syringe barrel for extrusion through a 400 μm nozzle to obtain a 3D structure on a glass substrate. The accepted 3D structure was further annealed to form 3D-G aerogel on which MnO_2 nanosheets were electrodeposited, as shown in Figure 17.8. The obtained product was subjected to electrochemical performance with the high mass loading of 45.2 mg/cm^2 on electrodes delivered an areal and volumetric capacitance of 115.5 F/cm^2 and 11.55 F/cm^3 at 0.5 mA/cm^2, demonstrating excellent electrochemical performance and rate capability. The 3D-G aerogel provided a large surface area due to the porosity and periodic arrangement of the pores allowing the fast diffusion of ions into the surface of the electrode. The multiple oxidation states of MnO_2 on the surface of 3D-G aerogel showed the hybrid capacitive behavior enhancing the performance of the real-time fabricated devices.

Iron (Fe) is another readily available material having three different oxidation states, allowing it to form multiple oxides (Fe_2O_3, Fe_3O_4, and FeO), which shows excellent behavior for electrochemically stored energy. In past, solution-based technologies were used to generate many iron oxides and composites, which were then used as electrode materials for electrochemical energy storage. However, iron oxide has relatively high conductivity, but a low specific capacitance compared to other metal oxides, limiting its practical usage. Researchers devote time to eliminating these drawbacks via incorporating high-conductive materials, increasing the surface area by designing the nanostructure, and combining with different materials to form Fe-based nanocomposites. Lai et al. [28] synthesized the nitrogen-doped carbon network (C) along with iron oxide (Fe_2O_3) to create the nanocomposites (N-C/Fe_2O_3). The required precursors for 3D printing were fabricated via various synthesis routes, and the prepared ink was used in 3D printing to obtain 3D-printed electrodes. The fabricated electrode exhibited 377 F/g at 2 mV/s and an energy density of 114 Wh/kg. The high electrochemical performance was attributed to the cross-coupled microstructures representing their homogeneous anchoring which helped to avoid the dead zones of the materials. The 3D printed material containing a carbon source showed an excellent ionic absorbing tendency due to the inhibition of the agglomeration of ions. The structural transmission of Fe_2O_3 exhibited the triangular GCD curves indicating its pure capacitive behavior and identical charging and discharging curves, resulting in symmetrical SCs. The nanocomposites show high performance, the anion species exhibited strong specific adsorption in all solvents, and the additional charge-storage

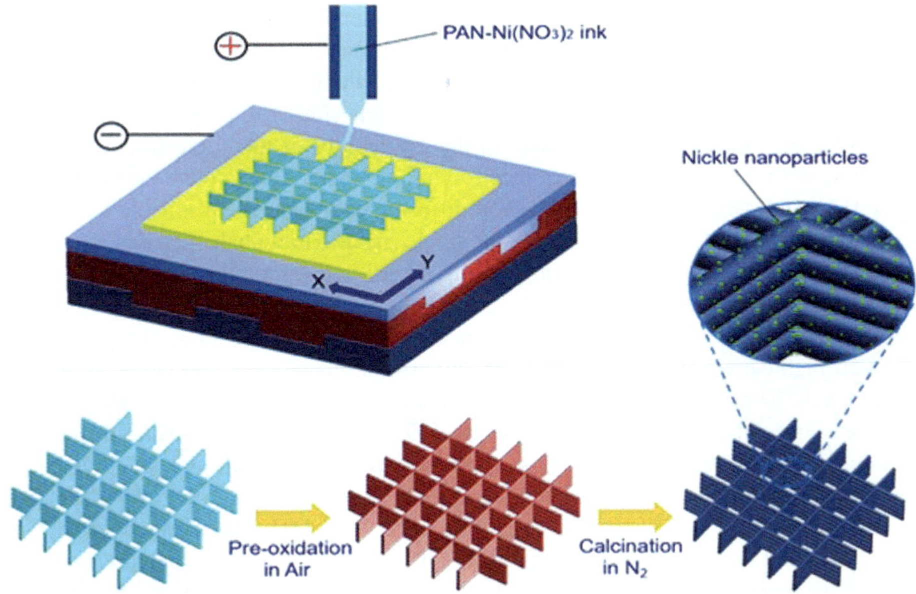

FIGURE 17.9 The visual diagram indicates the EHD 3D printing route for the Ni-based nanocomposites. Adapted with permission [29]. Copyright (2021), Elsevier.

capacity was due to the Fe_2O_3 coating. As Fe-based materials demonstrate sluggish kinetic reactions at a higher rate, therefore, researchers are looking forward to various types of nanocomposites which can elucidate the issue of low specific capacitance and poor stability.

Among multiple metals, nickel-based composites show better performance compared to other nanocomposites as they improve the charge storage capacity and form the nanopores due to the decomposition resulting in high conductivity and fast ionic transport through the electrolyte. Moreover, the C-doped nickel oxide (NiO) based nanocomposite was fabricated by Zhang et al. [29] via the facile electrohydrodynamic (EHD) and used for 3D printed C/NiO networked as SCs electrode. The schematic synthesis route is shown in Figure 17.9, indicating the array-like structure. The obtained electrodes in electrochemical characterization exhibited a specific capacitance of 79.1 F/g at 60 μm spacing between the fibers. The spacing between the fibers allows the interaction between the electrode and electrolyte interface resulting in fast and quasi-reversibility during the CV curve, exhibiting the contribution of both carbon and NiO. The NiO-based nanocomposites store charges via fast redox reactions at the electrode/electrolyte interface, indicating that they could be used as high-performance and low-cost electrode materials in SC devices. The pre-oxidation of $Ni(NO_3)_2$ ink onto a substrate forms the porous structure. This porous structure with open micropores help in the penetration of electrolyte and allows penetration to produce a fast redox reaction and a shorter ionic diffusion path, which plays a vital role in enhancing the capacitive behavior of these metal oxide-based nanocomposites.

17.3.3 Nanocomposites of Metal Sulfides for 3D-Printed SCs

Recently, metal dichalcogenides (MCs) have attracted researchers due to their 2D layered structure, high conductivity, and widespread applications in charge storage devices. These dichalcogenides form multiple layers, and their defective engineering allows researchers to tune the properties according to their applications. Most MCs are prepared using the selective synthesis approach allowing researchers to prepare these compounds widely. MC combines transition metal (TM) elements from Groups IV-VII B of the periodic table with elements from Group VIA, such as sulfur, selenium, and tellurium, to form two crystalline layers of material. The general formula of MCs is MX_2, where M is a transition metal element, and **X** is a chalcogen atom. The bandgap of MC type materials lies between ~ 0–2 eV, and the crystal structure is similar to pure graphene, having zero bandgaps. MCs are chemically reactive due to the availability of many electrochemically active surfaces, along with deeper penetration of electrolytes and metal ions into the electrode's intercalation sites. High electrical conductivity and short diffusion path make them suitable for applications in SCs. However, these 2D layered materials or single metal-based chalcogenides are unstable as their stability is generally phase-dependent, making it hard to integrate them into real SCs devices. These days, scientists are looking for stable and highly electrochemically active nanocomposites of metal sulfides, which can be achieved by using binary chalcogenides. The exceptional stability is due to their high electrical conductivity and more prosperous redox-active sites. Song et al. [30] take advantage of 3D printing and fabricated the reticulated lattice structure possessing mesoporous characteristics via octet-truss lattice (OTL) substrate. The OTL substrate was then subjected to electroless plating through a stereolithography process to fabricate the cobalt-nickel sulfide ($CoNi_2S_4$)/OTL and dip coating of carbon materials (CM-like CNT) for enhancing the cyclic stability. The process is described in Figure 17.10. While growing $CoNi_2S_4$ on OTL, the precursors have OH^- ions originating from H_2O decomposition in the anode. The obtained cascade structure maintains the continuity of the conductive materials and uniformity in the deposition of ions on the surface.

The BET test revealed the size of powdered $CoNi_2S_4$ of 1 nm, providing several routes for transmitting electrons and ions between the electrolyte and activated materials. The fabricated electrodes were further transferred for the electrochemical characterization where the CV curve indicated redox behavior of Co^{2+}/Co^{3+} and Ni^{2+}/Ni^{3+} based on Equations (17.3) and (17.4):

$$CoNi_2S_4 + 2OH^- \leftrightarrow CoS_{2x}OH + 2NiS_{2-x}OH + 2e^- \quad (17.3)$$

$$CoS_{2x}OH + OH^- \leftrightarrow CoS_{2x}O + H_2O + e^- \quad (17.4)$$

The GCD shows the symmetrical curves indicating the high coulombic efficiency at a lower electrodeposition time, but as soon as the electrodeposition time increases, the discharge time decreases, leading to the agglomeration of $CoNi_2S_4$ nanosheets referring to dead volumes on the surface of the electrode. Therefore, deposition timing plays a significant role in the performance of the SCs. A deposition time of 15 min is quantitively optimized for the SC electrodes indicating a specific capacitance of 3.17 F/cm^3 (or 1216 F/g) at 5 mA/cm^3. These excellent results tend researchers to fabricate the lattice asymmetric all-solid-state SC (LASC) using two electrodes, $CoNi_2S_4$/Ni/OTL and CM/Ni/OTL, as anode and cathode. The CM contains the CNT, whose amount decides the performance of LASCs as the amount of CNT increases the stacking phenomena, leading to a specific capacitance of 597.6 mF/cm^3 (or 157.6 F/g) at 1 mA/cm^3 with 40% capacitance retention. Still, two parameters remain challenging for 3D-printed SCs, i.e., a well-designed framework to provide a large surface area with high mass loading and enough toughness to provide repeated deformations. Therefore, Mei et al. [31] use photo-polymerization of liquid-sensitive materials through UV light, destroying the limit caused by other 3D-printing technology. This photo-polymerization allows for the development of the auxetic type of structure that contracts on compression and expands laterally during stretching which further provides strength to the overall system of ideal electrode framework for SC applications. The fabrication of $CoNi_2S_4$/NiCo-based nanocomposite was done after designing the physically developed layer. The dopping of Ni via electroless deposition is followed by the electrodeposition of NiCo on its surface and the secondary deposition of $CoNi_2S_4$ on the 3D-developed electrode. The nano-sized crystalline electrode having a high surface area was subjected to optimize the electrochemical performance. Ni and Co ion during electrochemical testing confirms the reversible redox activity and the anodic and cathodic peaks through CV testing. The

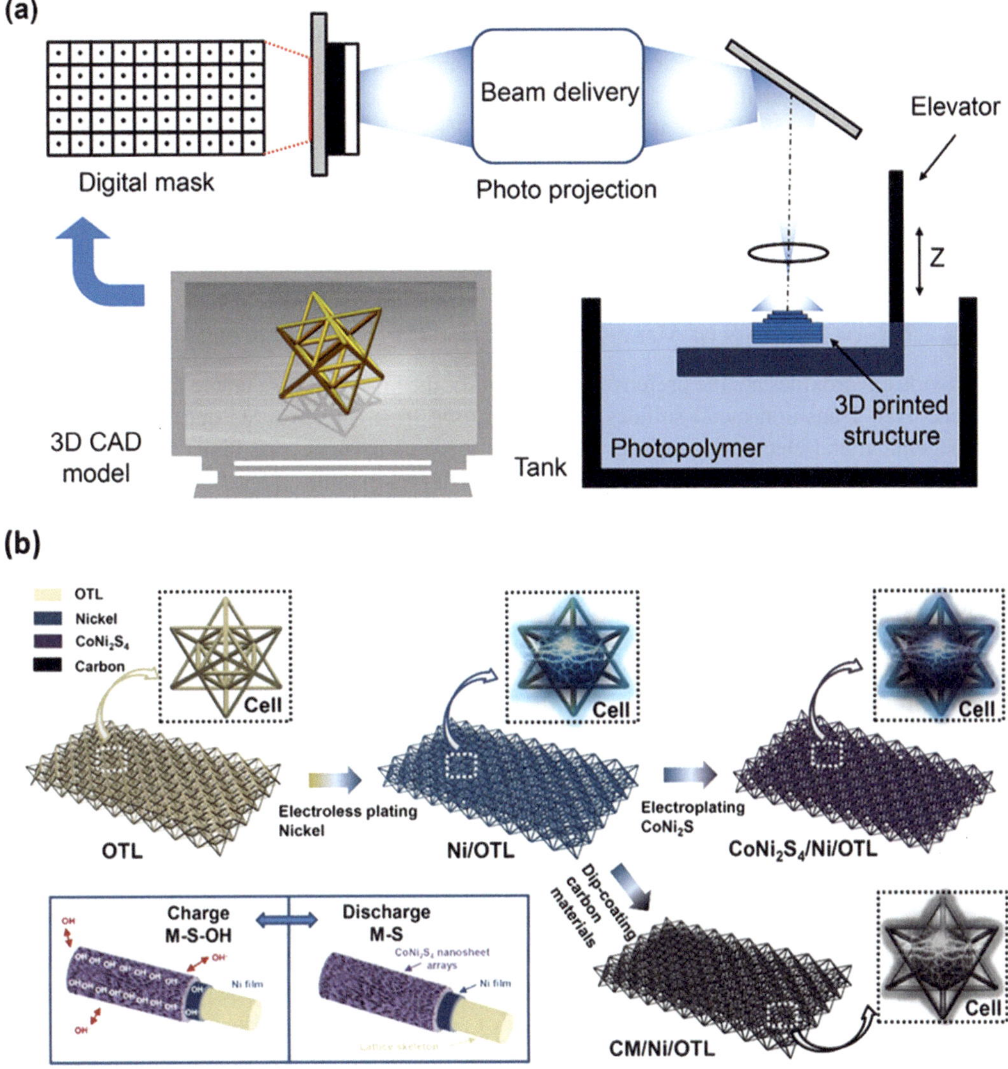

FIGURE 17.10 (a) The schematic illustration of the stereolithography process using the photosensitive layer structure. (b) The diagram indicates the fabrication route for the $CoNi_2S_4$/Ni/OTL and CM/Ni/OTL. Adapted with permission [30]. Copyright (2018), American Chemical Society.

stretchable synthesized electrode facilitates the transfer of ions and electrons between the electrolyte and electrode interface resulting in discharge volumetric C_{sp} of 28.5 F/cm^3 along with 92.2% capacitance retention. This implies strong and cohesive forces between the hierarchical nanocomposite and 3D current collectors. The lowest calculated resistance of 1.38 Ohm indicates the synergistic effect between these composites.

The other nanocomposite attracting researchers is copper (Cu)-based nanocomposite due to the naturally occurring state of Cu being corrosion-resistant to improve the SCs cyclability. The composite of Cu helps achieve high stability and large porosity for high-performance SCs. Tung et al. [32] use freeze gelation and 3D printing technology to fabricate rGO/copper-cobalt sulfide ($CuCo_2S_4$)-based nanocomposite electrode. They use the hydrothermal method for the fabrication of rGO/$CuCo_2S_4$, where rGO was prepared using a modified hummer's method. The formation of voids due to the presence of phenol is observed leading to an increase in the interfacial area and, hence, the capacitance value. Furthermore, the CV curve indicates the pseudocapacitive behavior by the redox peaks of both pairs Cu^{2+}/Cu^{+} and Co^{4+}/Co^{3+}, and the rectangular curve signifies the EDLC behavior. A specific capacitance of 1123 F/g at 5 mV/s using the CV curve while the GCD displayed a specific capacitance of 417 F/g at 125 A/g with 91.2% capacitance retention after 20,000 cycles. The reason for such a high specific capacitance was the low charge-transfer resistance (R_{ct} of 1.035 Ohm) offered by the electrode during the electrochemical activity. Moreover, the efficient approach for developing the metal sulfide for 3D-printed SC with a well-aligned structure helps outperform other nanocomposites.

17.3.4 Nanocomposites of Metal Phosphide for 3D-Printed SCs

Metal oxides/hydroxides are the preferred candidates for the kinetically occurring fast electron transport reactions. Transition metal sulfides are widely used in asymmetric SC due to their excellent electrochemical and electrical conductivity. Among these compounds, the bimetallic sulfide-based composites outrage the metallic sulfides, but their single-layer structure and parallel composition limit their specific capacity and electrochemical properties. The rational design of these nanostructured composites allows them to have a high specific surface area boosting their electrochemical properties. Nowadays, researchers focus on metal phosphide-based compounds and their nanocomposites due to their superior electrical conductivity and metalloid character. These metal phosphide-based composites facilitate elemental doping, and their nanostructure allows to fasten the kinetics of redox reactions. The foreign transition-metal ions modify the electronic structure and realign valence electrons to facilitate electron transfer; also, the large number of active sites provide a higher surface area leading to suitable composites for the 3D-printed SC devices. Generally, transition metal phosphide-based composites have high- theoretically specific capacitance only due to their better redox activity and ideal electrical conductivity. For example, the structure of Ni_{2p} is compared with Ni fcc structure, but when $Ni_{2p3/2}$ and $P_{2p3/2}$ are combined, their binding energies (BE) are substantially lower than that of oxide compounds, resulting in lower electric susceptibility $\Delta\chi$ leading the higher conductivity than oxides of the same element. Ni-P has different phases allowing them to form crystalline and amorphous morphology leading them towards the hybrid type of performance.

Xue et al. [33] used stereolithography-based technology due to its low cost, and ease of processability for the fabrication of the complex polymer-based structure. The octet-truss topology-based SC has been developed using the substrate of acrylic polyester, followed by the electroless deposition of nickel phosphide (NiP)-based compounds. The NiP was further targeted for the deposition of rGO to obtain NiP/rGO as shown in Figure 17.11. The obtained octet-truss system exhibits high mechanical strength and structural integrity after the coating of Ni_2P and could be easily scalable. The electrochemical characterization of the fabricated real-time SC reveals the high specific capacitance of 57.75 mF/cm^2 at 2 mA/cm^2 along with 70% capacitance retention when moving to 2–40 mA/cm^2 and 96% life cyclability after 5000 cycles.

The above study opens a new door for the detailed discussion on numerous research which provides information about the phosphide-based 3D-printed SCs. Among them, Yu et al. [34] and his team fabricated the bimetallic nickel-cobalt-phosphide (NiCoP)/MXene-based nanocomposite using the *in-situ* coupling. The 3D printed technology is used for the fine printing of electrodes where the NiCoP bimetallic compound along with MXene is prepared using the hydrothermal synthesis route, and the ink is prepared using CNT suspension. The developed electrodes are further subjected for application in SCs where they exhibit a specific capacitance of 1359 F/g at 1A/g as well as the areal and volumetric capacitance was calculated to 20 F/cm^2 and 137 F/cm^3 at 10 mA/cm^2 along with the 90% charge retention after 5000 cycles at 100 mA/cm^2 current density. The reason for the high specific capacitance lies behind the doping of MXene and phosphorus, creating a large number of vacancies and allowing the incorporation of electron/ion transport easier. The high

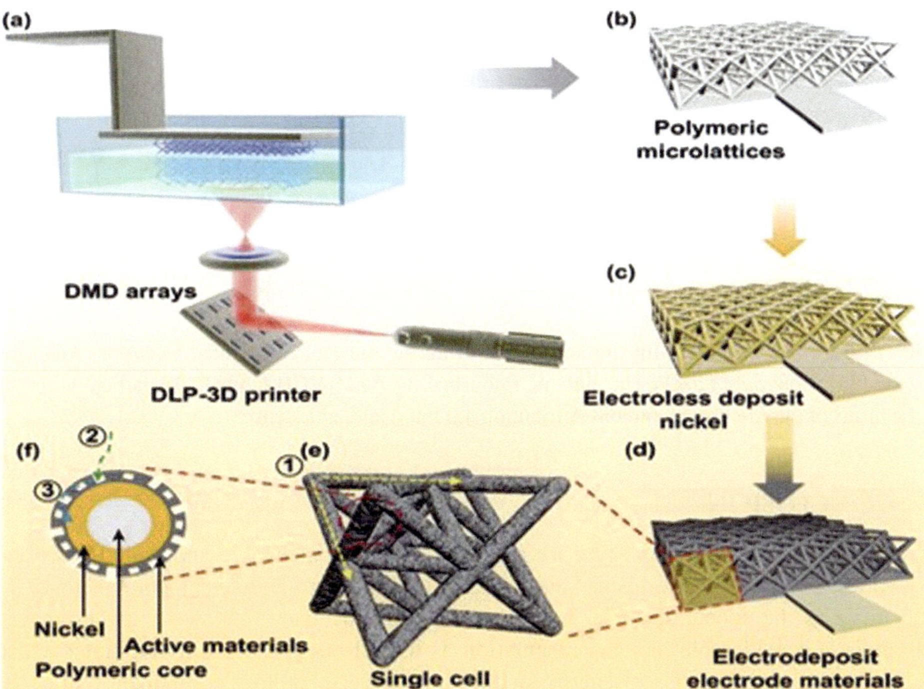

FIGURE 17.11 The schematic diagram (a–d) indicates the actual synthesis route for the preparation of 3D structure. (e–f) Depicts the magnified structure of composite. Adapted with permission [33]. Copyright (2019) The authors. Published by Springer. This article is distributed under the terms of the Creative Commons Attribution 4.0 International License.

stability of phosphide-based compounds allows the symmetrical CV curves to indicate the complete suppression of the available oxygen inside the real-time fabricated ASC. This ASC exhibits the non-linear CV curves indicating the hybrid type of performance, which further results in 3.29 F/cm^2 at 10 mA/cm^2 and capacitance retention of 88% after 5000 cycles. The as-synthesized ASC exhibits the areal and volumetric energy density of 0.89 mWh/cm^2 and 2.2 mWh/cm^3. The stable conductivity has been depicted in the high-frequency region along with high ionic diffusivity in the low-frequency region, which is demonstrated by the slope change lines during the cyclability. Among the variety of nanocomposites, molybdenum phosphide (MoP)-based nanocomposites are also suitable for charge storage devices due to their adjustable engineering and particle size. Zhong et al. [35] used surface anchoring for the diffusion of potassium (K^+) to reduce the volume expansion of the MoP during the charge/discharge process. They synthesized the bulk MoP using the hydrothermal method followed by 3D printed process which is depicted in Figure 17.12. In addition, they show the interconnected networks along with the mechanically stable structure allowing it to transport an adequate amount of charge.

Mo^{4+}/Mo^{6+} when incorporated with C, provides large mesoporous structures to the composite for the transport of ions and storing of more charges. The size of the MoP nanoparticle doped with C enhances the electrochemical properties. A specific capacitance of 256.1 mAh/g at 0.1 A/g was observed. The MoP@C nanocomposites indicate the synergistic effect between them due to the formation of uniform bonds. The nanoparticle size allows the crystallite to develop extra active sites for different energy levels capturing K^+ ions. As the particle size decreases, it converts the covalent bond between the MoP to the ionic, which further creates the heterojunction between the MoP@C nanocomposites, enhancing the electrolyte adsorption as the electronic conductivity.

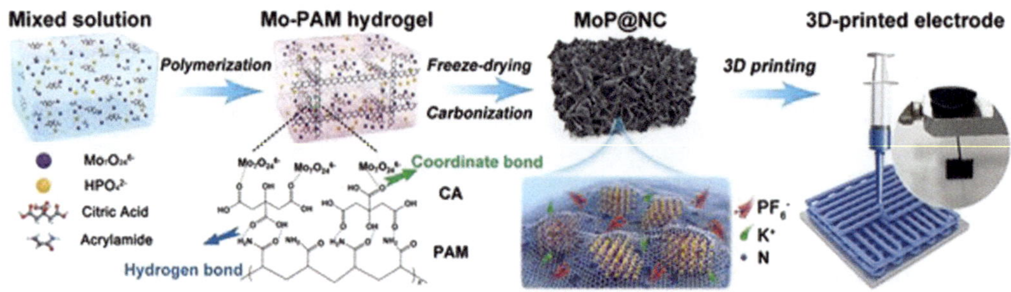

FIGURE 17.12 The schematic illustration for the synthesis route of 3D printed electrode. Adapted with permission [35]. Copyright (2021) The authors. Published by Wiley-VCH GmbH. This article is distributed under the terms of the Creative Commons Attribution 4.0 International License.

17.4 CONCLUSION

It is seen that energy capacity gadgets are moving towards an adaptable and scaled-down variant of conventional hardware. Alongside that, picking the most proficient nanocomposites is a significant piece of the supercapacitor's energy storage enhancement. Likewise, gathering various SCs assumes a considerable part in presentation as hybrid ones have exhibited better execution in contrast with different kinds of energy stockpiling gadgets. Another component that considers nanocomposite cathodes appealing candidates is their structure and morphology, which can be effectively integrated. Most of the reports portray the utilization of carbon-based nanomaterials to form nanocomposites. However, there is a rising number of logical articles effectively consolidating different materials, for example, metal oxides, progress metal dichalcogenides, MXenes, etc. Likewise, forming a composite with these nanomaterials has exhibited a critical improvement in the general electrochemical execution of their separate cathodes. There are still a few difficulties to conquer, like working on the centralization of electroactive composites, diminishing the utilization of expensive surfactants, and expounding procedures that can be performed with nothing or common misuse of parts. Despite these disadvantages, the opportunities for assembling electronic gadgets, in a perfect world in any state of 2D or 3D designs that can be bowed or wound somewhat, close by being more modest in size while conveying parallel execution as their bulkier and unbending partners it is the subsequent stage on mechanical development for these gadgets. Apart from this, efforts to widen the range of applications of these nanocomposite materials are being adapted for fuel cells, sensors, or biological applications, even though those are still in the infant phase. Through that, it is likely that a considerable part of research in the field of energy storage application and reasonable that a significant piece of the examination in a lot of energy will be committed to adaptable SCs.

REFERENCES

1 L.L. Zhang, X.S. Zhao, Carbon-based materials as supercapacitor electrodes, *Chem. Soc. Rev.* 38 (2009) 2520–2531.
2 F. Béguin, V. Presser, A. Balducci, E. Frackowiak, Carbons and electrolytes for advanced supercapacitors, *Adv. Mater.* 26 (2014) 2219–2251.
3 M. Winter, R.J. Brodd, What are batteries, fuel cells, and supercapacitors?, *Chem. Rev.* 104 (2004) 4245–4269.
4 S. Chu, A. Majumdar, Opportunities and challenges for a sustainable energy future, *Nature.* 488 (2012) 294–303.
5 C. Liu, F. Li, L.-P. Ma, H.-M. Cheng, Advanced materials for energy storage, *Adv. Mater.* 22 (2010) E28–E62.

6 T. Brousse, D. Bélanger, D. Guay, Asymmetric and hybrid devices in aqueous electrolytes, in: *Supercapacitors: Materials, Systems, and Applications*, F. Béguin, E. Frąckowiak (Eds.), Wiley-VCH Verlag GmbH & Co. KGaA, 2013: pp. 257–288.
7 O.M. Yaghi, H. Li, Hydrothermal synthesis of a metal-organic framework containing large rectangular channels, *Am. Chem. Soc.* 117 (1995) 10401.
8 Z. Wang, Q. Zhang, S. Long, Y. Luo, P. Yu, Z. Tan, J. Bai, B. Qu, Y. Yang, J. Shi, H. Zhou, Z.Y. Xiao, W. Hong, H. Bai, Three-dimensional printing of polyaniline/reduced graphene oxide composite for high-performance planar supercapacitor, *ACS Appl. Mater. Interfaces.* 10 (2018) 10437–10444.
9 P. Makkar, A. Malik, N.N. Ghosh, Biomass-derived porous carbon-anchoring $MnFe_2O_4$ hollow sphere and needle-like NiS for a flexible all-solid-state asymmetric supercapacitor, *ACS Appl. Energy Mater.* 4 (2021) 6015–6024.
10 D. Gogoi, P. Makkar, M.R. Das, N.N. Ghosh, $CoFe_2O_4$ nanoparticle decorated hierarchical biomass derived porous carbon based nanocomposites for high-performance all-solid-state flexible asymmetric supercapacitor devices, *ACS Appl. Electron. Mater.* 4 (2022) 795–806.
11 S. Bhardwaj, F. Martins de Souza, R.K. Gupta, 4 – High-performance lithium–sulfur batteries: role of nanotechnology and nanoengineering, in: *Lithium-Sulfur Batteries: Materials, Challenges and Applications*, R.K. Gupta, T.A. Nguyen, H. Song, G.B.T.-L.-S.B. Yasin (Eds.), Elsevier, 2022: pp. 57–73.
12 Y. Liu, Y. Wang, H. Wang, P. Zhao, H. Hou, L. Guo, Acetylene black enhancing the electrochemical performance of NiCo-MOF nanosheets for supercapacitor electrodes, *Appl. Surf. Sci.* 492 (2019) 455–463.
13 D.Y. Lee, S.J. Yoon, N.K. Shrestha, S.-H. Lee, H. Ahn, S.-H. Han, Unusual energy storage and charge retention in Co-based metal–organic-frameworks, *Microporous Mesoporous Mater.* 153 (2012) 163–165.
14 T. Sun, L. Yue, N. Wu, M. Xu, W. Yang, H. Guo, W. Yang, Isomorphism combined with intercalation methods to construct a hybrid electrode material for high-energy storage capacitors, *J. Mater. Chem. A.* 7 (2019) 25120–25131.
15 B. Anasori, M.R. Lukatskaya, Y. Gogotsi, 2D metal carbides and nitrides (MXenes) for energy storage, *Nat. Rev. Mater.* 2 (2017) 16098.
16 M. Naguib, O. Mashtalir, J. Carle, V. Presser, J. Lu, L. Hultman, Y. Gogotsi, M.W. Barsoum, Two-dimensional transition metal carbides, *ACS Nano.* 6 (2012) 1322–1331.
17 M.R. Lukatskaya, S. Kota, Z. Lin, M.- Q. Zhao, N. Shpigel, M.D. Levi, J. Halim, P.- L. Taberna, M.W. Barsoum, P. Simon, Y. Gogotsi, Ultra-high-rate pseudocapacitive energy storage in two-dimensional transition metal carbides, *Nat. Energy.* 2 (2017) 17105.
18 M. Ghidiu, M.R. Lukatskaya, M.- Q. Zhao, Y. Gogotsi, M.W. Barsoum, Conductive two-dimensional titanium carbide 'clay' with high volumetric capacitance, *Nature.* 516 (2014) 78–81.
19 C. Liu, Z. Yu, D. Neff, A. Zhamu, B.Z. Jang, Graphene-based supercapacitor with an ultrahigh energy density, *Nano Lett.* 10 (2010) 4863–4868.
20 C. Ma, R. Wang, H. Tetik, S. Gao, M. Wu, Z. Tang, D. Lin, D. Ding, W. Wu, Hybrid nanomanufacturing of mixed-dimensional manganese oxide/graphene aerogel macroporous hierarchy for ultralight efficient supercapacitor electrodes in self-powered ubiquitous nanosystems, *Nano Energy.* 66 (2019) 104124.
21 J.V. Vaghasiya, K. K-ípalová, S. Hermanová, C.C. Mayorga-Martinez, M. Pumera, real-time biomonitoring device Based on 2D black phosphorus and polyaniline nanocomposite flexible supercapacitors, *Small.* 17 (2021) 1–8.
22 X. Li, H. Li, X. Fan, X. Shi, J. Liang, 3D-printed stretchable micro-supercapacitor with remarkable areal performance, *Adv. Energy Mater.* 10 (2020) 1–12.
23 C. Yang, X. Wu, H. Xia, J. Zhou, Y. Wu, R. Yang, G. Zhou, L. Qiu, 3D printed template-assisted assembly of additive-free Ti_3C_2Tx MXene microlattices with customized structures toward high areal capacitance, *ACS Nano.* 16 (2022) 2699–2710.
24 J. Li, Z. Liu, Q. Zhang, Y. Cheng, B. Zhao, S. Dai, H.-H. Wu, K. Zhang, D. Ding, Y. Wu, M. Liu, M.-S. Wang, Anion and cation substitution in transition-metal oxides nanosheets for high-performance hybrid supercapacitors, *Nano Energy.* 57 (2019) 22–33.

25 C.D. Lokhande, D.P. Dubal, O.S. Joo, Metal oxide thin film based supercapacitors, *Curr. Appl. Phys.* 11 (2011) 255–270.

26 B. Rezaei, T.W. Hansen, S.S. Keller, Stereolithography-derived three-dimensional pyrolytic carbon/ Mn_3O_4 nanostructures for free-standing hybrid supercapacitor electrodes, *ACS Appl. Nano Mater.* 5 (2022) 1808–1819.

27 B. Yao, S. Chandrasekaran, J. Zhang, W. Xiao, F. Qian, C. Zhu, E.B. Duoss, C.M. Spadaccini, M.A. Worsley, Y. Li, Efficient 3D printed pseudocapacitive electrodes with ultrahigh MnO_2 loading, *Joule.* 3 (2019) 459–470.

28 F. Lai, C. Yang, R. Lian, K. Chu, J. Qin, W. Zong, D. Rao, J. Hofkens, X. Lu, T. Liu, Three-phase boundary in cross-coupled micro-mesoporous networks enabling 3D-printed and ionogel-based quasi-solid-state micro-supercapacitors, *Adv. Mater.* 32 (2020) 1–9.

29 B. Zhang, J. He, G. Zheng, Y. Huang, C. Wang, P. He, F. Sui, L. Meng, L. Lin, Electrohydrodynamic 3D printing of orderly carbon/nickel composite network as supercapacitor electrodes, *J. Mater. Sci. Technol.* 82 (2021) 135–143.

30 J. Song, Y. Chen, K. Cao, Y. Lu, J.H. Xin, X. Tao, Fully controllable design and fabrication of three-dimensional lattice supercapacitors, *ACS Appl. Mater. Interfaces.* 10 (2018) 39839–39850.

31 P. Chang, H. Mei, Y. Tan, Y. Zhao, W. Huang, L. Cheng, A 3D-printed stretchable structural supercapacitor with active stretchability/flexibility and remarkable volumetric capacitance, *J. Mater. Chem. A.* 8 (2020) 13646–13658.

32 D.T. Tung, L.T.T. Tam, H.T. Dung, N.T. Dung, P.N. Hong, H.M. Nguyet, N. Van-Quynh, N. Van Chuc, V.Q. Trung, L.T. Lu, P.N. Minh, Freeze gelation 3D printing of rGO-$CuCo_2S_4$ nanocomposite for high-performance supercapacitor electrodes, *Electrochim. Acta.* 392 (2021) 138992.

33 J. Xue, L. Gao, X. Hu, K. Cao, W. Zhou, W. Wang, Y. Lu, Stereolithographic 3D printing-based hierarchically cellular lattices for high-performance quasi-solid supercapacitor, *Nano-Micro Lett.* 11 (2019) 1–13.

34 L. Yu, W. Li, C. Wei, Q. Yang, Y. Shao, J. Sun, 3D Printing of NiCoP/Ti_3C_2 MXene architectures for energy storage devices with high areal and volumetric energy density, *Nano-Micro Lett.* 12 (2020) 1–13.

35 W. Zong, N. Chui, Z. Tian, Y. Li, C. Yang, D. Rao, W. Wang, J. Huang, J. Wang, F. Lai, T. Liu, Ultrafine MoP nanoparticle splotched nitrogen-doped carbon nanosheets enabling high-performance 3D-printed potassium-ion hybrid capacitors, *Adv. Sci.* 8 (2021) 1–11.

18 3D-Printed Carbon-Based Nanomaterials for Sensors

A.A.P.R. Perera[1,2], K.A.U. Madhushani[1,2], Buwanila T. Punchihewa[3], Felipe M. de Souza[2], and Ram K. Gupta[1,2]

[1]Department of Chemistry, Pittsburg State University, Pittsburg, Kansas 66762, USA

[2]National Institute for Materials Advancement, Pittsburg State University, Pittsburg, Kansas 66762, USA

[3]Department of Chemistry, University of Missouri-Kansas City, Missouri 64110, USA

*Corresponding author: ramguptamsu@gmail.com

CONTENTS

18.1 INTRODUCTION

The entire spectrum of human activities is full of sensors and their applications, from the switching of home electronics equipment to the manufacture of advanced military and healthcare devices. These sensors make it easier for people to live in the modern world [1]. By definition, a sensor is a device that detects or measures the variations or modifications of physical, chemical, or biological properties and gives the output as a readable electrical signal. Historically, sensor technology was founded in the nineteenth century, and later, thousands or more different types of sensors were developed and deployed with various successes. Over the years sensor technology has been used to improve the sensitivity, accuracy selectivity, precision of many automated electronic devices (AED), and resolution with rapid response times [2]. This is because the sensor is inevitably used as a monitoring unit in many AEDs, making it an important tool in the efficient operation of every household or industrial activity today as well as in the future. However, sensors have significant limitations such as expensive fabrication methods, instruments, need for trained operators, special materials, etc, which constrain their widespread use. Thus, new fabricating strategies are required to meet the expected goals. Thereby, the processes of sensor fabrication have evolved exponentially

DOI: 10.1201/9781003296676-18

over the past few decades. Laboratory and commercial-related sensors are being used to manufacture using 3D-printing techniques, which could be a turning point in this field [2]–[5].

The performance of a sensor is primarily determined by the electromechanical properties of its conductive materials. Sensors have now been developed using various conductive materials such as metals, conductive polymers, and carbon-based materials as sensing elements. However, metal electrodes are expensive, and in this context, much attention is focused on using them as carbon-based composite filaments to develop cost-effective sensing platforms. Also, the inclusion of nanomaterials in the sensor element has created a valuable opportunity to develop miniaturized sensors with superior sensor characteristics. Carbon-based nanomaterials, such as CNT and GR, have attracted the attention of scientists because of their ability to covalently binds with other materials. It leads to high strength, high density, and high hardness of the final material. Thus, CNTs, GR, and their composites have been extensively studied in recent years to determine their suitability for various sensing applications. Previous literature, therefore, indicates that CNT, GR, and their composites are acceptable alternatives to conventional materials due to their extremely high surface-volume ratio, chemical and thermal stability, mechanical strength, flexibility, high current carrying capacity with controllable bandgap, and cost-effectiveness [6]–[8]. Consumer demand for electronics, robotics, and internet applications is increasing day by day. The above-described features of carbon-based materials can fulfill the increasing demand for sensors. As a result, although this is already a widely studied field of research, the amendment of sensor fabrication using 3D-printed carbon-based nanomaterials will take the world to a new level. This chapter explores recent developments and limitations in 3D-printed carbon-based nanomaterials, highlighting that it is a promising candidate for the development of sensors.

18.2 WORKING PRINCIPLE OF SENSORS

Every sensor has its sensing elements and mechanisms with unique functions. A sensor is a module or chip of a complex AED, which is made to measure ambient conditions. Generally, the sensing device consists of a selective recognition phase and the transducer. Upon receipt of the input signal (mechanical, thermal, magnetic, electric, chemical, radiation, etc.) across the selective recognition phase, the sensor generates the output electrical signal. Upon receipt of the input signal across the selective recognition phase, the sensor generates the output electrical signal. In some cases, an output signal is followed through a signal conditioning unit to amplify the signal levels to meet the requirements of subsequent standard equipment or device. The resulting output signal can be displayed or recorded as numeric data, text, graphics, and diagrams (Figure 18.1a) [2], [9].

18.2.1 Types of Sensors

Sensors and sensing technologies are classified based on various criteria such as applications, materials used, and some production technologies. One of the most commonly used classification systems is a classification based on detection parameters at ambient conditions (Figure 18.1b). Many of the sensors show sensing mechanisms based on the capacitive, resistive, inductive, magnetic, and field-effect transistor properties of the sensing material [8], [10].

The capacitive sensing devices usually consist of two electrodes and a dielectric central layer. When functioning, external factors induce spatial change between the three layers, resulting in variation of capacitance with the current or voltage. The output signal relies on the altered spacing and area of electrode plates caused by physically stimulated deformation [2]. The capacitive sensing mechanism enables to detection of many types of sensors including pressure, strain, force, displacement, humidity, fluid level, acceleration, and biological molecules. In addition, capacitive sensors have the advantage of a simple device structure, low power consumption, low detection limit, fast dynamic response, and endurability which have been applied in a wide application range [11], [12].

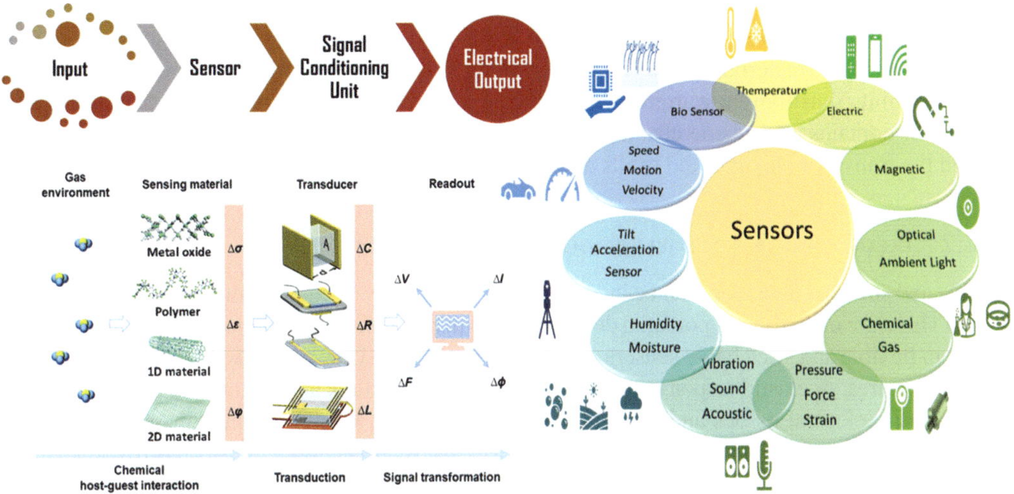

FIGURE 18.1 Schematic illustration of general sensing mechanism of a sensor with an example of an electrical gas sensing process and sensor classifications according to the detection properties. Adapted with permission [19]. Copyright (2022) American Chemical Society.

TABLE 18.1
Types of Resistive Sensors

Type of Sensor	Input Signal for the Resistance Change
Potentiometric sensors	Linear or angular position change
Resistive temperature sensors	Temperature variation (thermoresistive effect)
Photoresistive sensors	Light strikes a photo conductive material (photoresistive effect)
Piezoresistive sensors	Force is applied to a piezoresistive conductor (piezoresistive effect)
Magnetoresistive sensors	In the presence of an external magnetic field (magnetoresistive effect)
Chemoresistive sensors	Conductivity change in a material or solution due to chemical reactions that alter the number of electrons or concentration of ions
Bioresistive sensors	Bioresistance change in proteins or cells induced by structural variations and biological interactions

According to the definition, resistive sensors measure the change of electrical resistance in response to the external non-electricity quantity. Depending on the sensing mechanism and type of change, resistive sensors are further classified as potentiometric sensors, resistive temperature sensors, photoresistive sensors, piezoresistive sensors, magnetoresistive sensors, chemoresistive sensors, and bioresistive sensors [9]. Resistance changes that can happen due to various external stimuli are presented in Table 18.1.

Inductive sensors use electromagnetic induction to detect or measure electric and magnetic fields or other physical quantities such as displacement and pressure. These sensors do not require physical contact but can only detect metal targets. Non-metallic impurities (e.g., water, dirt) do not interact with the sensor. This feature allows inductive sensors to use in military equipment, robotics, aerospace, traffic lights, rail, car washes, and metal detectors [13]. Based on sensing principles, magnetic or electromagnetic sensors contain enormous amounts of sensors such as anisotropic magnetoresistance, giant magnetoresistance, tunneling magnetoresistance, the Hall effect, magnetoimpedance (MI) sensors, etc. Generally, a magnetic sensor operates based on the magnetic field, to detect magnetically active objects [14], [15]. Among the various types of sensors, research interest is increasing in the field-effect transistor (FET) for biosensing. A FET employs a semiconducting/conducting polymer material for the channel bridging two electrodes, the source and the drain. The channel exhibits sensitivity towards analyte concentration changes. For instance, a graphene-based FET sensor contains a graphene channel. By detecting the binding of the target analyte, the current through the transistor changes and sends an electrical signal [16]–[18].

18.2.2 Flexible and Stretchable Sensors

Flexible and stretchable sensors can be incorporated into a variety of equipment with intricate arbitrary shapes. These are often used as potential detection tools in biomedical applications. The sensor material present in flexible and stretchable sensors should be biomimetic. Low modulus, low bending stiffness, ultrathin features, or elastic response to strain deformations features are essential to provide intimate contact with biological tissues and irregular/deformed surfaces. Metal-contained materials presented excellent electrical and thermal conductivity. However, the relatively high density, corrosiveness, fragile conductive network, and cost of the metals restrict the practical production of wearable sensors. The main parameters for evaluating wearable sensor performance are sensitivity and stretchability [20]. Usually, the electronic circuit of the flexible sensor builds on top of the stretchable polymeric substrate such as polydimethylsiloxane, Ecoflex, polyurethane, polyethylene terephthalate, polyimide, and silicon/natural rubber [8]. These substrate materials can be 3D printed under ambient conditions to create relatively thin multi-material devices. However, the printed substrates are unable to maintain their geometric fidelity due to (i) the low elastic modulus of the pre- and post-cured material and (ii) deformation under gravity before curing [21]. In recent years, conductive carbon materials (GR and CNT) with chemical stability, mechanical strength, and electrical conductivity have received countless attention as conductive matrices for electronic circuits. Therefore, GR and CNT were introduced to the conductive network to be more stable during deformation in flexible sensors.

18.2.3 Figures of Merit of Sensors

Sensitivity to any parameter is liable for errors, and it is important to minimize those errors effectively. Accordingly, the accuracy, precision, sensitivity, linearity, and resolution characteristics of a sensor must be well-defined. In addition, the range, reproducibility, repeatability, response time, frequency range, impedance, and signal-to-noise ratio of a sensor should be appropriate for the applications [2], [9]. Thus, the specification or transfer characteristic (offset, sensitivity, or both) must be well-defined and stable for any new sensor.

18.2.4 Fabrication of Sensors via 3D Printing

There are two ways to develop a sensor using 3D-printing technology (i) embedding an existing sensor into a printed structure and (ii) printing the entire sensor [3], [4]. As per the previous discussion, sensors can be constructed to determine even non-electrical signals, by selecting the appropriate material. As a result, many 3D-printing approaches with different materials such as metals, conductive polymers, carbon-based materials, and their composites, have been suggested for the fabrication of sensors over the years. Particularly, the sensing platforms developed through 3D-printed one-dimensional CNT and two-dimensional GR-based compounds will discuss in the following sections.

18.3 ROLE OF 3D PRINT IN THE FABRICATION OF SENSORS

3D printing is a rapidly growing digital fabrication technology that has a critical role in modern interdisciplinary research and industry. In 3D printing, objects are created by adding material layer by layer with computer-controlled transformation stages according to the digital model files (software data). Unlike traditional methods that require molds, dies, or lithographic masks, the digital assembly of 3D-printing technology makes it possible to quickly create 3D prototypes using computer-aided designs (CAD) without wasting excess material. This revolutionary technology has already been applied to many industries. Similarly, a large number of new sensors thrived around the world in the late nineteenth century, based on the invention of 3D-printing techniques. Most of the time-consuming and expensive manufacturing techniques have been overshadowed by rapid, efficient, and cost-effective 3D-printing processes [22], [23].

The 3D-printing procedure consists of three important steps: (i) modeling of CAD, (ii) printing, and (iii) postprocessing. The design of the CAD model is a complex computational process. Firstly, the digital data is collected by a 3D scanner or simple digital camera with photometry software. The collected data is then converted to stereolithography (STL) file format. Before printing a 3D model from an STL file, should be examined for defects with holes, self-intersections, manifold errors, etc. that may occur during printing. Then, the STL file needs to be processed by a slicer (software), which converts the model into a series of thin layers and produces a file (G-code) containing instructions that fit into a specific 3D printer. Finally, with the use of this G-code file, the 3D printer prints materials layer by layer. The 3D-material layer is controlled by deposition rate as set by the 3D-printer operator, CAD file, and material properties (Figure 18.2). Most commonly high-quality materials of polymers, metal, and ceramics, their combination in the form of composites are used for 3D printing [3]. With the freedom of customization, the 3D-printing market has attracted the attention of a large audience. Recently scientists are trying to 3D print smart materials (materials that have the potential to alter the geometry and shape of the object, influenced by external conditions such as heat and water) for electronic skins, wearable electronics, and soft robotics manufacturing [23]. In addition, 3D bioprinting of implants and organs using bioinks – biocompatible materials and human cells – is an emerging technique in 3D printing to reduce the wait time for transplants [24]. Due to these special features, the 3D-printing technique is used to make various sensors mentioned in the above sections.

Mostly, sensors are fabricated through 3D-printing technology was done by printing the sensor components directly, printing molds for casting sensors, or by printing platforms integrated with commercial sensors. According to the standard terminology for 3D-printing techniques (ASTM F2792), there are seven main technologies: (i) binder jetting, (ii) material extrusion, (iii) directed energy deposition, (iv) material jetting, (v) sheet lamination, (vi) powder bed fusion, (vii) vat photopolymerization. Depending on the technical execution of the processes and material type or form, 3D-printing techniques can be further divided into various distinct categories as depicted in Figure 18.2. Binder jetting technique mainly uses binders and powders. Usually, the liquid binder is selectively deposited into a power bed. Subsequently, a fresh layer of powder spreads on the fused layer. The sequence is repeated until the desired 3D object is built. Similarly, inkjet printing devices selectively deposit drops of an adhesive or binder ink on a bed of leveled powder. Ink infiltrates into a thin layer (100–300 μm) of a freshly deposited powder bed. Stereolithography apparatus (SLA) and selective laser sintering (SLS) create 3D models using laser scanning. In SLA the starting material is a liquid photopolymer (mixture of monomer and oligomer components) with a viscosity range of ~100–2000 cP, which is rapidly solidified by a scanned UV laser beam under low operating temperatures. The SLS technique uses a bed of fine powder (metal and polymers), which can be sintered to solid layers with brief heating to a high temperature. The energy need to solidify is supplied by a focused IR laser beam. Digital light processing (DLP) is similar to SLA except for the energy source. In DLP projected light source is used to cure the photopolymer. Fused deposition modeling (FDM), fused filament fabrication (FFF), 3D dispensing, and 3D bioplotting fall into the material extrusion category. These techniques build 3D objects by heating thermoplastic materials and selectively depositing them in layers through a nozzle [4], [24]–[26].

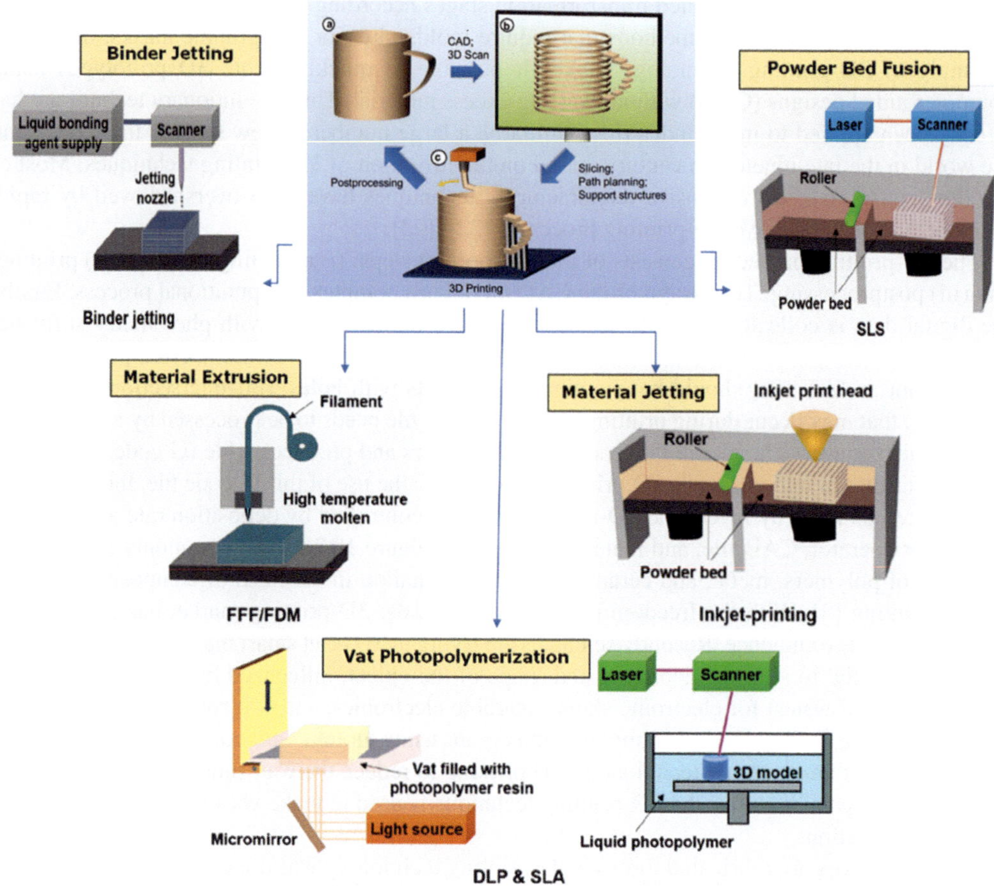

FIGURE 18.2 Basic principles of 3D printing techniques. Adapted with permission [26]. Copyright (2017) American Chemical Society. Adapted with permission [25]. Copyright (2021) American Chemical Society.

18.3.1 Graphene-Based 3D-Printed Sensors

GR is a material consisting of a ultrathin layer of sp^2 hybridized carbon atoms and is one of the strongest materials reported so far. In addition, flexibility, elasticity, impermeability, large theoretical specific surface area, lightweight, excellent thermal and electrical conductivity properties make it becomes a valuable and useful nanomaterial for countless practical applications. Due to the successful mechanical and electrical properties of GR, many studies have focused on the use of GR and GR-based composites in sensors as well (Figure 18.3). Owing to the 2D nanosheet structure, each carbon atom in GR is exposed to its environment, permitting it to detect minuscule changes in its surrounding. Furthermore, GR can be modified by chemical doping or aromatic factors and effectively functionalized with a receptor to detect target substances. It allows for the creation of GR-based micrometer size strain, gases, pressure, magnetic field, light, and temperature sensors [10], [27], [28]. In 2014, 5.6 wt% of the graphene-loaded composite was successfully 3D printed for the first time. In this study, graphene oxide (GO) was used as the starting filler material to minimize phase separation and clear graphene addition [29]. Infusion of GR-based materials into conventional metals, ceramics, and polymers allows for the production of strong, lightweight 3D-printed materials.

Inkjet printable GR offers a breakthrough in the fabrication of sensors. Many researchers and industries developed GR-based conductive inks for high-conductive and sensitive sensor films. The

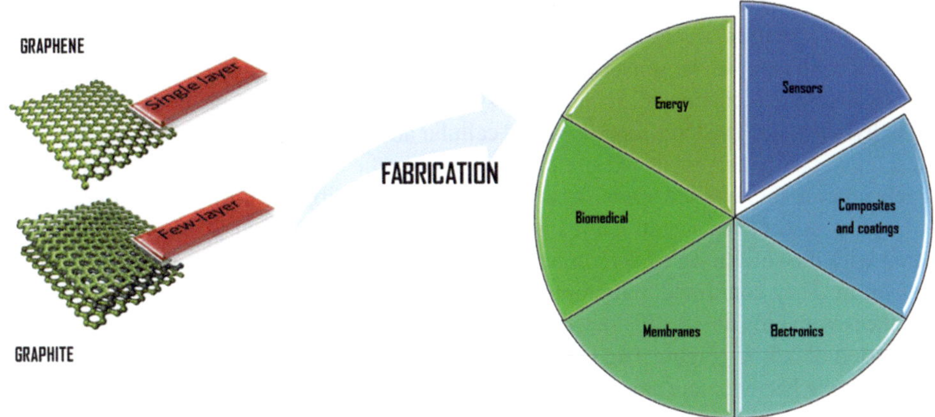

FIGURE 18.3 Molecular models of graphene; single-layer and graphite; few-layer (left), and promising applications of graphene (right). Adapted with permission [30]. Copyright (2011) John Wiley and Sons.

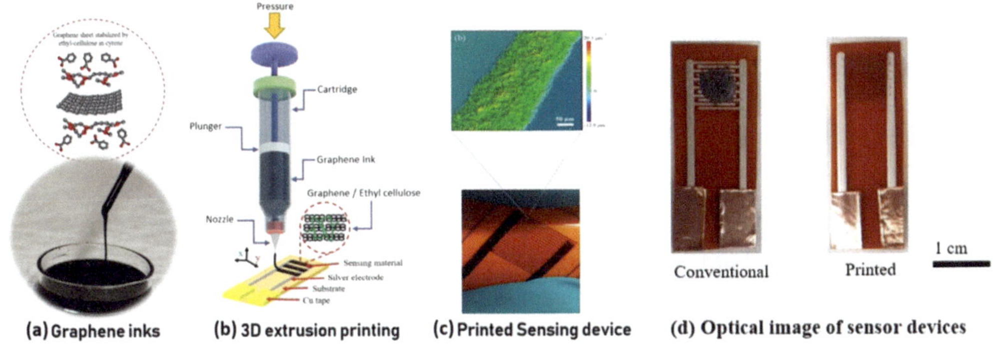

FIGURE 18.4 Schematic diagram of the fabrication process of 3D extrusion micro printing of chemo-resistive sensors for VOC detection using graphene-based ink. Adapted with permission [31]. Copyright (2021), Royal Society of Chemistry.

characteristics of inkjet-based printing are controlled by the particle size and structure of the ink. Unfortunately, the commercialization of sensors with GR-based inks is hampered by the non-uniform particle size and structures of GR. Pristine graphene inks commonly utilize polymer additives to maintain colloidal stability and controlled viscosity. In a study, a chemoresistive sensing device was fabricated through 3D inkjet printing of pristine graphene-ethyl cellulose-cyrene (pGR/EC/Cy) composite ink for volatile organic compounds(VOCs) detection at room temperature (Figure 18.4). The sensing device displayed a 10-fold improvement in the surface area/volume ratio compared to a conventional drop-casting method with high performances in terms of sensitivity and selectivity towards trace amount of VOC (e.g., linear detection range of 5–100 ppm with A_R ~ 0.45 sensitivity towards ethanol at room temperature). Compared to many of the graphene inks reported in the literature, the new pGR/EC/Cy ink provides a facile route to obtaining an environmentally friendly, sustainable, and low-cost chemo-resistive sensor [31], [32]. The main advantage of GR material-based gas sensors is that they can operate at room temperature or lower temperatures than gas sensors based on metal oxide semiconductors (operating temperature 100 to 1000 °C). Because it reduces the risk of explosion in the presence of combustible gases [33].

A high stretchable, durable sensor with extremely stable sensitivity was fabricated to monitor the temperature variation of a heated surface and human skin, overcoming the limited capability of a stretchable conductor on the sensor upon the large amplitude and frequent stretching. This sensor consists of stretchable conductive composites synthesized from the 3D-printed cellular graphene/polydimethylsiloxane. The long-range-ordered cellular architectures have a great potential for stable electrical conductivity and enhanced stretchability, due to the fine cellular architecture is capable of effectively sharing the external strain. In this study, the three different structural designs, including grid, triangular, and hexagonal structures were prepared using the direct ink writing technique and compared with the solid form to determine the most stable temperature sensing design under large external strain. They confirmed that graphene/PDMS composites with grid structure showed better performances with only about 15% sensitivity decrease at a large tensile strain of 20% and temperature resolution of 0.5 °C (Figure 18.5) [34].

The ambipolar behavior and nanoelectronic conductance properties of graphene play an important role in graphene-based field-effect transistors (GFETs). GFETs offer a versatile platform for chemical, physical, and biological sensing purposes. In particular, the application of GFET to biochemical sensing has been the focus of scientists recently [35]–[37]. However, a limited number of researchers have reported on GFET made using 3D-printed GR-based materials. Bhat et al. used the nozzle-jet printing method to print active silver/reduced graphene oxide (Ag/rGO) channel materials on a flexible and disposable polymer substrate for the chemical sensor. In this study, facile enzymeless phosphate ion detection in aqueous media is reported using a nozzle-jet-printed silver/reduced graphene oxide (Ag/rGO) composite-based field-effect transistor sensor on flexible and disposable poly(ethylene terephthalate) substrate. The nozzle-jet printing method is a simpler and modified form of the ink-jet printing technique. The sensor exhibits promising results in the detection of the concentration of phosphate ions with high performance, long-term stability, excellent reproducibility, and good selectivity in the presence of other interfering anions such as Ca^{2+}, Mg^{2+}, K^{+}, HCO^{3-}, and SO_4^{2-}. This sensor shows a wide linear range of 0.005–6.00 mM, high sensitivity of 62.2 μA/cm²/mM, and a low detection limit of 0.2 μM. The overall performance of the Ag/rGO FET sensors reveals better capability than that of the previously reported ones. Therefore, this method yields a low-cost, user-friendly sensing device for evaluating water eutrophication in natural water bodies [38].

18.3.2 CNT Based 3D-Printed Sensors

Carbon nanotubes were first evidenced in 1991 by the Japanese electron microscopist Sumio Iijima. According to the previous literature, CNT is one of the most widely studied nano-structured materials. CNT is a single-walled/multi-walled hollow carbon structure with a diameter of nanometer-scale and 1 to 100 microns in length. CNTs possess amazing mechanical and physical properties because of the well-ordered arrangement of carbon atoms connected via sp^2 bonds. Because of their properties, CNTs can be utilized in the field of energy storage, molecular electronics, thermal materials, structural materials, electrical conductivity, catalytic supports, biomedical, environmental, fabrics, fiber, etc [33], [39]. Hence it offers great promises for the next generation of sensors [33].

Similar to GR, the nanocomposites of CNTs are mainly used for sensor development in various fields because CNTs can enhance the mechanical, thermal, and electrical properties of the materials used for 3D printing [40]. Nonetheless, scientists are working to overcome the challenges in this field, including improving the dispersion of the carbon nanotube fillers with binders to prepare easily printable inks with stable sensing performance and stable electrical properties. Researchers have fabricated 3D-printing nanocomposite polymers using multi-walled CNTs (MWCNT) as fillers. Polymethylmethacrylate (PMMA), poly (ionic liquid) (PIL), ionic liquid (IL), and MWCNT were mixed to fabricate high conductive and flexible nanocomposite that used for 3D-printed electrochemical sensors. IL was used as a plasticizer and dopant for MWCNT. The composites that form after mixing IL and MWCNT have characteristic features including a long lifespan in air,

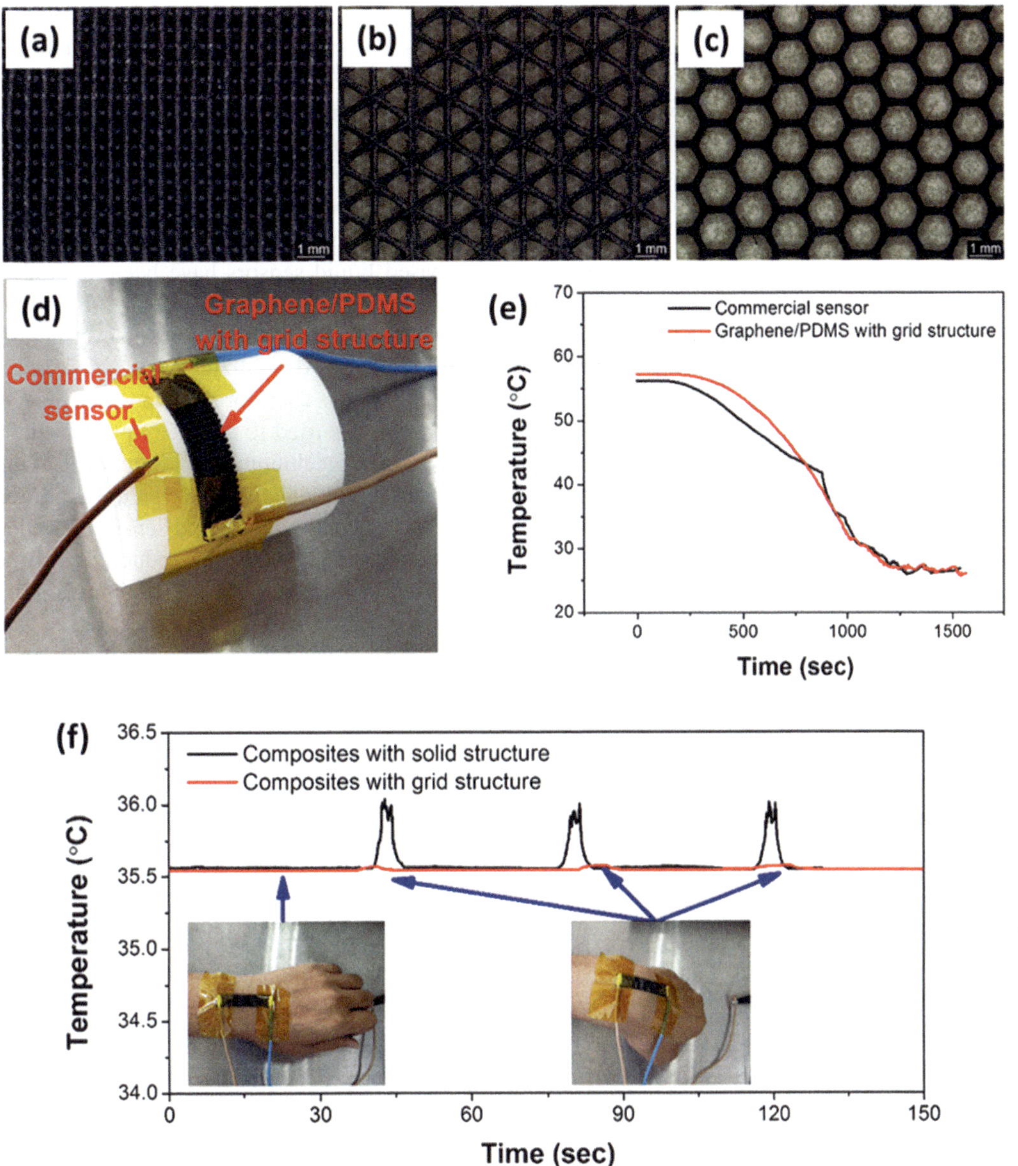

FIGURE 18.5 Optical images of the 3D-printed graphene/PDMS composites with (a) grid structure; (b) triangular porous structure; (c) hexagonal porous structure. Monitoring of the cooling process on a curved surface; (d) experimental setup; (e) comparison of the temperature performance between graphene/PDMS composites with a grid structure and commercial temperature sensor; and (f) simultaneous monitoring of wrist skin temperature and joint bending. Adapted with permission [34]. Copyright (2019), American Chemical Society.

low weight, and low operating voltage, which are ideal for the fabrication of sensors. PMMA and PIL were used to control the mechanical stiffness and ductility of the nanocomposite polymer. The highest conductivity of 520 S/m and enhanced mechanical properties were exhibited with 15 wt.% MWCNT. Nanocomposites with all compositions exhibited high thermal stability up to 340 °C and variable transition temperature depended on the contents of multi-walled CNT and IL [41].

Attractive properties of hydrogels including biocompatibility, shape-forming ability, swelling effect, and stretching ability directed the use of hydrogel as a promising material for flexible liquid sensor development. Due to its swelling effect, the hydrogel can provide a qualitative response to chemical liquids. However, above mentioned qualitative response can convert into quantitative when conductive fillers mix with hydrogel. CNTs and ionic compounds can use as conductive fillers and those materials improve the chemical sensitivity of pure hydrogel. The conductivity of composite hydrogel varies when it contacts liquid due to the swelling effect. The structure of a polymer matrix get expands when liquid is entering into a polymer matrix and expands the distance between conductive nanofillers. Composite hydrogel-based liquid sensors have been applied in different applications including the detection of the concentration of methanol in fuel cells and the detection of organic solvent leakage from gas stations, automobiles, and refineries. Nevertheless, traditional liquid sensors have some limitations including accurately recognizing the location or direction of stimuli and hard identifying micro-sized liquid leakages. Therefore, composite hydrogel based on poly (ethylene diacrylate) (PEGDA) and MWCNT was used to make 3D-printed bidirectional liquid sensors using a high-resolution stereolithography technique. A bidirectional liquid sensor was developed to identify microdroplet leakage and it contained three-layered 3D-printed artificial hair with PEGDA/MWCNT and pure PEGDA. The bidirectional liquid sensor identified liquid leakage separately from two sides with high efficiency. The electrical conductivity of the composite hydrogel is adjusted by controlling the swelling of the hydrogel. Absorbing of liquid molecules and the swelling effect decreased the resistance of pure PEGDA (top layer). Resistance of PEGDA/MWCNT (lower layer) increased with the effect of swelling due to the incremental distance between MWCNT. The 3D-printed mesh-shaped liquid sensor, which was developed to identify the position of liquid leakage, identified the position and volume of micro liquid leakage in a short time [42].

3D-printable conductive inks fabricated from nanocomposite of CNT and polylactic acid (PLA) were used for the fabrication of conductive scaffold structures applicable as liquid sensors. The ball mill mixing method was used to mix high concentrations of CNTs (up to 40 wt%) with PLA and scaffolds were fabricated using solvent-cast 3D-printing method with different structural parameters. The effects of structural parameters (inter-filament spacing (IFS), filament diameter, scaffold thickness, and structural patterns) on the sensitivity of liquid sensors were explored experimentally using 3D printing. The optimum liquid sensitivity was exhibited for IFS in the range of 0.5–1.5 mm and higher IFS exhibited lower sensitivity because scaffolds are less dense at higher IFS and lower total conductivity due to less material contact. The highest sensitivity was obtained at the lowest filament diameter (i.e., 128 μm) and sensitivity decreased when increasing the filament diameter. Low scaffold thickness provided higher sensitivity because a low number of printed layers have a low area of cover surface and since liquid can access the covered surface easily and provide higher sensitivity [43].

Generally, the FDM 3D-printing process uses to print thermoplastic material. Nowadays researchers are trying to expand the 3D-printing FDM applications using different composite materials. Two different approaches use when fabricating functional objects using FDM 3D printing; the first approach is to fabricate structural components using 3D printing and then attach prefabricated functional devices to the 3D-printed structure, and the second approach is to manufacture functional devices using fully composite materials. Kim et al. fabricated a multiaxial force sensor that is sensitive to forces in three orthogonal axes using FDM 3D printing. The sensing part of the multiaxial force sensor detects the mechanical inputs by changing their resistance and the structural part of the sensor provides a physical platform for this multiaxial force sensor. Nanocomposite composed of thermoplastic polyurethane (TPU) and CNT was used for the fabrication of the sensing part using FDM 3D printing. TPU/CNT has excellent electrical conductivity, piezoresistivity, and mechanical flexibility. The force sensor detects multidirectional force-sensing due to its 3D cubic cross structure, measuring the force exerted in three individual axes. Previously, multiaxial force sensor fabrication methods initially fabricated individual 1D force sensors followed by 3D assembly processes. However, the FDM 3D-printing method facilitates the fabrication of multiaxial force sensors without any additional assembly. This multiaxial force sensor is useful for different applications including electronic appliances and rehabilitation processes [44].

Many studies have been done on composites derived from GR and CNT for sensing applications. However, detailed comparative studies to evaluate the performance of these two materials are rarely reported. Li et al. attempted to systematically compare the electromechanical property of inkjet-printed CNT-polyimide (PI) and graphene nanoplate (GNP)-PI nanocomposite thin film devices as strain sensors. The details of the strain sensor fabricating showed that the thin-film sensing material was produced by inkjet printing of CNT/PI and GR/PI nanocomposite on flexible PI substrates. Inkjet printing technology has made it possible to control the geometry and thickness of thin films at the micron level. They observed that the electrical conductivity of both CNT-PI and GNP-PI thin film devices exhibits a significant increase in conductivity with low concentrations of CNT and GNP, while GNP-PIs have roughly two orders of magnitude higher conductivity than CNT-PIs. It confirms the 2D flat sheet shape of GNP can provide a more effective conducting path when mixing in a polymer matrix nanocomposite. The gauge factors for GNP-PI and CNT-PI sensor devices are 26 and 3.5, respectively. The relatively low gauge content of CNT indicates (i) a less effective electron tunneling effect and (ii) a mixing of CNTs with different chirality and alignment directions [6].

18.3.3 3D-Printed Carbon-Based Wearable Sensors

Wearable sensor technologies provide attractive solutions to our circadian rhythms. This is because wearable sensor technologies today have developed to provide unique features that go beyond everyday health monitoring. For example, wearable devices with a fall detection feature can detect sudden elevation changes. This information can save lives by automatically notifying authorities and emergency contacts through the mobile phone-connected instant alert system. Recent advances in wearable sensors focus on 3D-printed GR and CNT-based materials were surveyed in this section. Based on the unique mechanical and electrical characteristics of GR and CNT, more and more wearable sensors have been manufactured by scientists. However, the biocompatibility of GR and CNT compounds is still controversial [45]. Therefore, most of the existing literature was published in past years based on *in-vitro* studies. The main limitation of a wearable sensor is the incompact interface and the large impedance between the human body and sensors, which will decrease the signal quality. In addition, diversity and the high level of environmental noise of the input biosignals make it difficult to create an ideal wearable device. 3D printing is a revolutionary production technology that has led to the gradual development of sensor production due to the ease of production of complex conductor networks [5]. A common type of wearable sensor includes a flexible and wearable substrate material, with a sensing layer incorporated into or onto this material [46].

Polydimethylsiloxane (PDMS) and natural rubber (NR) is a widely used substrate polymer materials in flexible sensors and carbon black, CNTs, and GR have been used as conductive fillers [21]. Due to the lack of desirable mechanical properties of direct ink 3D-printed PDMS polymer, GR-based nanofillers are introduced into the ink formulation to improve the viscoelastic properties of the ink. A new 3D extrudable aqueous nanocomposite ink containing PDMS submicrobeads, graphene oxide (EGO) sheets, and PDMS prepolymer was developed in a study for piezoresistive strain sensor fabrication. PDMS/EGO nanocomposite ink is flowable at high shear stress and possesses high storage moduli to maintain its structure after extrusion. The author showed that the fabricated strain-sensing wearable patch is highly sensitive (capable of detecting human pulse) and robust (>1000 tensile cycle stability). Furthermore, this design demonstrated high linear sensitivity (gauge factor 20.3) at a high tensile strain of 40%, and a short response time (83 ms). This finding provided the basis for further direct ink 3D-printed sensor discoveries [47].

The inclusion of flexible, super-hydrophobic, multifunctional smart materials is the latest trend in the wearable sensing electronics industry. These materials have shown versatile applications in biomonitoring, environmental monitoring, haptic monitoring, self-cleaning, etc. [46]. Polymer-based conductive composites, which comprise conductive nanomaterials like GR, CNT, polymer-based elastic substrates like natural rubber, and silicone rubber are typically used to fabricate multifunctional superhydrophobic sensors. Subsequently, Liu et al. developed a multifunctional superhydrophobic silicone rubber (SR)/multi-walled carbon nanotubes (MWCNTs)/laser-induced graphene (LIG) composite strain sensor using laser direct writing for human body motion detection under complex and severe environmental conditions. It presented outstanding waterproof (water contact angle of 155^0 and sliding angle of 5^0), anti-corrosion ability (from pH = 1 to pH = 14), and photothermal anti-icing/deicing properties. Moreover, it exhibited a high sensitivity of 667, a large working range (up to 230%), and remarkable long-term stability (>2500 cycles). External ambient temperature affects stretchability, sensitivity, and long-term sensing stability. However, the MWCNTs / LIG cross-linked conductive network integrated with the sensor can maintain the sensitivity of the sensor in low-temperature environments [48].

Over the past ten years, 3D-printed electronics have revolutionalized the wearable sensor industry by producing cost-effective electronic sensors. PDMS/MWCNT nanocomposite is gaining high demand as a promising sensing material for capacitive and resistive pressure sensors due to its flexibility, stability, and sensitivity to conventional electronic materials. Some researchers have used MWCNTs nanoparticles primarily to achieve load-sensing capability based on piezoresistance with a sub-millimeter resolution. In a study, a piezoresistive sensor was produced by printing a PDMS polymer using modified extrusion-based 3D printing and then spreading the MWCNT over the wet PDMS matrix. The distribution of MWCNT in PDMS/MWCNT nanocomposite was analyzed using a scanning electron microscope and strain sensing ability was characterized using cyclic tensile loading at six different max strains. MWCNT repositioning and realignment at different tensile strains changed piezoresistivity and tunneling effects within the MWCNT were found to cause it. The gauge factor of the sensor was 4.3, confirming that PDMS/MWCNT nanocomposite sensor has two-time higher strain sensitivity than the metallic strain gauge. Human wrist bending at different speeds was monitored using 3D-printed sensor, and the results confirmed that 3D-printed PDMS/MWCNT strain sensor is potential enough to monitor human motions effectively [49]. Based on this research data authors further modified the sensing material fabrication method by dispersing MWNT in the PDMS base elastomer using the solvent-assisted ultrasonication method. The prepared MWNT/PDMS nanocomposites were printed on the pristine PDMS substrate. The new 3D-printed strain sensor demonstrated high flexibility, stretching to 146% strain before fracture, and linear piezoresistive response up to 70% strain with a gauge factor of 12.15 [21].

Fused deposition modeling (FDM) 3D printing has obtained more attention during the last decade due to its simple mechanical design and affordability. The lack of facial fabrication methods is the main limitation in sensor design fields. Therefore, FDM 3D printing based on highly elastic, electrically conductive nanocomposites is used to fabricate complex and multidirectional sensor patterns. Thermoplastic PU/MWCNT nanocomposite was used to fabricate a highly elastic strain sensor using FDM 3D printing. The material strength, electrical conductivity, and initial elastic modulus increased as the percentage of MWCNT filler loading percentage increased up to 5 wt%. Fabricated TPU/MWCNT exhibited excellent adhesion between the TPU layers, and there was no electrical conductivity degradation in through-layer and cross-layer directions after printing. Different percentages of MWCNT in TPU/MWCNT provided different ranges of sensitivity and flexibility that can use for different applications. The higher piezoresistivity gauge factors (high as 176) were obtained when applied a 100% large strain, and a highly repeatable resistance-strain response was received during cyclic loading. All the characteristic features mentioned above are confirming that FDM 3D printed TPU/ MWCNT nanocomposite can effectively apply to wearable electronics applications and soft robotics [50].

18.4 CONCLUSION AND PERSPECTIVES

3D-printing technology provides sustainable solutions to both the opportunities and challenges of emerging sensing technologies. Although different types of 3D-printing methods have been developed, DIW, FDM, and ink jetting are the most commonly used technologies for 3D-printed sensor fabrication using carbon-based nano-composite. Researchers have modified the traditional 3D-printing techniques to overcome the limitations and expand the potential for highly customizable low-cost sensors. Further, 3D-printable novel carbon-based composite materials with excellent sensing performances are formulated. Particularly, carbon-based composite conductive filaments formulated with PDMS, PU, SR, and NR are suggested for next-generation sensor fabrications due to the reduced cost and high sensitivity. The benefits of improvements made in sensors will enhance the quality and performance of the wearable electronics industry.

With the advent of the new material, advances in 3D modeling and design techniques have led to the development of smaller, lighter, and wearable sensor models. Because of the unique electrical, mechanical, and thermal properties many novel sensors have been developed using GR and CNTs are mostly used for strain, pressure, temperature, gas, chemical, and biosensing applications among all kinds of applications. GR-based novel ink materials endure extreme deformations by flexible sensors. These advances have thus led to the development of 3D-printable wearable devices that can fold, bend, and stretch while maintaining remarkable analytical performance. The development of wearable sensors is getting popular in the last decades due to its broad range of applications including health and wellness, early detection of disorders, force and motion tracking, home rehabilitation, etc. Most of the wearable sensors are combined into wearable objects or in direct contact with the body to monitor health conditions and collect clinically related data. Technologies for wearable sensors in recent years, there has been a significant shift from traditional wristband activity trackers to advanced sensor patches.

In addition to the traditional biosensor, the incorporation of environmental sensors, acoustic sensors, motion sensors, and pressure sensors into wearable sensors drives the demand for wearable devices. However, the fabrication of a multifunctional, cost-effective, miniaturized sensing device is challenging. Today, 3D-printing technology is creating new opportunities for various new wearable sensor technologies due to its suitability for versatile applications. Despite a large number of studies conducted on the 3D-printed conductive polymer and carbon-based nanocomposites, limited data are available for the direct 3D-printed carbon-based nanomaterials. Studies on improving the performance of direct 3D-printed carbon-based sensors are still ongoing. Scientists also expect to create carbon-based sensing materials with other functional properties, including magnetic capabilities. This chapter summarizes the results of recent research published related to 3D-printed carbon-based nanomaterials for sensors in terms of printing methods, material modifications, sensing mechanism, and application.

REFERENCES

1 S. Patel, H.-S. Park, P. Bonato, L. Chan, M. Rodgers, A review of wearable sensors and systems with application in rehabilitation, *J. Neuroeng. Rehabil.* 9 (2012) 21.

2 I. Sinclair, *Sensors and Transducers*, Elsevier, 2000.

3 M.R. Khosravani, T. Reinicke, 3D-printed sensors: Current progress and future challenges, *Sensors Actuators, A Phys.* 305 (2020) 111916.

4 M. Hofmann, 3D printing gets a boost and opportunities with polymer materials, ACS Macro *Lett.* 3 (2014) 382–386.

5 A. Kalkal, S. Kumar, P. Kumar, R. Pradhan, M. Willander, G. Packirisamy, S. Kumar, B.D. Malhotra, Recent advances in 3D printing technologies for wearable (bio)sensors, *Addit. Manuf.* 46 (2021) 102088.

6 Q. Li, S. Luo, Y. Wang, Q.M. Wang, Carbon based polyimide nanocomposites thin film strain sensors fabricated by ink-jet printing method, *Sensors Actuators, A Phys.* 300 (2019) 111664.

7 J. Huang, X. Yang, S.C. Her, Y.M. Liang, Carbon nanotube/graphene nanoplatelet hybrid film as a flexible multifunctional sensor, *Sensors (Switzerland).* 19 (2019) 317.

8 S. Li, X. Xiao, J. Hu, M. Dong, Y. Zhang, R. Xu, X. Wang, J. Islam, Recent advances of carbon-based flexible strain sensors in physiological signal monitoring, *ACS Appl. Electron. Mater.* 2 (2020) 2282–2300.

9 W.Y. Du, *Resistive, Capacitive, Inductive, and Magnetic Sensor Technologies*, CRC Press, 2014.

10 J. Zhu, X. Huang, W. Song, Physical and chemical sensors on the basis of laser-induced graphene: Mechanisms, applications, and perspectives, *ACS Nano.* 15 (2021) 18708–18741.

11 X. Ye, M. Tian, M. Li, H. Wang, Y. Shi, All-fabric-based flexible capacitive sensors with pressure detection and non-contact instruction capability, *Coatings.* 12 (2022) 1–10.

12 Q. Zhang, Y.L. Wang, Y. Xia, P.F. Zhang, T. V. Kirk, X.D. Chen, Textile-only capacitive sensors for facile fabric integration without compromise of wearability, *Adv. Mater. Technol.* 4 (2019) 1–11.

13 M. Jagiella, S. Fericean, Miniaturized inductive sensors for industrial applications, *Proc. IEEE Sensors.* 1 (2002) 771–778.

14 T. Dai, C. Chen, L. Huang, J. Jiang, L.M. Peng, Z. Zhang, Ultrasensitive magnetic sensors enabled by heterogeneous integration of graphene hall elements and silicon processing circuits, *ACS Nano.* 14 (2020) 17606–17614.

15 M. Hajiali, L. Jamilpanah, Z. Sheykhifard, M. Mokhtarzadeh, H. Yazdi, B. Tork, J.S.E Gharehbagh, B. Azizi, E. Roozmeh, G.R. Jafari, S.M. Mohseni, Controlling magnetization of Gr/Ni composite for application in high-performance magnetic sensors, *ACS Appl. Electron. Mater.* 1 (2019) 2502–2513.

16 K.M. Cheung, K.A. Yang, N. Nakatsuka, C. Zhao, M. Ye, M.E. Jung, H. Yang, P.S. Weiss, M.N. Stojanović, A.M. Andrews, Phenylalanine monitoring via aptamer-field-effect transistor sensors, *ACS Sensors.* 4 (2019) 3308–3317.

17 E. Wu, Y. Xie, B. Yuan, D. Hao, C. An, H. Zhang, S. Wu, X. Hu, J. Liu, D. Zhang, Specific and highly sensitive detection of ketone compounds based on p-type $MoTe_2$ under ultraviolet illumination, *ACS Appl. Mater. Interfaces.* 10 (2018) 35664–35669.

18 Y. Zhang, J. Clausmeyer, B. Babakinejad, A. López Córdoba, T. Ali, A. Shevchuk, Y. Takahashi, P. Novak, C. Edwards, M. Lab, S. Gopal, C. Chiappini, U. Anand, L. Magnani, R.C. Coombes, J. Gorelik, T. Matsue, W. Schuhmann, D. Klenerman, E. V. Sviderskaya, Y. Korchev, Spearhead nanometric field-effect transistor sensors for single-cell analysis, *ACS Nano.* 10 (2016) 3214–3221.

19 M. Khatib, H. Haick, Sensors for volatile organic compounds, *ACS Nano.* 16 (2022) 7080–7115.

20 H. Liu, H. Zhang, W. Han, H. Lin, R. Li, J. Zhu, W. Huang, 3D printed flexible strain sensors: From printing to devices and signals, *Adv. Mater.* 33 (2021) 2004782.

21 M. Abshirini, M. Charara, P. Marashizadeh, M.C. Saha, M.C. Altan, Y. Liu, Functional nanocomposites for 3D printing of stretchable and wearable sensors, *Appl. Nanosci.* 9 (2019) 2071–2083.

22 L.Y. Zhou, J. Fu, Y. He, A review of 3D printing technologies for soft polymer materials, *Adv. Funct. Mater.* 30 (2020) 1–38.

23 S. Mousavi, D. Howard, F. Zhang, J. Leng, C.H. Wang, Direct 3D printing of highly anisotropic, flexible, constriction-resistive sensors for multidirectional proprioception in soft robots, *ACS Appl. Mater. Interfaces.* 12 (2020) 15631–15643.

24 L.Y. Zhou, Q. Gao, J.F. Zhan, C.Q. Xie, J.Z. Fu, Y. He, Three-dimensional printed wearable sensors with liquid metals for detecting the pose of snakelike soft robots, *ACS Appl. Mater. Interfaces.* 10 (2018) 23208–23217.

25 K. Min, Y. Li, D. Wang, B. Chen, M. Ma, L. Hu, Q. Liu, G. Jiang, 3D printing-induced fine particle and volatile organic compound emission: An emerging health risk, *Environ. Sci. Technol. Lett.* 8 (2021) 616–625.

26 S.C. Ligon, R. Liska, J. Stampfl, M. Gurr, R. Mülhaupt, Polymers for 3D printing and customized additive manufacturing, *Chem. Rev.* 117 (2017) 10212–10290.

27 B. Nguyen, H. Nguyen Van, Promising applications of graphene and graphene-based nanostructures, *Adv. Nat. Sci. Nanosci. Nanotechnol.* 7 (2016) 23002.

28 K.F. Kelly, W.E. Billups, Synthesis of soluble graphite and graphene, *Acc. Chem. Res.* 46 (2013) 4–13.

29 X. Wei, D. Li, W. Jiang, Z. Gu, X. Wang, Z. Zhang, Z. Sun, 3D printable graphene composite, *Sci. Rep.* 5 (2015) 1–7.

30 M. Terrones, O. Martín, M. González, J. Pozuelo, B. Serrano, J.C. Cabanelas, S.M. Vega-Díaz, J. Baselga, Interphases in graphene polymer-based nanocomposites: Achievements and challenges, *Adv. Mater.* 23 (2011) 5302–5310.

31 K. Hassan, T.T. Tung, N. Stanley, P.L. Yap, F. Farivar, H. Rastin, M.J. Nine, D. Losic, Graphene ink for 3D extrusion micro printing of chemo-resistive sensing devices for volatile organic compound detection, *Nanoscale.* 13 (2021) 5356–5368.

32 W.J. Hyun, E.B. Secor, M.C. Hersam, White paper: Printable graphene inks stabilized with cellulosic polymers, *MRS Bull.* 43 (2018) 730–733.

33 J.S. Im, S.C. Kang, S.-H. Lee, Y.-S. Lee, Improved gas sensing of electrospun carbon fibers based on pore structure, conductivity and surface modification, *Carbon N. Y.* 48 (2010) 2573–2581.

34 Z. Wang, W. Gao, Q. Zhang, K. Zheng, J. Xu, W. Xu, E. Shang, J. Jiang, J. Zhang, Y. Liu, 3D-printed graphene/polydimethylsiloxane composites for stretchable and strain-insensitive temperature sensors, *ACS Appl. Mater. Interfaces.* 11 (2019) 1344–1352.

35 C. Huang, Z. Hao, T. Qi, Y. Pan, X. Zhao, An integrated flexible and reusable graphene field effect transistor nanosensor for monitoring glucose, *J. Mater.* 6 (2020) 308–314.

36 S. Wang, M.Z. Hossain, T. Han, K. Shinozuka, T. Suzuki, A. Kuwana, H. Kobayashi, Avidin–biotin technology in gold nanoparticle-decorated graphene field effect transistors for detection of biotinylated macromolecules with ultrahigh sensitivity and specificity, *ACS Omega.* 5 (2020) 30037–30046.

37 Z. Hao, C. Huang, C. Zhao, A. Kospan, Z. Wang, F. Li, H. Wang, X. Zhao, Y. Pan, S. Liu, Ultrasensitive graphene-based nanobiosensor for rapid detection of hemoglobin in undiluted biofluids, *ACS Appl. Bio Mater.* 5 (2022) 1624–1632.

38 K.S. Bhat, U.T. Nakate, J.Y. Yoo, Y. Wang, T. Mahmoudi, Y.B. Hahn, Nozzle-jet-printed silver/graphene composite-based field-effect transistor sensor for phosphate ion detection, *ACS Omega.* 4 (2019) 8373–8380.

39 N. Sinha, J. Ma, J.T.W. Yeow, Carbon nanotube-based sensors, *J. Nanosci. Nanotechnol.* 6 (2006) 573–590.

40 B. Podsiadły, P. Matuszewski, A. Skalski, M. Słoma, Carbon nanotube-based composite filaments for 3d printing of structural and conductive elements, *Appl. Sci.* 11 (2021) 1272.

41 K. Ahmed, M. Kawakami, A. Khosla, H. Furukawa, Soft, conductive nanocomposites based on ionic liquids/carbon nanotubes for 3D printing of flexible electronic devices, *Polym. J.* 51 (2019) 511–521.

42 X. Li, Y. Yang, B. Xie, M. Chu, H. Sun, S. Hao, Y. Chen, Y. Chen, 3D Printing of flexible liquid sensor based on swelling behavior of hydrogel with carbon nanotubes, *Adv. Mater. Technol.* 4 (2019) 1–9.

43 K. Chizari, M.A. Daoud, A.R. Ravindran, D. Therriault, 3D printing of highly conductive nanocomposites for the functional optimization of liquid sensors, *Small.* 12 (2016) 6076–6082.

44 K. Kim, J. Park, J. Hoon Suh, M. Kim, Y. Jeong, I. Park, 3D printing of multiaxial force sensors using carbon nanotube (CNT)/thermoplastic polyurethane (TPU) filaments, *Sensors Actuators, A Phys.* 263 (2017) 493–500.

45 H. Huang, S. Su, N. Wu, H. Wan, S. Wan, H. Bi, L. Sun, Graphene-based sensors for human health monitoring, *Front. Chem.* 7 (2019) 399.

46 J. Horne, L. McLoughlin, E. Bury, A.S. Koh, E.K. Wujcik, Interfacial phenomena of advanced composite materials toward weearable platforms for biological and environmental monitoring sensors, armor, and soft robotics, *Adv. Mater. Interfaces.* 7 (2020) 1–25.

47 G. Shi, S.E. Lowe, A.J.T. Teo, T.K. Dinh, S.H. Tan, J. Qin, Y. Zhang, Y.L. Zhong, H. Zhao, A versatile PDMS submicrobead/graphene oxide nanocomposite ink for the direct ink writing of wearable micron-scale tactile sensors, *Appl. Mater. Today.* 16 (2019) 482–492.

48 K. Liu, C. Yang, S. Zhang, Y. Wang, R. Zou, Q. Deng, N. Hu, Materials & Design Laser direct writing of a multifunctional superhydrophobic composite strain sensor with excellent corrosion resistance and Anti-icing/deicing performance, *Mater. Des.* 218 (2022) 110689.

49 M. Abshirini, M. Charara, Y. Liu, M. Saha, M.C. Altan, 3D printing of highly stretchable strain sensors based on carbon nanotube nanocomposites, *Adv. Eng. Mater.* 20 (2018) 1–9.

50 J.F. Christ, N. Aliheidari, A. Ameli, P. Pötschke, 3D printed highly elastic strain sensors of multiwalled carbon nanotube/thermoplastic polyurethane nanocomposites, *Mater. Des.* 131 (2017) 394–401.

19 3D-Printing of Carbon Nanotube-Based Nanocomposites for Sensors

*Nahal Aliheidari and Amir Ameli**
Department of Plastics Engineering, University of Massachusetts-Lowell, 1 University Ave, Lowell, MA, 01854, USA
*Corresponding author: amir_ameli@uml.edu

CONTENTS

19.1 INTRODUCTION AND BACKGROUND

19.1.1 Nanomaterials

Due to their outstanding electrical, optical, magnetic, chemical, and mechanical properties, nanomaterials have been employed in a vast range of applications in different areas, especially in the sensors field [1]. Nanofillers have been used over the years to enhance the properties of polymeric materials [1]–[4]. Nanofillers may enhance mechanical properties, electrical properties, thermal and barrier properties, and dimensional stability. Different features have been used to classify nanomaterials including dimension/size, morphology, state, and chemical composition (e.g., metallic, carbonaceous, etc.). Nanoscale materials based on dimensionality can be divided into 0D, 1D, 2D, 3D, heterostructured, and hierarchically nanostructured soft materials [2]–[5]. Figure 19.1 shows the schematic illustration based on dimensional classification [6]–[8].

19.1.2 Carbon Nanotubes

Carbon nanomaterials (CNs) contain diverse allotropes from graphene (GR) to buckminsterfullerene (C60). Carbon nanostructures are identified to consist of different low-dimension allotropes of carbon-containing graphite, activated carbon, carbon nanotubes, and the C60 family of buckyballs, polyaromatic molecules, and graphene. Over the last decades, nanotechnology has attracted significant attention due to its direct application to generate new materials with outstanding properties. While inorganic nanostructures such as MXenes are developing recent research interests, several

DOI: 10.1201/9781003296676-19

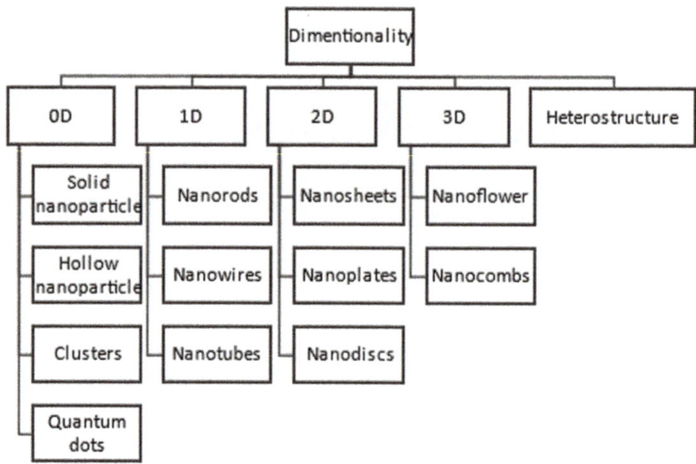

FIGURE 19.1 Schematic illustration of the classification of nanomaterials based on dimensionality.

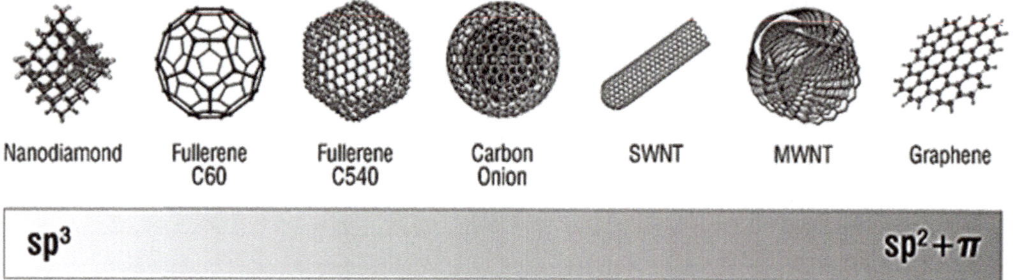

FIGURE 19.2 Schematic illustration of individual allotropes of CNTs. Adapted with permission [12], Copyright (2008), American Chemical Society.

carbonaceous nanostructures have been fully developed and commercialized. Many factors such as superior directionality, high surface area, and flexibility make carbonaceous nanostructures suitable for a wide range of applications [9]–[11]. The unique properties of carbonaceous nanomaterials are attributed to carbon's structural conformation and its specific state. Figure 19.2 summarizes some of the allotropes of CNs.

A very common allotrope of CN is a carbon nanotube (CNT). CNT, a microtubule of graphitic carbon with a nano-range diameter and micro-range length was discovered by Iijima in 1991 [13]. Since then, the development of more carbon structures attracted extensive attention, and high-quality CNTs have been produced by three major methods including discharge, laser ablation, and chemical vapor deposition (CVD) [14]. CNTs are identified by the number of walls to multi-wall CNTs (MWCNTs) and single-wall CNTs (SWCNTs). Surface-functionalized CNTs and doped CNTs are more advanced nanotubes that overcome the hydrophobic character of this nanomaterial and further expand their applications. CNTs offer extremely high modulus and strengths, large length-to-diameter aspect ratio, good chemical and environmental stabilities, and very high thermal and electrical conductivities [15].

19.1.3 Carbon Nanotube/Polymer Nanocomposites

In the past three decades, after the discovery of CNTs, due to their outstanding multifunctional properties compared to previous fillers, there has been an even more extensive research effort on a new class of polymer composites, i.e., CNT-filled polymer nanocomposites (CNT-PNC) [16]. The introduction of only a small amount of this nanomaterial can have a massive improvement in polymer properties such as transforming an insulator material into a conductive one. The remarkable performance and extraordinary properties have made CNT-PNCs favorable in many industries such as defense, automotive, marine, railway, civil engineering structures, sporting goods, and armor as structural components, sensors, actuators, semiconductors, etc. [17], [18]. CNT-PNCs have been used as strain sensors [19], thermal sensors [20], and electrochemical sensors [21]. There are also some reports on the electromagnetic interference (EMI) shielding capacity of CNT-PNCs [22]. Numerous studies have demonstrated the capability of CNT-PNCs in the sensing field. However, conventional manufacturing methods such as molding and casting still pose a challenge in fabricating complex, multifunctional, and embedded/integrated sensors. Innovative manufacturing technologies enabling complex geometries and design freedom need to be developed toward more successful integration of CNT-PNC-based sensors [23].

19.1.4 Additive Manufacturing of Polymer Nanocomposites

3D printing is a versatile, simple, and low-cost process that can make complex geometry not possible or costly to make using conventional methods [24]. 3D printing contains a broad range of technologies that can be classified based on process or materials. According to ASTM F42 [25] material jetting, binder jetting, material extrusion, vat polymerization, powder bed fusion, direct energy deposition, and sheet lamination are the seven major groups of additive manufacturing technologies. where, stereolithography (SLA), selective laser sintering (SLS), fused filament fabrication (FFF), and laminated object modeling (LOM) are the most well-known 3D-printing methods in the industry's applications. Another 3D-printing classification is based on the material type that is used as a feedstock, which includes metals, ceramics, and polymers. Among these, polymer-based has drawn tremendous attention from both industry and academia due to the low cost and being more user-friendly in practical experiences [26]. In FFF method, a feedstock (thermoplastic-based material) is heated above its melting temperature and is extruded through a nozzle at a specific flow rate to deposit the rasters on a build plate when the first 2D layer was completed the second layer will start depositing, the part is built by adding layers from the bottom to the top [27], [28].

To enhance the physical, mechanical, electrical, and/or chemical properties of FFF-printed parts, an effective strategy is to employ and adapt novel multifunctional feedstock materials that are developed based on polymer composites. In this case, the matrix thermoplastic polymer is loaded with additives. Similar to conventional polymer composites, the additives can be micro-sized reinforcements such as carbon fiber, glass fiber, and graphite fiber, [29]–[31] or nano-sized materials such as nanoclay, graphene, CNTs, montmorillonite (MT), graphite nanoplates, and TiO_2 and ZnO [32]–[34]. Nemourious studies reported enhancement of electrical and thermal conductivity, strength, and stiffness as well as achieving specific functional properties in 3D-printed nanocomposite parts, especially for various sensing applications [18]. The conventional sensor fabrication methods such as photolithography, evaporation deposition, lamination, and coating are complex, expensive, and time-consuming. However, by integrating the flexibility and design freedom of 3D printing and the multifunctionality of PNCs, a new generation of heterostructure PNC-based sensors that offer higher sensitivity and reliability can be achieved. PNC-based 3D-printed sensors possess the opportunity for novel, efficient, more customized, and multifunctional sensing platforms.

This book chapter aims to discuss the recent advances in 3D-printed PNC-based sensors. In particular, CNT-based nanocomposites 3D printed through FFF process are covered in more detail. Four different sensing capabilities, namely, piezoresistive, capacitive, liquid-vapor, and structural monitoring are covered. For each sensor, first, a brief introduction is provided on the working principles and governing mechanism and then the recent progress is presented followed by a more in detail discussion of some case studies.

19.2 3D-PRINTED PIEZORESISTIVE SENSORS

The electrical conductivity or resistivity of a polymer/conductive filler nanocomposite, such as CNT-based PNC can be divided into four stages. At a very small amount of CNT, the electrical conductivity is very low close to that of the insulative polymeric matrix (Figure 19.3, zone a). With a further increase of CNT content, the conductivity increases gradually due to the tunneling effect (Figure 19.3, zone b). In the third stage, (Figure 19.3, zone c), the insulation-conduction transition occurs and an electrically-conductive path is fully formed by the interconnection of the CNTs. The fourth and final stage is when the percolative network is completely connected and there is only a slight increase in the electrical conductivity by further increase of CNT content (Figure 19.3, zone d). The electrical conductivity of a percolative nanocomposite system in zones c and d can be described by the percolation theory:

$$\sigma = \sigma_0 (\rho - \rho_c)^t \tag{19.1}$$

where σ is the electrical conductivity of the nanocomposite, σ_0 is the scaling factor, ρ and ρ_c are the volume fraction of the filler corresponding to σ and the critical volume fraction (percolation threshold), respectively, and t is the exponent constant, which is close to 2 for CNTs.

The piezoresistive effect is a change in the electrical resistivity of a material, usually semiconductor or metallic when mechanical strain is applied. In monolithic materials such as copper wire, the piezoresistivity is governed by the change in the geometric configuration of the material due to mechanical loading. In conductive filler/polymer composites such as CNT-based PNC, the piezoresistive effect is however attributed to three major phenomena, namely, conductive path deformation within the nanocomposite, the inherent piezoresistivity of CNT, and the tunneling effect. The major governing factor is based on the mechanism that when the material undergoes mechanical deformation, the relative filler to filler configurations changes within the material and create new electric pathways, resulting in overall resistance change [36]. Depending on the nature of the mechanical loading, the interconnections between the conductive fillers within nanocomposite could be strengthened or weakened, resulting in a decrease or increase, respectively, in the material overall resistance.

Utilizing the high electrical conductivity of CNT, functional nanocomposites as piezoresistive or resistance-type strain sensors with high sensitivity [37] or large deformations have been reported.

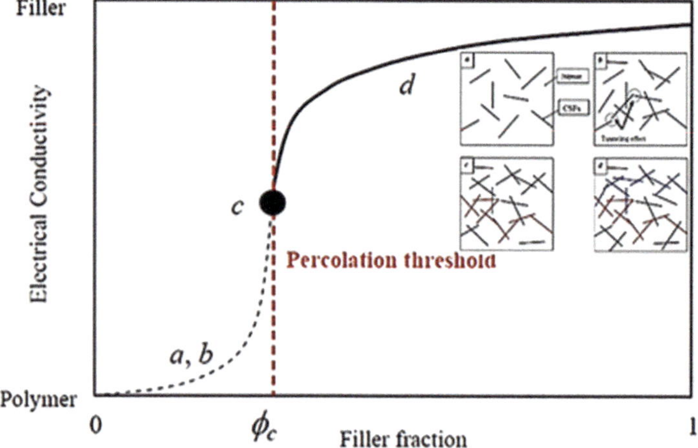

FIGURE 19.3 Electrical conductivity of conductive composites as a function of filler fraction. Adapted with permission [35], Copyright (2013), IOP Publishing.

In such sensors, the in-service conductivity will change with the strain applied. Small strain detection in this type of sensor has been explained using the tunneling effect, CNTs alignment, and agglomeration [38]. CNT-PNC-based piezoresistive sensors fabricated using conventional manufacturing methods have been extensively researched and widely employed in soft robotics, control devices, and electric skins [39], [40]. However, the more advanced and integrated applications of such sensors have been further enabled by coupling these multifunctional nanocomposites and additive manufacturing technologies. This is due to the capability of 3D-printing methods in delivering more complex, customized, embedded, and integrated sensors. The recent efforts on the development of 3D-printed piezoresistive sensors based on CNTs are discussed below.

Kim et al. [41] developed an FFF-made TPU/MWCNT system as a multiaxial force sensing to independently detect forces in x, y, and z directions. The system was based on the capability of the system for piezoresistive sensing. However, the force detection was heavily affected by the raster pattern. About the same time, Christ et al. [22] reported the fabrication of a highly elastic piezoresistive sensor using TPU/MWCNT which was employed to measure a wide range of strains. In another work, a strain sensor was fabricated using a modified FFF printer on a PDMS base and the sensor's behavior was shown to be linear up to 30% maximum strain with reliable reading [42]. This method has several steps and makes it complicated in comparison to the previous research. Kim et al. [43] also reported a sensor system of TPU/MWCNT. They claimed that the percolation threshold for the compound was observed at 3wt.% MWCNTs, the rest of the study was done on 5wt.% MWCNT and loaded up to 40% strain in both transverse and longitudinal directions. Both experimental results and modeling confirmed the isotropic evaluation of the resistivity. Very recently, Verma et al. [44] reported the fabrication of a piezoresistive sensor using polypropylene random copolymer and CNTs. The nanocomposite filament was produced using a corotating twin-screw extruder and printed by an FFF machine. Both mechanical and piezoresistance responses were investigated. The results indicate that both mechanical and piezoresistive responses depend on the interlayer adhesion during the FFF process. The sensors printed in a longitudinal direction showed a better response, compared to the transverse counterparts.

Christ et al. [45] reported the successful development of a TPU/MWCNT-based novel FFF feedstock that proves high elasticity and excellent piezoresistive properties. In this work, the TPU/MWCNT nanocomposite was prepared by melt extrusion using a twin-screw extruder. The nanocomposites with MWCNT content ranging from 0 to 5 wt% were fabricated using the optimized processing parameters. A rectangular sensor with dimensions of $1.6 \times 1.6 \times 10$ mm^3 (H × W × L) was printed using a MakerBot replicator.

The static and cyclic tensile performance and piezoresistivity behavior of TPU/MWCNT nanocomposite filaments and printed sensors were investigated. The quasi-static performance evaluation of the 3D-printed sensor was also performed. The strength of both filaments and printed samples increases by increasing the MWCNT content with an overall less increase in 3D-printed sensors. As seen in Figure 19.4, it was reported that the printed nanocomposites with 5wt.% MWCNT mechanically failed at ~ 60% strain and could not reach the targeted 100% strain due to the introduction of more defects, which act as local stress concentration locations. Also, at 2wt.% MWCNT, the sensors electrically failed at about 60–65% strain, due to significant loss of CNT interconnections and trending towards a more insulative characteristic. The other nanocomposites (3 and 4 wt.% MWCNT contents) were successfully stretched up to 100% strain with reliable measurement of resistance change. It was also shown that there was a modest decrease in the modulus of elasticity (~14%) of TPU/MWCNT samples after printing, which is attributed to the imperfect bonding at the interlayer. One desirable aspect of strain sensing materials is their ability to sense cyclic strains. As was mentioned before, for a sensor, it is crucial to keep the performance over the cyclic load. For both the strain-softening and Mullins effects, the most drastic change is observed within the first cycle and rapidly trended towards a steady-state within only a few cycles for both TPU and TPU/MWCNT. This rapid equilibrium demonstrates a high potential for cyclic strain sensing capabilities.

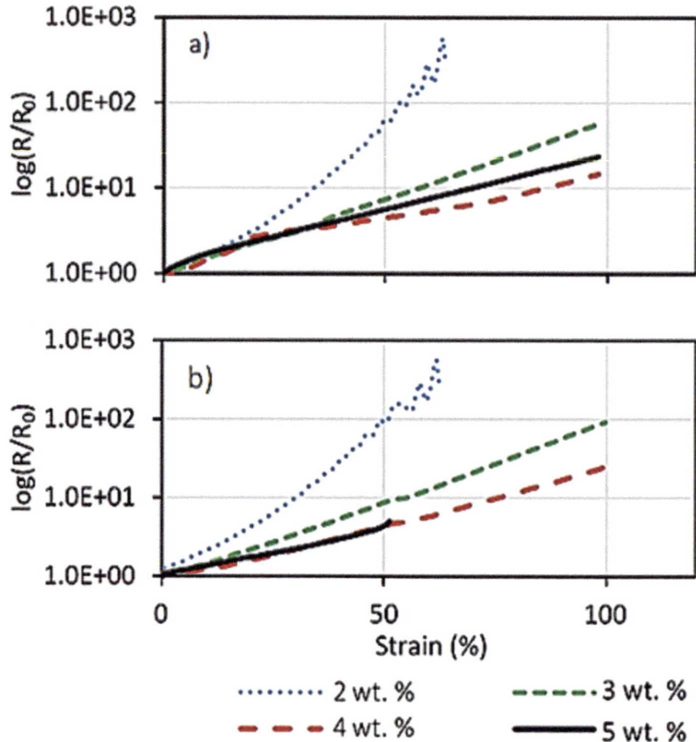

FIGURE 19.4 Logarithmic normalized resistance plots of (a) extruded filaments, and (b) FFF 3D printed TPU/MWCNT nanocomposites as a function of strain during loading. Adapted with permission [45], Copyright (2017), Elsevier.

The authors claim that the TPU/3 wt% MWCNT nanocomposite offers the most suitable combination of electrical and mechanical behaviors for sensing performance. It provides the highest sensitivity while remaining functional up to 100% strain as well as minimizing the amount of softening that occurs within the sample (Figure 19.4).

The developed nanocomposites in [45] were then employed to 3D print a range of embedded strain sensors [46] (Figure 19.5), in axial, transverse, and biaxial configurations. The sensors were then tested until 50% strain in the cyclic load. Figure 19.6 demonstrates the cyclic response for two types of sensors, i.e., linear axial and linear transverse. All the sensors showed a steady cyclic response after the first few initial cycles. It has to be noted that, in the first cycle, the resistance increased during both the loading and unloading phases, and a new higher level of resistance was established which can be attributed to the MWCNT network dissociation such that the conductive paths are broken, new paths are potentially created, and the conductive trace finds a new conductivity equilibrium during unloading [46]. It was also shown that simultaneous strain sensing in axial and transverse directions is possible when using linear biaxial, switchback biaxial, and sawtooth biaxial, and it was further demonstrated that with the change in the embedded sensor pattern (e.g., from the straight line to sawtooth), the sensitivity can be tuned in both directions.

19.3 3D-PRINTED CAPACITIVE SENSORS

Capacitive sensors have proven to be widely used in a broad range of applications owing to their ability to detect both metallic and nonmetallic parts, and position measurement in the sub-nanometer

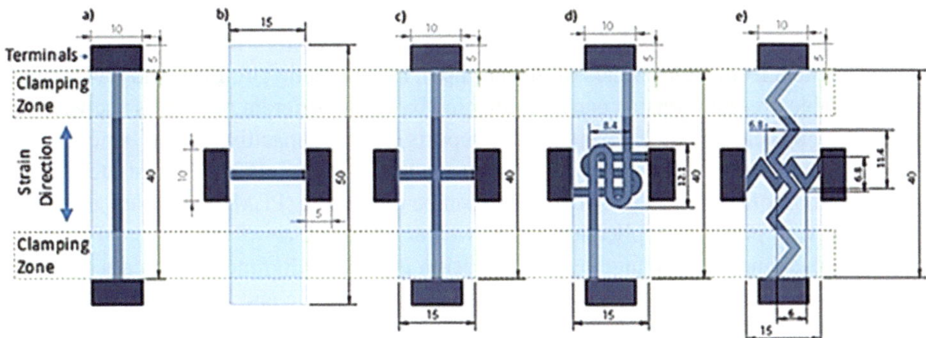

FIGURE 19.5 CAD drawings for (a) linear axial, (b) linear transverse, (c) linear biaxial, (d) switchback biaxial, and (e) sawtooth biaxial; All dimensions are in mm. Within each design, the black and light blue regions are TPU/MWCNT and pure TPU, respectively. The electrical connection terminals, clamping zone, and strain direction are also identified. The sample thickness in a-b is 1.0 mm and in c-e is 1.2 mm. The clamping zone size is 7.5 mm on each side, leaving a free length of 25 mm between the clamps. Adapted with permission [46], Copyright (2018), MDPI.

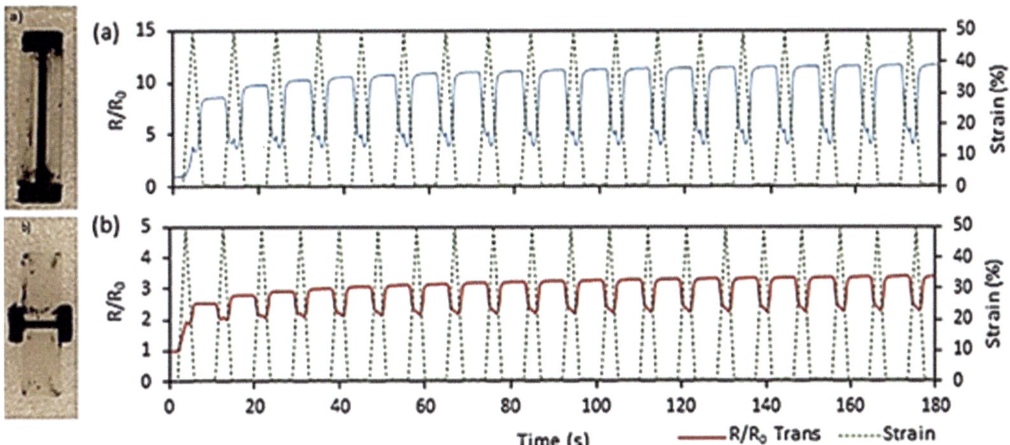

FIGURE 19.6 Cyclic relative resistance (R/R_0) plots for (a) linear axial sensor and (b) linear transverse sensor loaded to 50% maximum strain. Adapted with permission [46], Copyright (2018), MDPI.

range [47], [48]. Based on the principle of capacitive coupling, capacitive sensors can detect and measure objects that are conductive or have dielectric constant distinct from the air. Capacitive sensors can be used to measure position, humidity, touch, proximity, and void fraction [49], [50]. Conventional capacitance sensors are made of metallic foils which makes them not a good candidate for wearable devices due to their toxicity and low stretchability. A capacitive strain gauge can be made and modeled following a plate-like capacitor using a CNT film and silicone elastomer. When the device with an initial size of $l_0 * w_0 * t_0$ (initial length, width, and thickness) stretches by applied strain (λ), the capacitance of the stretched device can be calculated by [51]:

$$C = \varepsilon_0 \varepsilon_r \frac{\lambda * l * w_0}{t_0} = \lambda C_0 \tag{19.2}$$

where C_0, ε_0 *and* ε_r are the initial capacitance, the permittivity of vacuum, and the relative permittivity of silicone film, respectively. Other than this, for the capacitors with a finite size, the fringe effect, which results from the presence of the air gap in the magnetic circuit should be counted [52].

While wearable pressure sensors can be fabricated based on different principles such as resistive, capacitive, and piezoelectric mechanisms, some reports claim a capacitive sensor benefits from low power consumption and reliable operation. For instance, Choi et al. [53] demonstrate a method to produce a porous Ecoflex-multiwalled carbon nanotube composite (PEMC) structure as a capacitive pressure sensor with practical application on the wearable device.

Since the capacitance will change by sensor design and CNT alignment, fabrication of complex and small electrodes through conventional techniques is expensive and time-consuming, therefore employing advanced manufacturing methods such as 3D printing can be a solution. By that, 3D printing of the complex conductive structure and functional hollow structure has been easily possible which results in more sensitive sensors [48].

Hoheimer et al. [54] conducted a study on 3D-printed CNT nanocomposite sensors as robotic soft end effectors for grabbing delicate objects and examined the capability of 3D-printed built-in capacitive and piezoresistive sensors for such applications. In this work, to evaluate the capacitive and piezoresistive sensor's sensitivity, a pneumatically actuated soft end effector was 3D printed using FFF method as shown in Figure 19.7a. The three bottom layers for the base of the actuator were printed with electrically conductive TPU/3wt.%MWCNT nanocomposite, while the upper layers were printed with insulative pure TPU. When actuated with high-pressure air, the actuator bends towards this base layer, owing to its geometric design, until it contacts an object, which can be detected by a change in capacitance (Figure 19.7).

The mechanism is that once the sensor touches an object (such as an apple), the R-C delay is signified reaching a value of ~4000 from a low value of about 250. This significant rise can be used as an on-off logic control for touch sensing. Upon touching the sensor, the circuit's C increases, resulting in a longer delay before the signal is received, which is reflected as a larger R-C delay time. Further, as the touch sensing is based on the R-C time constant, the sensitivity, or magnitude of the response, can also be adjusted using a resistor with a lower or higher resistance in the circuit.

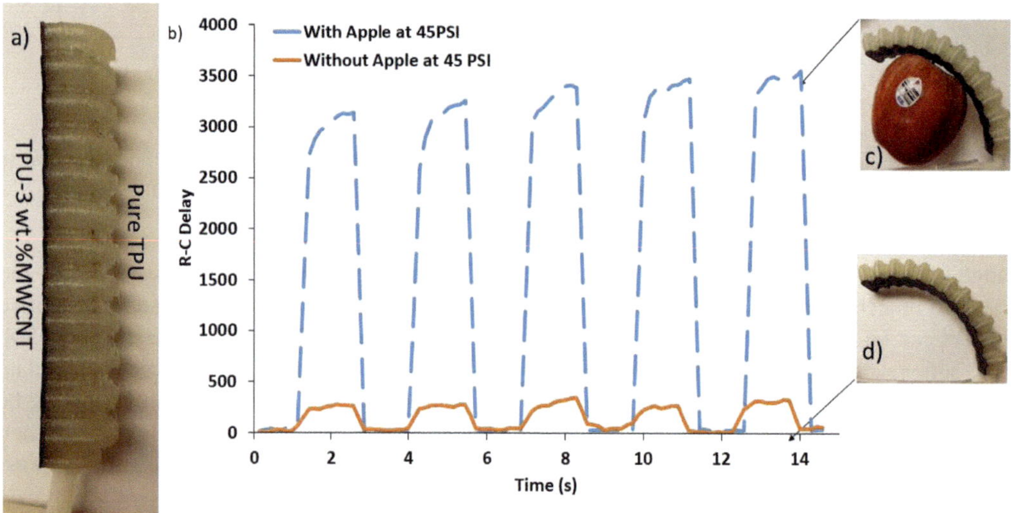

FIGURE 19.7 (a) 3D-printed actuator with a conductive base made of TPU-3wt.%MWCNT to act as a touch sensor upon actuation and (b) R-C time delay response of the actuator at ~310 kPa with (c) and without (d) physical contact of an object. Adapted with permission [54], Copyright (2020), Elsevier.

Recently, Liu et al. [55] reported the fabrication of a capacitive force sensor using a triple extruder FFF-machine with conductive filament to compare to the silicon-based microelectromechanical system (MEMS). The sensor was designed as a suspended beam-plate structure with HIPS as a sacrificing support layer. The sensor shows reliability for potential sensing and health monitoring when tested for 900 cycles. The sensor was attached to an adult neck and was able to detect the body pulse by capacitance changes due to the blood pulse. Loh et al. [56] also reported constructing flexible mechanical metamaterials programmed to have certain mechanical properties using the material design and 3D printing from a conductive polyurethane. The final sensor was employed as a human joint wearable device and was capable to measure force distribution and proximity detection which is crucial for safe gripping and identification of sensitive objects.

19.4 3D-PRINTED LIQUID/VAPOR SENSORS

A type of liquid sensor can be designed based on the volumetric swelling phenomenon. As the fluid diffuses into the polymer network, the volumetric expansion happens and disturbs the volumetric ratio of the conductive filler and the matrix in the virgin composite, causing a change in the interconnectivity of the conductive networks and thus the electrical resistivity changes. This change in resistance can be probed to detect or quantify fluid absorption [57]. The resistivity change is correlated to the local solvent uptake when the sensor is immersed in the solvent or exposed to its vapor. The swelling starts first on the surface and therefore inhomogeneous diffusion occurs through the cross-section of the sensor, if it is a conventional film-type sample [58]. The sensitivity and efficiency of liquid sensors depend on the type of sensing material, solvents, possible interactions between them, sensor geometrical features, temperature, and pressure [59].

The capability of FFF 3D printing to fabricate lattice structures and partially filled geometries from functional nanocomposites provides an opportunity to design liquid/vapor sensing platforms with highly-efficient diffusion paths. Recently, there have been some research efforts on the 3D-printed liquid sensor using electrically conductive nanocomposites. Aliheidari et al. [60], [61] reported a new U-shaped 3D-printed liquid sensor fabricated from highly elastic, electrically conductive TPU/MWCNT nanocomposites. The solvent selectivity of these sensors was investigated based on MWCNTs concentration, type of liquid, and infill density of 3D-printed specimens. Three different liquids, namely ethanol, acetone, and toluene were selected based on their widespread use in various industries.

Liquid (solvent) sensitivity of a polymer is related to the volumetric solvent uptake, which is attributed to Hansen solubility parameters [62] of the polymer and the solvent. Further, the polymer-liquid interaction is mainly determined by the Flory-Huggins interaction parameter $\chi 12$ which is calculated using [63]:

$$\chi 12 = \frac{V_{sol}(\delta_{pol} - \delta_{sol})^2}{RT} \tag{19.3}$$

where V_{sol} is the molar volume of the solvent, δ_{pol} and δ_{sol} are the solubility parameters of the polymer and the solvent, R is the gas constant (R=8.314 $JK^{-1}mol^{-1}$), and T is the absolute temperature.

The low values for $\chi 12$ indicate strong interactions between polymer and solvent and more solvent penetration. The low values of the Flory-Huggins parameter are obtained when the solubility parameters of the solvent and the polymer are as close as possible, indicating a good solubility of the polymer in the solvent (Table 19.1).

The results of [60], [61] revealed that increasing MWCNT content generated faster response and sense of solvents and caused a decrease in saturation time of the solvents. Based on Flory-Huggins interaction parameters for polyurethane/solvent, acetone is predicted to be the strongest and ethanol is the weakest solvent. This trend is in agreement with the measured liquid sensitivity of sensor/

TABLE 19.1
Characteristics of TPU/MWCNT and the Solvents Used for Sensing Experiments

Material	Hansen Solubility Parameter δ ($MPa^{1/2}$)	Boiling Point (°C)	Vapor Pressure KPa, 20 °C	Molar Volume (Cm^3mol^{-1})	Flory-Huggins Interaction Parameter (χ12)
Polyurethane	20.84	-	-	-	-
Acetone	19.9	56.2	24.6	73.4	0.026
Toluene	18.2	110.6	2.9	106.8	0.306
Ethanol	26.6	78.4	5.8	58.4	0.803

solvent pairs. The mechanical strength of the sensor structure and its electrical conductivity are the two important parameters, which have a crucial impact on designing an efficient and fast sensor. The sensor sensitivity enhances when the infill percentage is decreased. It is explainable that in a material with high packing of the printed layers, the sensor accessibility to solvent is limited and the solvent molecules could mostly become in contact with the sensor through its apparent surface area. Within the given short exposure time of 25 s, the solvent could only partially diffuse inside the sensor and thus caused limited swelling and less change in resistivity. However, as the infill density is decreased, the distance between the individual printed layers increases, helping them to be more accessible to the solvent. Therefore, the active surface area to volume ratio increases. This combination creates shortened diffusion paths, and the amount of solvent uptake increases for a given time.

Another type of vapor, gas or liquid sensing is based on the modification of doping level due to the redox interaction of analyte molecules which results in conductivity change. The PNCs usually are above the electrical percolation threshold. A recent work studied this type of vapor sensing strategy. Kennedy et al. [64] reported a PVDF-based 3D-printed nanocomposite to detect volatile organic compounds (VOC). The results revealed the potential capability of a freeform sensor that detects acetone vapor fast (20s) in a repeatable manner (25 cycles). Even the direction correlation was observed between the MWCNTs concentration and resistance change. Surprisingly, sensor thickness did not affect the relative increase in resistance with acetone exposure.

19.5 STRUCTURAL HEALTH MONITORING

Structural health monitoring (SHM) is a type of non-destructive evaluation (NDE) technique that essentially involves the strategic embedding of conductive filler into a structure such that the system is capable of self-assessment of structural health including damage detection and structural prognostics. Thaler et al. [65] investigated another class of smart nanocomposite with potential application in self-sensing. The ABS/CNT filaments having CNT contents of up to 10 wt% were first fabricated using a twin-screw extruder. The nanocomposites were then 3D printed in certain designs to investigate the tensile properties, fracture toughness, and critical strain energy release rate, as well as through-layer and in-layer conductivities. Results revealed very similar strength, stiffness, and strain-at-break of the printed pure ABS compared to those of the compression-molded samples. The electrical conductivity was investigated and it was observed that the compression-molded samples achieved higher conductivity values in comparison with 3D-printed ones. While the printed samples do not show the distinct percolation behavior and conductivity gradually increase as CNTs content increase.

Figure 19.8 presents the resistance change during the sample loading for different CNTs concentrations. A relatively linear relationship between the resistance and the strain was observed in the elastic region of the deformation. The small change at lower concentrations can be due to the tunneling effect, while at the higher CNT concentrations, with applying load the distances between

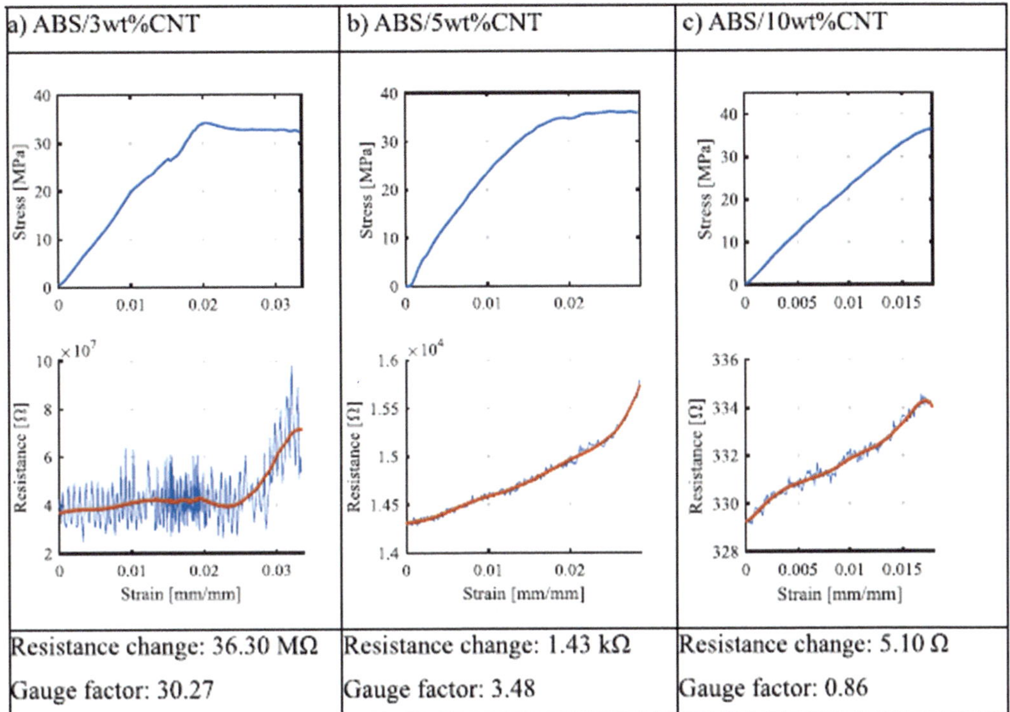

FIGURE 19.8 Stress-stress and the corresponding resistance-strain results of (a) ABS/3wt%CNT, (b) ABS/5wt% CNT, and (c) ABS/10wt%CNT nanocomposites. Adapted with permission [65], Copyright (2019), IOP Publishing Ltd.

CNTs increase, and therefore the resistance increases. In the printed samples with low CNT content, it is difficult to form conductive networks, but once the networks are formed, they show a better sensitivity to deformation. Therefore, measurable resistance change occurs in the printed ABS/CNT nanocomposite at higher concentrations once strained in tension. The results show that the strain or deformation in this nanocomposite can simply be measured by the resistance. Also, the onset of plastic deformation can potentially be detected. This nanocomposite may be printed as a self-sensing material for structural health monitoring purposes.

19.6 SUMMARY

3D printing of functional polymer composites for sensing applications was discussed. In particular, the fused filament fabrication (FFF) additive manufacturing method and carbon nanotube-based nanocomposites were focused and four main sensing types such as piezoresistance, capacitance, liquid/vapor, and structural health monitoring were discussed. Overall, CNT-based sensors exhibit strong potential due to their excellent mechanical, electrical, and functional properties. Higher sensitivity and reliability in sensors can be achieved via integrating the FFF additive manufacturing method and CNT-based nanocomposites. PNC-based 3D-printed sensors possess the opportunity for novel, efficient, more customized, and multifunctional sensing platforms. 3D-printed piezoresistive sensors perform with good repeatability and by changing the print pattern and sensor design, multi-directional sensing with tunable sensitivity is possible. Capacitive force sensors with a relatively complex structure can be fabricated in a one-step 3D-printing process without using any additional metallization process, alignment process, or assembling process. FFF methods show huge favor in increasing the sensitivity of the liquid sensors compared to the bulk compartment which is attributed to the elevation of the surface area and making the shorter path for diffusion. Structural health monitoring was another successful category for the 3D printing integrated with PNCs that could potentially be designed and integrated into structural applications.

REFERENCES

1. E. Sheikhzadeh, V. Beni, M. Zourob, Nanomaterial application in bio/sensors for the detection of infectious diseases, *Talanta*. 230 (2021) 122026.
2. J. Chen, A. Lotfi, P.J. Hesketh, S. Kumar, Carbon nanotube thin-film-transistors for gas identification, *Sensors Actuators, B Chem*. 281 (2019) 1080–1087.
3. B. Anasori, M.R. Lukatskaya, Y. Gogotsi, 2D metal carbides and nitrides (MXenes) for energy storage, *Nat. Rev. Mater.* 2 (2017) 1–17.
4. Z. Yue, F. Lisdat, W.J. Parak, S.G. Hickey, L. Tu, N. Sabir, D. Dorfs, N.C. Bigall, Quantum-dot-based photoelectrochemical sensors for chemical and biological detection, *ACS Appl. Mater. Interfaces*. 5 (2013) 2800–2814.
5. C. Wang, C. Wang, Z. Huang, S. Xu, Materials and structures toward soft electronics, *Adv. Mater*. 30 (2018) 1801368.
6. D. Rohilla, S. Chaudhary, A. Umar, An overview of advanced nanomaterials for sensor applications, *Eng. Sci.* 16 (2021) 47–70.
7. T.A. Saleh, Nanomaterials: Classification, properties, and environmental toxicities, *Environ. Technol. Innov*. 20 (2020) 101067.
8. C. Daulbayev, F. Sultanov, B. Bakbolat, O. Daulbayev, 0D, 1D and 2D nanomaterials for visible photoelectrochemical water splitting. *A Review, Int. J. Hydrogen Energy*. 45 (2020) 33325–33342.
9. N. Rao, R. Singh, L. Bashambu, Carbon-based nanomaterials: Synthesis and prospective applications, *Mater. Today Proc*. 44 (2021) 608–614.
10. J.N. Tiwari, V. Vij, K.C. Kemp, K.S. Kim, Engineered carbon-nanomaterial-based electrochemical sensors for biomolecules, *ACS Nano*. 10 (2016) 46–80.
11. Z. Wang, Z. Dai, Carbon nanomaterial-based electrochemical biosensors: An overview, *Nanoscale*. 7 (2015) 6420–6431.
12. M.S. Mauter, M. Elimelech, Environmental applications of carbon-based nanomaterials, *Environ. Sci. Technol*. 42 (2008) 5843–5859.
13. S. Iijima, Carbon nanotubes: Past, present, and future, *Phys. B Condens. Matter*. 323 (2002) 1–5.
14. Q. Zhang, J.Q. Huang, M.Q. Zhao, W.Z. Qian, F. Wei, Carbon nanotube mass production: Principles and processes, *ChemSusChem*. 4 (2011) 864–889.
15. N. Aliheidari, A. Ameli, Thermoplastic Polymers for Additive Manufacturing, in: Munmaya Mishra (Ed.), *Encycl. Polym. Appl*., Taylor & Francis Group, 2018, 486–500. www.taylorfrancis.com/books/9781351019415.
16. S. Wang, Y. Huang, E. Chang, C. Zhao, A. Ameli, H.E. Naguib, C.B. Park, Evaluation and modeling of electrical conductivity in conductive polymer nanocomposite foams with multiwalled carbon nanotube networks, *Chem. Eng. J*. 411 (2021) 128382.
17. P. Shukla, P. Saxena, Polymer nanocomposites in sensor applications: A review on present trends and future scope, *Chinese J. Polym. Sci. (English Ed.)* 39 (2021) 665–691.
18. A. Iqbal, A. Saeed, A. Ul-Hamid, A review featuring the fundamentals and advancements of polymer/CNT nanocomposite application in aerospace industry, *Polym. Bull*. 78 (2021) 539–557.
19. N. Hu, Y. Karube, M. Arai, T. Watanabe, C. Yan, Y. Li, Y. Liu, H. Fukunaga, Investigation on sensitivity of a polymer/carbon nanotube composite strain sensor, *Carbon N. Y.* 48 (2010) 680–687.
20. N. Sinha, J. Ma, J.T.W. Yeow, Carbon nanotube-based sensors, *J. Nanosci. Nanotechnol*. 6 (2006) 573–590.
21. H. Beitollahi, F. Movahedifar, S. Tajik, S. Jahani, A Review on the effects of introducing CNTs in the modification process of electrochemical sensors, *Electroanalysis*. 31 (2019) 1195–1203.
22. J. Christ, N. Aliheidari, A. Ameli, P. Pötschke, 3D printed highly elastic strain sensors of multiwalled carbon nanotube/thermoplastic nanocomposites, in: *Annu. Tech. Conf. – ANTEC, Conf. Proc*., 2017.

23 X. Sui, J.R. Downing, M.C. Hersam, J. Chen, Additive manufacturing and applications of nanomaterial-based sensors, *Mater. Today.* 48 (2021) 135–154.

24 H.G. Kim, S. Hajra, D. Oh, N. Kim, H.J. Kim, Additive manufacturing of high-performance carbon-composites: An integrated multi-axis pressure and temperature monitoring sensor, *Compos. Part B Eng.* 222 (2021) 109079.

25 ASTM Standard F2792 – 12a, Standard Terminology for Additive Manufacturing Technologies, ASTM Int. (2019).

26 J.W. Stansbury, M.J. Idacavage, 3D printing with polymers: Challenges among expanding options and opportunities, *Dent. Mater.* 32 (2016) 54–64.

27 N. Aliheidari, R. Tripuraneni, A. Ameli, S. Nadimpalli, Fracture resistance measurement of fused deposition modeling 3D printed polymers, *Polym. Test.* 60 (2017) 94–101.

28 N. Aliheidari, J. Christ, R. Tripuraneni, S. Nadimpalli, A. Ameli, Interlayer adhesion and fracture resistance of polymers printed through melt extrusion additive manufacturing process, *Mater. Des.* 156 (2018) 351–361.

29 X. Tian, T. Liu, C. Yang, Q. Wang, D. Li, Interface and performance of 3D printed continuous carbon fiber reinforced PLA composites, *Compos. Part A Appl. Sci. Manuf.* 88 (2016) 198–205.

30 B. Brenken, A. Favaloro, E. Barocio, N. Denardo, V.K. Pipes, R. Byron, Fused deposition modeling of fiber-reinforced thermoplastic polymers: Past progress and future needs, in: *Proc. Am. Soc. Compos. Thirty-First Tech. Conf.*, 2016.

31 F. Ning, W. Cong, J. Qiu, J. Wei, S. Wang, Additive manufacturing of carbon fiber reinforced thermoplastic composites using fused deposition modeling, Compos. *Part B Eng.* 80 (2015) 369–378.

32 V. Tambrallimath, R. Keshavamurthy, D. Saravanabavan, P.G. Koppad, G.S.P. Kumar, Thermal behavior of PC-ABS based graphene filled polymer nanocomposite synthesized by FDM process, Compos. *Commun.* 15 (2019) 129–134.

33 K.K. Sadasivuni, K. Deshmukh, M.A. Almaadeed, 3D and 4D printing of polymer nanocomposite materials: Processes, applications, and challenges, 3D 4D Print. *Polym. Nanocomposite Mater. Process. Appl. Challenges* Chapter 7, (2019) 191–229.

34 Z. Viskadourakis, M. Sevastaki, G. Kenanakis, 3D structured nanocomposites by FDM process: A novel approach for large-scale photocatalytic applications, *Appl. Phys. A Mater. Sci. Process.* 124 (2018) 1–8.

35 S. Xu, O. Rezvanian, K. Peters, M.A. Zikry, The viability and limitations of percolation theory in modeling the electrical behavior of carbon nanotube-polymer composites, *Nanotechnology.* 24 (2013) 155706.

36 M. Charara, M. Abshirini, M.C. Saha, M.C. Altan, Y. Liu, Highly sensitive compression sensors using three-dimensional printed polydimethylsiloxane/carbon nanotube nanocomposites, *J. Intell. Mater. Syst. Struct.* 30 (2019) 1216–1224.

37 Alamusi, N. Hu, H. Fukunaga, S. Atobe, Y. Liu, J. Li, Piezoresistive strain sensors made from carbon nanotubes based polymer nanocomposites, *Sensors.* 11 (2011) 10691–10723.

38 E. Chang, A. Ameli, A.R. Alian, L.H. Mark, K. Yu, S. Wang, C.B. Park, Percolation mechanism and effective conductivity of mechanically deformed 3-dimensional composite networks: Computational modeling and experimental verification, *Compos. Part B Eng.* 207 (2021) 108552.

39 C.M. Boutry, M. Negre, M. Jorda, O. Vardoulis, A. Chortos, O. Khatib, Z. Bao, A hierarchically patterned, bioinspired e-skin able to detect the direction of applied pressure for robotics, *Sci. Robot.* 3 (2018) eaau6914.

40 X. Fu, L. Wang, L. Zhao, Z. Yuan, Y. Zhang, D. Wang, D. Wang, J. Li, D. Li, V. Shulga, G. Shen, W. Han, Controlled assembly of MXene nanosheets as an electrode and active layer for high-performance electronic skin, *Adv. Funct. Mater.* 31 (2021) 2010533.

41 K. Kim, J. Park, J. hoon Suh, M. Kim, Y. Jeong, I. Park, 3D printing of multiaxial force sensors using carbon nanotube (CNT)/thermoplastic polyurethane (TPU) filaments, *Sensors Actuators, A Phys.* 263 (2017) 493–500.

42 M. Abshirini, M. Charara, Y. Liu, M. Saha, M.C. Altan, 3D printing of highly stretchable strain sensors based on carbon nanotube nanocomposites, *Adv. Eng. Mater.* 20 (2018) 1800425.

43 M. Kim, J. Jung, S. Jung, Y.H. Moon, D.H. Kim, J.H. Kim, Piezoresistive behaviour of additively manufactured multi-walled carbon nanotube/thermoplastic polyurethane nanocomposites, *Materials (Basel).* 12 (2019) 2613–2637.

44 P. Verma, J. Ubaid, K.M. Varadarajan, B.L. Wardle, S. Kumar, Synthesis and characterization of carbon nanotube-doped thermoplastic nanocomposites for the additive manufacturing of self-sensing piezoresistive materials, *ACS Appl. Mater. Interfaces*. 14 (2022) 8361–8372.
45 J.F. Christ, N. Aliheidari, A. Ameli, P. Pötschke, 3D printed highly elastic strain sensors of multiwalled carbon nanotube/thermoplastic polyurethane nanocomposites, *Mater. Des*. 131 (2017) 394–401.
46 J.F. Christ, N. Aliheidari, P. Pötschke, A. Ameli, Bidirectional and stretchable piezoresistive sensors enabled by multimaterial 3D printing of carbon nanotube/thermoplastic polyurethane nanocomposites, *Polymers (Basel)*. 11 (2018) 11–27.
47 H. Zangl, Capacitive sensors uncovered: Measurement, detection and classification in open environments, *Procedia Eng*. 5 (2010) 393–399.
48 S. Thangavel, S. Ponnusamy, Application of 3D printed polymer composite as capacitive sensor, *Sens. Rev*. 40 (2020) 54–61.
49 Y. Dai, S. Gao, A flexible multi-functional smart skin for force, touch position, proximity, and humidity sensing for humanoid robots, *IEEE Sens. J*. 21 (2021) 26355–26363.
50 A. Zolfagharian, A. Kaynak, A. Kouzani, Closed-loop 4D-printed soft robots, *Mater. Des*. 188 (2020) 108411.
51 L. Cai, L. Song, P. Luan, Q. Zhang, N. Zhang, Q. Gao, D. Zhao, X. Zhang, M. Tu, F. Yang, W. Zhou, Q. Fan, J. Luo, W. Zhou, P.M. Ajayan, S. Xie, Super-stretchable, transparent carbon nanotube-based capacitive strain sensors for human motion detection, *Sci. Rep*. 3 (2013) 1–9.
52 C.F. Hu, J.Y. Wang, Y.C. Liu, M.H. Tsai, W. Fang, Development of 3D carbon nanotube interdigitated finger electrodes on polymer substrate for flexible capacitive sensor application, *Nanotechnology*. 24 (2013) 444006.
53 J. Choi, D. Kwon, K. Kim, J. Park, D. Del Orbe, J. Gu, J. Ahn, I. Cho, Y. Jeong, Y. Oh, I. Park, Synergetic effect of porous elastomer and percolation of carbon nanotube filler toward high performance capacitive pressure sensors, *ACS Appl. Mater. Interfaces*. 12 (2020) 1698–1706.
54 C.J. Hohimer, G. Petrossian, A. Ameli, C. Mo, P. Pötschke, 3D printed conductive thermoplastic polyurethane/carbon nanotube composites for capacitive and piezoresistive sensing in soft pneumatic actuators, *Addit. Manuf*. 34 (2020) 101281.
55 G.D. Liu, C.H. Wang, Z.L. Jia, K.X. Wang, An Integrative 3D printing method for rapid additive manufacturing of a capacitive force sensor, *J. Micromechanics Microengineering*. 31 (2021) 065005.
56 L.Y.W. Loh, U. Gupta, Y. Wang, C.C. Foo, J. Zhu, W.F. Lu, 3D printed metamaterial capacitive sensing array for universal jamming gripper and human joint wearables, *Adv. Eng. Mater*. 23 (2021) 2001082.
57 K. Kobashi, T. Villmow, T. Andres, L. Häußler, P. Pötschke, Investigation of liquid sensing mechanism of poly(lactic acid)/multi-walled carbon nanotube composite films, *Smart Mater. Struct*. 18 (2009) 035008.
58 T. Villmow, S. Pegel, A. John, R. Rentenberger, P. Pötschke, Liquid sensing: Smart polymer/CNT composites, *Mater. Today*. 14 (2011) 340–345.
59 G. Postiglione, G. Natale, G. Griffini, M. Levi, S. Turri, Conductive 3D microstructures by direct 3D printing of polymer/carbon nanotube nanocomposites via liquid deposition modeling, *Compos. Part A Appl. Sci. Manuf*. 76 (2015) 110–114.
60 N. Aliheidari, P. Potschke, A. Ameli, Solvent sensitivity of smart 3D-printed nanocomposite liquid sensor, in: *Behav. Mech. Multifunct. Mater. Compos. XII, SPIE*, 2018: p. 26.
61 N. Aliheidari, C. Hohimer, A. Ameli, 3D-printed conductive nanocomposites for liquid sensing applications, in: Vol. 1 *Dev. Charact. Multifunct. Mater. Mech. Behav. Act. Mater. Bioinspired Smart Mater. Syst. Energy Harvest. Emerg. Technol., ASME*, 2017: p. V001T08A008.
62 C.M. Hansen, *Hansen Solubility Parameters: A User's Handbook*, CRC Press, 2007.
63 Y. Liu, B. Shi, Determination of Flory interaction parameters between polyimide and organic solvents by HSP theory and IGC, *Polym. Bull*. 61 (2008) 501–509.
64 Z.C. Kennedy, J.F. Christ, K.A. Evans, B.W. Arey, L.E. Sweet, M.G. Warner, R.L. Erikson, C.A. Barrett, 3D-printed poly(vinylidene fluoride)/carbon nanotube composites as a tunable, low-cost chemical vapour sensing platform, *Nanoscale*. 9 (2017) 5458–5466.
65 D. Thaler, N. Aliheidari, A. Ameli, Mechanical, electrical, and piezoresistivity behaviors of additively manufactured acrylonitrile butadiene styrene/carbon nanotube nanocomposites, *Smart Mater. Struct*. 28 (2019) 084004.

20 3D-Printed Metal-Organic Frameworks (MOFs) for Sensors

Vithyasaahar Sethumadhavan[1], *Preethichandra D. M. Gamage*[1,2], *and Prashant Sonar*[*1]
[1]Centre for Material Science and School of Chemistry and Physics, Faculty of Science, Queensland University of Technology – Brisbane.
[2]School of Engineering and Technology, Central Queensland University, Australia.
*Corresponding author: sonar.prashant@qut.edu.au

CONTENTS

20.1 INTRODUCTION

Metal-organic frameworks (MOFs) [1], [2] are composed of metal ions, porous nature materials, coordination polymers, and organic linkers. The functionality of MOFs can be tuned by changing their structures, chemistry, and material (functional) composites for their use in sensing activities. MOFs can be deposited using chemical vapor [3], electrochemical [4], liquid-gas phase [5], solvothermal [6], gel-vapour [7], and spray drying [8] deposition techniques. Among these deposition techniques, direct complex structures with high accuracy and less amount of material wastage can be achieved by 3D printing techniques. These techniques provide more control towards size, shape (symmetrical and asymmetrical), precision, cost-effectiveness, and surface morphology where it cannot be achieved using other conventional deposition technologies. The tunable 3D MOF's inherent (physical, chemical, and structural) properties have shown exceptional performance in ion adsorbents, fluorescent sensitizers, ligand interaction for sensing, and functional membranes [9]–[11]. The incorporation of polymers, functional materials, and metal ions improves the mechanical properties of MOFs which can be used to develop sensors for wearable [12] and flexible applications [13] (e.g., actuators, supercapacitors [14], implantable sensors, and optoelectronic sensors). 3D printed MOF materials are attracted towards sensing mainly because, the chemistry can be tuned between the interaction host-guest coordination sites, and their change in stable thermal, photochemical, and mechanical properties. MOFs can also be utilized in luminescent, calorimetric, and fluorescent sensing. The guest ions directly interact on the MOF functionalized surface, and it forms a coordination bond with metal sites. Strong coordination interaction improves the charge transfer or electron transfer between the analyte species and exhibits sensing phenomenon in MOFs [15], [16]. MOF

DOI: 10.1201/9781003296676-20

TABLE 20.1
Polymers for MOF to Improve the Chemical Stability and Mechanical Integrity [18]

Polymer (Active)	Printing Temp. (°C)	Substrate Temp (°C)	Biodegradability	Hygroscopicity
PLA (polylactic acid)	180–230 °C (~ 210 °C)	No	Yes, renewable sources	No
ABS (acrylonitrile butadiene styrene)	210–250 °C (Tg = 110 °C)	Yes, (80–100 °C)	No	No
PETG (polyethylene terephthalate)	220–250 °C (Tg = 80 °C)	Yes, (50–75 °C)	No	Yes
Nylon	240–260 °C	Yes, 70–100 °C	No	Yes
TPU	210–230 °C	Yes, 30–60 °C	No	Yes
Torlon	260–275 °C (Tg = 260 °C)	Yes	No	Yes

has a large surface area of ~8,000 m^2/g amongst other material composites which attracted sensing with the help of surface functionalization [17].

MOF composites can be deposited using various techniques such as Fused Filament Fabrication (FFF), Selective Laser Sintering (SLS), and Direct Ink Writing (DIW) for the fabrication of 2D and 3D structures. Various polymers for MOF and their properties are listed in below Table 20.1.

To improve the functionality of MOFs, possible approaches are listed below, and the pictorial workflow is shown in Figure 20.1:

1. organic linkers based on redox active materials: to enhance the charge (holes and electrons) transfer mechanism
2. use of transition metal ions to improve the coordination interaction between the ligands and metal sites
3. tuning the structural morphology and surface functionalization of MOFs
4. doping of MOFs to enhance the protons (h^+), and to improve the charge functionality for sensing.

20.2 3D-PRINTED MOF HYDROGELS AND IONOGELS FOR SENSING

3D-printed MOF hydrogels, and ionogels have attracted more attention towards the fabrication of *in situ* chemical and biological sensing fields due to their incredible physical and chemical properties. MOF materials are easily processable, tunable composites based on host-guest ion chemistry, highly stable based on the coordination sites, and porous to achieve adsorbing effect for sensing perspectives. MOF hydrogel displays tunable light emissive properties due to the combination of lanthanide ions with organic ligands. MOFs hydrogel is easily processable and it can be used in optoelectronic sensing of biological molecules.

Pal et al. proposed a 3D printable ink for MOF ionogel to detect the acidic compounds and human body motions using colorimetric sensing for wearable applications [19]. In this study, they used renewable nanomaterials like cellulose nanocrystals and dispersed them in choline chloride and ethylene glycol which forms a physical network in the composite as shown in Figure 20.2. After printing, the photopolymerization process occurred between the amides and it enhances the mechanical strength, stretchability, and ionic conductivity. The mechanical properties were examined for 0.1, 0.2 and 0.3 wt % of MOF showed relative increase and decrease in young's modulus and strain % of the composites. The ionic conductivity of ionogel with and without MOF was 10^{-1} and 10^{-2}

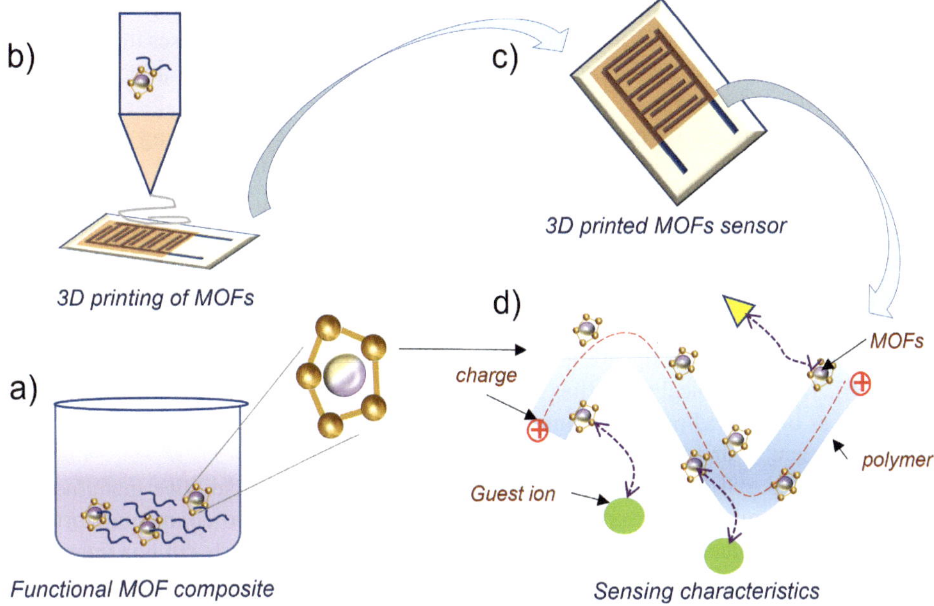

FIGURE 20.1 Schematic workflow of (a) Formulation of functionalized MOFs using metal ion sites, organic ligands, polymers, and solvents. (b) 3D printing of MOF layers using functionalized MOF materials. (c) 3D printed (interdigitated pattern) MOF sensor to sense gases and (d) interaction mechanism of analytes with 3D MOFs.

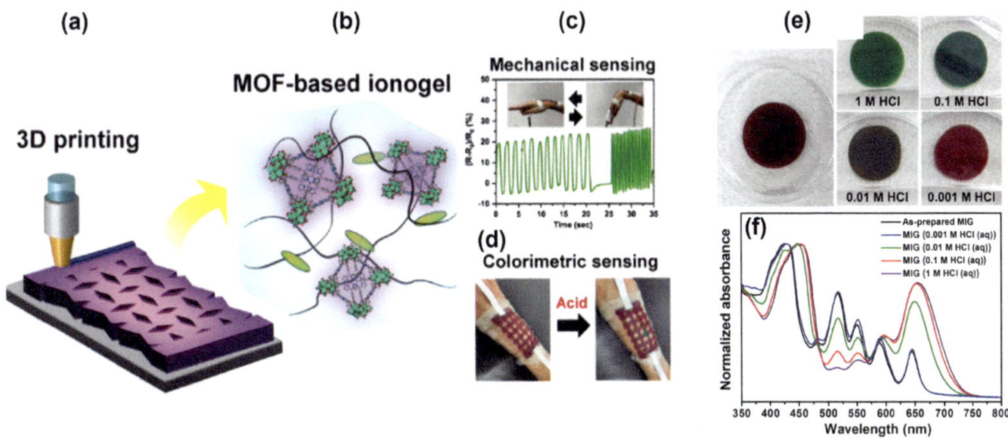

FIGURE 20.2 Pictorial representation of (a) 3D printing of ionogel (b) structure of MOF cellulose nanocrystals (c) sensing of human body vibration and their change in electrical response (d, e) sensing of acidic compounds (e.g., 0.01, 0.1, 1 M HCL) showed calorimetric response (f) change in absorbance vs. the concentration of HCl. Adapted with permission [22]. Copyright 2022, American Chemical Society.

S/cm was observed, the decrease of ionic conductivity due to the insulating behavior of MOF and restricts the movement of ions [20], [21]. Similar 3D printed ionogel showed brown color for pH = 2 and green for pH = 0 and 1 after being exposed to 0.01 M and 1 M of HCl as shown in Figure 20.2 (e, f). Upon increase in pH > 2, no color change was observed, and a similar result has been reported by Deibert et al., who they used porphyrinic zirconium (Zr-MOF) [22]. The change in color is due to the interaction of protons from acidic solutions with the MOF linker and it causes the delocalization of charge carries in the π - conjugation system. The change in charge transfer affects the band gap and the color shift is observed from acid to base.

Liu et al., demonstrated 3D-printed stretchable *in-situ* MOF hydrogel using monomer acrylamide, Irgacure 259 as photoinitiator, trimesic acid (H3BTC) as organic ligand, and triethylamine as a deprotonating agent. They formulated the MOF hydrogel using all the functional materials and then added hydroxyethyl cellulose (HEC) which acts as a shear-thinning agent for 3D printing. After 3D printing of various structures, UV curing was performed to initiate the free-radical polymerization in the printed network structures [23]. By following Lim and co-worker's work on iconic MOFs [24], printed structures were immersed into the Copper (II) nitrate Cu $(NO_3)_2$ solution to accelerate the ionic cross-linking with the ligands and it transformed from printed structure to transparent soft hydrogels [25]. The mechanical properties were examined, and it showed the stretchability of 400% with a toughness of 700 (kJ/m^3). By using this technique, 3D-printed stretchable MOFs hydrogel can be used in wearable and flexible sensor fields.

Huang et al. proposed a 3D assembly of luminescent MOFs using additive manufacturing as direct ink writing of lanthanum-based MOF structures. They have used two-component formulations composed of organic ligand and sodium alginate-based precursor ink which exhibits inter and intramolecular hydrogen bonding [26]. Europium (III) nitrate and Terbium (III) nitrate pentahydrate were used as an *in-situ* active material as a solution, where the 3D-printed structures were immersed in the solution for five minutes as shown in Figure 20.3.

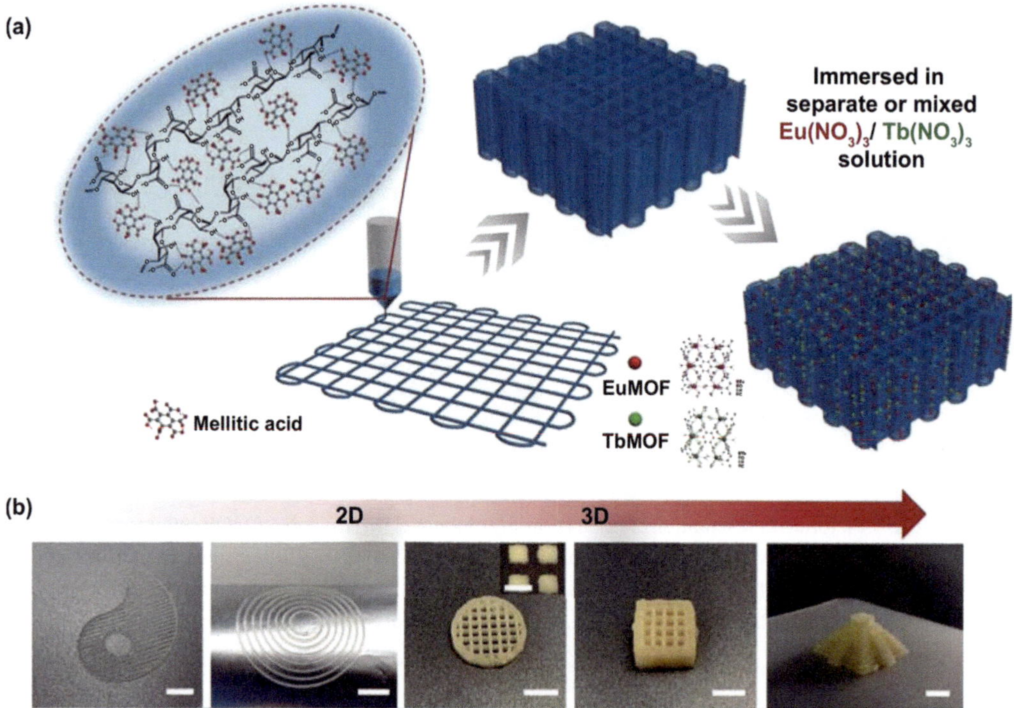

FIGURE 20.3 Schematic illustration of 3D printed MOFs (a) lanthanum (Eu and Tb) based MOFs in the precursor ink for 3D printing (b) 2D and 3D printed structures. Reproduced with permission [26]. Copyright 2021, Springer Nature.

Sodium alginate is mainly used to modify the rheological and shear thinning properties. After the post-treatment in lanthanide ions, 3D-printed structures of LnMOFs [27] exhibit fluorescence properties due to the coordination of lanthanide ions and chelation. Here in, Eu^{3+} and Tb^{3+} based lanthanum MOFs generate red and green fluorescence when it interacts with acetone [28]. The carboxylic acid from the metallic acid group delivers strong hydrogen bonding interactions and coordination with organic ligands. The study of Hu et al. prescribed that the organic ligands act as a sensitizer for the lanthanum-based ions because of the antenna effect. So, organic ligands can absorb the incident light and it shifts to Eu^{3+} and Tb^{3+} ions to exhibit fluorescence properties [29]. When the acetone concentration increases, the weakly coordinated molecules of water and ethanol were replaced by acetone molecules, and it results in fluorescence properties. The tuning of ions in MOFs for precursor ink to adsorb solvent molecules can be used for sensing ions in industrial applications.

Thakkar et al. developed 3D-printed MOF-based monoliths using solution-based methods where firstly cobalt (UTSA-16) and nickel (MOF-74) based powders were added with clay and ethanol of (80–20 wt %) [30]. On the other hand, polyvinyl alcohol (5 wt %) as a plasticizer, and deionized water with ethanol (5:95 vol %) were stirred vigorously for around 2.5 h using IKA RW 20 mixer. Later, both the solutions were loaded into the syringe with a diameter of 0.85 mm nozzle for the fabrication of (1.5 cm^2 circle) 3D-printed structures as shown in Figure 20.4.

The printed structures were cured thermally to dry completely using a conventional oven at 100 °C for 1 h. The mechanical properties were studied by compressing the sample using a 500 N loadcell at 0.0416 mm/s. 3D printed MOF monoliths adsorbed CO_2 / N_2 under the pressure of (1 bar) at 25, 50 and 75 °C showed a linear response with an increase in the concentration of adsorbed gases. The physical properties and mechanical integrity of 3D-printed monoliths showed stable performance and using these proofs-of-concept can be used for adsorbing various gases.

Wang et al. developed a MOF composite to fabricate 3D-printed ABS frameworks which can be used as an absorbent for dye absorption. They designed two kinds of frameworks (i) Printed skeleton-like structures using ABS polymers for scanning electron microscopy and (ii) ABS laminates for optical and structural characterization techniques (XRD, UV-Vis). Initially, ABS frameworks and laminates were washed with a 1:1 ratio of (EtOH: water) and then were immersed in 1,3,5-benzene tricarboxylic acid and triethylamine for four hours. Later, ABS components were immersed in copper (II) nitrate solution and it showed blue color which represents the ion exchange mechanism

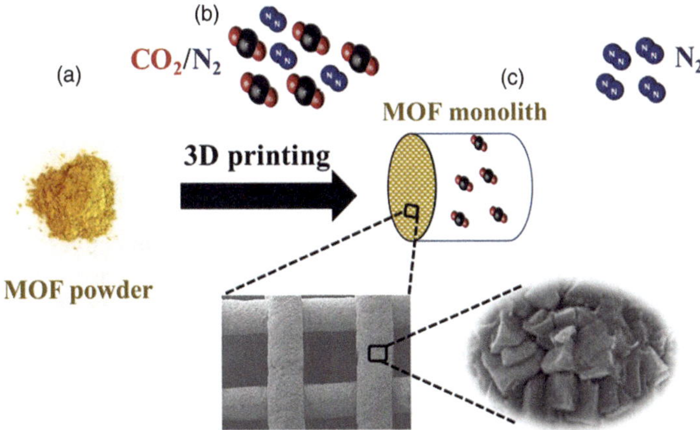

FIGURE 20.4 Schematic workflow of (a) Mixture of nickel and cobalt MOF powder (b) 3D printed MOF monoliths and the adsorption of CO_2/N_2 interaction (c) SEM image of 3D printed MOF monoliths and microporous flakes of nickel MOF-74 powder. Reproduced with permission [30]. Copyright 2017, American Chemical Society.

that occurred between ABS and copper nitrate solution [31]. From the UV-Vis analysis, the adsorption peak at 664 nm was observed because of the chemical adsorption of nitrogen atoms coordinated with the copper atom. Similar results were observed by Li et al. when the graphite oxide-based MOF was exposed to acetone, and it showed a blue color and it can be used for optical sensing using MOF composite materials [32]. Li et al. developed 3D-printed ZSM - 5 monolith catalysts by introducing a metal doping process to evaluate the selectivity of olefins [33]. To formulate the composite, metal dopants of Caesium (Ce), Chromium (Cr), Copper (Cu), Gallium (Ga), Lanthanum (La), Magnesium (Mg), Yttrium (Y), and Zinc (Zn) with nitrate as an anion as a precursor was added. The paste was then filled in the extruder to fabricate 3D-printed monoliths were ignited at 550 °C for 6 h to decay and remove the plasticizer. After this process, 3D printed monoliths exhibited micro-macro and mesoporous networks. Zn and Mg-based ZSM monolith showed an effective selectivity towards the light olefins amongst other metal dopants. This study prescribed the ion exchange of proton (h^+) for the hydroxyl group with the metal cation, and it can be used for sensing ion in future using MOF based composites with dopants.

20.3 BIOCHEMICAL AND BIOMEDICAL SENSORS

Biochemical sensing is the process of sensing chemical parameters in biofluids such as sweat, saliva, tears, urine, and blood to find the level of certain biological parameters of interest. Leland C. Clark Jr. is considered the pioneer in this field with his 1962 invention of the Clark electrode which measures oxygen in the blood. Since that time, the principle of biochemical sensing has been used to develop various biomedical sensors including the present-day blood glucose sensors which diabetes patients regularly use to check their blood glucose levels. The blood glucose sensors have pre-loaded enzymes (glucose oxidase) in them and produce H_2O_2 through a cyclic reaction. The enzyme participates in the reaction as a catalyst and H_2O_2 produced undergoes an electrolysis process where they produce oxygen proportional to the amount of glucose in the blood. An amperometric measurement of electrons produced in this electrolysis is measured and correlated with the amount of glucose in the blood. The sensor substrate morphology and surface area play a crucial role in enzyme binding, hence the performance of the biomedical sensors. There are many different approaches made by researchers to improve these two factors and using MOF is one of them.

Heavy metals such as Cd, As, Hg, etc. can sometimes be found in human food where most of the time they enter the food chain through illegal disposal of industrial waste. Detection of heavy metals in human tissues is very important to avoid long-term exposure of communities to contaminant food unknowingly, but the tests to detect heavy metals in human tissues are normally carried out in laboratories with the help of high-end analytical instruments. However, in the recent past researchers have developed sensors using MOF to monitor heavy metals easily and quickly. Kokinos *et. al* developed a 3D printed lab-in-a-syringe type of biochemical sensor to determine the Hg(II) [34] as shown in Figure 20.5. They have used Ca-MOF to modify the working electrode of graphene in different ratios between 5–20% (w/w).

Screen printing has been used as one of the easiest and most cost-effective 3D-printing methods for MOF working electrodes in biomedical sensors to develop MOF sensors. Some of them have used a single-layer screen print using the inherent 3D structure of the MOF used and the others have created 3D porous structures by printing multiple layers of the 2D MOFs. However, Lawson et al. have shown that the relatively large particle sizes have caused the printed MOF layers to be hydrophobic and special care need to be taken to bind them together in successive layer printing by fine-tuning the rheology of the binder [35]. They have developed an interesting way of growing MOF structures through a gel-print-grow process. Multilayer spin coating is another popular way of printing 3D MOF structures over pre-printed or pre-deposited sensor electrodes. Gupta et al. have developed a 3D MOF structure using this method to develop their biosensor for detecting *E. coli* [36]. They have used carbon-screen printed electrodes as the substrate and spin-coated MIL-53

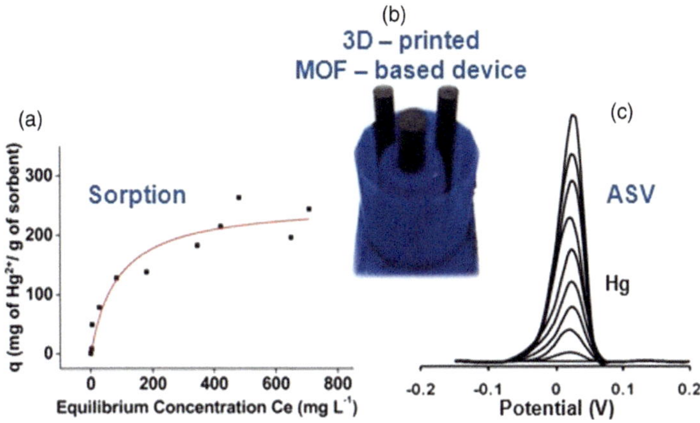

FIGURE 20.5 A pictorial representation of (a) isotherm data for Hg (II) sorption and fitted using Langmuir model (b) photograph of 3D printed Ce-MOF device (c) variable concentration of Hg(II) vs. applied potential using anodic stripping voltammetry. Reproduced with permission [34]. Copyright 2020, Elsevier.

(Fe) with poly(3,4-ethylenedioxytheophene) polystyrene sulphonate (PEDOT: PSS) as the sensing material. This sensor has shown a very wide detection range of 2.1×10^2 cfu/mL to 2.1×10^8 cfu/mL.

Kim et al. have developed a series of MOF-polymer composite inks suitable for pen-writing, spin coating, drop-casting, stencil printing, and 3D molding [37]. They have developed this class of inks by mixing methacrylate-terminated perfluoropolyether (PFPE) with different MOFs such as HKUST-1, MIL-101 (Cr), MOF-74 (Ni), ZIF-8, ZIF-67, etc. These inks can easily be used for printing sensors in various biomedical and biochemical sensor development work. There are many sensors developed using MOF inks with pre-printed sensor substrates and Zhao et al. have demonstrated a very successful biochemical sensor for detecting Fe(III) in water and human blood serum [38]. They have used MIL-53 (Fe)-$(OH)_2$ as the MOF to detect Fe (III) using the fluorescence quenching property of the MOF when exposed to Fe (III). The authors have attributed the fluorescence to 2,5-dihydroxyterepthelic acid (H4DOBDC) added to the solution. The MIL-Fe (III)- $(OH)_2$/ H4DOBC had a fluorescence quenching efficiency of 10.1% whereas the MIL-Fe (III)-$(OH)_2$/ H4DOBC exposed to Fe (III) has shown a fluorescence quenching efficiency of 26.9% and therefore a clear variation can be seen. They have also demonstrated that the developed sensor has very good selectivity for Fe^{+3} against various common cations.

L-tryptophan is a neutral amino acid available in the human body which contributes to protein synthesis. In addition, it is a precursor to several biologically active compounds found in the human body such as melatonin, quinolinic acid, kynurenic acid, and tryptamine. Recent studies have revealed that the quantitative detection of L-tryptophan can be used to diagnose several metabolic disorders. A 3D Ni-based conductive MOF has been proposed by Huang et. al for detecting the L-tryptophan levels [39] as shown in Figure 20.6. The surface area of Ni-MOFs was 0.218 cm^3/g, upon doping with Co it increased to 0.453 cm^3/g which has more porous and effective for sensing.

They have prepared Ni-MOFs and Cobalt (Co) doped Ni-MOFs using a single-pot synthesis method and the sensors were printed on glassy carbon electrodes by the drop-casting method. Cyclic voltammetry and impedance spectrometry methods have been employed to measure L-tryptophan levels in samples. XPS studies showed that some of the Ni^{2+} were substituted by Co^{2+} which increases the number of holes that can interact with negative ions. Their sensors have displayed a linear range of 10 nML^{-1} to 300 nML^{-1} with the lowest detection limit of 8.7 nML^{-1}. Fabricated sensors showed better chemical stability, selectivity, and reproducibility.

Human Immunoglobulin G (IgG) is the most common type of antibody in the blood, and it protects the human body from viruses, bacteria, and fungi. Measuring the level of IgG is important

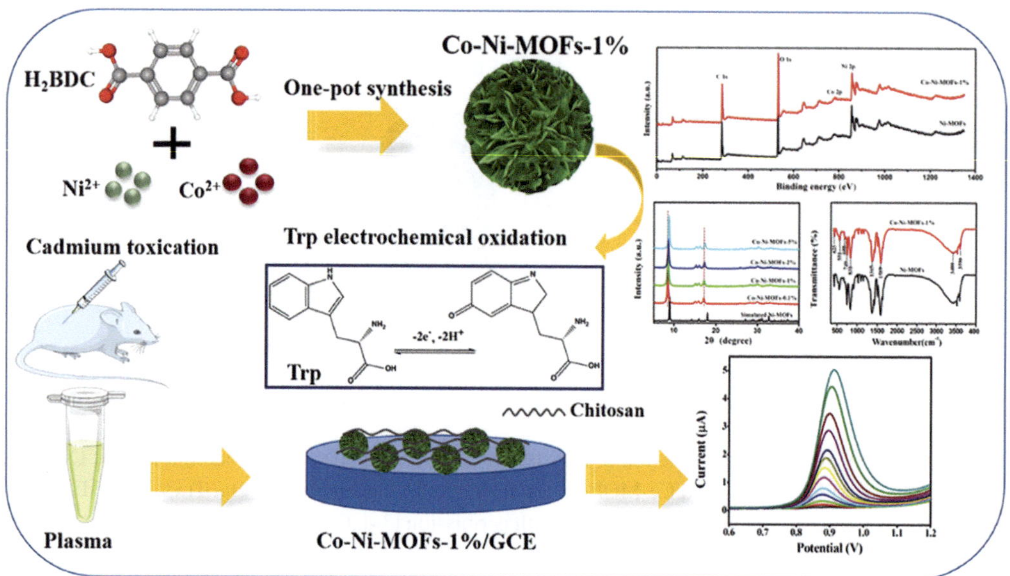

FIGURE 20.6 Schematic workflow of Co-Ni-MOF-1% and the sensing strategy for L-tryptophan (Trp). Reprinted with permission [39]. Copyright 2022, Elsevier.

for treating patients with different diseases, but it usually needs laboratory testing of blood samples. However, if IgG can be measured using a fast-response type clinical device, it would be a real convenience to medical clinicians in treating and managing patients. Liang et al. have developed a molecularly imprinted sensor based on Cu-MOF to successfully measure the IgG levels in human blood [40]. They propose a method of electrochemically depositing of Cu-MOFs on a glassy carbon electrode (GCE) and then followed by chitosan and glutaraldehyde to make high-binding loading sites for the IgG. Then this modified substrate was exposed to IgG and allowed it to bind the antibody and then a layer of polypyrrole was electropolymerized to cover the unoccupied sites. Next, they removed the IgG bound to the substrate leaving a cavity specific for IgG providing the selectivity for this protein. Consequently, non-electroactive glutaraldehyde species are associated with the amino group through aldehyde to form an imine bond with active species. Their CV measurements before and after the extraction of IgG clearly show that the molecular imprint has been successfully made. The sensor has shown a range of 0.01 to 10 ngmL-1 and a limit of detection of 3pgmL^{-1}.

Measuring the level of glucose in the blood is a common test to determine the health of a patient and measuring the glucose content in products is of vital importance in the food industry. Enzyme-based measurement of blood glucose is the default standard, but the shelf-lifetime of these sensors is a concern. In addressing this problem researchers have developed non-enzyme MOF-based glucose sensors where that can be stored on shelves for a long time as they do not have biomaterial in them. Meng *et al.* have developed a Co-based MOF (ZIF-67) with various Ag percentages (0.1%, 0.2%, 0.5%, and 0.8%) and printed the sensor electrode by drop casting these mixtures separately on polished glassy carbon electrodes [41]. Through cyclic voltammetry measurements, they have shown that there are two stages of redox peaks where one corresponds to the transition between Co(II) and Co(III), while the other peak corresponds to the transition between Co(III) and Co(IV). They have used amperometric measurements to determine the reaction current which is caused by the electrons released by glucose when it transits to gluconolactone under the provided potential. The range this sensor displayed is 2 – 1000 μM with a sensitivity of 0.379 μA μM^{-1} cm^{-2} and the limit of detection is 0.66 μM, where they are all suited for blood glucose measurement and industrial measurement as well. The amount of research work found in developing MOF-based biomedical and biochemical sensors has increased exponentially in the past decade. However, still there is more to investigate in developing and fine-tuning MOF parameters to match specific sensing applications.

20.4 3D PRINTED MOF CHEMICAL AND ELECTROCHEMICAL SENSORS

Electrochemical sensors are popular in environmental and biomedical fields as they can be fine-tuned to have a very good selectivity using a variety of techniques. The electrochemical sensors may be used in liquid phase or gas/vapor phase measurements where working with biofluids is common. Due to their relatively large size and structural properties, MOFs play a vital role in creating porous sensing material and increasing the active surface area for electrochemical reactions. The advantage of MOF-based electrochemical sensors is that due to the enhanced reaction area, they can provide significantly high reaction currents which can easily be monitored by microcontroller-based systems to develop low-cost instruments.

Organophosphates are neurotoxic chemicals commonly found in pesticides and nerve agents. Chlorpyrifos is an organophosphate banned by the UK in 2016, the EU in 2020, and the USA in 2021 as a dangerous chemical for agriculture due to its harmful effects. Therefore, detecting chlorpyrifos in water is important to monitor environmental health. Usually, it is detected by laboratory tests such as enzyme-linked immunoassay, high-performance liquid chromatography (HPLC), fluorimetry, surface-enhanced Raman scattering, etc. Janjani et al. have developed bimetallic MOF-modified screen-printed electrodes for electrochemical sensing of chlorpyrifos [42]. In their work pre- synthesized MnFe-MOF and Nafion mixture was drop cast on printed carbon electrodes. Nafion has been used as the binder for MnFe-MOS and has been mixed under sonication. They have achieved a linear range of 1 nM to 100 nM and a detection limit of 0.85 nM using this 3D printed MnFe-MOF sensor.

The presence of toxic heavy metals in waterways is a worldwide problem and authorities spend a huge sum of funding to monitor heavy metal contamination in water to prevent the public from being exposed to contaminated water. Normally water samples from waterways and reservoirs are taken into laboratories and analyzed using high-end instruments to measure traces of heavy metals, but this is an expensive process and takes time. There are in-situ measurement systems developed by researchers for this purpose using different techniques and MOF-based sensors becoming a promising approach due to their inherent large porosity and ease of fabrication. Both mercury (Hg) and thallium (Tl) are highly toxic heavy metals and therefore detecting their presence and measuring the quantities in natural water is very important. Bghayeri et al. developed a Cu-based MOF sensor that can determine traces of Tl(I) and Hg(I) [43]. The sensor was developed by drop casting the mixture of Cu-MOF, dimethylformamide, ethanol, and DI water on a polished GCE. They have used square wave anodic stripping voltammetry (SWASV) to measure the traces of these two heavy metals and the simultaneous measuring ranges are 0.5 – 700 ppb and 1–400 ppb for Tl and Hg, respectively. Chen et al. designed a platform to fabricate a portable, efficient and *in-situ* monitoring gas sensor using organolead halide-based MOF, which showed an immediate selectivity response to NH_3 at a very low detection rate of 12 ppm [44]. Herein, they used lead chloride-based MOF which has high porosity, and chemical stability and these materials exhibit photoluminescent, stable fluorescence characteristics towards sensing chemical gases. The high selectivity of NH_3 is due to the MOF porosity, where NH_3 molecules are small, and it is attributed to the fast charge (mass) transport and strong interaction (coordination) with Pb^{2+} ions. Long-term exposure to NH_3 will significantly affect the biological nervous system and it leads to blindness and neurological disorder. In aqueous-based MOF sensors, water molecules play a major role to enhance the strong coordination between the host and guest ion due to the nucleophilic substitution effect. Because NH_3 nucleophilicity is more than H_2O, tends to form a Pb^{2+}-NH_3. Various characterization techniques including X-ray photoelectron spectroscopy (XPS), Powder X-ray diffraction (PXRD), UV-Vis diffuse reflectance spectra

(UV-Vis), photoluminescence decay curves, and Density functional theory (DFT) showed a similar interaction site showed high selectivity towards NH_3 in compared with nitrogen, carbon dioxide, and oxygen. This study showed a better understanding of solid-state *in situ* gas sensors using various coordination sites. DMello et al. functionalized a MOF with an amine to detect acidic gases such as sulfur dioxide (SO_2), nitrogen dioxide (NO_2), and carbon dioxide (CO_2) with a detection limit of 1 to 10 ppm, 10 ppm, and 5000 ppm [45]. They have developed a composite based on metal oxide, Zirconium Zr-NH_2 benzene dicarboxylate (NH_2-UiO-66) at 150 °C and are semiconductive. Amine acts as a functionalization layer (linker), to increase the adsorption effect which allows the charge transfer between the functional complexes [46]. It showed a very good response over 1200 s by varying the concentration of gases over some time and it highlights the organic linker to sense other gases based on choosing suitable functional groups. From the understanding of the charge transfer phenomenon, the band position of the highest occupied crystal orbital (HOCO) at ~ -5 eV and lowest occupied crystal orbital (LUCO) at ~ -2.5 eV of MOF structure to the HOMO-LUMO of the guest (analyte) gases [47]. Further introduction of pore surface functionalization of MOF surface will increase the detection limit and operate at room temperature conditions.

Arsenic (As) is another heavy metal found in natural waters mainly through waste disposal and agricultural chemicals but also presents in soil by natural occurrences in some parts of the world. Arsenic can be found in the forms of As(III) and As(V) in natural waterways. Detecting and measuring Arsenic in water is like other heavy metal detection, where it needs expensive laboratory testing equipment. P. Xiao et al. have prepared Fe-MOF/MXene dispersed in DI water and drop cast it on polished GCE to make their sensor [48] as shown in Figure 20.7.

They employed the SAWSV method to determine the As (III) in water and showed good dual linear ranges of 1–10 ngL^{-1} and 10 -100 ngL^{-1} where the sensitivity of the first range was 8.94 $\mu AngL^{-1}$. Detection and measurement of volatile organic compounds (VOC) is an important part of air quality monitoring. This involves multiple sensors, mostly semiconductor or metal oxide-based ones. However, in the recent past, MOFs have shown that they are suitable for developing air quality monitoring sensors. Fernandez et al. have shown that an ionic liquid (IL) / MOF compound gives promising results for VOC detection [49]. The IL used was 1-Ethyl-3-methylimidazolium

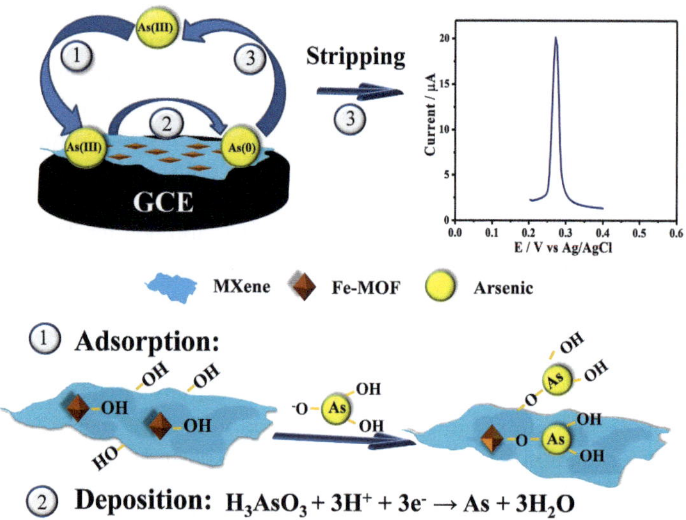

FIGURE 20.7 The schematic representation of adsorption and deposition assisted As (III) detection on Fe-MOF/MXene by square wave anodic stripping voltammetry. Reproduced with permission [48]. Copyright 2022, Elsevier.

bis(trifluoromethyl sulfonyl)imide and the MOF used was UiO-66-(OH)2. The 2D silver electrodes were screen printed and the sensor substrate was 3D printed using a spray printing method. This was done at 100 °C to promote rapid solvent evaporation. It could detect acetone and ethanol concentrations in the range of 10k to 100k ppm in less than a second. Acrylonitrile butadiene styrene (ABS) is extensively used in 3D-printing industries due to its viable mechanical, thermal, and chemical (binding) properties. Pellejero et al. functionalized ABS with MOF to sense toxic gases [50]. They have used ZIF-8 (Zeolitic Imidazolate Framework) as a MOF with ABS using post-synthesis surface modification techniques to improve the adsorption, wettability, chemical interaction, thermal stability, and mechanical integrity. By using the FFF technique, they extruded the circular mesh type structure using ABS-ZIF-8 composite and then deposited ZnO layer at 60 °C and 100 °C on 3D-printed structures and later exposed to diethyl zinc and it acts as a metal precursor. The adsorption studies were carried out by loading the MOF composites in the quartz column with a presence of dimethyl methylphosphonate vapor at 10 mL/min. Thermal studies showed the nucleation effect on the ZIF-8 surface due to the ZnO layer dissolution (rection of Zn^{2+}). From the understanding of metal oxide interaction, it is a new pathway to deposit on MOF surface or to bind with MOF to improve the functionalities in sensing toxic gases.

All these MOF-based electrochemical sensors show that there are opportunities for developing MOF-based sensors for environmental parameter measurement in the future.

20.5 CONCLUSION AND OUTLOOK

MOF materials and their intrinsic properties have extensively grown and started exploring 3D MOF composites. 3D-printed MOFs are showing more potential sensing materials because of their large surface area, surface functionalization by tuning the metal ions and ligands, and porosity to attract the host molecules (analytes). Some 3D-printed MOFs are ultra-sensitive to selective gases and chemically robust after several exposures in air and aqueous medium. In addition, 3D MOFs promote *in situ* sensing features to attract real-time adsorption of gases or ions for real-time applications. However, further improvements on 3D-printed MOF materials are still needed to use in real-time sensors such as high selectivity to other gases, reproducibility, response time, thermal stability, shelf life, and sensitivity. The binding of organic semiconductors and other functional materials to enhance the sensing functionalities and their real-world prototypes could be some of the innovative approaches for future applications. In the future, 3D-printed MOF-based sensors will grow rapidly by approaching practical sensing fields.

REFERENCES

1 A. E. Baumann, D. A. Burns, B. Liu, and V. S. Thoi, Metal-organic framework functionalization and design strategies for advanced electrochemical energy storage devices. *Communications Chemistry*, **2** (2019) 1–14.

2 S. Abednatanzi, P. G. Derakhshandeh, H. Depauw, F.-X. Coudert, H. Vrielinck, P. Van Der Voort, and K. Leus, Mixed-metal metal–organic frameworks. *Chemical Society Reviews*, **48** (2019) 2535–2565.

3 R. Haldar, L. Heinke, and C. Wöll, Advanced photoresponsive materials using the metal–organic framework approach. *Advanced Materials,* **32** (2020) 1905227.

4 M. Varsha and G. Nageswaran, Direct electrochemical synthesis of metal organic frameworks. *Journal of the Electrochemical Society,* **167** (2020) 155527.

5 C. Crivello, S. Sevim, O. Graniel, C. Franco, S. Pané, J. Puigmartí-Luis, and D. Muñoz-Rojas, Advanced technologies for the fabrication of MOF thin films. *Materials Horizons*, **8** (2021) 168–178.

6 M. Y. Zorainy, M. Sheashea, S. Kaliaguine, M. Gobara, and D. C. Boffito, Facile solvothermal synthesis of a MIL-47 (V) metal–organic framework for a high-performance Epoxy/MOF coating with improved anticorrosion properties. *RSC Advances,* **12** (2022) 9008–9022.

7 W. Li, P. Su, Z. Li, Z. Xu, F. Wang, H. Ou, J. Zhang, G. Zhang, and E. Zeng, Ultrathin metal–organic framework membrane production by gel–vapour deposition. *Nature Communications*, **8** (2017) 1–8.

8 A. Carné-Sánchez, I. Imaz, M. Cano-Sarabia, and D. Maspoch, A spray-drying strategy for synthesis of nanoscale metal–organic frameworks and their assembly into hollow superstructures. *Nature Chemistry,* **5** (2013) 203–211.

9 J.-R. Li, J. Sculley, and H.-C. Zhou, Metal–organic frameworks for separations. *Chemical reviews,* **112** (2012) 869–932.

10 K. S. Walton, Recognizing the unrecognizable. *Nature Chemistry*, **6** (2014) 277–278.

11 Z. Zhang, Z.-Z. Yao, S. Xiang, and B. Chen, Perspective of microporous metal–organic frameworks for CO_2 capture and separation. *Energy & Environmental Science,* **7** (2014) 2868–2899.

12 K. Rui, X. Wang, M. Du, Y. Zhang, Q. Wang, Z. Ma, Q. Zhang, D. Li, X. Huang, and G. Sun, Dual-function metal–organic framework-based wearable fibers for gas probing and energy storage. *ACS Applied Materials & Interfaces*, **10** (2018) 2837–2842.

13 J. Gong, Z. Xu, Z. Tang, J. Zhong, and L. Zhang, Highly compressible 3-D hierarchical porous carbon nanotube/metal organic framework/polyaniline hybrid sponges supercapacitors. *AIP Advances,* **9** (2019) 055032.

14 X. Xu, W. Shi, P. Li, S. Ye, C. Ye, H. Ye, T. Lu, A. Zheng, J. Zhu, and L. Xu, Facile fabrication of three-dimensional graphene and metal–organic framework composites and their derivatives for flexible all-solid-state supercapacitors. *Chemistry of Materials,* **29** (2017) 6058–6065.

15 S.-P. Yang, S.-R. Chen, S.-W. Liu, X.-Y. Tang, L. Qin, G.-H. Qiu, J.-X. Chen, and W.-H. Chen, Platforms formed from a three-dimensional Cu-based zwitterionic metal–organic framework and probe ss-DNA: Selective fluorescent biosensors for human immunodeficiency virus 1 ds-DNA and Sudan virus RNA sequences. *Analytical Chemistry,* **87** (2015) 12206–12214.

16 H.-S. Wang, J. Li, J.-Y. Li, K. Wang, Y. Ding, and X.-H. Xia, Lanthanide-based metal-organic framework nanosheets with unique fluorescence quenching properties for two-color intracellular adenosine imaging in living cells. *NPG Asia Materials*, **9** (2017) e354–e354.

17 H. Furukawa, K. E. Cordova, M. O'Keeffe, and O. M. Yaghi, The chemistry and applications of metal-organic frameworks. *Science*, **341** (2013) 1230444.

18 E. R. Kearns, R. Gillespie, and D. M. D'Alessandro, 3D Printing of Metal-Organic Frameworks for Clean Energy and Environmental Applications. *Journal of Materials Chemistry A*, **9** (2021) 27252 –27270.

19 S. Pal, Y.-Z. Su, Y.-W. Chen, C.-H. Yu, C.-W. Kung, and S.-S. Yu, 3D Printing of metal–organic framework-based ionogels: Wearable sensors with colorimetric and mechanical responses. *ACS Applied Materials & Interfaces*, **14** (2022) 28247–28257.

20 C.-W. Lai and S.-S. Yu, 3D printable strain sensors from deep eutectic solvents and cellulose nanocrystals. *ACS Applied Materials & Interfaces,* **12** (2020) 34235–34244.

21 D. J. Ramón and G. Guillena (Eds.), *Deep Eutectic Solvents: Synthesis, Properties, and Applications* (2020). ISBN: 978-3-527-34518-2, Wiley.

22 B. J. Deibert and J. Li, A distinct reversible colorimetric and fluorescent low pH response on a water-stable zirconium–porphyrin metal–organic framework. *Chemical Communications,* **50** (2014) 9636–9639.

23 W. Liu, O. Erol, and D. H. Gracias, 3D printing of an in situ grown MOF hydrogel with tunable mechanical properties. *ACS Applied Materials & Interfaces,* **12** (2020) 33267–33275.

24 J. Lim, E. J. Lee, J. S. Choi, and N. C. Jeong, Diffusion control in the in situ synthesis of iconic metal–organic frameworks within an ionic polymer matrix. *ACS Applied Materials & Interfaces,* **10** (2018) 3793–3800.

25 J.-Y. Sun, X. Zhao, W. R. Illeperuma, O. Chaudhuri, K. H. Oh, D. J. Mooney, J. J. Vlassak, and Z. Suo, Highly stretchable and tough hydrogels. *Nature*, **489** (2012) 133–136.

26 J. Huang and P. Wu, Controlled assembly of luminescent lanthanide-organic frameworks via post-treatment of 3D-printed objects. *Nano-Micro Letters*, **13** (2021) 1–14.

27 M. O. Rodrigues, F. A. A. Paz, R. O. Freire, G. F. de Sa, A. Galembeck, M. C. Montenegro, A. N. Araujo, and S. Alves Jr, Modeling, structural, and spectroscopic studies of lanthanide-organic frameworks. *The Journal of Physical Chemistry B*, **113** (2009) 12181–12188.

28 B. Chen, Y. Yang, F. Zapata, G. Lin, G. Qian, and E. B. Lobkovsky, Luminescent open metal sites within a metal–organic framework for sensing small molecules. *Advanced Materials*, **19** (2007) 1693–1696.

29 Z. Hu, B. J. Deibert, and J. Li, Luminescent metal–organic frameworks for chemical sensing and explosive detection. *Chemical Society Reviews,* **43** (2014) 5815–5840.

30 H. Thakkar, S. Eastman, Q. Al-Naddaf, A. A. Rownaghi, and F. Rezaei, 3D-printed metal–organic framework monoliths for gas adsorption processes. *ACS Applied Materials & Interfaces,* **9** (2017) 35908–35916.

31 Z. Wang, J. Wang, M. Li, K. Sun, and C.-J. Liu, Three-dimensional printed acrylonitrile butadiene styrene framework coated with Cu-BTC metal-organic frameworks for the removal of methylene blue. *Scientific Reports,* **4** (2014) 1–7.

32 L. Li, X. L. Liu, H. Y. Geng, B. Hu, G. W. Song, and Z. S. Xu, *A* MOF/graphite oxide hybrid (MOF: HKUST-1) material for the adsorption of methylene blue from aqueous solution. *Journal of Materials Chemistry A,* **1** (2013) 10292–10299.

33 X. Li, F. Rezaei, and A. A. Rownaghi, Methanol-to-olefin conversion on 3D-printed ZSM-5 monolith catalysts: Effects of metal doping, mesoporosity and acid strength. *Microporous and Mesoporous Materials*, **276** (2019) 1–12.

34 C. Kokkinos, A. Economou, A. Pournara, M. Manos, I. Spanopoulos, M. Kanatzidis, T. Tziotzi, V. Petkov, A. Margariti, and P. Oikonomopoulos, 3D-printed lab-in-a-syringe voltammetric cell based on a working electrode modified with a highly efficient Ca-MOF sorbent for the determination of Hg (II). *Sensors and Actuators B: Chemical*, **321** (2020) 128508.

35 S. Lawson, A.- A. Alwakwak, A. A. Rownaghi, and F. Rezaei, Gel–print–grow: A new way of 3D printing metal–organic frameworks. *ACS Applied Materials & Interfaces,* **12** (2020) 56108–56117.

36 A. Gupta, A. L. Sharma, and A. Deep, Sensitive impedimetric detection of E. coli with metal-organic framework (MIL-53)/polymer (PEDOT) composite modified screen-printed electrodes. *Journal of Environmental Chemical Engineering,* **9** (2021) 104925.

37 J.-O. Kim, J. Y. Kim, J.- C. Lee, S. Park, H. R. Moon, and D.- P. Kim, Versatile processing of metal–organic framework–fluoropolymer composite inks with chemical resistance and sensor applications. *ACS Applied Materials & Interfaces,* **11** (2019) 4385–4392.

38 Y. Zhao, H. Ouyang, S. Feng, Y. Luo, Q. Shi, C. Zhu, Y.- C. Chang, L. Li, D. Du, and H. Yang, Rapid and selective detection of Fe (III) by using a smartphone-based device as a portable detector and hydroxyl functionalized metal-organic frameworks as the fluorescence probe. *Analytica Chimica Acta,* **1077** (2019) 160–166.

39 W. Huang, Y. Chen, L. Wu, M. Long, Z. Lin, Q. Su, F. Zheng, S. Wu, H. Li, and G. Yu, 3D Co-doped Ni-based conductive MOFs modified electrochemical sensor for highly sensitive detection of l-tryptophan. *Talanta,* **247** (2022) 123596.

40 A. Liang, S. Tang, M. Liu, Y. Yi, B. Xie, H. Hou, and A. Luo, A molecularly imprinted electrochemical sensor with tunable electrosynthesized Cu-MOFs modification for ultrasensitive detection of human IgG. *Bioelectrochemistry*, **146** (2022) 108154.

41 W. Meng, Y. Wen, L. Dai, Z. He, and L. Wang, A novel electrochemical sensor for glucose detection based on Ag@ ZIF-67 nanocomposite. *Sensors and Actuators B: Chemical*, **260** (2018) 852–860.

42 P. Janjani, U. Bhardwaj, R. Gupta, and H. S. Kushwaha, Bimetallic Mn/Fe MOF modified screen-printed electrodes for non-enzymatic electrochemical sensing of organophosphate. *Analytica Chimica Acta,* **1202** (2022) 339676.

43 M. Baghayeri, A. Amiri, B. S. Moghaddam, and M. Nodehi, Cu-based MOF for simultaneous determination of trace Tl (I) and Hg (II) by stripping voltammetry. *Journal of the Electrochemical Society,* **167** (2020) 167522.

44 X. Chen, Y. Yu, C. Yang, J. Yin, X. Song, J. Li, and H. Fei, Fabrication of robust and porous lead chloride-based metal–organic frameworks toward a selective and sensitive smart NH_3 sensor. *ACS Applied Materials & Interfaces,* **13** (2021) 52765–52774.

45 M. E. DMello, N. G. Sundaram, A. Singh, A. K. Singh, and S. B. Kalidindi, An amine functionalized zirconium metal–organic framework as an effective chemiresistive sensor for acidic gases. *Chemical Communications*, **55** (2019) 349–352.

46 G.-Y. Lee, J. Lee, H. T. Vo, S. Kim, H. Lee, and T. Park, Amine-functionalized covalent organic framework for efficient SO_2 capture with high reversibility. *Scientific Reports*, **7** (2017) 1–10.

47 J. Long, S. Wang, Z. Ding, S. Wang, Y. Zhou, L. Huang, and X. Wang, Amine-functionalized zirconium metal–organic framework as efficient visible-light photocatalyst for aerobic organic transformations. *Chemical Communications*, **48** (2012) 11656–11658.

48 P. Xiao, G. Zhu, X. Shang, B. Hu, B. Zhang, Z. Tang, J. Yang, and J. Liu, An Fe-MOF/MXene-based ultra-sensitive electrochemical sensor for arsenic (III) measurement. *Journal of Electroanalytical Chemistry*, **916** (2022) 116382.

49 E. Fernandez, P. G. Saiz, N. Peřinka, S. Wuttke, and R. Fernandez de Luis, Printed capacitive sensors based on ionic liquid/metal-organic framework composites for volatile organic compounds detection. *Advanced Functional Materials*, **31** (2021) 2010703.

50 I. Pellejero, F. Almazán, M. Lafuente, M. A. Urbiztondo, M. Drobek, M. Bechelany, A. Julbe, and L. M. Gandía, Functionalization of 3D printed ABS filters with MOF for toxic gas removal. *Journal of Industrial and Engineering Chemistry*, **89** (2020) 194–203.

21 3D and 4D Printing for Biomedical Applications

Lorena Maria Dering[1,2], Beatriz Luci Fernandes[1], Matheus Kahakura Franco Pedro[1,2], André Giacomelli Leal[1,2], and Mauren Abreu de Souza[1]*
[1]Postgraduate Program on Health Technology – PPGTS
Pontifícia Universidade Católica do Paraná – PUCPR
Rua Imaculada Conceição, 1155 – Prado Velho – Postal code: 80215-901
Curitiba-PR/Brazil
[2]Instituto de Neurologia e Cardiologia de Curitiba – INC
Rua Jeremias Maciel Perretto, 300 – Ecoville – Postal Code: 81210-310
Curitiba-PR/Brazil
*Corresponding author: mauren.souza@pucpr.br

CONTENTS

21.1 INTRODUCTION

In addition to the various areas in which the 3D-printing technologies are used, this knowledge applied for biomedical purposes is gaining more space in several health applications. Going beyond the standard use for the production of medical equipment and implants, additive manufacturing can reproduce precise anatomic-specific 3D models for the visualization of organs and structures, usually called 3D biomodels. The use of biomodels has a wide range of advantages, including reduction of surgical costs, related to reducing the operating time due to previous surgical planning with the use of biomodels. This approach facilitates the communication between physician and patient, as the biomodels have proven to be adequate for educational tools, which facilitates the patients' understanding of their pathological process, thus allowing them to become better informed about the whole treatment process. Additionally, 3D printing is also used to produce personalized surgical guides, commonly used in the fields of orthopedics and maxillofacial surgery, optimizing and reducing surgery time. The most commonly used types of printers and processes for application in medical settings and surgical training are multijet modeling, binder jetting, selective laser sintering (SLS), stereolithography (STL), and fused deposition modeling (FDM) [1].

Therefore, this chapter aims to provide a broad overview of medical applications involving additive manufacturing. Different medical fields are covered, mainly those that more extensively use

DOI: 10.1201/9781003296676-21

3D-printing technologies. The chapter is divided into the following topics: 3D printing in neurosurgical planning and cardiology; prostheses and orthopedic surgeries; dentistry; 4D printing; and conclusion.

21.2 3D PRINTING IN NEUROSURGICAL PLANNING AND CARDIOLOGY

The continuous technological advances in medicine, including the visualization of internal organs provided by different imaging modalities, such as computerized tomography (CT), magnetic resonance imaging (MRI), and ultrasound (US)), and the possibility of transforming two-dimensional (2D) data into three-dimensional (3D) models, made possible the reproduction by additive manufacturing of patient-specific and precise anatomic models that allow for 3D visualization of organs and structures; these are called biomodels [2]. Some of these applications of additive manufacturing in the field of medical sciences were in neurosurgery, thus allowing for a broader range of research topics over the years and a more consolidated state-of-the-art knowledge.

These biomodels may be used for research, teaching, and surgical planning, which includes the visualization of the organs or tissues to be operated on, and the respective simulation of the surgical process. Therefore, it includes the use of equipment, in addition to demonstrating the process to the patient [1], [3], making the diagnosis, information, and treatment processes more personalized [4], since they facilitate communication between physician and patient. The use of biomodels is beneficial for the understanding of the disease by the patient, with studies dating back from 1999 onwards [5], [6]. Then, it consistently demonstrates that the tactile experience of holding a model of the affected organic structure (intracranial aneurysms), greatly adds to the overall information of the patient. Thus, it allows for the consent of eventual surgery to be truly based on an accurate understanding of their disease [7].

Other well-established usages of biomodels in neurosurgery include its role in the training of new medical professionals [3]. This is due since these biomodels have assisted in training for complex surgeries by young surgeons in cases of intracranial aneurysms [8]. These manufactured models are not only reproductions of the aneurysm and its surrounding vessels, but also includes the cranial structures surrounding them.

Yet another well-established advantage of the use of biomodels in neurosurgery is the construction of structures to visually facilitate comprehension of complex anatomical features. This is due to computer screens representing two-dimensional images, even if they are showing images of three-dimensional structures. Then, the understanding of the depth and other morphological features may be impaired, which leads to longer operating times. To overcome such problems, the use of biomodels has been extensively studied and showed to be of great assistance in contributing to the anatomical understanding of complex anatomies, especially in terms of depth. Therefore, biomodels have been shown to reduce operating times, which, in turn, leads to a reduction in surgical costs, since fewer resources (such as oxygen) are also used in surgeries performed with the aid of biomodels [9], [10]. In other terms, two-dimensional images may not be easily understood about other adjacent structures and complex anatomical irregularities, and also because the interpretation of three-dimensional virtual images depends on the interpreters' mental abilities to transform two-dimensional images into three-dimensional visualization. The printed biomodels are particularly useful to neurosurgeons, as the real models of the structures can bring a tactile expression with no need for the surgeon solely rely on two-dimensional images on a screen, since a tactile, real 3D biomodel is available.

Nowadays, many neurosurgical fields are being promoted from the use of 3D biomodels, including spinal surgery, neuro-oncology, and pediatric neurosurgery, among others. Perhaps more than any other field, though, neurovascular and endovascular surgery have benefited the most from the use of biomodels. The additive manufacture of biomodels for intracranial aneurysms has long been the prototype of the use of such models in neurosurgery since they represent relatively small lesions that are easily distinguished from the surrounding structures. An analogy may be established with the use of biomodels to plan for tumor resections, in which the lesion is deeply intertwined within health neuronal structures, which leads to difficult segmentation of medical images (using different imaging modalities, such as computed tomography or magnetic resonance imaging – even combined and fused), to generate a model solely on the structure of interest.

The aforementioned uses for biomodels in surgical planning, training of new surgeons, and patient information, have all been extensively tested in studies involving brain aneurysms, which have become one of the prototypical uses of biomodels within neurosurgery. The use of biomodels for surgical planning concerning intracranial aneurysms has followed what can be broadly considered as a pattern of evolutive, or incremental uses of the material. Many of the first models were produced using fused deposition modeling (FDM) printers to create, for instance, acrylonitrile butadiene styrene (ABS) models of the arteries, originally used for patient education purposes [6] and visualization of complex 3D vascular structures [5], sometimes with the integration of cranial structures [11]. Afterward, a new model emerged covering the ABS with silicone material, resulting in an integrated model Then, both are submerged into acetone, dissolving the ABS and generating a hollow model of the arterial wall [2], [8], [12]. More recent developments include the use of other modern materials for additive manufacturers, such as flexible resin, made with liquid-crystal display (LCD) 3D printers, to produce flexible and hollow aneurysms biomodels as well [13].

To illustrate some of the results, obtained from the design and manufacture of intracranial aneurysms in our research, Figure 21.1 shows the 3D-printed models of different regions and materials, as follows: (A) Internal carotid artery aneurysm, with a focus on the surrounding bone structures, produced in white resin in an LCD printer; the artery was painted afterward with acrylic paint, for better visualization purposes; (B) middle cerebral artery aneurysm, produced in resin in an LCD printer; (C) Internal carotid artery printed on an LCD printer, using flexible resin.

Still, in the neurovascular field, different studies have shown that 3D printing is a useful tool for surgical planning of even more complex lesions, such as arteriovenous malformations (AVMs). An example to be mentioned here is the process of prior training and procedural simulations using the manufactured biomodels. The main advantage of such an approach is the reduction in intraoperative time as previously reported in our studies involving aneurysms [2]. Figure 21.2 illustrates the 3D-printed biomodel being used for surgical planning, exactly when the surgery is happening

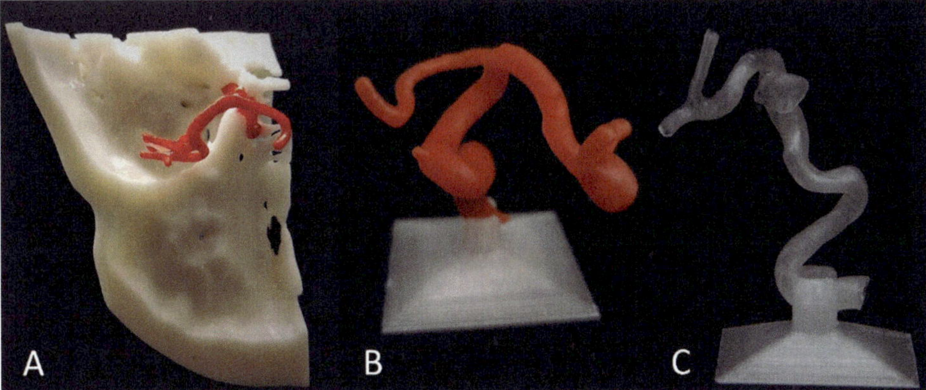

FIGURE 21.1 Aneurysm biomodels designed and manufactured for surgical planning: (A) Internal carotid artery aneurysm, with surrounding bone structures, produced in white resin in an LCD printer; the artery was painted afterward with acrylic paint, for better visualization purposes; (B) Middle cerebral artery aneurysm, produced in resin in an LCD printer; (C) Internal carotid artery printed on an LCD printer, using flexible resin.

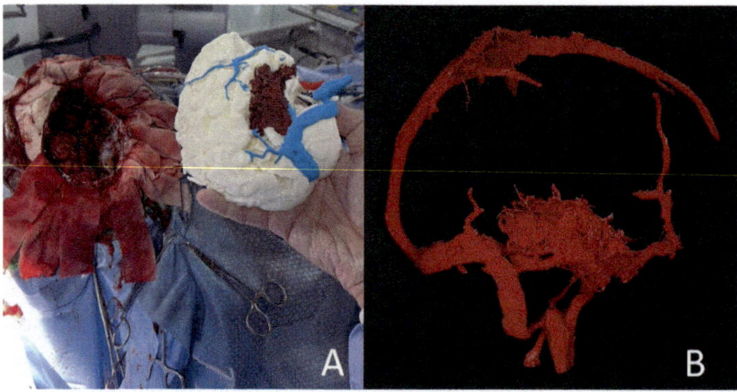

FIGURE 21.2 Biomodel of arteriovenous malformation (AVMs) used for surgical planning, showing: (A) 3D model of the same surgical region, compared with the 3D printed model (with posterior painting to discriminate arteries, veins, and AVM). (B) 3D model after printing the cerebral vessels using an FDM printer.

(Figure 21.2A). It is showing an arteriovenous malformation (AVMs) being compared with the 3D-printed model, representing both the AVM and the cerebral vessels (Figure 21.2B).

In the subfield of neuro-oncology, 3D biomodels have been used to assess the tumor infiltration area and its relationship with the surrounding healthy tissue, helping with the delineation of the resection borders. Thereby, 3D printing, associated with several medical imaging modalities, mostly magnetic resonance imaging (MRI), provides personalized surgical planning for patient care. Since it enables 2D images to be translated into personalized 3D biomodels that characterize the relationship between tumorous and healthy brain tissue, facilitating the anatomical understanding and resulting in the planning of the most effective treatment [14]. However, there are different levels of complexity, given the challenges in providing appropriate thresholding between the neoplasm and the healthy neuronal tissues, since they are often deeply infiltrated one into another. For the manufacturing of such complex models, it is possible to combine 3D printing with other techniques. For example, using the 3D printing of molds forming the brain parenchyma made of silicone (resembling the real structure) together with the tumor or other structures printed in different materials [15]; provides clearer interfaces between tumor and healthy tissues.

Moreover, biomodels can also be combined with traditional anatomy teaching methods, such as cadaveric pieces. The literature brings 3D biomodels representing vascular systems containing aneurysms, to be implanted in cadavers to conduct research for training and simulation of surgical scenarios and to study anatomical approaches [16]. Another branch of the circulatory system that is also subjected to studies on 3D printing is cardiology and cardiac surgery, which are contributing to more efficient planning of surgical interventions. For example, it is possible to print models through additive manufacturing of any cardiac compartment (ventricles and atria), or valve (aortic, mitral, pulmonary, or tricuspid). To illustrate, Figure 21.3 shows a 3D model of a left atrium, representing an enlarged atrial appendage, made with a flexible resin material.

The purposes that were described for other medical fields remain the same, such as: the models can be used for didactic purposes, i.e., for training new professionals and to inform and explain the procedures to the patients; also, it can be used as scaffolding in cell culture for replacing affected areas and for surgical simulations [17].

Cardiologic and pathological problems treated with the aid of additive manufacturing include aortic valve defects (such as stenosis due to calcification), atrial septal defect closure; complex arrhythmias (in which models of left atrium allow for better anatomical understanding), closure of enlarged left atrial appendages; pulmonary valve defects, and models of transcatheter caval valve implantation, among other cardiac requests [17]–[19].

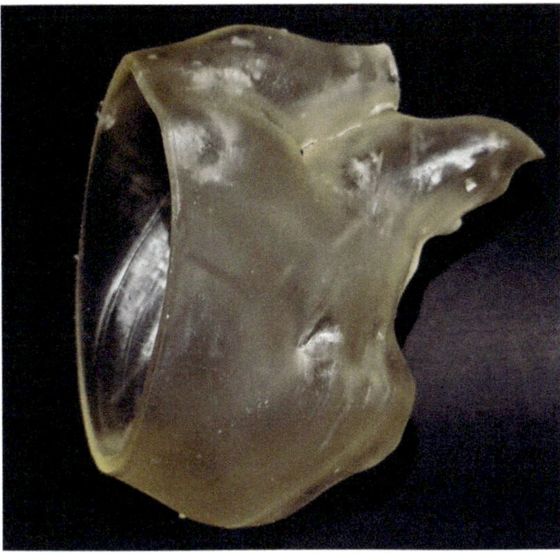

FIGURE 21.3 3D printed model of a left atrium with an enlarged atrial appendage, made with flexible resin.

21.3 PROSTHESES AND ORTHOPEDIC SURGERIES

Additive manufacturing has also been widely used in the field of prostheses. Prostheses refer to devices that are used to substitute a part of the body that is not functional anymore. One of the most widely applied uses of prostheses is in the field of cranioplasty surgeries, in which a synthetic plate is placed to correct a skull gap area, normally coming from previous surgery to decompress the skull. With 3D-printing techniques, these patient-specific plates are normally produced in two ways: (1) additive manufacture of the plate itself with biocompatible materials, or (2) additive manufacture of patient-specific molds, which will then be used to form the plate with conventional materials. The approach of 3D printing of the mold is the cheapest option, as it can be done with open-source software and desktop 3D printers. These molds are created from the imaging modalities available for the flaw region to be reconstructed (normally, employed in computerized tomography – CT), which is used to mold polymethylmethacrylate (PMMA) plates, using a material popularly known as 'bone cement'. Figure 21.4 illustrates a case study of the process of cranioplasty planning, which also used 3D printing. First, it is generated a digital 3D model of the affected region, i.e., the skull (A). Then it is created a skull plate, which is manufactured from the mold, in this case, made of PMMA (B). And finally, the implanted PMMA cranial plate is ready to be employed in the surgery and to be placed in the affected region, within the operating room. This technique can be used within hospitals, with the support of the surgical team, delivering excellent results, both in terms of mechanical properties and aesthetics [20], [21].

More recently, as an alternative to PMMA implants, prostheses made of synthetic materials, such as titanium or polyetheretherketone (PEEK) being manufactured within 3D printing have been developed. Titanium plates can be produced through the use of selective laser melting (SLM) 3D printers, corresponding to a powder bed fusion 3D-printing technology. Another option is printing plates with PEEK, a high-temperature thermoplastic polymer, on a fused filament fabrication (FFF) printer [20]. In orthopedic surgeries, the anatomical reconstructions of bone structures allow for better anatomical understanding by medical students and residents helping with teaching, training, and simulations [22]. As an example of such a medical application, Figure 21.5 illustrates a 3D-printed model of a knee, including several anatomical regions altogether, such as the patella, femur, and tibia, which were manufactured in ABS material.

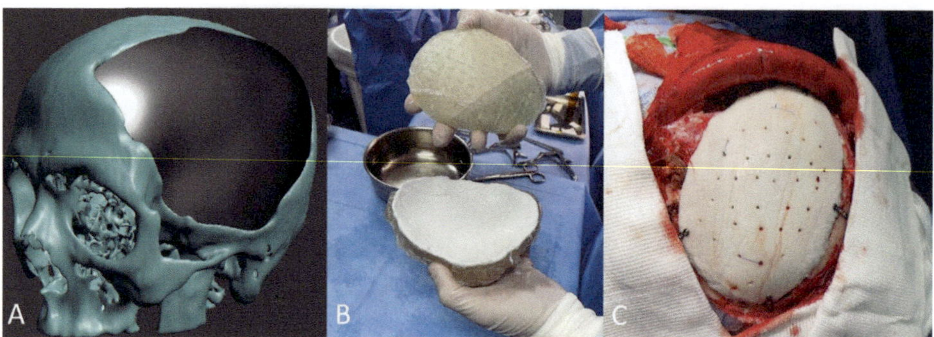

FIGURE 21.4 An example of cranioplasty planning, which was performed with 3D printing: (A) Digital 3D modeling of the cranial plate; (B) The skull plate made of PMMA, manufactured from the mold; (C) the implanted PMMA cranial plate placed within the affected region, in the operating room.

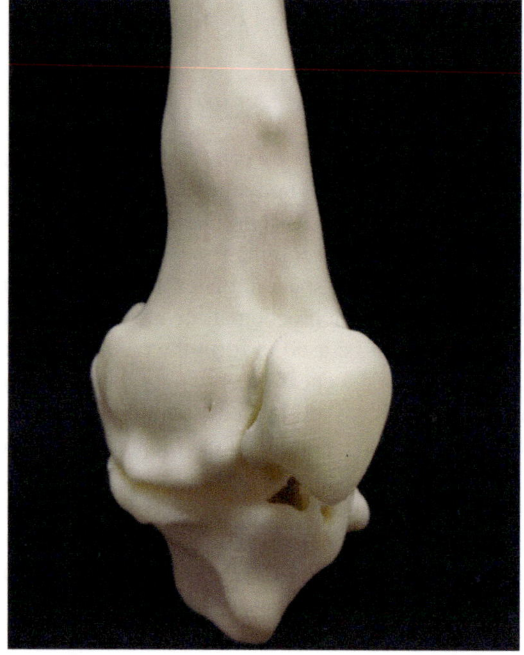

FIGURE 21.5 3D printed model of a knee, including patella, femur, and tibia, manufactured in ABS.

A more promising use of additive manufacturing in the orthopedic surgery field is the use of power bed fusion technologies to selectively melt and deposit layers of metal. The field is still growing, as the overall cost nowadays still exceeds the costs of traditional manufacturing of prosthetic materials. Titanium and chrome-cobalt alloys are particularly useful for the 3D printing of patient-specific prosthetic materials. The extremely porous surfaces of 3D-printed implants made with titanium facilitate bone growth and allow for fixation stability, as tested in acetabular cups for hip arthroplasty [23]. Other modifications, such as cell or drug loading, can contribute to lower infection rates and improve osseointegration. While costs may still be a concern, the overall spread of the technique will allow price reduction. 3D-printed orthopedic materials seem to have no increase in mechanical failure or aseptic loosening [24], which may widely encourage the adoption of such a technique.

21.4 DENTISTRY

In dentistry, additive manufacturing, combined with 3D digital modeling, is commonly used in the manufacture of dentures. The denture bases and denture teeth are 3D printed using methacrylate-based and light-curing resins. The teeth are printed separately and later joined using a light-curing adhesive. Due to the perfect digital processing, these dentures are more comfortable for the patient, corresponding accurately to their Bucco-maxilla anatomy.

Still, in dentistry, the common use of 3D printing is the manufacturing of crowns and bridges for implants in missing teeth regions. For this, the stereolithography (SLA) technique can be used, as it offers great efficiency and high levels of precision, which can produce parts with a resolution up to 0.05 mm [25]. Overall, various technologies in additive manufacturing present benefits in dentistry. Such 3D models can be used to study complex anatomic features, such as the structure presented in Figure 21.6. In this example, it is shown the model of a mandibular branch for understanding the relationship between the wisdom tooth and the nearby bony structures, is being compared to radiography.

Accurate implants can be made with low cost and superior strength [26], with the use of the binding jet technique. The same technique can be used to produce partial dentures using metallic powder. Contrary to neurosurgery, dentistry relies mostly on micro-computed tomography or intraoral 3D scanning to enable the digitalization of the buccal frame. Therefore, both approaches are responsible for providing the models for 3D printing (in STL format). Yet another advantage of the use of 3D printing is the ability to produce complex dental implants, meeting accurate shapes and sizes [27]. For dental restoration materials, comparisons have been made between additive manufacturing and the conventional subtractive manufacturing methods with zirconia and wax, leading to superior accuracy of the additive manufacturing techniques [26]. An additional advantage is the possibility of using digital reconstructions for inventory and reducing storage costs [28].

In summary, additive manufacturing in dentistry can be used to generate 3D-printed dental implants, with accuracy and increased comfort for the patients, helping in repairing and replacement of damaged teeth. The 3D models and their printed displays favor the following applications: construction of surgical guides; creation of orthodontic models for treatment of dental and mandibular

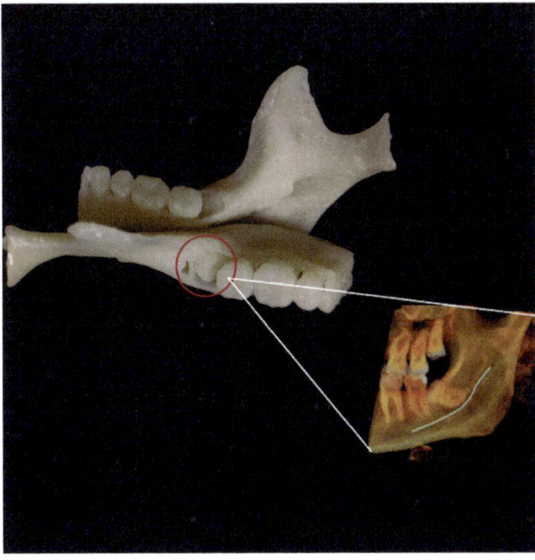

FIGURE 21.6 Polylactic acid model of the mandibular branch for the study of the relationship between the wisdom tooth and the nearby bony structures, as it is seen and compared to the radiography.

irregularities; printing of aligners; creation of patient-specific braces; and manufacturing of crowns and dentures. Therefore, such technology allows the manufacturing of dentistry devices in a fast, accurate, and less costly design of new surgical tools on-demand, as well as for educational purposes, considering the training of new professionals [26]. Then, in the dentistry field, it is expected that additive manufacturing will allow further cost reduction, including prosthodontic physical models and custom dental implants that are fully functional and with new biocompatible materials as well.

Still, for oral, maxillofacial surgeries and orthopedics, the use of 3D printing has become widespread, regarding the additional use of surgical guidance systems during complex and detailed procedures. These guides are based primarily on computed tomography, magnetic resonance imaging, and volumetric ultrasound. Additionally, these data may be supplemented by 3D digital scanners, especially in the context of maxillofacial surgeries, which is a branch of dentistry.

In surgeries for mandibular defects, titanium plates are widely used for mandibular reconstruction. Traditionally, these plates were molded during surgery, which resulted in time-consuming operations. With advances in additive manufacturing and the possibility of faithfully printing body parts, this process of modeling the plates can be done before surgery, using the 3D-printed mandible as a mold. This approach results in less surgery time, helping the patients with easier and faster recovery since they spend less time under anesthesia. Additionally, it brings benefits in terms of cost for the surgery and the health insurance as well [29].

21.5 4D PRINTING

The 3D-printing technology or additive manufacturing has been widely used in many industries and research in various fields, from pharmaceutics to textile. The success of this kind of manufacture is related to its characteristics such as low cost, versatility regarding the complex geometries, and material saving. 3D printing, although significant to speed up innovation, produces static structures useful, for example, in prototypes, static accessory parts, and decoration pieces [30]. However, if it is necessary to produce pieces that respond to a stimulus to perform a specific function, defined as a dynamic structure; 3D printing is not appropriate. Then 4D printing was developed and introduced by Tibbit Skylar [31]. Therefore, 4D printing is the 3D printing added with another stimulus in the dimension of time [31], [32]. It is possible due to the possibility of the structure changes its forms when some stimulus is applied, such as UV light, temperature, humidity, pressure, etc., as shown in Figure 21.7, allowing the generation of a smart dynamic structure.

Using additive manufacturing with two or more materials and the 4D-printing technique, it is possible to create a component where the primary shape printed can be transformed into another. In this sense, the structure can be programmed to change its unidimensional or 2D shape into a 3D form or even move when some external stimulus is imposed [31]. In this sense, the main purpose of 4D printing is to produce components that can be programmed according to their reaction to the controlled parameters of the environment. In some way, 4D printing emerged as an inspiration from nature. One example is the pinecone opening according to the humidity of the environment. In short, each pinecone scale is formed by active and passive fibers, which allow the pinecone to open to release seeds only when the humidity is favorable for germination [33]. Therefore, linking 4D printing to developing smart materials and metamaterials is possible.

Smart, functional, or active materials can respond to environmental stimuli, such as temperature, pressure, pH, light, and humidity [34], [35]. The response can lead to physical transformations such as shape and transparency, changes in mechanical properties such as stiffness or flexibility, or even degradation due to a chemical transformation. Among the smart materials used in 4D printing, the shape memory polymers (SMPs) stand out, changing their mechanical characteristic from stiff to rubbery due to their glass transition temperature (Tg) ranging from -70 to 100 °C [34]. Therefore, by printing different SMPs in the same piece, the differences in stiffness and flexibility cause tailored

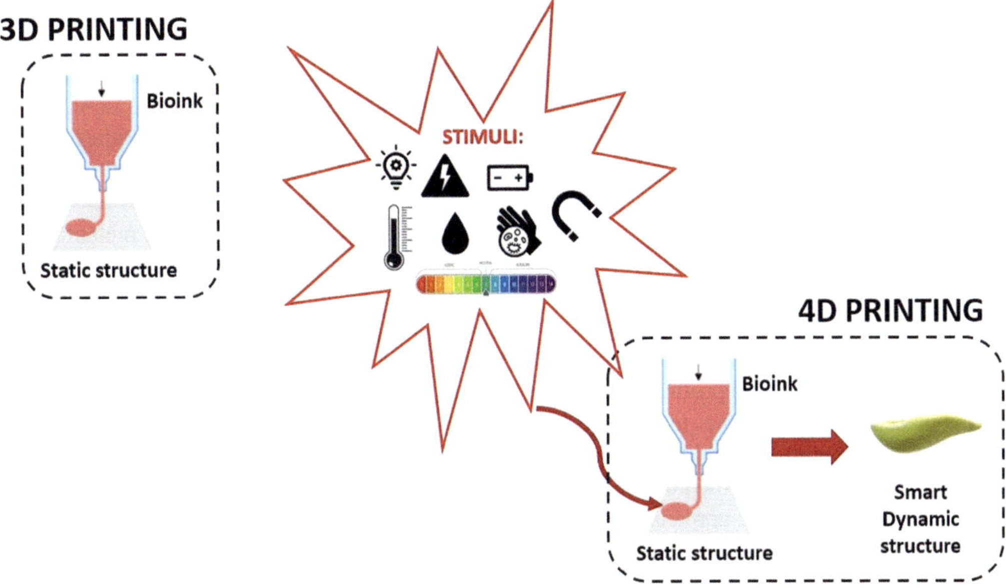

FIGURE 21.7 Representation of the differences between 3D and 4D printing. Focus is given to a variety of possible stimuli to be applied, such as UV light, temperature, humidity, pressure, magnetic field, voltage differences, etc – which allows the creation of a smart dynamic structure.

shapes to be formed. One example of a multi-material system in 4D printing is using SMP fibers with different Tg to create a composite that responds adversely to the environmental temperature. In their work, Wu et al. (2016 [36]) printed a system with three polymeric filaments, one used as the matrix and the other printed in subsequent layers. The printed system was subject to a longitudinal strain of 10% at 70 °C, then cooled to 0 °C, maintaining the stress. After uniformization of the 0 °C throughout the sample, the stress was released. The sample was heated gradually, showing the continuing curved shape recovery because of the differences in Tg of the fibers. However, due to the shape memory, the sample was completely straight when immersed in water at 70 °C [36]. It is worth noting that the SMPs have only temporary and permanent shapes, which means that their shape changes with the external stimulus. In the preceding example, if the environmental temperature varies, the sample changes its curvature. In other words, it is only possible to maintain the shape by controlling the temperature. Additionally, combining the materials to achieve the desired shape by applying a specific stimulus is a big challenge.

Other active or smart materials are among the hydrogels [37] that can be tailored to swell when hydrated or shrink when dried. Su et al. (2017 [38]) built 3D-printing patterns of a soft functional polymer based on swellable cyclopentanone entrapped in a non-swellable matrix. They printed a sample composed of active and passive parts. To produce the active region, the printed piece was exposed to UV light to polymerization, guaranteeing the retention of the swellable cyclopentanone in the matrix, followed by solidification through heating. The printed passive piece was created by heating to eliminate the cyclopentanone before the exposition to UV light to polymerization. Then, the shape-morphing sample was composed of a swellable active part that responded to acetone and a non-swellable passive part that did not respond to acetone creating a 3D structure from a planar one after immersion in acetone. The planar structure was recovered when the acetone was removed. This interesting work shows the versatility of 4D printing, which can be applied in many fields. For instance, this structure with active and passive parts, swelling and shrinking in response to an external stimulus, allows movements without robotics or motors.

Although the great advantage of 4D printing is to transform a one-dimensional component into a three-dimensional shape, a straightforward way to work with 4D printing is to create some anisotropy in the material during the printing process, as in the printing of metamaterials. The prefix 'meta' indicates that the characteristics of the material are beyond those we observe in nature. In other words, metamaterials are materials artificially manufactured to have their properties derived from their designed shape and size [39]. So, the metamaterial is shape-changing material with programmed architecture that changes its original shape to a new one after a stimulus and returns to its primary shape as soon as the stimulus is removed. In terms of 4D printing, metamaterials are mentioned as having shape-shifting or shape morphing behavior under the stimulus, useful in applications such as robots, medical devices, and actuators like in exoskeletons [40]. The literature brings some works where the main goal is shape transformation under stress to cause elastic deformations or use environmental temperature to control the degree of deformations.

Although 4D printing is a promising technology, there are some challenges mainly related to smart multi-material applications that must present printability, post-print stability, and be simultaneously resistant to the stimuli to be imposed. Some problems also need to be overcome, such as the low resolution of the pieces and the control in the speed of response to stimuli. The hydrogel constructs for cell scaffolds are future applications of 4D printing. The stimulus could physiologically modulate the molecular and cellular adhesion and improve the dynamic of cell migration, proliferation, and differentiation. Also, the future can build organs or constructs in reduced volume that reach their specific volume and shape inside the body due to programmed stimuli [41]. Stents can be a device printed in a smart polymer that changes its shape after being inserted into arteries stimulated by temperature, for instance. A great potential of 4D printing is in tissue regeneration, where there is a variation in mechanical properties according to the load applied, such as in muscles and bones [42]. In the repair of part of the cardiac muscle 4D printing can be used to produce biocompatible smart materials patches that can adjust to the curvature of the heart, enabling the appropriate regeneration of the tissue. Additionally, being foldable, the patches can be inserted with minimally invasive surgery. Some cardiovascular implants have been tested *in vitro* and *in vivo* in animal models [43].

21.6 CONCLUSION

As presented in this chapter, the 3D-printing technologies have many insertions in different medical applications, especially related to 3D printing in neurosurgical planning and cardiology, prostheses and orthopedic surgeries, and dentistry. Such technology is being applied to the teaching, training, and development of surgical techniques and medical devices among others. The advantages of the 3D-printing technique are the low-cost and versatility of the manufactured parts, as it is also well accepted in the medical field.

The use of biomaterials in medical devices is conditioned not only on the materials' properties but also on the manufacturing parameters that affect the quality and safety requirements for the desired application. Although recent literature has reported the versatility of 4D printing, some challenges must be overcome. The many types of biomaterials classified as smart or responsive materials, which respond to different stimuli in different ways, and can still be used together, provide a myriad of alternative architectures to suit the application of the medical device. However, 4D printing is expected to evolve rapidly in the coming years, allowing the inclusion of control over material responses to external stimuli.

REFERENCES

1 C. Y. Liaw and M. Guvendiren, "Current and emerging applications of 3D printing in medicine," *Biofabrication,* 2017;9.

2 A. Leal, M. Souza, and P. Nohama, "Additive manufacturing of 3D biomodels as adjuvant in intracranial aneurysm clipping," *Artificial Organs,* 2019;43:E9–15.

3 P. Low, J. Abdullah, A. Abdullah, S. Yahya, Z. Idris, and D. Mohamad, "Patient-specific reconstruction utilizing computer assisted three-dimensional modelling for partial bone flap defect in hybrid cranioplasty," *The Journal of Craniofacial Surgery,* vol.30, pp. e720–3, 2019.
4 A. Aimar, A. Palermo, and B. Innocenti, "The role of 3D printing in medical applications: A state of the art," *Journal of Healthcare Engineering,* vol.2019, pp.1–10, 2019.
5 P. D'Urso, R. Thompson, R. Atkinson, M. Weidmann, M Redmond, B. Hall, S. Jeavons, M. Benson, and W. Earwaker, "Cerebrovascular biomodelling: A technical note," *Surgical Neurology*, vol.52, no. 5, pp.490–500, 1999.
6 G. Wurm, M. Lehner, B. Tomancok, R. Kleiser, and K. Nussbaumer, "Cerebrovascular biomodeling for aneurysm surgery: Simulation-based training by means of rapid prototyping technologies," *Surgical Innovation,* vol.18, pp. 294–306, 2011.
7 A. G. Leal, L. B. Pagnan, R. T. Kondo, J. A. Foggiatto, G. J. Agnoletto, and R. Ramina, "Elastomers three-dimensional biomodels proven to be a trustworthy representation of the angiotomographic images," *Arquivos de Neuro-Psiquiatria*, vol.74, no. 9, FapUNIFESP (SciELO), pp.713–717, Sep-2016.
8 T. Mashiko, N. Kaneko, T. Konno, K. Otani, R. Nagayama, and E. Watanabe, "Training in cerebral aneurysm clipping using self-made 3-dimensional models," *Journal of Surgical Education*, vol.74, no. 4, Elsevier BV, pp.681–689, Jul-2017.
9 N. Martelli, C. Serrano, H. van den Brink, J. Pineau, P. Prognon, I. Borget, and S. El Batti, "Advantages and disadvantages of 3-dimensional printing in surgery: A systematic review," *Surgery*, vol.159, no. 6, Elsevier BV, pp.1485–1500, Jun-2016.
10 M. N. Nadagouda, V. Rastogi, and M. Ginn, "A review on 3D printing techniques for medical applications," *Current Opinion in Chemical Engineering,* vol.28, Elsevier BV, pp.152–157, Jun-2020.
11 A. Müller, K. G. Krishnan, E. Uhl, and G. Mast, "The application of rapid prototyping techniques in cranial reconstruction and preoperative planning in neurosurgery," *Journal of Craniofacial Surgery*, vol.14, no. 6, Ovid Technologies (Wolters Kluwer Health), pp.899–914, Nov-2003.
12 Kimura, A. Morita, K. Nishimura, H. Aiyama, H. Itoh, S. Fukaya, S. Sora, and C. Ochiai, "Simulation of and training for cerebral aneurysm clipping with 3-dimensional models," *Neurosurgery*, vol.65, no. 4, Ovid Technologies (Wolters Kluwer Health), pp.719–726, Oct-2009.
13 L. Dering, M. Pedro, and A. Leal, "Manufacture of 3D models of intracranial aneurysms for surgical training on a LCD 3D printer," *Brazilian Neurosurgery*, vol. 32, Brazilian Academy of *Neurosurgery*, pp. 213–214, 2021.
14 A. G. Leal, R. Ramina, P. H. P. de Aguiar, B. L. Fernandes, M. Abreu de Souza, P. Nohama "Clinical applications of additive manufacturing models in neurosurgery: A systematic review," *Brazilian Archives of Neurosurgery*, vol. 40, no. 4, Brazilian Society of Neurosurgery, e349–e360, 2021.
15 Y.-S. Dho, D. Lee, T. Ha, S. Y. Ji, K. M. Kim, H. Kang, M.-S. Kim, J. W. Kim, W.-S. Cho, Y. H. Kim, Y. G. Kim, S. J. Park, and C.-K. Park, "Clinical application of patient-specific 3D printing brain tumor model production system for neurosurgery," *Scientific Reports*, vol.11, no. 1, Springer Science and Business Media LLC, 26-Mar-2021.
16 A. Benet, J. Plata-Bello, A. A. Abla, G. Acevedo-Bolton, D. Saloner, and M. T. Lawton, "Implantation of 3D-printed patient-specific aneurysm models into cadaveric specimens: A new training paradigm to allow for improvements in cerebrovascular surgery and research," *BioMed Research International*, vol.2015, Hindawi Limited, pp.1–9, 2015.
17 A. Haleem, M. Javaid, and A. Saxena, "Additive manufacturing applications in cardiology: A review," *The Egyptian Heart Journal*, vol.70, no. 4, Springer Science and Business Media LLC, pp.433–441, Dec-2018.
18 M. Mathur, P. Patil, and A. Bove, "The role of 3D printing in structural heart disease," *JACC: Cardiovascular Imaging*, vol.8, no. 8, Elsevier BV, pp.987–988, Aug-2015.
19 B. O'Neill, D. D. Wang, M. Pantelic, T. Song, M. Guerrero, A. Greenbaum, and W. W. O'Neill, "Transcatheter caval valve implantation using multimodality imaging," *JACC: Cardiovascular Imaging,* vol.8, no. 2, Elsevier BV, pp.221–225, Feb-2015.
20 S. N. Schön, N. Skalicky, N. Sharma, D. W. Zumofen, and F. M. Thieringer, "3D-printer-assisted patient-specific polymethyl methacrylate cranioplasty: A case series of 16 consecutive patients," *World Neurosurgery*, vol.148, Elsevier BV, pp.e356–e362, Apr-2021.

21 J. A. Morales-Gómez, E. Garcia-Estrada, J. E. Leos-Bortoni, M. Delgado-Brito, L. E. Flores-Huerta, A. A. De La Cruz-Arriaga, L. J. Torres-Díaz, and Á. R. M.-P. de León, "Cranioplasty with a low-cost customized polymethylmethacrylate implant using a desktop 3D printer," *Journal of Neurosurgery*, vol.130, no. 5, *Journal of Neurosurgery Publishing Group (JNSPG),* pp.1721–1727, May-2019.

22 H. K. Gandapur and M. S. Amin, "Orthopaedics and additive manufacturing: The start of a new era," *Pakistan Journal of Medical Sciences*, vol.38, no. 3, *Pakistan Journal of Medical Sciences*, pp. 751–756, 17-Jan-2022.

23 L. Dall'Ava, H. Hothi, J. Henckel, A. Di Laura, P. Shearing, and A. Hart, "Characterization of dimensional, morphological and morphometric features of retrieved 3D-printed acetabular cups for hip arthroplasty," *Journal of Orthopaedic Surgery and Research*, vol.15, no. 1, Springer Science and Business Media LLC, pp. 1–12, 19-Apr-2020.

24 F. Pardo, B. Bordini, F. Castagnini, F. Giardina, C. Faldini, and F. Traina, "Are powder-technology-built stems safe? A midterm follow-up registry study," *Journal of Materials Science: Materials in Medicine,* vol.32, no. 1, *Springer Science and Business Media LLC,* pp. 1–6, Jan-2021.

25 D. Khorsandi, A. Fahimipour, P. Abasian, S. S. Saber, M. Seyedi, S. Ghanavati, A. Ahmad, A. A. De Stephanis, F. Taghavinezhaddilami, A. Leonova, R. Mohammadinejad, M. Shabani, B. Mazzolai, V. Mattoli, F. R. Tay, and P. Makvandi, "3D and 4D printing in dentistry and maxillofacial surgery: Printing techniques, materials, and applications," *Acta Biomaterialia,* vol.122, Elsevier BV, pp.26–49, Mar-2021.

26 M. Javaid and A. Haleem, "Current status and applications of additive manufacturing in dentistry: A literature-based review," *Journal of Oral Biology and Craniofacial Research*, vol.9, no. 3, Elsevier BV, pp.179–185, Jul-2019.

27 A. Mostafaei, E. L. Stevens, J. J. Ference, D. E. Schmidt, and M. Chmielus, "Binder jetting of a complex-shaped metal partial denture framework," *Additive Manufacturing,* vol.21, Elsevier BV, pp.63–68, May-2018.

28 J. W. Bae, D. Y. Lee, C. H. Pang, J. E. Kim, C.-K. Park, D. Lee, S. J. Park, and W.- S. Cho, "Clinical application of 3D virtual and printed models for cerebrovascular diseases," *Clinical Neurology and Neurosurgery,* vol.206, Elsevier BV, p.106719, Jul-2021.

29 I. Velasco, S. Vahdani, and H. Ramos, "Low-cost method for obtaining medical rapid prototyping using desktop 3D printing: A novel technique for mandibular reconstruction planning," *Journal of Clinical and Experimental Dentistry, Medicina Oral, S.L.*, vol. 9, pp.e1103–e1108, 2017.

30 M. Quanjin, M. R. M. Rejab, M. S. Idris, N. M. Kumar, M. H. Abdullah, and G. R. Reddy, "Recent 3D and 4D intelligent printing technologies: A comparative review and future perspective," *Procedia Computer Science*, vol.167, Elsevier BV, pp.1210–1219, 2020.

31 S. Tibbits, "4D printing: Multi-material shape change," *Architectural Design*, vol.84, no. 1, Wiley, pp.116–121, Jan-2014.

32 M. Invernizzi, S. Turri, M. Levi, and R. Suriano, "4D printed thermally activated self-healing and shape memory polycaprolactone-based polymers," *European Polymer Journal*, vol.101, Elsevier BV, pp.169–176, Apr-2018.

33 C. J. Eger, M. Horstmann, S. Poppinga, R. Sachse, R. Thierer, N. Nestle, B. Bruchmann, T. Speck, M. Bischoff, and J. Rühe, "The structural and mechanical basis for passive-hydraulic pine cone actuation," *Advanced Science*, vol.9, no. 20, Wiley, p.2200458, 14-May-2022.

34 S. Rebouillat and F. Pla, "A review: On smart materials based on some polysaccharides; within the contextual bigger data, insiders, 'improvisation' and said artificial intelligence trends," *Journal of Biomaterials and Nanobiotechnology*, vol.10, no. 02, Scientific Research Publishing, Inc., pp.41–77, 2019.

35 T. van Manen, S. Janbaz, K. M. B. Jansen, and A. A. Zadpoor, "4D printing of reconfigurable metamaterials and devices," *Communications Materials*, vol.2, no. 1, Springer Science and Business Media LLC, pp. 1–8, 03-Jun-2021.

36 J. Wu, C. Yuan, Z. Ding, M. Isakov, Y. Mao, T. Wang, M. L. Dunn, and H. J. Qi, "Multi-shape active composites by 3D printing of digital shape memory polymers," *Scientific Reports*, vol.6, no. 1, Springer Science and Business Media LLC, pp. 1–11, 13-Apr-2016.

37 S. Malekmohammadi, N. Sedghi Aminabad, A. Sabzi, A. Zarebkohan, M. Razavi, M. Vosough, M. Bodaghi, and H. Maleki, "Smart and biomimetic 3D and 4D printed composite hydrogels: Opportunities for different biomedical applications," *Biomedicines*, vol.9, no. 11, MDPI AG, p.1537, 26-Oct-2021.

38 J.-W. Su, X. Tao, H. Deng, C. Zhang, S. Jiang, Y. Lin, and J. Lin, "4D printing of a self-morphing polymer driven by a swellable guest medium," *Soft Matter*, vol.14, no. 5, Royal Society of Chemistry (RSC), pp.765–772, 2018.
39 S. Amin Yavari, S. M. Ahmadi, R. Wauthle, B. Pouran, J. Schrooten, H. Weinans, and A. A. Zadpoor, "Relationship between unit cell type and porosity and the fatigue behavior of selective laser melted meta-biomaterials," *Journal of the Mechanical Behavior of Biomedical Materials*, vol.43, Elsevier BV, pp.91–100, Mar-2015.
40 M. Bodaghi and W. H. Liao, "4D printed tunable mechanical metamaterials with shape memory operations," *Smart Materials and Structures*, vol.28, no. 4, IOP Publishing, p.045019, 20-Mar-2019.
41 M. Champeau, D. A. Heinze, T. N. Viana, E. R. de Souza, A. C. Chinellato, and S. Titotto, "4D printing of hydrogels: A review," *Advanced Functional Materials*, vol.30, no. 31, Wiley, p.1910606, 28-May-2020.
42 M. Javaid and A. Haleem, "4D printing applications in medical field: A brief review," *Clinical Epidemiology and Global Health*, vol.7, no. 3, Elsevier BV, pp.317–321, Sep-2019.
43 F. Kabirian, P. Mela, and R. Heying, "4D printing applications in the development of smart cardio-vascular implants," *Frontiers in Bioengineering and Biotechnology*, vol.10, Frontiers Media SA, pp. 1–11, 25-May-2022.

22 3D-Printed Carbon-Based Nanomaterials for Biomedical Applications

Farzaneh Jabbari[1], *Babak Akbari*[2*], *and Lobat Tayebi*[3]
[1]Nanotechnology and Advanced Materials Department, Materials and Energy Research Center (MERC), Imam Khomeini Blvd, Meshkin-Dasht, Karaj, Alborz, 31787-316, Iran
[2]Life Science Engineering Department, Faculty of New Sciences and Technologies, University of Tehran, North Kargar Street, Tehran, Tehran, 1439957131, Iran
[3]Department of Developmental Science, Marquette University School of Dentistry, 1801 West Wisconsin Ave., Milwaukee, WI, 53233, USA
*Corresponding author: babakbari@ut.ac.ir

CONTENTS

DOI: 10.1201/9781003296676-22

22.1 INTRODUCTION

Biomolecules and biomaterials have important roles in tissue engineering and biomedical applications. In recent years, biomaterials in the form of nanomaterials, due to their nanoscale-sized properties, have attracted much attention from biomedical researchers. Among these nanomaterials, CBNs have been developed as a novel type of biomaterial due to their unique physicochemical, mechanical, and biological properties such as excellent chemical and thermal stability, high mechanical strength, high electrical conductivity, ease of functionalization, and antimicrobial activity [1]. Many studies confirmed that CBNs can extensively adsorb various biomolecules through electrostatic interactions and can be developed as appropriate materials for biosensors and loading drugs. Each member of the carbon family presents unique properties and has been used in different biological applications. Despite the many advantages of the CBNs, their cytocompatibility is very complex and a severe challenge for scientists with biomaterials. The toxicity effects of CBNs originated from high dose usage; therefore, it is essential to determine the optimum dosage to minimize their toxicity [2]. CBNs need to be surface functionalized for biomedical applications to render them passive, non-toxic, and water-soluble. These serious challenges usually are overcome by surface modification with covalent approaches such as dehydrogenation, ozonolysis, oxidation, and plasma treatments [3]. In addition, functionalization is an appropriate and effective technique for producing multifunctional, multimodal, and high-performance biomaterials. For example, modified CBNs significantly increase the production of reactive oxygen species (ROS) in cancer cells, and ROS can inhibit cancer cell proliferation and kill them [4]. 3D printing is currently one of the easiest and most popular techniques for producing these biomaterials. This method can create suitable geometry, needs no dedicated clamp, provides a valuable tool for visualization, and 3D printing organs is possible, too [5]. This book chapter intends to study the properties and the performance of 3D-printed CBNs in tissue engineering and biomedical applications.

22.2 TYPES OF CBNS

The different types of CBNs are explained in the following sections.

22.2.1 Carbon Nanotubes (CNTs)

Iijima introduced CNTs at the beginning of the twenty-first century. They showed a tubular shape and large surface area, ultra-small size, high reactivity, and conjugation with therapeutic molecules are their privileged features [6]. They divide CNTs into two groups: single wall or multiple walls. Single-walled carbon nanotubes (SWCNTs) diameters are less than 1 nm, and they prefer to be curved. Multi-walled carbon nanotubes (MWCNTs) have different manner; the number of layers determines diameter. The outer diameter of MWCNTs ranges from 2 to 30 nm, and the inner diameter varies from 0.4 nm up to a few nanometers [3]. The properties of CNTs are indeed attractive; over the last 10 years, many studies have shown numerous applications of CNTs in regenerative medicine. The toxicity of CNTs is a crucial issue that has attracted the attention of many researchers and has not yet been answered correctly. More studies are needed to determine the optimal biological dose of CNTs [7].

22.2.2 Graphene

In 1947, physicist Philip R. Wallace conducted a theoretical study of graphene (G) to understand the electronic structure of graphite. In 1986, chemists Hanns-peter Boehm, Ralph Setton, and Eberhard Stumpp, coined the term *graphene* as a combination of the word graphite, referring to carbon in its ordered crystalline form, and the suffix '-ene' referring to graphite's polycyclic aromatic hydrocarbons, which organize the carbon atoms in a hexagonal, or six-sided, ring structure

[4]. Since G demonstrates high electrical conductivity, high surface area, and excellent thermal conductivity, many researchers are developing practical applications for drug delivery, as a material for building biosensors, as a potential antibacterial agent, and as a scaffold for tissue engineering [8].

22.2.3 Graphene Oxide and Reduced Graphene Oxide

Graphene oxide (GO) is an oxidized form of G, laced with oxygen-containing groups, making it easy to process since it is dispersible in water and other solvents. It can even be used to make G. Four basic methods are used to synthesize GO: Staudenmaier, Hofmann, Brodie, and Hummers. Variations are explored with an emphasis on achieving better results and cheaper processes. The effectiveness of oxidation processes is evaluated by the ratio of carbon/oxygen atoms of the GO. GO can easily be mixed with different polymers and other materials and enhance properties of composite materials like tensile strength, elasticity, conductivity, and more. GO's fluorescent appearance makes it appropriate for various medical applications. Biosensing and disease detection, drug carries, and antibacterial materials are some of the ways GO can be utilized in the biomedical field [4]. GO can be reduced to G using chemical, thermal or electrochemical methods. The material produced is called reduced GO (rGO). Reducing GO by chemical reduction is very scalable, but unfortunately, the produced rGO is typically poor. Thermally reducing GO requires temperatures of 1000 °C or higher which damages the structure of the G platelets. However, the overall quality of the produced rGO is quite good. Electrochemical methods have been shown to produce very high-quality rGO, almost identical to pristine G. However, the method still suffers from scalability issues (10). Furthermore, many researchers confirmed that G derivatives induce cells to differentiate to specific lineages such as adipogenesis, osteogenesis, chondrogenesis, etc. rGO improves the interactions with biomolecules, cells, and polymers. This manner of rGO was attributed to its structural defects [9].

22.2.4 Carbon Dots

Carbon dots (CDs) are optically active nanomaterials formed by 2–3-parallel G sheets. Excellent solubility in water, optical properties, biocompatibility, and chemical inertness are some of their properties. In addition, high brightness and photostability, low cost, small size, and controllable surface functions make CDs a promising tool in medicine and biomedical engineering. Compared with CDs and other CBNs, research on CDs is still in a primary step. It should be noted that there are no systematic and scalable synthesis methods to produce high-quality CDs with favorable structures [10].

22.3 TRENDS IN 3D-PRINTING METHODS FOR BIOMEDICAL APPLICATIONS

3D printing is also called additive manufacturing. This term accurately describes how this technology works to create objects. 'Additive' refers to the successive addition of thin layers to create an object. All 3D-printing technologies are similar, as they construct an object layer by layer to create complex shapes. It allows for the design and print of more complex designs than traditional manufacturing processes. Previous research classified 3D-printing methods according to the physical state of the main material, and it can be applied to ceramic, metal, polymer, and composites. Polymers are the first material group produced with this technique and, to date, constitute a wide range of materials. Natural and synthetic polymers are mainly used as inks [11], [12]. Chitosan, collagen, alginate, and hyaluronic acid are the most widely used natural polymers. Meanwhile, polylactic acid (PLA), polyglycolic acid (PGA), polylactic-co-glycolic acid (PLGA), polyurethane (PU), and polycaprolactone (PCL) are synthetic polymers that are attractive for biomedical applications [12].

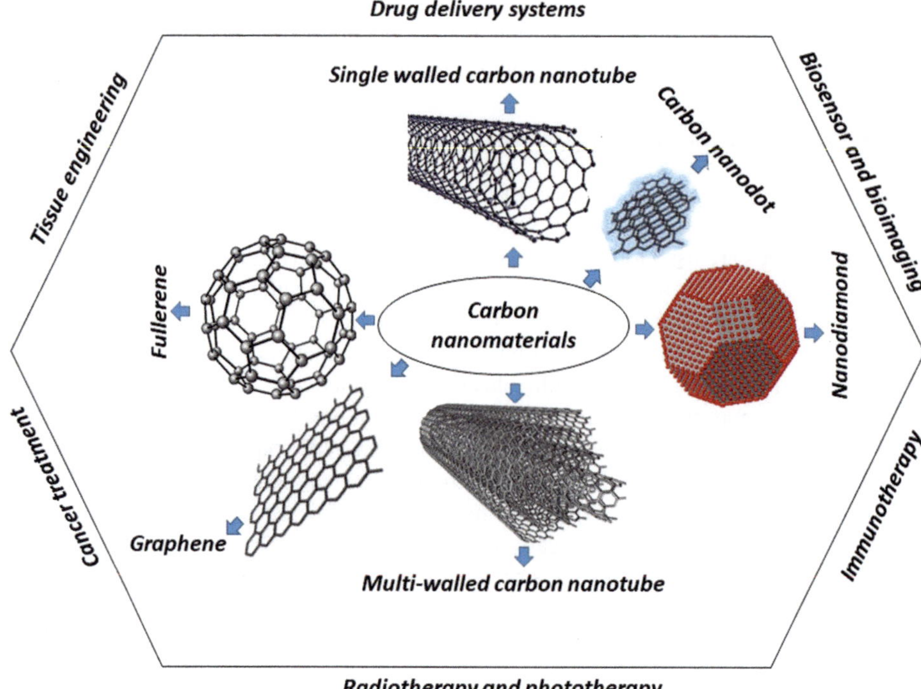

FIGURE 22.1 Schematic illustration for the biomedical applications of 3D printed carbon-based nanomaterials.

Biomedical applications of 3D printed CBNs have been discussed in the following sections. In Figure 22.1 some of the most important clinical applications of these materials are presented.

22.3.1 Drug Delivery

Innovations in 3D printing have led to more personalized medicine, with devices moving away from a one-size-fits-all model to a more custom approach. Conventional production methods would be costly, labor-intensive, and not exceptionally viable. By contrast, 3D printing can cut costs, save time, and open up a host of new applications for devices. Increasingly, the same applies to drug delivery. Over the last decade, 3D printing has gained significant attention in pharmaceutical formulations and drug delivery systems as a fantastic method to overcome the challenges of conventional manufacturing approaches. 3D-printed CBNs have gained much interest among various nanomaterials due to their unique and excellent properties and multifunctional application possibilities [8]. In an investigation, the anti-cancer drug methotrexate was loaded into GO. Both sides of GO layers adsorb methotrexate via non-covalent interactions. After loading the methotrexate onto GO as a carrier, the methotrexate–GO presented high specific cytotoxicity to the tumor cells compared to the normal cells [13]. In another study, GO nanoparticles were conjugated with polyethylene glycol, then loaded with the anti-cancer drug SN38 onto the GO surface. The complex showed high cytotoxicity for cancer cells, and the drug release from the GO surface was pH-dependent [8]. CNTs-based drug delivery systems are helpful in cancer therapy. Anti-cancer drugs loaded CNTs can gather drugs at the cancer sites. Both SWCNT and MWCNT are used in the drug delivery system. SWCNTs-based agents can kill human breast cancer cells without toxicity. Paclitaxel-loaded polyethylene glycol-graft-SWCNT presented an acceptable performance for cancer therapy by extensively killing breast cancer cells. Prostate cancer is the deadliest cancer among men in the world. PC3 and DU145 (prostate cancer cell lines) have different body functions. PC3 cells have high metastatic activity compared to DU145 cells with a moderate metastatic manner. Cis-diamminedichloroplatinum was implanted in SWCNTs, and this complex released the cisplatin and killed cancer cells. In another study, SWNT-integrin alpha (v) beta (3) monoclonal antibody system (SWNT-PEG-mAb) was used against SH-SY5Y and U87 (brain cancer cell lines). The results demonstrated that SWNT-PEG-mAb could be a good candidate for cancer drug delivery. SWCNT showed prevention of the proliferation of human embryonic kidney 293 cells (HEK 293 cells) in a time and dose-dependent manner. It can be concluded that conventional cancer treatment methods can harm healthy tissues. Therefore, CBNs, as drug delivery systems, must be developed to overcome many clinical challenges [11].

22.3.2 Gene Delivery

Gene therapy is a modern and helpful method to treat many diseases caused by genetic disorders. A gene that is inserted directly into a cell usually does not function. Instead, a vector carrier is genetically engineered to deliver the gene. One of the critical challenges of gene therapy is the absence of safe gene vectors. The use of viral vectors is highly controversial. The ability to carry a limited amount of genetic material and immune responses in patients are the most important reasons that limit their clinical applications. In recent years, the most commonly used non-viral vectors are lipids, inorganic particles, and polymers due to their non-toxicity and low immunogenicity effects. CBNs have been applied extensively due to their unique properties and high loading capacity. CNTs improve gene diffusion via the cell membrane without interfering with the endocytosis process of cells. In a study, an SWCNT linked with siRNA from Caspase3 was applied to treat cardiovascular diseases. Another research showed that the combination of polyethyleneimine and GO provides the cytomegalovirus promoter with a plasmid that controls luciferase gene expression. In many studies, polyethyleneimine-GO complex, chitosan- GO composite, polyethyleneimine-GO- polyethylene glycol (PEG) complex were used to improve the efficiency of plasmid DNA transfection. Preparation of low-molecular-weight polyethyleneimine-rGO- PEG gene delivery carrier increases the efficacy of the gene transfection since it is light-responsive. PEGylated rGO can be used to deliver single-stranded RNA (ssRNA) by a covalent bond of PEG to GO and converting PEG-GO to PEG-rGO. PEG-rGO complex can absorb a significant amount of ssRNA. Although various studies about CBNs as gene carriers are currently in progress by many researchers, only a small number of studies have been done for in vivo applications. To overcome the current limits of gene therapies and open the possibility of treating new diseases, researchers are trying to improve the delivery process so that the genes get to the suitable organs, without necessarily going through the liver, and targeting so that the genes get to the right cells [8], [14].

22.3.3 Biosensing

3D-printed CBNs, due to their unique and excellent mechanical, electrical and optical properties, have been used significantly as biosensors in the field of biosensing technology. High conductivity, acceptable chemical stability and sensitivity, excellent electrocatalytic activity, and high electron transferring activity make them a promising tool for biosensing applications. For example, G and CNTs can be used to sense glucose. The glucose level detection in diabetic patients is a critical indicator for clinical treatments. CNTs-based biosensors can sense nitric oxide and epinephrine. In a study, 3D printed G/PLA composite filaments were applied for detecting picric acid and ascorbic acid in various concentrations. Also, they detected glucose in blood plasma, but the detection efficiency was very low. Surface modification of these filaments enhanced their electrochemical properties and increased detection efficiency.

G-based sensors are very thin with high mechanical flexibility, making them suitable for cases where there is a possibility of deformation. These sensors receive high-quality signals and have close contact with the body. A recent approach in 3D-printed G-based sensors was the detection of SARS-CoV-2 in patients and virus spike protein from nasal swabs. Many studies confirmed that SWCNT could detect arginase-1, ctDNA, infectious diseases, proteins, carcinoembryonic marker and squamous cancer antigen, biomolecules such as thrombin, amino acid, etc. [10]. Rocha and coworkers fabricated a PLA/carbon black (CB) 3D-printed electrode to detect metal ions in biological samples. The electrode has been activated by immersion in 0.5MNaOH to create more conductive sites. The activated electrodes have been tested in urine and saliva. The results showed that the detection efficiency changed to the following values, 93% to 108% and 96% to 112% for Cd^{2+} and Pb^{2+}, respectively. Another fantastic study presented PLA/rGO 3D-printed electrodes to detect catechol. By solvent treatments, the electrodes were activated, and the tyrosinase enzyme was immobilized at its surface. The activated electrode showed acceptable performance. By considering all of these studies, it can be concluded that 3D-printed CBNs have a promising future in the biosensing field, but further in vivo studies should be performed [15].

22.3.4 Bioimaging

Bioimaging is a technique that can non-invasively visualize biological processes in real-time via probes and detectors. The goal of bioimaging is to not interfere with life processes. It is helpful to provide images of the whole body, organs, tissues, and cells. Among the promising biomaterials, 3D-printed CBNs have specific properties that have been extensively selected for biological imaging. G and G-containing materials, due to their intrinsic photoluminescence, are extensively used for fluorescence (FL) imaging, two-photon FL imaging (TPFI), and Raman imaging. GOs as Raman tags show naturally strong D and G peaks and adding metal nanoparticles can improve them. For tracking the labeled materials in vivo with high sensitivity, radionuclide-based imaging methods such as positron emission tomography (PET) and single-photon emission computed tomography (SPECT) are helpful. The presence of GO in the tracker complex has great tissue/cell imaging and is fast removed from the body. Magnetic resonance imaging (MRI) is another medical imaging method with high resolution. Manganese and gadolinium, paramagnetic toxic metals, are MRI contrast agents. GO with oxygen-containing functional groups can be easily coordinated with these ions and decrease their toxicity. Furthermore, GO-grafted gadolinium can be used to deliver chemotherapeutic drugs and microRNAs. Photoacoustic imaging (PAI) is another method based on the PA effect. PAI allows the delivery of light energy absorbed by tissues causing an expansion. This expansion then generates ultrasound waves detected by the transducer and produces images of optical absorption contrast within tissues. rGO is an ultrahigh sensitive PA contrast agent and absorbs NIR light significantly. Based on the studies, it can be concluded that, for using CBNs in the bioimaging field, it is necessary to control their size, surface coatings, and components. Another important issue that needs special attention is the long-term toxicity of these materials for in vivo applications. Furthermore, more information about their biological properties is useful to understand their biocompatibility and benefits to promoting their bioimaging performances. Surface modification and functionalization of CBNs effectively cross blood-brain barriers, mucus, etc. It is believed that these materials will play notable roles in bioimaging and determine the fate of the treatments used [8], [16].

22.3.5 Antimicrobial

For in vivo biomaterials and biomedical device applications, bacterial infection at the implantation site is a severe concern in tissue engineering. Nanomaterials, an excellent alternative to antibiotics, can cross cell membranes without challenges and destroy bacterial cells. Recently, CBNs, due to their antibacterial property, have received significant attention. There are several mechanisms of antibacterial property of these materials, but each type of CBN improves one or more of these mechanisms. For example, CNTs stick to bacterial membranes, damage the membrane, and release DNA content. The antibacterial effect of SWCNTs is stronger compared to MWCNTs. Due to their smaller diameter and greater surface area, SWCNTs can easily penetrate the cell wall. G, GO, and rGO act based on membrane stress. Cell deposition on these materials and contact with sharp nanosheets cause membrane stress, as well as GO, can destroy cell membrane by reactive oxygen species (ROS) groups. Many researchers have confirmed that the type of bacteria, size, and shape are effective parameters for antimicrobial activity. Gram-positive bacteria are more resistant to nanoparticles' bactericidal effect than gram-negative since they can easily change their structure. The concentration of CBNs is another factor that influences antibacterial activity. Based on the studies, the threshold concentrations of GO and rGO are 80 μg/mL and 100 μg/mL, respectively. CBNs, due to their antibacterial property, have been used extensively in biomedical applications such as wound dressing and bandages, surface coatings, medical implants, innovative packaging, and hydrogels. Several challenges about these materials must be resolved before using them. To conclude, expanding alternative antibiotics or advanced antibacterial strategies is immediately required to eradicate the infection created by antibiotic-resistance bacteria. Antibacterial strategies based on 3D-printed CBNs are not mature for in vivo applications, and most are passing clinical trial phases. The number of publications about these materials must be accelerated in the future to combat bacterial infections [1], [17].

22.3.6 Tissue Engineering

In recent years, 3D printing has received remarkable attention in tissue engineering as a promising tool for better tissue regeneration, replacement, and repair performance. Figure 22.2 shows some of these applications. Osteoconductivity, antimicrobial activity, high surface area, electronic and thermal conductivity, mechanical strength, and biocompatibility are the most important features of tissue engineering. Among a wide range of materials, CBNs have most of these properties and are attractive for many biomedical applications. For example, CBN's electrical and optical properties are amazing for biosensors and medical imaging. In addition, their high aspect ratio is useful for drug/gene/therapeutic agent delivery. In tissue engineering, incorporating these materials into a polymer matrix shows an excellent structure with acceptable physicochemical properties, high mechanical strength, biocompatibility, and high gene expression ability for tissue regeneration. For tissue regeneration, these materials are so attractive; for example, CNTs support neuron attachment and encourage cell differentiation.

Polymeric scaffolds do not have electrical conductivity and suitable tensile strength. Therefore, CNTs overcome these challenges and are extremely useful in neuronal tissue engineering. CNTs promote intercalated disc formation in cardiac tissue engineering. Biological results confirmed that CNT-containing scaffolds increased the gene expression of cardiac markers and promoted excitation conduction velocities. Bone is a specialized form of connective tissue and can remodel and repair itself in slight fractures, but in pathological fractures, bone loss, and bone tumors, it cannot heal itself. There are many therapeutic methods for bone regeneration, and each method has its limitations. Bone tissue shows piezoelectric properties that promote bone regeneration via electrical stimulation. CNTs as a tool for bone regeneration present high alkaline phosphatase (ALP) activity, high protein absorption, high cell proliferation, high osteoconduction, and high compressive strength. Although these materials are widely used in bone, cardiac, cartilage, and nerve tissue engineering, many serious challenges still exist, and further studies are needed to overcome them. Developing cytocompatible and green CBNs for tissue engineering is a serious problem. The researchers demonstrated that with the increasing concentration of CBNs in composites, physiochemical and mechanical properties are promoted, and toxicity effects increase accordingly. It is better to function these materials with amazing functional groups such as –COOH, –OH, C=O, and

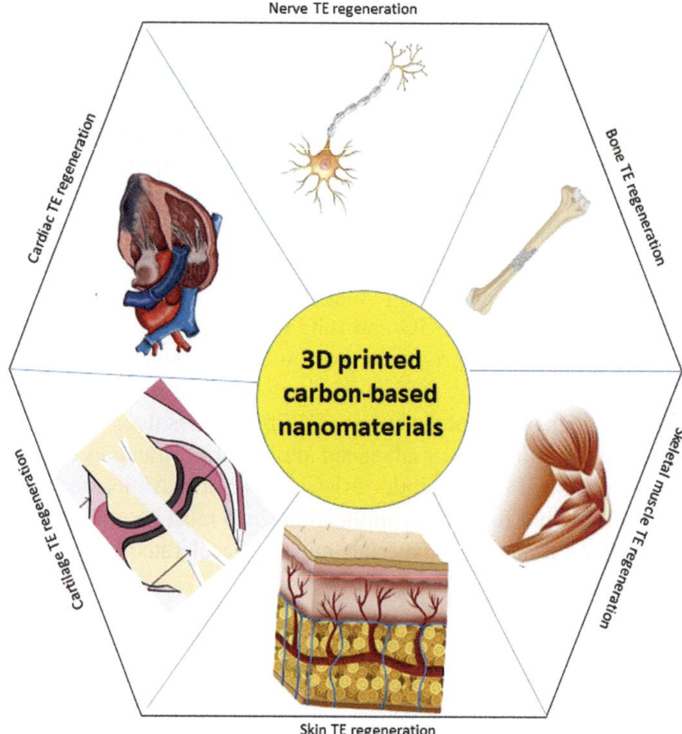

FIGURE 22.2 3D printed carbon-based nanomaterials as a promising tool for tissue regeneration.

esters to minimize the toxicity. Oxidative stresses are the main factors for the toxic effects of these materials, which can be attributed to their shape, dimensions, synthesis method, and dosage and surface parameters [8], [9].

22.3.6.1 Cardiac Tissue Engineering

Myocardial infarction (MI) is one of the most common diseases worldwide that causes cell death and heart function loss. Damaged cardiomyocytes have a poor ability to repair themselves. Cell therapy, heart transplantation, and tissue engineering have been introduced as excellent methods for the treatment of MI. Among these methods, tissue engineering has attracted much attention by using a wide range of materials. Many synthetic and natural materials have been introduced for cardiac tissue regeneration in recent years. 3D-printed CBNs are interesting for repairing heart tissue due to their thermal and electrical conductivity, high mechanical strength, and high surface area properties. For example, GO-containing composite scaffolds enhance cell signaling, cell adhesion and propagation, cardiac gene expression, and conductivity. It has been confirmed that GO and its family can stimulate cell differentiation to cardiomyocytes and improve cardiogenic phenotype without external electrical stimulation (12). In a study, rGO was incorporated into gelatin methacrylyol (GelMA) to increase electrical conductivity and mechanical strength for heart tissue regeneration. The results showed that rGO remarkably enhanced electrical conductivity and mechanical strength, proliferation rate, and contractility. Moreover, the viability of cardiomyocytes cultured in the electroactive hydrogel significantly increased. rGO has angiogenic properties at lower doses, which is necessary for cardiovascular applications, and is anti-angiogenic at higher doses ($\geq$100 ng mL $^{-1}$). Many published studies showed that rGO and GO at high concentrations induced cell death via the formation of ROS groups. ROS can activate many signaling cascades such as the epidermal growth factor receptor, nuclear factor-KB, transcription factor activator protein-1 (AP-1), and participate in cell growth, differentiation, and proliferation. CNTs are anisotropic materials that improve the alignment of cardiomyocytes. CNTs encourage cardiomyocytes to act in a way that promotes cardiac tissue regeneration. CNTs have excellent interaction with cells and promote the expression of α-actinin and myosin heavy chain proteins. Mechanical properties of CNTs impress the cardiomyocytes phenotype and induce the ventricular-like phenotype, while GO and rGO induce atrial-like phenotype and mixed phenotype, respectively. In a study, GO was added to oligo (poly (ethylene glycol) fumarate) (OPF) hydrogels to increase the mechanical strength and electrical conductivity of remaining healthy cells in a damaged heart. The composite hydrogel enhanced cardiac fibroblast attachment and increased the expression of cardiac genes. Therefore, the addition of GO into composites improves cell viability, and differentiation and induces angiogenesis for in vivo applications. It can be concluded that there are a variety of novel materials and amazing technologies which can be used for heart tissue engineering. CBNs possess many unique properties, making them fantastic for heart tissue regeneration. As mentioned, one or more severe challenges need to be overcome for clinical applications. The biocompatibility of these materials and their validation in various in vitro and animal models need to be studied. Generally, more experiments and studies are required to confirm the safety of CBNs for in vivo applications [18], [19].

22.3.6.2 Skeletal Muscle Tissue Engineering

Skeletal muscle tissues provide movement, heat production, stability, expansion, contraction of orifices, and glycemic control. A significant percentage of muscle injuries occur during athletic activities. Current clinical treatments of using autologous muscle flaps or allografts have many challenges. Therefore, stem cells and nanomaterials-based therapies are an amazing approach to promoting skeletal muscle tissue regeneration. According to this, CBNs are extensively used due to their high electric conductance, high surface area, and excellent physical and mechanical properties. GO-based scaffolds promote attachment, proliferation, differentiation, and migration of skeletal muscle precursor cells. These structures can mimic the behaviors of natural muscles more than other carbon nanomaterials. Combining GO with natural polymers such as collagen, chitosan, and gelatin induces cell differentiation. rGO nanoparticles present better electrical conductivity than other materials if other carbon family members connect to their surface. A study related to GO and rGO showed that GO has more activity in inducing myogenic protein/gene expression and multinucleate myotube formation since GO can easily absorb serum proteins via its oxygen-containing functional groups. Subsequently, rGO improves proliferation and myogenic differentiation and is firmly electronically stimulable. The electrically conductive biological medium can increase muscle cells' electrical interactions and immediately increase myogenic differentiation. Electrical stimulation (ES) has remarkable effects on cell alignment and helps irregular cells to have a neat arrangement. It must be noted that the orientation of the cells is proportional to the direction of the applied force. ES plays an important role in cell migration. The path of cell migration due to ES depends on the cell type. The intensity of ES is tolerable for most cells and does not affect cell phenotype, cell viability, and cell differentiation. Although the use of ES in tissue engineering has significantly increased in recent years, many types of research are needed for clinical applications. In vivo degradation rate of G/GO/rGO-containing structures depends on the synthesis process, surface functionalization/modification, and dimensions. It is confirmed that H_2O_2 can degrade G-based materials, and the degradation rate can be sped up by modification with O_2 plasma.

Due to their appropriate surface roughness and wettability, CNT-containing scaffolds also presented similar improvements in myoblast adhesion, proliferation, and differentiation. While these structures promote muscle cell functions, their safety needs further investigation. As a result, based on many studies, CNTs (SWCNTs and MWCNTs) show a different levels of toxicity. Surface modification of CNTs increases biocompatibility. These materials can be functionalized with many biomolecules via covalent and noncovalent bindings. Functionalized CNTs are outstanding candidates for producing high-quality nanocomposites. Due to their essential role in tissue engineering and medical science, further in vitro and in vivo studies should be performed to evaluate these materials' safety [18], [20].

22.3.6.3 Nerve Tissue Engineering

Neuronal tissue regeneration has been a controversial clinical challenge due to the complex structures of the nervous system. Furthermore, the central nervous system has a poor ability to repair itself. Transplantation of stem cells and neural cells is an important therapeutic strategy for repairing damaged tissue and restoring its functions. Direct cell transplantation has many limitations and disadvantages. 3D-printed CBNs have emerged as excellent materials for repairing damaged nervous tissues. In vitro use of these materials in neural substrates and scaffolds composition has many benefits such as high neural cell viability, cell adhesion, proliferation, differentiation, and migration. CNTs, due to their mechanical, electrical, and physical properties, are extensively used in neural tissue regeneration and can revitalize damaged tissue functional activities. CNTs are so stable in biological environments for long-term applications and are flexible. The interaction between neural cells and CNTs promotes neural cell signaling. G, GO, and rGO are also attractive materials for neural tissue engineering. GO/rGO–containing culture media promote glial differentiation of adipose stem cells and regulate the expression of neurotrophin and filament proteins. The presence of nerve regenerative growth factors in GO/rGO containing composite scaffolds protect neural stem cells from H_2O_2-induced oxidative damage and promotes cell signaling in vitro. Combining CNTs with hydrogels is an excellent strategy for neural tissue regeneration. In a study, G-gelatin composite scaffolds containing mesenchymal stem cells demonstrated Schwann cell marker. With electrical stimulation to the scaffolds, the secretion of nerve growth factor increased accordingly. In another study, SWCNT was added to the hydrogel scaffold, and neurite outgrowth significantly increased since SWCNT, and electrical stimulation has synergistic effects in boosting cell growth. The best strategy to increase CNTs solubility in a biological medium is to incorporate hydrophilic polymers. This complex improves neuron growth due to an excellent electrostatic interaction between these materials and the negatively charged cell membranes. There are two methods for controlling neural cell functions via biofunctionalized CNTs. In the first method, CNTs are added to the neuronal cell culture medium, and carbon structures directly interact with nerve culture. In the second method, CNTs are modifiers of other biomaterials to promote their neurofunctionality. It is possible to form more membrane/material tight junctions and indirect interactions between these materials and neurites. CNTs, containing scaffolds, act as a reservoir of adsorbed proteins and have a dynamic role in promoting neuronal electrical actions. It can be concluded that CBNs, as a promising tool for nerve tissue engineering, have several serious challenges. The toxicity effects in direct contact with neural cells limit their use in neuronal applications. In this regard, more research is required about the relationship between toxicity and surface area, size, dimension, functional groups, and release of CBNs. 3D-printing technology is attractive for producing CBNs with a unique design, high cytocompatibility, and proper functionality, thus promoting their application for nervous regeneration [18], [21].

22.3.6.4 Cartilage Tissue Engineering

Articular cartilage cannot repair itself, so repairing and regeneration are severe issues in the biomedical field. Tissue engineering is an excellent new treatment for cartilage tissue regeneration. By using CBNs in cartilage tissue scaffolds and other structures, these materials' conductivity and mechanical properties can be significantly improved. Many studies used GO/G/rGO and their biocomposites for the chondrogenic differentiation of stem cells. G and its family members in scaffold composition promote chondrogenic differentiation and proliferation rates since these materials have excellent electrostatic interactions with transforming growth factor-β and stimulating new tissue formation. The combination of GO with other nanomaterials significantly enhances mechanical strength; GO-chitosan (CS) composite scaffold is an example of this combination in which the mechanical strength of the scaffold, due to the presence of GO nanoparticles, increases. Furthermore, CS/Polyvinyl alcohol (PVA)/GO composite showed promising results for cartilage tissue engineering. Hyaluronic acid (HA)-GO composite can promote cartilage tissue repair. HA can improve polysaccharide-based scaffolds' mechanical, physical, thermal, and surface properties. HA-GO composites are used as a lubricant and have a regulatory role in the joint cavity microenvironment during cartilage regeneration. In a study, polycaprolactone (PCL) /gelatin/ MWCNTs were used for cartilage tissue repair. MWCNTs increased the hydrophilicity, bioactivity, and mechanical strength of the composite. The degradation rate of the PCL-gelatin /MWCNTs was decreased since the strong bonds between NH_3^+ (gelatin) and COO^- (MWCNTs) were formed, and the chemical stability of the composite was increased. The simulated body fluid (SBF) converted COOH groups to COO^- and promoted calcium and phosphorous deposition and increased bioactivity of the structure. In another study, GO was added to alginate hydrogel cross-linked with $CaCl_2$, and the composite was used as a fantastic platform for cartilage tissue regeneration. The addition of GO into the hydrogel structure increased compressive elastic modulus, and cell viability was also increased. This composite hydrogel acts as an artificial ECM for cell growth and differentiation. One of the most attractive benefits of CBNs is that due to the large surface area of these materials, only a tiny amount of growth factor is required compared to the large amounts of it needed during external application. The addition of CBNs with favorable properties into other natural or synthetic materials makes them appropriate tools for cartilage tissue engineering. While many in vitro studies have confirmed the positive effects of CBNs-containing scaffolds for cartilage repairing, several in vivo studies are needed to be done before clinical applications of them. Furthermore, the physiological, pathophysiological, and histological differences should be paid attention to when the animal data and results are attributed to the human clinical setting. To find effective methods for tissue regeneration while overcoming all limitations, a new outlook must be created for tissue engineering to be a promising tool for many needy people around the world [18], [22].

22.3.6.5 Bone Tissue Engineering

Bone loss or devitalization occurs due to many reasons such as traffic accidents, delayed union, rarefaction of bone, corrective osteotomies, and bone tumor resection the most important. Bone tissue engineering has been introduced as a practical approach to regenerating bone defects. In the past decade, 3D printing has advanced the rapid fabricating of synthetic substitutes for bone tissue regeneration. This technology unifies engineering and biological sciences to produce bone repair materials, often combined with living cells and growth factors to overcome the limitations of traditional methods. Many studies have demonstrated the use of conventional biomaterials such as metals, polymers, ceramics, and their composites for 3D bioprinting, but their microporous structure has limited their applications [23]. CBNs, due to high surface area, excellent mechanical and antibacterial properties, fatigue resistance, non-cytotoxic effect, good interaction with biological molecules, and nanoscale size for imitating the native bone structure, are a great alternative to conventional materials. In a rat calvarial defect model, Cui et al. used the 3D printed polyion complex (PIC)/MWCNT biohybrid hydrogels. The scaffolds presented high mineralized bone volume, excellent biocompatibility with rat bone marrow stem cells, and sufficient porosity for cell growth [24]. Researchers confirmed that 3D printed GO-containing scaffolds significantly improved new bone formation since they can stimulate the osteogenic differentiation of cells by adjusting bone-related gene expression [24], [25]. In another study, PCL-G scaffolds were printed and associated with microcurrent therapy to treat bone defects in rats.

After 120 days of implantation of composite scaffolds, the osteoprotegerin levels increased and RANKL levels decreased. This promoted tissue formation and modulated the bone remodeling phase. An important issue of 3D printed CBNs in bone tissue engineering is their antibacterial activity. Results on antibacterial properties are so challenging. It is believed that the antimicrobial activity of CBNs depends on many factors related to both the bacterial and nanomaterials. Many studies are required to clearly understand these materials' antibacterial activity and their correlation with the human innate immune system [25].

22.3.6.6 Skin Tissue Engineering

As the body's largest organ, skin protects against germs, regulates body temperature, and enables touch sensations. Skin tissue engineering is so important and necessary for advanced wound healing. It was initially proposed to solve problems such as inflammation and tissue infection. In this way, 3D printing is an excellent method that has attracted positive opinions in the medical and healthcare fields. The advent of this technology provides impressive new facilities to reconstruct the injured skin. In terms of materials, materials with a high surface area similar to the extracellular matrix (ECM) of tissues, micro-inflammation characteristics, and non-toxic catabolites are effective for skin regeneration. Accordingly, 3D printed CBNs are extensively used in this field. Soliman et al. showed that GO encouraged endothelial cell migration and caused a remarkable improvement in a full-thickness skin wound healing model in rats. It can also be related to GO's ability to clean ROS in the wound area. The other ROS can damage cell membrane and DNA, which causes tissue regeneration to stop. GO can eliminate ROS and regularize inflammation to speed up wound regeneration [26]. 3D printed GO has also been used as filler in wound healing hydrogels. For example, rGO-GelMA hydrogels promote endothelial, keratinocyte, and fibroblast migration and angiogenesis. The researchers speculate that rGO increases intracellular ROS levels. A profound and fundamental problem for wound healing hydrogel is their bacterial infection. Adding Ag nanoparticles to hydrogel can inhibit bacterial growth and increase the healing rate [27]. In another study, PCL/GO/silver (Ag)/arginine (Arg) accelerated the wound healing process due to its angiogenic and antimicrobial properties. Based on the studies, GO has pro-angiogenic features at low doses (1–50 ng/mL) while it presents anti-angiogenic properties at high doses (≥100 ng/mL). The physical-chemical properties of GOs, such as size, aggregation states, surface charge, functionalization, and purity, can also regularize their toxicity [27].

22.3.7 Dentistry

In recent years, polyether ether ketone (PEEK) has been widely used in oral and craniomaxillofacial surgery. In addition, PEEK can be applied in skull implants, dental implants, and osteosynthesis plates. Many studies showed that incorporating carbon fibers can enhance the mechanical strength of PEEK. Typically, carbon fiber reinforced PEEK (CFR-PEEK) has an elastic modulus close to human cortical bone and can be considered a promising tool to replace metallic materials. Fused deposition modeling (FDM), the fastest-growing 3D-printing method, has significantly been used to produce the CFR-PEEK [28]. Recently, G-printed scaffolds have been used in dentistry. Ma and colleagues showed that 3D-printed GO-PVA β-tricalcium phosphate (TCP) scaffolds accelerate the new bone formation and have anti-cancer properties [28]. Glass ionomer (GI) is widely used in restorative dentistry. 3D-printed G, when combined with GI, significantly promotes the mechanical, biological, and antibacterial properties of GIs. This combined material also decreases microcracks in the internal structure and protects it from erosion and disintegration by microbial invasion [29]. In endodontics, bioactive cement is used to manage perforation, pulp capping, and retrograde root filling. 3D-printed G improves the mechanical property of the bio-cement and significantly reduces the setting time. 3D-printed G seems to be engaging in dental implants as a platform to enhance implants' osseointegration and new bone formation. It possesses an osteogenic property that increases the expression of osteogenic related genes such as RUNX2, COL-I, and ALP, and, consequently, improves the deposition of the mineralized matrix. Teeth can become discolored on their outer surfaces when people use colored foods and drinks. As a result, many people prefer non-invasive whitening treatments that bleach the teeth. Many studies confirmed that using rGO and

H_2O_2 under photoactivation significantly increased the whitening effect of H_2O_2 and decreased the treatment time [29].

22.3.8 Diagnosis

Early detection is vital for effective therapy. Recently, 3D-printed CNTs, due to their excellent thermal and electronic conductivity, suitable optical property, and high drug loading capacity, have been extensively applied for cancer therapy and diagnosis. CNTs can serve as drug carriers to deliver great anti-cancer drugs and function as excellent mediators in phototherapy because of their intrinsic optical properties. On the other hand, they can be used as an effective tool for detecting the expression of cancer genes [30]. In recent years, CNTs have been utilized to diagnose ovarian cancer. To improve cancer detection and diagnosis by 3D-printed CNTs, the researchers attached an antibody specific to a biomarker (a protein that appears in the blood of ovarian cancer) allowing it to recognize and bind to the marker. Once the nanotubes are bound to the protein, the light they give off changes color, picked up on by sensors outside the body [31]. Many studies confirmed that 3D-printed CNTs, G, and GO, could also be used to fight against the COVID-19 pandemic. The interactions between these materials and the SARS-CoV-2 virus were studied, and hydrophobic and strong electrostatic interactions were detected. Hence, it can be concluded that CBNs are such helpful tools in preparing diagnostic kits and personal protective equipment. Although numerous in vitro studies have been done, clinical and in vivo are still scarce [32].

22.4 FUTURE DIRECTIONS AND CHALLENGES

Innovations and new technologies are changing the world and the daily lives of every one of us. Many things that were mere visions of the future yesterday are now a reality. 3D-printing technology, which has attracted much attention in many fields, presents a great outlook for the future. It is optimistic that it can solve many problems we have experienced lately. The coronavirus pandemic is a clear example to illustrate this point. During pandemic crises, meeting the demand for face masks has become one of the main issues for people fighting the pandemic. As an effective solution to this problem, 3D-printing technology is applied to produce masks. Subsequently, protective face masks are made in a shorter time. In addition, this technology is widely used in the biomedical field. Based on the studies, CBNs are shown to be a helpful tool with unique advantages such as excellent optical property, high drug loading capacity, thermal and electronic conductivity, and easy functionalization ability, among others. 3D-printed CBNs have wonderful potential for various medical applications, from tissue engineering to cancer therapy and diagnostic applications, but the challenges we face are severe and must prevail. One of the most critical challenges is a deep understanding of CBNs-cell (or tissue/organ) interactions. This knowledge will help expand the application of these materials in clinical fields. Another primary concern is the toxicity of CBNs at in vitro and in vivo levels. Although the cytotoxicity effects of these materials are still extremely debatable, however, global interest and attention are growing. We believe that the increasing expansion of these materials soon will bring incredible innovations to the biomedical field and can maintain human health.

ACKNOWLEDGMENT

Part of the research reported in this paper was supported by the National Institute of Dental & Craniofacial Research of the National Institutes of Health under award numbers R15DE027533, 1R56 DE029191-01A1, and 3R15DE027533-01A1W1.

The content is solely the responsibility of the authors and does not necessarily represent the official views of the National Institutes of Health.

REFERENCES

1 S. Lopez de Armentia, J.C. Del Real, E. Paz, N. Dunne, Advances in biodegradable 3D printed scaffolds with carbon-based nanomaterials for bone regeneration, *Materials* 13(22) (2020) 5083.

2 R. Madannejad, N. Shoaie, F. Jahanpeyma, M.H. Darvishi, M. Azimzadeh, H. Javadi, Toxicity of carbon-based nanomaterials: Reviewing recent reports in medical and biological systems, *Chemico-Biological Interactions* 307 (2019) 206–222.

3 H. Li, S.I. Song, G.Y. Song, I. Kim, Non-covalently functionalized carbon nanostructures for synthesizing carbon-based hybrid nanomaterials, *Journal of Nanoscience and Nanotechnology* 14(2) (2014) 1425–1440.

4 T.A. Tabish, S. Zhang, P.G. Winyard, Developing the next generation of graphene-based platforms for cancer therapeutics: The potential role of reactive oxygen species, *Redox Biology* 15 (2018) 34–40.

5 W. Liu, Y. Li, J. Liu, X. Niu, Y. Wang, D. Li, Application and performance of 3D printing in nanobiomaterials, *Journal of Nanomaterials 2013* (2013) 1–7.

6 D. Maiti, X. Tong, X. Mou, K. Yang, Carbon-based nanomaterials for biomedical applications: A recent study, *Frontiers in Pharmacology* 9 (2019) 1401.

7 J.E. Contreras-Naranjo, V.H. Perez-Gonzalez, M.A. Mata-Gómez, O. Aguilar, 3D-printed hybrid-carbon-based electrodes for electroanalytical sensing applications, *Electrochemistry Communications* 130 (2021) 107098.

8 H. Shen, L. Zhang, M. Liu, Z. Zhang, Biomedical applications of graphene, *Theranostics* 2(3) (2012) 283.

9 G. Rajakumar, X.-H. Zhang, T. Gomathi, S.-F. Wang, M. Azam Ansari, G. Mydhili, G. Nirmala, M.A. Alzohairy, I.-M. Chung, Current use of carbon-based materials for biomedical applications—A prospective and review, *Processes* 8(3) (2020) 355.

10 J. Liu, R. Li, B. Yang, Carbon dots: A new type of carbon-based nanomaterial with wide applications, *ACS Central Science* 6(12) (2020) 2179–2195.

11 Y. Bozkurt, E. Karayel, 3D printing technology; methods, biomedical applications, future opportunities and trends, *Journal of Materials Research and Technology* 14 (2021) 1430–1450.

12 A. Ghilan, A.P. Chiriac, L.E. Nita, A.G. Rusu, I. Neamtu, V.M. Chiriac, Trends in 3D printing processes for biomedical field: Opportunities and challenges, *Journal of Polymers and the Environment* 28(5) (2020) 1345–1367.

13 H.N. Abdelhamid, K.H. Hussein, Graphene oxide as a carrier for drug delivery of methotrexate, *Biointerface Res. Appl. Chem* 11 (2021) 14726–14735.

14 A. Gholami, S.A. Hashemi, K. Yousefi, S.M. Mousavi, W.-H. Chiang, S. Ramakrishna, S. Mazraedoost, A. Alizadeh, N. Omidifar, G. Behbudi, 3D nanostructures for tissue engineering, cancer therapy, and gene delivery, *Journal of Nanomaterials* 2020 (2020) 1–24.

15 M.H. Omar, K.A. Razak, M.N. Ab Wahab, H.H. Hamzah, Recent progress of conductive 3D-printed electrodes based upon polymers/carbon nanomaterials using a fused deposition modelling (FDM) method as emerging electrochemical sensing devices, *RSC Advances* 11(27) (2021) 16557–16571.

16 J. Lin, Y. Huang, P. Huang, Chapter 9 – Graphene-based nanomaterials in bioimaging. In *Biomedical Applications of Functionalized Nanomaterials*, Sarmento, B., das Neves, J., (Eds.). Elsevier: Amsterdam, The Netherlands (2018) 247–287.

17 S. Yaragalla, K.B. Bhavitha, A. Athanassiou, A review on graphene based materials and their antimicrobial properties, *Coatings* 11(10) (2021) 1197.

18 M. Maleki, R. Zarezadeh, M. Nouri, A.R. Sadigh, F. Pouremamali, Z. Asemi, H.S. Kafil, F. Alemi, B. Yousefi, Graphene oxide: A promising material for regenerative medicine and tissue engineering, *Biomolecular Concepts* 11(1) (2020) 182–200.

19 K.N. Alagarsamy, S. Mathan, W. Yan, A. Rafieerad, S. Sekaran, H. Manego, S. Dhingra, Carbon nanomaterials for cardiovascular theranostics: Promises and challenges, *Bioactive Materials* 6(8) (2021) 2261–2280.

20 C. Chen, X. Bai, Y. Ding, I.-S. Lee, Electrical stimulation as a novel tool for regulating cell behavior in tissue engineering, *Biomaterials Research* 23(1) (2019) 1–12.

21 S.-J. Lee, W. Zhu, M. Nowicki, G. Lee, D.N. Heo, J. Kim, Y.Y. Zuo, L.G. Zhang, 3D printing nano conductive multi-walled carbon nanotube scaffolds for nerve regeneration, *Journal of Neural Engineering* 15(1) (2018) 016018.

22 N. Amiryaghoubi, M. Fathi, A. Barzegari, J. Barar, H. Omidian, Y. Omidi, Recent advances in polymeric scaffolds containing carbon nanotube and graphene oxide for cartilage and bone regeneration, *Materials Today Communications* 26 (2021) 102097.

23 P. Memarian, F. Sartor, E. Bernardo, H. Elsayed, B. Ercan, L.G. Delogu, B. Zavan, M. Isola, Osteogenic properties of 3D-printed silica-carbon-calcite composite scaffolds: Novel approach for personalized bone tissue regeneration, *International Journal of Molecular Sciences* 22(2) (2021) 475.

24 L. Cheng, K. Shoma Suresh, H. He, R.S. Rajput, Q. Feng, S. Ramesh, Y. Wang, S. Krishnan, S. Ostrovidov, G. Camci-Unal, 3D printing of micro-and nanoscale bone substitutes: A review on technical and translational perspectives, *International Journal of Nanomedicine* 16 (2021) 4289.

25 V. Palmieri, W. Lattanzi, G. Perini, A. Augello, M. Papi, M. De Spirito, 3D-printed graphene for bone reconstruction, *2D Materials* 7(2) (2020) 022004.

26 M. Soliman, A.A. Sadek, H.N. Abdelhamid, K. Hussein, Graphene oxide-cellulose nanocomposite accelerates skin wound healing, *Research in Veterinary Science* 137 (2021) 262–273.

27 P. Bellet, M. Gasparotto, S. Pressi, A. Fortunato, G. Scapin, M. Mba, E. Menna, F. Filippini, Graphene-based scaffolds for regenerative medicine, *Nanomaterials* 11(2) (2021) 404.

28 X. Han, D. Yang, C. Yang, S. Spintzyk, L. Scheideler, P. Li, D. Li, J. Geis-Gerstorfer, F. Rupp, Carbon fiber reinforced PEEK composites based on 3D-printing technology for orthopedic and dental applications, *Journal of Clinical Medicine* 8(2) (2019) 240.

29 R. Guazzo, C. Gardin, G. Bellin, L. Sbricoli, L. Ferroni, F.S. Ludovichetti, A. Piattelli, I. Antoniac, E. Bressan, B. Zavan, Graphene-based nanomaterials for tissue engineering in the dental field, *Nanomaterials* 8(5) (2018) 349.

30 L. Tang, Q. Xiao, Y. Mei, S. He, Z. Zhang, R. Wang, W. Wang, Insights on functionalized carbon nanotubes for cancer theranostics, *Journal of Nanobiotechnology* 19(1) (2021) 1–28.

31 M. Barani, M. Bilal, F. Sabir, A. Rahdar, G.Z. Kyzas, Nanotechnology in ovarian cancer: Diagnosis and treatment, *Life Sciences* 266 (2021) 118914.

32 S. Mallakpour, E. Azadi, C.M. Hussain, Fight against COVID-19 pandemic with the help of carbon-based nanomaterials, *New Journal of Chemistry* 45(20) (2021) 8832–8846.

23 3D-Printed Graphene for Biomedical Applications

Sara Lopez de Armentia[1*], *Juan Carlos del Real*[1], *Nicholas Dunne*[2], *and Eva Paz*[1]
[1]Institute for Research in Technology/Mechanical Engineering Dept., Universidad Pontificia Comillas, Alberto Aguilera 25, Madrid, 28015, Spain
[2]School of Mechanical and Manufacturing Engineering, Dublin City University, Collins Avenue, Dublin 9, Ireland; Centre for Medical Engineering Research, School of Mechanical and Manufacturing Engineering, Dublin City University, Stokes Building, Collins Avenue, Dublin 9, Ireland; School of Pharmacy, Queen's University of Belfast, 97 Lisburn Road, Belfast BT9 7BL, UK; Trinity Centre for Biomedical Engineering, Trinity Biomedical Sciences Institute, Trinity College Dublin, Dublin 2, Ireland; Department of Mechanical and Manufacturing Engineering, School of Engineering, Trinity College Dublin 2, Ireland; Advanced Materials and Bioengineering Research Centre (AMBER), Royal College of Surgeons in Ireland and Trinity College Dublin, Dublin 2, Ireland; Advanced Manufacturing Research Centre (I-Form), School of Mechanical and Manufacturing Engineering, Dublin City University, Dublin 9, Ireland; Advanced Processing Technology Research Centre, Dublin City University, Dublin 9, Ireland; Biodesign Europe, Dublin City University, Dublin 9, Ireland
*Corresponding author: sara.lopez@comillas.edu

CONTENTS

23.1 INTRODUCTION

The term *graphene* appeared for the first time in 1986 when Boehm et al. [1] described graphene as a single two-dimensional carbon sheet. It is the basic structure of all graphitic materials and atoms are arranged in a honeycomb lattice structure. The word *graphene* derives from graphite with the suffix '-ene', which is used for fused polycyclic aromatic hydrocarbons. In 2004, graphene was

DOI: 10.1201/9781003296676-23

isolated for the first time by Professor Andre Geim and Professor Konstantin Novoselov [2]. From that point, the extraordinary properties of this nanomaterial have been explored in depth for a wide variety of applications. It presents a very high surface area – 2,630–2,640 m^2/g, excellent thermal conductivity – 3,000–5,000 W/m·K, good mechanical properties – around 42 N/m of break strength, 130.5 GPa of intrinsic tensile strength with Young's modulus of around 1 TPa, and high electrical conductivity – 10^4–5.9 × 10^5 S/m approximately [3], [4].

Despite its superior properties, the use of graphene is limited by the difficulty and high cost to obtain a single defect-free carbon sheet. Therefore, graphene derivatives are a good alternative to graphene. Graphene oxide (GO) is produced through the oxidation of graphite in an acidic medium and it is produced more easily than graphene. Besides, if most oxygen-containing functional groups are eliminated, reduced graphene oxide (rGO) is obtained. By the reduction of GO, electrical conductivity is partially recovered. Both, GO and rGO, have oxygenated functional groups, which potentiate covalent bonding and strengthen the interface with the polymers or provide reactivity with biomolecules. Graphene nanoplatelets (GNP) consist of small stacks of graphene nanosheets, its production much more cost-effective than graphene. They exhibit interesting properties such as good electrical and thermal conductivity, low density, high aspect ratio, planar morphology, and mechanical toughness, at low cost. The characteristics of graphene derivatives are still excellent, but lower than graphene. The properties of graphene and its derivatives are shown in Table 23.1 [5], [6]. The structure of the most common graphene-based nanomaterials (GBN) is shown in Figure 23.1.

In addition to these excellent properties, graphene and its derivatives present many specific properties useful in the field of Biomedical Engineering. They stimulate cell attachment, proliferation, and differentiation, decrease the immune response, and present antimicrobial activity and osteoconductivity [9], [10]. However, there are some issues regarding the cytotoxicity of graphene-based nanomaterials that are not fully understood. Thanks to the increasing interest in GBN in biomedical engineering applications, the number of studies regarding its biocompatibility and cytotoxicity has increased in the last years. However, there is no scientific consensus about the cytotoxicity and biocompatibility of these nanomaterials yet.

Major advances have been made in recent years not only in terms of materials but also in manufacturing technologies. Additive Manufacturing (AM) is a technology that is based on the manufacturing of parts layer by layer. By contrast, conventional manufacturing technologies are subtractive. In AM, objects are defined by a computer-aided-design (CAD) software and they are 'sliced' into thin layers of some microns to create STL files. ISO/ASTM 52900:2021 [11] classifies AM technologies into seven primary processes: (i) Vat Polymerization, (ii) Material Jetting, (iii) Binder Jetting, (iv) Material Extrusion, (v) Powder Bed Fusion, (vi) Sheet Lamination, and (vii) Directed Energy Deposition. In recent decades, the application of AM has experienced an important raise in many research fields. One of these fields is Biomedical Engineering, where AM opens up the possibility of producing tailor-made designs, which can be customized for each patient and specific case. Besides, AM presents other advantages compared to conventional manufacturing technologies, like rapid prototyping and manufacturing of complex structures. These structures manufactured by AM have controllable and reproducible geometries, which used to be a drawback of conventional technologies.

TABLE 23.1
Properties of Graphene and Its Derivatives

	Graphene	GNP	GO	rGO
Thermal conductivity [$W{\cdot}m^{-1}{\cdot}K^{-1}$]	3,000–5,000	≈ 5,000	≈ 3,000	30–250
Young's Modulus [GPa]	≈ 1,000	≈ 1,000	200–220	≈ 250
Electrical conductivity [$S{\cdot}m^{-1}$]	10^4–$5.9{\cdot}10^5$	≈ 10^5	Insulator	200–35,000

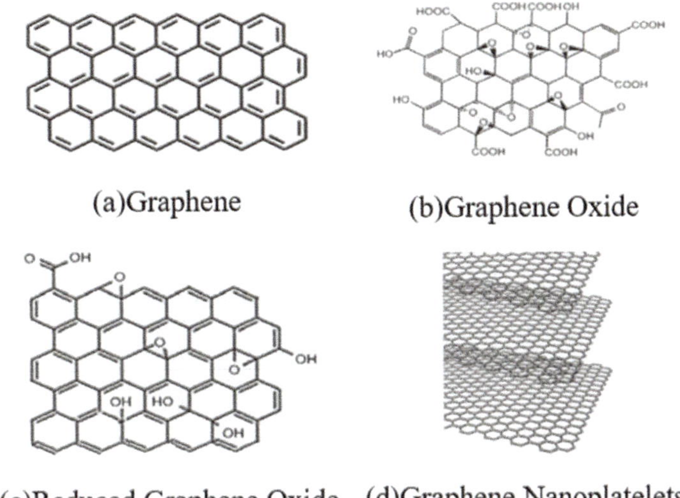

FIGURE 23.1 (a) Graphene is a single layer of carbon atoms arranged in a honeycomb lattice structure. (b) GO has a similar structure, but it presents oxidized groups (carboxyl, hydroxyl, and epoxy). (c) rGO shows much less oxidized groups than GO. (d) GNP has more than one carbon layer, without oxidized groups. (a–c) adapted with permission [7], Copyright (2021), Elsevier. (d) adapted with permission [8], Copyright (2021), Elsevier.

To obtain three-dimensional (3D) printed structures that present the good properties of graphene and its derivatives, it is necessary to use a material that supports the nanomaterials. The most common matrix is polymer, which can be in different states: solid thermoplastics, photocurable resins, inks, etc. Depending on the AM technology, the state of the polymer and the GBN addition method change. However, some studies used ceramics as the matrix, which are usually in form of powder. Finally, some researchers used polymer as the matrix, and GBN together with ceramics as fillers. For researchers, it is challenging to develop nanocomposites with GBN with excellent printability, good biological performance, and adequate mechanical properties. Therefore, the method to prepare the raw material to feed the printer, as well as the printing parameters, need to be optimized for each printing technology, GBN type, and specific application.

23.2 APPLICATIONS OF 3D GRAPHENE-CONTAINING STRUCTURES FOR BIOMEDICAL ENGINEERING

23.2.1 Drug and/or Gene Delivery

Drugs and active molecules are introduced into the body via different administration routes, i.e., buccal/sublingual, nasal, ocular, oral, pulmonary, anal/vaginal, transdermal, and parenteral drug delivery. However, recently, different drug delivery systems based on organic and inorganic micro- and nanoparticulated systems have been developed, resulting in an increase in effectiveness and reduction of side effects compared to conventional administration routes [12].

Graphene presents characteristics that are interesting to be explored. It is a perfect platform for binding protein molecules, besides, it protects proteins from proteolysis, which is a hydrolysis reaction that results in the breakdown of proteins [13]. GBN was explored as a drug delivery system for the first time by Liu et al. [14], who attached a water-insoluble, aromatic drug to PEGylated GO sheets. GBN structure, i.e., sp^2-carbons with two-dimensional structure, facilitates the binding of different substances, like biological molecules, drugs, or fluorescent probes (an example is showed

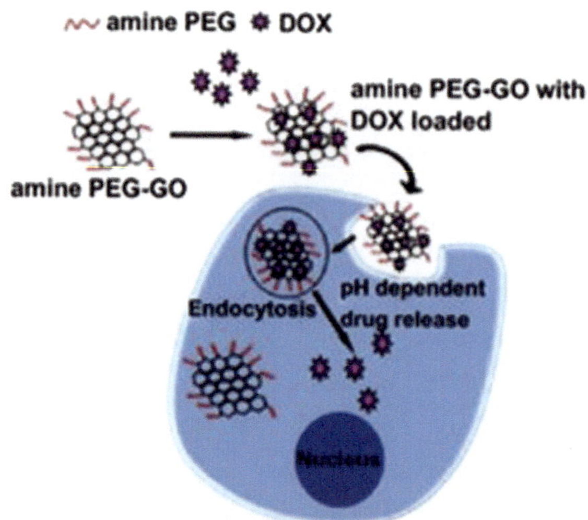

FIGURE 23.2 GO functionalized with amine PEG is loaded with doxotubicin (DOX) and the drug is released at a certain pH. Adapted with permission [13]. Copyright S, Syama; P.V. Mohanan, some rights reserved; exclusive licensee Springer. Distributed under a Creative Commons Attribution License 4.0 (CC BY).

in Figure 23.2). Besides, they present high drug-loading capacity with a controlled release thanks to their large specific surface area and strong ion exchange capacity [15]. The advantage of GBN when compared to other nanomaterial-based systems is their high surface area, which allows for drug loading on both sides of the planar structure [16].

Amongst graphene and its derivatives, graphene presents some restrictions that limit its interaction with other molecules because it has an unstable chemical structure and insufficient active sites. However, GO shows a higher ability to absorb biomolecules thanks to carboxyl, epoxy, and hydroxyl groups [17]. In terms of fabrication, the use of AM has paved the way for the development personalized approach to medicine. For example, by combining 3D scanning with 3D printing, personalized salicylic acid-loaded patches for acne treatment that adapt perfectly to the nose anatomy of the patient were created [18]. The combination of 3D printing with graphene and its derivatives allows us to obtain outstanding results. For example, in cartilage protection, the use of AM gives the opportunity of mimicking the natural tissue and the presence of graphene in a collagen-chitosan matrix leads to control of the release of biological factors [15]. In the case of coronary interventions, polycaprolactone (PCL) stents loaded with GNP obtained by fused-deposition modeling (FDM) printer allows having personalized stenting with the controllable release of anticoagulation and anti-restenosis agents and improved mechanical properties due to the use of GNP [19].

23.2.2 Biosensing and Bioimaging

GBN are attractive to be used in the development of electrochemical sensors and biosensors due to their excellent electrical conductivity, large surface area, and high electron transfer potential. The use of graphene derivatives to create electrically conductive nanocomposites allows obtaining electrodes and circuits by AM, which can enable the electrochemical detection of organic compounds and biologically active molecules. Biosensors typically are based on oxidoreductase enzymes coupled with conductive electrodes that transduce the biochemical interactions into a readable electronic signal. Currently, the production of biosensors is a complex and expensive process that generally implies the necessity of designed facilities, trained personnel and specialized materials [20]. AM has been used to produce metallic electrodes for sensing by selective laser sintering (SLS), but it has a high production cost. Conversely, 3D graphene-polymer electrodes are cheaper to produce by FDM, present higher stability, and have a wide range of working potentials for electrochemical applications [21].

Researchers have developed graphene-based field effect transistor biosensors that allow to detection of biomolecules like nucleic acids, proteins, and growth factors, monitoring the changes in an electrical signal [13]. Since the development of novel plastic matrices, such as polylactic acid (PLA) or acrylonitrile butadiene styrene (ABS), with conductive carbon materials (e.g., PLA-graphene, ABS-graphene) AM, especially FDM and inkjet printing, has been explored as a cheap manufacturing technology to obtain printed biosensors with graphene-loaded polymers. Amongst the biosensors, it is possible to find in literature biosensors to analyze blood plasma and detect nitrite and uric acid [22], picric acid, ascorbic acid, glucose, hydrogen peroxide, etc. [5]. Figure 23.3 shows an example of PLA-graphene biosensors manufactured using FDM.

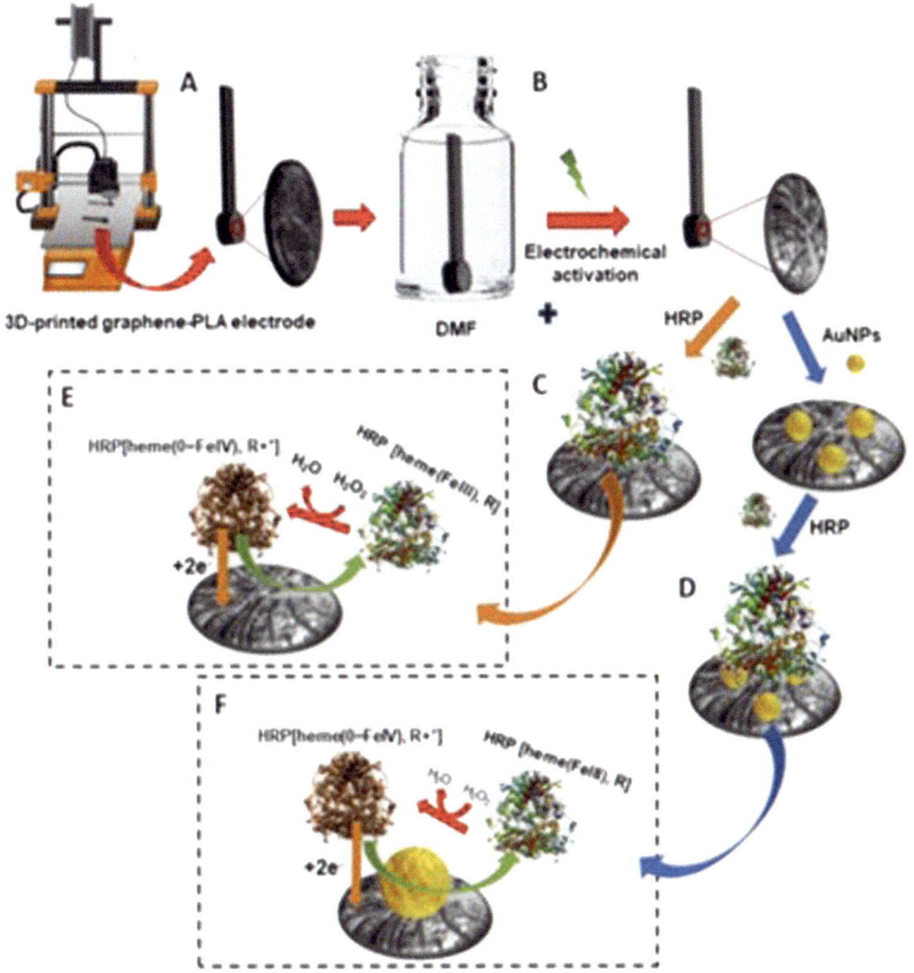

FIGURE 23.3 Biosensors were manufactured by FDM with PLA loaded with graphene. After printing, they were electrochemically activated in DMF and modified following two different paths: (1) modification of electrodes with HRP enzyme and (2) modification with gold nanoparticles followed by HRP enzyme. These components were attached to the surface of the electrode. The modified electrode allows for the detection of H_2O_2. Adapted with permission [21], Copyright (2019), Elsevier.

For fluorescence imaging, dyes, photosensitizers, quantum dots, gold nanoclusters, or up-converting nanoparticles functionalized GO/rGO have been widely investigated [23]. GO has strong absorbance and fluorescence properties in the near-infrared region, which allows it to be used in bioimaging applications. By changing the synthesis conditions such as the pH, rate of reduction, and size, fluorescence can be induced in GO. Functional groups of GO can be conjugated with fluorescent dyes for bioimaging [13]. There is another graphene derivative that does not need to be functionalized to be used in bioimaging applications. Graphene quantum dots (GQD) are nano-sized quantum dots with excellent optical properties for bioimaging and phototherapies. GQD are formed by cutting G or GO and they show fluorescent properties and are blue-luminescent. Compared with other quantum dots, GQDs possess high stability, excellent biocompatibility, good solubility, and low cytotoxicity. They were synthesized via hydrothermal route for the first time in 2010 by Pan et al. [24]. Gomes [25] explored cells embedded in a 3D-printed scaffold with GQD. They demonstrated that the quantum dots can enable deep tissue imaging under biomimetic conditions.

23.2.3 Tissue Engineering and Regenerative Medicine

Successful tissue engineering depends on the biocompatible substrates that offer cells to attach, grow and proliferate. Scaffolds are highly porous structures that act as an extracellular matrix to allow cells to proliferate and differentiate to repair hard or soft tissue. Amongst others, porosity and pore size are the main factors that define the performance of a scaffold. Scaffolds have been conventionally manufactured using techniques like solvent casting, freeze-drying, and salt leaching. However, these technologies allow to obtain of porous structures, but their geometry, porosity, and pore size are not controlled. This control can be achieved by AM technologies. AM allows for an imitation of natural tissue or organs and controls the special arrangement among cells, which is crucial for regenerative medicine. PLA-graphene scaffolds obtained by FDM were explored [26] and it was found that adequate topography may promote the alignment and orientation of neuronal, fibroblast, and myoblast cells, whilst graphene may stimulate cell differentiation.

Amongst AM technologies, bioprinting can use not only biomaterials but also living cells and/or other biological components, like proteins or growth factors. It enables the creation of biological structures that can regenerate or augment lost or damaged tissues and organs. Bioprinting can be performed by two different approaches: (i) pre-seeding or direct if cells are added before the printing, and (ii) post-seeding or indirect if printed structures are co-cultured with the proper cells [5]. For instance, post-seeded 3D-printed scaffolds with GO have shown high chondrocyte proliferation and activation of cell apoptosis when it was used for cartilage proliferation. The addition of GO resulted in thicker cartilage formation compared to the created by scaffolds without GO [27]. Graphene-loaded inks have also been studied to be used as scaffolds with improved human mesenchymal stem cell (hMSC) adhesion, viability, proliferation, and neurogenic differentiation with significant upregulation of glial and neuronal genes [28]. Some ceramics, like bioactive glass and diopside, showed the good mechanical and biological performance to be used as scaffolds when they are reinforced with GBN. 3D-printed scaffolds of ceramics reinforced with GBN are obtained by SLS [29], [30]. Figure 23.4 shows a bioactive glass scaffold for bone regeneration reinforced with GO obtained by SLS.

Pre-seeded structures were also developed to be produced by extrusion, inkjet, vat polymerization, and laser-assisted technologies [31]. Zhu et al. [32] combined pre-seeded bioprinting by stereolithography (SLA) with the addition of GNP to target nerve tissue regeneration.

23.3 DESIGN AND FABRICATION

AM offers high design freedom because the only information that needs to create a structure is its 3D digital model. In one manufacturing step, complex geometries can be easily and directly transformed

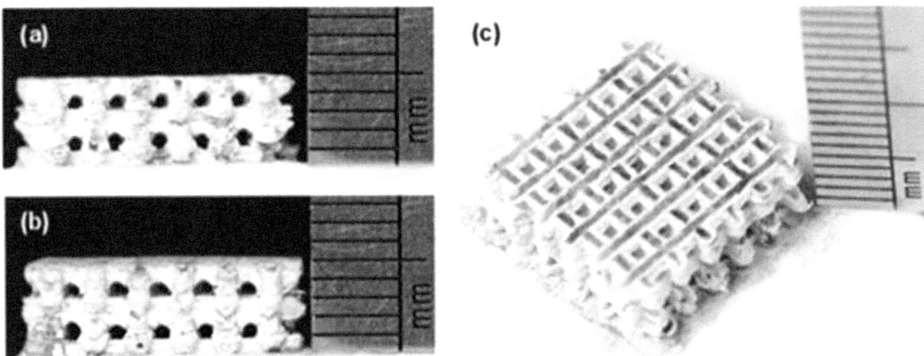

FIGURE 23.4 Scaffolds are highly porous 3D structures. In this example, the scaffold has square porous that is done in the three dimensions with high interconnectivity. Pores are approximately 0.8 μm in size and isotropic in their distribution. Adapted with permission [29]. Copyright C, Gao; T. Liu; C. Shuai; S. Peng, some rights reserved; exclusive licensee Nature Research. Distributed under a Creative Commons Attribution License 3.0 (CC BY).

into end-usable parts. To create these 3D digital models, it is possible to use reverse engineering with some medical tools, like computer tomography or magnetic resonance imaging. This aspect is very attractive in the medical field since it allows to obtain of 3D models of customized parts that match the patient's anatomy without extra cost [33]. Due to the potential risk to the human health of GBN through different mechanisms, like the binding of key signaling molecules, it is imperative to establish a proactive approach for the safe design of this kind of material [16]. In this regard, Safe-by-Design (SbD) concepts are focused on identifying and reducing potential risks and uncertainties in terms of human health and environmental safety since the early stages of product development, including the risk to health resulting from the potential airborne emission of nanoparticles and/or agglomerates during the manufacturing of medical devices [34]. In the last years, there is an increasing interest in applying SbD in nanomedicine. For instance, the GoNanoBioMat project elaborated a methodological SbD approach for nanomedicines with a focus on polymeric nanobiomaterials used for drug delivery [35].

Depending on the final application, the preferred AM technology to create structures with GBN changes, because they offer different characteristics of the raw material and different accuracy. The addition of GBN also could define the technology used. Mainly, five different technologies are reported in the literature to be used with this kind of material, but the most commonly used with graphene nanocomposites are those based on extrusion:

- **Fused Deposition Modeling (FDM):** The raw material is typically a thermoplastic polymer, e.g., polylactic acid, polycaprolactone, etc. It is melted in a heated nozzle, and it is deposited in the melted state by extrusion through the nozzle, building the structure which cools down and solidifies after extrusion. When GBN are added to the thermoplastic, the resulting nanocomposite must flow and present flexibility to achieve good printability. Therefore, the concentration of GBN within filament is usually low (< 10%wt.) [36].
- **Direct Ink Writing (DIW):** In this case, the raw material is liquid because it uses a solvent and a polymer, e.g. DCM (dichloromethane) and PLGA (poly(lactic-co-glycolic acid)) [28]. The polymer act as a binder, which gives consistency to the printed part. In this technology, the ink is extruded from a nozzle that rapidly solidifies into a defined structure. The solidification of the structure is driven via the evaporation of the solvent. The colloidal ink must maintain its shape during the solidification or drying process. This technology admits high

concentrations of fillers, achieving 60% vol. and often GBN are combined with ceramics to achieve good properties [28].

- **Selective Laser Sintering (SLS):** This technology uses powder as raw materials. The most common polymer powder used in SLS is polyamide, but ceramics can also be used, like bioactive glass or diopside [29], [30]. The addition of GBN can be facilitated using a solvent to favor the mixing process or by mechanical mixing with the powder.
- **Stereolithography (SLA) and Digital Light Processing (DLP):** These technologies are based on Vat Polymerization and they differ in terms of the light source. SLA produces the curing of the resin by a laser, whilst DLP does this process by a digital projector, curing the whole layer at the same time. This technology is used for sensors and tissue engineering and the addition of GBN is done at low concentrations, usually lower than 2% wt. [37], since it can affect the curing process or act as light scattering, resulting in low printability. In these technologies, there is a drawback that has to be considered for Biomedical Engineering applications. Resins are a mix of oligomers, monomers, photoinitiators, and additives. The polymerization process occurs during the printing process and it is important to assure that the polymerization is complete, typically via postprocessing, since residual monomer and photoinitiator usually show cytotoxicity.

Pristine graphene sheets cannot form stable homogeneous dispersions, which can be solved when the graphene is oxidized or functionalized. GO is soluble in water (hydrophilic) due to its carboxyl, epoxy, and hydroxyl functional groups [13]. In any case, it is important to use an adequate method to disperse GBN to achieve good dispersion. The method used to produce the nanocomposite to feed the printer differs depending on the technology and the kind of raw material [38].

In the case of solid materials, those processed using FDM and SLS, usually a solvent is added to disperse GBN by ultrasonication or mechanical mixing. Besides, if the raw material is easily melted, GBN can be added to the melted matrix and there is no need to use any solvent. Conversely, DIW and SLA/DLP technologies use liquid raw materials, which facilitates the addition and dispersion of GBN. In this case, ultrasonication and/or mechanical mixing can be done without any solvent, achieving good dispersion. However, these liquid raw materials present often high viscosity, which makes it difficult to have a good dispersion of GBN. When it occurs, a solvent can be added to promote the dispersibility of nanofillers.

When solvents are used to disperse GBN, it is important to remove them completely because any trace of solvent could result in a reduction of mechanical properties of the structure and/or possible reduction of the biocompatibility. After all, usually, the solvents used are toxic. The raw materials, GBN addition method, and advantages and disadvantages of each technology are shown in Table 23.2.

In the case of drug delivery devices, SLA is one of the most common AM technologies [45]. Conversely, to manufacture biosensors, FDM [22] and inkjet printing [20] have been explored. For *the manufacturing of scaffolds, the most explored technology is FDM, but some studies with DIW,* SLA, DLP, and SLS technologies can be found [5].

23.4 BIOLOGICAL FUNCTIONALITY

Some studies showed that bare graphene and GBN exhibit potential adverse effects and unsatisfied interaction at the biological and cellular interfaces [46]. The hydrophobic nature of graphene often impedes its use in the Biomedical Engineering field. However, the functionalization of GBN can result in better biological performance. Hydrophobicity is one of the factors that are responsible for cell mortality since they interfere with the hydrophobic protein-protein interaction [5]. Pristine graphene presents low chemical reactivity and solubility, which is partially solved by oxidizing it to produce GO. It contains hydroxyl and epoxide groups which are polar and hydrophilic. However,

TABLE 23.2
AM Technologies Used in Biomedical Engineering Applications

	Raw Material	GBN Addition	Advantages	Disadvantages	Ref
FDM	Thermoplastic	- Solvent dissolution - Melt mixing by extrusion	- Low cost - Not using toxic solvents	- High anisotropy - Low resolution - Flowability and flexibility limits GBN concentration	[36], [39]
DIW	Liquid	- Solvent mixing - Centrifuge mixing - Ultrasonication	- Low viscosity materials - Cells can survive during printing	- Formulation of inks	[40], [41]
SLS	Powder	- Melt mixing by extrusion - Dissolution-precipitation - Physical mixing	- No supports required - Facility to produce complex geometries - No solvent required	- Medium resolution - Post-processing required	[38], [42]
SLA	Liquid resin	- Solvent mixing - Ultrasonication	- High resolution - Liquid raw material, which facilitates GBN addition	- Cytotoxic residual monomer and/or photoinitiator - Viscosity limits GBN concentration - GBN can affect polymerization and/or light behavior	[42], [43]
DLP	Liquid resin	- Solvent mixing - Ultrasonication	- High resolution - Liquid raw material, which facilitates GBN addition - Faster than SLA	- Toxic residual monomer and/or photoinitiator - Viscosity limits GBN concentration - GBN can affect polymerization and/or light behavior	[38], [44]

there are still some regions of the basal plane that are unmodified and remain with the same properties as graphene [16].

One of the main reasons why the functionalization of graphene leads to a reduction in cellular toxicity is because it reduces the hydrophobicity of these nanomaterials. Graphene, GO and rGO can be functionalized with biomolecules, e.g., gelatin, cellulose, DNA, enzymes, hydroxyapatite, etc, to increase solubility in biological solutions, biocompatibility, response towards stimuli, the catalytic activity of enzymes, promoting cell growth, amongst others [47]. For instance, dextran-functionalized GO showed to be cleared from the reticuloendothelial system of mice within a week without significant toxicity [48].

There are different strategies to functionalize the surface: covalent and non-covalent functionalization, plasma hydrogenation, and nanoparticle functionalization [16]. For Biomedical Engineering applications, GBN are covalently functionalized with DNA, proteins, and polymers. The objective is to enhance biocompatibility or introduce additional specific properties, like targeting a precise location in cells and organisms for diagnosis or disease treatment [49]. In the case of non-covalent functionalization, π–π stacking and/or electrostatic interaction is produced between the graphene surface and the functionalities [50]. For example, in the Biomedical Engineering field, graphene was functionalized with Fe_3O_4 and Au nanoparticles to be used as an electrochemiluminescence marker for biosensing applications [51]. In the case of plasma treatment, the activation of hydrogen gases is performed by radio frequency or electric discharge to form plasma. When graphene is exposed to this plasma, the water molecules physisorbed on its surface are fragmented producing graphene hydrogenation. With this surface functionalization, a good level of hydrophilicity is obtained [52], [53].

23.5 CLINICAL TRANSLATION

In general, the application of nanotechnology in the Biomedical Engineering field (nanomedicine) showed interesting results, but it carries new barriers due to the difficulty to predict potential adverse effects on human health and the environment because of the complexity of nanobiomaterials. The unpredictability of the interaction of nanomaterials with biological systems makes it difficult to bring these to the market [35]. Undoubtedly, graphene and its derivatives have shown promising results in the Biomedical Engineering field. However, there are still some problems to solve to achieve their clinical translation. One of the main issues is the lack of standardization in the production of graphene-based nanomaterials. Besides, there is a lack of guidelines, standards, and tools adapted to nanomedicine for assessing their risks.

To complete clinical assays, it is important to control the lateral size, purity, aggregation, and oxidative state of these nanomaterials [49]. The importance of controlling these factors is not only standardization to know the properties of the nanomaterials, but also to control their toxicity, since the number of layers, lateral size, shapes, and chemistry affect the interaction with the cells. For this reason, cytotoxicity studies found in the literature present inconclusive results because it depends on these factors, which are not completely controlled and standardized, as previously explained. Therefore, the lack of standardization of both, production and risk evaluation of GBN results in an under-exploitation of their potential benefits in medicine. To facilitate clinical translation, it is mandatory to satisfy strict regulatory requirements of the US-FDA (Food and Drug Administration) and EU-MDR (Medical Device Regulation). The critical quality attributes (CQAs) should capture the attributes that could affect the quality, safety, or efficacy of the final product, and they should be reported as part of the premarket application: chemical composition, average particle size, particle size distribution, aspect ratio, physical and chemical stability. However, FDA has not established regulatory definitions of 'nanotechnology', 'nanomaterial', 'nanoscale' or other terms related to the use of nanoparticles yet [54].

For the regulatory approval by the US-FDA of a new therapeutic entity, four different pathways can be followed: tissues, biological products, drugs, and medical devices. Depending on the application, the therapeutic entity will match on a pathway. However, some approaches may fall into more than one category, even all four of these categories. It implies an important difference in times; bringing a medical device to market takes an average of three to seven years, compared with an average of 12 years for drugs [38]. Even with these issues, some graphene-based materials for Biomedical Engineering applications have already been commercialized. For example, graphene-polymer disks for dental applications [55], graphene-microalgae cosmetic skincare [56], and graphene-based glucose sensors [57] were developed by Graphenano®. However, to the best of our knowledge, 3D-printed graphene-based biomedical devices have not still been commercialized.

23.6 FUTURE PERSPECTIVES AND CHALLENGES

In this chapter, the use of 3D-printed graphene nanocomposites in the Biomedical Engineering field has been analyzed. The combination of AM and GBN has offered promising results that support its usage in different domains of Biomedical Engineering like Drug and Gene delivery, Biosensing, Bioimaging, Tissue Engineering, and Regenerative Medicine. Despite the good results obtained, clinical translation of 3D-printed devices made of graphene nanocomposites is not still a global reality because of a lack of standardization in terms of manufacturing and risk assessment. Therefore, the way forward in this research field should be directed towards that end.

REFERENCES

1 H.P. Boehm, R. Setton, E. Stumpp, Nomenclature and terminology of graphite intercalation compounds. *Carbon N. Y.* 24 (1986) 241–245.

2 K.S. Novoselov, A.K. Geim, S.V. Morozov, D. Jiang, Y. Zhang, S.V. Dubonos, I.V. Grigorieva, A.A. Firsov, Electric field effect in atomically thin carbon films. *Science*, 306 (2004) 666–669.

3 S.K. Tiwari, V. Kumar, A. Huczko, R. Oraon, A. Adhikari, G.C. Nayak Magical allotropes of carbon: Prospects and applications. *Crit. Rev. Solid State Mater. Sci.* 41 (2016) 257–317.

4 P. Kamedulski, M. Skorupska, P. Binkowski, W. Arendarska, A. Ilnicka, J.P. Lukaszewicz, High surface area micro-mesoporous graphene for electrochemical applications. *Sci. Rep* 11 (2021) 22054.

5 M. Silva, I.S. Pinho, J.A. Covas, N.M. Alves, M.C. Paiva, 3D printing of graphene-based polymeric nanocomposites for biomedical applications. *Funct. Compos. Mater*. 2 (2021) 8.

6 T.P. Dasari Shareena, D. McShan, A.K. Dasmahapatra, P.B. Tchounwou, A review on graphene-based nanomaterials in biomedical applications and risks in environment and health. *Nano-Micro Lett.* 10 (2018) 53.

7 K. Sreeja, T. Naresh Kumar, Effect of graphene oxide on fresh, hardened and mechanical properties of cement mortar. *Mater. Today Proc.* 46 (2021) 2235–2239.

8 Z. Jiang, O. Sevim, O.E. Ozbulut, Mechanical properties of graphene nanoplatelets-reinforced concrete prepared with different dispersion techniques. *Constr. Build. Mater*. 303 (2021) 124472.

9 W. Wang, J.R.P. Junior, P.R.L. Nalesso, D. Musson, J. Cornish, F. Mendonça, G.F. Caetano, P. Bártolo, Engineered 3D printed poly(ε-caprolactone)/graphene scaffolds for bone tissue engineering. *Mater. Sci. Eng. C*. 100 (2019) 759–770.

10 H. Elkhenany, S. Bourdo, S. Hecht, R. Donnell, D. Gerard, R. Abdelwahed, A. Lafont, K. Alghazali, F. Watanabe, A.S. Biris, Graphene nanoparticles as osteoinductive and osteoconductive platform for stem cell and bone regeneration. *Nanomedicine Nanotechnology, Biol. Med.* 13 (2017) 2117–2126.

11 International Organization for Standarization ISO/ASTM 52900:2021 Additive manufacturing—General principles—Fundamentals and vocabulary. 2021.

12 T. Limongi, F. Susa, M. Allione, E.Di Fabrizio, Drug delivery applications of three-dimensional printed (3DP) mesoporous scaffolds. *Pharmaceutics* 12 (2020) 085.

13 S. Syama, P.V. Mohanan, Comprehensive application of graphene: Emphasis on biomedical concerns. *Nano-Micro Lett.* 11 (2019) 6.

14 Z. Liu, J.T. Robinson, X. Sun, H. Dai, PEGylated nanographene oxide for delivery of water-insoluble cancer drugs. *J. Am. Chem. Soc.* 130 (2008) 10876–10877.

15 Z. Cheng, B. Landish, Z. Chi, C. Nannan, D. Jingyu, L. Sen, L. Xiangjin, 3D printing hydrogel with graphene oxide is functional in cartilage protection by influencing the signal pathway of Rank/Rankl/OPG. *Mater. Sci. Eng. C*. 82 (2018) 244–252.

16 Z. Guo, S. Chakraborty, F.A. Monikh, D.D.Varsou, A.J. Chetwynd, A. Afantitis, I. Lynch, P. Zhang, Surface functionalization of graphene-based materials: Biological behavior, toxicology, and safe-by-design aspects. *Adv. Biol.* 5 (2021) 2100637.

17 M. Aleemardani, P. Zare, A. Seifalian, Z. Bagher, A.M. Seifalian, Graphene-based materials prove to be a promising candidate for nerve regeneration following peripheral nerve injury. *Biomedicines* 10 (2022) 73.

18 A. Goyanes, U. Det-Amornrat, J. Wang, A.W. Basit, S. Gaisford, 3D scanning and 3D printing as innovative technologies for fabricating personalized topical drug delivery systems. *J. Control. Release*. 234 (2016) 41–48, doi:10.1016/j.jconrel.2016.05.034.

19 S.K. Misra, F. Ostadhossein, R. Babu, J. Kus, D. Tankasala, A. Sutrisno, K.A. Walsh, C.R. Bromfield, D, Pan, 3D-printed multidrug-eluting stent from graphene-nanoplatelet-doped biodegradable polymer composite. *Adv. Healthc. Mater*. 6 (2017) 1–14.

20 G. Rosati, M. Urban, L. Zhao, Q. Yang, C. De Carvalho, S. Bonaldo, C. Parolo, E.P. Nguyen, G. Ortega, P. Fornasiero. A plug, print & play inkjet printing and impedance-based biosensing technology operating through a smartphone for clinical diagnostics. *Biosens. Bioelectron.* 196 (2022) 113737.

21 A.M. López Marzo, A.C. Mayorga-Martinez, M. Pumera, 3D-printed graphene direct electron transfer enzyme biosensors. *Biosens. Bioelectron.* 151 (2020) 111980.

22 R.M. Cardoso, P.R.L. Silva, A.P. Lima, D.P. Rocha, T.C. Oliveira, M. Thiago, E.L. Fava, O. Fatibello-filho, E,M. Richter, R.A.A. Muñoz, Sensors and actuators B: Chemical 3D-printed graphene/polylactic acid electrode for bioanalysis: Biosensing of glucose and simultaneous determination of uric acid and nitrite in biological fluids. *Sensors Actuators B. Chem.* 307 (2020) 127621.

23 J. Lin, X. Chen, P. Huang, Graphene-based nanomaterials for bioimaging. *Adv. Drug Deliv. Rev.* 105 (2016) 242–254.

24 D. Pan, J. Zhang, Z. Li, M. Wu, Hydrothermal route for cutting graphene sheets into blue-luminescent graphene quantum dots. *Adv. Mater.* 22 (2010) 734–738.

25 V. Gomes, Graphene quantum dots for multiphoton imaging and drug delivery. Graphene 2017 Conf. 2017.

26 M. Gasparotto, P. Bellet, G. Scapin, R. Busetto, C. Rampazzo, L. Vitiello, D.I. Shah, F. Filippini, 3D printed graphene-PLA scaffolds promote cell alignment and differentiation. *Int. J. Mol. Sci.* 23 (2022) 1736.

27 Z. Cheng, L. Xigong, D. Weiyi, H. Jingen, W. Shuo, L. Xiangjin, W. Junsong, Potential use of 3D-printed graphene oxide scaffold for construction of the cartilage layer. *J. Nanobiotechnology* 18 (2020) 97.

28 A.E. Jakus, E.B. Secor, A.L. Rutz, S.W. Jordan, M.C. Hersam, R.N. Shah, Three-dimensional printing of high-content graphene scaffolds for electronic and biomedical applications. *ACS Nano.* 9 (2015) 4636–4648.

29 C. Gao, T. Liu, C. Shuai, S. Peng, Enhancement mechanisms of graphene in nano-58S bioactive glass scaffold: Mechanical and biological performance. *Sci. Rep.* 4 (2014) 4712.

30 C. Shuai, T. Liu, C. Gao, P. Feng, T. Xiao, K. Yu, S. Peng, Mechanical and structural characterization of diopside scaffolds reinforced with graphene. *J. Alloys Compd.* 655 (2016) 86–92.

31 P. Rider, Z.P. Kačarević, S. Alkildani, S. Retnasingh, M. Barbeck, Bioprinting of tissue engineering scaffolds. *J. Tissue Eng.* 9 (2018) 1–16.

32 W. Zhu, B. Harris, L. Zhang, Gelatin methacrylamide hydrogel with graphene nanoplatelets for neural cell-laden 3D bioprinting. *Annu. Int. Conf. IEEE Eng. Med. Biol. Soc.* 2016 (2016) 4185–4188.

33 F. Calignano, M. Galati, L. Iuliano, P. Minetola, Design of additively manufactured structures for biomedical applications: A review of the additive manufacturing processes applied to the biomedical sector. *J. Healthc. Eng.* 2019 (2019) 9748212.

34 J.M. López De Ipiña, C. Vaquero, A. Egizabal, A. Patelli, L. Moroni, Safe-by-design strategies applied to scaffold hybrid manufacturing. *J. Phys. Conf. Ser.* 1953 (2021) 012009.

35 M. Schmutz, O. Borges, S. Jesus, G. Borchard, G. Perale, M. Zinn, Ä.A.J.A.M. Sips, L.G. Soeteman-Hernandez, P. Wick, C. Som, A methodological safe-by-design approach for the development of nanomedicines. *Front. Bioeng. Biotechnol.* 8 (2020) 1–7.

36 S.F. Melo, S.C. Neves, A.T. Pereira, I. Borges, P.L. Granja, F.D. Magalhães, L.C. Gonçalves, Incorporation of graphene oxide into poly(ε-caprolactone) 3D printed fibrous scaffolds improves their antimicrobial properties. *Mater. Sci. Eng. C.* 109 (2020) 110537.

37 M.M. Hanon, A. Ghaly, L. Zsidai, Z. Szakál, I. Szabó, L. Kátai, Investigations of the mechanical properties of DLP 3d printed graphene/resin composites. *Acta Polytech. Hungarica.* 18 (2021) 143–161.

38 S. Lopez de Armentia, J.C. del Real, E. Paz, N. Dunne, Advances in biodegradable 3D printed scaffolds with carbon-based nanomaterials for bone regeneration. *Materials (Basel).* 13 (2020) 5083.

39 X. Wei, D. Li, W. Jiang, Z. Gu, X. Wang, Z. Zhang, Z. Sun, 3D printable graphene composite. *Sci. Rep.* 5 (2015) 11181.

40 G. Shi, S.E. Lowe, A.J.T. Teo, T.K. Dinh, S.H. Tan, J. Qin, Y. Zhang, Y.L. Zhong, H. Zhao, A versatile PDMS submicrobead/graphene oxide nanocomposite ink for the direct ink writing of wearable micron-scale tactile sensors. *Appl. Mater. Today.* 16 (2019) 482–492.

41 K. Chizari, M.A. Daoud, A.R. Ravindran, D. Therriault, 3D printing of highly conductive nanocomposites for the functional optimization of liquid sensors. *Small*, 12 (2016) 6076–6082.

42 H. Qu, Additive manufacturing for bone tissue engineering scaffolds. *Mater. Today Commun.* 24 (2020) 101024.

43 Z. Feng, Y. Li, L. Hao, Y. Yang, T. Tang, D. Tang, W. Xiong, Graphene-reinforced biodegradable resin composites for stereolithographic 3D printing of bone structure scaffolds. *J. Nanomater.* 2019 (2019) 1–13.

44 Z. Feng, Y. Li, C. Xin, D. Tang, W. Xiong, H. Zhang, Fabrication of graphene-reinforced nanocomposites with improved fracture toughness in net shape for complex 3D structures via digital light processing. *J. Carbon Res.* 5 (2019) c5020025.

45 J. Wang, Y. Zhang, N.H. Aghda, A.R. Pillai, R. Thakkar, A. Nokhodchi, M. Maniruzzaman, Emerging 3D printing technologies for drug delivery devices: Current status and future perspective. *Adv. Drug Deliv. Rev.* 174 (2021) 294–316.

46 K.H. Liao, Y.S. Lin, C.W. MacOsko, C.L. Haynes, cytotoxicity of graphene oxide and graphene in human erythrocytes and skin fibroblasts. *ACS Appl. Mater. Interfaces.* 3 (2011) 2607–2615.

47 C. Nie, L. Ma, S. Li, X. Fan, Y. Yang, C. Cheng, W. Zhao, C. Zhao, Recent progresses in graphene based bio-functional nanostructures for advanced biological and cellular interfaces. *Nano Today.* 26 (2019) 57–97.

48 S. Zhang, K. Yang, L. Feng, Z. Liu, In vitro and in vivo behaviors of dextran functionalized graphene. *Carbon N. Y.* 49 (2011) 4040–4049.

49 G. Reina, J.M. González-Domínguez, A. Criado, E. Vázquez, A. Bianco, M. Prato, Promises, facts and challenges for graphene in biomedical applications. *Chem. Soc. Rev.* 46 (2017) 4400–4416.

50 V. Georgakilas, J.N. Tiwari, K.C. Kemp, J.A. Perman, A.B. Bourlinos, K.A. Kim, R. Zboril. Noncovalent functionalization of graphene and graphene oxide for energy materials, biosensing, catalytic, and biomedical applications. *Chem. Rev.* 116 (2016) 5464–5519.

51 W. Gu, X. Deng, X. Gu, X. Jia, B. Lou, X. Zhang, J. Li, Wang, E. Stabilized, Superparamagnetic functionalized graphene_Fe_3O_4@Au nanocomposites for a magnetically-controlled solid-state electrochemiluminescence biosensing application. *Anal. Chem.* 87 (2015) 1876–1881.

52 D. Jones, W. Hoffmann, A. Jesseph, C. Morris, G. Verbeck, J. Perez, On the mechanism for plasma hydrogenation of graphene. *Appl. Phys. Lett.* 97 (2010) 233104.

53 M. Moschetta, J.Y. Lee, J. Rodrigues, A. Podestà, O. Varvicchio, J. Son, Y. Lee, K. Kim, G.H. Lee, F. Benfenati, M. Bramini, A. Capasso, Hydrogenated graphene improves neuronal network maturation and excitatory transmission. *Adv. Biol.* 5 (2021) 2000177.

54 U.S. Department of Health and Human Services; Food and Drug Administration; Center for Drug Evaluation and Research; Center for Biologics Evaluation and Research. Drug Products, Including Biological Products, that Contain Nanomaterials 2022.

55 Graphenano Nanotechnologies Graphenano Dental. Available online: www.graphenano.com/graphenano-dental/.

56 AquaGraph AcquaGraph Cosmetics. Available online: www.acquagraph.com/.

57 Graphenano Nanotechnologies Graphenano Sensors. Available online: www.graphenano.com/graphenano-sensors/.

24 3D-Printed Metal Oxides for Biomedical Applications

Dipesh Kumar Mishra[1*], *Rudranarayan Kandi*[2*], *Dayanidhi Krishana Pathak*[3], *and Pawan Sharma*[4]
[1]Department of Manufacturing Technology and Automation, Center for Advanced Studies, AKTU Lucknow, 226031, India
[2]Assistant professor, Department of Industrial and Production Technology, Dr B R Ambedkar National Institute of Technology Jalandhar, Punjab.
[3]Department of Mechanical and Automation Engineering, G. B. Pant Government Engineering College, New Delhi, 110020, India
[4]Department of Mechanical Engineering, Indian Institute of Technology, (BHU) Varanasi, 221005, India
[*]Corresponding author: mishradipesh14@gmail.com, rudranarayan07@gmail.com

CONTENTS

24.1 INTRODUCTION

Metal oxides are compounds that are abundantly found in nature. These are formed by the reaction of metal and oxygen. Metal oxides have a vital role in various chemical reactions and physical areas. In the last decades, the critical uses of metal oxides, such as application in gas sensing, making in transistors, catalysis, fabrication in biosensors, microelectronics, and many more, have attracted researchers. Metal oxides exhibit promising electronic and magnetic properties, which can be inferred from Table 24.1. The available metal anion and oxide anion held together by electrostatic force in crystalline metal oxides induce higher carrier mobility, optical transparency, and tolerance to mechanical stress, making them more efficient for electronic applications such as the fabrication of solar cells and semiconductors. Furthermore, metal oxide-based nanostructures have unique physical and chemical properties because of their regular and limited size, which affect the basic interaction properties of materials. Therefore, a large variety of metal oxides such as nanowires,

DOI: 10.1201/9781003296676-24

TABLE 24.1
Application of Metal Oxides in Various Fields [1]

Types of Oxides	Properties	Applications
BiO_3	(a) Optoelectronicronic, (b) Semiconductor with photocatalytic activity	Water treatment plants
Co_3O_4	(a) Optical, (b) magnetic, and (c) electrochemical properties	Energy storage plants
Fe_2O_3/Fe_3O_4	(a) Magnetic properties	Biomedical uses like MRI and Drug delivery.
Sb_2O_3	(a) Semiconductor material	Use as a Chemical catalysis
SiO_2	(a) Biocompatible properties	Cosmetic and biomedical
TiO_2	(a) Electronic properties	Environment preservation and biomedical application.
UnO_2	(a) Electronic properties	Nuclear
ZnO	(a) Electronic properties	Decontamination of gases
ZrO_2	(a) Electro-optical, (b) piezoelectric and (c) dielectric material	Catalysis
SnO_2	(a) Optical and (b) electronic	Biomedical

nanoporous structures, and nanotubes are widely used for fulfilling the need available in different energy and engineering sectors. Moreover, the ferro, ferri, and antiferromagnetic behavior of metal oxides make them more suitable for biomedical applications (i.e., drug delivery, magnetic resonance imaging (MRI), and hyperthermia [1]).

Currently, the metal oxides nanoparticles (NPs) have been extensively utilized in different biomedical applications owing to their remarkable properties like selective oxidation, dehydration, photocatalysis, electrolysis, higher surface area, better magnetic properties, and enhanced catalytic activity. Interestingly, the biomedical properties of NPs cannot be compared with the properties of bulk particles corresponding to macroscopic size particles. The extensively used metal-based NPs for biomedical applications are TiO_2, ZrO_2, and ceria (CeO_2), and more detailed applications of different metal oxides are presented in Table 24.1. These mentioned materials have promising antioxidant, bacterial activities, mechanical stability, and biocompatibility properties that made them incomparable for biomedical appliances such as neurochemical monitoring and drug delivery instrument [2]. Furthermore, few metal oxides can create a favorable environment for drug delivery by developing oxides' monolayer surface carriers [3].

The shape and size of NPs have strongly impacted their applications in various fields. Generally, the range of NPs lies between 1 to 100 nm. As compared to bulk or macro-sized metal oxides, NPs exhibit distinct features like they can go in the cell membrane or internal cell organelles, and may even ease to surpass the blood-brain hurdle. In a recent study, it was found that the 20 nm sized NPs are extra capable to cross the internal cell organelles. Even, these tiny-sized NPs can pierce the bacterial cells and release their toxic effects by dissolution. The major parameters that mainly help to improve the characteristics of metal oxides are large surface area and proportionate effects. The larger surface area of NPs per unit mass makes the metal oxides more effective by constantly increasing their diffusion mechanism. Thereby, the magnetic, optics, and electrical properties of NPs are largely enhanced [4].

The important challenge of metal oxides highlighted by researchers in the context of biomedical applications is toxicity. From the prior study, it was found that the size and loading of nanoparticles over the surface affected the toxicity [5]. To investigate the toxicity of metal oxides, the delicate cellular changes in DNA damage and oxidative damage on human tissue are vital to examine. Additionally, chemical inertness is another important factor that is being used for artificial implants.

In this context, iron oxide, zinc oxide, and titanium oxides are popular metal oxides that are widely used. The nanoparticles of these metal oxides were found to meet the challenges associated with metal oxides for biomedical applications. The NPs are analyzed where NPs with proper purposeful sets permit the clinging with medicines, antibodies and lingands. The operationalization of the metal oxides set can be assisted by non-covalent connections between the hydroxyl group of metal oxides and ligands [6]. The following features that are associated with NPs of metal oxides are (i) non-toxic, (ii) chemically stable, (iii) biocompatible, and (iv) resistance to wear and scratch [7], [8]. Presently, the usages of metal oxide groups in different fields are recognized by international standard groups (like ISO) [9].

24.2 BIOMEDICAL APPLICATIONS OF METAL OXIDES

The potential features of metal oxides have gained their application in the biomedical field such as used for drug delivery, artificial implant fabrication, cancer therapy, etc.

24.2.1 Drug Delivery and Theranostic Applications

The designing of drug delivery systems is planned in such a way that delivers the relaxing chemicals to the intended locations of the human body. In this regard, cerium oxide, zinc oxide, and superparamagnetic iron oxide nanoparticles (SPIONs) have been broadly utilized as metal oxide-made carriers for drug delivery. The main function of the drug delivery system is that it does not comprise the functionality of drugs and the same can be unloaded as prompted by stimuli [10].

Theranostic is a new era of medicine that used to target the site with combine specific therapy and diagnostic tests. Theranostics is used by specific biomarkers for obtaining the diagnostic images from inside the human body and further, the therapeutic dose of radiations is provided to the patient on the behalf of diagnostic images obtain from them. Therefore, in the recent trends, different nanomedicine-based technology and imaging contrast agents are widely using a theranostic system that has the facility of therapeutics and diagnostic functionality in a single platform. In this context, zinc oxide NPs have been used.

The application of SPIONs is popularly known as the contrast agent for MRI (according to FDA). It was observed from the literature review, that the DOX-loaded SPIONs were efficiently used to resist the breast cancer cells by increasing the efficacy of the drug delivery system. Further, this effect reduces loads of medicine dose and its side effects [11].

24.2.2 Cancer Therapy

The better thermal property of SPIONs helps to develop localized heating near tumor areas when the suitable magnetic field is activated. As the temperature increases beyond 40 °C, localized hyperthermia assists to release the drug that further caused cancer cell death.

Titanium oxide (TiO_2) based NPs have been widely used as a therapy for killing cancer cells. These NPs produced cytotoxic radical oxygen species (ROS) as they were exposed to ultraviolet (UV) light. Hence, the patient suffered hair loss and baldness. A primary issue with this technique is the straight exposure against UV lights to live cells. In this regard, Shrestha et al. [12] reported that the toxicity of TiO_{2-}based NPs can be reduced by mixing with SPIONs. In addition, cerium oxide-based NPs have exhibited as a vital tool for providing resistance against ROS-induced toxicity [12]. Further, the additional biological control feature of cerium oxide-based NPs can also use as a radio sensitizing agent in the context of controlling the damage that occurred due to ROS toxicity. The latest study reported that the cerium oxide's toxicity was against the cancerous cell, not normal living cells. Therefore, it can be used as a toxic-free agent for normal cells. The radio sensitizing property of cerium oxide assists to make it a nongenotoxic agent. Thereby, it can control the damage to DNA. Although, the *in-vivo* study has presented a reduced tumor weight and volume in nude xenografted mice after treatment with cerium oxide [5].

24.2.3 Protection of Implants

Mostly, the cases of tumors were found in the patients due to bacterial adhesion and its multiplication in the implants. Consequently, bacterial adhesion leads to bone resorption and the implant has removed. The latest study has reported that the use of antibacterial coating can limit the result of bacterial infection. The parameters like biocompatibility, tenability, and resistance against mechanical stress are found vital characteristics of implants that are responsible for the effective working of antibacterial adhesion. In this context, titania is often used as a coating material over bio-implants for excellent cell adhesion and proliferation [13]. Moreover, Al_2O_3, ZnO, and CuO NPs are better coating materials for the hindrance of bacterial infection [5]. Figure 24.1 shows the antibacterial properties of the artificial bone implant.

24.2.4 Control of Bacterial Effect and Wound Healing

The low concentration of metal oxides has shown a better influence on antibacterial activity. Minimum repressive and antibacterial concentrations are found in ZnO NPs in comparison with zinc acetate. In the prior study, Azam et al. proposed the antibacterial exercise of ZnO, CuO, and Fe_2O_3 NPs contrary to the different bacteria such as Gram positive and Gram negative [13]. This study revealed that ZnO possesses large energy in response to Gram-positive bacterial than Gram-negative bacterial strains. Small particle-sized NPs are higher in surface to volume ratio, which leads to induced hindrance against bacterial growth. In addition, ZnO NPs have been used to produce ROS that is further utilized to kill cancer cells [13].

The aim of wound healing research helps decrease the risk of bacterial infection and accelerate the speed of recovery. In this regard, ZnO NPs are widely used as wound healing agents because they have better antibacterial and antifungal properties. The addition of ZnO-based NPs for wound dressing has been found to have improved the wound therapeutic process. The result of the latest

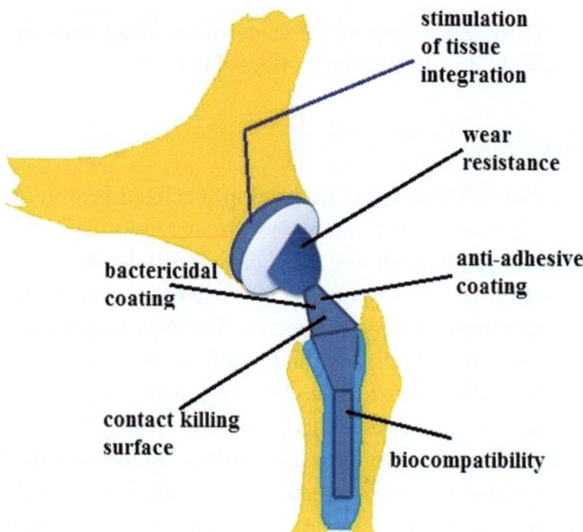

FIGURE 24.1 Schematic of artificial bone prostheses showing antibacterial properties. Adapted with permission [5].

studies has shown that the effect of metal oxide nanoparticles against bacteria was increased significantly if the particle size was found less than 100nm [14]. Moreover, TiO_2 NPs were also obtained to be more efficacious in the wound therapeutic process. When UV light is exposed to TiO_2 NPs then titanium ions were released, which then leads to resistance the bacterial infections and significantly increased wound healing [3].

The process used for fabricating the artificial implants made from metal oxides also induced a vital impact on its successful applications. In this context, three-dimensional (3D) printing techniques have found many advantages over other conventional processing techniques. This emerging processing technique has gathered great concern among not only manufacturers for fabricating medical equipment but also inducing promising properties for its successful applications. The same can be studied in the next section in a detailed manner [15].

24.3 3D PRINTING OF METAL OXIDES

3D printing is a versatile manufacturing technique that covers a large range of structures and intricate geometry from the 3D CAD model data. This technique follows layer-by-layer deposition steps for the fabrication of any products. Initially, this technique was invented by Charles Hull in 1986 under the name of Stereolithography (SLA) which was followed by subsequent growth such as fused deposition modeling (FDM), powder bed fusion (PBF), direct energy deposition (DED), inkjet printing and many more. 3D printing has grown over the years, and the advantages such as less wastage, freedom of design, and automation make it capable to induce change in the manufacturing system. Therefore, it is widely used in different fields, including construction, prototyping and biomedical. Especially, in the last decade, this latest technology is extensively preferred for the fabrication of medical-specific devices. Applications using biocompatible materials such as the development of tissue without damaging the surrounding living cells, designing of drug delivery systems, blood delivery systems, dental implants, and special medical devices used for cancer therapy, and protection of medical implants [16].

Several 3D-printing methods have been attempted to fabricate parts with metal oxides/polymer and metal oxide/metal powder in various biomedical applications. The manufacturing method includes fused deposition modeling (FDM), resin-based 3D printing, selective laser sintering/melting (SLS and SLM), solvent-based extrusion 3D printing, two-photon-based 3D printing, and binder jet printing, and selective laser gelling (SLG). Further, the chapter examines various applications of 3D-printed metal oxides in the biomedical field.

24.3.1 3D Printing of Iron Oxides

A broad range of material combinations with iron oxides has been attempted to suit various biomedical applications. The selection of various materials with metal oxides is based on both the application and type of implants. To date, various types of 3D-printing methods have been adapted to implement the utilities of iron oxides in the biomedical field. Table 24.2 presents a list of materials with iron oxide for various biomedical applications.

To overcome the limitations such as poor load-bearing capacity and low specific gravity, iron oxide has been incorporated into polylactic acid (PLA) to improve both the mechanical and biocompatibility of the composite [17]. The investigation mentioned the addition of F_3O_4 significantly improved the radiopacity and the location of the bone screw implant was clear in the micro-CT scan without any blooming effect. Figure 24.2 shows the typical scanning images of the neat polymeric bone screw and 20% Fe_3O_4/polymeric bone screw at the implant site. Fe_3O_4 nanoparticles exhibited excellent osteogenic effects [18]–[20]. Mixing unmodified iron oxides in biocompatible polymers improves biocompatibility with the induction of magnetism [21]. The inductive coupling magnetism can promote the growth, differentiation, and mineralization of the bone cells on the scaffold. A report

TABLE 24.2
Various Biomedical Applications of Iron Oxides Using 3D Printing

Application	Materials	Fabrication Method	Product
Bone Implant/repair/Bone tissue engineering	Polylactic acid + Fe_3O_4	Fused deposition method (FDM)	Bone screw [17]
	Chitosan + Fe_3O_4	Solvent-based extrusion printing	Bone Scaffold [21]
	Ti-6Al-4V + Fe_3O_4 + polydopamine (PDA)	Selective laser melting (SLM)	Bone Scaffold [23]
	Silk fibroin + Fe_3O_4	3D printing	3D printed frame for tissue culture plate with magnets [22]
Drug Delivery	Chitosan + Fe_2O_3	Two photon printing	Micro swimmers [24]
	Glycol chitosan + oxidized hyaluronate (OHA) + Iron oxide	Extrusion 3D printing	3D printed constructs [25]
Soft robotics/Bionic devices	GelMA + Iron oxides	Solvent based extrusion printing + UV curing	Bio-inspired soft robotic systems [26]
Regenerative/Restoration application	Polyglycolic acid (PLGA) + GelMA + Iron oxide	Direct writing 3D printing	Bile duct scaffold [27]
	Silk fibroin + Iron oxide	Digital light printing (DLP)	Magnetic Bioreactor [28]

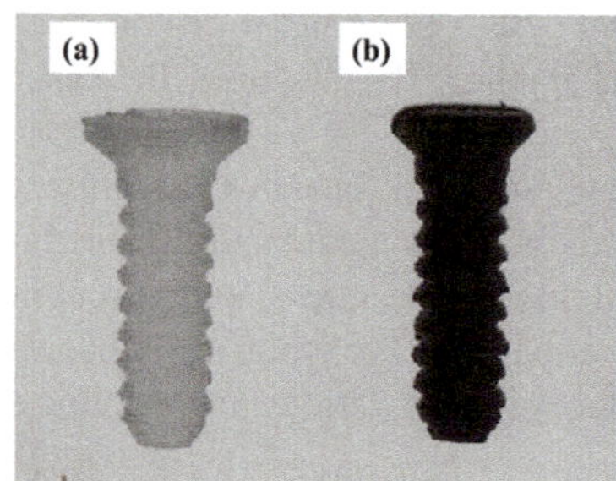

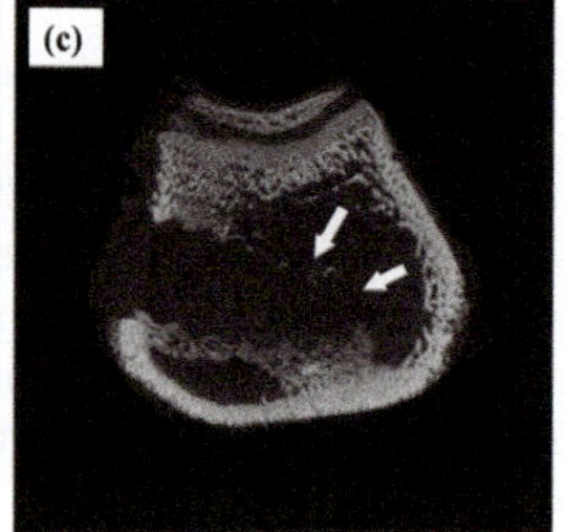

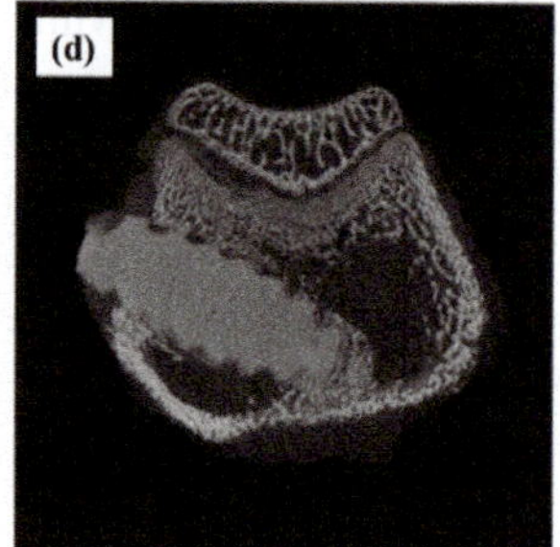

FIGURE 24.2 Illustration of screw implant made of (a) polymer, (b) polymer/Fe_3O_4 composite, micro-CT of implanted, (c) polymeric screw, and (d) polymer/Fe_3O_4 composite screw in rabbit bone site. Adapted with permission [17]. Copyright: The Authors, some rights reserved; exclusive licensee [MDPI]. Distributed under a Creative Commons Attribution License 4.0 (CC BY).

checked the effects of the inductive coupling magnetism on the scaffold biocompatibility and the effect was positive on the osteogenic responses with higher cell proliferation and collagen production. Similar observations were reported where a low static magnetic field positively impacted the osteogenic differentiation potential of the cells inside the biomimetic magnetic scaffold [22]. 3D-printed titanium bone implants coated with iron oxides are suitable options for large bone defects at load-bearing sites. Huang and the group [23] developed a 3D-printed titanium bone implant fabricated using selective laser melting (SLM). To improve cell proliferation and osteogenic differentiation, magnetic Fe_3O_4 mixed with polydopamine (PDA) was coated over the titanium implant.. Post-implantation in the rabbit bone revealed that the coated implant in the presence of a slow magnetic field (SMF) could overcome the poor osteogenic activity enhancing cell proliferation, attachment, and differentiation.

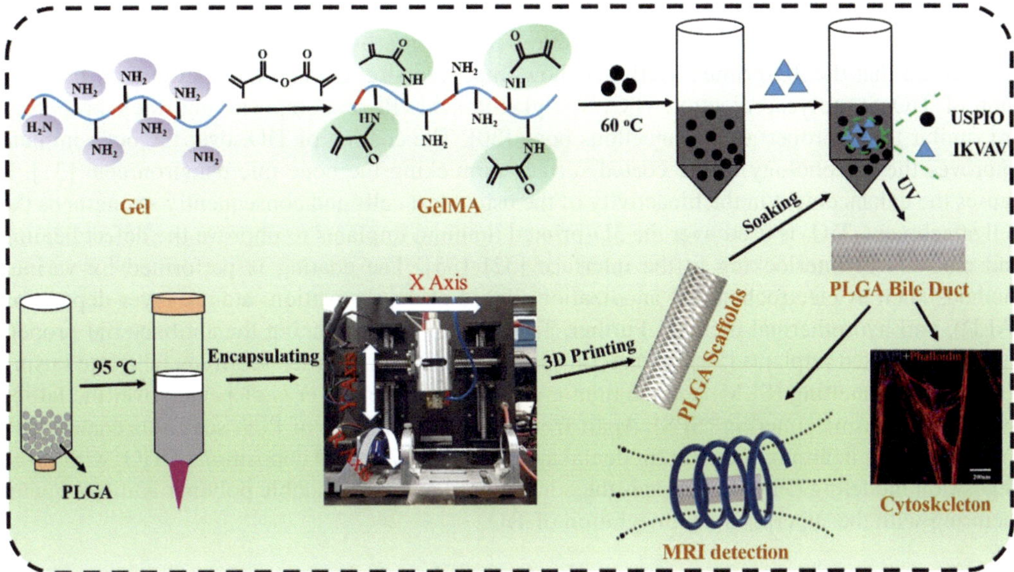

FIGURE 24.3 Schematic synthesis GelMA and fabrication of artificial tubular scaffold of PCL for bile duct using a 3D printing method. Adapted with permission [27]. Copyright: The Authors, some rights reserved; exclusive licensee [Frontiers]. Distributed under a Creative Commons Attribution License 4.0 (CC BY).

Apart from bone implant and bone tissue repair, iron oxides have been used for drug delivery applications. Magnetically actuated biocompatible and biodegradable chitosan/iron oxide-based micro-swimmers are fabricated using two photon-based 3D printing [24]. The magnetic field helps the micro-swimmers steer in the medium to reach the target site. Further, the micro-swimmers are activated using light induction to release the drug at the target site with high precision and efficacy. In another report, magnetic 3D constructs of Glycol chitosan+ oxidized hyaluronate (OHA)+ Iron oxide are 3D printed using solvent extrusion 3D printing for bioinspired soft robotic applications [25]. Iron oxides have shown great potential for a wide array of applications in fabricating magneto-responsive soft robots and bionic devices. Gelatin methacryloyl matrix (GelMA)/ iron oxide is used to fabricate starfish soft robots and is actuated using a magnetic field [26]. Additionally, iron oxides are used in tissue regenerative applications such as bile duct regeneration and cardiac muscle generation [27], [28]. Figure 24.3 shows the fabrication of a tubular scaffold of poly (lactic-co-glycolic acid) PLGA/GelMA/ IKVAV laminin peptide/iron oxide using a direct writing method of 3D printing. The composite artificial bile duct shows excellent MRI imaging function due to the presence of iron oxide for real-time non-invasive detection of tissue repair.

24.3.2 3D Printing of Titania (TiO_2)

Titanium oxide continues to be one of the most evolved biomaterials used in biomedical applications. It has been a suitable candidate for bone tissue engineering and controlled drug delivery applications. A report demonstrates the fabrication of bone implant of polymer/ceramic oxide composite using novel aerosol-based 3D printing and TiO_2 is well dispersed over the 3D-printed implant using an atomizer [29]. The well-dispersed TiO_2 improves the mechanical and biocompatibility of the implant by improving the load-bearing capacity, improved osteoblast adhesion, and collagen deposition compared to the implant with agglomerated TiO_2. Further, it reduced the acidic pH changes at the local site and improves the mechanical integrity of the implant. TiO_2 is known for its biocompatibility, non-toxicity, and antibacterial properties. The nanoparticles upon mixed with biodegradable polymer improve the stability of the polymer. It is observed that the 3D-printed scaffolds show improved strength and elongation with the addition of TiO_2 in polycaprolactone (PCL)/polylactic acid (PLA) composite and their properties are similar to the properties of cancellous bone [30]. The coating of TiO_2 over the bone implant improves the morphology of the coated surface mimicking the bone microenvironment [31]. It causes the enhancement in the bioactivity of the osteoblast cells and consequently strengthens the cell attachment. TiO_2 is used over the 3D-printed titanium implants to improve the defect healing and mechanical interlocking at the interface [32]–[35]. The coating is performed by various methods such as electrochemical anodization, liquid phase deposition, atomic layer deposition (ALD), and hydrothermal method. Further, TiO_2 is used for enhancing the antibacterial properties of 3D-printed implants [36]. Figure 24.4 shows the titanium lattice scaffolds fabricated using selective laser melting (SLM) and the titania-silver composite power is dispersed over the lattice using spark plasma sintering (SPS). Apart from these, 3D printing of PLA scaffolds coated with TiO_2 is used for dentin fabrication for dental applications [37]. ALD deposition of TiO_2 with fused deposition modeling (FDM) combines the convenience of biodegradable polymer additive manufacturing with the superior osseointegration of TiO_2.

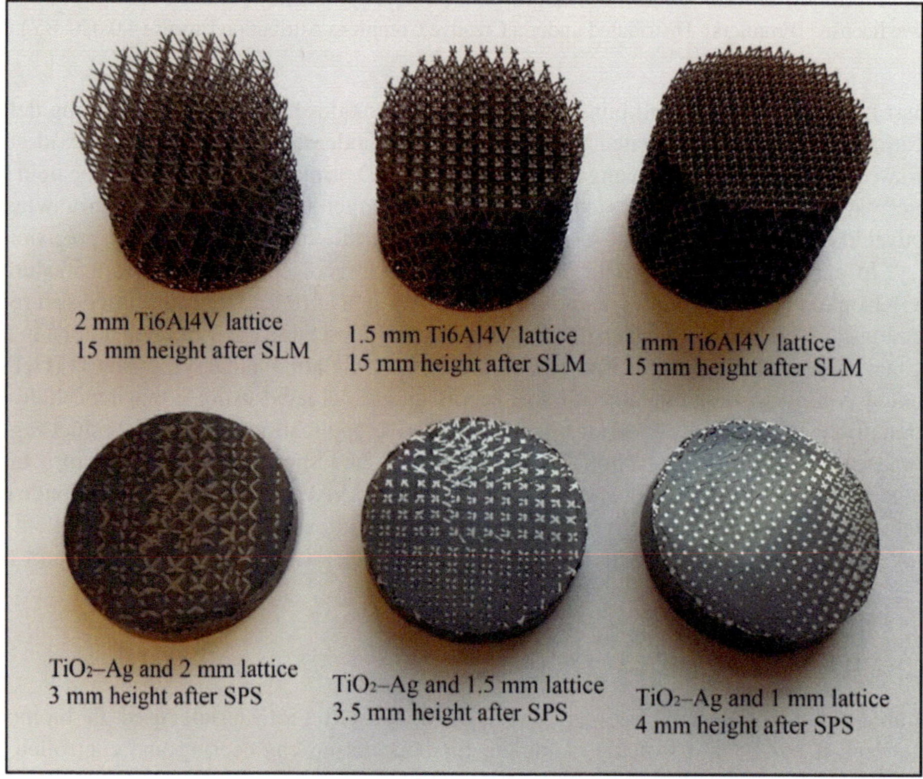

FIGURE 24.4 Top row: Illustrative images of Ti6Al4V lattice structures using Selective laser melting (SLM); Bottom row: Illustrative images of sintered TiO_2–2.5% Ag embedded in the Ti6Al4V lattice structures using Spark plasma sintering (SPS). Adapted with permission [36].

24.3.3 3D Printing of Zirconia (ZrO_2)

Zirconia (ZrO_2) has been successfully used in clinics due to its toughest property in oxide ceramics and the excellent biocompatibility endows the ZrO_2 scaffolds with bone tissue engineering. The introduction of ZrO_2 into pure ceramics may probably regulate the degradation rate and create a more stable microenvironment for related cell activities, eventually promoting new bone formation. Potential manufacturing of polyamide/ ZrO_2 is studied using fused deposition modeling (FDM) 3D printing to fabricate composite with comparable strength to biomedical implants [38]. 3D printed ceramic scaffold incorporated with ZrO_2 is one of the promising candidates for bone tissue engineering. An increase in ZrO_2 in the 3D printed β-Ca2SiO4 scaffolds improves the compressive strength and stimulates cell attachment, proliferation, and differentiation of osteoblast [39]. Importantly, post-implantation in the rat defect site promotes new bone generation compared to the scaffold without ZrO_2. Novel 3D-printing methods such as fusion 3D-printing process (FPG) and selective laser gelling (SLG) have been adopted incorporating ZrO_2 to improve the bone regeneration property of the scaffolds. In the fusion 3D-printing process, [40] ZrO_2 is added to melted biodegradable polymer (PCL) in a heatable barrel at around 105 °C and extrusion of PCL/ZrO_2 is performed over a flat platform with the help of compressed air. ZrO_2 improves the surface wettability of the overall scaffolds by improving the hydrophilicity nature conducive to cell adhesion and water uptake, beneficial for nutrient exchange. In selective laser gelling (SLG) [41], ZrO_2 and SiO_2 powders (fillers) are mixed with the SiO_2 solution (binder), and the slurry is used for the 3D printing as shown in Figure 24.5.

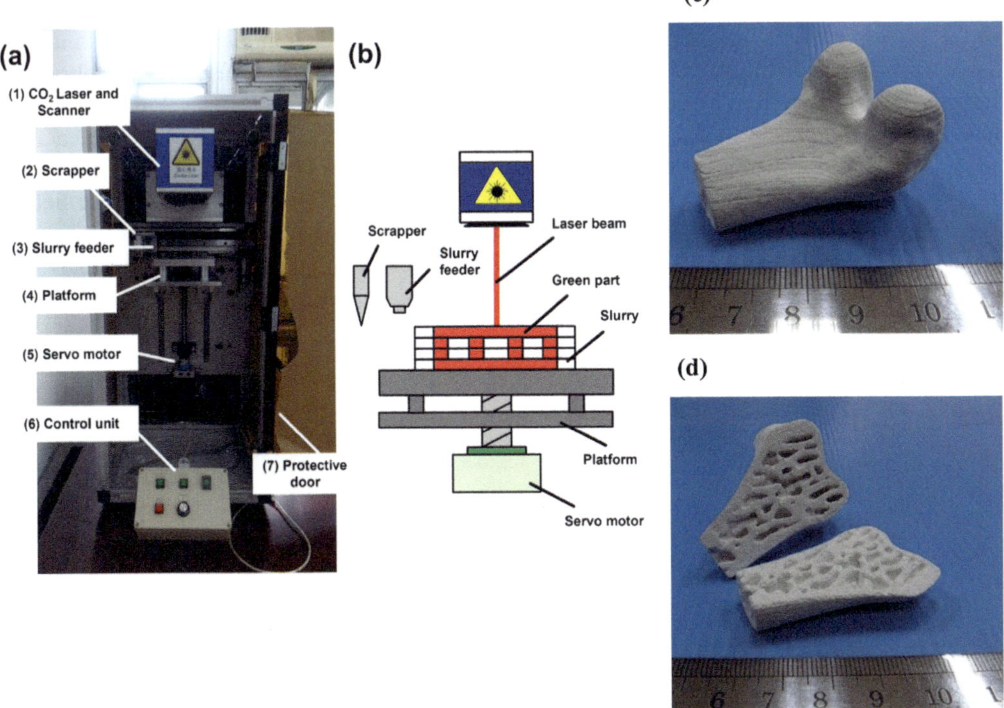

FIGURE 24.5 Illustrative image of (a) Selective laser gelling (SLG) based 3D printing machine, (b) schematic figure of developed SLG(c) fabricated part for a bone implant, and (d) cross-sectional view of the bone implant. Adapted with permission [41].

The method possesses a CO_2 laser, scanner, slurry feeder, scrapper, servomotor, and a platform. The laser selectively solidifies the solution to form the green part. The green part is later heat treated at 1300 °C to generate the ceramic bone implant. The scaffold with ZrO_2 possesses mechanical properties and cell affinity more comparable to those of natural human bone. ZrO_2 is used as the filler material to improve the antibacterial property of the implant. Gel casting, a novel manufacturing method is used where both resin 3D printing, gel casting, and sintering are performed to fabricate ceramic hip prostheses of ZrO_2 [42]. The ZrO_2 shows excellent antibacterial property that effectively kills the pathogen bacteria like *Escherichia coli* and *Staphylococcus aureus*. However, the limitations of ZrO_2 are the brittleness of the fabricated scaffolds. Further, ZrO_2 improves the radiopacity and biocompatibility of regenerative devices. Polymers are radiolucent and are not visible under X-ray imaging and CT scanning. The incorporation of radiopaque material into the polymer improves the exact position of the implant device during X-ray imaging. The incorporation of radiopaque material into the polymer improves the exact position of the implant device during X-ray imaging [43]. The image for the 3D-printed part doped with ZrO_2 depicts all the bones of the hand, unlike the part without radiopaque ZrO_2.

24.3.4 3D Printing of Zinc Oxide (ZnO)

Zinc oxide (ZnO) offers potential applications in the field of bone tissue engineering, photocatalysts, and antibacterial activities. The addition of silica with ZnO improves the compressive strength of 3D-printed tricalcium phosphate scaffolds for bone regeneration [44]. Further, it improves the hydrophilicity and cellular proliferation over the bone scaffolds suitable for bone tissue regeneration application [45]. With the increase of ZnO content up to 0.5 % in 3D-printed (powder-based Polyjet 3D printing) calcium sulfate hemihydrate (CSHH) scaffolds improve the Young's modulus and compressive strength of the scaffold [46]. Apart from these, ZnO is used in the fabrication of photocatalysts to remove or degrade pharmaceutical contaminations in water. Fused deposition modeling (FDM) is used to manufacture photocatalysts of polymer/ZnO and the photocatalytic ability is improved to the efficiency of ~98% after 60 mins under UV treatment [47]. 3D-printed ZnO/ABS/TPU/CaSiO33D using fused deposition modeling (FDM) offers an alternative field of wastewater treatment in large chemical equipment and plants [48]. Figure 24.6 shows the fabrication of photocatalysis of polylactic acid coated with ZnO using fused deposition modeling (FDM)

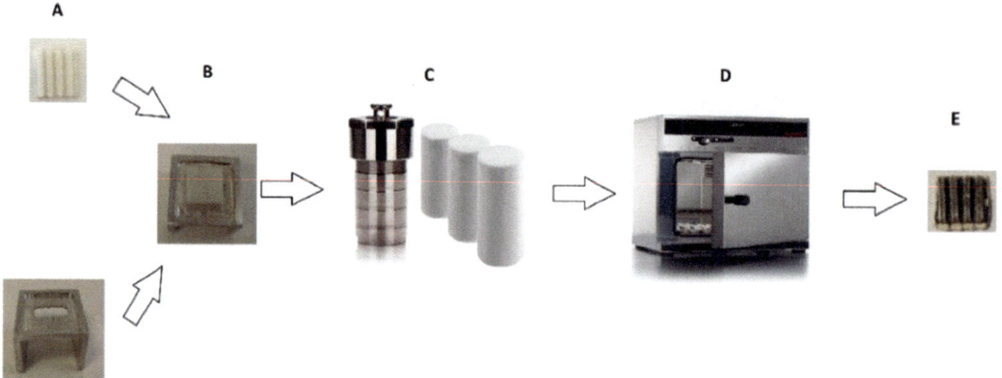

FIGURE 24.6 Illustrative image of zinc oxide coated PLA (a) fabrication of PLA part using 3D printing, (b) placement of PLA part over the metal base, (c) Stainless steel (ss) autoclaves, (d) coating process inside an oven for 2 h, and (e) 3D printed PLA/ZnO part. Adapted with permission [49]. Copyright: The Authors, some rights reserved; exclusive licensee [MDPI]. Distributed under a Creative Commons Attribution License 4.0 (CC BY).

to remove pharmaceutical contaminants in water [49]. Apart from these, ZnO is used with polymeric biomaterials for wood healing [50] and to improve the antibacterial properties of the scaffolds [51].

24.4 CONCLUSIONS

1. In this chapter, various 3D printing methods in combination with suitable biomaterials (polymer and metal) has been reviewed for different applications comprehensively.
2. Bioengineered scaffold fabrication processes, including material, application, and manufacturing methods are described briefly.
3. Major applications of 3D-printed metal oxides such as orthopedic application, improving the antibacterial properties of bone implants, wound healing, drug delivery, and soft bionic and photocatalytic applications have been highlighted.
4. Another growing area in the 3D printing of metal oxide polymer composite is the improvement in the radiopacity of the biomedical devices for clear visibility in X-ray imaging and micro-CT scanning.
5. Several 3D-printing methods to improve the radiopacity of devices are explained. Further, the combination of 3D-printed metal oxide composite and sintering method to generate bone implants are thoroughly covered.
6. A few research works have been performed on the 4D printing of metal oxide composites. Hence, this area needs to be investigated taking the utilities of metal oxide composites.

REFERENCES

1 S. Laurent, S. Boutry, R.N. Muller, *Metal Oxide Particles and Their Prospects for Applications*, Elsevier Ltd., 2018.

2 W.E. Frazier, Metal additive manufacturing: A review, *J. Mater. Eng. Perform.*, 23 (2014) 1917–1928.

3 S. Murthy, P. Effiong, C.C. Fei, 11 – Metal oxide nanoparticles in biomedical applications, in: Y. Al-Douri (Ed.), *Met. Oxide Powder Technol.*, Elsevier, 2020: pp. 233–251.

4 S. Laurent, S. Boutry, R.N. Muller, Chapter 1 – Metal oxide particles and their prospects for applications, in: M. Mahmoudi, S. Laurent (Eds.), *Iron Oxide Nanoparticles Biomed. Appl.*, Elsevier, 2018: pp. 3–42.

5 L.D. Duceac, S. Straticiuc, E. Hanganu, L. Stafie, G. Calin, S.L. Gavrilescu, Preventing bacterial infections using metal oxides nanocoatings on bone implant, *IOP Conf. Ser. Mater. Sci. Eng.*, 209 (2017).

6 Y.X. Hou, H. Abdullah, D.H. Kuo, S.J. Leu, N.S. Gultom, C.H. Su, A comparison study of SiO_2/nano metal oxide composite sphere for antibacterial application, *Compos. Part B Eng.*, 133 (2018) 166–176.

7 B. Shivaramakrishnan, B. Gurumurthy, A. Balasubramanian, Potential biomedical applications of metallic nanobiomaterials: A review, *Int. J. Pharm. Sci. Res.*, 8 (2017) 985–1000.

8 S.M. Dizaj, F. Lotfipour, M. Barzegar-Jalali, M.H. Zarrintan, K. Adibkia, Antimicrobial activity of the metals and metal oxide nanoparticles, *Mater. Sci. Eng. C*, 44 (2014) 278–284.

9 A.-M. Brezoiu, M. Deaconu, I. Nicu, E. Vasile, R.- A. Mitran, C. Matei, D. Berger, Heteroatom modified MCM-41-silica carriers for Lomefloxacin delivery systems, *Microporous Mesoporous Mater.*, 275 (2019).

10 R. Langer, New methods of drug delivery, *Science*, 249 (1990) 1527–1533.

11 Y. Wang, F. Yang, H.X. Zhang, X.Y. Zi, X.H. Pan, F. Chen, W.D. Luo, J.X. Li, H.Y. Zhu, Y.P. Hu, Cuprous oxide nanoparticles inhibit the growth and metastasis of melanoma by targeting mitochondria, *Cell Death Dis.*, 4 (2013).

12 M.P. Vinardell, M. Mitjans, Antitumor activities of metal oxide nanoparticles, *Nanomaterials,* 5 (2015) 1004–1021.

13 H.A. Jeng, J. Swanson, Toxicity of metal oxide nanoparticles in mammalian cells, *J. Environ. Sci. Heal. Part A, Toxic/Hazardous Subst. Environ. Eng.*, 41 (2006) 2699–2711.

14 A. Mohandas, S. Deepthi, R. Biswas, R. Jayakumar, Chitosan based metallic nanocomposite scaffolds as antimicrobial wound dressings, *Bioact. Mater.*, 3 (2018) 267–277.

15 T. Abudula, R.O. Qurban, S.O. Bolarinwa, A.A. Mirza, M. Pasovic, A. Memic, 3D printing of metal/metal oxide incorporated thermoplastic nanocomposites with antimicrobial properties, *Front. Bioeng. Biotechnol.*, 8 (2020) 1–8.

16 T.D. Ngo, A. Kashani, G. Imbalzano, K.T.Q. Nguyen, D. Hui, Additive manufacturing (3D printing): A review of materials, methods, applications and challenges, *Compos. Part B Eng.*, 143 (2018) 172–196.

17 H.T. Wang, P.C. Chiang, J.J. Tzeng, T.L. Wu, Y.H. Pan, W.J. Chang, H.M. Huang, In vitro biocompatibility, radiopacity, and physical property tests of nano-Fe_3O_4 incorporated poly-L-lactide bone screws, *Polymers (Basel).*, 9 (2017) 1–11.

18 Y. Wu, W. Jiang, X. Wen, B. He, X. Zeng, G. Wang, Z. Gu, A novel calcium phosphate ceramic-magnetic nanoparticle composite as a potential bone substitute, *Biomed. Mater.*, 5 (2010) 15001.

19 J. Meng, Y. Zhang, X. Qi, H. Kong, C. Wang, Z. Xu, S. Xie, N. Gu, H. Xu, Paramagnetic nanofibrous composite films enhance the osteogenic responses of pre-osteoblast cells, *Nanoscale*, 2 (2010) 2565–2569.

20 D. Shan, Y. Shi, S. Duan, Y. Wei, Q. Cai, X. Yang, Electrospun magnetic poly (L-lactide)(PLLA) nanofibers by incorporating PLLA-stabilized Fe_3O_4 nanoparticles, *Mater. Sci. Eng. C*, 33 (2013) 3498–3505.

21 U. Bozuyuk, O. Yasa, I.C. Yasa, H. Ceylan, S. Kizilel, M. Sitti, Light-triggered drug release from 3D-printed magnetic chitosan microswimmers, *ACS Nano,* 12 (2018) 9617–9625.

22 E. Tanasa, C. Zaharia, A. Hudita, I.C. Radu, M. Costache, B. Galateanu, Impact of the magnetic field on 3T3-E1 preosteoblasts inside SMART silk fibroin-based scaffolds decorated with magnetic nanoparticles, *Mater. Sci. Eng. C*, 110 (2020) 1–13.

23 Z. Huang, Y. He, X. Chang, J. Liu, L. Yu, Y. Wu, Y. Li, J. Tian, L. Kang, D. Wu, H. Wang, Z. Wu, G. Qiu, A magnetic iron oxide/polydopamine coating can improve osteogenesis of 3D-printed porous titanium scaffolds with a static magnetic field by upregulating the TGFβ-smads pathway, *Adv. Healthc. Mater.*, 9 (2020) 1–13.

24 H.Y. Lin, H.Y. Huang, S.J. Shiue, J.K. Cheng, Osteogenic effects of inductive coupling magnetism from magnetic 3D printed hydrogel scaffold, *J. Magn. Magn. Mater.*, 504 (2020) 166680.

25 E.S. Ko, C. Kim, Y. Choi, K.Y. Lee, 3D printing of self-healing ferrogel prepared from glycol chitosan, oxidized hyaluronate, and iron oxide nanoparticles, *Carbohydr. Polym.*, 245 (2020) 116496.

26 R. Tognato, A.R. Armiento, V. Bonfrate, R. Levato, J. Malda, M. Alini, D. Eglin, G. Giancane, T. Serra, A stimuli-responsive nanocomposite for 3D anisotropic cell-guidance and magnetic soft robotics, *Adv. Funct. Mater.*, 29 (2019) 1–10.

27 Y. Xiang, W. Wang, Y. Gao, J. Zhang, J. Zhang, Z. Bai, S. Zhang, Y. Yang, Production and characterization of an integrated multi-layer 3D printed PLGA/GelMA scaffold aimed for bile duct restoration and detection, *Front. Bioeng. Biotechnol.*, 8 (2020) 1–14.

28 O. Ajiteru, K.Y. Choi, T.H. Lim, D.Y. Kim, H. Hong, Y.J. Lee, J.S. Lee, H. Lee, Y.J. Suh, M.T. Sultan, O.J. Lee, S.H. Kim, C.H. Park, A digital light processing 3D printed magnetic bioreactor system using silk magnetic bioink, *Biofabrication*, 13 (2021).

29 H. Liu, T.J. Webster, Enhanced biological and mechanical properties of well-dispersed nanophase ceramics in polymer composites: From 2D to 3D printed structures, *Mater. Sci. Eng. C*, 31 (2011) 77–89.

30 S.E.. Nájera, M. Michel, N.-S. Kim, 3D printed PLA/PCL/TiO_2 composite for bone replacement and grafting, *MRS Adv.*, 2 (2017) 3865–3872.

31 S. Maher, A. Mazinani, M.R. Barati, D. Losic, Engineered titanium implants for localized drug delivery: Recent advances and perspectives of Titania nanotubes arrays, *Expert Opin. Drug Deliv.*, 15 (2018) 1021–1037.

32 S. Bose, D. Banerjee, A. Shivaram, S. Tarafder, A. Bandyopadhyay, Calcium phosphate coated 3D printed porous titanium with nanoscale surface modification for orthopedic and dental applications, *Mater. Des.*, 151 (2018) 102–112.

33 F. Xiao, C. Zong, W. Wang, X. Zhu Liu, A. Osaka, X. Chun Ma, Low-temperature fabrication of titania layer on 3D-printed 316L stainless steel for enhancing biocompatibility, *Surf. Coatings Technol.*, 367 (2019) 91–99.

34 J. Qin, D. Yang, S. Maher, L. Lima-Marques, Y. Zhou, Y. Chen, G.J. Atkins, D. Losic, Micro- and nano-structured 3D printed titanium implants with a hydroxyapatite coating for improved osseointegration, *J. Mater. Chem. B*, 6 (2018) 3136–3144.
35 F. Xiao, Y. Zhai, Y. Zhou, X. Xu, Y. Liu, X. Ma, X. Gu, W. Wang, Low-temperature fabrication of titania layer on 3D-printed PEKK for enhancing biocompatibility, *Surf. Coatings Technol.*, 416 (2021) 2–11.
36 R. Rahmani, M. Rosenberg, A. Ivask, L. Kollo, Comparison of mechanical and antibacterial properties of TiO_2/Ag Ceramics and Ti6Al4V-TiO_2/Ag composite materials using combined SLM-SPS techniques, *Metals (Basel).*, 9 (2019) 1–13.
37 K.C. Feng, J. Li, L. Wang, Y.C. Chuang, H. Liu, A. Pinkas-Sarafova, C.C. Chang, C.Y. Nam, M. Simon, M. Rafailovich, Combination of 3D printing and ALD for dentin fabrication from dental pulp stem cell culture, *ACS Appl. Bio Mater.*, 4 (2021) 7422–7430.
38 T.N.A.T. Rahim, A.M. Abdullah, H.M. Akil, D. Mohamad, Z.A. Rajion, Preparation and characterization of a newly developed polyamide composite utilising an affordable 3D printer, *J. Reinf. Plast. Compos.*, 34 (2015) 1628–1638.
39 S.Y. Fu, B. Yu, H.F. Ding, G.D. Shi, Y.F. Zhu, Zirconia Incorporation in 3D Printed β-Ca_2SiO_4 Scaffolds on their physicochemical and biological property, *J. Inorg. Mater.*, 34 (2019) 444–454.
40 Q. Wang, Z. Ma, Y. Wang, L. Zhong, W. Xie, Fabrication and characterization of 3D printed biocomposite scaffolds based on PCL and zirconia nanoparticles, *Bio-Design Manuf.*, 4 (2021) 60–71.
41 C.H. Chang, C.Y. Lin, C.H. Chang, F.H. Liu, Y.T. Huang, Y.S. Liao, Enhanced biomedical applicability of ZrO_2–SiO_2 ceramic composites in 3D printed bone scaffolds, *Sci. Rep.*, 12 (2022) 1–11.
42 Y. Zhu, K. Liu, J. Deng, J. Ye, F. Ai, H. Ouyang, T. Wu, J. Jia, X. Cheng, X. Wang, 3D printed zirconia ceramic hip joint with precise structure and broad-spectrum antibacterial properties, *Int. J. Nanomedicine*, 14 (2019) 5977–5987.
43 A. Shannon, A. O'Connell, A. O'Sullivan, M. Byrne, S. Clifford, K.J. O'Sullivan, L. O'Sullivan, A radiopaque nanoparticle-based ink using polyjet 3D printing for medical applications, *3D Print. Addit. Manuf.*, 7 (2020) 259–268.
44 G.A. Fielding, A. Bandyopadhyay, S. Bose, Effects of silica and zinc oxide doping on mechanical and biological properties of 3D printed tricalcium phosphate tissue engineering scaffolds, *Dent. Mater.,* 28 (2012) 113–122.
45 G. Fielding, S. Bose, SiO_2 and ZnO Dopants in 3D Printed TCP Scaffolds Enhances Osteogenesis and Angiogenesis in vivo, *Bone*, 23 (2014) 1–7.
46 B. Aldemir Dikici, S. Dikici, O. Karaman, H. Oflaz, The effect of zinc oxide doping on mechanical and biological properties of 3D printed calcium sulfate based scaffolds, *Biocybern. Biomed. Eng.*, 37 (2017) 733–741.
47 Z. Viskadourakis, M. Sevastaki, G. Kenanakis, 3D structured nanocomposites by FDM process: A novel approach for large-scale photocatalytic applications, *Appl. Phys. A Mater. Sci. Process.*, 124 (2018) 1–8.
48 M. Zhang, X. Xia, C. Cao, H. Xue, Y. Yang, W. Li, Q. Chen, L. Xiao, Q. Qian, A ZnO@ABS/TPU/$CaSiO_3$ 3D skeleton and its adsorption/photocatalysis properties for dye contaminant removal, *RSC Adv.*, 10 (2020) 41272–41282.
49 M. Sevastaki, V.M. Papadakis, C. Romanitan, M.P. Suchea, G. Kenanakis, Photocatalytic properties of eco-friendly ZnO nanostructures on 3d-printed polylactic acid scaffolds, *Nanomaterials*, 11 (2021) 1–13.
50 L. Siebert, E. Luna-Cerón, L.E. García-Rivera, J. Oh, J. Jang, D.A. Rosas-Gómez, M.D. Pérez-Gómez, G. Maschkowitz, H. Fickenscher, D. Oceguera-Cuevas, C.G. Holguín-León, B. Byambaa, M.A. Hussain, E. Enciso-Martínez, M. Cho, Y. Lee, N. Sobahi, A. Hasan, D.P. Orgill, Y.K. Mishra, et al., Smart wound scaffolds: Light-controlled growth factors release on tetrapodal ZnO-incorporated 3D-printed hydrogels for developing smart wound scaffold, *Adv. Funct. Mater.*, 31 (2021) 2170154.
51 R.C. Nonato, L.H.I. Mei, B.C. Bonse, C. V. Leal, C.E. Levy, F.A. Oliveira, C. Delarmelina, M.C.T. Duarte, A.R. Morales, Nanocomposites of PLA/ZnO nanofibers for medical applications: Antimicrobial effect, thermal, and mechanical behavior under cyclic stress, *Polym. Eng. Sci.*, 62 (2022) 1147–1155.

25 3D-Printed MXenes for Biomedical Applications

Sirawit Pruksawan[1], Heng Li Chee[1], Suppanat Puangpathumanond[2], and FuKe Wang[1]*
[1]Institute of Materials Research and Engineering (IMRE), Agency for Science, Technology and Research (A*STAR), 2 Fusionopolis Way, 138634 Singapore
[2]College of Design and Engineering, National University of Singapore (NUS), 4 Engineering Drive 3, 117583 Singapore
*Corresponding author: wangf@imre.a-star.edu.sg

CONTENTS

25.1 SYNTHESIS AND STRUCTURE OF MXENE MATERIALS

MXenes with 2D structures are typically synthesized through selective removal of 'A' layers from their parent MAX phase using etching solutions such as hydrofluoric acid (HF). The precursor MAX phases are ternary carbides and nitrides with laminated hexagonal structures, with the general formula $M_{n+1}AX_n$ (n = 1 to 3). "M" is typically an early transition metal (e.g., Sc, Ti, V), 'A' denotes an A-group element (e.g., Al, Si, P) and 'X' is either C and/or N [1], [2] (Figure 25.1a). By using strong etching solutions such as HF and ammonium bifluoride (NH_4HF_2), elements in the 'A' layers are replaced by surface termination groups such as fluorine (-F) and hydroxyl (-OH), forming 2D multi-layered MXenes with the general formula of $M_{n+1}X_nT_x$ (n = 1 to 4) where the 'T' is the surface termination group are produced [3] (Figure 25.1b). The multi-layered MXenes are held together by van der Waals attraction or hydrogen bonds which can be broken through delamination or exfoliation, to physically separate the MXene sheets into a monolayer of MXene. (Figure 25.1b) [4]. Synthesis MXene from non-MAX phases is also used sometimes [5] and one type is Mo_2CT_x, which was prepared from a non-MAX phase of Mo_xGa_2C.

MXenes have a large assortment of alloy compositions as the MAX phases can be comprised of a wide variety of different 'M', 'A' and 'X' atom combinations. Despite this, few of them, however, are stable. The first MXene, Ti_3C_2, was synthesized in 2011 by etching the Ti_3AlC_2 MAX phase in concentrated HF. Etching conditions (e.g., time and acid concentration) depend largely on the characteristics of MAX phases such as structure and chemistry, with MXenes carrying larger n

DOI: 10.1201/9781003296676-25

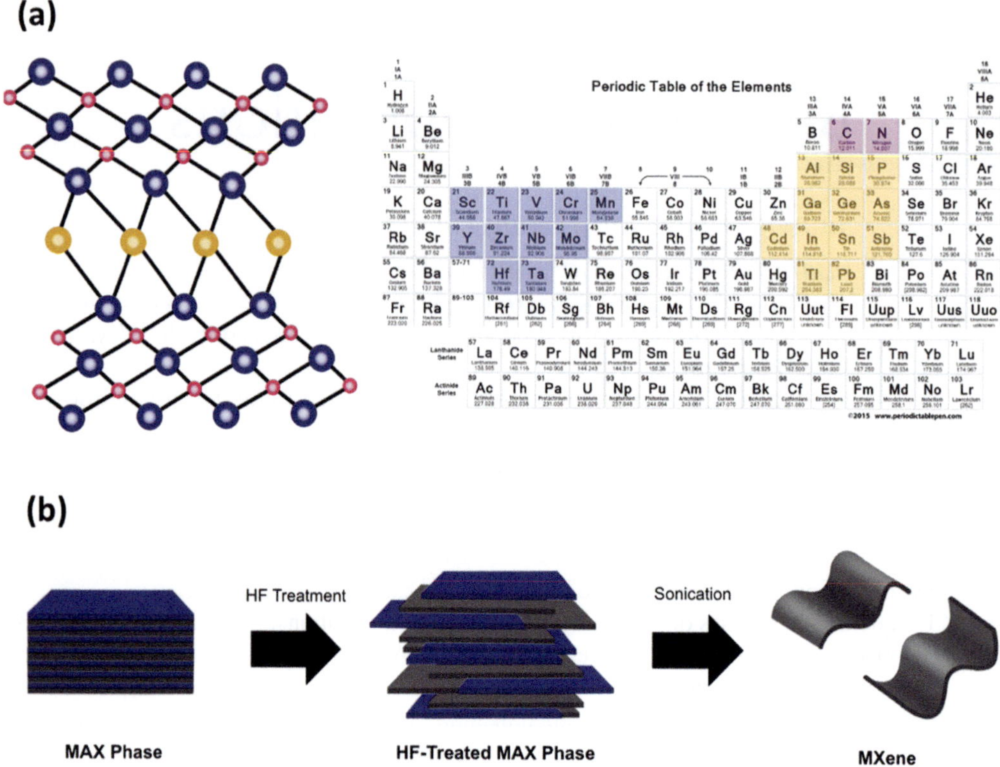

FIGURE 25.1 Synthesis and structure of MXene materials. (a) Periodic table of the elements showing compositions of MXene materials. (b) Illustration of MXene synthesis from their MAX phases.

values requiring more severe etching conditions [6]. As the use of HF is detrimental to the environment, several 'cleaner' alternative synthesis methods were reported including alkali etching [7], electrochemical etching [8], and water-free etching [9]. Etching by strong alkali treatment is suitable in the case of MAX phases with basic/amphoteric 'A' layers (e.g., S in Zr_2SC), in which 'A' atoms in the precursor MAX phases are oxidized and dissolved in high-concentration NaOH at high temperature, resulting in 2D multi-layered MXenes. Electrochemical etching involves the selective removal of 'A'''layers at a certain electrical voltage using a MAX phase electrode in electrolytes such as hydrochloric acid (HCl). Water-free etching of MXenes was achieved using a mixture of ammonium dihydrogen fluoride and a polar solvent. This method allows the use of MXene materials in water-sensitive applications, for example, structural polymer composite materials, energy conversion, and some biomedical devices. The properties of MXenes vary greatly depending on the synthesis way and processing conditions, and thus, a deeper understanding of the synthesis of MXene materials is desired to advance the field.

Another crucial step in MXene synthesis is the exfoliation of multi-layered MXenes. The exfoliation process of MXene is different and more complex as compared to other 2D structure materials like graphene and phosphorene, mainly because the interleaved 'A' atoms in MXense bind each layer via M–A metallic bonds which are stronger than van der Waals interaction in the other 2D-structure materials. To date, two well-established approaches are commonly used for the exfoliation of MXenes: (1) mechanical exfoliation and (2) chemical exfoliation. The mechanical exfoliation of multi-layered MXenes is achieved through the adhesive tape method, where MXenes 2D flakes are physically removed using adhesive tape [10]. This approach, however, can only be used when the

TABLE 25.1
Examples of Experimentally Synthesized MXenes, Classified Based on Structure Type

MXene Structure Type		
M_2X	M_3X_2	M_4X_3
Ti_2C	Ti_3C_2	V_4C_3
V_2C	Zr_3C_2	Nb_4C_3
Nb_2C	Hf_3C_2	Ta_4C_3
Mo_2C	Mn_3N_2	Ti_4N_3
Ti_2N	Ti_3CN	$(Nb_{0.8}Ti_{0.2})_4C_3$
V_2N	Mo_2TiC_2	$(Nb_{0.8}Zr_{0.2})_4C_3$
Mo_2N	Cr_2TiC_2	$Mo_2Ti_2C_3$
$(V_{1-x}Ti_x)_2C$	Mo_2ScC_2	
$(Ti_{0.5}Nb_{0.5})_2C$	$(V_{0.5}Cr_{0.5})_3C_2$	

M-A bonds in the precursor MAX phases are relatively weak. For the chemical exfoliation process, interactions between the individual MXene layers are suppressed by the chemical intercalation of organic molecules into MXenes such as tetrabutylammonium hydroxide (TBAOH) and dimethyl sulfoxide (DMSO). Chemical exfoliation is the most widely used to obtain MXenes from MAX phases due to its high exfoliation efficiency [11]. Electrochemical exfoliation is also used occasionally by applying electrochemical potentials to intercalate ions into interlayers of the bulk material, resulting in the weakening of interactions between the layers [12].

A high-throughput screening method for the identification of possible MXene candidates was introduced to identify MXene alloy compositions that have the lowest formation energy [13]. In the high-throughput screening method, a multiscale computational model is used to compute the relative stability of different possible MXene alloy configurations. The thermodynamic stability of different MXene alloy configurations is computed based on their structural arrangements and compositions, which can be used to guide the synthesis of MXene alloys. To date, at least 80 ternary MAX phases have been discovered experimentally, 20 of which have been successfully etched into 2D MXene sheets [14]. It should be noted that most of them were synthesized by etching the Al layer of the MAX phases. Typical examples of synthesized MXenes are shown in Table 25.1. In general, MXenes can be classified into three major structures: (1) mono-M elements (M_2X), (2) solid solutions (M_3X_2), and (3) ordered double-M elements [15]. Mono-M element MXenes contain only one kind of transition metal such as Sc_2C and Ti_2C, while MXene solid solutions are comprised of two kinds of transition metals in the M layers, like $(Ti,V)_3C_2$ and $(Cr,V)_3C_2$. In the case of ordered double-M element MXenes, single or bilayers transition metal are sandwiched between the other transition metal type such as Mo_2TiC_2 and $Mo_2Ti_2C_3$. The MXene structure can be observed experimentally using atomic-resolution scanning transmission electron microscopy (STEM) and electron energy-loss spectroscopy (EELS).

25.2 UNIQUE CHARACTERISTICS OF MXENES FOR BIOMEDICAL APPLICATIONS

Biomedical applications such as tissue engineering and drug delivery require particular material properties such as hydrophilicity, biocompatibility, electrical properties, and mechanical properties. The unique 2D layered structure of MXenes and their excellent mechanical performance, easy surface chemical modification, and multifunction including electrical, optical, and magnetic properties make Mxenes the ideal candidates for biomedical applications [16].

A typical MXene structure is shown schematically in Figure 25.1b, where the layered structures make MXenes an important family in 2D materials. Some of the layers of MXenes are stacked together through Van der Waals forces and hydrogen bonds, which can be separated through the chemical stabilization of individual layers in solution or printing ink. The unique layered structure of MXenes provides a large specific surface area and numerous anchorage sites for drug molecules, which improves drug loading, sustained release, and stability of MXene. Unlike another important 2D material, graphene, which has only the π-π interaction, MXenes have many different functional groups such as –H, –OH, –F on the surface, which can graft with various functional groups [17]. Performance characteristics of MXenes such as mechanical stability, electrical properties, and functions can be further enhanced by surface modification of MXenes, providing a good strategy for MXenes in biomedicine applications. Typically, the surface of MXenes can be modified with functional small molecules, polymers, and inorganic particles. Various small molecules MXene surface modifications have been reported to improve the performance and process of MXenes. Successful examples include the surface modification with hexaethylene glycol monododecyl ether ($C_{12}E_6$) and diazonium salts. The composites of $C_{12}E_6$@Ti_3C_2 showed increased molecular interactions of MXene nanosheets, which can function as a non-ionic surfactant. Diazonium ions were used for the in situ modification of the Ti_3C_2 surface to improve its electrical properties [18]. Other molecules such as bismuth ferrite [19], sodium aluminum hydride [20], and (3-Aminopropyl)tiethoxysilane [21] have been applied to add functional groups on the MXene's surface as a platform to immobilize different biomolecules.

The surface modification of MXene with polyethylene glycol (PEG) had showed excellent stability and biocompatibility, and great application potential in fields such as biomedicine to target tumor accumulation [22], [23]. The self-initiated photographing-photopolymerization (SIPGP) of (dimethylamino)ethyl methacrylate (DMAEMA) to the surface of MXenes gave a CO_2 and temperature dual responsible smart MXene system for biological application [24]. The surface of MXene can also be modified with inorganic nanomaterials to obtain multifunctional nanocomposites to open new opportunities for biomedical applications. M_nO_x nanoparticles have been grown on the surface of the Ti_3C_2 MXene through a 'redox-induced growth' method [25]. The resulting nanocomposites showed unique T1-weighted MRI capabilities and effective ablation of the tumor [25]. The nanocomposite of polyoxometalates (POMs)/Ti_3C_2 MXenes showed high biocompatibility and high photothermal ablation performance for cancer cells [26]. While the combination of mesoporous silica nanoparticles (mMSNs) with MXenes allows the development of an efficient drug delivery platform. The mMSN/Ti_3C_2 composites prepared by using the nanopore engineering strategy exhibit enhanced dispersibility and biocompatibility, together with multiple surface chemistry for targeted modification (RGD coupling) [27]. While mMSN/Nb_2C composites prepared by using the sol-gel technique showed a reduction in the overexpressed integrin $\alpha_v\beta_3$ on the cancer cell membrane through RGD coupling and selective identification. Furthermore, the high photothermal conversion efficiency of Nb_2C gave the resulting nanocomposites a stronger therapeutic effect on brain cancer cells at both the intracellular and subcutaneous levels [28]. Direct growth of metal nanoparticles on the surface provides another approach to surface modification of MXenes. Gold nanoparticles' growth on the surface of Ti_3C_2 MXene showed powerful PTT under the NIR-I and NIR-II biological windows [29] while the silver particles modified MXenes showed fast and effective light-triggered therapy function [30].

The hydrophilic behavior of MXenes offers the possibility for surface functionalization and antibacterial characteristics, which is highly desired in the development and fabrication of medical devices. Additionally, hydrophilic MXenes are a promising candidate for targeted delivery systems, owing to a combination of high hydrophilicity and the large surface area which grants the MXenes drug-loading capacities. The biocompatibility of MXene is greatly dependent on its size and surface properties. Some Ti-based MXenes like Ti_3AlC_2 were reported to be biocompatible [16], however apart from the latter, there are very limited studies reporting the biocompatibility of MXenes to date.

Electrical properties of MXenes are probably the most studied characteristic because all bare MXenes monolayers are metallic, while the surface groups such as OH and F tune the metal-semiconductors transition. The electrical conductivity of MXene is also affected by its elemental compositions, structures, and surface termination groups. Due to the existence of transition metal carbide/nitride and surface termination groups, MXenes generally exhibit high metallic conductivity. The electron mobilities achieved by MXenes can reach as high as ~10^6 cm^2 V^{-1} s^{-1}, which is greater compared to that of other 2D structure materials like graphene (~10^5 cm^2 V^{-1} s^{-1}) [14]. MXenes also exhibit good mechanical properties with moduli of up to ~300 GPa, which is sufficient to meet the reliability requirements of conventional biomedical products [31]. The mechanical properties of MXenes are mainly based on the mass of the transition metal and the surface termination groups [32]. In addition, MXene nanosheets inherently possess excellent photothermal converting properties that can be further developed as photothermal nanoagents for cancer hyperthermia when exposed to external near-infrared (NIR) laser irradiation. In recent years, MXenes have been applied as agents for photo-thermal therapy (PTT), as a bioimaging technology, for drug delivery, and as an antibacterial substance and electro-conductive base in biomaterial applications.

25.3 3D-PRINTING TECHNIQUES OF MXENES IN BIOMEDICAL APPLICATIONS

3D-printing technologies (also known as Additive Manufacturing (AM)) allow people to design and fabricate parts with complex shapes and features unachievable using traditional manufacturing methods such as molding and machining. 3D printing is particularly promising in biomedical applications where parts or products are highly complex and require a high degree of customization and personalization. Currently, there is a broad range of materials used for 3D printing such as metals, ceramics, and polymers, with 2D materials being among its latest and most promising additions [33], [34]. The 3D printing of 2D materials offers high design flexibility at low costs while, at the same time, retaining the 2D materials' intrinsic properties. Compared to 2D inks, 3D printing inks usually require mechanically self-supporting when depositing onto a substrate. To date, the 3D printing of several 2D materials, including graphene, MXenes, transition-metal dichalcogenides (TMDs), and black phosphorus (BP) has been explored. Among these 2D materials, MXenes are one of the most extensively used. The most common printing techniques in the 3D printing of MXenes for biomedical applications include (1) 3D extrusion-based printing, (2) 3D inkjet printing, and (3) 3D screen printing (Figure 25.2). The inks for 3D printing are composed of the active component (or pigment), a solvent, and an ink binder. Solvent helps to improve the homogeneity of the mixture while the binder is used to ensure cohesion between particles and to tune the viscosity of the ink [14]. A typical component of MXene inks is as follows: 12–20 wt% active component, 20–30 wt% solvent, 45–65 wt% binders, and 1–5 wt% additives [35]. Examples of solvents for MXenes include water, isopropyl alcohol (IPA), and ethyl alcohol, while the binder is generally a polymer. In some cases where ink viscosity is not critical, binders and additives can be removed from the MXene inks [36].

3D extrusion-based printing operates based on the use of pneumatic air pressure or mechanical screw systems to extrude the materials/filaments via a nozzle to build an object layer-by-layer. There are two main types of extrusion-based printing technologies: (1) direct ink writing (DIW) and (2) fused deposition modeling (FDM). Extrusion-based printing of 2D materials like MXenes uses mainly the DIW technique while FDM is limited to thermoplastic polymers such as acrylonitrile butadiene styrene (ABS). During the DIW printing process, colloidal inks are extruded in a layer-based fashion onto a substrate through a nozzle to construct objects. The rheological behavior of the MXene inks play a very crucial role in the printing process. In most cases, additives such as secondary solvents and/or surfactants are added to tune the rheological behaviors of the ink to facilitate DIW. The extruded colloidal ink solidifies instantaneously, making DIW an extremely fast and

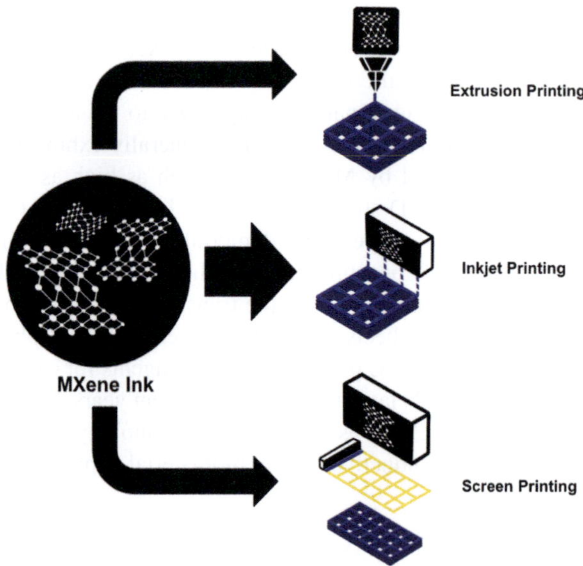

FIGURE 25.2 Illustration of 3D MXene ink printing: Extrusion-based printing, inkjet printing, and screen printing.

convenient prototyping tool [34]. Extrusion-based printing of MXenes is thus commonly used to manufacture printed flexible devices, health monitoring sensors, and implantable medical devices [37], [38].

3D inkjet printing is another major technique used for printing MXenes. Unlike traditional 2D inkjet printing, 3D inkjet printing constructs a 3D structure consisting of multiple printed layers. The advantages of inkjet printing include high precision, low cost, and the ability to print multi-materials. 3D inkjet printing is a non-contact printing approach that is ideal for biomedical immobilization because cross-contamination with the surface being printed on can be avoided. In the inkjet printing process, ink droplets are deposited in rapid succession onto the substrate to manufacture designed 3D objects. Two main droplet jetting mechanisms are used: continuous inkjet (CIJ) and drop-on-demand inkjet (DOD). The CIJ printer creates a stream of ink droplets out of an ink reservoir while the DOD printer produces droplets only when and where they are needed [39]. One limitation of inkjet printing is that the inkjet-printed structure usually lacks structural integrity and stiffness. The print quality of any inkjet printing process mainly depends on both the viscosity and interfacial surface tension of the ink. A fluid parameter known as the Ohnesorge number (Oh), is used to monitor drop formation in the inkjet printing process. The Oh indicates the tendency for a droplet formation to either stay together or disperse during its descent, by comparing viscous forces to surface tension forces. The Jetting of droplets without satellite droplets is possible if Oh is in the range of 0.1 to 1. There are several reported works on $Ti_3C_2T_x$ MXene-based inkjet inks [40]. Although inkjet printing of MXenes is mainly used for energy storage and harvesting applications, it could also be applied in biomedicine such as smart bioelectronics devices, where easy-to-integrate components are preferred.

3D screen printing is a high-throughput, mature printing method where a mesh (e.g., metal thread, synthetic fiber, or fabric) is used to transfer ink onto a substrate, except in areas made impermeable to the ink by a screen stencil. The 3D screen printing process can be considered as repeated traditional 2D screen printing to manufacture 3D objects through a layer-by-layer building in the z-axis direction. During the 3D screen printing process, the screen stencil is placed on top of the substrate, and the ink is poured on one end of the screen stencil. After that, the ink is scraped across the screen stencil with the aid of a squeegee. Subsequently, the ink dries and forms a single layer of the previously printed structure. The process is then repeated until the construction of the entire object consisting of many screen-printed layers is complete [41]. After printing, the screen stencil is removed, and the solvent in the ink is volatilized [42]. Viscosity plays an important role in the screen printing process – the inks should be viscous (between 1000 and 10,000 mPa s) and lose their viscosity if stress is applied. Using screen-printing MXene inks, various flexible devices such as wearable sensors can be fabricated for biomedical applications [4]. 3D inkjet printing is also possible for large-scale manufacturing [43].

Table 25.2 summarizes the major characteristics of each printing technology for MXene printing. Although significant progress in 3D printing of MXenes has been made in recent years, several key challenges need to solve: (1) The intrinsic properties of 2D Mxenes materials may be lost during the 3D-printing process, (2) it is difficult to balance the trade-off between rheological properties of MXene inks and other properties such as surface tension or dispersity.

25.4 BIOMEDICAL APPLICATIONS OF 3D-PRINTED MXENES

To date, MXenes have been utilized in a wide variety of fields, from optical to energy storage, electronics, and catalysis, with environmental, structural, and biomedical applications on the rise [44]. Specifically, MXenes play a crucial role in diagnostics, therapeutics, theranostics, biosensors, antimicrobials, tissue engineering, and regenerative medicine owing to their unique 2D layered structure and outstanding chemical and physical properties, such as biocompatibility, large surface area, mechanical flexibility, antibacterial activity, electrical conductivity and light-to-heat conversion performance as discussed above [45]–[47].

The unique and tunable properties of 2D MXenes combined with 3D printing have helped to advance developments in all the aforementioned fields. AM techniques such as DIW, inkjet printing, and stereolithography (SLA) have been successfully used to fabricate 3D MXene structures, thus granting the MXenes a whole new level of versatility. For instance, the 3D printing of MXenes enables the manufacturing of highly customized biomedical devices with complex geometries that is unachievable by traditional manufacturing methods, together with fringe benefits such as design freedom and flexibility, manufacturing on demand, and optimized material usage [33], [47]. Despite these advantages, 3D-printing MXene-based materials are a relatively unexplored field, much unlike its 2D counterpart [46]. In the following paragraphs, we aim to shed light on some of the most recent and impactful advancements in 3DP MXenes for biomedical applications. A summary of these advancements is recorded in Figure 25.3 and Table 25.3.

25.4.1 Sensors

Apart from the well-established advantages of MXenes over other two-dimensional materials, they are also known to have exceptional electrochemical properties [61], which allow using them as active sensing elements or a supporting substrate in MXene-based sensors in biomedical applications. Overall, the MXene-based sensors can roughly be classified into three major groups, namely, chemical, physical, and biosensors [33].

25.4.1.1 Biosensors

Biosensors are used to detect target molecules within a living system and mainly consist of (i) a sensing element (an immobilized biomolecule, for instance) that identifies its complementary analyte; (ii) a transducer that transforms the biochemical signal to other signal types such as optical and electrical signals, and (iii) a data interpretation unit [45]. Other than the benefits listed above, MXenes also possess a large number of surface functional groups, allowing for spontaneous interactions with other ions and molecules [47], making them suitable candidate materials for the fabrication of

TABLE 25.2
Typical Compositions, Viscosity Properties, Resolution, Speed, Setup, and Example Biomedical Applications of MXene Inks for Each 3D Printing Technology

	Ink Composition (wt%)								
3D Printing Techniques	**Pigment**	**Solvent**	**Binder**	**Additive**	**Example MXene Inks**	**Viscosity (mPa s)**	**Resolution (μm)**	**Speed**	**Setup Cost**
Extrusion-based printing	10–15	25–35	1,000–10,000	1–5	$Ti_3C_2T_x$	1,000–10,000	50–200	Average	Average
Inkjet printing	5–10	65–95	4–30	1–5	$Ti_3C_2T_x$	4–30	20–100	Low	High
Screen printing	12–20	20–30	1,000–10,000	1–5	Ti_3C_2Tx	1,000–10,000	30–100	High	Average

Source: Adapted and modified with permission [33]. Copyright Gómez et al., some rights reserved; exclusive licensee MDPI. Distributed under a Creative Commons Attribution License 4.0 (CC BY).

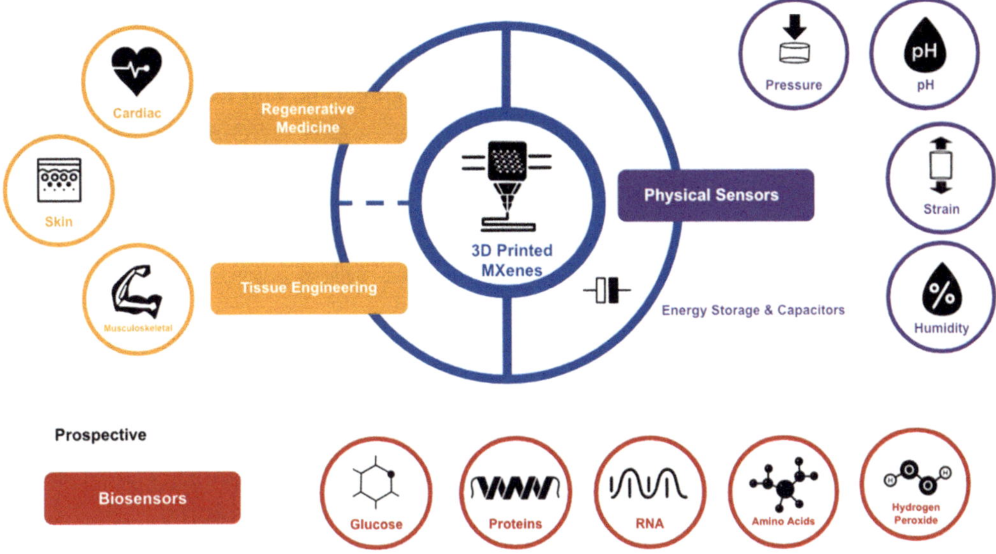

FIGURE 25.3 Summary of currently explored application fields for 3DP Mxenes.

TABLE 25.3
Mxene Ink Composition, 3D Printing Technique, and Their Respective Potential Applications in the Field of Physical Sensors, Tissue Engineering, and Regenerative Medicine

Ink Composition	Fabrication Method	Demonstrated/Potential Application	Reference
		Physical Sensors	
$Ti_3C_2T_x$ nanosheets in Di water	3D extrusion printing	Smart electronics, sensors, electromagnetic shielding, antennas and other applications.	[48], [49]
$Ti_3C_2T_x$ nanosheets in Di water	3D extrusion printing	Electronic devices, electromagnetic shielding, sensors, antennas, and biomedical applications.	[50]
TEMPO (2,2,6,6-tetramethylpiperidine-1-oxylradical)-mediated oxidized cellulose nanofibrils (TOCNFs)/ Ti_3C_2 MXene hybrid inks	3D extrusion printing	Wearable heating textiles, human health monitoring, and human–machine interfaces	[51]
$Ti_3C_2T_x$ nanosheets in Di water	3D freeze-Printing	Flexible electronics, sensors, energy storage and pressure sensors	[12]
N-$Ti_3C_2T_x$ ink (made of CNTs and MXene) in water	3D extrusion printing	Energy storage devices	[52]
NCPM/CNT ink(NiCoP/MXene+ CNT)	3D freeze-Printing	High-energy-density energy storage systems.	[53]
Crumpled MXene with AC, CNT, and GO	3D extrusion printing	Energy storage systems	[54]

(continued)

TABLE 25.3 (Continued)
Mxene Ink Composition, 3D Printing Technique, and Their Respective Potential Applications in the Field of Physical Sensors, Tissue Engineering, and Regenerative Medicine

Ink Composition	Fabrication Method	Demonstrated/Potential Application	Reference
		Tissue Engineering and Regenerative Medicine	
MXene in Di water	Aerosol jet 3D printing	Cardiac tissue engineering applications	[55]
MXene nanosheets, hyaluronic acid, alginate	3D extrusion printing	Neural tissue engineering and other biomedical applications	[56]
GelMA, TCP, sodium alginate and MXene	3D extrusion printing	Bone regeneration and other biomedical applications	[57]
GelMA/MXene with and without C2C12 cells	3D extrusion printing	tissue engineering applications, specifically as biocompatible and biomimetic bioinks with enhanced printability and conductivity.	[58]
MXene,Hydroxyapatite, Sodium Alginate	3D extrusion printing	Bone regeneration and other biomedical applications	[59]
MXene, collagen,silk fibroin, hydroxyapatite	3D extrusion printing	Bone regeneration and multifunctional treatment of maxillofacial tumours	[60]

Note: CNT, carbon nanotubes; NCPM, NiCoP/MXene; NiCoP, nickel cobalt phosphate; AC, activated carbon; GO, graphene oxide; GelMA, gelatin methacrylate; TCP, beta-tricalcium phosphate.

direct electrochemical biosensors [46]. Thus far, the research done on MXenes in this field focuses primarily on the 2D MXenes' role in formulations used for 2D inkjet and screen printing or the fabrication of (synthesis) composite biomaterials that have been successfully used to detect and monitor electrical and optical signals. Examples of electrical biosensing analytes include proteins, amino acids, RNA, and more commonly, Hydrogen Peroxide and Glucose, all of which are used for the early diagnosis of diseases [45]. Optical signal applications can be further classified into two groups based on the principle of illumination, which is electrochemiluminescence (ECL) and surface plasmon resonance (SPR). Both groups are used to detect biological analytes such as enzymes, DNA, and proteins for clinical diagnosis which are pertinent research tools for pharmaceuticals and life science [45]. Despite the potential of MXenes in developing extremely robust biosensing technologies, there are currently no reports on research being carried out to incorporate 2D MXenes into AM in the field. This could be due to the challenges faced in creating a reliable MXene-biological analyte resin/bioink formulation in addition to restrictions presented by 3D-printing techniques. However, with research advancements in MXene development and AM technologies progressing at a fast pace, opportunities to develop 3D-printed MXene-based biosensors are expected to be on the rise in the near future.

25.4.1.2 Physical Sensors

Apart from biomolecules, pressure and tensional forces could also induce changes in electrical signals output by MXenes as a result of their excellent mechanical and electrical properties. Taking advantage of this fact, MXene-based wearable sensors that detect changes in pressure, strain, humidity, or pH can be designed to monitor physiological signals from human activity, making them suitable for personal health monitoring and clinical diagnosis [45], [62]. This is especially the case when they are integrated into the prospective body skin textile substrates, which are an attempt at seamless integration of sensors into clothing. Though numerous developments of wearable MXene sensors have been reported [63], work on 3D-printed MXenes-based wearable sensors is promising but limited. Notably, Cao et al. reported the successful fabrication of 3D-printed multifunctional flexible smart textiles and fibers using extrusion-based AM at room temperature. MXene Ti_3C_2 nanosheets were fabricated, mixed with TEMPO (2,2,6,6-tetramethylpiperidine-1-oxylradi-cal)-mediated oxidized cellulose nanofibrils (TOCNFs) to form hybrid inks, 3D printed and immersed in an ethanol coagulation bath. The developed MXene-based inks exhibited astounding rheological characteristics, allowing the rapid fabrication of dimensionally precise structures. It is also worth noting that the 3D-printed TOCNFs@Ti_3C_2 smart fibers demonstrated excellent responsiveness to external electrical, mechanical, and photonic stimuli as compared to traditional synthetic fibers, thus allowing them to be processed into highly sensitive strain sensors. These smart fibers and textiles are promising in a multitude of biomedical applications such as next-generation healthcare electronics that include wearable heating and sensing systems and human health monitoring [4], [51], [64].

Next-generation physical sensors incorporated into wearable electronics for biomedical applications are expected to be wireless and powered by reliable energy storage devices that can produce large and prolonged energy output with good energy densities [65]. The success of this would be indirectly dependent on capacitors and their energy density and storage capability. With potential applications of 3D-printed MXenes targeted towards the development of such high-powered sensors, constant work is being done to develop energy storage devices that can deliver high energy density and power within a limited carbon footprint area [12], [48]–[50], [52], [54]. For example, Orangi et al. reported on the development of 3D-printed MXene-based micro-super capacitors with ultra-high energy densities. Water-based $Ti_3C_2T_x$ MXene ink with compatible rheological properties for extrusion-based 3D printing was fabricated without the addition of additives or high-temperature drying. The MXene ink was then printed using a pneumatic fluid dispenser with a print head speed of approximately 4 psi and 3 mm/s, respectively. Their results showed that outstanding areal capacities could be achieved by increasing the number of deposition layers and electrode height of the prints. The maximum energy density reported was significantly higher than previously reported MXene-based energy storage devices. Additionally, the orientations of MXene flakes can be controlled by modifying ink characteristics and printing parameters in extrusion-based 3D printing (Figure 25.4a). With this, 3D-printed nanosheets can be stacked and aligned horizontally (Figure 25.4c–d). Flexion tests involving MSCs resulted in no signs of cracks or detachment from the substrate, making them suitable for use in flexible electronics (Figure 25.4b). More DIW-based printing has been developed to improve the electrical properties of MXene-based physical sensors and biomedical devices [50], [66]. Zhang et al. formulated and incorporated $Ti_3C_2T_x$ MXene inks (Figure 25.4e) into Micro-Supercapacitors (MSC), conductive tracks, and ohmic resistors (Figure 25.4f) through DIW. Scanning electron microscopy images of the 3D-printed MXene-based structures displayed interconnected, stacked MXene nanosheets creating a continuous film, with a line gap ~120 um between two paths (Figure 25.4g–h). Additional conductive agents and current collectors were unnecessary due to the metallic conductivity achieved through compacted printed lines (Figure 25.4i). Furthermore, the printouts displayed strong adhesion between stacked MXene sheets through hydrogen bonding, thus eliminating the need for polymeric binders (Figure 25.4i). The authors reported that increases in energy density and volumetric capacitance could be achieved by synergizing the pseudocapacitive behavior and high electrical conductivity of MXenes with the high printing resolution and spatial uniformity that DIW offers. As a result, MSCs developed using this technique were reported to have astounding electrochemical performance, including volumetric capacitance of up to 562 F/cm^3 and energy densities as high as 0.32 μWh/cm^2. These developments look extremely promising and can likely be applied to biomedical smart electronics and sensors [47], [49], [66]. Using a different approach, a straightforward 3D Freeze Printing (3DFP) approach developed by combining inkjet-based 3D printing and unidirectional freeze casting was used to fabricate macroscopic porous aerogels with vertically aligned $Ti_3C_2T_x$ nanosheets. It was found that ink concentration could be tuned to control the gels' mechanical properties. The 3DFP aerogels

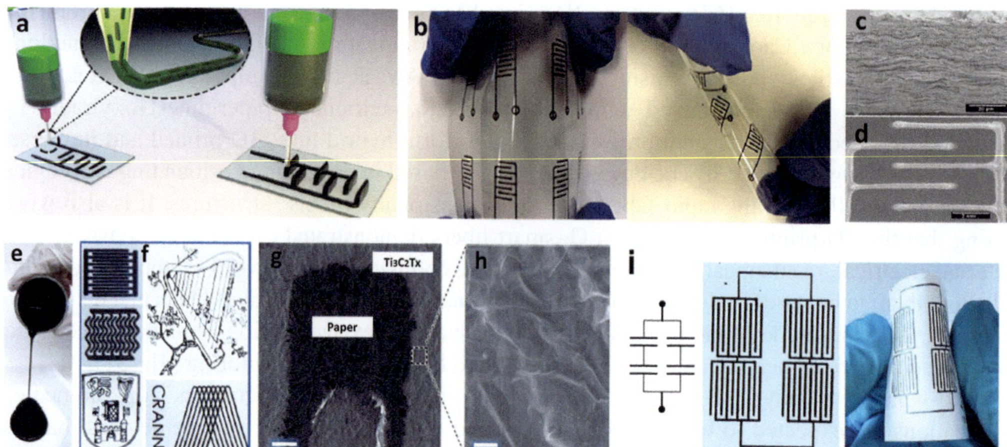

FIGURE 25.4 3D Printed MXene structures for physical sensors. (a) MXene ink printable mechanism for extrusion-based 3D printing. (b) Flexibility testing of printed MSCs demonstrates excellent adhesion of printouts to a substrate during repetitious twisting and bending. (c) A cross-sectional SEM image of MSC displays MXene flakes stacked compactly with horizontal alignment. (d) Top-view SEM of MSC device printed on polymer substrate. Figures 25.4a–d are adapted with permission [50]. Copyright Orangi et al., some rights reserved; exclusive licensee American Chemical Society. Distributed under a Creative Commons Attribution License 4.0 (CC BY). (e) Optical image of MXene aqueous ink, showing its viscous nature, used for 3D extrusion-based printing. (f) Optical images of 3D printed MXene patterns. (g) SEM low magnification image of 3D printed MXene utilized as MSCs. Scale bar = 200 μm. (h) SEM high magnification image of MXene MSC for the framed area in (g), displaying interconnected, stacked MXene nanosheets forming a continuous film. Scale bar = 500 nm. (i) 3D extrusion printed tandem devices on a paper substrate, comprising of two serial MSCs and two parallel MSCs show good flexibility when printed. Figures 4e–i are adapted with permission [36]. Copyright Zhang et al., some rights reserved; exclusive licensee Springer Nature. Distributed under a Creative Commons Attribution License 4.0 (CC BY).

were also found to demonstrate structural resilience and stable electrical conductivity during repeated compression cycles, with the ability to withstand nearly 50% compression before recovery. Furthermore, rate capability and capacitive responses were improved with all-MXene-based MSCs, reaffirming the importance of microstructure engineering and its effect on electrochemical performance [12]. The demonstration of success with 3DFP MXene-based devices paves the way for the research community to develop flexible electronics and physical sensors for various applications, particularly those used in the biomedical field.

25.4.2 Tissue Engineering and Regenerative Medicine

Regenerative medicine is defined as a process of replacing, regenerating, or engineering human tissues, cells, or organs to establish or restore normal function. It is a diverse, interdisciplinary field that includes nanotechnology, biomaterials, gene therapy, and tissue engineering. Research endeavors in this field are mainly focused on developing strategies and therapies that mimic the healing mechanisms present in the human body [67].

Besides possessing unique physiochemical properties, 2D MXenes' excellent surface functionality, photothermal properties, mechanical plasticity and resistance, biocompatibility and biodegradability make them an exciting prospect in tissue engineering and regenerative medicine. Numerous studies have been done on 2D MXenes in relation to tissue engineering and regenerative medicine

[33], [38], [68], specifically, the integration of 2D MXenes into 3D-printed Bioactive Glass (BG) scaffolds through surface dip-coating takes advantage of their photothermal properties for phototherapy [68]–[71]. With MXenes being one of the emerging nanomaterials in the biomedical field, 3D-printed MXenes for tissue engineering and regenerative medicine have recently started to gain traction as well.

In recent years, 3D-printed MXenes have been developed to aid in bone regeneration by improving the poor-self healing capability of bone defects above a critical size. In a recent study, Nie et al. reported successful 3D printing of a composite hydrogel scaffold composed of MXene, GelMA, β-TCP, and sodium alginate (Sr^{2+}). Synthesized MXene-composite bioinks comprising different MXene concentrations of 0.1 mg/ml, 0.3 mg/ml or 0.5 mg/ml were printed at 20°C using an extrusion-based bioprinter through a 200 μm diameter nozzle. In vitro testing conducted with rat bone marrow mesenchymal stem cells and in vivo testing with rats concluded that the material's photothermal antibacterial properties towards both gram (+) and gram (-) bacteria and bone tissue regeneration functions allow playing a synergistic role in antibacterial and osteogenic effects to treat infected mandibular defects. These findings not only provide an effective strategy for the individualized treatment of infected bone defects but also broadens the field of application for 2D MXenes in biomedicine [57]. In another study, Xue et al. evaluated the MXenes' potential for bone regeneration applications by fabricating MXene composite scaffolds using extrusion-based 3D printing. Ti_3C_2 MXenes were synthesized and mixed with hydroxyapatite (HA) and sodium alginate (SA) to form a homogeneous MXene/HA/SA bioink. Printed samples were then crosslinked by soaking in 5 wt% $CaCl_2$ solutions for 30 minutes. 3D printouts showed that greater MXene concentration resulted in a darker colored scaffold (Figure 25.5a). In vitro cellular experiments and in-vivo animal testing confirmed cell proliferation and differentiation, owing to the scaffold's excellent biocompatibility and its ability to enhance bone regeneration of calvarial bone defects. Promotion of osteogenic differentiation with MXene-based scaffolds was confirmed through an increase Alazarin Red (AR) staining (Figure 25.5b) as compared to that of control and HA/SA scaffolds. These results demonstrated the MXene-composite 3D-printed scaffolds' osteoinductivity and their desirability for bone regeneration and reconstruction [58], [59].

With a different approach, Li et al. used extrusion-based 3D printing followed by freeze drying to fabricate scaffolds for maxillofacial bone regeneration. Collagen, silk fibroin, and hydroxyapatite were mixed to form a homogeneous solution before the addition of MXenes in various concentrations (0.4, 0.8, 1.2, and 1.6 mg/g). Thereafter, the mixture was used to print grid-shaped scaffolds. Post-processing involved vacuum freeze drying of the scaffolds for eight hours to achieve a stable 3D printed structure. SEM imaging of MXene composite scaffolds displayed cubic structures with regular microporous structures (Figure 25.5c–d). The MXene composite scaffolds were found to present excellent photothermal performance which could be controlled using near-infrared (NIR) radiation. SEM analysis (Figure 25.5e and f) and Phalloidine/DAPI staining results (Figure 25.5g and h) showed that the CAL27 cell population was significantly reduced after laser treatment and residual cells had irregular morphology and few pseudopods. Live/dead staining (Figure 25.5i–j) also showed red fluorescence after laser treatment, indicating that most of the CAL27 cells were killed. In addition, the printed scaffolds showed successful adhesion and proliferation of mouse embryonic osteogenic precursor cells (MC3T3-E1). Cellular experiments also established the killing efficiency of squamous cell carcinoma cells (CAL-27) and inhibition of tumor growth in-vivo. Furthermore, the scaffolds were also able to induce MC3T3-E1 protein expression in-vitro and stimulate maxillofacial bone formation in-vivo. Overall, this study provides a promising clinical therapy strategy for maxillofacial bone repair [60].

Novel well-characterized MXene-based bioinks were also developed to address the growing need for customizable conductive bioinks for the 3D bioprinting of specific tissue types. Boularaoui et al. fabricated hydrogels comprising gelatin methacryloyl (GelMA) with either Gold Nanoparticles (GNR) or 2D MXene nanosheets by 3D extrusion-based bioprinting. GNRs and 2D MXenes were

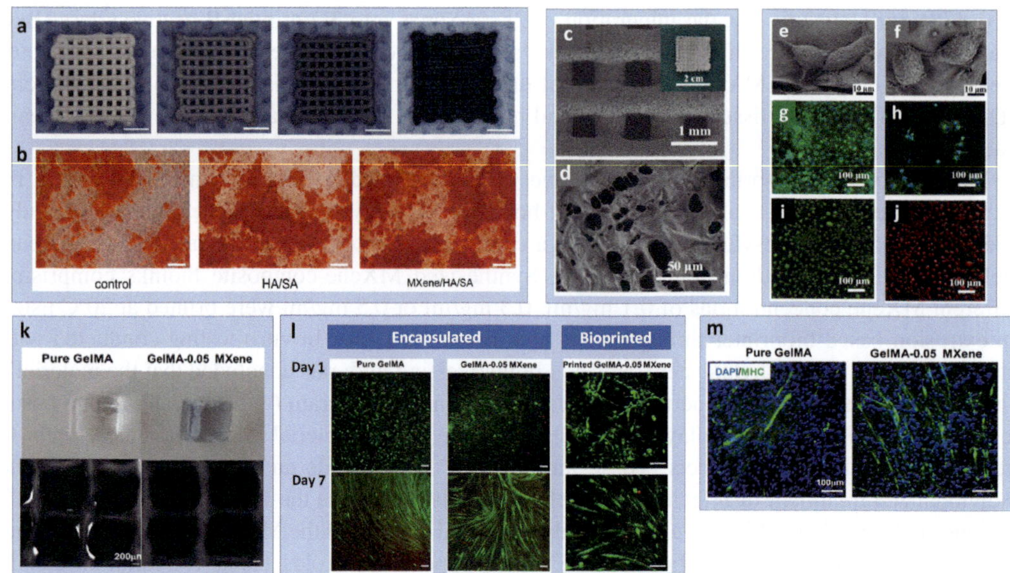

FIGURE 25.5 3D Printed MXene structures for tissue engineering and regenerative medicine. (a) Digital photographs of 3D-printed MXene composite scaffolds with increasing MXene concentration. (b) Alazarin Red (AR) staining of control, HA/SA, and MXene/HA/SA (1 mg/ml) at day 14 (scale bar: 100 μm). Figures 5a–b are adapted with permission [59]. Copyright Mi et al., some rights reserved; exclusive licensee IOP Publishing (United Kingdom). Distributed under a Creative Commons Attribution License 4.0 (CC BY). SEM images of collagen, silk, and hydroxyapatite (CSH) functionalized with MXene nanosheets (M-CSH) at 30x (c) and 800x (d) magnification. SEM images (e, f), FITC-Phalloidin/DAPI staining (g, h) and live/dead staining (i, j) of CAL-27 cells cultured on M-CSH scaffolds (e, g, i) before and (f, h, j) after laser treatment. Figures 5c–j are adapted with permission [60]. Copyright Li et al., some rights reserved; exclusive licensee Oxford University Press. Distributed under a Creative Commons Attribution License 4.0 (CC BY). (k) Optical image of cylindrical-shaped pure GelMA and MXene-based hydrogels together with their respective top-view SEM (bottom) of mesh constructs achieved through 3D extrusion printing. (l) Fluorescent microscopy images of C2C12 cells either encapsulated or bioprinted with pure GelMA and/or GelMA-0.05 Mxene and cultured over 1 day and 7 days. (m) Fluorescent microscopy images of pure GelMA and GelMA-0.05 MXene hydrogels stained for myosin heavy chain (MHC) to determine C2C12 cell differentiation. Figures 5k–m are adapted with permission [58]. Copyright Boularaoui et al., some rights reserved; exclusive licensee American Chemical Society. Distributed under a Creative Commons Attribution License 4.0 (CC BY).

added to GelMA to improve the bioink's printability and conductivity. Specifically, for GelMA/MXene hydrogel samples, 2% of GelMA was mixed with various MXene concentrations (0.05, 0.1, 0.5, 1, and 3 mg/mL) to test its effects on printability, conductivity, rheological characteristics, mechanical properties, and biocompatibility. Bioink containing 0.1 and 0.5 mg/ml of MXene was also mixed with skeletal muscle cells C2C12 and subsequently, 3D printed. As compared to pure GelMA hydrogel, the GelMA-MXene printed hydrogel displayed a darker contrast with better-formed fibers, signifying better printability (Figure 25.5k). Biocompatibility of MXene-based scaffolds and bioinks was tested and compared with that of pure GelMA bioinks for 7 days. Live/dead staining was conducted with C2C12 cells and captured through fluorescent images (Figure 25.5l). Results showed that the encapsulated C2C12 cells did not display any visible cytotoxicity when grown on GelMA-0.05 MXene scaffolds and had viability results similar to that of pure GelMA. Additionally, GelMA-0.05 MXene bioinks containing C2C12 cells were also assessed to be able to maintain high cell viability for seven days. To understand MXene's effect on C2C12 differentiation, C2C12 cells were encapsulated in pure GelMA and GelMA-0.05 Mxene samples, stained for myosin heavy chain (MHC) after seven days, and imaged through fluorescence imaging (Figure 25.5m). Results concluded that the incorporation of MXene additives into GelMA samples enhanced the skeletal muscle differentiation capability of C2C12 cells. This study demonstrated that the addition of conductive MXene nanosheets enhanced GelMA's electrical signal conduction and promoted differentiation in C2C12 cells even without electrical stimulation. Furthermore, the addition of MXene did not severely alter the mechanical stiffness of GelMA, thus allowing cells to retain their regular morphology. These results provide a stepping stone for the research community to explore the fabrication of 3D conductive tissue constructs with physiological relevance [58]. Similarly, Rastin et al. demonstrated the utility of 2D MXene nanosheets with Hyaluronic Acid (HA) and Alginate (Alg) in developing conductive cell-laden bioinks through the 3D printing bio application. 5 wt% of HA and 1 wt% of Alg were mixed with either 1mg/ml or 5mg/ml of MXene to form a homogenous bioink that was printed through a 3D extrusion bioprinter. Thereafter, bioink containing 1mg/ml of MXene and Human Embryonic Kidney 293 (HEK-293) cells was extruded to form cell-ladened 3D conductive hydrogels which were evaluated based on biocompatibility and cell viability. Results concluded that the MXene-based hydrogels displayed impressive rheological characteristics, thus allowing the fabrication of multilayered structures with good shape retention and high resolution. Additionally, the introduction of electrical conductivity to the 3D-printed hydrogel scaffolds via the addition of MXene resolves the longstanding issue of current bioinks' incompatibility with tissues. Along with a >95% cell viability, these findings stand to establish 3D printing as a viable means of assembling functional scaffolds for the regeneration of damaged tissues, thus paving the way for future developments of customized conductive MXene-based bioinks for tissue and neural engineering [56]. In a different approach, Basara et al. developed a new composite construct through Aerosol Jet 3D-printing (AJP) of MXene-polyethylene glycol (PEG) bioink for cardiac patch fabrication. 3D MXene-PEG hydrogels were prepared by pre-molding PEG on a polydimethylsiloxane (PDMS) mold, followed by AJP of the MXene bioink onto the PEG hydrogel at a cell-level resolution. The composite hydrogel was seeded with human-induced pluripotent stem cell-derived cardiomyocytes (iCMs) and cultured for a week to make cardiac patches. A significant increase in MYH7, SERCA2, and TNNT2 protein expression together with an improvement in conduction velocity and beating synchronicity was observed, suggesting MXenes play a vital role in iCM alignment. Again, this work indicates that the MXene-PEG composite provides topographical as well as conductive cues for iCMs, making it physiologically relevant as a cardiac patch for myocardial infarction treatment and other cardiac tissue engineering applications [46], [55].

25.5 CONCLUSIONS AND PERSPECTIVE

Remarkable advancements in MXene research have been made in recent years to prove their utility in a wide range of fields such as energy storage, electronics, catalysis, optical, environmental, structural, and biomedical applications. This is a direct result of the MXenes' excellent chemical and physical properties, biocompatibility, mechanical flexibility, electrical conductivity, and photothermal characteristics. Advancements in the 3D printing of MXenes have also been made in hopes of synergizing the advantageous characteristics of both MXenes and 3D printing to develop highly functional and performance biomedical devices. The exploration of 3D-printed MXenes in the biomedical field, however, remains surprisingly limited despite the prevalence of its 2D counterpart, while 3D-printed MXenes have amazing potential in tissue engineering, regenerative medicine, and biosensing applications, making them extremely exciting prospects in the biomedical field. However, there remain gaps in to use MXenes for biomedical devices 3D-printing manufacturing. First of all, resin viscosity plays a critical influence on most of the available 3D-printing technologies for MXenes printing as high viscosity resin generally decreases the printing resolution or leads to a fail printing. Hence, low loading of MXenes is used in most current MXene 3D printing because high loading of MXenes leads to high viscosity. Surface modification of MXene with a low energy interface is required for the high-loading of 2D MXene for practical applications. In addition, in most biomedical applications, most MXenes are modified with hydrophilic polymers such as PEG and PVP through non-covalent and physical adsorption interactions. However, the non-covalent interactions are not stable enough during the printing process, which often leads to the separation of MXenes during the printing. Furthermore, as a newly developed 2D material, there are limited reports on the biocompatibility of MXenes, and most of the studies ended with the cell toxicity test. Systematic impact factors of surface modification and concentration of MXene loading on biocompatibility are important for most biomedical applications but little work has been done. To move forward with the practical application of 3D-printed MXenes for biomedical applicators, various efforts such as long-term toxicity, biocompatibility in animal models, and ecological environment should be considered.

REFERENCES

1 N.C. Ghosh and S.P. Harimkar, 3 – Consolidation and Synthesis of MAX Phases by Spark Plasma Sintering (SPS): A Review, in *Advances in Science and Technology of Mn+1AXn Phases*, I.M. Low, ed., Woodhead, Sawston, United Kingdom, 2012, pp. 47–80.

2 K. Deshmukh, A. Muzaffar, T. Kovářík, M.B. Ahamed and S.K.K. Pasha, Chapter 1 – Introduction to 2D MXenes: Fundamental Aspects, MAX Phases and MXene Derivatives, Current Challenges, and Future Prospects, in *Mxenes and Their Composites*, K.K. Sadasivuni, K. Deshmukh, S.K.K. Pasha and T. Kovářík, eds., Elsevier, Amsterdam, Netherlands, 2022, pp. 1–47.

3 C. Zhou, X. Zhao, Y. Xiong, Y. Tang, X. Ma, Q. Tao, C. Sun and W. Xu, A Review of Etching Methods of MXene and Applications of MXene Conductive Hydrogels, *Eur. Polym. J.* 167 (2022), 111063.

4 S.P. Sreenilayam, I. Ul Ahad, V. Nicolosi and D. Brabazon, MXene Materials Based Printed Flexible Devices for Healthcare, Biomedical and Energy Storage Applications, *Mater. Today* 43 (2021), 99–131.

5 J. Zhou, X. Zha, F.Y. Chen, Q. Ye, P. Eklund, S. Du and Q. Huang, A Two-Dimensional Zirconium Carbide by Selective Etching of Al_3C_3 from Nanolaminated $Zr_3Al_3C_5$, *Angew. Chem. Int. Ed.* 55 (2016), 5008–5013.

6 Y. Gogotsi and B. Anasori, The Rise of MXenes, *ACS Nano* 13 (2019), 8491–8494.

7 T. Li, L. Yao, Q. Liu, J. Gu, R. Luo, J. Li, X. Yan, W. Wang, P. Liu, B. Chen, W. Zhang, W. Abbas, R. Naz and D. Zhang, Fluorine-Free Synthesis of High-Purity $Ti_3C_2T_x$ (T=OH, O) via Alkali Treatment, *Angew. Chem. Int. Ed.* 57 (2018), 6115–6119.

8 W. Sun, S.A. Shah, Y. Chen, Z. Tan, H. Gao, T. Habib, M. Radovic and M.J. Green, Electrochemical Etching of Ti_2AlC to Ti_2CTx (MXene) in Low-concentration Hydrochloric Acid Solution, *J. Mater. Chem. A* 5 (2017), 21663–21668.

9 X. Zhao, M. Radovic and M.J. Green, Synthesizing MXene Nanosheets by Water-free Etching, *Chem* 6 (2020), 544–546.

10 M. Khazaei, A. Ranjbar, K. Esfarjani, D. Bogdanovski, R. Dronskowski and S. Yunoki, Insights into Exfoliation Possibility of MAX Phases to MXenes, *Phys. Chem. Chem. Phys.* 20 (2018), 8579–8592.

11 P. Srivastava, A. Mishra, H. Mizuseki, K.-R. Lee and A.K. Singh, Mechanistic Insight into the Chemical Exfoliation and Functionalization of Ti_3C_2 MXene, *ACS Appl. Mater. Interfaces* 8 (2016), 24256–24264.

12 H. Tetik, J. Orangi, G. Yang, K. Zhao, S.B. Mujib, G. Singh, M. Beidaghi and D. Lin, 3D Printed MXene Aerogels with Truly 3D Macrostructure and Highly Engineered Microstructure for Enhanced Electrical and Electrochemical Performance, *Adv. Mater.* 34 (2022), e2104980.

13 T.L. Tan, H.M. Jin, M.B. Sullivan, B. Anasori and Y. Gogotsi, High-Throughput Survey of Ordering Configurations in MXene Alloys Across Compositions and Temperatures, *ACS Nano* 11 (2017), 4407–4418.

14 O. Salim, K.A. Mahmoud, K.K. Pant and R.K. Joshi, Introduction to MXenes: Synthesis and Characteristics, *Mater. Today Chem.* 14 (2019), 100191.

15 Z. Li and Y. Wu, 2D Early Transition Metal Carbides (MXenes) for Catalysis, *Small* 15 (2019), 1804736.

16 B. Lu, Z. Zhu, B. Ma, W. Wang, R. Zhu and J. Zhang, 2D MXene Nanomaterials for Versatile Biomedical Applications: Current Trends and Future Prospects, *Small* 17 (2021), 2100946.

17 X. Han, J. Huang, H. Lin, Z. Wang, P. Li and Y. Chen, 2D Ultrathin MXene-Based Drug-Delivery Nanoplatform for Synergistic Photothermal Ablation and Chemotherapy of Cancer, *Adv. Healthcare Mater.* 7 (2018), 1701394.

18 H. Wang, J. Zhang, Y. Wu, H. Huang and Q. Jiang, Chemically Functionalized Two-Dimensional Titanium Carbide MXene by in situ Grafting-Intercalating with Diazonium Ions to Enhance Supercapacitive Performance, *J. Phys. Chem. Solids* 115 (2018), 172–179.

19 M.A. Iqbal, S.I. Ali, F. Amin, A. Tariq, M.Z. Iqbal and S. Rizwan, La- and Mn-Codoped Bismuth Ferrite/Ti_3C_2 MXene Composites for Efficient Photocatalytic Degradation of Congo Red Dye, *ACS Omega* 4 (2019), 8661–8668.

20 R. Wu, H. Du, Z. Wang, M. Gao, H. Pan and Y. Liu, Remarkably Improved Hydrogen Storage Properties of $NaAlH_4$ Doped with 2D Titanium Carbide, *J. Power Sources* 327 (2016), 519–525.

21 S. Kumar, Y. Lei, N.H. Alshareef, M.A. Quevedo-Lopez and K.N. Salama, Biofunctionalized Two-dimensional Ti_3C_2 MXenes for Ultrasensitive Detection of Cancer Biomarker, *Biosens. Bioelectron.* 121 (2018), 243–249.

22 X. Han, X. Jing, D. Yang, H. Lin, Z. Wang, H. Ran, P. Li and Y. Chen, Therapeutic Mesopore Construction on 2D Nb_2C MXenes for Targeted and Enhanced Chemo-photothermal Cancer Therapy in NIR-II Biowindow, *Theranostics* 8 (2018), 4491–4508.

23 B. Rashid, A. Anwar, S. Shahabuddin, G. Mohan, R. Saidur, N. Aslfattahi and N. Sridewi, A Comparative Study of Cytotoxicity of PPG and PEG Surface-Modified 2-D Ti_3C_2 MXene Flakes on Human Cancer Cells and Their Photothermal Response, *Materials* 14 (2021), 4370.

24 J. Chen, K. Chen, D. Tong, Y. Huang, J. Zhang, J. Xue, Q. Huang and T. Chen, CO_2 and Temperature Dual Responsive "Smart" MXene Phases, *Chem. Commun.* 51 (2015), 314–317.

25 C. Dai, H. Lin, G. Xu, Z. Liu, R. Wu and Y. Chen, Biocompatible 2D Titanium Carbide (MXenes) Composite Nanosheets for pH-Responsive MRI-Guided Tumor Hyperthermia, *Chem. Mater.* 29 (2017), 8637–8652.

26 L. Zong, H. Wu, H. Lin and Y. Chen, A Polyoxometalate-functionalized Two-dimensional Titanium Carbide Composite MXene for Effective Cancer Theranostics, *Nano Res.* 11 (2018), 4149–4168.

27 Z. Li, H. Zhang, J. Han, Y. Chen, H. Lin and T. Yang, Surface Nanopore Engineering of 2D MXenes for Targeted and Synergistic Multitherapies of Hepatocellular Carcinoma, *Adv. Mater.* 30 (2018), 1706981.

28 H. Lin, Y. Chen and J. Shi, Insights into 2D MXenes for Versatile Biomedical Applications: Current Advances and Challenges Ahead, *Adv. Sci.* 5 (2018), 1800518.

29 W. Tang, Z. Dong, R. Zhang, X. Yi, K. Yang, M. Jin, C. Yuan, Z. Xiao, Z. Liu and L. Cheng, Multifunctional Two-Dimensional Core–Shell MXene@Gold Nanocomposites for Enhanced Photo–Radio Combined Therapy in the Second Biological Window, *ACS Nano* 13 (2019), 284–294.

30 X. Fan, Y. Ding, Y. Liu, J. Liang and Y. Chen, Plasmonic $Ti_3C_2T_x$ MXene Enables Highly Efficient Photothermal Conversion for Healable and Transparent Wearable Device, *ACS Nano* 13 (2019), 8124–8134.

31 L. Alexey, L. Haidong, A. Mohamed, A. Babak, G. Alexei, G. Yury and S. Alexander, Elastic Properties of 2D $Ti_3C_2T_x$ MXene Monolayers and Bilayers, *Adv. Sci.* 4 (2022), eaat0491.

32 B.C. Wyatt, A. Rosenkranz and B. Anasori, 2D MXenes: Tunable Mechanical and Tribological Properties, *Adv. Mater.* 33 (2021), 2007973.

33 I.J. Gómez, N. Alegret, A. Dominguez-Alfaro and M. Vázquez Sulleiro, Recent Advances on 2D Materials towards 3D Printing, *Chemistry (Easton)* 3 (2021), 1314–1343.

34 K. Hassan, M.J. Nine, T.T. Tung, N. Stanley, P.L. Yap, H. Rastin, L. Yu and D. Losic, Functional Inks and Extrusion-Based 3D Printing of 2D Materials: A Review of Current Research and Applications, *Nanoscale* 12 (2020), 19007–19042.

35 J. Azadmanjiri, T.N. Reddy, B. Khezri, L. Děkanovský, A.K. Parameswaran, B. Pal, S. Ashtiani, S. Wei and Z. Sofer, Prospective Advances in MXene inks: Screen Printable Sediments for Flexible Micro-supercapacitor Applications, *J. Mater. Chem. A* 10 (2022), 4533–4557.

36 C. Zhang, L. McKeon, M.P. Kremer, S.-H. Park, O. Ronan, A. Seral-Ascaso, S. Barwich, C.Ó. Coileáin, N. McEvoy, H.C. Nerl, B. Anasori, J.N. Coleman, Y. Gogotsi and V. Nicolosi, Additive-free MXene Inks and Direct Printing of Micro-supercapacitors, *Nat. Commun.* 10 (2019), 1795.

37 B. Zazoum, A. Bachri and J. Nayfeh, Functional 2D MXene Inks for Wearable Electronics, *Materials* 14 (2021), 6603.

38 X. Lin, Z. Li, J. Qiu, Q. Wang, J. Wang, H. Zhang and T. Chen, Fascinating MXene Nanomaterials: Emerging Opportunities in the Biomedical Field, *Biomater. Sci.* 9 (2021), 5437–5471.

39 R. Yang, J. Zhou, C. Yang, L. Qiu and H. Cheng, Recent Progress in 3D Printing of 2D Material-Based Macrostructures, *Adv. Mater. Technol.* 5 (2020), 1901066.

40 M. Vural, A. Pena-Francesch, J. Bars-Pomes, H. Jung, H. Gudapati, C.B. Hatter, B.D. Allen, B. Anasori, I.T. Ozbolat, Y. Gogotsi and M.C. Demirel, Inkjet Printing of Self-Assembled 2D Titanium Carbide and Protein Electrodes for Stimuli-Responsive Electromagnetic Shielding, *Adv. Funct. Mater.* 28 (2018), 1801972.

41 E. Jabari, F. Ahmed, F. Liravi, E.B. Secor, L. Lin and E. Toyserkani, 2D Printing of Graphene: A Review, *2D Materials* 6 (2019), 042004.

42 S. Harada, K. Kanao, Y. Yamamoto, T. Arie, S. Akita and K. Takei, Fully Printed Flexible Fingerprint-like Three-Axis Tactile and Slip Force and Temperature Sensors for Artificial Skin, *ACS Nano* 8 (2014), 12851–12857.

43 T. Sekine, R. Sugano, T. Tashiro, J. Sato, Y. Takeda, H. Matsui, D. Kumaki, F. Domingues Dos Santos, A. Miyabo and S. Tokito, Fully Printed Wearable Vital Sensor for Human Pulse Rate Monitoring using Ferroelectric Polymer, *Sci. Rep.* 8 (2018), 4442.

44 A. Bhat, S. Anwer, K.S. Bhat, M.I.H. Mohideen, K. Liao and A. Qurashi, Prospects Challenges and Stability of 2D MXenes for Clean Energy Conversion and Storage applications, *npj 2D Mater. Appl.* 5 (2021), 61.

45 J. Huang, Z. Li, Y. Mao and Z. Li, Progress and Biomedical Applications of MXenes, *Nano Select* 2 (2021), 1480–1508.

46 R. Vankayala, S. Thangudu, N. Kuthala and P. Kalluru, MXenes and Their Composites for Medical and Biomedical Applications, in *Mxenes and Their Composites*, D.D. Sadasivuni, S.K.K Pasha, K. Deshmukh and T. Kovářík, eds., Elsevier, Amsterdam, Netherlands, 2022, pp. 499–524.

47 Y.Z. Zhang, Y. Wang, Q. Jiang, J.K. El-Demellawi, H. Kim and H.N. Alshareef, MXene Printing and Patterned Coating for Device Applications, *Adv. Mater.* 32 (2020), e1908486.

48 J.D. Cain, A. Azizi, K. Maleski, B. Anasori, E.C. Glazer, P.Y. Kim, Y. Gogotsi, B.A. Helms, T.P. Russell and A. Zettl, Sculpting Liquids with Two-Dimensional Materials: The Assembly of $Ti_3C_2T_x$ MXene Sheets at Liquid-Liquid Interfaces, *ACS Nano* 13 (2019), 12385–12392.

49 C.J. Zhang, L. McKeon, M.P. Kremer, S.H. Park, O. Ronan, A. Seral-Ascaso, S. Barwich, C.O. Coileain, N. McEvoy, H.C. Nerl, B. Anasori, J.N. Coleman, Y. Gogotsi and V. Nicolosi, Additive-free MXene Inks and Direct Printing of Micro-supercapacitors, *Nat. Commun.* 10 (2019), 1795.

50 J. Orangi, F. Hamade, V.A. Davis and M. Beidaghi, 3D Printing of Additive-Free 2D $Ti_3C_2T_x$ (MXene) Ink for Fabrication of Micro-Supercapacitors with Ultra-High Energy Densities, *ACS Nano* 14 (2020), 640–650.

51 W. Cao, C. Ma, D. Mao, J. Zhang, M. Ma and F. Chen, MXene-Reinforced Cellulose Nanofibril Inks for 3D-Printed Smart Fibres and Textiles, *Adv. Funct. Mater.* 29 (2019), 1905898.

52 Z. Fan, C. Wei, L. Yu, Z. Xia, J. Cai, Z. Tian, G. Zou, S.X. Dou and J. Sun, 3D Printing of Porous Nitrogen-Doped Ti_3C_2 MXene Scaffolds for High-Performance Sodium-Ion Hybrid Capacitors, *ACS Nano* 14 (2020), 867–876.

53 L. Yu, W. Li, C. Wei, Q. Yang, Y. Shao and J. Sun, 3D Printing of NiCoP/Ti_3C_2 MXene Architectures for Energy Storage Devices with High Areal and Volumetric Energy Density, *Nano-Micro Lett.* 12 (2020), 143.

54 L. Yu, Z. Fan, Y. Shao, Z. Tian, J. Sun and Z. Liu, Versatile N-Doped MXene Ink for Printed Electrochemical Energy Storage Application, *Adv. Energy Mater.* 9 (2019), 1901839.

55 G. Basara, M. Saeidi-Javash, X. Ren, G. Bahcecioglu, B.C. Wyatt, B. Anasori, Y. Zhang and P. Zorlutuna, Electrically Conductive 3D Printed $Ti_3C_2T_x$ MXene-PEG Composite Constructs for Cardiac Tissue Engineering, *Acta Biomater* 139 (2022), 179–189.

56 H. Rastin, B. Zhang, A. Mazinani, K. Hassan, J. Bi, T.T. Tung and D. Losic, 3D Bioprinting of Cell-laden Electroconductive MXene Nanocomposite Bioinks, *Nanoscale* 12 (2020), 16069–16080.

57 R. Nie, Y. Sun, H. Lv, M. Lu, H. Huangfu, Y. Li, Y. Zhang, D. Wang, L. Wang and Y. Zhou, 3D Printing of MXene Composite Hydrogel Scaffolds for Photothermal Antibacterial Activity and Bone Regeneration in Infected Bone Defect Models, *Nanoscale* 14 (2022), 8112–8129.

58 S. Boularaoui, A. Shanti, M. Lanotte, S. Luo, S. Bawazir, S. Lee, N. Christoforou, K.A. Khan and C. Stefanini, Nanocomposite Conductive Bioinks Based on Low-Concentration GelMA and MXene Nanosheets/Gold Nanoparticles Providing Enhanced Printability of Functional Skeletal Muscle Tissues, *ACS Biomater. Sci. Eng.* 7 (2021), 5810–5822.

59 X. Mi, Z. Su, Y. Fu, S. Li and A. Mo, 3D Printing of Ti_3C_2-MXene-Incorporated Composite Scaffolds for Accelerated Bone Regeneration, *Biomed. Mater.* 17 (2022), 035002.

60 F. Li, Y. Yan, Y. Wang, Y. Fan, H. Zou, H. Liu, R. Luo, R. Li and H. Liu, A Bifunctional MXene-modified Scaffold for Photothermal Therapy and Maxillofacial Tissue Regeneration, *Regener. Biomater.* 8 (2021), rbab057.

61 A. Saleh, S. Wustoni, E. Bihar, J.K. El-Demellawi, Y. Zhang, A. Hama, V. Druet, A. Yudhanto, G. Lubineau, H.N. Alshareef and S. Inal, Inkjet-printed $Ti_3C_2T_x$ MXene Electrodes for Multimodal Cutaneous Biosensing, *J. Phys.: Mater.* 3 (2020), 044004.

62 Y. Guo, M. Zhong, Z. Fang, P. Wan and G. Yu, A Wearable Transient Pressure Sensor Made with MXene Nanosheets for Sensitive Broad-Range Human-Machine Interfacing, *Nano Lett.* 19 (2019), 1143–1150.

63 M. Xin, J. Li, Z. Ma, L. Pan and Y. Shi, MXenes and Their Applications in Wearable Sensors, *Front. Chem.* 8 (2020), 297.

64 S.S. Athukorala, T.S. Tran, R. Balu, V.K. Truong, J. Chapman, N.K. Dutta and N. Roy Choudhury, 3D Printable Electrically Conductive Hydrogel Scaffolds for Biomedical Applications: A Review, *Polymers (Basel)* 13 (2021), 474.

65 C. Li, S. Cong, Z. Tian, Y. Song, L. Yu, C. Lu, Y. Shao, J. Li, G. Zou, M.H. Rümmeli, S. Dou, J. Sun and Z. Liu, Flexible Perovskite Dolar Cell-driven Photo-rechargeable Lithium-ion Capacitor for Self-powered Wearable Strain Sensors, *Nano Energy* 60 (2019), 247–256.

66 M. Saadi, A. Maguire, N.T. Pottackal, M.S.H. Thakur, M.M. Ikram, A.J. Hart, P.M. Ajayan and M.M. Rahman, Direct Ink Writing: A 3D Printing Technology for Diverse Materials, *Adv. Mater.* 34 (2022), e2108855.

67 H. Huang, C. Dong, W. Feng, Y. Wang, B. Huang and Y. Chen, Biomedical Engineering of Two-dimensional MXenes, *Adv. Drug Delivery Rev.* 184 (2022), 114178.

68 C. He, L. Yu, H. Yao, Y. Chen and Y. Hao, Combinatorial Photothermal 3D-Printing Scaffold and Checkpoint Blockade Inhibits Growth/Metastasis of Breast Cancer to Bone and Accelerates Osteogenesis, *Adv. Funct. Mater.* 31 (2020), 2006214.

69 S. Pan, J. Yin, L. Yu, C. Zhang, Y. Zhu, Y. Gao and Y. Chen, 2D MXene-Integrated 3D-Printing Scaffolds for Augmented Osteosarcoma Phototherapy and Accelerated Tissue Reconstruction, *Adv. Sci.* 7 (2020), 1901511.

70 Q. Yang, H. Yin, T. Xu, D. Zhu, J. Yin, Y. Chen, X. Yu, J. Gao, C. Zhang, Y. Chen and Y. Gao, Engineering 2D Mesoporous Silica@MXene-Integrated 3D-Printing Scaffolds for Combinatory Osteosarcoma Therapy and NO-Augmented Bone Regeneration, *Small* 16 (2020), e1906814.

71 J. Yin, S. Pan, X. Guo, Y. Gao, D. Zhu, Q. Yang, J. Gao, C. Zhang and Y. Chen, Nb_2C MXene-Functionalized Scaffolds Enables Osteosarcoma Phototherapy and Angiogenesis/Osteogenesis of Bone Defects, *Nano-Micro Lett.* 13 (2021), 30.

26 The Application of 3D Print in the Formulation of Novel Pharmaceutical Dosage Forms

*Touraj Ehtezazi**
School of Pharmacy and Biomolecular Sciences, Liverpool John Moores University, Liverpool, L3 3AF, UK
*Corresponding author: t.ehtezazi@ljmu.ac.uk

CONTENTS

26.1 INTRODUCTION

Pharmaceutical dosage forms are normally manufactured in large size batches such as 100,000 pieces [1]. It is expected the batch size for solid pharmaceutical dosage forms to be at least 100,000 units unless it is for orphan medicinal products [2]. Batch size depends on the demand and also the time of product lifetime [3]. For an abbreviated new drug application (ANDA), the exhibit batch size should be at least one-tenth (1/10) of the commercial batch size or 100,000, whichever is greater[1]. The current manufacturing establishments and regulatory aspects have met the public demand well. However, there are new emerging healthcare aspects such as over seven billion population, and an increase in average age, with a median age of 42 for more developed regions (28.5 years in 1950 and

DOI: 10.1201/9781003296676-26

42 years in 2020). In general, the median age has increased from 23.6 years in 1950 to 30.9 years in 2020 in the world [4]. This new demographic population requires a different type of supply chain and production. In addition, technological advancements provide novel opportunities that make medicines more efficient. For example, elderly people or patients may need to take multiple medicines every day. Taking several dosage forms (polypharmacy) is a common practice among elderly patients, but there are risks of drug interactions, medication errors, and contra-indications, leading to a lack of patient compliance [5]. Therefore, combining these into fewer dosage forms would make life much easier for them and followed with better compliance. However, mixing a large number of active ingredients could lead to chemical incompatibilities such as aspirin, statin, and blood-pressure-lowering drugs [6].

The prevalence of cancer has increased from 9% in medieval times to 50% in modern Britain, at the time of death. This could be attributed to modern carcinogens, the spread of viruses that trigger malignancy, industrial pollutants, and longer life expectancy. Since cancer is a heterogeneous disease, then standard treatments are effective only for a subset of patients. Thus, the intrinsic variability of cancer needs precision and personalized medicine. For example, traditional therapies such as mAbs are effective against certain cancer cells and these drugs will not be effective against mutant cancer cells. Therefore, a new line of mAbs is required. Thus, by increasing the number of cancer patients, more precision and personalized medicine are needed, something that established conventional pharmaceutical products cannot meet.

Rare diseases are defined as health-related problems that are heterogeneous, and geographically disparate. Most of them are chronic and result in early death, although only a few of them are preventable. Rare diseases are defined as conditions whose prevalence is not more than 50 per 100,000. It has been reported that there are more than 300 million people with rare diseases worldwide and these patients need personalized medicine.

Therefore, the well-established pharmaceutical manufacturing process faces challenges to meet the new demands. Three-dimensional printing (3DP) has been employed for the development of novel pharmaceutical dosage forms. These include tablets, capsules, nose patches, filaments, core-shell tablets, gastro-floating tablets, hollow cylinders, dual compartmental dosage units, multi-compartment capsular devices, orodispersible films, and fast dissolving oral films, and liquid capsules. Currently, there is only one 3DP pharmaceutical product approved by regulatory bodies, which is Spritam. It just provides extraordinary fast disintegrating tablets, which may be convenient for the treatment of patients with epilepsy. However, this product is not meant to meet the new demands for pharmaceutical dosage forms. This chapter aims to evaluate whether the application of 3DP can meet the demands for new and novel pharmaceutical dosage forms.

26.2 WHAT IS PERSONALIZED MEDICINE?

Personalized medicine is defined as tailoring therapy for a patient based on individual needs [7]. Personalized medicine plays important role in the prevention, diagnosis, treatment, and prognosis of cancer. Choosing a suitable antibody is one aspect of personalized medicine in cancer therapy. Personalized medicine is also identified through genome sequencing; for example, it can identify hypersensitivity to abacavir in HIV treatment. Personalized medicine is applied for transfusion medicine, matching donors and recipients, for example, avoiding antibody-mediated hemolysis due to passenger lymphocyte syndrome through genetic testing. It is expected that personalized medicine represent better value for money, as patients will receive the correct treatment right from the beginning. For example, the warfarin dose should be adjusted based on the naturally occurring genetic variations in both the VKORC1 and CYP2C9 genes. Personalized medicine may be applied to achieve the right immunotherapy combination for acute myeloid leukemia treatment. The phase three clinical trial data did not show a reduction in the rate of moderate or severe exacerbations in a patient with asthma taking fevipiprant chewable tablets, and it has been suggested that further efforts are needed to personalize treatments to the patient's need. In addition, personalized medicine is recommended for the treatment of psoriatic arthritis, based on characteristic phenotypes of peripheral T helper cells and administering correct antibodies. To achieve effective treatment, or a personalized treatment for diseases like rheumatoid arthritis, serum biomarkers, synovial biopsy, and gene expressions should be obtained before commencing the treatment. This would achieve much quicker clinical outcomes than conventional treatments. For example, the patient's treatment starts with methotrexate and glucocorticoids, and when a response is not observed then it is switched to tumor necrosis factor inhibitor, and this process may take about nine months. Personalized medicine changes the drug development process. There is an early stage of bench-to-bedside investigations, where clinical data is acquired. This is called forward translation. Then, there is a reverse-translation process, where drug candidates are further optimized based on biomarkers. Nanoparticles may be used for personalized medicine when a wide range of nanoparticle formulations are available, and a specific task is needed from each formulation. For example, a nanoparticle formulation may be employed for drug delivery to the brain, but another formulation may be employed for gene editing [8].

26.3 WHY IS PERSONALIZED MEDICINE IMPORTANT?

UK National Health Service (NHS) published a document in 2016 (www.england.nhs.uk/wp-content/uploads/2016/09/improving-outcomes-personalised-medicine.pdf), which states four 'Ps' for personalized medicine

1. Prediction
2. Prevention
3. Precision
4. Participatory.

The prediction and prevention will come through genomic characterization. For example, familial hypercholesterolemia causes an increase in blood cholesterol levels and the risk of developing heart attacks and cardiac events [9]. Early interventions can lead to a substantial prevention of cardiac disease and death [9]. Next-generation sequencing is now developed, which allows meeting the demand for sequencing millions of DNA or RNA sequences with higher sensitivity to detect low-frequency variants, and high throughput as fast as 6.5 h turnaround. In addition, next-generation plasma protein profiling has been developed allowing 1463 proteins to be analyzed compared to 794 proteins analyzed by the convention methods.

Precision stands for precise diagnosis of the disease. This will also depend on the patient's molecular and cellular characteristics. Certainly, a precise diagnosis will allow clinicians in choosing the optimum medicine and dose, for example, warfarin.

The participatory helps targeted discussions between clinicians and individuals based on individual's genomic characteristics and lifestyle. For example, there is certain Type-1 diabetes that can be better controlled by an optimum combination of insulin and medical nutrition therapy [10]. The international Consortium for Personalised Medicine anticipates that there will be next-generation healthcare by 2030. One of the key questions is 'Will scientists, innovators, healthcare providers, and others be able to provide the most suitable medicine, at the right dose, for the right person, at the right time, at a reasonable cost?' [11] Therefore, the pharmaceutical industry needs to prepare itself for this imminent transition in the healthcare system.

26.4 WHAT ARE THE CURRENT METHODS OF PERSONALIZED MEDICINE?

Manufacturing tablets with different strengths is the desired approach for personalized medicine. For example, different warfarin doses are needed based patient's genotype [12]. As a result, there are warfarin tablets with strengths of 0.5/1/2/3/4/5 mg. Splitting tables are the most common method of personalized medicine to only adjust the dose. This approach assumes that the drug is uniformly distributed in the tablet. Certainly, this method applies to low potent drugs. Tablet splitting can be done by the patient or caregiver (such as a nurse, or parents). Tablet splitting can be achieved by tools available at home such as a kitchen knife [13], [14]. This method can be suitably applied to scored tablets. Splitting tablet by kitchen knife can be performed in the following steps: by placing the tablet on a hard surface such as a bench top, then by placing the sharp side of the blade of a clean knife along the score-line, pressing the knife or non-sharp end of the knife by one hand, and ensuring holding the handle by other hands. It is recommended that the knife blade be made of stainless steel, with a length of 8 cm, and a thickness of 1 mm. The handle length should be 12 cm.

Tablet splitters may be used for splitting tablets (Figure 26.1). These are usually made of plastic with a stainless-steel blade inside. Typical dimensions are length 9 cm, width 5 cm, and height 3 cm. Table splitters can hold tablets in the size of 6–20 mm. The tablet splitter is opened, and a tablet is placed inside a designated area of the splitter, which is parallel to the base of the device. Then the lid is closed which brings the blade into contact with the tablet and breaks into two similar size sections.

Grinders or tablet crushers may also be used to produce drug powders from uncoated tablets. Tablet crushers typically are cylindrical devices with two compartments, which are separated by a base. The base may have a rough surface or a slit. The upper compartment has a lid that screws down to the vicinity of the base. A tablet is placed in the upper compartment and on the base. Then, the screw cap is replaced on the top compartment and fully twisted. By this action, the gap between the screw cap and the base is reduced, and the tablet is crushed in between. Normally, tablet powder is passed through the slit of the base plate and collected in the lower compartment of the crusher. This type of device is not normally used for adjusting drug doses, but to make it easier for swallowing by patients that have difficulty swallowing tablets such as children.

Cutting transdermal patches are also used for adjusting the dose for pediatric use. This applies to those patches that do not have a reservoir. A pair of scissors typically are used to adjust the size of the patch. The assumption is that the drug is uniformly distributed in the patch. As cutting the transdermal patches mat damages the integrity of the device, an occlusive dressing has been suggested to be inserted between the patch and the skin, which can block the contact area of the skin. Adjusting the area of the absorption surface provides means of adjusting the dose from a transdermal patch.

Unlicensed oral formulations are another method of personalized medicine. Unlicensed oral suspensions are prepared by dispersing drug powders that are obtained either from emptying drug capsules or crushing drug tablets in a liquid vehicle.

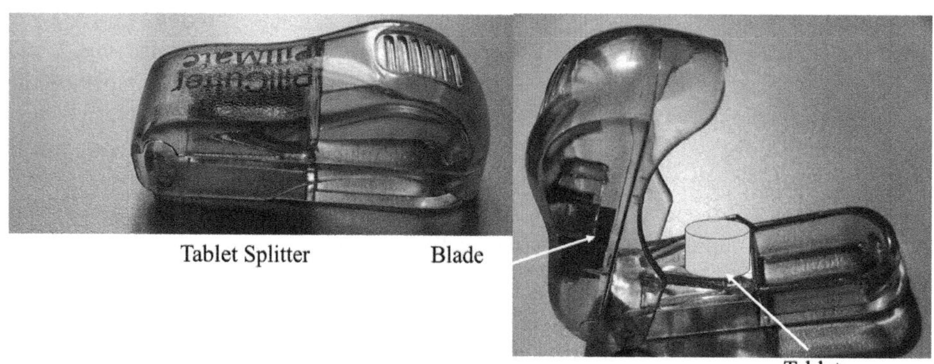

FIGURE 26.1 Photographs of a typical tablet splitter.

3D modeling of anatomical parts is also considered personalized medicine for planning surgery [15]. 3DP allows for reducing implant surgery time, intraoperative blood loss, and postoperative complications.

26.5 WHAT ARE THE PROBLEMS WITH CURRENT PERSONALIZED MEDICINE?

The splitting tools play an important role in producing parts with close masses. For example, a kitchen knife splits tablets with a large variation in weight compared to other tools such as Pilomat and Vitality tablet splitters [13]. The tablet splitting method should comply with regulatory requirements. This means that tablet parts should be within 85–115% and all parts within 75–125% of the theoretical weight of a tablet part. It was found that not all the time this regulatory requirement is met. Furthermore, there are significant variations in intra and inter-accuracy of tablet splitting tools [16]. Certain brands meet the regulatory requirement all the time, but some brands do not meet the requirement all the time.

A tablet crusher also may be used for preparing oral unlicensed oral suspension to achieve desired doses. However, tablet crushing could change the drug efficacy by incomplete dosing, raise safety concerns related to operators exposed to airborne particles from crushed tablets, lead to significant drug loss by crushers, change the drug absorption rate, cause adverse events, and reduce palatability. Grinders may cause drug loss as high as 14% [13]. In cutting the transdermal patches, again, it is assumed that the drug is uniformly distributed in the patch, while this depends on the formulation. It should be noted that cutting a transdermal patch may affect the delivery of the drug or affect the adhesive properties of the patch. Furthermore, the patch manufacturer may not recommend cutting the patch. The use of approved oral suspensions is more appropriate for adjusting the dose for personalized medicine compared to unlicensed suspensions made by emptying capsules. However, unlicensed oral suspensions may present limitations such as drug chemical stability, formulation physical stability to deliver uniform doses and microbial contamination of the product during compounding [17].

26.6 WHAT ARE THE ALTERNATIVE METHODS OF ACHIEVING PERSONALIZED MEDICINE?

3DP is currently the main emerging technology as an alternative method for personalized medicine [18]. Alternatively, induced pluripotent stem cells (iPSCe) provide a unique tool for personalized medicine by deriving disease-specific stem cells. iPSCs can self-renew and differentiate into many cell types [19]. In this approach, somatic cells are obtained from the patient such as fibroblasts (obtained from skin biopsies) or peripheral blood cells, which is minimally invasive. A viral vector (such as Sendai virus vector SeVdp-302L) is used to deliver transcription factors: *KLF4, OCT4, SOX2,* and *c-MYC* to reprogramme the cells. The generated iPSCs may be pushed to precardiac mesodermal lineage (KDR^+, $MESP1^+$) via GSK-3β inhibition, and then differentiated to cardiac progenitor cells. Autologous bone-marrow-derived mesenchymal stem cells were employed for the treatment of multiple sclerosis. As another example, neuronal precursors were obtained from embryonic and mesenchymal stem cells, which were delivered intravenously for the treatment of Parkinson's disease. Viral vectors may be employed for gene delivery in the context of personalised medicine. For example, adenoviral vectors have been widely for gene therapy in clinical trials [20]. Recently, baculoviruses have been considered for gene therapy [21]. As these viral vectors are safe (not causing genetic mutations or being immunogenic), easy to scale up, with transient gene expression [21].

26.7 WHY DOES 3DP SEEM BETTER THAN OTHER METHODS?

The advances in 3DP, availability and low cost make this technology appealing, in particular the extrusion-based methods (semisolid printing, pellet/powder extrusion, and FDM) [22]. 3DP can be useful for providing medicines with desired doses or combining several active ingredients in a single dosage form. This is a major demand in the healthcare sector. 3DP is also can be employed for gene delivery and gene editing. This is an exciting and novel application of 3DP. Gene delivery can be achieved by formulating viral or non-viral vectors, containing the desired nucleic acids (siRNA or DNA). This is followed by dispersing the gene delivery vectors in a biogel/bioink and using a bioprinter to produce a scaffold with desired size and shape, or the scaffold can provide micron-sized pockets to hold gene delivery vectors. Then, the scaffold is placed in the target area of the body, such as defected bones to promote bone generation. In another approach, a composite was 3D printed, which contained stem cells and PLGA particles encapsulating a transcription factor. Then the scaffold can be printed in the shape of a personalized lumber vertebra. recently, 4D printing has been developed. These are referred to as objects that are created in 3D (even 1D or 2D), but these respond to environmental stimuli (i.e., smart materials) [23]. For example, Melocchi et al. (2019) employed 3D printing and PVA-based materials to produce 3D-printed cylindrical-shaped coils for oral administration, which swell in the stomach and avoid excretion through open pylorus [24].

26.8 WHAT IS 3D PRINTING?

3DP is an iterative manufacturing process, where each time a 2D object is created. The layers are joined to each other, either due to adhesion properties of the molten layer, the use of adhesives added by droplet generators, or due to chemical reactions. 3DP production originated in 1974 by Wyn Kelly Swainson, followed by Charles W. Hull in 1984.

3DP required a computer-aided design (CAD) model or a digital scan. There are several software programs are available such as SolidWorks (Dassault Systèmes), and Autocad (Autodesk, California, US). The CAD model is saved in the Standard Tessellation Language (STL) format. However, a slicer program is needed to instruct the printer on the operation and specify the operational parameters. The printing parameters depend on the type of 3D printing. For example, fused deposition modeling (FDM) requires setting the temperature of the nozzle, the build-plate temperature, the travel speed of the printer head, each layer thickness, and infill percentage. Following completion of this layer, the printer makes another layer (usually on top of that). This iteration process is continued until the object is materialized in full-size. 3D objects materialize on a platform either in the air or inside a liquid. The thickness of each layer varies between 100 nm [25] for two-photon polymerization 3DP to 600 μm for FDM 3DP. This permits the construction of highly detailed 3D physical objects on a nano-scale. For example, a detailed boat was 3D printed at microscales by applying the two-photon absorption polymerization technique [26]. Nano-printing may provide opportunities for developing nano-drug delivery systems.

26.9 WHAT CAN NOVELTIES 3D PRINTING PROVIDE TO PHARMACEUTICAL SCIENCES?

3DP has been employed extensively to prepare novel pharmaceutical dosage forms. This approach allows the formulation of a variety of dosage forms that cannot be prepared by conventional methods. For example, Ehtezazi et al. (2018) showed the fabrication of mesh or plain films by 3D printing, which allows faster disintegration compared to plain films (Figure 26.2) [27]. A Hilbert's film was also prepared by 3DP (Figure 26.2). 3DP provides the opportunity for preparing personalized medicine [18, 28], such as chewable printlets for children with maple syrup urine disease, or extended-release tablets of prednisolone with varying doses in the range of 2–10 mg. 3DP facilitates the production of polypills with defined release profiles. Robles-Martinez et al. (2019) employed stereolithographic (SLA) 3DP to produce tablets with six different layers, each containing a different drug [29].

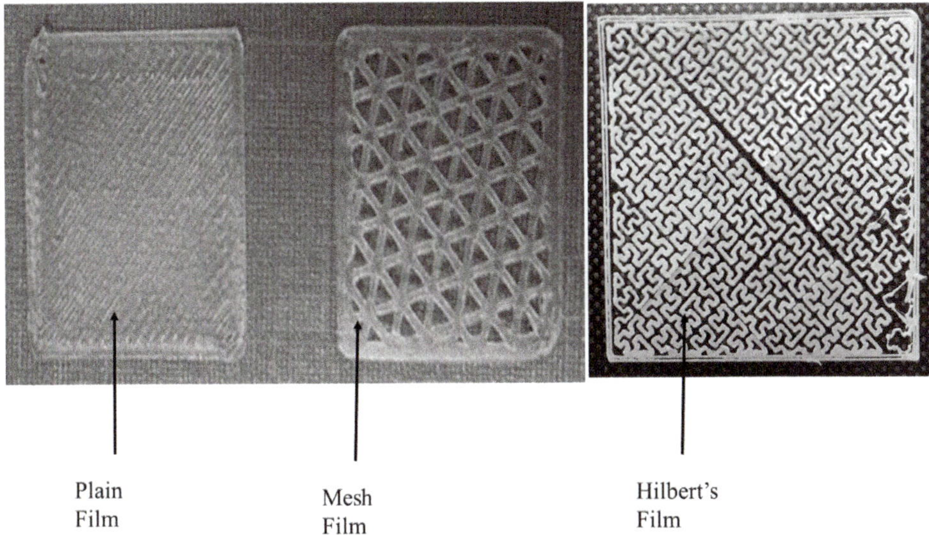

FIGURE 26.2 Fabrication of fast dissolving oral films by the fused deposition modeling 3DP.

3DP has been utilized to manufacture Spritam tablets, which disintegrate within two seconds. This rapid disintegration time is outstanding and has not been met by conventional tablet manufacturing methods. Furthermore, 3D-printed tablets of amitriptyline hydrochloride achieved >80% of drug release within a few minutes after immersing in the release media. While sugar-coated amitriptyline hydrochloride released 80% of the content over 10 minutes. Furthermore, 3D printing allowed producing oral films with higher loading of indomethacin nanocrystals compared to the traditional solvent casting method. 3DP has been employed to produce gummy drug formulations with different shapes and colors to improve adherence to medication in children (Figure 26.3). In another approach, compartmentalized capsules were manufactured by 3DP, where each compartment disintegrated at different time points, allowing a burst release followed by a sustained release drug delivery [30].

26.10 WHAT HAVE 3DP METHODS BEEN DEVELOPED RECENTLY?

There are several 3DP techniques have been developed, and most of these have been investigated for pharmaceutical applications. These are described below.

26.10.1 Inkjet 3DP

Inkjet 3DP employs a liquid dispenser (an inkjet device), which produces droplets of uniform size (Figure 26.4A). The ink contains a binder solution and like normal 2D printing, the inkjet device deposits droplets on the powder (instead of paper) on a platform. The powder particles adhere to each other due to the presence of the binder in the droplets and form a thin object. The platform level is lowered by a fraction of a millimeter and another powder layer is spread over the 2D object, and the process of depositing binder solution starts again. The 3D-printed tablet is embedded inside a powder bed and it is recovered. The tablets require a drying process before packaging. Katstra et al. (2000) employed inkjet 3DP to produce tablets [31]. As the droplets were estimated to be 80–90 μm from a 45 μm nozzle, then continuous inkjet printing should have been utilized. This was a wise decision, as smaller inkjet nozzles tend to block. Inkjet 3DP is the only technique so far that has been

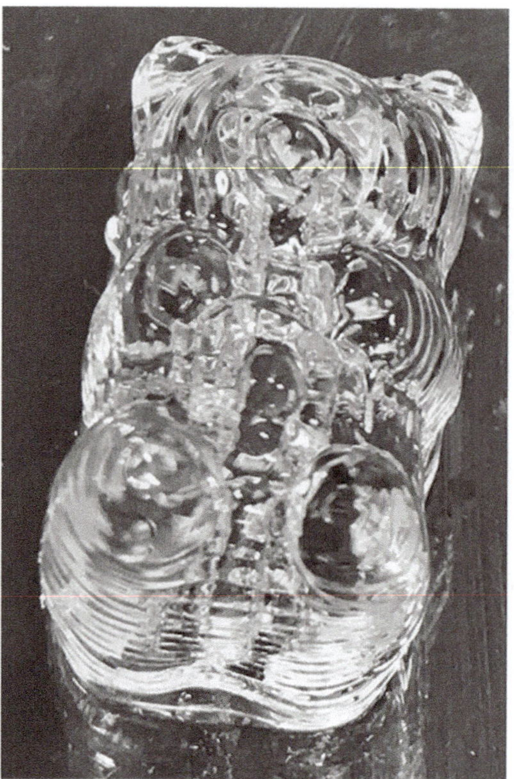

FIGURE 26.3 3D printed transparent gummy bear.

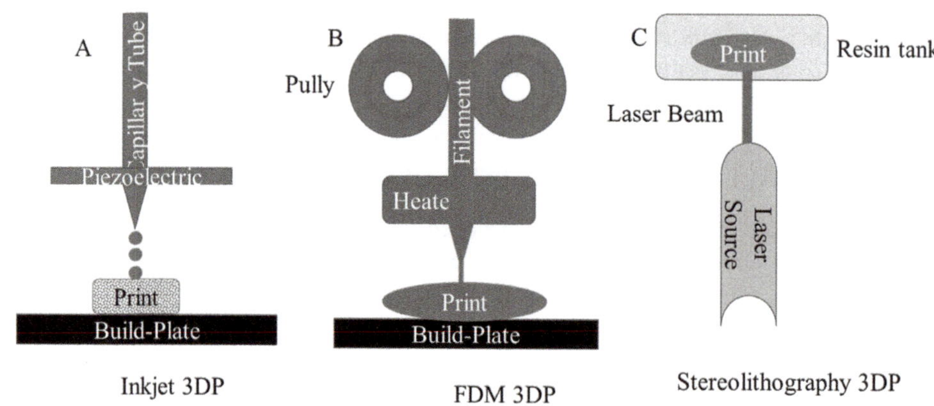

FIGURE 26.4 Schematic examples of different types of 3DP: (A) inkjet, (B) FDM, (C) Stereolithography.

used to produce tablets at an industrial scale (Spirtam tablets by the Aprecia's patented ZipDose® Technology). Inkjet printing has also been employed for the formulating of tablets with an internal honeycomb structure to control drug release [28]. Sen et al. (2020) reported that increasing the surface tension of API ink leads to less variation in the content and weight uniformity of printed tablets [32].

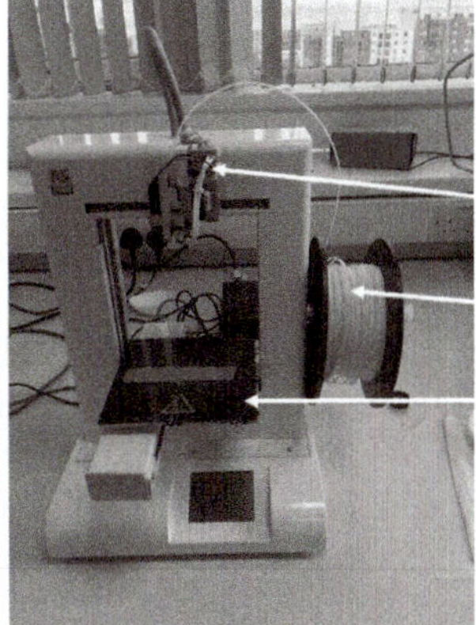

FIGURE 26.5 Different types of FDM 3D printers. (Left) Direct 3DP that does not require filament. (Right) a typical FDM that uses filament for printing.

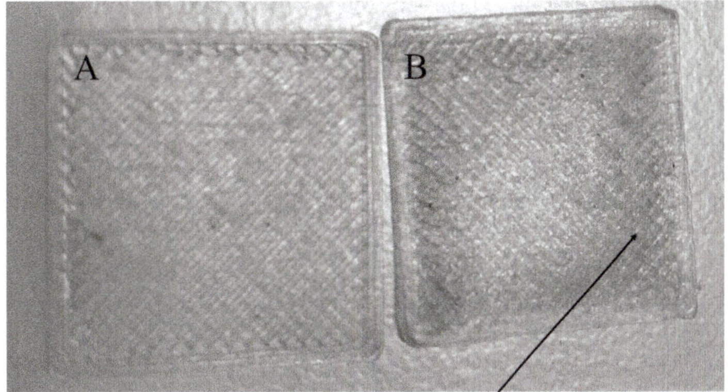

FIGURE 26.6 Degradation of material during FDM 3DP. (A) Formulation without scorching. (B) The formulation contained a disintegrant that decomposes at high temperatures, which can be recognized by its darkened color.

26.10.2 Fused Deposition Modelling

Fused deposition modeling (FDM) 3DP is the most commonly used technique for the formulation of pharmaceutical dosage forms. This method is based on hot melt extrusion (Figure 26.4B). This means that a powder mixture or a filament is heated to melt in the printer head (Figure 26.5). This is followed by the raster movement of the printer head or the print bed (platform), which leads to the deposition of a 2D object on the print bed. After completion of one layer, either the printer head or print bed change moves in the *z*-direction (height) and another layer is deposited. The printer head can be heated up to 230 °C, and the print bed may be heated up to 80°C to ensure the formation of a 2D object without defects. FDM has been applied for the formulation of sustained released tablets, immediate-release tablets, capsules, implants, oral films, and radiator-like oral dosage forms. Polymers have been used for FDM including polyvinyl alcohol, hydroxypropyl methylcellulose, poloxamer 407, polyethylene oxide, and thermoplastic polyurethanes [33]. Thermoplastic polymers are desired for FDM printing. The polymer melts or softens at elevated temperatures. Drug or excipient degradation at high temperatures is the main disadvantage of FDM (Figure 26.6, author's unpublished data). In addition, certain rheological properties of molten polymer should be met for printing by FDM. It was found that at the share rate of the nozzle (1000 s^{-1}), the molten polymer should exhibit viscosity in the range of 100–1000 Pa.s to be printable [34]. The lowest filament stiffness requirement was 80 kg/mm^2 for the Prusa I3 MK3S 3D printer [35].

26.10.3 Digital Light Processing

Digital light processing (DLP) employs ultraviolet light to form a 2D object. In contrast to FDM and inkjet printing, the 2D layer forms inside a tank that contains a photoreactive resin. The ultraviolet light initiates polymerization, but the polymerization is continued as long as there is UV light illumination. The DLP printer uses a digital micromirror device (DLD). This is a mirror (2 x 3 cm) made of millions of pixels, such as 20 megapixels, like digital cameras. The DLD device is electronically connected to a computer, and DLD can present an image visible to the naked eye. This is simply achieved through binary data. This means that an image is converted to a two-dimensional array, and each cell of the array can be either '0' or '1'. By '0', the array will be black (not reflecting light) and by '1', the array will be white (reflecting light). If the light is shone on the DLD, the mirror will reflect the light and the image on the DLD will appear on the wall/paper/platform standing opposite to the mirror. Normally, an optical lens focuses the light on the wall/paper/platform.

Depending on the type of the printer, a layer of resin is spread over the print bed, and with the help of DLD and UV light the resin is cured, and a 2D object forms. Then the platform position is changed (may be lowered) and another resin layer is deposited on the 2D object and the curation process re-starts. As photo-polymerization occurs simultaneously in all directions of the platform, DLP printing is much faster than inkjet printing or FDM. DLP is suitable for thermolabile drugs. Photo-reactive polymers have been employed for the pharmaceutical application of DLP, such as poly(ethylene glycol) diacrylate (PEGDA) and poly(ethylene glycol) dimethacrylate (PEGDMA) [36]. A photoinitiator is required for DLP printing such as 2-Hydroxy-4′-(2-hydroxyethoxy)-2-methylpropiophenone. Kadry et al. 2019 employed DLP printing for the formulation of theophylline tablets containing chambers with the width of 500 μm [36].

26.10.4 Stereolithographic 3D Printing

Stereolithographic 3D Printing (SLA 3DP) uses a UV-laser beam to cure a resin. Depending on the printer, a layer of resin is dispersed in a tank and then the print bed is lowered into the tank until a thin layer of the resin stays between the platform and the bottom of the tank (Figure 26.4C). A UV-laser beam is shone from the bottom of the tank to solidify the resin on the platform. The laser movement is dictated by the CAD file in a computer. After the formation of the first layer, the platform is raised and another layer of fresh resin is added to the tank, then the platform is lowered, and a thin layer of the resin stays between the bottom of the tank and the first printed layer. The curation process starts again. SLA 3DP is known for producing objects with extreme details. SLA 3DP uses photocrosslinkable polymers such PEGDA, PEGDMA, poly (2-hydroxyethyl methacrylate) (pHEMA), and poly(-propylene fumarate)/diethyl fumarate (PPF/DEF). Currently, there is no regulatory approved photocrosslinkable polymer for pharmaceutical applications [37]. SLA 3DP has been employed for the formulation of scaffolds containing the drug for tissue engineering, [38] and a bladder device for intravesical drug delivery [39].

26.10.5 Selective Laser Sintering

Selective laser sintering is similar to inkjet printing, but with the difference that a laser beam fuses the powder particles rather than liquid droplets containing a binder. The powder-bed temperature could reach 110 °C [40]. Interesting FDA-approved polymers could be used for SLS 3DP, such as

Kollicoat IR and Eudragit L100-55. A laser absorbing material is required for SLS 3DP such as Candurin® Gold Sheen, which is a food ingredient. SLS 3DP was employed to produce tablets with controlled porosity, allowing preparing immediate or sustained release tablets [40]. Allahham et al. (2020) employed SLS 3DP to produce orodispersible tablets disintegrating within 15 s [41].

26.10.6 Semisolid Extrusion

Semisolid extrusion (SSE) 3DP is based on enclosing a paste inside a chamber (a syringe) and applying pneumatic or mechanical pressure on the paste, forcing the material to flow through a nozzle. This causes the formation of a narrow filament. The raster movement of the chamber/nozzle will deposit a 2D pattern on the print bed. Following the completion of one layer, the printer head moves up and another layer is deposited. Typical air pressure is 20–103 kPa. Typical excipients are gelatine, starch, xanthan gum, and HPMC. SSE has been employed for the formulation of orodispersible films. This was achieved by printing the paste on the heated bed (70 °C), which led to the solidification of the layer(s). Films achieved a disintegration time of 40 s [42]. SSE may be used for printing formulations that increase adherence to medication in children (Figure 26.3).

26.11 WHAT HAVE 3DP METHODS BEEN IN THE MARKET/CLINICAL EVALUATIONS?

Currently, the production of Spritam tablets is the only pharmaceutical product on the market. An inkjet 3DP is utilized for the production of the tablets. Spritam tablets contain levetiracetam as the active ingredient, which is an antiepileptic drug. The formulation is available as 250 mg, 500 mg, 750 mg, and 1000 mg tablets. The tablet is round, white to off-white color with a spearmint flavor. The cost of Spritam 250 mg tablets is around $642 for a supply of 60.

Candy 3DP was employed to produce chewable printlets of isoleucine for delivery of personalized doses (50–200 mg) in children with maple syrup urine disease. Chewable printlets were made in different colors and flavors, which were welcomed by patients [43]. Currently, M3DIMAKER™ is the world's first GMP 3D printer for personalized medicine [44]. This 3DP has not been used yet in clinical trials.

26.12 WHAT ARE THE CHALLENGES FOR THE NOVEL 3DP METHODS TO GET TO THE MARKET?

Novel 3D methods are continuously developing for the formulation of pharmaceutical dosage forms. These need to reach patients to improve their quality of life. It is estimated that the market for 3D-printed pharmaceuticals will reach £1.68b by 2027 [45]. However, there are barriers to these novel technologies compared to traditional and established methods. These are explained in the following.

First, the 3D printers should comply with cGMP requirements, i.e., they must be qualifiable. Although there are no specific details about this term, generally, 3D printers should meet the following requirements:

1. The design of 3D printers should allow easy cleaning.
2. 3D printers must not have a negative impact on product quality.
3. 3D printers must comply with technical rules, i.e., any relevant pharmacopeia specifications, or any relevant ISO requirements.
4. 3D printers must be suitable for their purposes. This means that 3D printers should produce printlets with the desired weight uniformity, content uniformity, predetermined disintegration time (if relevant), and predetermined level of impurities (comes from the degradation of the product during printing or any contamination from the printer head or print bed).

Second, from a manufacturer's point of view, three major risk areas have been identified related to the 3D printing of pharmaceuticals:[46]

1. Product liability risk; product fits its purposes all the time and is free from printing defects [47] if it is locally produced.
2. Counterfeit risk (fake products cannot be copied from the original product).
3. Safety and efficacy risk (the manufacturing process does not put the health of patients at further risk and produced product has desired efficacy).

It appears each 3D-printing method will require separate regulatory requirements due to considerable variations in the 3D-printing methods.

26.13 USE OF 3DP IN PRE-FORMULATION STUDIES

3DP has already been employed in the pre-formulation studies. 3DP FDM has been utilized to manufacture oropharyngeal models to understand the performance of inhalers in patients [48]. The live oropharyngeal models were prepared from subjects with different anatomical sizes [49]. 3D printing allowed to manufacture of the oropharyngeal models in four different sections and evaluate the drug deposition distribution in the mouth and throat. In addition, 3D printing has been employed for the manufacture of nasal casts for pharmaceutical products intended for nose-to-brain delivery [50]. 3D printing allows modifying the nasal casts, for example, to erase the sinuses and make them simpler without significant deviation from global drug distribution in the nasal cavity. Generally, there are two types of casts: standard and individual. The standard casts are usually made of stainless steel, and these tend to represent an average person. These types of casts are robust and may be manufactured through the mold injection process. The production number should be high to reduce the cost of each model. However, these models do not represent the inter-subject variation. While the availability of digital data for a selection of subjects allows researchers around the world either print the models or add new data to the existing collection.

It has been suggested that 3DP could be used in other aspects of pre-formulation studies. For example, 3DP could be employed for the fabrication of miniature reaction vessels for high throughput drug discovery. As 3DP allows for the preparation of a wide range of doses, then 3DP could enable the identification of clinical go/no-go decisions at the early stages of the drug development process. Inkjet 3DP of solid oral dosage forms would have an immediate indication in these early pre-formulation studies, by considering the regulatory approval of Spritam. If inkjet 3DP is found not suitable, then FDM 3DP may be employed for the production of tablets with desired hardness.

26.14 CONCLUSION AND FUTURE TRENDS

3DP of pharmaceuticals may meet the new requirements from the medicines, in terms of providing the right dose to a patient and reducing polypills. Extensive research has been conducted for understanding and optimization of 3DP for the formulation of pharmaceutical dosage forms. Certainly, 3DP can provide novel dosage forms that are hard to achieve by the conventional pharmaceutical manufacturing methods. This could lead to better patient compliance and treatment outcomes. The future direction needs immediate attention in the following areas for the 3DP of pharmaceutical dosage forms.

1. **Material:**
   Currently, a limited number of excipients is available for pharmaceutical 3DP. Most of the excipients are not approved by regulatory bodies for use in 3DP of pharmaceutical dosage forms. Therefore, extensive research is required to build a library of excipients, which can be used with a variety of 3DP methods.

2. **Stability:**
   API or excipients may degrade during the 3DP process. The chemical changes should be identified for both excipients and APIs in the process of 3DP. Most excipients are in the solid state at room temperature, but APIs and excipients may interact when they melt during 3DP.
3. **Speed:**
   3DP owes its novelty partly due to the formation of highly detailed 2D objects at a time. This comes with the cost of slowing the production process. Although it may be suggested that several 3D printers would solve the problem, companies avoid this option, firstly due to the capital investment, and secondly the maintenance and troubleshooting that each printer may require. Therefore, 3D printers should be developed that print several objects at the same time.
4. **Robustness:**
   3D printers should be designed to be robust, i.e., this equipment should tolerate expected variation in environmental conditions, raw material, and human factors. For example, the heated print bed may be required for FDM 3DP. However, variation in the environmental temperature may lead to the separation of the printlet from the print bed and prevent the formation of a complete object.

REFERENCES

1 N.A. Charoo, Z. Rahman, Integrating QbD tools for flexible scale-up batch size selection for solid dosage forms, *Journal of Pharmaceutical Sciences*, 109 (2020) 1223–1230.

2 EMA, Guideline on manufacture of the finished dosage form (2017).

3 K. Matsunami, T. Miyano, H. Arai, H. Nakagawa, M. Hirao, H. Sugiyama, Decision support method for the choice between batch and continuous technologies in solid drug product manufacturing, *Industrial and Engineering Chemistry Research*, 57 (2018) 9798–9809.

4 WHO, Median age population https://population.un.org/wpp/Download/Standard/Population/, 2020.

5 S. Patris, Polymedication among elderly patients Preventing drug related problems and inappropriate medication, *Journal de Pharmacie de Belgique* 3 (2016) 4–9.

6 N. Keikhosravi, S.Z. Mirdamadian, J. Varshosaz, A. Taheri, Preparation and characterization of polypills containing aspirin and simvastatin using 3D printing technology for the prevention of cardiovascular diseases, *Drug Development and Industrial Pharmacy*, 46 (2020) 1665–1675.

7 F.R. Vogenberg, C. Isaacson Barash, M. Pursel, Personalized medicine: part 1: evolution and development into theranostics, *Pharmacy & Therapeutics,* 35 (2010) 560–576.

8 M.J. Mitchell, M.M. Billingsley, R.M. Haley, M.E. Wechsler, N.A. Peppas, R. Langer, Engineering precision nanoparticles for drug delivery, *Nature Reviews Drug Discovery*, 20 (2021) 101–124.

9 V.E. Bouhairie, A.C. Goldberg, Familial hypercholesterolemia, *Cardiology Clinics*, 33 (2015) 169–179.

10 A.A. Akil, E. Yassin, A. Al-Maraghi, E. Aliyev, K. Al-Malki, K.A. Fakhro, Diagnosis and treatment of type 1 diabetes at the dawn of the personalized medicine era, *Journal of Transnational Medicine*, 19 (2021) 137.

11 A.M. Vicente, W. Ballensiefen, J.I. Jönsson, How personalised medicine will transform healthcare by 2030: the ICPerMed vision, *Journal of Transnational Medicine*, 18 (2020) 180.

12 K. Ravvaz, J.A. Weissert, C.T. Ruff, C.-L. Chi, P.J. Tonellato, Personalized anticoagulation: Optimizing warfarin management using genetics and simulated clinical trials, *Circulation: Cardiovascular Genetics,* 10 (2017) e001804.

13 H.J. Woerdenbag, J.C. Visser, M. Leferink Op Reinink, R.R. van Orsoy, A.C. Eissens, P. Hagedoorn, H. Dijkstra, D.P. Allersma, S.W. Ng, O. Smeets, H.W. Frijlink, Performance of tablet splitters, crushers, and grinders in relation to personalised medication with tablets, *Pharmaceutics,* 14 (2022) 320.

14 S.F. Gharaibeh, L. Tahaineh, Effect of different splitting techniques on the characteristics of divided tablets of five commonly split drug products in Jordan, *Pharmacy Practice (Granada)*, 18 (2020) 1776.

15 P. Foundation, *The personalised medicine technology landscape* (2018).

16 D.A. van Riet-Nales, M.E. Doeve, A.E. Nicia, S. Teerenstra, K. Notenboom, Y.A. Hekster, B.J.F. van den Bemt, The accuracy, precision and sustainability of different techniques for tablet subdivision: Breaking by hand and the use of tablet splitters or a kitchen knife, *International Journal of Pharmaceutics*, 466 (2014) 44–51.

17 G. Binson, C. Sanchez, K. Waton, A. Chanat, M. Di Maio, K. Beuzit, A. Dupuis, Accuracy of dose administered to children using off-labelled or unlicensed oral dosage forms, *Pharmaceutics*, 13 (2021) 1014.

18 V.M. Vaz, L. Kumar, 3D printing as a promising tool in personalized medicine, *AAPS PharmSciTech*, 22 (2021) 49.

19 Y.S. Chun, K. Byun, B. Lee, Induced pluripotent stem cells and personalized medicine: current progress and future perspectives, *Anatomy & Cell Biology*, 44 (2011) 245–255.

20 C.S. Lee, E.S. Bishop, R. Zhang, X. Yu, E.M. Farina, S. Yan, C. Zhao, Z. Zheng, Y. Shu, X. Wu, J. Lei, Y. Li, W. Zhang, C. Yang, K. Wu, Y. Wu, S. Ho, A. Athiviraham, M.J. Lee, J.M. Wolf, R.R. Reid, T.C. He, Adenovirus-mediated gene delivery: Potential applications for gene and cell-based therapies in the new era of personalized medicine, *Genes & Diseases*, 4 (2017) 43–63.

21 S. Schaly, M. Ghebretatios, S. Prakash, Baculoviruses in gene therapy and personalized medicine, *Biologics: Targets & Therapy*, 15 (2021) 115–132.

22 A. Basit, Recent innovations in 3D-printed personalized medicines: an interview with Abdul Basit, *Journal of 3D Printing in Medicine,* 4 (2020) 5–7.

23 N.G.A. Willemen, M.A.J. Morsink, D. Veerman, C.F. da Silva, J.C. Cardoso, E.B. Souto, P. Severino, From oral formulations to drug-eluting implants: using 3D and 4D printing to develop drug delivery systems and personalized medicine, *Bio-Design and Manufacturing*, 5 (2022) 85–106.

24 A. Melocchi, M. Uboldi, N. Inverardi, F. Briatico-Vangosa, F. Baldi, S. Pandini, G. Scalet, F. Auricchio, M. Cerea, A. Foppoli, A. Maroni, L. Zema, A. Gazzaniga, Expandable drug delivery system for gastric retention based on shape memory polymers: Development via 4D printing and extrusion, *International Journal of Pharmaceutics*, 571 (2019) 118700.

25 Z. Faraji Rad, P.D. Prewett, G.J. Davies, High-resolution two-photon polymerization: the most versatile technique for the fabrication of microneedle arrays, *Microsystems & Nanoengineering*, 7 (2021) 71.

26 V. Hahn, T. Messer, N.M. Bojanowski, E.R. Curticean, I. Wacker, R.R. Schröder, E. Blasco, M. Wegener, Two-step absorption instead of two-photon absorption in 3D nanoprinting, *Nature Photonics*, 15 (2021) 932–938.

27 T. Ehtezazi, M. Algellay, Y. Islam, M. Roberts, N.M. Dempster, S.D. Sarker, The application of 3D printing in the formulation of multilayered fast dissolving oral films, *Journal of Pharmaceutical Sciences,* 107 (2018) 1076–1085.

28 M. Kyobula, A. Adedeji, M.R. Alexander, E. Saleh, R. Wildman, I. Ashcroft, P.R. Gellert, C.J. Roberts, 3D inkjet printing of tablets exploiting bespoke complex geometries for controlled and tuneable drug release, *Journal of Controlled Release*, 261 (2017) 207–215.

29 P. Robles-Martinez, X. Xu, S.J. Trenfield, A. Awad, A. Goyanes, R. Telford, A.W. Basit, S. Gaisford, 3D Printing of a multi-layered polypill containing six drugs using a novel stereolithographic method, *Pharmaceutics*, 11 (2019) 274.

30 J. Li, M. Wu, W. Chen, H. Liu, D. Tan, S. Shen, Y. Lei, L. Xue, 3D printing of bioinspired compartmentalized capsular structure for controlled drug release, *Journal of Zhejiang University Science B*, 22 (2021) 1022–1033.

31 W.E. Katstra, R.D. Palazzolo, C.W. Rowe, B. Giritlioglu, P. Teung, M.J. Cima, Oral dosage forms fabricated by Three Dimensional Printing™, *Journal of Controlled Release*, 66 (2000) 1–9.

32 K. Sen, A. Manchanda, T. Mehta, A.W.K. Ma, B. Chaudhuri, Formulation design for inkjet-based 3D printed tablets, *International Journal of Pharmaceutics*, 584 (2020) 119430.

33 C. Parulski, O. Jennotte, A. Lechanteur, B. Evrard, Challenges of fused deposition modeling 3D printing in pharmaceutical applications: Where are we now?, *Advanced Drug Delivery Reviews*, 175 (2021) 113810.

34 M. Elbadawi, T. Gustaffson, S. Gaisford, A.W. Basit, 3D printing tablets: Predicting printability and drug dissolution from rheological data, *International Journal of Pharmaceutics*, 590 (2020) 119868.

35 P. Xu, J. Li, A. Meda, F. Osei-Yeboah, M.L. Peterson, M. Repka, X. Zhan, Development of a quantitative method to evaluate the printability of filaments for fused deposition modeling 3D printing, *International Journal of Pharmaceutics*, 588 (2020) 119760.

36 H. Kadry, S. Wadnap, C. Xu, F. Ahsan, Digital light processing (DLP) 3D-printing technology and photoreactive polymers in fabrication of modified-release tablets, *European Journal of Pharmaceutical Sciences: Official Journal of the European Federation for Pharmaceutical Sciences*, 135 (2019) 60–67.

37 S. Deshmane, P. Kendre, H. Mahajan, S. Jain, Stereolithography 3D printing technology in pharmaceuticals: a review, *Drug Development and Industrial Pharmacy*, 47 (2021) 1362–1372.

38 M. Vehse, S. Petersen, K. Sternberg, K.-P. Schmitz, H. Seitz, Drug delivery from Poly(ethylene glycol) diacrylate scaffolds produced by DLC based micro-stereolithography, *Macromolecular Symposia*, 346 (2014) 43–47.

39 X. Xu, A. Goyanes, S.J. Trenfield, L. Diaz-Gomez, C. Alvarez-Lorenzo, S. Gaisford, A.W. Basit, Stereolithography (SLA) 3D printing of a bladder device for intravesical drug delivery, *Materials Science and Engineering: C*, 120 (2021) 111773.

40 F. Fina, A. Goyanes, S. Gaisford, A.W. Basit, Selective laser sintering (SLS) 3D printing of medicines, *International Journal of Pharmaceutics*, 529 (2017) 285–293.

41 N. Allahham, F. Fina, C. Marcuta, L. Kraschew, W. Mohr, S. Gaisford, A.W. Basit, A. Goyanes, Selective laser sintering 3D printing of orally disintegrating printlets containing ondansetron, *Pharmaceutics,* 12 (2020) 110.

42 J. Elbl, J. Gajdziok, J. Kolarczyk, 3D printing of multilayered orodispersible films with in-process drying, *International Journal of Pharmaceutics*, 575 (2020) 118883.

43 A. Goyanes, C.M. Madla, A. Umerji, G. Duran Piñeiro, J.M. Giraldez Montero, M.J. Lamas Diaz, M. Gonzalez Barcia, F. Taherali, P. Sánchez-Pintos, M.L. Couce, S. Gaisford, A.W. Basit, Automated therapy preparation of isoleucine formulations using 3D printing for the treatment of MSUD: First single-centre, prospective, crossover study in patients, *International Journal of Pharmaceutics*, 567 (2019) 118497.

44 J.J. Ong, A. Awad, A. Martorana, S. Gaisford, E. Stoyanov, A.W. Basit, A. Goyanes, 3D printed opioid medicines with alcohol-resistant and abuse-deterrent properties, *International Journal of Pharmaceutics*, 579 (2020) 119169.

45 S. Mohapatra, R.K. Kar, P.K. Biswal, S. Bindhani, Approaches of 3D printing in current drug delivery, *Sensors International*, 3 (2022) 100146.

46 T. Ehtezazi, M. Algellay, A. Hardy, Next steps in 3D printing of fast dissolving oral films for commercial production, *Recent Patents on Drug Delivery & Formulation*, 14 (2020) 5–20.

47 I.-B. Dumitrescu, D. Lupuliasa, C.M. Drăgoi, A.C. Nicolae, A. Pop, G. Şaramet, D. Drăgănescu, The age of pharmaceutical 3d printing. Technological and therapeutical implications of additive manufacturing, *Farmacia*, 66 (2018) 365–389.

48 T. Ehtezazi, K.W. Southern, D. Allanson, I. Jenkinson, C. O'Callaghan, Suitability of the upper airway models obtained from MRI studies in simulating drug lung deposition from inhalers, *Pharmaceutical Research*, 22 (2005) 166–170.

49 T. Ehtezazi, I. Saleem, I. Shrubb, D.R. Allanson, I.D. Jenkinson, C. O'Callaghan, The interaction between the oropharyngeal geometry and aerosols via pressurised metered dose inhalers, *Pharmaceutical Research*, 27 (2010) 175–186.

50 L. Deruyver, C. Rigaut, P. Lambert, B. Haut, J. Goole, The importance of pre-formulation studies and of 3D-printed nasal casts in the success of a pharmaceutical product intended for nose-to-brain delivery, *Advanced Drug Delivery Reviews*, 175 (2021) 113826.

27 Materials and Challenges of 3D Printing for Regenerative Medicine Applications

Shailly H. Jariwala[1,2*], *Richard C. Steiner*[1,2], *Jack T. Buchen*[1,2], *and Luis M. Alvarez*[2,3]
[1]The Henry M. Jackson Foundation for the Advancement of Military Medicine, Inc., 6720-A Rockledge Drive, Suite 100, Bethesda, MD 20817, USA
[2]The Center for Rehabilitation Sciences Research, Department of Physical Medicine and Rehabilitation, Uniformed Services University of Health Sciences, 4301 Jones Bridge Rd, Bethesda, MD 20814, USA
[3]Lung Biotechnology PBC, 1040 Spring Street, Silver Spring, MD 20910, USA
*Corresponding author: shailly.jariwala.ctr@usuhs.edu

CONTENTS

27.1 INTRODUCTION

Three-dimensional printing (3DP) is defined as the process of depositing materials layer by layer and then fusing these layers before the next layer is added, as a bottom-up approach to manufacturing. The term 3D printing is often used interchangeably with additive manufacturing (AM) and was first suggested as a potential means for rapid manufacturing as early as the 1970s when Wyn Kelly Swainson in 1977 filed a patent for a device that could 3D print using lasers to crosslink liquid

DOI: 10.1201/9781003296676-27

monomers into a 3D design. To date, 3D printing and similar AM processes within the scope of regenerative medicine applications have focused on the creation of cell-free scaffolds for implantation in surgery with the hope of successful integration and propagation of new tissue into the scaffold's matrix [1]. Investigation into AM technology that incorporates and can assemble living cells, biomaterials, and biochemicals in functional tissue-like structures begins to take hold. We see the first example of 3D printing of cells into a 2D/3D format by Klebe R. J. using cytoscribing technology in an inkjet printer in 1988 to print cells into specific 2D patterns that can be built one layer after another into 3D structures [2]. It is here that 3D printing has evolved to differentiate from the conventional process of 3D printing scaffolds followed by seeding them with cells, to a simultaneous process that creates 3D-bioprinted matrix and cells concurrently, what we now term as 3D bioprinting (3DBP). In the following years, we have seen the introduction of many new bioprinting products such as the artificial liver and cartilage in 2012, coaxial printing in 2015, rapid continuous optical printing using 3D digital light processing (DLP) in 2016, and collagen human hearts based on the freeform reversible embedding of suspended hydrogels (FRESH) technology in 2020 [3]. Considering the great progress already made in this field, the prospect of utilizing 3DBP for creating functional tissues and organs is now a real possibility in the near future.

3D bioprinting is the emerging subfield of 3D printing. The term 'bioprinting' has been broadly used to refer to the printing of either biologically inert materials or biomaterials that incorporate living cells and other biologics in 3D constructs using AM processes [4]. The concept of using living components to create 3D structures is what fundamentally differentiates it from other types of 3D printing. While 3D printing and bioprinting use the same AM techniques to build 3D objects, there is a vastly different set of conditions, standards, and requirements that need to be addressed for 3D bioprinting to be successful compared to a typical 3D-print design. The critical component of 3DBP is the bioink, which is important for the development of functional tissue structures. The key characteristics for the selection of suitable materials for bioprinting involve the printability, mechanical, biocompatibility, biodegradability, and sterility capabilities of the material. Bioinks are printable materials that can be composed of either only cells (scaffold-free bioinks) or a combination of cells and scaffold biomaterials (scaffold-based bioinks). Scaffold-free bioinks are printed directly to create a 3D tissue in a process that imitates embryonic development, while scaffold-based bioinks are primarily made of soft biodegradable hydrogels allowing the encapsulated cells to grow and form desired 3D structures. The selection of biomaterials for bioinks is thus very critical for bioprinting.

In this chapter, the various 3DBP modalities used currently for regenerative applications are defined, elaborated, and compared. The advantages and limitations of each modality are discussed. Next, we have summarized the use of 3DBP in the generation of various tissue constructs and addressed the existing limitations of 3DBP. Finally, we will conclude with the future outlook of 3DBP and the evolution of 4D bioprinting for the fabrication of tissues for regenerative applications.

27.2 3D BIOPRINTING MODALITIES

Several 3DBP modalities exist: inkjet, extrusion, digital light processing, and laser-assisted bioprinting, as shown in Figure 27.1. In the following sections, we will discuss each modality in depth as they relate to scaffold fabrication and contrast their advantages and disadvantages.

27.2.1 Inkjet Bioprinting

Inkjet bioprinting is a 3DBP modality built upon the technology of inkjets, described by Lord Rayleigh in 1878. From Rayleigh's theory, continuous inkjet printing technology was developed wherein a droplet flow of controllable size can be produced. Shortly after, drop-on-demand (DOD) technology was developed, allowing the ejection of droplets only when stimulated by a digital signal. DOD technology was later commercialized to produce inkjet office printers, and, in recent years, has been adapted by researchers to allow the printing of bioinks, both with and without cells, with extreme precision [5].

Continuous inkjet printing is not currently utilized for 3DBP as the devices are complex, cannot be precisely controlled, and introduce opportunities for ink contamination [5]. However, DOD inkjet bioprinting has several variations which prove extremely effective for 3DBP. DOD inkjet bioprinting can be subdivided into thermal, piezoelectric, electrostatic, and electrohydrodynamic (Figure 27.1A).

Thermal inkjet bioprinters operate by heating a localized area of ink rapidly for a short time, thereby causing the formation of heat bubbles, and through the subsequent expansion, pushing ink through the inkjet outlet [6]. The heating element warms to 250–350 °C, which may cause cell death for cells in the immediate vicinity, however, the overall ink temperature does not raise more than

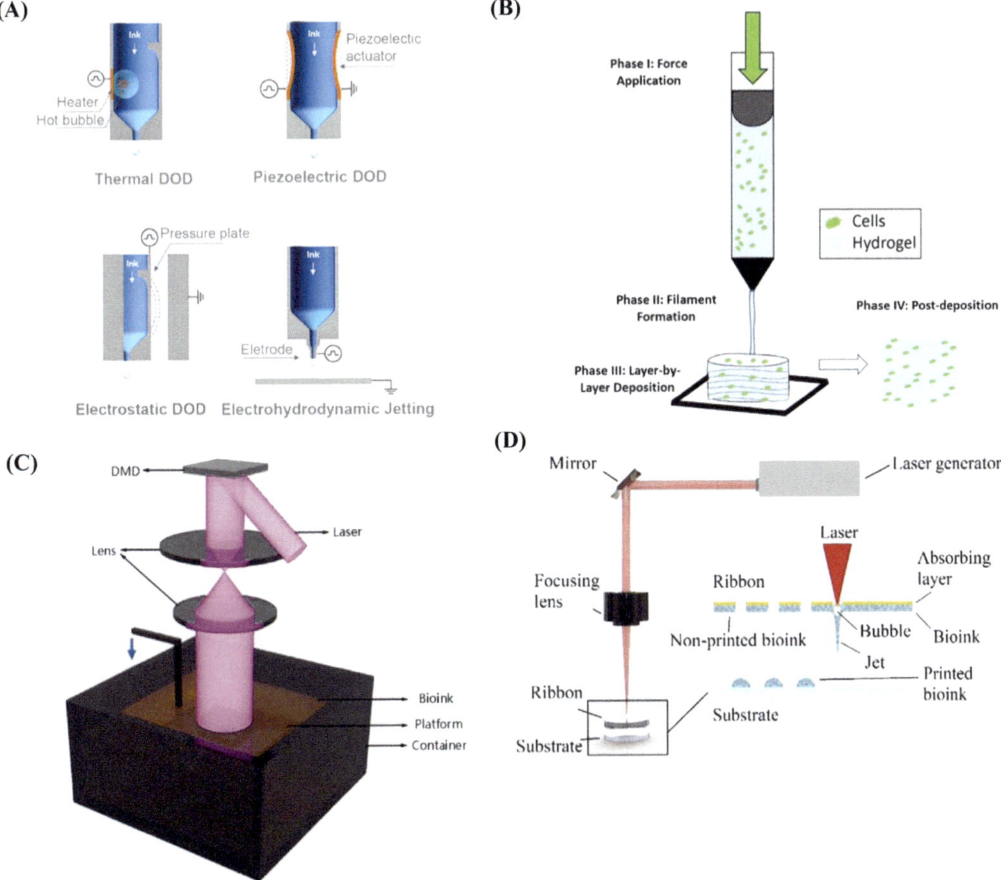

FIGURE 27.1 A schematic of various 3DBP modalities. (A) Thermal, piezoelectric, electrostatic, and electrohydrodynamic inkjet bioprinting modalities. Reprinted with permission [5], Copyright 2020, American Chemical Society. (B) Extrusion bioprinting, various phases of extrusion printing process are highlighted. Reprinted with permission [8], Copyright 2021, Elsevier. (C) DLP bioprinting. Adapted with permission [3]. Copyright Zeming Gu, Jianzhong Fu, Hui Lin, Yong He, some rights reserved; exclusive licensee Elsevier. Distributed under a Creative Commons Attribution License 4.0 (CC BY). (D) Laser-assisted bioprinting. Reprinted with permission [9], Copyright 2021, John Wiley & Sons.

10 °C, allowing for cell viability of around 90% [5]. The diameter of the inkjet outlet is typically about 50 μm, a comparable size to many cells, which may cause shear stress during printing. This being said, the nozzle size allows for the printing of droplets ranging from 30–80 μm, resulting in high resolution [5]. Thermal inkjet bioprinting is favorable for its low cost, relative simplicity, and high-speed, but can be difficult to tune due to the changes in bioink viscosity resulting from rapid heating and vaporization [6].

Another variation is piezoelectric inkjet bioprinting, which operates by utilizing piezoelectric ceramics, generating a precise displacement in the bioink reservoir corresponding to the applied voltage [7]. Piezoelectric 3DBP can have a customized printhead size of 19–120 μm, allowing for control of drop size from 50–100 μm [6]. The viability of cell-laden bioinks printed using the piezoelectric method may be less than that of the thermal method due to ultrasonic-induced cell lysis, or excessive force applied during ink ejection. Still, piezoelectric inkjet bioprinting is often able to achieve cell viability of 70–90% [5].

Electrostatic inkjet bioprinting utilizes an ink-chamber deformation to induce an increase in pressure. In this method, however, the sidewall of the ink reservoir is displaced by an electrode plate that attracts the sidewall through static electricity. This increases in volume, and a subsequent vacuum, drawing ink from an external reservoir to fill the void. When the electrode plate is deactivated, the ink reservoir sidewall returns to its original position, causing an increase in pressure [5] Electrostatic inkjet bioprinters offer precise control, offering nozzle diameters of 8–160 picoliters, and resolutions of up to 2880 x 1140 dpi [5]. Due to the small nozzle size, shear stress must be considered when printing with cell-laden bioinks, however, complications introduced by other inkjet methods, such as heat and sonication, are not limiting factors.

Electrohydrodynamic jet bioprinting is the final variation of the inkjet bioprinting method. It utilizes an applied pressure to produce a Taylor cone at the end of the inkjet outlet, to which a voltage is applied, producing a precise jet of bioink droplets. This modality is similar in its application to that of electrospinning, as the voltage is applied to the area containing the Taylor cone, and the receiving substrate is grounded to encourage the collection of the bioink [5]. The main advantage of electrohydrodynamic jet bioprinting is that the jet can be smaller in diameter than the nozzle of the ink reservoir, as the jet is produced from the tip of the Taylor cone, rather than the nozzle. This allows for extremely precise printing, with resolution reaching 2–60 μm in diameter, including some high viscosity inks which require a larger nozzle size [5].

In general, inkjet bioprinters hold several key advantages over other 3DBP modalities: high-throughput, small drop size, drop-on-demand (DOD), and non-contact printing [5], [6]. Inkjet prints are highly efficient and capable of printing from a virtually unlimited quantity of nozzles simultaneously, allowing for rapid, high-volume printing. In addition, they allow for high resolution, to the extent that researchers can customize drop size to that of a single cell [5]. DOD technology allows inkjet bioprinters the capability of reproducing highly complex and organized structures by deploying multiple bioinks one drop at a time with a resolution as low as 2 μm [5]. Unlike other bioprinting modalities, inkjet bioprinters do not require substrate contact or manipulation during printing, improving resolution and preventing additional contamination.

27.2.2 Extrusion Bioprinting

Extrusion-based printing or bioprinting is a 3DP modality where biomaterials (cell-laden and/or non-cell laden) are extruded as viscous fluids through a syringe using a pneumatic or mechanical pump at precise locations on receiving substrates (e.g., a culture dish, growth medium, or support gel), for fabrication of 3D scaffolds (Figure 27.1B) [3]. It is the most widely used bioprinting modality due to its versatility and affordability. The 3DP biomaterials used for extrusion printing can be cured using ultraviolet radiation, time-dependent solidification, and chemical or photo-crosslinking agent [3]. Printing parameters such as nozzle diameter, extrusion pressure, temperature, and speed influence the final bioprinted structures. Extrusion bioprinting also allows for use of single or multiple syringes containing a combination of biomaterials and/or cell types, enabling the formation of a more compositionally complex scaffold. This grants integration of cells and biomaterials which results in the replication of a more physiologically relevant construct. Pneumatic-driven extrusion systems use compressed air to dispense liquid from the syringes, while piston-driven systems use micro-infusion pumps to push the bioink out of the nozzle. Piston-driven methods provide higher resolution and better printability with semi-solid or solid biomaterials in comparison to pneumatic-driven methods [10]. However, piston-driven methods are volumetrically limited, cost more, and are difficult to clean and disinfect [10].

Currently, numerous extrusion-based bioprinters are available commercially, such as 3D Bioplotter® (EnvisionTEC), NovoGen MMX Bioprinter™ (Organovo), 3DDiscovery (RegenHU), BIO V1 (REGEMAT3D), INKCREDIBLE (CELLINK), BIOBOT ™ and BIOASSEMBLYBOT® (Advanced Solutions), ALLEVI3 (Allevi, 3D Systems). Due to its flexibility, inexpensiveness, and ability to print porous constructs, extrusion printing has been employed by several researchers globally. It has already facilitated bioprinting of cells, tissue constructs, and tissue/organ-on-chip devices, with the expectation of printing functional scaled-up organs in the future. Several tissue constructs have been successfully engineered using extrusion printing, including but not limited to cartilage, blood vessels, bone, skin, liver, nerve, brain, and cardiac tissue constructs [3], [11].

Extrusion-based bioprinting has several advantages and disadvantages with respect to other printing modalities such as inkjet and laser-based technology. A wide variety of bioinks, such as hydrogels, tissue spheroids, micro-carriers, and cell pellets [10] can be used with extrusion printing in comparison to other techniques that can only facilitate printing cell-laden hydrogels. This is possible due to the flexibility in nozzle tip design, larger nozzle diameters utilized, and the ability to extrude bioink in near solid-state (also within a fugitive liquid delivery medium). Additionally, high cell density bioprinting is currently feasible only with extrusion printing with a reasonably lower process-induced cell damage and injury. Most importantly, extrusion printing enables printing anatomically accurate porous structures, which is very difficult using other printing modalities (except with a modified stereolithography technique). It is also the only bioprinting process that is scalable and has the potential to print human-sized organs.

Despite these advantages, extrusion printing does have a few limitations. Foremost, the resolution offered by this technology is limited, with the minimum feature size being greater than 100 μm [11]. This is considerably lower than the resolution offered by other printing modalities. In addition, bioinks must be shear thinning and have a higher viscosity to begin with in order to be extruded in the form of filaments [8]. However, higher viscosities can trigger clogging inside the nozzle tips and require optimization for the diameter of the nozzle tips being utilized. Furthermore, the shear stress experienced by the cells at the walls of the nozzle tip during extrusion induces a significant drop in cell viability within the printed construct [11]. Clogged nozzle tips also demand a higher printing pressure which increases the shear stress experienced by the cells being extruded and subsequently compromises cell viability. Cell viability is also affected in this modality as biomaterials are not printed directly into cell culture media, thus exposing cells to dehydration and lack of nutrients while the entire construct is being printed. Lastly, requirements for gelation during the extrusion process limit the hydrogels that can be used with extrusion bioprinting. Hence, maintaining a fine balance between using a high viscosity ink and cell viability/printability is very critical for extrusion printing. Therefore, the development of new bioinks that are well suited for extrusion printing is still a great need.

27.2.3 Digital Light Processing (DLP)

Digital light processing 3DP is another type of vat polymerization technique, similar to stereolithography (SLA), but uses a laser beam. The DLP is a bottom-to-top process that uses a digital photomask to project a 2D image onto a photocurable liquid resin (bioink) as shown in Figure 27.1 C. DLP systems use a digital photomask which is usually a digital mirror device (DMD), an array of micro-mirrors, to control the curing laser beam [12]. Using DMD allows for a complete layer of resin to be cured at once, making DLP relatively faster than SLA [3]. There are three main components involved in DLP printing: i) a container filled with bioink which is typically a photocurable hydrogel or photosensitive resin which can solidify on exposure to a laser of a specific wavelength (usually ultraviolet (UV) light) or a photoinitiator, ii) a moving platform, and iii) an imaging system above the container (Figure 27.1 C). Starting with the bottom layer first, each new layer is printed and crosslinked above the previous one.

The DLP process requires bioinks with a lower viscosity allowing a faster flow of the monomer solution and a faster relaxation process, resulting in a better quality of prints. Having said that, when live cells are incorporated in DLP printing bioinks must be adequately viscous to prevent cell settling or precipitation during the printing process, to avoid inhomogeneous distribution of cells within the printed construct. Amongst the bioinks used with the DLP platform, polyethylene glycol diacrylate (PEGDA) and gelatin methacryloyl (GelMA) are the most commonly used bioinks for bioprinting tissue models. Compared to PEGDA, printing with GelMA typically leads to a lower printing resolution due to a lower concentration of polymer content. Although, an improved cell attachment with GelMA enables the utilization of a DLP bioprinter for printing constructs with live cells [12]. Additionally, a combination of PEGDA and GelMA bioinks has resulted in achieving an optimal printing resolution as well as cell attachment in the microtissues printed using a DLP platform.

In comparison to SLA, extrusion, and inkjet printing, DLP has a considerable advantage in printing speed ($\sim 10^3$ mm^3/s), a superior printing resolution (down to 10 μm), and easier control over the mechanical properties of bioinks [12]. Additionally, unlike the interfaces present between printed layers in inkjet printing (between droplets) or extrusion printing (between adjacent fibers), DLP technology can fabricate much smoother 3D structures, resulting in enhanced structural integrity. Despite the high resolution attainable by DLP printers, the limited projection size mandatory for achieving this high precision restricts printing to only small-sized objects. Moreover, DLP systems are very expensive and due to their squared voxel produce a distinctive 'boxy' surface finish. The optical transparency of bioinks and their interaction with UV light has also been a bottleneck for creating high-resolution constructs. Thus, the next logical progression for DLP printing would be to improve the optical properties of bioink formulations by the addition of efficient photoinitiators and photo-absorbing molecules.

27.2.4 Laser-Assisted Bioprinting (LaBP)

Laser-assisted bioprinting (LaBP) is a 3DBP modality that consists of a pulse-emitting laser, a refocusing element to refine the laser pulse, and an absorbing medium to convert the laser pulse into a bioink ejection (Figure 27.1 D). LaBP is based on laser-induced forward transfer (LIFT) technology, which is a direct-writing technique utilizing laser irradiation to induce material transfer with high precision and accuracy [13]. The laser is typically emitted perpendicular to the printing substrate, and then reflected 90° by a mirror. Before the laser pulse reaches the absorbing medium, it is passed through a lens, allowing for beam focusing to increase resolution. Upon contact with the absorbing medium, a bubble forms that force the ejection of bioink droplets onto the printing substrate [9], [13]. The absorbing medium can be either absorbing film-assisted LIFT (AFA-LIFT) or matrix-assisted pulse laser evaporation direct writing (MAPLE-DW) [9]. AFA-LIFT is composed of three layers, in the following order: a transparent layer (typically made of glass or quartz), a thin layer of absorbing metal (such as titanium or gold), and the bioink layer [9], [13]. When irradiated, the absorbing metal is vaporized, resulting in the bubble and subsequent droplet jet of bioink necessary for LaBP [13]. MAPLE-DW is composed of only two layers, in the following order: the transparent layer and the bioink. MAPLE-DW typically targets bioink proteins for energy absorption, requiring the use of an ultraviolet (UV) laser. Due to cell viability issues associated with UV exposure, MAPLE-DW LaBP is not well suited to cell-laden bioinks [9].

More recent advancement in LaBP is that of laser-induced backward transfer (LIBT). LIBT was developed by Doucastella et al. [9] to address limiting factors in other LaBP methods: namely issues with print consistency caused by surface tension and evaporation acting on the small layer of bioink in the absorbing medium of AFA-LIFT and MAPLE-DW LaBPs. In this method, there is no energy absorbing layer, and the bioink and printing substrate are switched. Bioink is held in a vat, which mitigates evaporative and surface tension issues caused by low ink volume. The bioink is irradiated, either from below or from above through a transparent substrate, with a femtosecond infrared laser pulse, which forces a small quantity of bioink to squirt up and deposit on the printing substrate [9].

For all forms of LaBP, selection of the laser emission source, as well as the pulse time, may be varied depending on the intended applications. As previously mentioned, infrared (IR) lasers are often preferred as they are less damaging to most bioinks and present less issues of cell viability. In addition, it's essential to consider the thermal impact of various wavelengths and emission times, and to identify a balance when selecting emission parameters. Though IR laser sources are preferred for their prevention of DNA double helix damage to cell-laden bioinks, excessive exposure times may result in cell death from overheating [9].

LaBPs are capable of printing single-cell droplets of bioink, as well as achieving high levels of resolution down to the micron level [9]. In addition, LaBP is capable of high throughput printing and is adaptable in technique and laser parameters. Cell viability is one of the strongest advantages of LaBP, as viability is typically 80–100%. Bioink and cell viability must be considered and can be improved by carefully selecting laser sources and emission time to reduce thermal and radiative cell destruction. The other main source of cell destruction during LaBP is that of shear stress resulting from the rapid acceleration when ejected from the absorbing medium, and negative acceleration when collected on the printing substrate. During printing, the velocity of the bioink jet can reach 500–1000 m/s, and the resulting acceleration from ejection and deposition can range from 10^5–10^9 g [9]. Shear stress from the negative acceleration when collecting on the printing substrate can be mitigated by creating a softer impact for bioinks, thereby improving cell viability. Researchers have accomplished this in the past by coating the printing substrate with a thin layer of liquid or gel. With current methods, there is an inverse relationship between the coating thickness and cell viability, as a thicker coating will reduce impact force to a greater extent, but will also isolate cells and prevent necessary nutrient absorption [9].

27.3 3D BIOPRINTING FOR TISSUE REGENERATION

Being a relatively newer aspect of tissue engineering, 3DBP has unlocked the possibility of creating exceptional biomimicry in the form and function of fabricated tissue substitutes, which has great significance in regenerative medicine. 3DBP is expanding considerably into an enormous industry due to its diverse applications. Broadly, the applications of bioprinting have been divided into two categories: i) tissue regeneration, such as printing of blood vessels, heart valves, nerves, musculoskeletal tissues, liver, and skin; and ii) biomedical applications, including drug screening. In this chapter, we will be focusing on tissue regeneration applications of bioprinting.

27.3.1 3D Bioprinting: Neural Regeneration

The nervous system is often organized by regions such as the central nervous system (CNS), and the peripheral nervous system (PNS). As such, nerve injuries are categorized by their region, etiology, severity, and potential for a return to normal functionality. Due to the complexity of nerve tissue inter-communication, specific nerve fiber connection/orientation, and the body's endogenous response to nerve injury, it is extremely difficult to achieve full recovery from nerve injuries even with the medical intervention [14]. The mechanisms by which nerve injuries can occur are extremely varied with etiologies including systemic illnesses (e.g., diabetes), toxicity, hereditary, autoimmune, infection, and physical trauma [15]. Peripheral nerve injury (PNI) has a higher chance of axonal repair and a return to acceptable neuronal function but only if the injury is small and not greater than 3 cm will it be possible to achieve adequate neuroplasticity [16].

Given the intrinsic limitations for the natural repair of nerve tissue, there are therapeutic approaches that can significantly improve the outcome of successful nerve repair. In neuro regenerative applications, bioprinted artificial scaffolds can be implanted at the site of injury in the hopes of fully encompassing the injury gap/defect and fully integrating with the native tissue to guide new tissue growth, in effect increasing the rate of nerve repair. Neural guide conduits (NGCs) are often created with either synthetic or natural biomaterials and are constructed using a variety of different 3DBP techniques (Figure 27.2). Depending on the material and manufacturing process used it is often very difficult to create exact biomimetic cues in the scaffold architecture and then ensure the design fulfills the criteria for successful conduit performance such as biocompatibility, mechanical stability, degradability, cell/biomolecule support, and mass-transport capable. 3DBP manufacturing has received great interest in developing new neural scaffolds due to the high degree of control it has for exact biomaterial placement in 3D environments.

Hydrogel bioinks based on natural polymers such as collagen, fibrin, gelatin methacrylate (GelMA), alginate, chitosan, and silk have had great success as neural scaffold matrices [18] Excellent biocompatibility, biodegradability, and low inflammatory responses give natural polysaccharides an advantage when supporting neural cell lines while presenting limitations in scaffold matrix mechanical stability unless strengthened by the addition of physical or chemical crosslinkers. Synthetic polymers have also shown much potential in 3D printing hydrogels with successful designs for neural applications having been reported using materials such as polyethylene oxide (PEO), polyurethane (PU), polyethylene glycol diacrylate (PEGDA), and polyethylene glycol dimethacrylate (PEGDMA), polypropylene fumarate (PPF), and polypropylene maleate (PPM). Synthetic polymers have a distinct advantage over natural polymers in that their mechanical and degradation profiles can be easily tunable in ink form to meet neural conduit design requirements, allowing for a greater

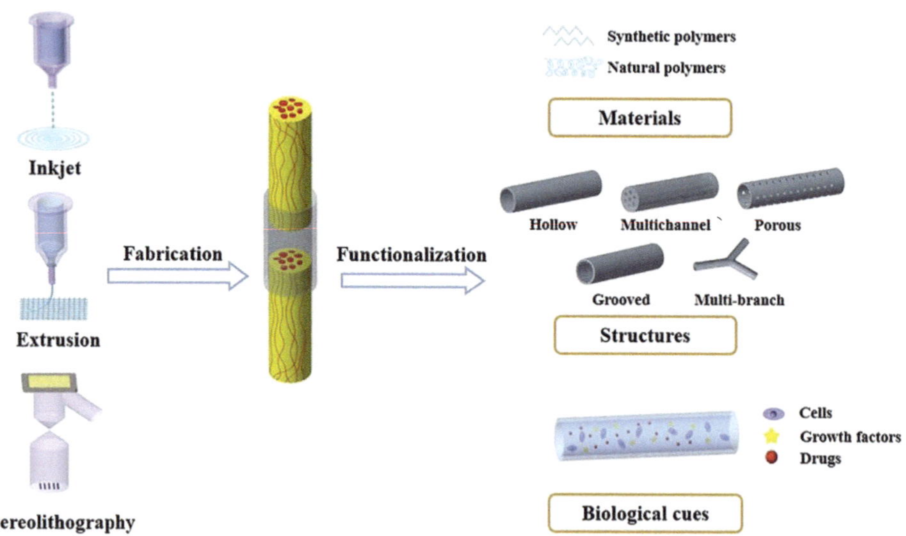

FIGURE 27.2 Schematic diagram of 3D printing technologies for NGCs fabrication and the strategies for the construction of functional NGCs. Reprinted with permission [17], Copyright 2021, Oxford University Press.

degree of control when 3D printing neural conduits [19]. Pure synthetic polymer bioinks however can present an issue for neural conduits designed to support specific cell lines and for prolonged nerve repair applications, as they are usually not as biocompatible as natural polymers and can present further complications by inducing chronic inflammation and foreign body reactions at the site of the implant. By combining both natural and synthetic polymers in ink form 3D-printed hydrogels are capable of incorporating the advantages of both natural and synthetic polymers while addressing these limitations for improved neural conduit designs. Examples of successful composite hydrogel bioinks used in neural conduit designs include gelatin-alginate, PEGDA-GelMA, PEG-PEGDA, and GelMA-chitosan microspheres.

An additive approach to further improving 3DBP neural conduit designs is the incorporation of materials that have conductive properties so that action potential or induced electrical stimulation can propagate across the nerve gap. Polypyrrole (PPy), polyaniline (PANI), and poly (3,4-ethylenedioxythiophene) (PEDOT) are currently the most promising conductive polymers that have shown success in neural conduit designs composed of hydrogels and synthetic polymers [20]. Carbon nanomaterials have also been widely used in the manufacturing of 3D-printed NCs due to their unique electrical, mechanical, and biological properties. Currently, the most utilized forms of carbon used in bioinks have been graphene, single-walled carbon nanotubes (SWCNT), and multi-walled carbon nanotubes (MWCNT). Most often these materials are included in composite ink formulations to improve the electrophysiological properties of conductively inert links, many of which have been discussed in the previous sections [21].

Inkjet printing has advantages in resolution, and studies have shown that the precise positioning of materials by inkjet can promote micropatterned topographies that when incorporated into NGC designs can provide necessary structural cues for guiding nerve growth [22]. Ink jetting has also shown to be compatible with 3DBP applications for nerve regeneration, as demonstrated by the ability to print neural stem cells (NGC) and Schwan cell suspensions while retaining excellent cell viability after piezoelectric jetting [23]. The amount of biocompatible bioinks that can be utilized with inkjet printing is substantial, and we see that a number of those inks have also shown success in NGC designs. Such examples include Poly(e-caprolactone) (PCL) in combination with different materials for improved material properties including poly(acrylic acid) (PAA), glacial acetic acid (GAA), and reduced graphene O]oxide (rGO) [24]. While collagen and fibrin inks have demonstrated improved growth factor and cell support [25].

Extrusion-based printing is one of the most widely used technologies to construct 3DBP neural conduits. The variations in ink viscosities give extrusion printing an advantage over ink jetting when creating precise print patterns, and guidance channels that have been shown to greatly improve neural cell development on the conduit surface. Examples of successful extrusion bioinks include synthetic polymers like PU, PCL, PLGA, and PLA while hydrogel designs have used gelatin-alginate, and GelMA/GC-MS. There are also examples of extrusion bioink composites that use carbon-based materials such as graphene + gelatin hydrogel/PLA, and graphene + PLGA [21]. These bioinks have also shown the ability to support a multitude of neural-based cell lines and maintain high viability during and after the bioprinting step, including Schwann cells, NSCs (Neural Stem Cells), induced pluripotent stem cells (iPSC), and human fibroblast [26].

Light-based forms of 3DP such as SLA and DLP are not as frequently utilized for neural conduits as compared to ink jetting and extrusion modalities but are quickly gaining great interest in 3DP neural applications. While they are limited in maintaining excellent cell viability due to the photoinitiators and light sources needed for crosslinking, they have shown further improvements in print fidelity and resolution, with conduit topographic features reaching details at the 100 μm scale allowing for even greater control in cell manipulation through biomimicry and chemotaxis. SLA has been demonstrated successfully in neural conduit designs including bioinks composed of PEGDA, and PEG-PEGDA hydrogels [25], [27]. SLA bioinks have also been demonstrated as composites with conductive polymers and carbon-based materials for neural conduit designs such as PEDOT +

GelMA/PEGDA hydrogel, PPy + PEGDA hydrogel, and MWCNT + PEGDA hydrogel [26]. DLP bioinks have also reported successful neural conduit designs including PEG-PPF, and GelMA multichannel neural conduits co-cultured the neural conduit with PC12 cells [28], [29].

27.3.2 3D Bioprinting: Osteochondral Tissue

Bone regeneration involves investigating methods and techniques to replace damaged bone tissue or improve the rate/quality of bone repair for bone defects caused by bone fractures, osteodegenerative, and tumor diseases. Current techniques for repairing bone defects are based on grafting autologous, allografts, and synthetic grafts. However, site morbidity, variable immunoreaction, and a lack of control in tissue organization and mechanical properties limit these techniques for establishing stable quality bone tissue. With the increased use of grafts for the repair of bone defects and the minimal success of the adequate return of quality bone tissue, there is a need to investigate alternative methods for bone repair [30]. Bioprinting technology has been shown to deeply improve graft performance and the availability of effective synthetic-bone substitutes (Figure 27.3(i) and (ii)). The primary use of 3D-printing modalities in osteogenic repair is in the manufacturing of 3D bone scaffolds that can successfully cover bone defects and integrate them into the native tissue bordering the site, to support new bone tissue growth. These scaffolds are designed with a critical focus on the composition of the scaffold being compatible with the native bone/cartilage tissue and having to withstand a more frequent and greater mechanically stressed microenvironment. Currently, 3D-printed bone scaffolds are primarily created using inkjetting, extrusion, stereolithography, and DLP methods with a few lesser-known methods under investigation such as phase separation, cryogenic, indirect, and powder printing [30].

Bone-like structures have been produced using ink jetting technology. Examples include osteo conduits composed of hydroxyapatite (HA) and tricalcium phosphate droplets deposition onto powders. Ink jetting has also demonstrated cell viability in scaffolds produced using poly(ethylene glycol) dimethacrylate supplemented with HA and osteoinductive ceramics co-printed with bone marrow-derived human mesenchymal stem cells (BMSCs). Extrusion printing has a relatively poor resolution for microscale bone scaffolds, but it is particularly suitable for bioinks with high viscosities and high cell densities [32]. This feature enables the production of 3D bone scaffolds that better resemble the cell density of the native tissue. This method also allows for higher structural integrity by providing a wider range of speeds at which the ink can be deposited. This flexibility in printing parameters gives extrusion printing greater compatibility with different bioinks, including both scaffold-free and scaffold-based inks and inks designed to incorporate live cells that can obtain high levels of structural and functional complexities, such as native cartilages and bone tissue. Such recent examples include (PLA) + acrylonitrile butadiene styrene (ABS), Periosteal derived cells alginate hydrogel + collagen I, II, RGD alginate hydrogels, and HA-GelMA [30].

SLA and DLP printing methods have not been as extensively used for bone scaffold designs compared to extruder/inkjet printing methods. However, light-based printing methods have demonstrated printing resolutions that conventional extruder printers are incapable of replicating, allowing for greater replication of the bone tissue architecture. SLA has demonstrated its ability to create osteochondral (OC) scaffold at the micron scale to support bone tissue using materials such as GelMA + nanocrystalline HA for breast cancer bone defects, PEG hydrogel + nanocrystalline HA, and hydrogel resins (PEG, PEG-diacrylate) [30]. However, based on the potentiator and laser wavelength used, SLA bioinks can be detrimental to the health of osteo-based cell lines making cell-infused prints very difficult. DLP in comparison has less chance for cell disruption and apoptosis due to the use of visible light for photo-crosslinking as exemplified by the use of DLP to produce radially orientated hydrogel cartilage constructs assisted with cell infiltration [33]. DLP has also been used to create the entire OC scaffold as demonstrated by the use of PEG materials combined with bovine cartilage ECM [34].

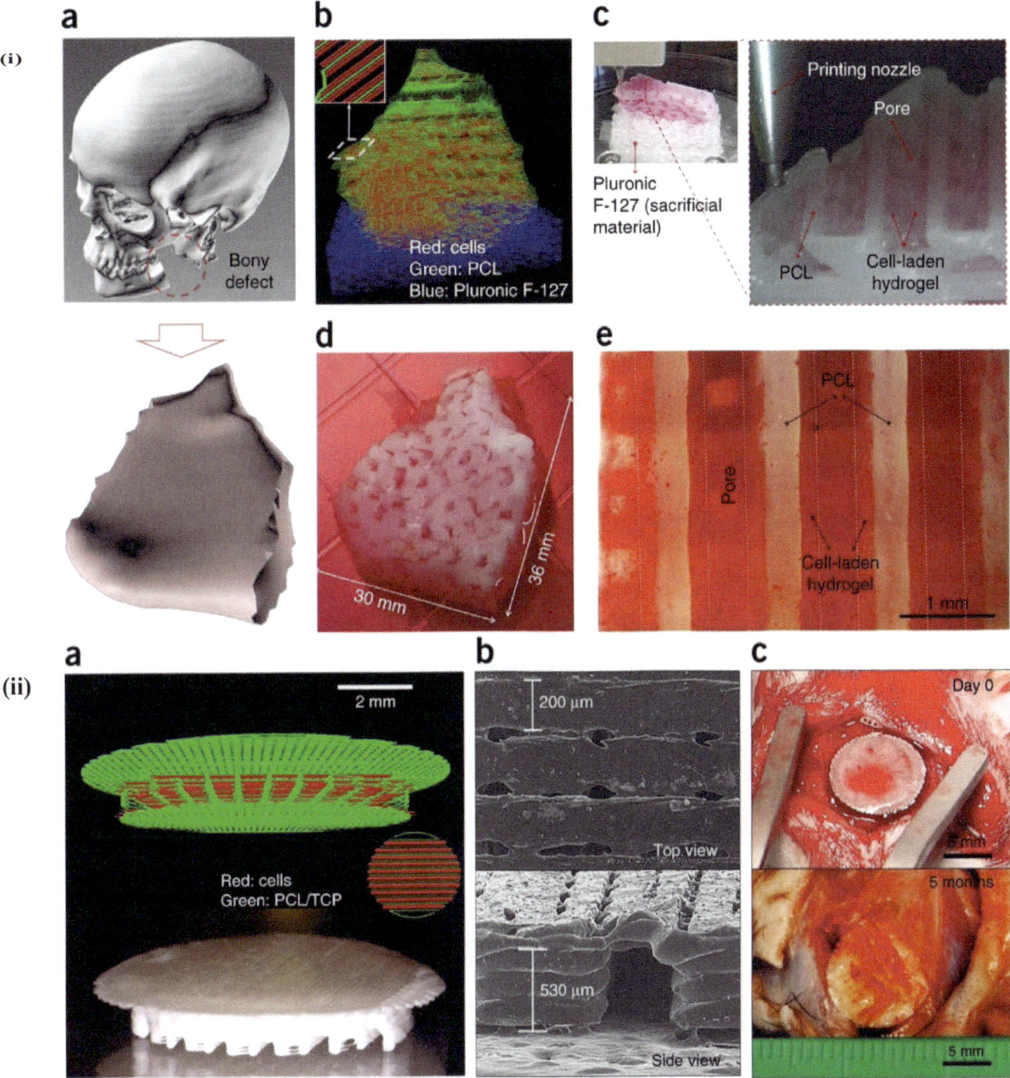

FIGURE 27.3 (i). Example mandible bone reconstruction. (a) 3D CAD model recognized a mandible bony defect from human CT image data. (b) Visualized motion program was generated to construct a 3D architecture of the mandible bone defect using CAM software developed by our laboratory. Lines indicate the dispensing paths of PCL, Pluronic F-127, and cell-laden hydrogel, respectively. (c) 3D printing process using the integrated organ printing system. The image shows the patterning of a layer of the construct. (d) Photograph of the 3D printed mandible bone defect construct, which was cultured in osteogenic medium for 28 d. (e) Osteogenic differentiation of hAFSCs in the printed construct was confirmed by Alizarin Red S staining, indicating calcium deposition. Reprinted with permission [31], Copyright 2016, Oxford University Press. (ii). Example calvarial bone reconstruction. (a) Visualized motion program (top) used to print a 3D architecture of calvarial bone construct. Lines indicate the dispensing paths of the PCL/TCP mixture and cell-laden hydrogel, respectively. Photograph of the printed calvarial bone construct (bottom). (b) Scanning electron microscope images of the printed bone constructs. (c) Photographs of the printed bone constructs at day 0 (top) and 5 months (bottom) after implantation. Reprinted with permission [31], Copyright 2016, Oxford University Press.

27.3.3 3D Bioprinting: Cardiac Tissue

Cardiovascular diseases are the main cause of death worldwide, and because damage to the myocardium is irreversible, advancing 3DBP of cardiac tissues is vital to improve morbidity and mortality [35]. The heart is composed of three layers, each of which contains a complex, specifically organized structure of cardiomyocytes and various other non-myocyte cell types, as well as capillaries, arteries, and an extracellular matrix. Each of these components requires a high level of precision when 3DBP to avoid misalignment and poor tissue function. Cardiac scaffolds should feature porosity that allows for cell infiltration and nutrient exchange, as well as mechanical support of surrounding tissues, cell survival, proliferation and differentiation, promotion of vascularization, cell-cell and cell-matrix interactions [35]. The three main methods employed for cardiac 3DBP are extrusion, inkjet, and laser or projection-based bioprinting.

To date, 3DBP has been used for several different cardiac applications: i) heart patches for the repair of myocardial scar tissue, as well as long-term drug delivery, ii) ex-vivo cardiac models for the study of drug interactions and physiological function, iii) engineered-heart components, including valves, ventricles, and some neonatal-scale hearts (Figure 27.4). Extrusion-based 3DBP is the most commonly utilized 3DBP modality, providing a fast, efficient, and cost-effective option, however,

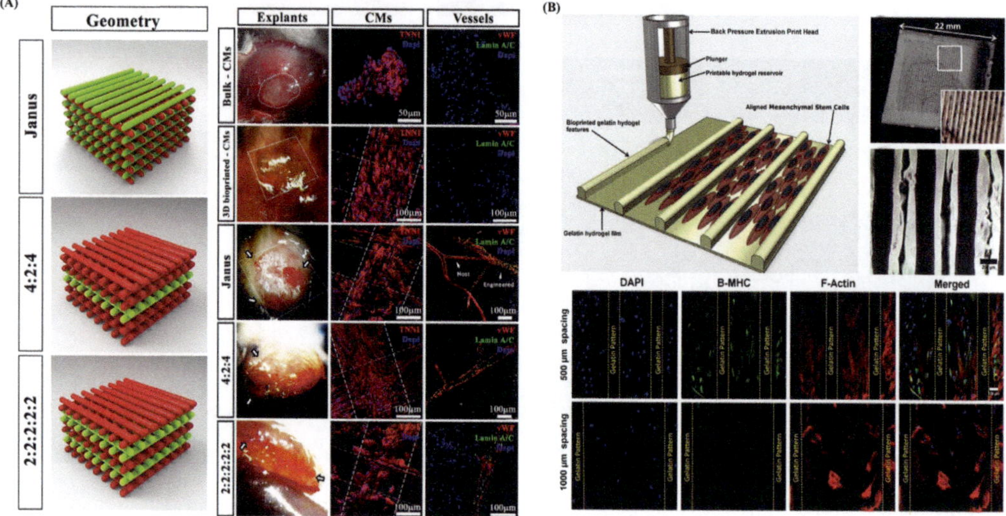

FIGURE 27.4 Applications of cardiac 3DBP in a heart patch, and an ex-vivo cardiac muscle model. (A) Left: A schematic illustrating the geometry of 3DBP, vascularized heart patches, using induced pluripotent stem cell cardiomyocyte laden bioink, and human umbilical vein endothelial cell laden bioink. Right: Images of patches after 7 days in culture, displaying an expression of troponin I and connexin 43 in cardiomyocytes and von Willebrand factor in human umbilical vein endothelial cells, showing growth of vasculature in 3DBP patch. Adapted with permission [37], (B) Top Left: Schematic of 3DBP microchannel hydrogels with various widths. Top Right: Scanning electron microscope images of microchannel hydrogels. Bottom: Fluorescence microscopy images, taken after 5 days in culture, illustrating the effect of various microchannel widths on the alignment, elongation, and differentiation of human mesenchymal stem cells. Adapted with permission [35],

this method struggles to offer the high level of accuracy required for cardiac 3DBP. When extrusion bioprinting cardiac tissues, the freeform reversible embedding of suspended hydrogels (FRESH) method is often utilized, which was developed and demonstrated by Lee et al. [36] when printing a tri-leaflet heart valve, a neonatal-scale collagen heart, and a human cardiac ventricle model.

Inkjet 3DBP is commonly utilized for its increased precision when fine-tuning porosity, as well as biomechanical and cytotoxic cardiac tissue properties. As previously discussed, inkjet bioprinters provide for a high level of cell viability. This being said, inkjet printers are not able to accurately print high-density, cell-laden bioinks, which is a critical characteristic for the creation of cardiac scaffolds [35]. DLP 3DBP provides for quick, high-resolution printing while avoiding shear stress to cells. Though DLP has been utilized by Liu et al. [38] to create scaffolds laden with neonatal mouse ventricular cardiomyocytes, several limiting factors need to be addressed to create complex cardiac scaffolds: (i) the use of multiple bioinks is challenging, which limits complex scaffold fabrication, (ii) cell settling within the bioink reservoir needs to be addressed to achieve uniform distribution, (iii) photo-crosslinking processes may present a challenge when printing with certain cell-laden bioinks [35].

27.3.4 3D Bioprinting: Vasculature

Organs and tissues require vasculature to maintain viability and metabolic activity. The vasculature is critical in supplying oxygen and nutrients to cells as well as in the removal of metabolic waste products, otherwise, cells located more than 200 μm away from their nearest capillaries will die [39]. This in turn compromises tissue viability and functionality. It is even more significant for bioprinted tissues that have a higher oxygen-consumption rate (e.g., heart, liver) to have vasculature integrated within the 3D printed scaffold to maintain long-term viability. One strategy is to rely on the controlled release of angiogenic factors for induction of blood vessel growth in 3D printed tissues, while the second strategy involves direct printing of vascularized scaffolds [40]. To date, 3DBP has been used to build two types of vascular constructs: i) vascularized tissue constructs with integrated channels or vessels, and ii) standalone tubular vascular structures. Predominantly, both direct (extrusion, inkjet, DLP) and indirect (sacrificial bioprinting) printing have been employed in vascularized tissue engineering (Figure 27.5A). Direct printing allows for the fabrication of structures with concurrent cell seeding, while indirect printing is more efficient in fabricating complex architectures [39]. Hence, 3DBP modalities should be employed selectively to achieve the successful fabrication of vascular constructs based on desirable applications.

As reported by previous literature and ongoing studies, extrusion, inkjet, and laser-based printing have been employed for bioprinting vasculature. In extrusion-based methods, cells (endothelial and/or smooth muscle cells) are suspended in a hydrogel (alginate, fibrin, PEG, GelMA) and the scaffold undergoes crosslinking (physical or chemical) to maintain the desired structure [40]. Hollow tubular constructs have been printed around a solid mandrel or using coaxial printing (Figure 27.5B), or a sacrificial hydrogel which is removed post-printing. Pi et al. [41] utilized a modified extrusion system equipped with multilayered coaxial nozzles to fabricate highly organized perfusable vascular constructs (Figure 27.5B). Sacrificial materials such as Pluronic F127, gelatin, agarose, and alginate, with reversible crosslinking mechanisms, are typically used as the fugitive bioinks.

DLP bioprinting can eliminate the need for a sacrificial material. Zhu et al. [43] developed pre-vascularized tissues using a continuous DLP technique called microscale continuous optical bioprinting (μCOB), without any sacrificial material. They encapsulated multiple cell types, mimicking vascular cell composition, directly into hydrogels and observed spontaneous formation of lumen-like structures by endothelial cells *in vitro*. Similarly, Grigoryan et al [42] have used DLP printing to generate pre-vascularized hepatic hydrogel carriers using endothelial cells (Figure 27.5C). In summary, a detailed review of various 3D printing techniques employed for fabricating vasculature to date can be found in the Chen and Zhu et al. articles [40], [39].

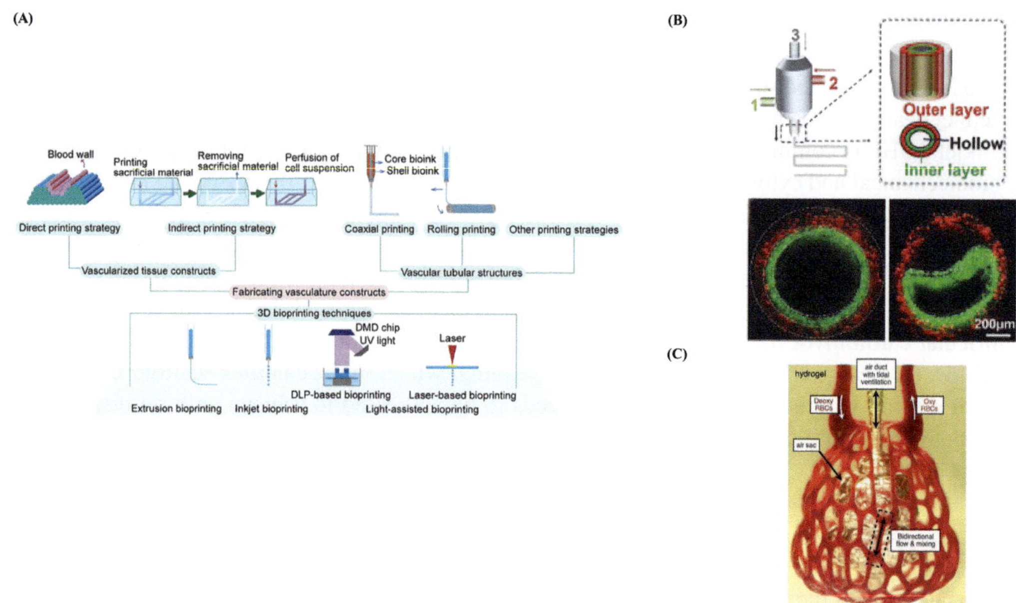

FIGURE 27.5 (A) 3DBP strategies employed for printing vascular constructs (Reprinted with permission [39], Copyright 2021, John Wiley & Sons). Both, direct and indirect printing strategies are utilized in the literature. Vascularized constructs fabricated by (B) Extrusion printing, coaxial printing of multilayered tubular structures using multichannel coaxial extrusion system. Reprinted with permission [41], Copyright 2018, John Wiley & Sons. (C) DLP printing, tidal ventilation, and oxygenation in hydrogels with vascularized alveolar model topologies. Reprinted with permission [42], Copyright 2019, American Association for the Advancement of Science.

27.4 CONCLUSIONS AND FUTURE DIRECTIONS

Bioprinting techniques have advanced remarkably in the past decade, from proof-of-concept prints to complex multi-material tissue structures. Nevertheless, as a relatively new technology, certain limitations exist that need to be overcome for it to achieve its ultimate goal - the fabrication of fully functional human organs. The challenges pertaining to bioprinting techniques, cellular sources, and biomaterial selection remain crucial for the future development of 3DBP. From a technological perspective, there is a need for increased printing speed, resolution, and compatibility for various biomaterials. Improvement of resolution to the nanoscale (subcellular/molecular) is required for better control of physical guidance provided by the microarchitecture. Similarly, shortened printing times are required not only for economic and productivity reasons in large-scale production of tissues for clinical use but also for situations wherein extended working times would be detrimental to sustaining bioink properties, along with the viability of partially printed structures. On the other hand, the selection of bioinks remains a significant challenge, as the materials need to be biocompatible, printable, and possess mechanical properties to support and maintain cell viability and function. Multi-material and multi-head printing may have the potential to resolve some of these issues moving forward, as it is highly unlikely that a single bioink would create an artificial microenvironment to mimic an actual *in vivo* scenario.

Besides challenges from technological and material standpoints, integration of a vascular network within the printed structures is one of the biggest hurdles in the field of regeneration. Although there are several bioprinting techniques used thus far to build vascular networks, each method has its limitations, and coaxial bioprinting may solve some issues of vascularization in scale-up tissues.

Furthermore, with increasing personalized patient-specific 3DBP, there are ethical and regulatory concerns as well. It raises questions regarding testing a bioprinted organ using a patient's cells on another patient first, and regarding developing a regulatory framework where a patient would serve as his/her testing subject. Another significant limitation of 3DBP is that it considers only the initial static condition of the printed object and assumes it to be inanimate. In contrast, natural tissue regeneration involves dynamic functional changes caused in response to intrinsic and/or extrinsic stimuli, which cannot be reproduced by bioprinting. Thus, the 3DBP field still has a lot of room for improvement.

27.4.1 4D Bioprinting

The term 4D printing was first conceptualized by Dr. Skylar Tibbits in 2012, where he established the broad definition of 4D printing as '3D printing but with time' as a fourth dimension [44]. Further understanding of this manufacturing process has evolved the definition to more clearly describe the dynamic changes in 3D print shape, structure, and function, to create constructs with self-assembly, deformation, and self-repair potential. Currently, the best definition to describe 4D printing is the evolution in the shape, property, and functionality of a 3D-printed structure with time when it is exposed to an external and/or an internal stimulus (Figure 27.6A) [45]. The two main strategies of 4D bioprinting are changes in shape and functionalities of printed constructs over time. By using stimuli-responsive materials, 4D bioprinting can generate 3D structures that are capable of dynamic transformation in response to desired stimuli over time, overcoming the limitations of 3DBP.

Presently, various stimuli-responsive shape-recovery polymers, known as shape-responsive materials (SRMs), are being extensively studied for use with 4D printing. Stimuli applied can be physical (thermal, magnetic, acoustic, electric, etc.), chemical (pH-responsive), or biological (enzymes, glucose-sensitive) (Figure 27.6B) [46], [48]. These SRMs are broadly classified into the following four categories: self-assembly, self-actuation, self-sensing, and self-healing materials [47]. To date, compression, torsion, rolling, stretching, and bending (or folding), are the five basic types of shape deformation processes that can be actuated by the 4D-printed structures [47]. The

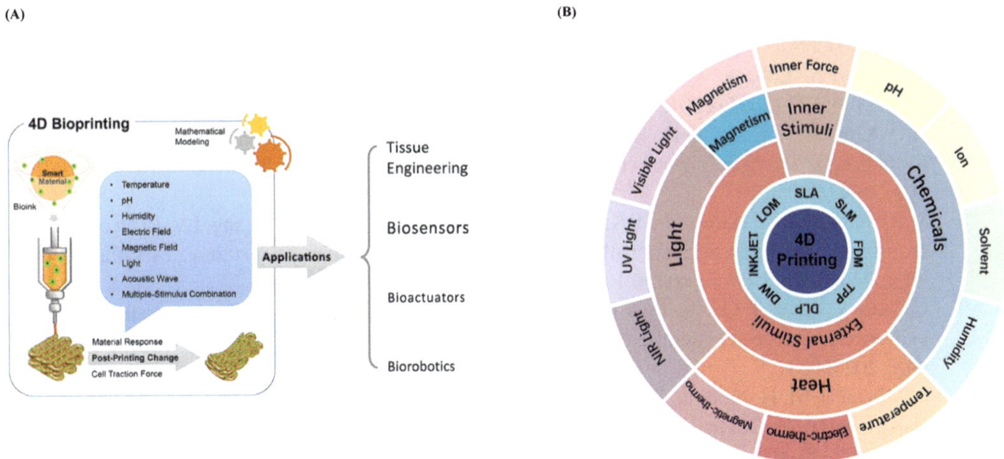

FIGURE 27.6 (A) Schematic depicting 4D bioprinting process. Reprinted with permission [46], Copyright 2018, John Wiley & Sons. (B) Stimulation sources for 4D bioprinting sorted into two types: external stimuli and internal stimuli, and subtypes under five subcategories: inner-, magnetism-, light-, heat-, and chemicals-stimuli. Reprinted with permission [47], Copyright 2022, John Wiley & Sons.

recent studies on 4D -printed tissue constructs are summarized in depth in a review article by Wang et al. [47].

4D bioprinting could also benefit pediatric patients through the production of self-growing constructs that would grow with the patient over time. Nevertheless, this technology is in its infancy and still not widely understood. There are challenges associated with 4D printing such as, the number of materials suitable for 4D printing, simple design of the programmed/patterned structures, and shortage of reversible transformation in printable materials, which are currently being researched. In conclusion, 4D bioprinting holds great potential for personalized treatments and precision medicine which are of utmost importance to the field of regenerative sciences.

27.5 DISCLAIMER

The opinions and assertions expressed herein are those of the authors and do not necessarily reflect the official policy or position of the Uniformed Services University (USU) or the Department of Defense. The opinions and assertions expressed herein are those of the authors and do not necessarily reflect the official policy or position of the Henry M. Jackson Foundation for the Advancement of Military Medicine, Inc.

REFERENCES

1 G. Saini, N. Segaran, J.L. Mayer, A. Saini, H. Albadawi, R. Oklu, Applications of 3d bioprinting in tissue engineering and regenerative medicine, *Journal of Clinical Medicine*. 10 (4966) (2021) 1–19. https://doi.org/10.3390/jcm10214966.

2 R.J. Klebe, Cytoscribing: A method for micropositioning cells and the construction of two- and three-dimensional synthetic tissues, *Experimental Cell Research*, 179 (2) (1988), 362–373.

3 Z. Gu, J. Fu, H. Lin, Y. He, Development of 3D bioprinting: From printing methods to biomedical applications, *Asian Journal of Pharmaceutical Sciences*. 15 (2020) 529–557. https://doi.org/10.1016/j.ajps.2019.11.003.

4 L. Moroni, T. Boland, J.A. Burdick, C. de Maria, B. Derby, J.J. Yoo, G. Vozzi, Biofabrication: A guide to technology and terminology, *Trends in Biotechnology*. 36 (2017) 19. https://doi.org/10.1016/j.tibtech.2017.10.015.

5 X. Li, B. Liu, B. Pei, J. Chen, D. Zhou, J. Peng, X. Zhang, W. Jia, T. Xu, Inkjet bioprinting of biomaterials, *Chemical Reviews*. 120 (2020) 10793–10833. https://doi.org/10.1021/ACS.CHEMREV.0C00008.

6 P. Kumar, S. Ebbens, X. Zhao, Inkjet printing of mammalian cells – Theory and applications, *Bioprinting*. 23 (2021) e00157. https://doi.org/https://doi.org/10.1016/j.bprint.2021.e00157.

7 H.P. Kim, W.S. Kang, C.H. Hong, G.J. Lee, G. Choi, J. Ryu, W. Jo, Piezoelectrics, *Advanced Ceramics for Energy Conversion and Storage*. (2019) 157–206. https://doi.org/10.1016/B978-0-08-102726-4.00005-3.

8 S. Ramesh, O.L.A. Harrysson, P.K. Rao, A. Tamayol, D.R. Cormier, Y. Zhang, I.V. Rivero, Extrusion bioprinting: Recent progress, challenges, and future opportunities, *Bioprinting*. 21 (2021) e00116. https://doi.org/https://doi.org/10.1016/j.bprint.2020.e00116.

9 C. Dou, V. Perez, J. Qu, A. Tsin, B. Xu, J. Li, A state-of-the-art review of laser-assisted bioprinting and its future research trends, *ChemBioEng Reviews*. 8 (2021) 517–534. https://doi.org/10.1002/cben.202000037.

10 I.T. Ozbolat, M. Hospodiuk, Current advances and future perspectives in extrusion-based bioprinting, *Biomaterials*. 76 (2016) 321–343. https://doi.org/10.1016/j.biomaterials.2015.10.076.

11 K. Shinkar, K. Rhode, Could 3D extrusion bioprinting serve to be a real alternative to organ transplantation in the future?, *Annals of 3D Printed Medicine*. 7 (2022) 100066. https://doi.org/10.1016/j.stlm.2022.100066.

12 H. Goodarzi Hosseinabadi, E. Dogan, A.K. Miri, L. Ionov, Digital Light processing bioprinting advances for microtissue models, *ACS Biomaterials Science and Engineering*. 8 (2022) 1381–1395. https://doi.org/10.1021/acsbiomaterials.1c01509.

13 R.D. Ventura, An overview of laser-assisted bioprinting (LAB) in tissue engineering applications, *Medical Lasers*. 10 (2021) 76–81. https://doi.org/10.25289/ml.2021.10.2.76.

14 R.Y. Tam, T. Fuehrmann, N. Mitrousis, M.S. Shoichet, Regenerative therapies for central nervous system diseases: A biomaterials approach, *Neuropsychopharmacology*. 39 (2014) 169–188. https://doi.org/10.1038/npp.2013.237.

15 C. Hammi, B. Yeung, *Neuropathy*, 2022. StatPearls Publishing.

16 P.G. Nagappan, H. Chen, D.Y. Wang, Neuroregeneration and plasticity: A review of the physiological mechanisms for achieving functional recovery postinjury, *Military Medical Research*. 7 (2020) 1–16. https://doi.org/10.1186/s40779-020-00259-3.

17 Y. Huang, W. Wu, H. Liu, Y. Chen, B. Li, Z. Gou, X. Li, M. Gou, 3D printing of functional nerve guide conduits, *Burns and Trauma*. 9 (2021) 1–11. https://doi.org/10.1093/burnst/tkab011.

18 J. Li, C. Wu, P.K. Chu, M. Gelinsky, 3D printing of hydrogels: Rational design strategies and emerging biomedical applications, *Materials Science and Engineering R: Reports*. 140 (2020) 1–76. https://doi.org/10.1016/j.mser.2020.100543.

19 D.A. Gyles, L.D. Castro, J.O.C. Silva, R.M. Ribeiro-Costa, A review of the designs and prominent biomedical advances of natural and synthetic hydrogel formulations, *European Polymer Journal*. 88 (2017) 373–392. https://doi.org/10.1016/j.eurpolymj.2017.01.027.

20 D.N. Heo, S.J. Lee, R. Timsina, X. Qiu, N.J. Castro, L.G. Zhang, Development of 3D printable conductive hydrogel with crystallized PEDOT:PSS for neural tissue engineering, *Materials Science and Engineering C*. 99 (2019) 582–590. https://doi.org/10.1016/j.msec.2019.02.008.

21 M. Uz, M. Donta, M. Mededovic, D.S. Sakaguchi, S.K. Mallapragada, Development of gelatin and graphene-based nerve regeneration conduits using three-dimensional (3D) printing strategies for electrical transdifferentiation of mesenchymal stem cells, *Industrial & Engineering Chemistry Research*. 58 (2019) 7421–7427. https://doi.org/10.1021/acs.iecr.8b05537.

22 J.M. Unagolla, A.C. Jayasuriya, Hydrogel-based 3D bioprinting: A comprehensive review on cell-laden hydrogels, bioink formulations, and future perspectives, *Applied Materials Today*. 18 (2020) 100479. https://doi.org/10.1016/j.apmt.2019.100479.

23 C. Tse, R. Whiteley, T. Yu, J. Stringer, S. MacNeil, J.W. Haycock, P.J. Smith, Inkjet printing Schwann cells and neuronal analogue NG108-15 cells, *Biofabrication*. 8 (2016) 015017. https://doi.org/10.1088/1758-5090/8/1/015017.

24 K. Liu, L. Yan, R. Li, Z. Song, J. Ding, B. Liu, X. Chen, 3D printed personalized nerve guide conduits for precision repair of peripheral nerve defects, *Advanced Science*. 9 (2022) 1–22. https://doi.org/10.1002/advs.202103875.

25 Y.-B. Lee, S. Polio, W. Lee, G. Dai, L. Menon, R.S. Carroll, S.-S. Yoo, Bio-printing of collagen and VEGF-releasing fibrin gel scaffolds for neural stem cell culture, *Experimental Neurology*. 223 (2010) 645–652. https://doi.org/10.1016/j.expneurol.2010.02.014.

26 X. Yu, T. Zhang, Y. Li, 3D printing and bioprinting nerve conduits for neural tissue engineering., *Polymers (Basel)*. 12 (2020) 1–27. https://doi.org/10.3390/polym12081637.

27 C.J. Pateman, A.J. Harding, A. Glen, C.S. Taylor, C.R. Christmas, P.P. Robinson, S. Rimmer, F.M. Boissonade, F. Claeyssens, J.W. Haycock, Nerve guides manufactured from photocurable polymers to aid peripheral nerve repair, *Biomaterials*. 49 (2015) 77–89. https://doi.org/10.1016/j.biomaterials.2015.01.055.

28 R.A. Dilla, C.M.M. Motta, S.R. Snyder, J.A. Wilson, C. Wesdemiotis, M.L. Becker, Synthesis and 3D printing of PEG–poly(propylene fumarate) diblock and triblock copolymer hydrogels, *ACS Macro Letters*. 7 (2018) 1254–1260. https://doi.org/10.1021/acsmacrolett.8b00720.

29 W. Zhu, K.R. Tringale, S.A. Woller, S. You, S. Johnson, H. Shen, J. Schimelman, M. Whitney, J. Steinauer, W. Xu, T.L. Yaksh, Q.T. Nguyen, S. Chen, Rapid continuous 3D printing of customizable peripheral nerve guidance conduits, *Materials Today*. 21 (2018) 951–959. https://doi.org/10.1016/j.mattod.2018.04.001.

30 T. Genova, I. Roato, M. Carossa, C. Motta, D. Cavagnetto, F. Mussano, Advances on bone substitutes through 3d bioprinting, *International Journal of Molecular Sciences*. 21 (2020) 1–28. https://doi.org/10.3390/ijms21197012.

31 H.W. Kang, S.J. Lee, I.K. Ko, C. Kengla, J.J. Yoo, A. Atala, A 3D bioprinting system to produce human-scale tissue constructs with structural integrity, *Nature Biotechnology*. 34 (2016) 312–319. https://doi.org/10.1038/nbt.3413.

32 C. Mandrycky, Z. Wang, K. Kim, D.H. Kim, 3D bioprinting for engineering complex tissues, *Biotechnology Advances*. 34 (2016) 422–434. https://doi.org/10.1016/j.biotechadv.2015.12.011.

33 L. Gong, J. Li, J. Zhang, Z. Pan, Y. Liu, F. Zhou, Y. Hong, Y. Hu, Y. Gu, H. Ouyang, X. Zou, S. Zhang, An interleukin-4-loaded bi-layer 3D printed scaffold promotes osteochondral regeneration, *Acta Biomaterialia*. 117 (2020) 246–260. https://doi.org/10.1016/j.actbio.2020.09.039.

34 S. Zhu, P. Chen, Y. Chen, M. Li, C. Chen, H. Lu, 3D-printed extracellular matrix/polyethylene glycol diacrylate hydrogel incorporating the anti-inflammatory phytomolecule honokiol for regeneration of osteochondral defects, *The American Journal of Sports Medicine*. 48 (2020) 2808–2818. https://doi.org/10.1177/0363546520941842.

35 Z. Wang, L. Wang, T. Li, S. Liu, B. Guo, W. Huang, Y. Wu, 3D bioprinting in cardiac tissue engineering, *Theranostics*. 11 (2021) 7948–7969. https://doi.org/10.7150/THNO.61621.

36 A. Lee, A.R. Hudson, D.J. Shiwarski, J.W. Tashman, T.J. Hinton, S. Yerneni, J.M. Bliley, P.G. Campbell, A.W. Feinberg, 3D bioprinting of collagen to rebuild components of the human heart, *Science*. 365 (6452), 482–487.

37 F. Maiullari, M. Costantini, M. Milan, V. Pace, M. Chirivì, S. Maiullari, A. Rainer, D. Baci, H.E.S. Marei, D. Seliktar, C. Gargioli, C. Bearzi, R. Rizzi, A multi-cellular 3D bioprinting approach for vascularized heart tissue engineering based on HUVECs and iPSC-derived cardiomyocytes, *Scientific Reports*. 8 (2018) 1–15. https://doi.org/10.1038/s41598-018-31848-x.

38 F. Liu, X. Wang, Synthetic polymers for organ 3D printing, *Polymers (Basel)*. 12 (2020). https://doi.org/10.3390/polym12081765.

39 J. Zhu, Y. Wang, L. Zhong, F. Pan, J. Wang, Advances in tissue engineering of vasculature through three-dimensional bioprinting, *Developmental Dynamics*. 250 (2021) 1717–1738. https://doi.org/10.1002/dvdy.385.

40 E.P. Chen, Z. Toksoy, B.A. Davis, J.P. Geibel, 3D Bioprinting of vascularized tissues for in vitro and in vivo applications, *Frontiers in Bioengineering and Biotechnology*. 9 (2021) 1–22. https://doi.org/10.3389/fbioe.2021.664188.

41 Q. Pi, S. Maharjan, X. Yan, X. Liu, B. Singh, A.M. van Genderen, F. Robledo-Padilla, R. Parra-Saldivar, N. Hu, W. Jia, C. Xu, J. Kang, S. Hassan, H. Cheng, X. Hou, A. Khademhosseini, Y.S. Zhang, Digitally tunable microfluidic bioprinting of multilayered cannular tissues, *Advanced Materials*. 30 (2018) 1706913. https://doi.org/10.1002/adma.201706913.

42 B. Grigoryan, S.J. Paulsen, D.C. Corbett, D.W. Sazer, C.L. Fortin, A.J. Zaita, P.T. Greenfield, N.J. Calafat, J.P. Gounley, A.H. Ta, F. Johansson, A. Randles, J.E. Rosenkrantz, J.D. Louis-Rosenberg, P.A. Galie, K.R. Stevens, J.S. Miller, Multivascular networks and functional intravascular topologies within biocompatible hydrogels, *Science* (1979). 364 (2019) 458–464. https://doi.org/10.1126/science.aav9750.

43 W. Zhu, X. Qu, J. Zhu, X. Ma, S. Patel, J. Liu, P. Wang, C.S.E. Lai, M. Gou, Y. Xu, K. Zhang, S. Chen, Direct 3D bioprinting of prevascularized tissue constructs with complex microarchitecture, *Biomaterials*. 124 (2017) 106–115. https://doi.org/10.1016/j.biomaterials.2017.01.042.

44 Skylar Tibbits, The emergence of "4D printing," TED Conference (2013).

45 A. Ahmed, S. Arya, V. Gupta, H. Furukawa, A. Khosla, 4D printing: Fundamentals, materials, applications and challenges, *Polymer (Guildf)*. 228 (2021) 1–25. https://doi.org/10.1016/j.polymer.2021.123926.

46 N. Ashammakhi, S. Ahadian, F. Zengjie, K. Suthiwanich, F. Lorestani, G. Orive, S. Ostrovidov, A. Khademhosseini, Advances and Future Perspectives in 4D Bioprinting, *Biotechnology Journal*. 13 (2018) 1–21. https://doi.org/10.1002/biot.201800148.

47 Y. Wang, H. Cui, T. Esworthy, D. Mei, Y. Wang, L.G. Zhang, Emerging 4D printing strategies for next-generation tissue regeneration and medical devices, *Advanced Materials*. 34 (2022) 2109198. https://doi.org/10.1002/adma.202109198.

48 Z. Wan, P. Zhang, Y. Liu, L. Lv, Y. Zhou, Four-dimensional bioprinting: Current developments and applications in bone tissue engineering, *Acta Biomaterialia*. 101 (2020) 26–42. https://doi.org/https://doi.org/10.1016/j.actbio.2019.10.038.

28 Analysis of the Use of Hydrogels in Bioprinting

Jesús M. Rodríguez Rego[*1], *Laura Mendoza Cerezo*[1], *Antonio Macías García*[2], *and Alfonso C. Marcos Romero*[1]
[1]Department of Graphic Expression. School of Industrial Engineering. University of Extremadura. Avenida de Elvas, s/n. 06006-Badajoz. Spain
[2]Department of Mechanical, Energy and Materials Engineering. School of Industrial Engineering. University of Extremadura. Avenida de Elvas, s/n. 06006-Badajoz. Spain
*Corresponding autor: jesusrodriguezrego@unex.es

CONTENTS

28.1 INTRODUCTION

Bioprinting is a booming tool that promises to offer important results in fields such as science, medicine, and biotechnology. Mostly aimed at use in medicine, specifically in the regeneration of organs and tissues and in the study of diseases, there are many studies to be carried out to achieve its normal implementation in hospitals and clinical laboratories, as it is still at an incipient stage. Many discoveries have led bioprinting to achieve its current advances. Achieving its first results in the 1990s, with cell growth on a prefabricated three-dimensional surface, biodegradable and suitable for cell survival, bioprinting has gradually undergone a great evolution over the years, with the most promising results beginning to be achieved in the 2010s. From bioabsorbable splints to heterogeneous aortic valves, vascular channels, different types of tissues, functional thyroid glands, ovarian tissue, and tumor reconstructions (tumor-on-a-chip) to facilitate and optimize the study of effective treatments, the possibilities of bioprinting are increasing every year thanks to new research (Figure 28.1).

To achieve all these advances, it has been necessary to study not only bioprinting methods and the development of new bioprinters, but also the hydrogels that make up the bioinks loaded with cellular material. These hydrogels are the matrix that will maintain the integrity of the bioprinted structure while causing the least possible damage to the cells during and after bioprinting, as well as allowing angiogenesis and the exchange of substances to ensure cell proliferation and maintenance. For this reason, it is essential to know the characteristics of the hydrogels that make up bioinks

DOI: 10.1201/9781003296676-28

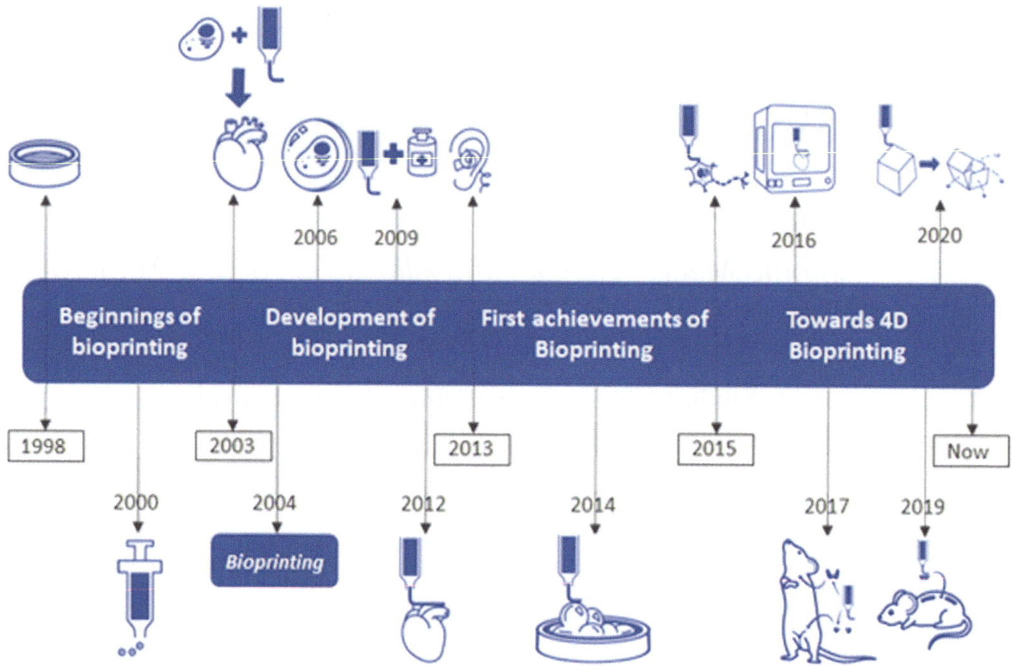

FIGURE 28.1 Relevant milestones in the evolution of bioprinting.

and how to use them to obtain an overview of how they work and their possibilities. Thus, in this chapter, we study the characteristics, rheological properties, composition, types of bioink, and ways of crosslinking hydrogels, together with the different existing biofabrication technologies. All this will allow us to understand the complexity of these compounds and the need to continue studying them, investigate new combinations, and incorporate new additives that improve their structural properties, viability, and mimicry of human body tissues for their optimal clinical use.

28.2 CHARACTERISTICS OF HYDROGELS

Tissue engineering plays a fundamental role in the evolution of medicine. It deals with the study of the regeneration, repair, or construction of tissues or functional organs similar to human organs. The advancement of tissue engineering over the years, together with 3D printing technologies, has made the development of bioprinting possible. Bioprinting is defined as the fabrication of three-dimensional structures composed of biological materials. To this end, it uses hydrogels as a printing matrix because they possess the mechanical and biocompatibility qualities necessary to carry out the process of generating tissues and organs with an organization similar to those present in the human body.

Hydrogels are defined as hydrophilic and cross-linked polymers, which can absorb and swell in water and biofluids, and transform into insoluble 3D networks [1]. They are composed of synthetic, natural, or semi-synthetic/hybrid materials, and can be used either to realize a cell-engineered biological construct (e.g., synthetic medium/biological medium) and reconstruct a damaged element of an organism or to simulate the pathophysiology of tissue and study the molecular processes behind it. The structure that hydrogels possess is due to the formation of an insoluble network in the biofluid they are made of, through a cross-linking process. This cross-linking process is what confers elastic properties to the hydrogel when it is subjected to stress (during bioprinting) (Figure 28.2). Within the aqueous environment in which the network is formed, an equilibrium is maintained between the

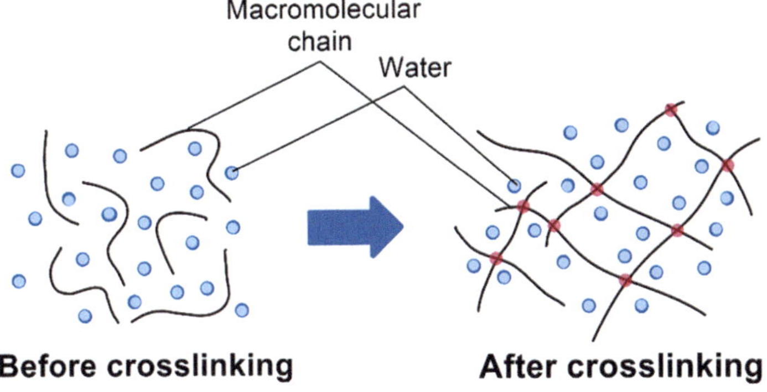

FIGURE 28.2 Hydrogel crosslinking process.

osmotic forces of the liquid and the elastic forces of the crosslinked polymer, and it is the composition of the chemical structure and the crosslink density that defines the swelling rate and permeability of the structure. They are capable of absorbing an amount of water equivalent to between 10 and 1000 times their dry weight [2].

These properties give the formed structure great elasticity and water retention capacity, which allows its use in different biomedical applications due to its resemblance to biological tissues. For a hydrogel to be considered suitable for use, it must be compatible with the biological material it is to carry, and it must be able to acquire a structure that conforms to the tissue it is intended to mimic. Hydrogels must also maintain proximity to tissue with minimal adhesive effect. In addition, synthetic matrices must be able to envelop cells and promote cell proliferation without damaging them under the effect of osmotic pressure. Therefore, synthetic matrices should be highly porous to promote the diffusion of nutrients and metabolites between cells and the surrounding environment [3].

28.3 RHEOLOGICAL PROPERTIES OF HYDROGELS

The rheological properties of bioinks greatly influence the integrity and fidelity of the printed structure and the viability of cells. The main rheological properties affecting the final characteristics of 3D bioprinted tissues and biological constructs include flow behavior, viscosity, shear stress, and viscoelasticity [4].

- **Flow behavior**: indicates the resistance to shear deformation of a fluid. It is characterized by the interaction between viscosity and shear rate. Depending on this flow, the behavior is generally classified as Newtonian or non-Newtonian. Hydrogel-based bioinks exhibit preferably non-Newtonian behavior in addition to shear thinning to flow smoothly and avoid clogging of the bioprinter nozzle. Moreover, these characteristics improve the printing fidelity and stability of 3D bioprinted structures [5].
- **Viscosity**: In bioinks, higher viscosity implies improved structural stability, while decreasing cell viability. Conversely, a lower viscosity decreases the structural stability of the print while increasing cell viability [1]. Too high a viscosity can cause occlusions in the nozzle, so it is necessary to match the viscosity of the bioink to the shape and size of the nozzle and vice versa, bearing in mind that adjusting both parameters can also result in higher cell viability. In the case of bioink formulations, viscosity can be controlled by adjusting molecular weight, polymer concentration, additive mass, temperature, and pre-crosslinking [6].
- **Shear stress**: is determined by the viscosity of the bioink, and negatively influences cell viability, since the higher the shear stress, the higher the rate of cell damage (Figure 28.3). Shear

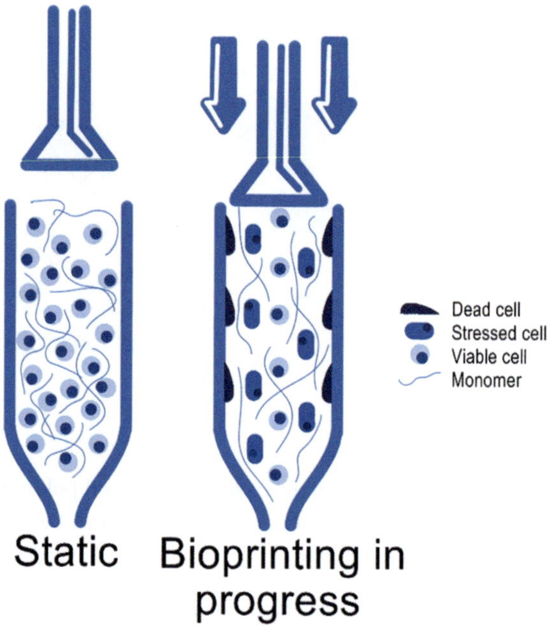

FIGURE 28.3 Effect of shear stress during bioprinting on cells.

stress acts on the tube wall during flow to oppose the driving force due to pressure. Away from the wall, velocity increases and the faster layers apply shear stress on the adjacent slower layers to increase their velocity. At the same time, the slower layers attempt to restrain the faster layers with a force opposite to the direction of flow as the frictional effects propagate through the fluid [7].

- **Viscoelasticity**: the behavior of materials with viscous and elastic properties when deformed. This property depends on the type of hydrogel, its concentration, and the cross-linking method applied to it, whether physical or chemical and affects cell proliferation and differentiation. Viscoelasticity is based on the measurement of the storage modulus (elastic part) and heat loss (viscous part) of stored energy. Thus, a high storage modulus results in bioinks with higher structural stability while a higher loss modulus will result in bioinks with lower structural stability [8]. It is necessary to adapt this rheological property to the needs of bioprinting.

28.4 COMPOSITION OF HYDROGELS USED IN BIOPRINTING

The composition of hydrogels used in bioprinting can be as follows:

- **Natural hydrogels** are composed of biomaterials that have inherent functions within the human body itself, such as collagen, fibrin, hyaluronic acid (HA), or Matrigel® [9]. They are best suited to achieve high initial cell viability because, being derived from biological tissues, they better mimic the natural microenvironment of the organism, enhancing cell proliferation and differentiation and replicating with greater fidelity those processes, functions, and reactions that occur in vivo. Despite this, they can present some problems, since they may be immunogenic or carry viruses and diseases from the donor when grafted.
- **Synthetic hydrogels** are composed of materials such as polyethylene glycol (PEG), polyacrylic acid (PAA), or polyhydroxymethylmethacrylate (pHEMA), among others. Like natural ones, they possess high biocompatibility and can be specially functionalized with cell

adhesion groups and growth factors, as well as chemically and physically modified with bioactive molecules to enhance growth [10]. In addition, they can function as a substrate and can be customized according to the requirements for each type of bioprinting, achieving better mechanical and biological properties.

- **Semi-synthetic/hybrid hydrogels** are formed by natural hydrogels that have been chemically modified or blended with synthetic polymers, acquiring new properties that can solve some of the problems that natural hydrogels may present, such as structural weakness, rapid degradation or variability of the source of origin [11]. As natural hydrogels form part of their formulation, they may even present the same immunogenicity and incompatibility problems as natural hydrogels. In addition, the use of bioinks containing other biomaterials can provide additional mechanical support for bioprinted cells, helping them to organize, migrate and differentiate autonomously to form functional tissues.

28.5 TYPES OF BIOINKS BASED ON HYDROGELS

Numerous types of hydrogels exhibit good bioprinting properties due to their biocompatibility and cell integration characteristics. In addition, the modified forms of some of them allow their immediate cross-linking after exposure to UV light, increasing their structural integrity as printing occurs.

According to their composition, hydrogels can be classified as follows:

- Protein-based bioinks:
  - **Collagen:** is the main structural protein of the extracellular matrix of mammalian cells. Thus, collagen possesses tissue-like physicochemical properties, along with in vitro/in vivo biocompatibility, and has been widely used in biomedical applications. Collagen can be combined with alginate for use as a composite bioink [12] and is capable of maintaining intercellular communications between different cell types. It can be modified with photopolymerizable methacrylate groups to give rise to ColMA, thus allowing its physical cross-linking in the presence of UV light.
  - **Gelatin:** it is produced by the denaturation of collagen. It is derived from mammalian skin, tendons, or bones, and has a low cost. Its properties confer it the ability to be biocompatible, biodegradable, and thermosensitive, being able to form hydrogels at low temperatures. Moreover, it is possible to perform numerous studies by varying its concentration, and its modified forms, which can be chemically or physically cross-linked, have also been adapted for bioprinting, such as gelatin methacrylate (GelMA) [13]. This GelMA is achieved by its modification with photopolymerizable methacrylate groups, allowing covalent crosslinking. Physical gelation arises from the assembly of intermolecular triple helices that possess a collagen-like structure [14]. Gelatin can be combined with alginate to improve the mechanical properties of bioprinted soft tissues. It can also be combined with hyaluronic acid to create more robust structures.
  - **Fibrinogen:** is a large, fibrous, soluble glycoprotein involved in blood clot formation, where it is converted into an insoluble fibrin molecule by thrombin in the presence of Ca^{2+} through intermolecular interactions [15]. In tissue engineering, it is mainly used for the reconstruction of damaged tissues because of its null immunogenicity and its biocompatibility, and biodegradability, as well as inducing the formation of an extracellular matrix. It can be combined, for example, with collagen to improve mechanical properties in cartilage tissue engineering or with hyaluronic acid to form vascular networks [16].
  - **Silk:** a biomaterial with high elasticity and slow biodegradation, low immunogenicity, and biocompatibility, which has important applications in tissue engineering as it is traditionally used as a support for cell growth.

- Bioinks based on polysaccharides:
    - **Alginate:** is a natural anionic polysaccharide obtained from brown algae, and is very similar to the glycosaminoglycans present in the extracellular matrix present in the human body. Moreover, it exhibits biocompatibility, low cytotoxicity, a gentle gelling process, and low cost. Due to its rapid gelation without the formation of harmful by-products, it is widely used in bioprinting as a bioink. Physical crosslinking and solidification of alginate are easily accomplished by the addition of divalent cations such as Ca^{2+} [17]. It is the most widely used bioink because its rapid gelation implies good printability, and it mixes easily with other bioinks or additives. Despite this, as alginate is a biologically inert material, it has low cell attachment, so it must be modified by an oxidation process and subsequent addition of RGD peptides to promote cell adhesion [18].
    - **Gellan gum:** is a hydrophilic anionic polysaccharide produced by bacteria and used in the food and pharmaceutical industry as a gelling agent. Like alginate, it forms a hydrogel at low temperatures when mixed with monovalent or divalent cations. It can be combined with other bioinks to provide mechanical stability to the bioprinted structure, increasing its viscosity and not adversely affecting cell viability.
    - **Hyaluronic acid:** is a non-sulfated glycosaminoglycan found in the natural extracellular matrix. It plays an important role in synovial fluid, vitreous humor, and hyaline cartilage. It is a polysaccharide suitable for bioimpressions that require high viscosity and good rheological properties, since the higher the molecular weight and concentration of hyaluronic acid, the higher the viscosity [19]. As with gelatin and collagen, it can be modified by the addition of photopolymerizable methacrylate groups, obtaining HAMA that can be crosslinked in the presence of UV light.
    - **Dextran:** a natural hydrophilic complex polysaccharide that is synthesized by acid lactic acid bacteria from sucrose, and can be degraded by the body by dextranase enzymes, making it biodegradable [20]. The combination of hyaluronic acid hydrogels and methacrylated dextran exhibit viscoelastic and pseudoplastic characteristics as bioinks and can be used in various bioprinting applications.
    - **Agarose:** polysaccharide extracted from marine algae with good gelling properties that can be used to form bioinks in conjunction with other bioinks.
    - **Chitosan:** a polysaccharide obtained from the deacetylation of chitin present in the exoskeletons of crustaceans and insects and fungi, green algae, and yeasts. It has antimicrobial and wound healing properties and can be combined with other bioinks, such as agarose [21].
    - **Konjac**: Konjac glucomannan (KGM) is a natural polysaccharide consisting of β D-glucose and β D-mannose linked by β-1,4-glycosidic bonds [22]. The properties that make it suitable for bioprinting are its high water retention, high viscosity and easy gelatinization. Few studies have been found in the literature on this polysaccharide and its behavior during bioprinting processes, so more research is needed with different concentrations, combinations with other bioinks and different bioprinting parameters.
- Bioink based on decellularized extracellularized matrix (dECM):
    - **Matrigel:** matrix composed from solubilized basement membrane secreted by mouse sarcoma cells, rich in laminin and collagen IV, and used in cell culture. It serves as a natural biomimetic extracellular matrix and is widely used in cell and tissue culture, where it undergoes thermal cross-linking at higher temperatures [23]. Other bioinks, such as alginate, can be added to enhance its bioprincibility, in addition to other growth factors and components that aid cell growth and proliferation. Although such a combination with other bioinks may result in interesting bioprinting properties, their transfer to the clinical setting may be limited due to their origin from mouse sarcoma.

- Bioinks based on synthetic polymers:
    - **PEG (polyethylene glycol):** is a synthetic polymer synthesized by oxidation of ethylene oxide, which does not present immunogenicity or cytotoxicity, is bioinert, and presents appropriate properties for use in bioprinting because its mechanical properties favor the integrity of the printed structures. Since it is a bioinert material, cells cannot easily adhere to its surface, so for its correct use as a bioink it must be mixed with other biologically active inks. PEG can be modified to contain reactive groups (PEGX), to be photoreticulable (PEGDA), and to be combined with other components such as gelatin, fibrinogen, atelocollagen [6], or hyaluronic acid, among others.
    - **Pluronic (poloxamer):** is a block copolymer consisting of a central block of poly (propylene propylene oxide) (PPO) flanked by blocks of poly (ethylene oxide) (PEO) [24]. It is a thermosensitive polymer that possesses inverse thermogelling properties (it gels with increasing temperature), which allows its easy handling in the liquid state and high viscosity at low temperatures to carry out its mixing with the cellular material and biopolymers that are added to incorporate biological signals. Thus, the gelation temperature depends on the concentration and type of Pluronic. These gels show shear thinning behavior and good shear recovery, important for good precision when printing. It can be modified to include photoreticulable groups in its structure. The main problem with the use of this compound as a bioink is that it requires high concentrations to produce the gelation and rheological behavior required for extrusion bioprinting, negatively affecting long-term cell viability, although for a short period cells can be encapsulated in high concentrations of Pluronic without any detrimental effect on their viability [25].

28.6 CROSS-LINKING OF HYDROGELS

When the hydrogels used in bioprinting do not possess the necessary physical characteristics to meet the requirements of the part to be reproduced, compounds that allow crosslinking for greater structural stability must be used. For this purpose, both chemical and physical modifications can be used to induce gelation, provided that the encapsulated cells survive and proliferate afterward (Figure 28.4).

Cross-linking in hydrogels can occur either physically or chemically. In physically activated gels, light or temperature stimulates the occurrence of cross-linking within hydrogels, whereas, in chemically activated gels, covalent bonds or coordinated bonds between polymer chains create a stable hydrogel network through molecular or ionic crosslinking agents, respectively [26]. Physically activated hydrogels have multiple advantages for clinical application, such as a mild gelling temperature and low-toxicity cross-linking reaction in the absence of cross-linking chemicals, making them the perfect solution for fractures and bone defects with small holes. In contrast, hydrogels that are chemically activated can be used in hard and large bone defects through mono, dual, or multiple covalent cross-linking hydrogels [27].

- **Cross-linked hydrogels by physical conditions:** they are shaped by physical cross-linking interactions, which include ionic interactions, temperature, and self-assembly.
    - Temperature-activated hydrogels: temperature-activated hydrogels can respond to different thermal stimuli through changes in their shape or internal structure. Therefore, the addition of hydrogels of polymers with small thermosensitive behaviors used in biomedicine confers them the ability to be thermosensitive. Some of the thermosensitive polymers used are PLGA-PEG (lactic acid coglycolic acid-PEG) [28], poly(N,N-diethylacrylamide) [29], PNIPAAm (poly(N-isopropylacrylamide)) [30] and soluplus [31]. Of these polymers, the most widely employed is PNIPAAm, as it is a typical monomer with excellent temperature sensitivity, and has become the most popular material due to its sensitive phase

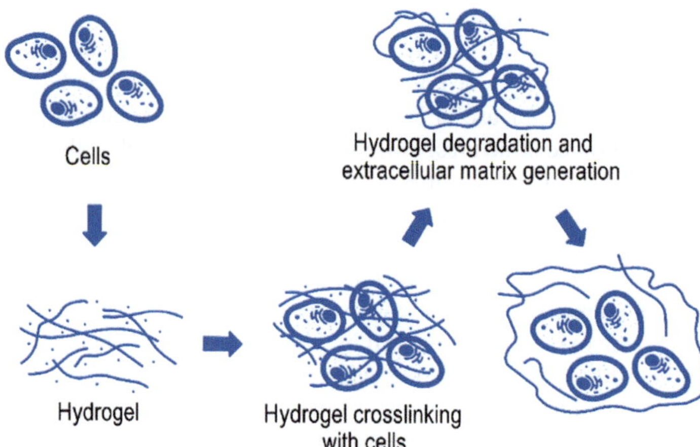

FIGURE 28.4 Crosslinking process of hydrogels together with cells for subsequent tissue formation.

translation ability when the temperature reaches 32 °C (lower critical solution temperature). The volume-phase transition temperature of a temperature-sensitive hydrogel can be adjusted by changing the amount of anionic monomer or the ratio of hydrophilic/hydrophobic groups in the gel [32].

- Hydrogels activated by ionic crosslinking agents: most natural polysaccharides derived from algae, such as agarose and carrageenan, and some proteins of animal origin, such as gelatin, elastin, and collagen, form thermal hydrogels. The hydrogel can be formed by chelation or by electrostatic interaction, facilitating spontaneous physical gelation due to the presence of electrically charged species [33]. These polymers are known as polyelectrolytes: they contain a net charge along the polymer backbone and crosslink to form insoluble complexes when combined with multivalent cations or anions depending on the charge of the polymer. Ionic crosslinking agents result in the formation of ionic intermolecular bonds, forming a three-dimensional molecular network. The introduction of Ca^{2+} and Cu^{2+} ions into the system to form coordination bonds improves the mechanical strength and tensile properties of hydrogels [34]. In addition, the addition of mineral ions often benefits cell migration, proliferation, and differentiation, showing benefits beyond the physical and mechanical properties of hydrogels. The most important of the ionic crosslinking method is its self-healing ability so that the gel network can be destroyed with high shear force and repaired once the shear force is removed. However, the poor mechanical property seriously limits the application of hydrogels in bone and other hard tissues due to the physical crosslinking method of ionic interaction [27].
- Self-assembly-activated hydrogels: self-assembly is driven by non-covalent weak-bonding mechanisms, such as hydrogen bonds, hydrophobic interactions, and electrostatic interactions, and occurs mainly in peptide- and protein-based hydrogels [35].

- **Chemically crosslinked hydrogels**: they have better mechanical properties than physically crosslinked hydrogels, and are formed by the addition of polymers that have been modified and the subsequent formation of covalent bonds between their chains. Crosslinking can occur by photoinduced crosslinking, enzymatically induced crosslinking, or small molecule crosslinking:
  - Photoinduced crosslinking: hydrogels that can be photoactivated induce the formation of a three-dimensional network and morphological changes of the gel precursor upon activation

by light at a specific wavelength, either UV light or visible light. Both natural and synthetic hydrogels and hybrids can be modified by the addition of photoreticulable components or photoinitiators to achieve in situ gelations. Depending on the reaction mechanism, photoinitiators are divided into two groups:

- Type I or excision: UV light generates the excision of the initiator radicals. The most commonly used excision photoinitiators in bioprinting are I2959 and LAP. LAP is a single-component initiation system with high thermal stability and good solubility and is colorless; however, it is activated by blue visible light, which was shown to have some level of toxicity [36].
- Type II or hydrogen extraction: they require less excitation energy than type I photoinitiators. The most commonly used are riboflavin and eosin Y. Riboflavin, also known as vitamin B2, is reduced by light, which releases free radicals, so it can be useful in radical polymerization [37]. Eosin Y, on the other hand, is a two-component initiation system, in which eosin Y acts as a photosensitizer. It is used in combination with triethanolamine (initiator) and N-vinylcaprolactam (comonomer to accelerate gelation kinetics) to produce the photoinduced cross-linking reaction under visible light [38]. When the absorption of eosin Y overlaps with the irradiated light spectrum, it absorbs the light and goes into an excited state. The excited eosin Y extracts hydrogen radical from triethanolamine, generating triethanolamine radical, which serves as the initiator of crosslinking [39].

- Enzymatically induced crosslinking: due to the cellular toxicity of chemical crosslinking agents, the use of enzymatic crosslinking hydrogels is proposed to improve material properties, including fast gelation time and excellent biocompatibility. One of the components used for enzymatic crosslinking of hydrogels is horseradish peroxidase, which is capable of forming covalent bonds between tyramine-modified polymers in the presence of hydrogen peroxide [40]. Another enzymatic method of hydrogel crosslinking is the use of transglutaminase factor XIII in chondroitin sulfate-PEG hybrid hydrogels. The addition of functional groups to chondroitin sulfate along with specific substrate sequences of transglutaminase factor XIII is used to enable crosslinking of chondroitin sulfate with hydrogel precursors [41].
- Small molecule crosslinking: small molecules of specific crosslinking are often used to carry out crosslinking of different polymers, which are introduced into their networks to improve their performance. Some of these molecules are genipin, caffeic acid or dopamine, and tannic acid.
  - Genipin: obtained from plants such as Genipa Americana or Gardenia jasminoides, it is used as a high-performance crosslinking agent to connect polymers with the amino group. This molecule can maintain injectability at room temperature, ensure favorable compatibility and obtain an anti-inflammatory effect. It cross-links proteins, collagen, gelatin, and chitosan [42], being able to repair tissues.
  - Dopamine and caffeic acid: catecholic compounds. They possess catechol functional groups that exhibit binding capacity. Caffeic acid has many advantages over dopamine, as it is more readily available, has lower cost and desirable aesthetics, and exhibits internal biological characteristics such as antibacterial, anti-inflammatory, and antioxidant effects [43].
  - Tannic acid: derived from green tea, grapes, and wine, among others. The abundant phenolic compounds in tannic acid have a strong adhesive capacity on organic or inorganic surfaces, allowing its development as a potential crosslinking agent for the engineering of functional hydrogels. In addition, due to the large amount of catechol groups present in the structure, their free radical scavenging ability can be used in materials for anticancer and anti-inflammatory therapy [27].

28.7 BIOFABRICATION TECHNOLOGIES

Studies of new biofabrication methodologies are increasingly perfecting the final finish of bioprinted structures, making it possible to increase their resemblance to the parts they imitate with each advance. In addition, they also ensure greater cell viability and more efficient bioprinting, bringing researchers closer to the goal of achieving the creation of organs and biological structures ready for transplantation. The following is a description of the bioprinting methods most commonly used today:

- **Inkjet bioprinting:** inkjet bioprinting is an accurate, fast and low-cost method, which is based on the process of microelectromechanical systems with thermal bubbles or piezoelectric injection microdroplet molding. It is divided into on-demand droplet bioprinting and continuous inkjetting:
  - Drop-on-demand inkjet bioprinting: drop-on-demand inkjet bioprinting can be divided into thermal, piezoelectric, and electrohydrodynamic jet bioprinting.
    - In thermal or piezoelectric bioprinting, the bioink is converted into droplet form by applying a voltage to a piezoelectric crystal transducer to vibrate the materials or by heating the formulation to above-boiling temperature, thus creating the droplets [44] (Figure 28.5). The advantages associated with this type of bioprinting are its low cost and its ability to print rapidly while retaining high cell viability, and the disadvantages are its poor material selectivity, temperature variations during the printing process, and frequent clogging of the printheads.
    - Electrohydrodynamic jet bioprinting involves the application of an electric field in the bioprinting nozzle so that the bioink shoots out and the cells experience a very low electric current that has no negative effects on cell integrity, viability, and proliferation [45].
  - Continuous inkjet bioprinting: this bioprinting technique, droplets are formed by a transducer or a droplet loading device that produces a continuous flow of droplets. The droplets are then directed to an electrically charged element to obtain the desired charge, and finally, the formed droplets reach the substrate and create the 3D product [46]. Requiring conductive bioinks, it is not well suited for bioprinting. In addition, this method has a recirculation flow that makes the bioink very susceptible to contamination [1].
- **Extrusion bioprinting:** this is the most widely used method due to its ease of use, the availability of a wide range of material selection (cast polymer, hydrogel, dECM, nanoclay, etc.), and low application cost. The process begins with loading the bioink to be printed into a suitable cartridge or syringe for the bioprinter, along with the placement of the desired nozzle. Through a pneumatic or mechanical mechanism, the bioink is deposited on a platform forming the previously digitally designed structure. The result is that different structures are built through the extrusion of bioink to form continuous fibers (Figure 28.6). This method

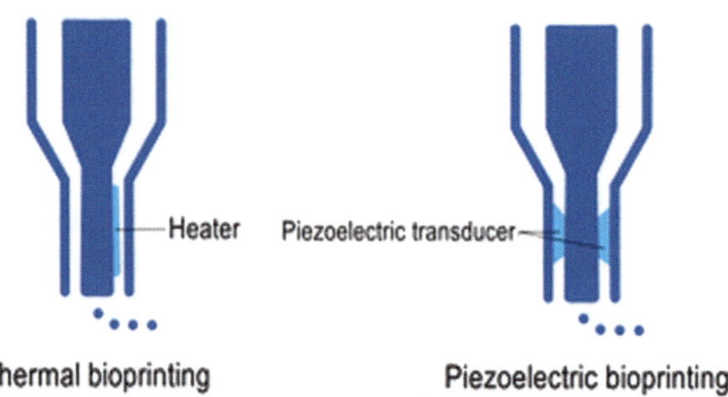

FIGURE 28.5 Thermal bioprinting and piezoelectric bioprinting.

FIGURE 28.6 Extrusion-based bioprinting.

has several disadvantages, such as the high rate of cell death due to the shear stress that the cells undergo during their flow through the syringe until they exit through the nozzle, and a low printing resolution [46]. The latter disadvantage can be improved by the addition of nanoparticles that confer higher printability. Furthermore, modification of the size and shape of the nozzle can also help to decrease the cellular stress produced during bioprinting, achieving an increase in viable cells after printing.

- **Laser-assisted bioprinting (LAB):** this bioprinting technique uses a laser as an energy source to deposit biomaterials on a substrate, and consists of three parts: a pulsed laser source, a ribbon coated with liquid biological materials that are deposited on the metal film, and a receiving substrate. The physical principle of LAB is based on the generation of a cavitation-like bubble, deep in the bioink film, whose expansion and collapse induce the formation of a jet and thus the transfer of the bioink from the ribbon to the substrate, forming a microdroplet [47] (Figure 28.7).
- **Digital light processing (DLP) bioprinting:** works by projecting product cross-sectional graphics onto the surface of liquid photosensitive resin using digital micromirror elements to project, which allows the irradiated resin to photopolymerize layer by layer (Figure 28.8), resulting in a relatively fast printing speed. DLP bioprinting can build products with complex geometric features by depositing materials according to digital files. With this bioprinting technology, artificial tissues and organs are constructed with precise bionic structure and high cell viability [48].

28.8 CONCLUSIONS

A wide variety of hydrogels are currently available for use in bioprinting, with very different properties and characteristics. Each of them must meet certain minimum requirements, such as maintaining high cell viability after the bioprinting process, achieving good structural stability, adequately mimicking the tissue to be mimicked, and achieving an internal structure that allows the formation of blood vessels and the exchange of molecules, as well as not showing toxicity after implantation in the organism and being able to biodegrade. Furthermore, depending on the type of hydrogel, it will be more or less suitable for mimicking a specific part of the organism.

In addition to the types of hydrogels, there is a wide variety of types of bioprinting, each with advantages and disadvantages depending on the bioink used and the structure to be bioprinted. Due to the variety presented by both bioinks and bioprinting methods, together with the discoveries that are being made every day in both concepts, it is important to continue studying their properties and combinations to standardize methods that make it possible to reproduce specific parts of human

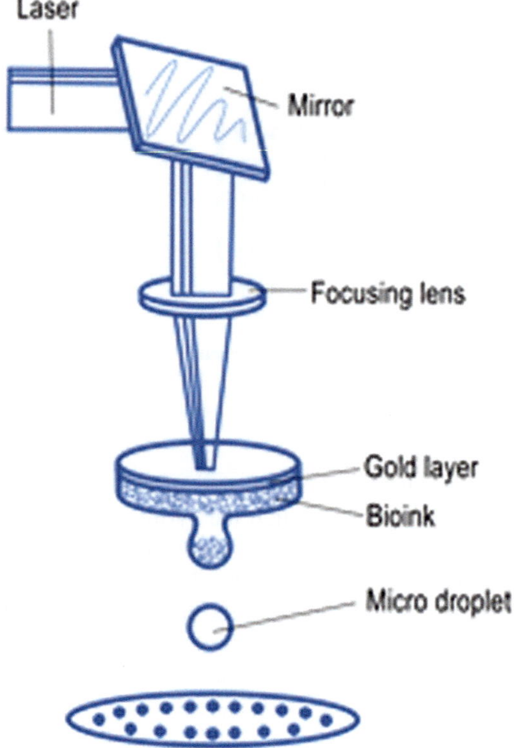

FIGURE 28.7 Laser-assisted bioprinting.

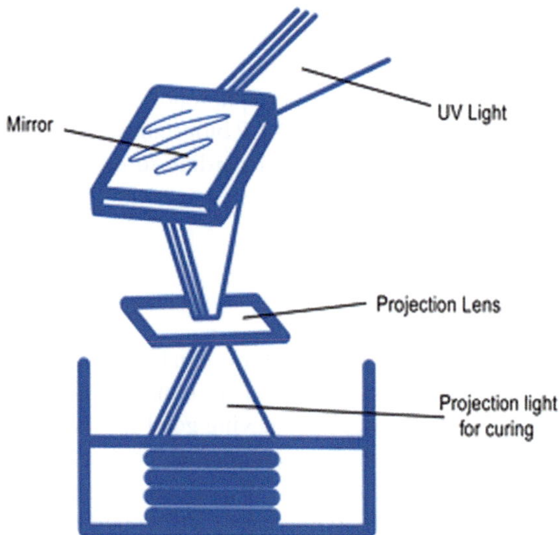

FIGURE 28.8 Digital Light Processing bioprinting.

tissues with the highest possible success rates, bringing bioprinting ever closer to its standardized application in laboratories and hospitals as a method for generating personalized transplants.

ACKNOWLEDGMENTS

This research has been funded by the European Regional Development Fund (FEDER) in the framework of the project (BIOSIMPRO) with code 2021/00110/001.

REFERENCES

1 A. Fatimi, O. V. Okoro, D. Podstawczyk, J. Siminska-Stanny, and A. Shavandi, "Natural hydrogel-based bio-inks for 3D bioprinting in tissue engineering: A review," *Gels*, vol. 8, no. 3, p. 179, Mar. 2022.

2 A. S. Hoffman, "Hydrogels for biomedical applications," *Advanced Drug Delivery Reviews,* vol. 64, no. SUPPL., pp. 18–23, Dec. 2012.

3 N. A. Peppas, P. Bures, W. Leobandung, and H. Ichikawa, "Hydrogels in pharmaceutical formulations," *European Journal of Pharmaceutics and Biopharmaceutics: Official Journal of Arbeitsgemeinschaft fur Pharmazeutische Verfahrenstechnik e.V*, vol. 50, no. 1, pp. 27–46, Jul. 2000.

4 A. Ribeiro, M. M. Blokzijl, R. Levato, C. W. Visser, M. Castilho, W. E. Hennink, T. Vermonden, and J. Malda, "Assessing bioink shape fidelity to aid material development in 3D bioprinting," *Biofabrication*, vol. 10, no. 1, p. 014102, Jan. 2017.

5 L. Ning, H. Sun, T. Lelong, R. Guilloteau, N. Zhu, D. J. Schreyer and X. Chen, "3D bioprinting of scaffolds with living Schwann cells for potential nerve tissue engineering applications," *Biofabrication*, vol. 10, no. 3, p. 035014, Jun. 2018.

6 A. L. Rutz, K. E. Hyland, A. E. Jakus, W. R. Burghardt, and R. N. Shah, "A multimaterial bioink method for 3D printing tunable, cell-compatible hydrogels," *Advanced Materials,* vol. 27, no. 9, pp. 1607–1614, Mar. 2015.

7 K. M. Foster, D. V. Papavassiliou, and E. A. O'rear, "Elongational stresses and cells," *Cells,* vol. 10, no. 9, p. 2352, Sep. 2021.

8 A. Schwab, R. Levato, M. D'Este, S. Piluso, D. Eglin, and J. Malda, "Printability and shape fidelity of bioinks in 3D bioprinting," *Chemical Reviews,* vol. 120, no. 19, pp. 11028–11055, Oct. 2020.

9 J. Yang, X. Sun, Y. Zhang, and Y. Chen, "The application of natural polymer-based hydrogels in tissue engineering," In Y. Chen (Ed.), *Hydrogels Based on Natural Polymers*, 1st ed., vol. 1, pp. 273–307, Jan. 2019. Elsevier. https://doi.org/10.1016/B978-0-12-816421-1.00010-0

10 J. Thakor, S. Ahadian, A. Niakan, E. Banton, F. Nasrollahi, M. M. Hasani-Sadrabadi and A. Khademhosseini, "Engineered hydrogels for brain tumor culture and therapy," *Bio-Design and Manufacturing,* vol. 3, no. 3, pp. 203–226, Sep. 2020.

11 E. Garcîa-Gareta, *Biomaterials for Skin Repair and Regeneration*, Synthetic Polymers, 1st Edition, 2019. Woodhead.

12 P. S. Gungor-Ozkerim, I. Inci, Y. S. Zhang, A. Khademhosseini, and M. R. Dokmeci, "Bioinks for 3D bioprinting: An overview," *Biomaterials Science,* vol. 6, no. 5, pp. 915–946, May 2018.

13 J. W. Nichol, S. T. Koshy, H. Bae, C. M. Hwang, S. Yamanlar, and A. Khademhosseini, "Cell-laden microengineered gelatin methacrylate hydrogels," *Biomaterials,* vol. 31, no. 21, pp. 5536–5544, Jul. 2010.

14 V. Normand, S. Muller, J. C. Ravey, and A. Parker, "Gelation kinetics of gelatin: A master curve and network modeling," *Macromolecules,* vol. 33, no. 3, pp. 1063–1071, Feb. 2000.

15 T. Rajangam and S. S. A. An, "Fibrinogen and fibrin based micro and nano scaffolds incorporated with drugs, proteins, cells and genes for therapeutic biomedical applications," *International Journal of Nanomedicine,* vol. 8, pp. 3641–3662, Sep. 2013.

16 M. Gruene, M. Pflaum, C. Hess, S. Diamantouros, S. Schlie, A. Deiwick, L. Koch, M. Wilhelmi, S. Jockenhoevel, A. Haverich and B. Chichkov, "Laser printing of three-dimensional multicellular arrays for studies of cell-cell and cell-environment interactions," *Tissue Engineering – Part C: Methods,* vol. 17, no. 10, pp. 973–982, Oct. 2011.

17 K. Yamamoto, Y. Yuguchi, B. T. Stokke, P. Sikorski, and D. C. Bassett, "Local structure of Ca^{2+} alginate hydrogels gelled via competitive ligand exchange and measured by small angle x-ray scattering," *Gels*, vol. 5, no. 1, Mar. 2019.

18 J. Jia, D. Richards, S. Pollard, Y. Tan, J. Rodriguez, R. Visconti, T. Trusk, M. Yost, H. Yao, R. Markwald and Y. Mei, "Engineering alginate as bioink for bioprinting," *Acta Biomaterialia,* vol. 10, no. 10, pp. 4323–4331, Oct. 2014.

19 M. K. Cowman, T. A. Schmidt, P. Raghavan, and A. Stecco, "Viscoelastic properties of hyaluronan in physiological conditions," *F1000Res*, vol. 4, p. 622, 2015.

20 G. Sun and J. J. Mao, "Engineering dextran-based scaffolds for drug delivery and tissue repair," *Nanomedicine (Lond),* vol. 7, no. 11, pp. 1771–1784, Nov. 2012.

21 D. F. D. Campos, A. Blaeser, A. Korsten, S. Neuss, J. Jäkel, M. Vogt and H. Fischer, "The stiffness and structure of three-dimensional printed hydrogels direct the differentiation of mesenchymal stromal cells toward adipogenic and osteogenic lineages," *Tissue Engineering Part A,* vol. 21, no. 3–4, pp. 740–756, Feb. 2015.

22 Y. Li, R. Deng, N. Chen, J. Pan, and J. Pang, "Review of konjac glucomannan: Isolation, structure, chain conformation and bioactivities," *Journal of Single Molecule Research*, vol. 1, no. 1, p. 7, 2013.

23 L. Horvath, Y. Umehara, C. Jud, F. Blank, A. Petri-Fink, and B. Rothen-Rutishauser, "Engineering an in vitro air-blood barrier by 3D bioprinting," *Scientific Reports*, vol. 5, no. 1, pp. 1–8, Jan. 2015.

24 M. Müller, J. Becher, M. Schnabelrauch, and M. Zenobi-Wong, "Nanostructured Pluronic hydrogels as bioinks for 3D bioprinting," *Biofabrication,* vol. 7, no. 3, p. 035006, Aug. 2015.

25 N. E. Fedorovich, I. Swennen, J. Girones, L. Moroni, C. A. van Blitterswijk, E. Schacht, J. Alblas and W. J. A. Dhert, "Evaluation of photocrosslinked Lutrol hydrogel for tissue printing applications," *Biomacromolecules,* vol. 10, no. 7, pp. 1689–1696, Jul. 2009.

26 L. Voorhaar and R. Hoogenboom, "Supramolecular polymer networks: Hydrogels and bulk materials," *Chemical Society Reviews,* vol. 45, no. 14, pp. 4013–4031, Jul. 2016.

27 X. Xue, Y. Hu, S. Wang, X. Chen, Y. Jiang, and J. Su, "Fabrication of physical and chemical crosslinked hydrogels for bone tissue engineering," *Bioactive Materials,* vol. 12, pp. 327–339, Jun. 2022.

28 X. Li, L. Chen, H. Lin, L. Cao, J. Cheng, J. Dong, L. Yu and J. Ding, "Efficacy of poly(D,L-lactic acid-co-glycolic acid)-poly(ethylene glycol)-poly(D,L-lactic acid-co-glycolic acid) thermogel as a barrier to prevent spinal epidural fibrosis in a postlaminectomy rat model," *Clinical Spine Surgery*, vol. 30, no. 3, pp. E283–E290, 2017.

29 L. Hanyková, I. Krakovský, E. Šestáková, J. Šťastná, and J. Labuta, "Poly(N,N'-diethylacrylamide)-based thermoresponsive hydrogels with double network structure," *Polymers (Basel),* vol. 12, no. 11, pp. 1–19, Nov. 2020.

30 B. L. Ekerdt, C. M. Fuentes, Y. Lei, M. M. Adil, A. Ramasubramanian, R. A. Segalman and D. V. Schaffer, "Thermoreversible hyaluronic acid-PNIPAAm hydrogel systems for 3D stem cell culture," *Advanced Healthcare Materials*, vol. 7, no. 12, p. 1800225 Jun. 2018.

31 H. Wu, K. Wang, H. Wang, F. Chen, W. Huang, Y. Chen, J. Chen, J. Tao, X. Wen and S. Xiong, "Novel self-assembled tacrolimus nanoparticles cross-linking thermosensitive hydrogels for local rheumatoid arthritis therapy," *Colloids Surf B Biointerfaces,* vol. 149, pp. 97–104, Jan. 2017.

32 L. Zhou, Z.- Y. Lu, X. Zhang, and H. Dai, "Studies on the temperature sensitive of N-isopropylacrylamide copolymer," *Gaofenzi Cailiao Kexue Yu Gongcheng/Polymeric Materials Science and Engineering,* vol. 22, pp. 165–168, Mar. 2006.

33 J. M. Unagolla and A. C. Jayasuriya, "Hydrogel-based 3D bioprinting: A comprehensive review on cell-laden hydrogels, bioink formulations, and future perspectives," *Applied Materials Today,* vol. 18, p. 100479, Mar. 2020.

34 R. das Mahapatra, K. B. C. Imani, and J. Yoon, "Integration of macro-cross-linker and metal coordination: A super stretchable hydrogel with high toughness," *ACS Applied Materials and Interfaces,* vol. 12, no. 36, pp. 40786–40793, Sep. 2020.

35 S. Zhang, "Fabrication of novel biomaterials through molecular self-assembly," *Nature Biotechnology,* vol. 21, no. 10, pp. 1171–1178, Sep. 2003.

36 V. Marek, A. Potey, A. Réaux-Le Goazigo, E. Reboussin, A. Charbonnier, T. Villette, C. Baudouin, W. Rostène, A. Denoyer and S. M. Parsadaniantz, "Blue light exposure in vitro causes toxicity to trigeminal neurons and glia through increased superoxide and hydrogen peroxide generation," *Free Radical Biology Medicine,* vol. 131, pp. 27–39, Feb. 2019.

37 J. L. Covre, P. C. Cristovam, R. R. Loureiro, R. M. Hazarbassanov, M. Campos, É. H. Sato and J. Á. P. Gomes, "The effects of riboflavin and ultraviolet light on keratocytes cultured in vitro," *Arquivos Brasileiros de Oftalmologia*, vol. 79, no. 3, pp. 180–185, 2016.

38 C. S. Bahney, T. J. Lujan, C. W. Hsu, M. Bottlang, J. L. West, and B. Johnstone, "Visible light photoinitiation of mesenchymal stem cell-laden bioresponsive hydrogels," *European Cells and Materials*, vol. 22, pp. 43–55, 2011.

39 S. Sharifi, H. Sharifi, A. Akbari, and J. Chodosh, "Systematic optimization of visible light-induced crosslinking conditions of gelatin methacryloyl (GelMA)," *Scientific Reports*, vol. 11, no. 1, p. 23276, Dec. 2021.

40 Y. Zhang,, Y. Cao, H. Zhao, L. Zhang, T. Ni, Y. Liu, Z. An, M. Liu and R. Pei, "An injectable BMSC-laden enzyme-catalyzed crosslinking collagen-hyaluronic acid hydrogel for cartilage repair and regeneration," *Journal of Materials Chemistry B,* vol. 8, no. 19, pp. 4237–4244, May 2020.

41 F. Anjum, P. S. Lienemann, S. Metzger, J. Biernaskie, M. S. Kallos, and M. Ehrbar, "Enzyme responsive GAG-based natural-synthetic hybrid hydrogel for tunable growth factor delivery and stem cell differentiation," *Biomaterials,* vol. 87, pp. 104–117, May 2016.

42 A. Lauto, L. J. R. Foster, L. Ferris, A. Avolio, N. Zwaneveld, and L. A. Poole-Warren, "Albumin-genipin solder for laser tissue repair," *Lasers in Surgery and Medicine*, vol. 35, no. 2, pp. 140–145, 2004.

43 J. H. Ryu, P. B. Messersmith, and H. Lee, "Polydopamine surface chemistry: A decade of discovery," *ACS Applied Materials & Interfaces,* vol. 10, no. 9, pp. 7523–7540, Mar. 2018.

44 N. Samiei, "Recent trends on applications of 3D printing technology on the design and manufacture of pharmaceutical oral formulation: A mini review," *Beni-Suef University Journal of Basic and Applied Sciences*, vol. 9, no. 1, p. 12, Dec. 2020.

45 S. L. Sampson, L. Saraiva, K. Gustafsson, S. N. Jayasinghe, and B. D. Robertson, "Cell electrospinning: An in vitro and in vivo study," *Small,* vol. 10, no. 1, pp. 78–82, Jan. 2014.

46 J. Gong, C. C. L. Schuurmans, A. M. van Genderen, X. Cao, W. Li, F. Cheng, J. J. He, A. López, V. Huerta, J. Manríquez, R. Li, H. Li, C. Delavaux, S. Sebastian, P. E. Capendale, H. Wang, J. Xie, M. Yu, R. Masereeuw, T. Vermonden and Y. S. Zhang, "Complexation-induced resolution enhancement of 3D-printed hydrogel constructs," *Nature Communications*, vol. 11, no. 1, Dec. 2020.

47 V. Keriquel, H. Oliveira, M. Rémy, S. Ziane, S. Delmond, B. Rousseau, S. Rey, S. Catros, J. Amédée, F. Guillemot and J. C. Fricain, "In situ printing of mesenchymal stromal cells, by laser-assisted bioprinting, for in vivo bone regeneration applications," *Scientific Reports*, vol. 7, no. 1, pp. 1–10, May 2017.

48 J. Zhang, Q. Hu, S. Wang, J. Tao, and M. Gou, "Digital light processing based three-dimensional printing for medical applications," *International Journal of Bioprinting*, vol. 6, no. 1, pp. 12–27, 2020.

29 Additive Manufacturing in the Automotive Industry

Panagiotis Stavropoulos, Harry Bikas, Thanassis Souflas, Konstantinos Tzimanis, Christos Papaioannou, and Nikolas Porevopoulos*
Laboratory for Manufacturing Systems & Automation, Department of Mechanical Engineering & Aeronautics, University of Patras, University Campus, Patras, 26504, Greece
*Corresponding author: pstavr@lms.mech.upatras.gr

CONTENTS

29.1 INTRODUCTION

Additive Manufacturing (AM), more commonly known as '3D printing' refers to a wide family of manufacturing processes, capable of processing various materials. ASTM standards define AM as 'the process of joining materials to make objects from 3D model data, usually layer upon layer, as opposed to subtractive manufacturing methodologies, such as traditional machining' [1]. This layer-by-layer material deposition is what makes AM processes unique, providing design freedom and manufacturing flexibility, enabling customization, efficient use of resources, multi-functional parts, and end parts with high stiffness to weight ratio.

The wide spectrum of AM processes available nowadays can be categorized into seven main process families, based on their physical process mechanism. According to ASTM, these process families are Powder Bed Fusion (PBF), Directed Energy Deposition (DED), Vat Polymerization (VP), Material Extrusion (MEx), Material Jetting (MJ), Binder Jetting (BJT), and Sheet Lamination (SL). Each process family may comprise multiple process variants, using different energy sources or slight variations of the same main process mechanism. Different material families can be used as raw materials, such as metals, ceramics, elastomers, polymers, composites, etc. The raw material

DOI: 10.1201/9781003296676-29

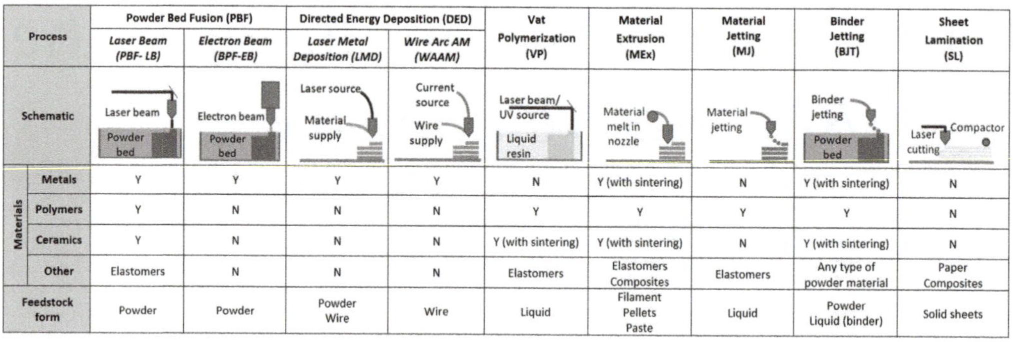

Process		Powder Bed Fusion (PBF)		Directed Energy Deposition (DED)		Vat Polymerization (VP)	Material Extrusion (MEx)	Material Jetting (MJ)	Binder Jetting (BJT)	Sheet Lamination (SL)
		Laser Beam (PBF- LB)	*Electron Beam (BPF-EB)*	*Laser Metal Deposition (LMD)*	*Wire Arc AM (WAAM)*					
Schematic		Laser beam Powder bed	Electron beam Powder bed	Laser source Material supply	Current source Wire supply	Laser beam/ UV source Liquid resin	Material melt in nozzle	Material jetting	Binder jetting Powder bed	Laser cutting Compactor
Materials	Metals	Y	Y	Y	Y	N	Y (with sintering)	N	Y (with sintering)	N
	Polymers	Y	N	N	N	Y	Y	Y	Y	N
	Ceramics	Y	N	N	N	Y (with sintering)	Y (with sintering)	N	Y (with sintering)	N
	Other	Elastomers	N	N	N	Elastomers	Elastomers Composites	Elastomers	Any type of powder material	Paper Composites
Feedstock form		Powder	Powder	Powder Wire	Wire	Liquid	Filament Pellets Paste	Liquid	Powder Liquid (binder)	Solid sheets

FIGURE 29.1 AM process families, processable materials, and feedstock forms. Adapted with permission [2]. Copyright (2015) The Authors, some rights reserved; exclusive licensee Springer Nature. Distributed under a Creative Commons Attribution License 4.0 (CC BY).

feedstock form can be wire, powder, liquid, and sheets. Figure 29.1 introduces the different AM process families, alongside their processable material families and feedstock form.

AM benefits can be realized by using the technology in different stages of product development and production and different application areas. In general, AM is already used in the product development phase since the mid-1990s (where the term Rapid Prototyping originated from), to create proof of concepts, placeholders in larger assemblies, or even functional prototypes of the final product. Advances in AM are making it more and more relevant for the production phase though, where application areas can be distinguished between using AM indirectly or directly to produce end-use parts. Due to the unique characteristics of AM already summarized, end-use parts could be either high-performance components, personalized/mass customized parts, spare parts that can be produced on-demand, or aftermarket components and accessories. Bridge production (introducing AM for initial production scale-up before transitioning to another technology) is also a promising application field, that could help reduce time to market, react faster to demand changes or supply chain disruptions, reduce cost and make manufacturing more resilient.

The aforementioned potential AM application areas for the automotive industry are summarized in Figure 29.2 and analyzed in the following Sections.

Owing to the unique characteristics compared to conventional manufacturing processes, AM processes have significantly disrupted the manufacturing sector worldwide over the last decade. This is also reflected in the rapid continuous growth of the AM market worldwide, reaching $13bn in 2020, a 20% growth compared to 2019 despite the global effects of the COVID-19 crisis [3]. This trend is expected to continue, with a forecasted compound annual growth rate (CAGR) of 17% over the next 3 years, while average predictions show that the AM market is foreseen to double by 2026, reaching $37bn. Specifically, regarding the automotive sector, AM has generated $1.39bn in 2019 and is estimated to grow to $9bn by 2029 from the production of end-use parts alone [4]. MEx is the predominant technology used, representing more than 50% of the sales, and pointing towards the fact that the main use of AM in the automotive sector is for prototyping and tooling purposes (60% according to [5]).

29.2 AM IN AUTOMOTIVE PRODUCT DEVELOPMENT

During the design process of a vehicle, several parts need to be designed and developed by different teams of people working on creating different systems, that are assembled to form the final product. AM has proven a strong tool for the rapid deployment of prototypes, ranging from proof of concept of a particular design (allowing ease of visualization and testing ergonomics for example)

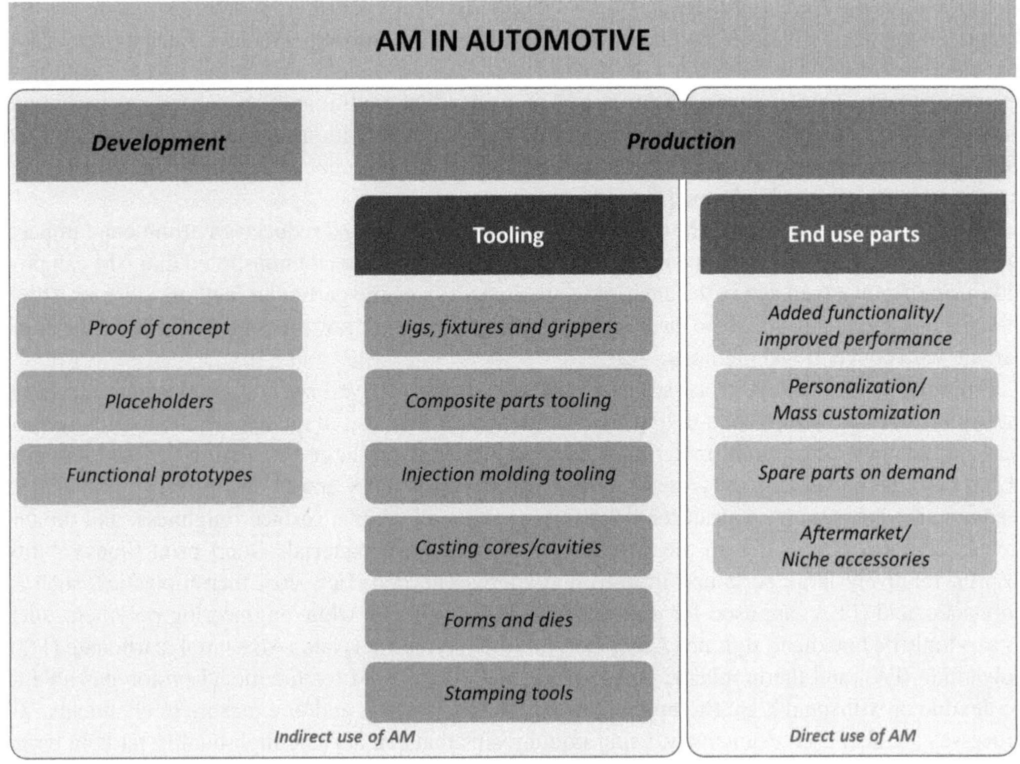

FIGURE 29.2 Application areas for AM in the automotive industry.

to placeholders used for handling volumetric constraints and ensuring the assembly of different components, up to functional prototypes that can be used for functional testing and operational verification of complete systems.

AM offers a series of unique opportunities during the product development phase, especially when compared with currently utilized manufacturing processes, such as machining. AM technologies make it possible to manufacture prototypes more rapidly and with fewer resources than would otherwise be needed for production using conventional methods. Furthermore, since the designers utilize the same equipment to print many models rather than buying specific tooling needed for subtractive manufacturing, AM lowers the cost of prototyping by removing the tooling expenditures. This enables automotive engineers to make more design adjustments in a shorter amount of time, which in turn decreases the time to market. Additionally, AM shortens the lead time of tool manufacturing to allow for rapid design modifications while using existing equipment. The use of AM allows companies to validate their designs before investing in costly tooling and production setup/changeovers. As such, automobile producers can shorten the product development cycle. Indicative examples of how the automotive industry has successfully used AM during the vehicle development phase are presented below.

BMW already uses AM in various areas, and mainly in the pre-development, vehicle validation, and testing phase, such as, in the case of BMW i8 vehicles, where no predecessor vehicles existed, the prototype vehicles have been produced primarily with the use of AM [6]. SEAT makes use of AM, in the prototype development, in different stages, to evaluate how a product assembly works, mostly through the use of PBF. Examples include the replacement of missing elements during a prototype assembly with printed ones, to simulate fast and cheap the volume of it and the overall assembly procedure [7]. In 2016, Chrysler developed and manufactured the use of VP, transparent resin differential housings to analyze the lubricant distribution and flow inside the differential components in different conditions [8]. Audi, in the Audi Pre-Series Center in Ingolstadt, manufactures prototypes with the use of AM, to accelerate design verification and evaluate new design concepts. One example is the tail-light covers that traditionally had been manufactured through milling or injection molding, where AM enabled the creation of transparent multi-colored tail lights over a single print, while these previously have been produced separately and assembled, significantly decreasing lead time [9].

A very important benefit of AM is sustainability improvement and reduced environmental impact. During the research of a prototype urban vehicle by [10], it has been demonstrated that AM can provide a significant advantage in the environmental impact since the particular method achieves a high material usage efficiency and the electric energy consumption for low production series can be comparable with conventional methods.

From the variety of AM processes available, only a few are well suited to the product development phase, being mature enough to provide end parts of the desired quality with consistency, process a wide range of different materials, and having a relatively large production rate and low cost [11], [12]. The most commonly used family of processes is MEx, mainly for prototyping small to medium size items with no strict requirements about small details, surface roughness, and dimensional accuracy. This is due to the low cost of machines and materials, short print times, ability to print relatively large parts and little to no expertise needed. Low-cost thermoplastics, such as polylactic acid (PLA) are used for prototypes and placeholders, while engineering polymers, such as acrylonitrile butadiene styrene (ABS), acrylonitrile styrene acrylate (ASA), polycarbonate (PC), polyamide (PA), and thermoplastic polyurethane (TPU) are used for functional prototypes able to be flexible or withstand higher temperature, mechanical loading, and/or exposure to chemicals. VP processes are also used extensively, using liquid resins that can achieve high-quality parts in terms of surface roughness, dimensional accuracy, and fine features. Additionally, the outputs of the VP process can be flexible, transparent, high temperature resistant, and with good mechanical strength (depending on the resin system used), covering a wide range of applications for product development. However, the process mechanism of VP processes leads to longer print time compared to MEx, while the attainable build volume is considerably smaller. BJT is also extensively used, due to its unique ability to process ceramics as well as produce full-color prints; a characteristic useful for prototypes intended for visual validation of design. Most commonly used materials include PA and polypropylene (PP). The last process family applicable in the product development phase is PBF; mostly used for functional prototypes as it can produce mechanically robust parts with fine details and good dimensional accuracy. Materials for prototyping are limited to various grades of PA (including aluminum-, glass- and carbon-filled variants).

29.3 AM IN AUTOMOTIVE PRODUCTION

29.3.1 Indirect Use of AM

Indirect use of AM refers to using AM processes for manufacturing components and tools that will not be directly used in the final product sold to the customer; instead, they will be used to assist the production (manufacturing and assembly) phase of the product. The main value proposition of AM in this family of applications is the reduction in lead time and cost (linked with supply chain simplification), in addition to the inherent flexibility of AM allowing the users to create unique tools tailored to their specific requirements. Typical applications in this category include the manufacturing of jigs, fixtures, and grippers, as well as tooling and molds.

29.3.1.1 Jigs, Fixtures, and Grippers

Jigs and fixtures are devices used to assist in part positioning and assembly. Grippers are mechanisms that are electrically, pneumatically, or hydraulically actuated and used to grasp and handle parts in a production environment. Conventionally, these devices are often machined from metal (or sometimes polymers) to achieve the required accuracy. Their function often dictates complex geometries, which leads to high lead times and costs.

As such, additively manufactured jigs and fixtures use is increasing, as it allows rapid revisions based on operator feedback and satisfaction. AM also allows for the reduced lead time when switching between different products, or when the product design is revised, creating 'tooling on demand' and leading to significant cost savings. Another driver for AM adoption in this application area is lightweighting. AM allows complex, optimized structures to be built with minimal effort, making the tools more ergonomic for human operators, but at the same time reducing the weight of grippers used by machines and/or robots, which allows lower payload robots to be used.

A recent trend in this application category is metal replacement; the redesign or remanufacturing of existing components and systems, so that the metallic raw material is replaced with engineering polymers [13]. The key driver for metal replacement is the reduced resource consumption for raw material manufacturing and the significantly reduced manufacturing cost, compared to metallic alternatives. However, the new, polymer-based components should have at least equal properties to the original metallic ones. Modern engineering polymers can achieve equal or superior properties to their metal counterparts in terms of strength to weight ratio and can provide tailored, superior properties regarding heat and chemical resistance in several cases. AM opens up new possibilities for the development of reinforced polymer materials with tailored properties, also known as super-polymers. The additional advantage of using AM for metal replacement in the manufacturing of automotive tooling is that further weight reduction can be achieved through the higher strength to weight ratio of engineering polymers compared to their metallic counterparts. This is also often accompanied by a further cost reduction, as both processes and materials are lower cost than their metallic counterparts.

In terms of application examples, Volkswagen Autoeuropa switched to AM saving over 90% in tool development costs (from $800 to $21 per tool) and time (from 56 to 10 days). Aside from the cost reductions, AM tools are more ergonomic and result in higher operator involvement since feedback is easier to include in the design [14].

A case study has been executed by Ford in their Sharonville transmission plant, regarding pucks and work holding fingers for robotic arms. These pucks are used to carry parts down the assembly line and are one of the most common tooling components to exhibit failure, leading to a high overall replacement cost. By remanufacturing these parts with polymer AM (MEx), and utilizing in-house machines, Ford was able to reduce $140,000 per year production costs. The second case study coming from Ford refers to work holding fingers for robotic arms. Previously made of metal, these fingers resulted in scratching and nicking of the parts that were handled by the robot, introducing a cost of more than $1 million per year for Ford. By replacing them with Polyethylene terephthalate glycol (PETG) manufactured by MEx, the damages to the parts were eliminated.

A similar case study on cobot grippers, spacers, and end effectors was presented by Thyssenkrupp Bilstein. Such components need to be replaced often, due to the different parts that were handled with the cobots, with a cost of more than $800 for a replacement. These cobot components were replaced with polymer ones, manufactured by PLA or TPU (where a soft, flexible material is needed), resulting in a cost per part of less than $5. Some of these components have been in service for more than one year, demonstrating the durability of these materials and the equally good performance that can be achieved by metal replacement, while minimizing production costs for an automotive factory [15].

Typical materials that can be utilized for replacing metal tools, jigs, and fixtures in automotive industries through AM include PLA and ABS where temperature resistance and toughness are not critical, engineering polymers such as polyamides and polyethylene, or even high-performance engineering polymers, such as reinforced polyamides or polyetherimides, suitable for high load bearing applications.

29.3.1.2 Tooling and Molds

Another important potential area for the application of AM is the manufacturing of molds and tooling. These can be used for different purposes (such as manufacturing composite parts, injection molding, casting, and forming/deforming). Depending on the intended use, different materials and hence AM processes can be used.

In terms of tooling for composite material parts, the main requirement is a smooth, watertight surface in the mold cavity, without any pores that would allow leaks in the mold. As such, the most promising process family for this particular application is VP. Since the curing process to produce the parts is typically exothermal (it generates heat) and the parts need to be cured at a high temperature, suitable resins should be used. This has been demonstrated through the manufacturing of a mold for fabricating a Carbon Fiber Reinforced Polymer (CFRP) steering wheel, where the mold cost has been reduced from $900 to $610 and the mold production time has been reduced from four to six weeks to two days, compared with an outsourced machined mold [16].

Another promising application is using AM to create injection molding cavities and cores. There are two main approaches regarding AM in this application; creating tooling that is better performing or creating tooling rapidly and at a low cost, allowing fast and economic small batch production of injection molded parts. Aiming to improve cooling performance, a mold was manufactured by DED and subsequently machined to obtain the required accuracy and surface finish. The use of DED enabled the manufacturing of conformal grid cooling channels, which resulted in a 45% reduction in temperature difference between the mold and coolant when compared with the conventional mold; this in turn significantly lowers the cycle times that can be achieved for the injection molding process [17]. Aiming to reduce the cost of small batch injection molding, a polymer AM process (in this case VP) has been successfully used to manufacture a thin cavity mold [18].

A relatively economic way of producing large complex metal parts is by using AM for printing casting molds. The predominant process for this particular application is BJT, where a specially formulated binder suitable for casting is used in combination with casting sand [19]. ExOne has been marketing such systems for a few years now and can demonstrate numerous applications. BMW is now utilizing this approach (printing sand via a binder jetting process) to produce casting cores for its S58 engine cylinder head, achieving geometrical complexity that would not be possible using conventional manufacturing processes [20].

Last but not least, tooling for low-volume forming and deforming processes can be produced via AM. The use of polymer tools and dies has been demonstrated as an effective alternative where only a few parts are needed. The same logic of 'metal replacement' applies here as well, with engineering-grade polymers being able to withstand the mechanical loads required. A characteristic example involves the design, manufacturing, and testing of a PA-based die, with embedded carbon in the filament and metallic inserts for supports for a small production batch of stamped metal parts [21].

Even more interesting is the application of metal AM processes, capable of manufacturing full-scale tooling used in large volume serial production. Due to the size of tooling required by the automotive industry, the predominantly used process family is DED (both LMD and WAAM). The direct costs of AM tooling regarding stamping processes are higher than conventional methods, mainly due to equipment costs and post-processing steps [22]. However, AM tooling can provide a net economic benefit to the manufacturer, due to fewer process steps required to manufacture it, simplified supply chains, and significantly reduced lead time [23] compared to conventional manufacturing methods. Using AM allows the tooling manufacturer to build the vast majority of the volume of the tool in a relatively low-cost metal, whereas AM can be used to deposit high-performance alloys on top of it, that can exhibit increased hardness, meeting the performance requirements of the hot extrusion die for automotive parts [24]. In addition, the inherent flexibility of the process allows the integration of geometrically complex heating/cooling channels within the tool, improving its performance. Hot stamping dies have been manufactured through LMD and milling [25], to illustrate the potential of conformal cooling channels on the die. This enabled an improved temperature distribution, leading to lower cycle times, thus increasing the efficiency of the hot stamping process.

DED processes can also be used to extend the service life of tooling, by removing damaged areas and depositing new material on them. Bilsing Automation has been offering repair of automotive stamping dies through the LMD process, without having to reproduce the malfunctioning part [26]. Upon post-processing (needed to achieve the required tolerances and surface quality characteristics), the tooling can enter service again, with significant environmental and economic sustainability benefits for the end user.

29.3.2 Direct Use of AM

While AM use during the product development phase, and also the production phase through tooling applications, is slowly becoming established practices in the industry, end-use parts are still hampered in terms of AM uptake. Disruptive technologies (such as AM) are typically first undertaken, demonstrated, and tested in an environment that can support the required high expenses for research and development activities (R&D) [27]. In motorsport, race teams collaborate with automakers to examine innovative solutions which can offer a competitive advantage. However, what is suitable for industry interested in making a few cars per year is rarely applicable for mass production. Nevertheless, the technological progress of R&D departments and their close link with automakers (Renault-Renault F1, McLaren Automotive-McLaren F1) accelerates the adoption of advanced technologies in automotive industries with limited edition and mass production vehicles. Indicative examples of this technological exchange are the use of carbon fiber as structural material [27], [28] and the Kinetic Energy Recovery System (KERS) which have been developed and fully adopted into the motorsport environment before adoption from the automotive industry [29], [30]. A similar timeline and technology exchange is expected in AM.

AM can offer a series of production-related benefits to manufacturers, including simplification of the supply chain, increased resilience, reduced time to market, and reduced costs. Most of the AM production benefits are listed below.

- They enable geometrically complex, customized designs to be realized, leading to improved performance (heat exchangers, seat support structures, etc.) of the respective components.
- They produce lightweight parts due to unique process characteristics that enable to use of material where is needed, creating hollow structures/cavities that cannot be made with conventional processes.
- They provide optimized parts in terms of load paths to avoid excessive use of material, leading to less raw material needed and less waste.
- They allow parts consolidation, creating complex geometries in one piece instead of using assembly and fixation parts/methods (screws, retaining rings, welding, or adhesives) reducing the number of parts for a product with certain functionalities, thus reducing time and cost.
- They provide flexibility in production lines since different designs and materials can be produced using the same equipment.
- They enable on-demand production since AM processes do not require significant machine setup before the process.
- They contribute to reduced energy consumption and emissions during the operation of the vehicle, due to lower weight and reduced moment of inertia that can be achieved for rotating parts.
- They contribute to the increased lifecycle of components (bearing, mounts, tires, brake pad) due to reduced weight of the vehicle, contributing to sustainability.

A field of particular interest is the use of AM technologies in combination with advanced engineering polymers to replace conventionally metallic components, as it has significant potential for cost reduction. However, while metal replacement in automotive tooling is quite straightforward, when it comes to end-use parts directly used in automobiles, several attributes need to be considered for part conformity. First and foremost, the strength (either ultimate or fatigue) of the polymer counterparts should be equal or superior to the original metal components. Additionally, high performance in elevated temperatures is required to ensure part conformity, since automotive parts are very often subjected to high thermal loads. Furthermore, additional aspects of polymer materials, and more importantly the ability to create polymer materials with tailored properties, are related to their chemical resistance and durability in harsh environments (e.g., contact with an engine or transmission oils, corrosion resistance, etc.). The aforementioned attributes dictate the available materials that can be utilized. In general, high-performance engineering polymers are preferred for automotive parts, such as polyether ether ketone (PEEK), polyether ketone (PEKK), PA, etc. Another very important attribute that enables metal replacement through AM for automotive parts is their ability to be reinforced. Carbon reinforcement is a very typical option for metal replacement in the automotive sector, as it can provide material properties very close to aluminum, which is one of the key materials constituting a vehicle. Furthermore, the use of polymer materials for automotive parts can provide superior properties, compared to metals, in other aspects that are equally important for the optimal operation and ride quality of the vehicle. AM polymers can improve the insulation characteristics of a part, reduce Noise, Vibration and Harshness (NVH) of the vehicle, and provide abrasion resistance and self-lubricating functions, due to their very low friction coefficients.

One of the most promising applications for metal replacement through AM in automotive parts is the use of PEEK to manufacture polymer gears. PEEK is a high-performance polymer with excellent properties for its use in gears manufacturing. Such properties include high strength, self-lubrication capabilities, low coefficient of friction, abrasion resistance, high service temperature (260 °C), and chemical resistance to oils and petrol. Due to these properties, PEEK gears can achieve significantly lower inertia and effectively reduce the vibration and NVH during their operation, compared to their metal counterparts. As such, AM PEEK gears have been used in automotive parts, such as seat mechanisms and internal mechanical components [31]. Additionally, some case studies have been presented by Roboze, where Carbon filled PA and PEEK were used to replace metal to manufacture supporting brackets and covers. The ability of these materials to withstand engine operating temperatures and contact with petrol and oils, along with their high strength, rendered them as suitable alternatives for their metal counterparts that were usually manufactured from aluminum [32]. The sectors where direct use of AM process is identified alongside the related benefits are described in the following sections.

29.3.2.1 Parts with Improved Performance

The complex internal structures, optimized load paths, and the case-dependent infill percentage/part density make the AM end parts have improved performance, equipping mostly high-performance, limited edition, and low production volume cars with lightweight and high-quality components aiming to improve the overall performance and efficiency of the vehicle.

Fiat Chrysler Automobiles (FCA) with the aid of the Fraunhofer Institute provided a very interesting concept for the development of a wheel carrier with an integrated brake caliper [33]. This part will be made with a PBF-LB process. Different metals alloys have been tested aiming to provide high structural and thermal behavior and reduced weight. This unique part aims to combine an assembly of 12 parts (including a wheel carrier, a heat shield, brake caliper, etc.) into one, reducing not only the manufacturing time of the individual components but also the total weight (36% weight reduction) considering that multiple joints and screws will be avoided.

Porsche and Bugatti are also interested in the application of AM in their high-performance cars. Porsche has used PBF-LB alongside a high-strength alloy M174+ aluminum alloy to create a topology-optimized piston with internal cooling channels, for the engine of its flagship Porsche 911 [34]. This design with the integrated cooling duct into the component's 'crown' has contributed to 10% weight reduction, increased engine speed, reduced temperature load, and optimized ignition that has led to an increase of 30 BHP to the engine's horsepower compared to conventional forged pistons.

Porsche has also developed an AM aluminum housing for the complete electric drive unit, integrating it into a two-speed gearbox. The process selected was PBF-LB, and the housing achieved approximately a 10% reduction in the overall weight while increasing up to double the stiffness in the high-stress areas, in comparison with the conventional casting process. The use of AM enabled additional design freedom, which reduced the number of parts needed to be assembled to form the final unit [35].

Bugatti has created additively manufactured titanium exhaust tailpipes and brake calipers. Both parts are made through PBF-LB in a titanium alloy, and they are placed in Bugatti Chiron, which is a limited-edition sports car [36], [37]. Some of the functionalities are common for both parts; high-temperature resistance and minimized weight are significant aspects considering their applications. The brake caliper handles the high temperature of brake rotors and pad while being part of the unsprang mass of the vehicle which increases the importance of minimized weight. Bugatti says that the AM brake calipers weigh 40% less (4.9 kg to 2.9 kg) compared to the alternative solution made with conventional machining while they offer higher stiffness than before due to topology optimized design. On the other hand, the tailpipes should manage the exhaust heat and their weight should be reduced.

Another application includes the design of a topology optimized engine mount for a diesel engine that can serve also as an arching pulley that provides tension to the belt driving the vehicle cooling system, produced by PBF-LB [38]. The optimized part reduced the weight and the volume of the components needed to serve these two functionalities.

In a similar ethos, an engine subframe produced previously by the casting of AlSi10Mg alloy has been redesigned using topology optimization, exploring the design freedom enabled by using AM processes [39]. The process used was PBF-LB with powder material of similar chemical composition of Aluminum AlSi10Mg. The final weight reduction of the part was more than 20%; from 2.2 kg to 1.71 kg.

Managing temperatures in the engine head/block and their parts becomes relevant through the turbocharging era of engines, which leads to high exhaust temperatures. AM allows hollow engine valves to be manufactured, using the cavity of the valve for cooling through internal liquid sodium. Hollowing also enables mass reduction, reducing parasitic losses and improving the valvetrain dynamic response for overall improved engine performance and efficiency [40]. The valve was printed using PBF-LB in Inconel 718, thanks to its good mechanical properties across high temperatures. A weight reduction of 29.4% was achieved while only increasing the stresses by 12%. Nevertheless, AM could only be suitable for motorsport and low production volume applications due to the increased cost.

29.3.2.2 Personalization and Mass Customization

Due to the geometrical flexibility, fully digitalized process, and minimal (or even zero when no material change is needed) changeover time and cost, AM is ideal for lot-size one production. This is a key enabler of personalized and mass customized production, one of the hot topics in the automotive industry.

Porsche provides additively manufactured personalized bucket seat inserts, made from TPU powder using a BJT process [41]. This fully customized part comes along with a lightweight structure and a minimized environmental footprint aiming also to link the motorsport nature of the brand with more conventional vehicles. Moreover, the customer can select the desired firmness levels: soft, medium, and hard, ensuring a perfect fit for the customer's body.

Stellantis is offering accessories that can be individualized and produced via AM, through the new Peugeot 308 series. The material used for printing is a flexible polymer while the selected AM process is BJT. A range of accessories is offered, such as sunglasses, cup or phone holders. The use of AM enhanced design freedom, as also the agility of the production toward the different consumer demands [42].

29.3.2.3 Spare Parts on Demand

Spare or replacement parts for older vehicles are often discontinued, as maintaining them in stock comes with a significant overhead cost for automotive companies. AM processes can provide parts on demand, without any change/modification of the machine set up. Only the 3D CAD model and the part requirements are required to produce spare parts for old cars that have been out of stock. On the other hand, significant time and resources are needed to re-produce these parts conventionally. Therefore, AM seems to be perfect for producing spare parts on demand, contributing also to inventory reduction while releasing machine tools and tooling for series production.

BMW has been able to manufacture spare parts for restoring the door handle and window lever of the legendary BMW 507, with the use of AM [6]. Graphite AM company has produced together with Eagle, a company that restores E-Type Jaguars, a heater/air conditioning duct that fits underneath the dash top that has better quality and fits better than the cheaply produced OEM part, which is hard to find due to the age of the vehicle [43]. Wolfs Engineering has also developed a heat and fuel-resistant air intake manifold for a 1970s Classic Mini which was hard to find on the market and the second-hand parts were too expensive. They have also developed velocity stacks for the intake that improved the power of the engine by almost 10%. The products are now sold to other Mini enthusiasts, while the company also teamed up with a Mazda car restoration specialist to further expand their business [44].

Porsche runs a program for cost-effective part replacement by creating spare parts for rare and classic cars with AM processes, to ensure customer satisfaction. For the provided parts they use the PBF-LB process for metal parts and polymer parts, alongside suitable materials for each specific case [45]. Each part is inspected with Non-Destructive Testing techniques, to ensure the high quality and performance that is expected from Porsche. Among the parts that are made with AM, processes are the clutch of Porsche 959, a filler cap, a crank arm, and a variety of brake parts. Similarly, AUDI provides water connectors for the W12 engine that are made from Aluminum alloys using the PBF-LB process [46].

29.3.2.4 Aftermarket/Niche Accessories

This category could fit under the mass customization one; however, aftermarket accessories are typically standardized products, exhibiting higher performance, and sold individually in much smaller quantities. AM is slowly increasing in presence in this field as well, mainly due to design freedom and suitability to small lot sizes, allowing for components to be optimized performance (including reduced weight) compared to the ones they replace, as well as for individualization/personalization of products. This category includes both polymer and metallic components.

In terms of polymer components, the primary target market is manufacturing personalized interior car accessories. These can be used to retrofit new technology to older generation cars (in-car infotainment systems for example), but also to individualize and personalize the vehicle interior. This concept is slowly endorsed by OEMs as well; as an example, Ford has released CAD files that allow owners and aftermarket companies to design and manufacture via AM their accessories [47].

In terms of metallic components, the main goal is to produce parts with increased performance. Most applications are currently focusing on flow improvements (such as engine intake and exhaust systems [48]); this is possible due to the design flexibility enabled by AM. In addition, AM allows for light-weighting (and thus further increased performance of the components), as it can produce much thinner walls without pores when compared with casting processes which would be the alternative for such components. Coupled with topology optimization, AM can also enable the manufacturing of lightweight structural components, such as wheels [49].

29.4 DISCUSSION

As it is evident from the previous section, AM processes are very well positioned for several uses within the automotive manufacturing environment. This is particularly true when used as production assistive technologies. Nevertheless, it can be deduced that in terms of end-use parts, automakers have decided to adopt the AM processes mostly in concept -limited edition cars. The main reasons for this are:

- **High investment costs.** The investment cost for AM machines and post-processing steps, as well as the cost of design services (topology optimization) for optimized load paths and conformity to Design for AM guidelines, are factors that contribute to increased AM costs. On top of that, the low maturity of AM processes dictates an iterative approach until a successful result is achieved. In addition, post-processing steps to achieve the desired surface quality, dimensional accuracy, and material properties are case-dependent, which means that these steps are required each time a new concept comes to the design office. This leads to the conclusion that significant funds are required, which can be covered only by increasing the selling price of parts.
- **Low production rates**. The multiple steps that are required after the AM process until the part be ready for use, alongside the significant time that is required for each one of these processes make them suitable for small batches that can be used either in limited edition cars or as spare parts for classic vehicles. In addition, they can be used in fully customized vehicles that always need extended delivery time or need to be developed based on the customer requirements.
- **Low maturity of processes**: Many of the parts that are manufactured through AM are prototypes or used in prototype vehicles aiming to prove the capabilities of this advanced manufacturing technology. To this end, automotive industries prefer mainly cars with short and strictly specified maintenance schedules, examining extensively these parts. It is true to say that industries want to take advantage of this technology more in the future so they will find the proper way to standardize all the required aspects and speed up the adoption of AM processes in a wider range of applications in the automotive industry.
- **Small working envelope**. Automotive companies are increasingly investing in modular and multi-functional concepts that can enable them to use the same basic platform for as a large variety of car models as possible, without significant needs for hardware reconfiguration in the production line. This flexibility could be provided through the use of AM, as demonstrated by a sports car that uses chassis components made through AM [50], enabling customized suspension behavior by selecting suspension type (push rod, pull rod, etc.) or different wishbone designs/materials. However, large-scale AM chassis components are not a viable option with the currently available AM technologies and machines.

A critical factor in the successful implementation of AM in an automotive environment is understanding the opportunities and limitations of available AM technologies, and following a structured approach for selecting candidate parts, materials and processes.

The first step involves selecting parts from the portfolio of the company that could be candidates for transitioning to manufacturing using AM. Selection can be done based on technical or economic criteria, or most commonly a techno-economic approach containing both. Successful implementation examples given herein could act as a starting point for companies. A list of criteria that should be examined and taken into account when deciding on potential parts that can be produced via AM should be created. This will act as an initial checklist, ensuring that the users will consider the correct factors and will not overlook any aspects in this initial decision-making process. Parts that exhibit the following characteristics are in general good candidates for production with AM and should be prioritized:

- Parts with high geometric and functional complexity. This includes parts whose performance could be improved by increasing their complexity.
- Parts that reducing weight is important.
- Parts with frequent design revisions, both for operational reasons but also due to customization/personalization potential.
- Parts whose tooling costs represent a large percentage of the total cost of the part.
- Parts with elaborate and time-consuming assembly and joining procedures (typically fabricated assemblies).
- Parts with high purchase value or manufacturing costs can be prioritized.
- Parts with high inventory cost.
- Parts with relatively low production volume and/or low demand rate. This implies on the cycle time; parts with short cycle times should be avoided.
- Parts that normally have a high MOQ from the supplier (or a very elaborate and costly process to set up for internal production).
- Parts with a normally long lead time.
- Customer critical parts (high-demand parts that are critical to the business operation) should normally be avoided. Nevertheless, in times of supply chain disruption, AM should be examined as a potential solution for reducing lead time for CCPs.
- Obsolete parts/spare parts on demand.

Upon selecting candidate parts, their operating conditions and environment (in terms of loads, temperature, electrical, thermal, and chemical properties) should be defined, select suitable materials for the application. AM processes are limited in terms of processable materials, so a general knowledge of the material families available can be beneficial. It is worth mentioning that AM part properties are generally anisotropic (with lower values on the z-axis), which however can be mitigated through post-processing. In addition, the material properties of AM components are generally lower compared to machined and/or forged components made from the same material.

The main factors that affect the selection of a particular AM process are the size and accuracy requirements for the part. In general, parts with low to medium precision requirements (tolerances, surface roughness) should be considered to be produced with AM. Moreover, AM machines typically have a smaller working volume compared to their conventional counterparts. The selection of appropriate material(s) can also influence the AM process to be used, as not all process families are capable of processing all materials. A general guideline (including typical build volume sizes and processable materials) is given in Table 29.1.

29.5 CONCLUSIONS

Due to their unique characteristics, AM processes find an increasing number of applications in the automotive environment; from product development to production phases, either for the creation of prototypes, tooling, or direct production of parts. Each application area has its requirements; as such different kinds of AM processes and materials are applicable in each one of those, leveraging different aspects/competitive advantages granted by the technology.

A booming application field that is going to significantly increase in the coming years is the indirect use of AM. Placeholders and prototypes built with AM are becoming commonplace, with polymer AM tooling (jigs, fixtures, and grippers) also gaining traction. Nevertheless, these are relatively low-cost, low-risk, low-return investments. One of the most promising applications that

TABLE 29.1
Summary of AM Process Families, Materials, Build Volumes, Prost Processing Needs, Target Applications and Competitive Advantages

Process, Mechanism, Build Volume (mm^3)	Typical Materials	Post-Processing Needs	Target Applications	Competitive Advantage
PBF Melting and solidification of powder particles in a powder bed **Typical:** *250x250x250* **Maximum:** *750x550x550*	• Polymers (PA6, PA11, PA12, Glass/carbon-filled PA, PP, PEEK) • Metal alloys (Steel, Titanium, Tungsten-based superalloys, Nickel-based superalloys, Cobalt chrome, Copper, Precious metals)	• Heat treatment • Supports removal • Machining to meet tolerances • Surface finishing	• Direct metal and polymer part production • Conformally cooled tooling • End-use parts with good mechanical properties	• Low cost from a material point of view • Strong and durable parts • Minimal /no support material required for polymer PBF • Good dimensional accuracy
DED Melting and solidification of powder or wire directly in the process area **Typical:** *800x800x600* **Maximum:** *6000x2000x2000*	• Weld-able metals available in powder or wire form (Steel alloys, Titanium alloys, Aluminum alloys, Nickel-based superalloys)	• Heat treatment • Machining to final shape • Surface finishing	• Parts where a near net shape is acceptably machined as a secondary operation • Repairing damaged or worn parts • Large scale tooling	• High deposition rate • Large build volume • Good mechanical properties • Suitable for repairs
VP Photo-polymerization of liquid resins **Typical:** *145x145x175* **Maximum:** *1500x750x550*	• UV Curable photopolymer resins (PC-like, PP-like, ABS-like) • Ceramic filled resins	• UV curing • Supports removal • Surface finishing/ treatment	• Functional transparent prototypes • Composite parts tooling • Short-run injection molding tooling • Intricate parts with small details • Replacement of injection molded parts	• Very high accuracy • Best surface finish • Can produce transparent parts • Isotropy in properties • No porosity • Wide range of material properties depending on the resin system used
MEx Melting, extrusion, deposition, and solidification of the material **Typical:** *200x200x200* **Maximum:** *1000x1000x1000*	• General purpose polymers (PLA) • Engineering polymers (ABS, PETG, ASA, PP, PC) • High-performance polymers (PPSF, PPSU, PEEK, PEKK) • Particle or fiber-reinforced polymers • Metal filled polymers	• Supports removal • Debinding & sintering (for metallic parts) • Surface finishing	• Proof of concept prototypes • Placeholders • Jigs, fixtures, and grippers • Forms and dies	• Most affordable machines • Prints in standard engineering thermoplastics • Low-cost materials available • Easy to use • Versatile in terms of size

(*continued*)

TABLE 29.1 (Continued)
Summary of AM Process Families, Materials, Build Volumes, Prost Processing Needs, Target Applications and Competitive Advantages

Process, Mechanism, Build Volume (mm^3)	Typical Materials	Post-Processing Needs	Target Applications	Competitive Advantage
MJ Melting, jetting, and solidification **Typical:** *380x250x200* **Maximum:** *1000x800x500*	• UV-photosensitive resins • Acrylic photopolymers (thermoset) • Polypropylene, HDPE, PS, PMMA, PC, ABS, HIPS, EDP	• Supports removal • UV curing • Surface finishing	• Proof of concept prototypes • Investment casting patterns • Short-run injection molding tools	• Good surface finish • Can produce transparent parts • Can produce multi-material and full-color parts • Can make wax parts for investment casting
BJT Joining of powder particles through binder (adhesive) **Typical:** *800x500x400* **Maximum:** *2200x1200x600*	• Metals • Ceramics • Polymers	• Surface finishing • Sintering (for metallic parts)	• Proof of concept prototypes • Functional prototypes • Sand-molds for casting • Ceramic parts • Metal part production (with sintering)	• Fast process without the need for support • Wide variety of powder materials • Can produce full-color parts • Can make metal parts (after sintering)
SL Cutting sheets of material per layer and adhesive bonding **Typical:** *260x380x380* **Maximum:** *800x500x500*	• Paper • Polymer film • Composites	• Surrounding material removal • Surface finishing	• Very limited prototyping applications	• Low cost • Environmentally friendly • Ease of material handling

would significantly disrupt the industry is the use of AM to produce tooling for series production; either conformally cooled injection molding cavities, or hot stamping forms and dies, as it can offer massive benefits in terms of cycle time reduction and process performance. Low-cost AM tooling for small volume production is expected to also increase in use, albeit with limited effects on the industry at large.

In terms of end-use parts, expecting AM technologies to replace conventional manufacturing processes for mass-produced components is unrealistic. However, AM is going to be increasingly used in niche market vehicles due to increased performance. This is expected to be further supported by decreasing costs and increased robustness of available AM machines, combined with the emergence of low-cost engineering polymers with much improved properties, that could be used to replace conventionally metallic components, even in semi-structural applications. In addition, the flexibility granted by AM processes makes them ideal for mass customization applications, an ever-increasing trend in the automotive sector. On-demand production with minimal need for the reconfiguration will boost AM's use for producing obsolete spare parts for classic cars. This is a benefit that will also lead to increased adoption of AM in the aftermarket sector, where increased performance and the need for personalization are key, and production volumes and/or rates are significantly lower.

REFERENCES

1 ASTM International (2012). *Standard Terminology for Additive Manufacturing Technologies: Designation F2792-12a*. West Conshohocken, PA: ASTM International.
2 H. Bikas, P. Stavropoulos, G. Chryssolouris, Additive manufacturing methods and modeling approaches: A critical review. *International Journal of Advanced Manufacturing Technology*. 83 (2016) 389–405.
3 3D Printing Trend Report 2022, www.hubs.com/get/trends/, Retrieved July 28, 2022.
4 SmarTech Analysis (2019). Additive Manufacturing Opportunities In Automotive, www.smartechanalysis.com/reports/additive-manufacturing-automotive-part-production-2019/, Retrieved July 27, 2022.
5 3D Printing in Automotive Market Size, By Component, By Technology, By Application, Industry Analysis Report, Regional Outlook, Growth Potential, Competitive Market Share & Forecast, 2018–2024. Global Market Insights Inc. from www.gminsights.com/industry-analysis/3d-printing-in-automotive-market, Retrieved July 25, 2022.
6 Additive Manufacturing: 3D Printing to Perfection, www.bmw.com/en/innovation/3d-print.html, Retrieved July 28, 2022.
7 SEAT and 3D Printing: The Next Big Thing in Automotive Manufacturing?, www.bcn3d.com/seat-and-3d-printing-the-next-big-thing-in-automotive-manufacturing/, Retrieved July 24, 2022.
8 Chrysler Engineers See Clearly with 3D Printed Transparent Differential Housings, www.makepartsfast.com/chrysler-engineers-see-clearly-3d-printed-transparent-differential-housings/, Retrieved July 26, 2022.
9 How Is 3D Printing Transforming the Automotive Industry? (2021), https://amfg.ai/2018/06/13/3d-printing-transforming-automotive-industry/, Retrieved July 29, 2022.
10 V. Lunetto, A. R. Catalano, P. C. Priarone, A. Salmi, E. Atzeni, S. Moos, L. Iuliano, L. Settineri, Additive manufacturing for an urban vehicle prototype: Re-design and sustainability implications. *Procedia CIRP*. 99 (2021) 364–369.
11 Additive Manufacturing in the Automotive Industry, www.futurebridge.com/blog/additive-manufacturing-in-the-automotive-industry/?print=print, Retrieved July 27, 2022.
12 Additive Manufacturing in Product Development Processes, https://industrytoday.com/additive-manufacturing-in-product-development-processes/, Retrieved July 28, 2022.
13 Metal Replacement Market by End-Use Industry (Automotive, Aerospace & Defense, Construction, Healthcare, Others), Type (Engineering Plastics, Composites), and Region – Global Forecast to 2021, www.marketsandmarkets.com/Market-Reports/metal-replacement-market-87266199.html, Retrieved July 28, 2022.

14 Volkswagen Autoeuropa: Maximizing Production Efficiency with 3D Printed Tools, Jigs, and Fixtures, https://ultimaker.com/learn/volkswagen-autoeuropa-maximizing-production-efficiency-with-3d-printed, Retrieved August 02, 2022.

15 10 Examples of 3D-Printed Tooling, www.additivemanufacturing.media/articles/big-ideas-in-am-10-examples-of-3d-printed-tooling, Retrieved July 28, 2022.

16 Carbon Fiber Parts Manufacturing with 3D Printed Molds, https://3d.formlabs.com/white-paper-carbon-fiber-parts-manufacturing-with-3d-printed-molds/#form, Retrieved July 21, 2022.

17 M. Soshi, J. Ring, C. Young, Y. Oda, M. Mori, Innovative grid molding and cooling using an additive and subtractive hybrid CNC machine tool. *CIRP Annals – Manufacturing Technology*. 66 (2017) 401–404.

18 R. Surace, V. Basile, V. Bellantone, F. Modica, I. Fassi, Micro injection molding of thin cavities using stereolithography for mold fabrication. *Polymers*. 13 (2021) 1848.

19 Sand Mold Casting Applications, www.exone.com/en-US/Resources/case-studies/Sand-Mold-Casting-Applications, Retrieved June 03, 2022.

20 Production in 3D Printing Process of the Sand Core for the Water Jacket of the All-New High-Performance M TwinPower Turbo 6-Cylinder Petrol Engine, www.press.bmwgroup.com/global/photo/detail/P90335316/Production-in-3D-printing-process-of-the-sand-core-for-the-water-jacket-of-the-all-new-High, Retrieved July 26, 2022.

21 Stress Testing 3D Printed Parts on a Trumpf Press Brake, www.centerlinees.com/blog/2018/1/26/stress-testing-3d-printed, Retrieved July 21, 2022.

22 Getting to Production Faster with 3D-Printed Press Brake Tooling, www.additivemanufacturing.media/blog/post/getting-to-production-faster-with-3d-printed-press-brake-tooling, Retrieved July 22, 2022.

23 R. Leal, F. Barreiros, M. Alves, L. Romeiro, F. Vasco, J. C. Santos, Additive manufacturing tooling for the automotive industry. *International Journal of Advanced Manufacturing Technology*. 92 (2017) 5–8.

24 Z. Yang, H. Hao, Q. Gao, Y. Cao, R. Han, H. Qi, Strengthening mechanism and high-temperature properties of H13 + WC/Y2O3 laser-cladding coatings. *Surface and Coatings Technology*. 405 (2021) 126544–126555.

25 M. Cortina, J. I. Arrizubieta, A. Calleja, E. Ukar, A. Alberdi, Case study to illustrate the potential of conformal cooling channels for hot stamping dies manufactured using hybrid process of laser metal deposition (LMD) and milling. *Metals*. 8 (2018) 102.

26 Bilsing Adds Laser Metal Deposition Repair Services for Aero Components and Stamping Dies, www.metalformingmagazine.com/article/?/additive-manufacturing/additive-processes/bilsing-adds-laser-metal-deposition-repair-services-for-aero-components-and-stamping-dies, Retrieved July 26, 2022.

27 Innovation in Automotive, https://raeng.org.uk/media/at5b35sq/innovation_in_automotive_report.pdf, Retrieved June 09, 2022.

28 The Future of Automotive and Mobility, www.forbes.com/sites/sap/2021/05/05/the-future-of-automotive-and-mobility/?sh=31e6634d59d5, Retrieved June 09, 2022.

29 What Is KERS? How It Is Used in Formula One?, www.mechead.com/kers-used-formula-1/#:~:text=The%20acronym%20KERS%20stands%20for,called%20upon%20to%20boost%20acceleration, Retrieved June 09, 2022.

30 KERS Technology Races to Le Mans, www.electronicsweekly.com/market-sectors/power/kers-technology-races-to-le-mans-2012-06/, Retrieved June 09, 2022.

31 Dematerialization of the Warehouse and on Demand Manufacturing: All the Advantages of Roboze 3D Printing, www.roboze.com/en/resources/roboze-3d-printing-a-new-opportunity-to-produce-peek-gears.html, Retrieved June 09, 2022.

32 Metal Replacement and Application Solutions: The Motor Crankcase, www.roboze.com/en/resources/metal-replacement-and-application-solutions-the-motor-crankcase.html, Retrieved June 09, 2022.

33 And Suddenly the Printed Wheel Carrier Brakes, www.iapt.fraunhofer.de/en/press-media/Press_releases/Press_Release_Fiat_Chrysler.html, Retrieved May 29, 2022.

34 Porsche Partners with Mahle and Trumpf to 3D Print Pistons for Its 911 Supercar, https://3dprintingindustry.com/news/porsche-partners-with-mahle-and-trumpf-to-3d-print-pistons-for-its-911-supercar-173852/, Retrieved May 29, 2022.

35 Prototype for Small-Series Production: Electric Drive Housing from a 3D Printer, https://newsroom.porsche.com/en_US/technology/porsche-protoype-small-production-electric-drive-housing-3d-printer-23277.html, Retrieved July 28, 2022.

36 Apworks 3D Prints Titanium Part for New Bugatti Chiron Pur Sport, https://3dprintingindustry.com/news/apworks-3d-prints-titanium-part-for-new-bugatti-chiron-pur-sport-170374/, Retrieved May 29, 2022.

37 World Premiere Brake Caliper from 3D printer, www.bugatti.com/media/news/2018/world-premiere-brake-caliper-from-3-d-printer/, Retrieved May 29, 2022.

38 T. R. Marchesi, R. D. Lahuerta, E. C. N. Silva, M. S. G. Tsuzuki, T. C. Martins, A. Barari, I. Wood, Topologically optimized diesel engine support manufactured with additive manufacturing. *IFAC-PapersOnLine*. 28 (2015) 2333–2338.

39 A. Merulla, A. Gatto, E. Bassoli, S. I. Munteanu, B. Gheorghiu, M. A. Pop, T. Bedo, D. Munteanu, Weight reduction by topology optimization of an engine subframe mount, designed for additive manufacturing production. *Materials Today: Proceedings*. 19 (2019) 1014–1018.

40 D. Cooper, J. Thornby, N. Blundell, R. Henrys, M. A. Williams, G. Gibbons, Design and manufacture of high performance hollow engine valves by Additive Layer Manufacturing. *Materials and Design*. 69 (2015) 44–55.

41 Porsche Set to Launch 3D Printed Personalized Bucket Seats, https://3dprintingindustry.com/news/porsche-set-to-launch-3d-printed-personalized-bucket-seats-170320/, Retrieved May 29, 2022.

42 New Technology: 3D Printing in the New Peugeot 308, www.media.stellantis.com/em-en/peugeot/press/new-technology-3d-printing-in-the-new-peugeot-308?adobe_mc_ref=, Retrieved July 19, 2022.

43 Case Study: Graphite AM & Eagle, www.graphite-am.co.uk/case-studies/graphite-and-eagle/, Retrieved July 287, 2022.

44 How a Classic Car Enthusiast Solved the Spare Parts Dilemma, https://am.covestro.com/content/dam/dsm/additive-manufacturing/en_US/documents/covestro/am-casestudy-novamid-wolfs-engineering-en-digital.pdf, Retrieved July 22, 2022.

45 Porsche Green-Lights 3D Printed Spare Parts for Classic Car, https://3dprintingindustry.com/news/porsche-green-lights-3d-printed-spare-parts-classic-cars-128792/, Retrieved May 31, 2022.

46 Additive Manufacturing in the Automotive Industry, www.futurebridge.com/blog/additive-manufacturing-in-the-automotive-industry/, Retrieved May 29, 2022

47 How Do I Use 3D Printing to Customize My Ford Maverick Pickup?, www.ford.com/support/how-tos/owner-resources/vehicle-documents/how-do-i-use-3d-printing-to-customize-my-ford-maverick-pickup/, Retrieved July 19, 2022

48 BBi 991/997.2 Intake Manifold, https://bbiautosport.com/products/bbi-991-9972-dfi-intake-manifold-system, Retrieved July 19, 2022

49 The future of wheel technology, www.hrewheels.com/wheels/concepts/hre3d, Retrieved July 19, 2022

50 Metal AM in automotive: How the Czinger 21C is redefining next-generation car manufacturing, www.metal-am.com/articles/metal-3d-printing-in-automotive-how-the-czinger-21c-is-redefining-next-generation-car-manufacturing/, Retrieved May 29, 2022.

30 Materials and Challenges of 3D Printing for Defense Applications and Humanitarian Actions

*Rafael L. Germscheidt, Mariana B. Silva, Evandro Datti, and Juliano A. Bonacin**
Institute of Chemistry, University of Campinas, PO Box 6154, ZIP Code 13083-970, Campinas, SP, Brazil.
*Corresponding authors: jbonacin@unicamp.br

CONTENTS

30.1 INTRODUCTION TO 3D PRINTING AND THE IMPORTANCE OF ON AND OFF-SITE PRODUCTION

The additive manufacturing (AM) process, or 3D printing, builds an object by depositing successive layers of the printing material on top of each other. 3D printing is versatile and customizable, does not require molding tools, allows high design flexibility, and enables free-form fabrication, making it a technology that can be applied in many sectors of industrial production [1], [2]. Additionally, new applications for 3D printing are continuously emerging as new technologies and materials are established. As 3D printers have become more affordable and popular, it has been common to see them in spaces such as laboratories, houses, and schools being used for prototyping, the printing of educational items, individual objects, and simple maintenance parts[3], [4]. Its use in the construction industry has also been developed and is a great example to demonstrate how 3D printing can also be used for the production of complex structures from different types of materials. Whereas 3D printing technology was once restricted to designers and engineers, recent advances and improvements have made 3D printing more affordable and even more versatile.

Currently, there is a wide range of AM techniques that comprise different types of materials. Although polymers are the most popular type of material for 3D printing, especially for prototyping, it is also possible to obtain printed parts from ceramics, metals, concrete, and composite materials, as well as hybrid and functional materials[3], [4]. These materials are employed as solids, liquids,

DOI: 10.1201/9781003296676-30

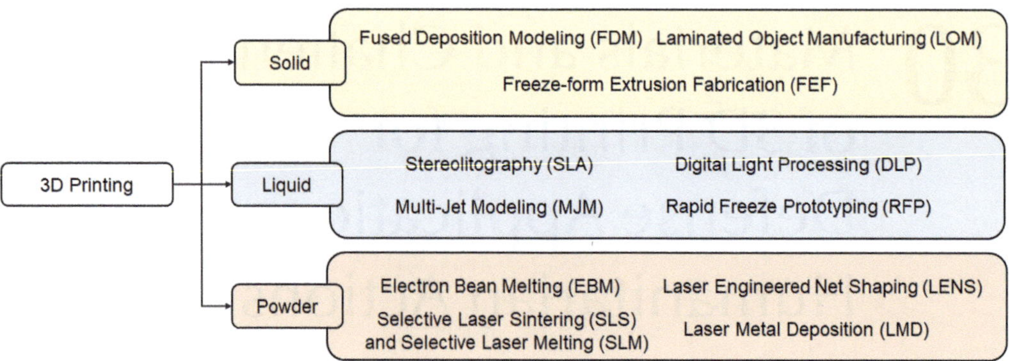

FIGURE 30.1 Examples of 3D printing processes classified according to the physical state of the precursor material.

or powders, depending on the processing technique applied [1]. AM techniques can be classified according to the type of material used, as summarized in Figure 30.1.

The most common method for 3D printing is Fused Deposition Modelling (FDM) due to its accessibility and ease of use. This method employs a filament wire made of a thermoplastic polymer that is heated to reach a semi-liquid state and then extruded through a size-controlled nozzle [5]. Besides FDM, which is an extrusion technique, contour crafting (CC), stereolithography (SLA), digital light processing (DLP), powder bed fusion processes, inkjet printing, direct energy deposition (DED) and laminated object manufacturing (LOM) are currently used technologies for 3D printing [3]. The main AM methods used are summarized and briefly explained in Table 30.1, based on the technology they apply.

As discussed above, the versatility and flexibility of 3D printing, as well as the varied available techniques and the possibility of applying different types of materials are characteristics that make 3D printing a suitable tool for different applications. The flexibility of 3D printing is especially interesting when we consider areas with constrained access for different reasons, such as war zones, areas affected by natural disasters, or blocked access due to political conflicts. In such cases, developing and fabricating necessary objects locally or as close as possible to the application place with speed and flexibility is essential to supply medical, protection, defense, and maintenance tools.

In this sense, it is important to highlight the concept of on and off-site production, as well as the role of 3D printing in these scenarios. On-site production means that the required object or structure is fabricated at the same place as it will be applied, while off-site production has the required part fabricated in one place, such as a production or distribution center, and then transferred to where it will be applied. When considering remote locations or places with constrained access, 3D printing can be used to perform both on and off-site production of required objects, as it presents flexibility, fast production and many 3D printers can be easily transported. Off-site 3D printing has been widely employed in the construction industry, for example, where robust printing machines are required to obtain more complex structures, which are then transported to their place of application. However, 3D-printing technologies such as FDM and SLA can be transported to remote locations, as they are usually small machines that do not require additional equipment.

In this chapter, different applications of AM are explored, always considering its flexibility and potential to assist in remote locations or restricted regions. Given this scenario, applications for the defense industry are addressed, considering military and machine maintenance, as well as possible applications for humanitarian actions including medical and protection equipment, the printing of individual objects, and resources for water treatment and purification. Figure 30.2 presents a scheme of the possible applications of AM that will be explored.

TABLE 30.1
Summary of Main Additive Manufacturing Methods, Separated Based on the Technology They Apply [1], [3], [4], [6]

	AM Technique	Process	Materials
Material extrusion	Fused deposition modeling (FDM)	The solid material is extruded through a nozzle at appropriate temperature, deposited onto a heated platform and then layer-by-layer	Thermoplastic filaments, composites (solids)
	Contour Crafting (CC)	Applying high pressures and large nozzles allows for the extrusion of material for large structures	Concrete or soil, as a concentrated dispersion of particles (ink or paste)
Photopolymerization	Stereolithography (SLA)	A photocurable resin is hardened with a laser as a platform moves downwards after each layer of liquid resin is cured	Liquid photopolymers
	Digital Light Processing (DLP)	The resin is cured with a projector	Liquid photopolymers
Powder bed fusion	Selective Laser Melting (SLM) and Selective Laser Sintering (SLS)	A laser source is used to melt or sinter a fine powder bed which is then fused on top of a previous layer	Compact fine powders of metals, alloys, and some polymers (solids)
Material jetting	Inkjet printing (3DP)	A stable material suspension is pumped and deposited as droplets via the injection nozzle onto a substrate	Ceramics, metals, polymers
Direct Energy Deposition	Laser Engineered Net Shaping (LENS)	A laser is directed to a small region of the substrate and used to melt the feedstock material simultaneously. The melted material is deposited onto the melted substrate and then solidified after the laser is moved	Metal wire or powder (solid)
	Electron Beam Melting (EBM)	A metal wire is melted by using an electron beam and forms an object layer by layer	Metal wire (solid)
Sheet lamination	Laminated Object Manufacturing (LOM)	Layers in sheet form are cut with a mechanical cutter or laser, and then linked together	Metal-filled tapes, polymer composites, ceramics (solid)

30.2 MATERIALS AND STRATEGIES IN 3D PRINTING FOR DEFENSE APPLICATIONS AND HUMANITARIAN ACTIONS

As previously mentioned, the flexibility and speed of 3D printing make it a suitable technique to supply required objects in areas with limited access, where the traditional manufacturing process would find obstacles such as complicated logistics and shortage of raw materials. In this sense, 3D printing is an important tool for defense applications and to help humanitarian actions in times of crisis, such as in war zones and places affected by conflicts and natural disasters.

3D printing has been increasingly adopted by the defense industry worldwide, comprising a sector that is expected to be worth 1.7 billion dollars by 2027 [7]. This impressive growth is expected as governments and armed forces around the world have recognized the potential of AM to shorten supply chains, reduce costs, and promote military readiness by 3D printing all types of parts, from

FIGURE 30.2 Applications of AM: 3D printing can be applied in the defense and aerospace industry, individual protective equipment as well as diverse individual objects, water treatment, and purification, parts for maintenance of machines, and medical and humanitarian applications.

individual objects to be used in the field, such as rifle grips, to engine parts and bunkers to support soldiers in the field [8].

Equipment for soldiers' protection is one example of how 3D printing can be used for defense applications. The US army and the 3D design company General Lattice have been working on a 3D-printed helmet with advanced lattice geometries that intends to provide better impact absorption and improve soldier protection in the field [9]. The development of materials with outstanding physical and mechanical properties is also required to obtain efficient protection equipment. Carbon-based composites, for example, have been widely explored for their applications in the defense industry. In this sense, some works have demonstrated the potential of 3D printing to obtain carbon-based composites with great mechanical properties [10], [11]. Tian et al., for example, have presented a continuous fiber reinforced thermoplastic composite (CFRTPC) based on continuous carbon fiber and polylactic acid (PLA), obtained by fused deposition modeling, with easy control of fiber content and advanced mechanical performance [11].

In addition to individual protection equipment, AM can also be used to produce weapons parts in a short period of time, allowing for fast repairing of damaged weapons. In the field, the possibility of repairing weapons based only on the damaged part, without the need for transportation of spare parts, could save time, and improve combat power [12], [13].

AM can also be used for transportation and shelter of military units. A 3D-printed runway has been developed for military expeditionary airfields using direct metal laser sintering technology. A runway is a component used on weaker ground surfaces to allow military aircraft to land and take off safely [7]. Ships also benefited from 3D printing: a ship's propeller has been produced, this time using wire arc AM process and mainly bronze material, greatly reducing the construction time and minimizing the amount of material used [14]. In a more ambitious project, the US military intends to 3D print the structure of combat vehicles in one piece, which will significantly reduce the cost and weight of vehicles. The type of structure aimed by the US military, called the monolithic hull, is not a new idea, however, the available manufacturing technologies so far did not prove cost-effective nor suitable for large-scale production. With AM the production of monolithic hulls can be faster and provide a better design that will make the structure more durable [8].

AM can also be used to fabricate barracks and bunkers for military shelter, as well as more complex structures such as bridges to facilitate the movement of military units. The manufacturer Icon has demonstrated the construction of a concrete structure for a bunker able to hide a truck-mounted multiple rocket launcher system in just 36 hours, as well as training barracks. The system uses robotic 3D printers and quick-drying cement-based material to fabricate large structures that are then assembled into the desired layout [7], [8].

Aside from defense applications, 3D printing in areas with limited access can also greatly benefit humanitarian actions by providing required supplies in times of need. A great example of how AM can be applied to help the population in restricted areas due to wartime is currently found in Ukraine, where an invasion from Russia started in February 2022: a small number of printers and printing filament were sent to Lviv, in western Ukraine, not long after the conflict started, and then a bigger team has joined the initiative to gather 3D printer owners across Europe available to print needed items. This initiative has been able to produce and send to Ukraine protective gear, tourniquets, medical equipment such as periscopes, and drones [15]. Another recent example is the COVID-19 pandemic, which had its outbreak in March 2020 and, due to the intense and sudden demand, many medical and individual protection equipment were in dangerously short supply. 3D printing technology was widely used to provide local and fast replacements for these items. Furthermore, the mobilization that took place at this time has shown many emerging applications for 3D printing and greatly strengthened the outreach of AM [16].

Humanitarian assistance can be defined as the actions taken to save lives, alleviate suffering and maintain human dignity during crises caused by man or natural hazards, as well as actions to prevent and strengthen vulnerable regions against such situations [17]. In this sense, humanitarian actions include the provision of resources such as food, water, basic hygiene items, individual protection equipment, and medical supplies, as well as safe shelter when needed.

AM can have a great impact on humanitarian aid actions as it provides a versatile and fast alternative to fabricating parts. It can be used to provide missing or replacement parts that would take a long time to arrive, allowing workers to design and fabricate needed parts on demand. This can be applied for both infrastructure demands, such as transportation and housing, and medical equipment, to keep them functional during the crisis [18].

It can also be applied to provide housing infrastructure [19], individual objects, as will be further discussed, and to produce different types of devices, such as medical and biomedical equipment (for example, stethoscopes, umbilical cord clamps, microscopes, and defibrillators) [20], miniature diagnostic kits and different electrochemical devices, such as sensors for different applications [21].

30.3 3D PRINTING FOR INDIVIDUAL OBJECTS

Applications of 3D printing for fabricating individual objects are numerous and varied. Firstly, this technology is continuously employed and improved by the army to enhance the capability to produce smaller and lighter precision weaponry. Researchers from the US Army Aviation and Missile Research, Development, and Engineering Center (AMRDEC), for example, are creating nano-based structures and components for advanced weaponry, aviation, and autonomous air/ground systems applications [22].

As aforementioned, AMRDEC keeps investigating methods linked to 3D printing for direct fabrication of antennas, interconnects, printed circuit boards, sensors, and other devices. There are also some research programs focused on developing stretchable materials using 3D printing to obtain form-fitting objects that are amenable for wearable applications, which include smart physical assisting and therapeutic wearable devices that can be adjusted to a certain patient condition and the body for thermotherapy [22].

A noteworthy research and development program called PRIntable Materials with embedded Electronics (PRIME2) was initiated by the US Army AMRDEC. The goal of this program was to create new fabrication capabilities that integrate Radio Frequency (RF) and electronic components into AM processes, aiming to reduce the size, weight, and overall cost of weaponry components and subsystems. One of the main objectives of this project was the use of a one-step process to print an entire wiring board with embedded passive components and integrated RF structures [22].

RF structures and electronic components manufactured by the conventional method involve piecemeal assembly techniques, which takes a lot of time and presents drawbacks associated with the manufacturing process for fabricating antenna elements, RF structures, and connections. On the other hand, AM provides the reduction of complexity, elimination of the requirement for assembly parts, and other advantages, meaning that it has the potential to revolutionize existing weaponry systems by significantly reducing the size, weight, and cost [22].

In addition to military applications, 3D printing has also found use in healthcare, automotive, defense industries, aerospace, pharmaceutical, and several other fields that will be briefly discussed in this text. A particular area in which 3D printing utilization is increasing a lot is medicine, providing the possibility of creating customized implants, prostheses, medical models, and medical devices [23].

The use of AM allows the manufacturing of three-dimensional solid objects of any desired shape, thus allowing its application in different areas in a fast and easy way. Several personal objects can be 3D printed by using this technique, such as hearing aids, prostheses, stethoscopes, eye lenses, drugs, dental implants, medical face shields, and masks. In the case of hearing aids, for example, its production usually takes more than a week, with the introduction of 3D printing into this domain, it can now take less than a day [23].

To further emphasize the benefits of using AM for individual object design, 3D printed prostheses can be designed according to individual preference and cost much less than traditional alternatives. Besides, they are more comfortable than conventional prostheses and can be fabricated in a day [23].

Therefore, one of the advantages related to this technology is the exclusion of the traditional measurement methods and the reduction of waiting times. 3D printing is often preferred in knee joints, tibia and femur bone, fibula bone implants, and dental implants since it enables precise manufacturing specific to the patient. Moreover, successful surgical imaging of the patient is obtained by employing this process, avoiding complications and mistakes that might arise in the fixture of the implant [2].

An interesting approach based on 3D printing that resulted from recent developments is called bioprinting. It allows biocompatible materials to be 3D printed on living tissues, in other words, it can be used to produce tissues, bone structures, and organs proper for organ transplantation. This technology has also been used for drug discovery, and toxicology, among others [2].

With respect to pharmaceutical applications of AM, which is a relatively new area and it's currently receiving close attention, there are three major subjects in these applications. One of them is a drug delivery system that includes controlled release, polypills, gastrofloating, orodispersible, and microneedles. Moreover, 3D printing is also applied in the development of pharmaceutical devices, including pharmacy dispensing aids and drug-eluting devices [24].

Making the comparison between traditional manufacturing methods and AM in this area, the first one does not provide personalized patient dosing as it is not cost-effective and impractical while 3D printing has high accuracy, a highly adaptable nature, and can be used as an alternative manufacture tool. For instance, a dose can easily be varied by altering the tablet size by using the last method and the resultant tablet can have an accurate dosage for a specific patient. Thus, AM technology may be capable of producing medications on-demand and increasing the accessibility to medicines for those living in remote areas or even in risk areas, such as war zones [24].

Another important aspect related to this issue is that this method can also be utilized to fabricate complex geometries, which has made the production of certain oral dosage forms and medical devices possible. Thereby, high-dose tablets of a given drug can be printed with the use of AM, which is not feasible by using conventional manufacturing methods as a result of the limitations in material blending and tableting compression. In addition, certain types of AM technologies can create highly precise products, making it possible to produce microscale drug delivery systems, such as microneedles (MNs) [24].

Levetiracetam was the first 3D-printed drug produced and has obtained U.S. Food and Drug Administration (FDA) approval. 3D printing enables the creation of very porous pills, obtaining high drug doses in a single pill. It dissolves quickly and can be ingested easily [23].

Furthermore, a great example that highlights the effectiveness of 3D printing in solving problems faced by mankind is the production of masks and other medical devices to protect us from the coronavirus pandemic. Due to the unavailability of supplies during the current COVID-19 pandemic, 3D printing has been increasingly employed to print spare parts for medical devices and protective gear. The benefits of AM over conventional manufacturing are faster production, digital storage and traceability of part files, reduction in delivery time, and the ability to produce components regardless of the complexity of part geometry [25].

As well known, the COVID-19 pandemic has caused a rapid increase in the demand for medical supplies and spare parts, which posed a high risk of materials shortage. In this scenario, 3D printed masks have been manufactured especially for healthcare professionals and, in addition, can potentially be applied to fabricate time-critical parts on demand, like nasal test swabs, face shields, gloves, goggles, respirators, and spares for ventilators [2], [25]. Any type of mask can be designed using CAD-CAM programs within 10 minutes and printed in a short time. Unlike traditional methods, production can be fast and with lower labor costs, 3D printing also prevents the generation of waste through manufacturing on demand, furnishing a greener and more environmentally friendly future [2].

30.4 3D PRINTING FOR OBJECTS FOR MACHINE MAINTENANCE

Machine maintenance regarding the repair work and the replacement of worn or nonfunctional parts is a critical issue since spare parts for products that are at the end of their life cycles are logistically complicated once they are not normally stored in the central warehouse. Thus, these uncommon spare parts occupy valuable space in smaller inventories and take a long time to be transported to the point where their use is needed, leading to a delay in the repair process [26].

Production of machine spare parts with AM on demand and storing them on a server can shorten the repair cycle by making logistics easier. Therefore, the introduction of 3D printing in the repair supply chain decreases the number of items that have to be reimbursed to the client and also reduces the number of objects to be stored [26].

Apart from that, the spare parts provider would benefit from the application of 3D printing in this area, due to the reduced warehousing and logistics costs since the number of units to keep in stock would be reduced, allowing them to get rid of parts that stay a long time in storage without a demand [26].

Considering some cases associated with this matter, the US Navy for instance recently released a plan to connect suppliers who are not able to meet the increasing demand for submarine parts with 3D printing companies that can print the metal parts without stopping to boost supply [27]. The utilization of AM to print spare parts is not something new in the US military, 3D printed parts have been used in critical aircraft engines, on tanks and submarines as described above, and also on the soldiers themselves. This is because 3D printing has offered the Department of Defense incredible supply chain agility and it allows the army to produce new objects rapidly and cost-effectively, on-demand and on-site. Additionally, it can improve the lifespan of aircraft and vehicles that may otherwise be out of service [27]. In another example, Chekurov et al. [10] showed an application of AM to produce a consumer electronic device spare part in offsite repair, which was a memory cover of a Dell portable computer that can break during the repair process. This spare part was successfully fabricated, it fitted without any problem and the computer worked normally.

AM has also been extensively applied in the aerospace industry with regard to the production of highly complex and lightweight parts, besides it can be used for repairing complex components such as engine blades, combustion chambers, and other structures [28]. Durable, light and strong components are required in airplane fabrication and the 3D printing technology can offer the aerospace industry great possibilities to fulfill these challenges. AM has been incorporated by this sector from concept design to end-use parts and repairs, aiming to get rapid prototyping of components over the design phase, manufacture complex shape metal parts, and repair damaged objects rather than scrapping or replacing the broken ones [28]. It is worth mentioning that laser metal deposition (LMD) technology is the most suitable process to repair aerospace parts. It is a type of AM in which the metal powder is directly fed onto the damaged portion of the part and laser cured, restoring the original strength of the component related to the case of LMD for repair [28].

The production of aerospace parts using a conventional manufacturing process generates a high amount of waste of extremely expensive material, while only a little percentage of the original material is part of the finished object. However, when 3D printing is applied, maximum utilization of the material is achieved. This technology, therefore, has caused a breakthrough in the fabrication process of several engine components and other objects and their repair [28].

GE Aviation, for instance, has successfully produced a Leap Engine Fuel Nozzle made of cobalt and chrome by using a laser AM process, certified for being employed on civil aircraft. The mentioned part has substituted an assembly of 20 components with reduced cost, weight, without joints, and improved performance [28]. Therefore, 3D printing offers an important opportunity for tool repairing with time and money saving. Instead of discarding an old tool and replacing it, is possible to repair the worn-out parts in a tool by using metal additive technologies. They can be employed to fix tool inserts, dies, or molds and for the repair of aerospace components with high flexibility and reliability as already described [29].

SLM, a type of powder-bed fusion AM, is a particularly suitable technique for repairing damaged mold inserts. It enables a fast repair only in the damaged areas, simplifying the process, and decreasing the downtime and costs linked to the maintenance and repair compared to conventional methods [29].

In addition to that, 3D printing has also played a crucial role in the production of ventilator parts during the COVID-19 pandemic. The lack of these parts in the hospitals was so severe that there weren't enough ventilators for every patient; doctors then had to make decisions about which one would receive their support. To respond to this shortage, a certain company started to print a respiratory valve, which is a necessary part to connect respiratory tubing with the ventilator and ensure flow in the right direction for oxygen input and outgoing. Moreover, it made its design available for anyone who wants to print the valve in case of need [30].

Besides the respiratory valve, another company designed and printed T-connector and Y-connector to use one ventilator for two patients at the same time. These parts were distributed to local hospitals to meet the high demand for ventilators during the coronavirus pandemic [30]. All the alternatives mentioned show AM, in ways of 3D printing can be widely used and applied in different areas, playing a crucial role in remote and risk areas, helping both the population and the soldiers in a war zone.

30.5 WATER TREATMENT USING 3D PRINTING

Water and alimentation are basic needs for human survival; however, in war zones and in times of crisis, access to good and healthy food and clean water sources may be the biggest challenge for the population. Restricting access to water and food sources is often used as a tactic of war, although, most of the time, water sources in these areas tend to be scarce, and even when it is possible to find some, they are probably contaminated by either chemical and/or biological matter, making them unfit for human consumption. Some people even dare to affirm that having access to water is increasingly a matter of survival in conflict zones. Thus, the development of easy, cheap, and accessible water treatment technologies is key to overcoming this humanitarian issue in both war zones and also areas with strict access to drinkable water [31], [32].

As already discussed in the previous sections, 3D printing is a versatile technique, bringing a lot of advantages in war zones and during humanitarian actions. In the same way, 3D printing can be crucial to facilitating access to clean and drinkable water even when it is scarce or has been restricted as a conflict tactic. 3D printing can easily be used to print water bottles or canteens, to facilitate the transportation of water by the soldiers, or even to help a better distribution of this resource to the population of a risk area. In 2017, Marine Corps Captain Justin Carrasco designed a 3D water bottle with 0.5 and 1 L capacities, minimizing the logistical footprint of packaged water bottles while they are being transported in military convoys, in a way to maximize logistical functionality while maintaining usability [33]. Besides, by using the different printing technologies, such as FDM, SLA, and SLS, these recipients can be printed in different designs, sizes, shapes, and materials, making them adaptable for their application in the best possible way [34].

Besides being useful for fast and cheap prototyping of water recipients, 3D printing can be used to facilitate and promote on-site water treatment, in this way, if the water source available close by is not suitable for drinking, soldiers or the population living nearby, can easily get this water and make it drinkable. Research in this area is focused on the development of materials for membrane separation, water treatment, and purification process applications, aimed at the design of spacers, membranes, adsorbents, photocatalysts, ceramic filters, modules, capsules, etc [35–37].

The biggest challenge faced when designing 3D-printed membranes is the determination of porous sizes. Since they need to be very small (the order of um or even nm) to achieve better filtration, the 3D printer usually must be adapted to create these nanoporous, with good accuracy. Furthermore, it is also important to highlight that different printing technologies (SLA, FDM or SLS) will directly affect how the membrane will be produced, also changing the membrane performance and fouling properties. Tan et al. [38] compared different 3D printed membranes (varying the printing technology) with a commercial membrane, and they showed that the 3D-printed spacers maintained superior mass transfer performance when compared to the commercial spacer. In terms of printing accurateness, their tests showed a preference for polyjet > SLS > FDM, showing that the printing technology affects the membrane and separator performance.

These membranes can be also further modified with chemical or biological agents working to improve the membrane functioning and allow other types of separation. Lv et al. [39] reported the fabrication of a superhydrophobic porous membrane for oil-water separation by 3D printing with nanosilica-filled polydimethylsiloxane (PDMS) ink. The modified ink was 3D printed and the obtained membrane showed a maximum oil-water separation efficiency of 99.6% with pore sizes of 0.37 mm, also exhibiting a high flux of approximately 23700 L/m^2.h, showing the versatility and facility of this technique.

Another strategy also used is the development of a high-performance and low-cost solar collector for water treatment manufactured by 3D-printing technology. Recently, Mai et al. [37] used 3D printing to produce water treatment equipment that could efficiently treat wastewater quickly, conveniently, and inexpensively. The composite filaments with a diameter of 1.75 ± 0.05 mm were prepared by a melt blending method, containing 10 wt% of modified TiO_2 and 90 wt% of PLA. The results then showed that those modified particles were uniformly dispersed in the PLA matrix and were stable enough to perform the proposed photocatalytic degradation, showing a good result for waste-water treatment, with also improved after adding 5 wt% of silver.

Since the 3D printing process provides flexible materials with porous and open structures, they can also be used to directly adsorb organic dyes with high efficiency, and can potentially be recycled afterward. In this sense, Wang et al. [40] proposed the utilization of 3D printing for the fabrication of ABS filament coated with porous Cu-Benzene tricarboxylic acid adsorbents for methylene blue (MB) removal. The Cu-BTC/ABS composite was designed and printed in different shapes, and the structure was able to enhance surface wettability, which in turn helps to improve the adsorption of metals and linkers. The MB removal efficiency for solutions with concentrations of 10 and 5 mg/L was 93.3% and 98.3%, respectively, within 10 min. After MB adsorption, the composite could easily be recovered without the need for centrifugation or filtration, making it easily reusable. This could then easily be used in war zones with no access to clean water sources, making the water suitable for drinking in an easy and efficient way.

Another option that can be used is the modification of the filaments with Ag or Cu to produce on-site disinfecting filters, able to provide disinfection directly in the water bottle or recipient, making it suitable for drinking. Luukkonen et al. [41] recently compared the main characteristics of highly porous geopolymer components for water treatment applications manufactured by 3D printing. In their research, they were able to fabricate components that possessed mesoporosity, suitable mechanical strength, and water permeability. Furthermore, it was possible to prepare filters with low metal leaching between a pH of 3–7, in a way that the released Ag and Cu concentrations would be within drinking water standards, presenting another great alternative for easy, cheap, and accessible water treatment.

Furthermore, wastewater or water non-suitable for drinking can be used as a feedstock for the water splitting (WS) process, providing on-site treatment for water recycling and reuse, along with the production of energy, in the form of hydrogen, that could be produced and stored for further applications. This is an excellent alternative to obtaining drinking water in areas where it is scarce, especially since the H_2 produced during this process can be used as an energy source, helping both poor areas and war zones. Microbial electrolysis cells (MECs) or wastewater electrolysis cells (WECs), the ones used to produce H_2 from water non-suitable for drinking, can easily be designed and produced using 3D-printing technologies facilitating, even more, access in remote areas [42], [43].

All these alternatives for water treatment can be very useful in war zones and during humanitarian actions. Furthermore, using 3D printing for these alternatives can provide easy, fast, and cheap on-site water treatment that could help to solve water access problems faced by people in remote areas or even during conflict times.

30.6 FINAL REMARKS AND PERSPECTIVES FOR NEXT YEARS

In summary, in this chapter, we have shown how 3D printing can be a versatile technique since it is easy, cheap, and can be used for on-site production according to local needs. Therefore, 3D printing can be very useful for different applications, especially in remote areas, where access to clean water, food, medical supporting materials, parts for machine maintenance, and other supplies can be more difficult. Furthermore, it can be very useful for defense applications, in war zones and places affected by conflicts and natural disasters. The low cost and versatility of 3D-printing techniques allow for easy and fast transportation to any area in need, facilitating and speeding up access.

Besides being very useful during war times, allowing fast and cheap weaponry design and fabrication, it is also useful for prototyping individual protection equipment, medical supplies, and even parts for machine maintenance, for on-site repairing of general machines, war machines, and vehicles or even weapons of all kinds. Additionally, it can also be used to help with clean water and food access in risky areas, either by providing 3D-printed food or even to make 3D-printed water bottles with printed membranes for water purification and treatment.

Since conflicts and humanitarian crises are a constant in our society, the development of AM as an alternative technology to assist in such situations presents great potential and tends to become even more evident in the next years, as 3D-printing technologies become even more affordable and new materials are established.

As previously discussed, many companies, military, humanitarian and governmental organizations are already looking into it and developing research related to these applications. Thus, we believe for the next years a great number of new technologies will be developed and applied in these areas, assisting soldiers and populations in risky areas to have access to a better quality of life.

REFERENCES

1 O. Abdulhameed, A. Al-Ahmari, W. Ameen, and S. H. Mian, "Additive manufacturing: Challenges, trends, and applications," *Advances in Mechanical Engineering*, vol. 11, no. 2, p. 1687814018822880, Feb. 2019.

2 Y. Bozkurt and E. Karayel, "3D printing technology; methods, biomedical applications, future opportunities and trends," *Journal of Materials Research and Technology*, vol. 14, pp. 1430–1450, Sep. 2021.

3 T. D. Ngo, A. Kashani, G. Imbalzano, K. T. Q. Nguyen, and D. Hui, "Additive manufacturing (3D printing): A review of materials, methods, applications and challenges," *Composites Part B: Engineering*, vol. 143, pp. 172–196, Jun. 2018.

4 B. A. Praveena, N. Lokesh, A. Buradi, N. Santhosh, B. L. Praveena, and R. Vignesh, "A comprehensive review of emerging additive manufacturing (3D printing technology): Methods, materials, applications, challenges, trends and future potential," *Materials Today: Proceedings*, 2022, vol. 52, pp. 1309–1313.

5 N. A. S. Mohd Pu'ad, R. H. Abdul Haq, H. Mohd Noh, H. Z. Abdullah, M. I. Idris, and T. C. Lee, "Review on the fabrication of fused deposition modelling (FDM) composite filament for biomedical applications," *Materials Today: Proceedings*, vol. 29, pp. 228–232, Jan. 2020.

6 A. Mitchell, U. Lafont, M. Hołyńska, and C. Semprimoschnig, "Additive manufacturing—A review of 4D printing and future applications," *Additive Manufacturing*, vol. 24. pp. 606–626, Dec. 01, 2018.

7 M. Clemens, "The Use of Additive Manufacturing in the Defense Sector," *3Dnatives*. Jun. 2022. [Online]. Available: www.3dnatives.com/en/the-use-additive-manufacturing-defense-sector30 0620224/

8 A. Cottingham, "Military Applications of 3D Printing," *All3DP Pro*. Jul. 2021. [Online]. Available: https://all3dp.com/1/3d-printing-military-applications/

9 M. L., "General Lattice Awarded Contract to Improve Combat Helmet Impact Absorption for United States Army," *3Dnatives*. Sep. 2021. [Online]. Available: www.3dnatives.com/en/general-lattice-com bat-helmet-united-states-army-230920214/

10 R. Matsuzaki, M. Ueda, M. Namiki, T. Jeong, H. Asahara, K. Horiguchi, T. Nakamura, A. Todoroki, and Y. Hirano, "Three-dimensional printing of continuous-fiber composites by in-nozzle impregnation," *Scientific Reports*, vol. 6, no. 1, p. 23058, Mar. 2016.

11 X. Tian, T. Liu, C. Yang, Q. Wang, and D. Li, "Interface and performance of 3D printed continuous carbon fiber reinforced PLA composites," *Composites Part A: Applied Science and Manufacturing*, vol. 88, pp. 198–205, Sep. 2016.

12 S. C. Kim, M. Kim, and N. Ahn, "3D printer scheduling for shortest time production of weapon parts," *Procedia Manufacturing*, vol. 39, pp. 439–446, Jan. 2019.

13 M. Kim, S. Kim, and N. Ahn, "Study of rifle maintenance and parts supply via 3D printing technology during wartime," *Procedia Manufacturing*, vol. 39, pp. 1510–1516, Jan. 2019.

14 "Ship-shape: The World's First 3D Printed Marine Propeller," *The Engineer*. [Online]. Available: www.theengineer.co.uk/content/advanced-manufacturing/ship-shape-the-world-s-first-3d-printed-marine-propeller/

15 A. Feldman, "Putting 3D Printers To Work In Ukraine's War Zone," *Forbes*. [Online]. Available: www.forbes.com/sites/amyfeldman/2022/03/31/putting-3d-printers-to-work-in-ukraines-war-zone/

16 A. Ahmed, A. Azam, M. M. Aslam Bhutta, F. A. Khan, R. Aslam, and Z. Tahir, "Discovering the technology evolution pathways for 3D printing (3DP) using bibliometric investigation and emerging applications of 3DP during COVID-19," *Cleaner Environmental Systems*, vol. 3, p. 100042, Dec. 2021.

17 "Defining Humanitarian Assistance." Nov. 2017. [Online]. Available: https://web.archive.org/web/20171102215158/; www.globalhumanitarianassistance.org/data-guides/defining-humanitarian-aid

18 J. Owen, "Positive Ways 3D Printing Impacts Humanitarian Aid," *3D Universe*. Dec. 2019. [Online]. Available: http://3duniverse.org/2019/12/02/positive-ways-3d-printing-impacts-humanitarian-aid/

19 H. Amelia, "The Role of Construction 3D Printing in Humanitarian Aid," *3Dnatives*. Jan. 2022. [Online]. Available: www.3dnatives.com/en/the-role-of-construction-3d-printing-in-humanitarian-aid-080620214/

20 "Doing Good with 3D Printing: Humanitarian Aid, Prosthetics & More Projects – 3DSourced." Jan. 2022. [Online]. Available: www.3dsourced.com/editors-picks/doing-good-3d-printing-humanitarian-aid-projects/

21 R. M. Cardoso, C. Kalinke, R. G. Rocha, P. L. dos Santos, D. P. Rocha, P. R. Oliveira, B. C. Janegitz, J. A. Bonacin, E. M. Richter and R. A. A. Munoz, "Additive-manufactured (3D-printed) electrochemical sensors: A critical review," *Analytica Chimica Acta*, vol. 1118. pp. 73–91, Jun. 29, 2020.

22 J. Booth, E. Edwards, M. Whitley, M. Kranz, M. Seif, and P. Ruffin, "Military comparison of 3D printed vs commercial components," *Nano-, Bio-, Info-Tech Sensors, and 3D Systems II*, Mar. 2018, vol. 10597, pp. 52–66.

23 H. Dodziuk, "Applications of 3D printing in healthcare," *Kardiochirurgia i Torakochirurgia Polska = Polish Journal of Cardio-Thoracic Surgery*, vol. 13, no. 3, pp. 283–293, Sep. 2016.

24 G. Chen, Y. Xu, P. Chi Lip Kwok, and L. Kang, "Pharmaceutical applications of 3D printing," *Additive Manufacturing*, vol. 34, p. 101209, Aug. 2020.

25 M. Salmi, J. S. Akmal, E. Pei, J. Wolff, A. Jaribion, and S. H. Khajavi, "3D printing in COVID-19: Productivity estimation of the most promising open source solutions in emergency situations," *Applied Sciences*, vol. 10, no. 11, p. 4004, Jan. 2020.

26 S. Chekurov and M. Salmi, "Additive manufacturing in offsite repair of consumer electronics," *Physics Procedia*, vol. 89, pp. 23–30, Jan. 2017.

27 C. Schwaar, "U.S. Military to 3D Print Its Way Out of Supply Chain Woes," *Forbes*. [Online]. Feb. 2022. Available: www.forbes.com/sites/carolynschwaar/2022/02/27/us-military-to-3d-print-its-way-out-of-supply-chain-woes/

28 L. J. Kumar and C. G. Krishnadas Nair, "Current Trends of Additive Manufacturing in the Aerospace Industry," in *Advances in 3D Printing & Additive Manufacturing Technologies*, D. I. Wimpenny, P. M. Pandey, and L. J. Kumar, Eds. Singapore: Springer, 2017, pp. 39–54.

29 E. Yasa, Ö. Poyraz, N. Cizicioğlu, and S. Pilatin, "Repair and Manufacturing of High Performance Tools by Additive Manufacturing," Jun. 2015.

30 Md. S. Tareq, T. Rahman, M. Hossain, and P. Dorrington, "Additive manufacturing and the COVID-19 challenges: An in-depth study," *Journal of Manufacturing Systems*, vol. 60, pp. 787–798, Jul. 2021.

31 "Having access to water is increasingly a matter of survival in conflict zones – World ReliefWeb." [Online]. Mar. 2022. Available: https://reliefweb.int/report/world/having-access-water-increasingly-matter-survival-conflict-zones

32 N. M. Abu-Lohom, Y. Konishi, Y. Mumssen, B. Zabara, and S. M. Moore, *Water Supply in a War Zone*. Washington, DC: World Bank, 2018.

33 S. Saunders, "Marine Corps Captain 3D Prints Water Bottle Prototype, Designed to Reduce Logistical Footprint in Military Convoys," *3DPrint.com The Voice of 3D Printing/Additive Manufacturing*. Feb. 2017. [Online]. Available: https://3dprint.com/165210/3d-printed-military-water-bottle/

34 "Clean Currents: 3D-printed water bottles and more from ocean plastic," *plasticstoday.com*. Jul. 2018. [Online]. Available: www.plasticstoday.com/packaging/clean-currents-3d-printed-water-bottles-and-more-ocean-plastic

35 L. D. Tijing, J. R. C. Dizon, I. Ibrahim, A. R. N. Nisay, H. K. Shon, and R. C. Advincula, "3D printing for membrane separation, desalination and water treatment," *Applied Materials Today*, vol. 18, p. 100486, Mar. 2020.

36 N. Yanar, P. Kallem, M. Son, H. Park, S. Kang, and H. Choi, "A New era of water treatment technologies: 3D printing for membranes," *Journal of Industrial and Engineering Chemistry*, vol. 91, pp. 1–14, Nov. 2020.

37 Z. Mai, D. Liu, Z. Chen, D. Lin, W. Zheng, X. Dong, Q. Gao, and W. Zhou, "Fabrication and application of photocatalytic composites and water treatment Facility based on 3D printing technology," *Polymers (Basel)*, vol. 13, no. 13, p. 2196, Jan. 2021.
38 W. S. Tan, S. R. Suwarno, J. An, C. K. Chua, A. G. Fane, and T. H. Chong, "Comparison of solid, liquid and powder forms of 3D printing techniques in membrane spacer fabrication," *Journal of Membrane Science*, vol. 537, pp. 283–296, Sep. 2017.
39 J. Lv, Z. Gong, Z. He, J. Yang, Y. Chen, C. Tang, Y. Liu, M. Fan, and W. Lau, "3D printing of a mechanically durable superhydrophobic porous membrane for oil–water separation," *Journal of Materials Chemistry A*, vol. 5, no. 24, pp. 12435–12444, Jun. 2017.
40 Z. Wang, J. Wang, M. Li, K. Sun, and C. Liu, "Three-dimensional printed acrylonitrile butadiene styrene framework coated with Cu-BTC metal-organic frameworks for the removal of methylene blue," *Scientific Reports*, vol. 4, no. 1, p. 5939, Aug. 2014.
41 T. Luukkonen, J. Yliniemi, H. Sreenivasan, K. Ohenoja, M. Finnila, G. Franchin, and P. Colombo, "Ag- or Cu-modified geopolymer filters for water treatment manufactured by 3D printing, direct foaming, or granulation," *Scientific Reports,* vol. 10, no. 1, p. 7233, Apr. 2020.
42 R. L. Germscheidt, D. E. B. Moreira, R. G. Yoshimura, N. P. Gasbarro, E. Datti, P. L. dos Santos, and J. A. Bonacin, "Hydrogen environmental benefits depend on the way of production: An overview of the main processes production and challenges by 2050," *Advanced Energy and Sustainability Research*, vol. 2, no. 10, p. 2100093, Oct. 2021.
43 J. P. Hughes, P. L. dos Santos, M. P. Down, C. W. Foster, J. A. Bonacin, E. M, Keefe, S. J. Rowley-Neale and C. E. Banks, "Single step additive manufacturing (3D printing) of electrocatalytic anodes and cathodes for efficient water splitting," *Sustainable Energy & Fuels*, vol. 4, no. 1, pp. 302–311, Dec. 2019.

Index

F

G

H

I

J

K

L

M

N

O